INTRODUCTION TO
Organic and Biochemistry

SEVENTH EDITION

Frederick A. Bettelheim

William H. Brown
Beloit College

Mary K. Campbell
Mount Holyoke College

Shawn O. Farrell
Olympic Training Center

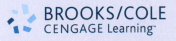
BROOKS/COLE
CENGAGE Learning™

Australia • Brazil • Japan • Korea • Mexico • Singapore • Spain • United Kingdom • United States

Introduction to Organic and Biochemistry, Seventh Edition
Frederick A. Bettelheim, William H. Brown, Mary K. Campbell, Shawn O. Farrell

Publisher: Mary Finch

Senior Acquisitions Editor: Lisa Lockwood

Senior Development Editor: Sandra Kiselica

Assistant Editor: Brandi Kirksey

Editorial Assistant: Elizabeth Woods

Media Editor: Lisa Weber

Marketing Manager: Nicole Hamm

Marketing Assistant: Elizabeth Wong

Marketing Communications Manager: Linda Yip

Content Project Manager, Editorial Production: Teresa L Trego

Creative Director: Rob Hugel

Art Director: John Walker

Print Buyer: Judy Inouye

Permissions Editor: Mollika Basu

Production Service: Prepress PMG

Text Designer: Ellen Pettengell

Photo Researcher: Marcy Lunetta/Page to Page

Copy Editor: Kevin Gleason

Illustrator: Greg Gambino/2064 Designs

OWL Producers: Stephen Battisti, Cindy Stein, David Hart (Center for Educational Software Development, University of Massachusetts, Amherst)

Cover Designer: John Walker

Cover Image: Lester Lefkowitz/CORBIS

Compositor: Prepress PMG

Library of Congress Control Number: 2008941327

ISBN-13: 978-0-495-39116-6

ISBN-10: 0-495-39116-6

Brooks/Cole
10 Davis Drive
Belmont, CA 94002-3098
USA

Cengage Learning is a leading provider of customized learning solutions with office locations around the globe, including Singapore, the United Kingdom, Australia, Mexico, Brazil, and Japan. Locate your local office at:
international.cengage.com/region

Cengage Learning products are represented in Canada by Nelson Education, Ltd.

For your course and learning solutions, visit **academic.cengage.com**

Purchase any of our products at your local college store or at our preferred online store **www.ichapters.com**

Printed in Canada
1 2 3 4 5 6 7 12 10 09 08

SN 276905

To my lovely wife, Courtney — Between textbook revisions, a

full-time job, and school, I have been little more than a ghost

around the house, hiding in my study writing. Courtney held

the family together, taking care of our children and our

home while maintaining her own writing schedule. None of this

would have been possible without her love, support, and

tireless effort. —SF

To my grandchildren for the love and joy they bring

to my life: Emily, Sophia, and Oscar; Amanda and Laura;

Rachel; Gabrielle and Max. —WB

To Andrew, Christian, and Sasha — Thank you for the rewards

of being your mom. And to Bill, Mary, and Shawn. — It is

always a pleasure to work with you. —SK

Contents in Brief

Organic Chemistry

Biochemistry

Contents

Chapter 6 Chirality: The Handedness of Molecules 114

Chapter 7 Acids and Bases 135

Chapter 8 Amines 171

Chapter 9 Aldehydes and Ketones 187

Chapter 10 Carboxylic Acids 205

Chapter 16 Chemical Communications: Neurotransmitters and Hormones 369

Chapter 17 Nucleotides, Nucleic Acids, and Heredity 395

Chapter 18 Gene Expression and Protein Synthesis 425

Chapter 19 Bioenergetics: How the Body Converts Food to Energy 456

Chapter 20 Specific Catabolic Pathways: Carbohydrate, Lipid, and Protein Metabolism 477

Preface

Perceiving order in nature of the world is a deep-seated human need. It is our primary aim to convey the relationship among facts and thereby present a totality of the scientific edifice built over the centuries. In this process we marvel at the unity of laws that govern everything in the ever-exploding dimensions: from photons to protons, from hydrogen to water, from carbon to DNA, from genome to intelligence, from our planet to the galaxy and to the known universe. Unity in all diversity.

As we prepare the ninth edition of our textbook, we cannot help but be struck by the changes that have taken place in the last 30 years. From the slogan of the '70s, "Better living through chemistry" to today's saying "Life by chemistry" one is able to sample the change in the focus. Chemistry helps to provide the amenities of good life but is at the core of our concept and preoccupation of life itself. This shift in emphasis demands that our textbook designed primarily for the education of future practitioners of health sciences should attempt to provide both the basics and the scope of the horizon within which chemistry touches our life.

The increasing use of our textbook made this new edition possible and we wish to thank our colleagues who adopted the previous editions for their courses. Testimony from colleagues and students indicates that we managed to convey our enthusiasm for the subject to students, who find this book to be a great help in studying difficult concepts.

Therefore, in the new edition we strive further to present an easily readable and understandable text. At the same time we emphasized the inclusion of new relevant concepts and examples in this fast-growing discipline especially in the biochemistry chapters. We maintain an integrated view of chemistry. From the very beginning in the general chemistry part we include organic compounds and biochemical substances to illustrate the principles. The progress is ascension from the simple to the complex. We urge our colleagues to advance to the chapters of biochemistry as fast as possible, because there lies most of the material that is relevant to the future professions of our students.

Dealing with such a giant field in one course, and possibly the only course in which our students get an exposure to chemistry, creates the selection of the material an overarching enterprise. We are aware that even though we tried to keep the book to a manageable size and proportion, we included more topics than could be covered in a two-semester course. Our aim was to give enough material from which the instructor can select the topics he or she deems important. We organized the sections so that each of them can stand independently and, therefore, leaving out sections or even chapters will not cause fundamental cracks in the total edifice.

We have increased the number of topics covered and provided a wealth of new problems, many of them challenging and thought-provoking.

"To see the world in a grain of sand And heaven in a wild flower Hold infinity in the palm of your hand And eternity in an hour."
WILLIAM BLAKE:
Auguries of Innocence

The cure for boredom is curiosity There is no cure for curiosity.
DOROTHY PARKER

Audience

Like the previous editions, we intend this book for non chemistry majors, mainly those entering health sciences and related fields, such as nursing, medical technology, physical therapy, and nutrition. It can also be used by students in environmental studies. In its entirety, it can be used for a one-year (two-semester or three-quarter) course in chemistry, or parts of the book can be used in a one-term chemistry course.

We assume that the students using this book have little or no background in chemistry. Therefore, we introduce the basic concepts slowly at the beginning and increase the tempo and the level of sophistication as we go on. We progress from the basic tenets of general chemistry to organic and finally to biochemistry. We consider this progress as an ascent in terms of both practical importance and sophistication. Throughout we integrate the three parts by keeping a unified view of chemistry. We do not consider general chemistry sections to be the exclusive domain of inorganic compounds, so we frequently use organic and biological substances to illustrate general principles.

While teaching the chemistry of the human body is our ultimate goal, we try to show that each subsection of chemistry is important in its own right, besides being required for future understanding.

Chemical Connections (Medical and Other Applications of Chemical Principles)

The Chemical Connections boxes contain applications of the principles discussed in the text. Comments from users of the earlier editions indicate that these boxes have been especially well received, and provide a much-requested relevance to the text. Up-to-date topics appear here, including coverage of anti-inflammatory drugs such as Vioxx and Celebrex (Chemical Connections 13H). Another example is the coverage of novel wound dressings based on polysaccharides obtained from shrimp shells (Chemical Connections 12E). In Chapter 22, which deals with nutrition, students can get a new look at the Food Guide Pyramid (Chemical Connections 22A). The ever-present issues related dieting are described in Chemical Connections 22B. In Chapter 23, students can learn important implications for the use of antibiotics (Chemical Connections 23D) and a detailed explanation of the important and often controversial topic of stem cell research (Chemical Connections 23E).

The presence of Chemical Connections allows a considerably degree of flexibility. If an instructor wants to assign only the main text, the Chemical Connections do not interrupt continuity, and the essential material will be covered. However, because they enhance core material, most instructors will probably wish to assign at least some of the Chemical Connections. In our experience, students are eager to read the relevant Chemical Connections, without assignments and they do with discrimination. From such a large number of boxes, an instructor can select those that best fit the particular needs of the course. So students can test their knowledge, we provide problems at the end of each chapter for all of the Chemical Connections.

Metabolism: Color Code

The biological functions of chemical compounds are explained in each of the biochemistry chapters and in many of the organic chapters. Emphasis is placed on chemistry rather than physiology. Because we have received much positive feedback regarding the way in which we have organized the topic of metabolism (Chapters 19, 20, and 21), we have maintained this organization.

First we introduce the common metabolic pathway through which all food will be utilized (the citric acid cycle and oxidative phosphorylation), and only then do we discuss the specific pathways leading to the common pathway. We find this a useful pedagogic device, and it enables us to sum the caloric values of each type of food because its utilization through the common pathway has already been learned. Finally, we separate the catabolic pathways from the anabolic pathways by treating them in different chapters, emphasizing the different ways the body breaks down and builds up different molecules.

The topic of metabolism is a difficult one for most students, and we have tried to explain it as clearly as possible. As in the previous edition, we enhance the clarity of presentation by the use of a color code for the most important biological compounds discussed in Chapters 19, 20, and 21. Each type of compound is screened in a specific color, which remains the same throughout the three chapters. These colors are as follows:

ATP and other nucleoside triphosphates

ADP and other nucleoside diphosphates

The oxidized coenzymes NAD^+ and FAD

The reduced coenzymes NADH and $FADH_2$

Acetyl coenzyme A

In figures showing metabolic pathways, we display the numbers of the various steps in yellow. In addition to this main use of a color code, other figures in various parts of the book are color coded so that the same color is used for the same entity throughout. For example, in all Chapter 15 figures that show enzyme-substrate interactions, enzymes are always shown in blue and substrates in orange.

Features

- **[NEW] Problem-Solving Strategies** The in-text examples now include a description of the strategy used to arrive at a solution. This will help students organize the information in order to solve the problem.

- **[NEW] Visual Impact** We have introduced illustrations with heightened pedagogical impact. These include ones that show the microscopic and macroscopic aspects of a topic under discussion. The Chemical Connections essays have been enhanced further with more photos that illustrate each topic.

- **Key Questions** We use a Key Questions framework to emphasize key chemical concepts. This focused approach guides students through each chapter by using section head questions.

- **[UPDATED] Chemical Connections** Over 150 essays describe applications of chemical concepts presented in the text, linking the chemistry to their real uses. Many new application boxes on diverse topics are added, such as carbohydrate bandages, organic foods, and monoclonal antibodies.

- **Summary of Key Reactions** In each organic chemistry chapter (1–11) an annotated summary presents reactions introduced in the chapter, identifies the section in which each is introduced, and gives an example of each reaction.

- **[UPDATED] Chapter Summaries** Summaries reflect the Key Questions framework. At the end of each chapter, the Key Questions are restated and the summary paragraphs that follow highlight the concepts associated with the questions. New to this edition are links between the summaries and the end-of-chapter problems that are assignable in OWL.

- **[UPDATED] GOB OWL Problems** End-of-chapter problems that can be assignable in GOB OWL, the web-based homework system that accompanies this book, are marked with a blue square.

- **[UPDATED] Looking Ahead Problems** At the end of most chapters are challenge problems designed to show the application of principles in the chapter to material in following chapters.

- **[UPDATED] Tying It Together and Challenge Problems** At the end of most chapters are problems that build on past material as well as problems that test students' knowledge of the material. In response to reviewer feedback, the number of these problems has increased in this edition.

- **[UPDATED] How To Boxes** This edition marks an increase in the number of boxes that emphasize the skills students need to master the material. These boxes are available in OWL in an interactive form.

- **Molecular Models** Ball-and-stick models, space-filling models, and electron density maps are used throughout the text where appropriate as aids to visualizing molecular properties and interactions.

- **Margin Definitions** Many terms are also defined in the margin to help students learn terminology. By skimming the chapter for these definitions, students will have a quick summary of its contents.

- **Margin Notes** Additional bits of information, such as historical notes, reminders, and so forth complement nearby text.

- **Answers to all in-text and odd-numbered end-of-chapter problems** Answers to selected problems are provided at the end of the book. Detailed worked-out solutions to these same problems are provided in the Student Solutions Manual.

- **Glossary** The glossary at the back of the book gives a definition of each new term along with the number of the section in which the term is introduced.

ORGANIZATION AND UPDATES

Organic Chemistry (Chapters 1–11)

- **Chapter 1, Organic Chemistry,** introduces the characteristics of organic compounds and the most important organic functional groups.

- In **Chapter 2, Alkanes,** we introduce the concept of a line-angle formula and continue using these formulas throughout the organic chapters (Chapters 2–11). These structures are easier to draw than the usual condensed structural formulas and are easier to visualize.

- In **Chapter 3, Alkenes and Alkynes,** we introduce the concept of a reaction mechanism through the hydrohalogenation and acid-catalyzed hydration of alkenes. In addition, we present a mechanism for catalytic hydrogenation of alkenes and later, in Chapter 10, show how the reversibility of catalytic hydrogenation leads to the formation of "*trans* fats." The purpose of this introduction to reaction mechanisms is to demonstrate to students that chemists are interested not only in what happens in a chemical reaction, but also in how it happens.

- **Chapter 4, Benzene and Its Derivatives,** follows immediately after the treatment of alkenes and alkynes. Our discussion of phenols includes phenols and antioxidants.

- **Chapter 5, Alcohols, Ethers, and Thiols,** discusses the structure, names, and properties of alcohols first, and then gives a similar treatment to ethers, and finally to thiols.

- In **Chapter 6, Chirality: The Handedness of Molecules,** the concept of a stereocenter and enantiomerism is slowly introduced using 2-butanol as a prototype. We then treat molecules with two or more stereocenters and show how to predict the number of stereoisomers possible for a particular molecule. We also explain R,S convention for assigning absolute configuration to a tetrahedral stereocenter.

- **Chapter 7, Acids and Bases,** introduces the use of curved arrows to show the flow of electrons in organic reactions. Specifically, we use them here to show the flow of electrons in proton-transfer reactions. The major theme in this chapter is the application of acid–base buffers and the Henderson-Hasselbalch equation.

- In **Chapter 8, Amines,** we trace the development of new asthma medications from epinephrine as a lead drug to albuterol (Proventil).

- **Chapter 9, Aldehydes and Ketones,** has a discussion of $NaBH_4$ as a carbonyl-reducing agent with emphasis on it as a hydride transfer reagent. We then make the parallel to NADH as a carbonyl-reducing agent and hydride transfer agent.

- The chemistry of carboxylic acids and their derivatives are divided into two chapters.

- **Chapter 10, Carboxylic Acids,** focuses on the chemistry and physical properties of carboxylic acids themselves. We briefly discuss *trans* fatty acids and omega-3 fatty acids and the significance of their presence in our diets.

- **Chapter 11, Carboxylic Anhydrides, Esters, and Amides,** describes the chemistry of these three important functional groups with emphasis on their acid-catalyzed and base-promoted hydrolysis, and reactions with amines and alcohols.

Biochemistry (Chapters 12–23)

- **Chapter 12, Carbohydrates,** begins with the structure and nomenclature of monosaccharides, their oxidation and reduction, and the formation of glycosides, and concludes with a discussion of the structure of disaccharides, polysaccharides, and acidic polysaccharides. A new Chemical Connections box addresses *Life-Saving Carbohydrate Bandages*.

- **Chapter 13, Lipids,** covers the most important features of lipid biochemistry, including membrane structure and the structures and functions of steroids. New information on steroid use and Olympic sprinter Marion Jones has been added.

- **Chapter 14, Proteins,** covers the many facets of protein structure and function. It gives an overview of how proteins are organized beginning with the nature of individual amino acids, and describes how this organization leads to their many functions, giving the student the basics needed to lead into the sections on enzymes and metabolism. A new Chemical Connections box discusses *Aspartame, the Sweet Peptide*.

- **Chapter 15, Enzymes,** covers the important topic of enzyme catalysis and regulation. The focus is on how the structure of an enzyme leads to the vast increases in reaction rates observed with enzyme-catalyzed reactions. Specific medical applications of enzyme inhibition are included as well as an introduction to the fascinating topic of transition-state analogs and their use as potent inhibitors. A new Chemical Connections explores *Enzymes and Memory*.

- In **Chapter 16, Chemical Communications,** we see the biochemistry of hormones and neurotransmitters. The health-related implications of how these substances act in the body are a main focus of this chapter. New information on the possible causes of Alzheimer's disease is explored.

- In **Chapter 17, Nucleotides, Nucleic Acids, and Heredity,** introduces DNA and the processes surrounding its replication and repair. How nucleotides are linked together and the flow of genetic information that occurs due to the unique properties of these molecules are emphasized. The sections on the types of RNA have been greatly expanded as our knowledge increases daily about these important nucleic acids. The uniqueness of an individual's DNA is described with a Chemical Connections box that introduces *DNA Fingerprinting* and how forensic science relies on DNA for positive identification.

- **Chapter 18, Gene Expression and Protein Synthesis,** shows how the information contained in the DNA blueprint of the cell is used to produce RNA and eventually protein. The focus is on how organisms control the expression of genes through transcription and translation. The chapter ends with the timely and important topic of gene therapy, our attempt to cure genetic diseases by giving an individual a gene he or she was missing. New Chemical Connections boxes describe *Human Diversity and Transcription Factors* and *Silent Mutations*.

- **Chapter 19, Bioenergetics,** is an introduction to metabolism that focuses strongly on the central pathways, namely the citric acid cycle, electron transport, and oxidative phosphorylation.

- In **Chapter 20, Specific Catabolic Pathways,** we address the details of carbohydrate, lipid, and protein breakdown, concentrating on the energy yield.

- **Chapter 21, Biosynthetic Pathways,** starts with a general consideration of anabolism and proceeds to carbohydrate biosynthesis in both plants and animals. Lipid biosynthesis is linked to production of membranes, and the chapter concludes with an account of amino acid biosynthesis.

- In **Chapter 22, Nutrition,** we take a biochemical approach to understanding nutrition concepts. Along the way, we look at a revised version of the Food Guide Pyramid, and debunk some of the myths about carbohydrates and fats. Chemical Connections boxes expand on two topics that are often important to students—dieting and enhancement of sports performance through proper nutrition. New Chemical Connections boxes discussing *Iron: An Example of a Mineral Requirement* and *Organic Food—Hope or Hype?* have been added.

- **Chapter 23, Immunochemistry,** covers the basics of our immune system and how we protect ourselves from foreign invading organisms. Considerable time is spent on the acquired immunity system. No chapter on immunology would be complete without a description of the human immunodeficiency virus. The chapter ends with a description of the controversial topic of stem cell research—our hopes for its potential and concerns for the potential downsides. Added is a new Chemical Connection box *Monoclonal Antibodies Wage War on Breast Cancer.*

- **Body Fluids,** can be found on the companion website at **www.cengage.com/chemistry/bettelheim.**

Support Package

For Instructors

OWL (Online Web-based Learning)
Instant Access to OWL (two semesters)
ISBN-10: 0-495-11105-8; ISBN-13: 978-0-495-11105-4
Instant Access to OWL e-Book (two semesters)
ISBN-10: 0-495-39123-9; ISBN-13: 978-0-495-39123-4

Online Web-based
Learning

Authored by Roberta Day, Beatrice Botch and David Gross of the University of Massachusetts, Amherst; William Vining of the State University of New York at Oneonta; and Susan Young of Hartwick College. Developed at the University of Massachusetts, Amherst, and class tested by more than a million chemistry students, OWL is a fully customizable and flexible web-based learning system. OWL supports mastery learning and offers numerical, chemical, and contextual parameterization to produce thousands of problems correlated to this text. The OWL system also features a database of simulations, tutorials, and exercises, as well as end-of-chapter problems from the text. With OWL, you get the most widely used online learning system available for chemistry with unsurpassed reliability and dedicated training and support. For Bettelheim's seventh edition, OWL includes parameterized end-of-chapter questions from the text (marked in the text with ■) and tutorials based on How To boxes in the text.

The optional **e-Book in OWL** includes the complete electronic version of the text, fully integrated and linked to OWL homework problems. Most e-Books in OWL are interactive and offer highlighting, notetaking, and bookmarking features that can all be saved. A fee-based access code is required for OWL. To view an OWL demo and for more information, visit **www.cengage.com/owl** or contact your Cengage Learning Brooks/Cole representative.

Instructor's Manual, by Mark Erickson (Hartwick College), Shawn Farrell, and Courtney Farrell contains worked-out solutions to all in-text and end-of-chapter problems. ISBN-10: 0-495-39115-8; ISBN-13: 978-0-495-39115-9.

Power Lecture This dual platform, one-stop digital library and presentation tool includes:

PowerLecture™

- Prepared Microsoft® PowerPoint® Lecture Slides by William H. Brown that cover all key points from the text in a convenient format that you can enhance with your own materials or with additional interactive video and animations from the CD-ROM for personalized, media-enhanced lectures.

- Image Libraries in PowerPoint and in JPEG format that provide electronic files for all text art, most photographs, and all numbered tables in the text. These files can be used to print transparencies or to create your own PowerPoint lectures.

- Electronic files for the Instructor's Manual and Test Bank.

- Sample chapters from the Student Solutions Manual and Study Guide.

- ExamView testing software, with all test items from the printed Test Bank in electronic format, enables you to create customized tests of up to 250 items in print or online.

- JoinIn™ clicker questions authored by Frederick A. Bettelheim and Joseph M. Landsberg specifically for this text, for use with the classroom response system of your choice. Assess student progress with instant quizzes and polls, and display student answers seamlessly within the Microsoft PowerPoint slides of your own lecture questions. Please consult your Brooks/Cole representative for more details. ISBN-10: 0-495-39114-X; ISBN-13: 978-0-495-39114-2.

Test Bank on eBank, by Stephen Z. Goldberg (Adelphi University), contains approximately 50 multiple-choice questions per chapter for each of the 23 chapters in this text. To access, contact your Brooks/Cole representative.

Transparency Acetates One hundred fifty full-color overhead transparencies include key figures and tables from the text. ISBN: 0-495-39117-4; ISBN-13: 978-0-495-39117-3

Instructor's Manual for Laboratory Experiments, Seventh Edition This manual will help instructors in grading the answers to questions and in assessing the range of experimental results obtained by students. The Instructor's Manual also contains important notes for professors to tell students and details on how to handle the disposal of waste chemicals. ISBN: 0-495-39197-2; ISBN-13: 978-0-495-39197-5.

Faculty Companion Website Accessible from **www.cengage.com/chemistry/ bettelheim**, this website provides downloadable files for the Instructor's Manuals for the text and for the lab manuals as well as WebCT and Blackboard versions of the Test Bank. Students will find the online chapter as well as tutorial quizzes and interactive forms of the Active Figures and How To boxes from the text.

For Students

- **Student Study Guide,** by William Scovell of Bowling Green State University includes reviews of chapter objectives, important terms and comparisons, focused reviews of concepts, and self-tests. ISBN 0-495-39118-2; ISBN-13: 978-0-495-39118-0.

- **Student Solutions Manual,** by Mark Erickson (Hartwick College), Shawn Farrell, and Courtney Farrell. This ancillary contains complete worked-out solutions to all in-text and odd-numbered end-of-chapter problems. ISBN 0-495-39119-0; ISBN-13: 978-0-495-39119-7.

- **OWL (Online Web-based Learning System) for the GOB Course** See the description above in the "For Instructors" section.

- *Laboratory Experiments for General, Organic, and Biochemistry,* **Seventh Edition,** by Frederick A. Bettelheim and Joseph M. Landesberg. Forty-eight experiments illustrate important concepts and principles in general, organic and biochemistry. Includes 11 organic chemistry experiments, 17 biochemistry experiments, and 20 general chemistry experiments. Many experiments have been revised and a new addition is an experiment on the properties of enzymes. All experiments have new Pre- and Post-lab Questions. The large number of experiments allows sufficient flexibility for the instructor. ISBN: 0-4953-9196-4; ISBN-13: 978-0-495-39196-8.

Student Companion Website

Accessible from **www.cengage.com/chemistry/bettelheim**, this website provides the online chapter as well as tutorial quizzes and interactive forms of the Active Figures and How To boxes from the text.

Acknowledgments

The publication of a book such as this requires the efforts of many more people than merely the authors. We would like to thank the following professors who offered many valuable suggestions for this new edition:

We are especially grateful to Garon Smith, University of Montana; Paul Sampson, Kent State University and Francis Jenney, Philadelphia College of Osteopathic Medicine, who read page proofs with eyes for accuracy. As reviewers, they also confirmed the accuracy of the answer section in the book.

We give special thanks to Sandi Kiselica, our Senior Development Editor, who has been a rock of support through the entire revision process. We appreciate her constant encouragement as we worked to meet deadlines; she has also been a valuable resource person. We appreciate the help of our other colleagues at Brooks/Cole: Executive Editor Lisa Lockwood, Production Manager Teresa Trego, Associate Editor Brandi Kirksey, Media Editor Lisa Weber, and Patrick Franzen of Pre-Press PMG.

We so appreciate the time and expertise of our reviewers who have read our manuscript and given us helpful comments. They include:

Allison J. Dobson, *Georgia Southern University*

Sara M. Hein, *Winona State University*

Peter Jurs, *The Pennsylvania State University*

Delores B. Lamb, *Greenville Technical College*

James W. Long, *University of Oregon*

Richard L. Nafshun, *Oregon State University*

David Reinhold, *Western Michigan University*

Paul Sampson, *Kent State University*

Garon C. Smith, *University of Montana*

Steven M. Socol, *McHenry County College*

Health-Related Topics

Key

ChemConn = Chemical Connections Box number
Sect. = Section number
Prob. = Problem number

Organic Chemistry

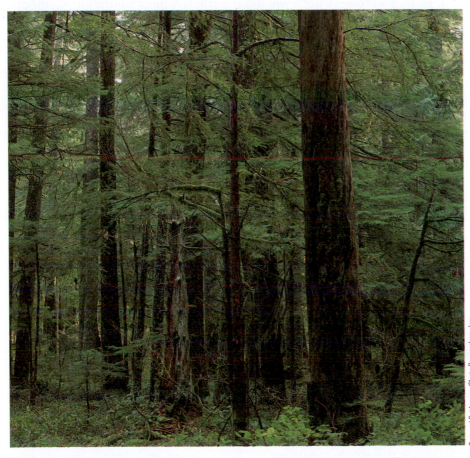

Tom and Pat Leeson/Photo Researchers, Inc.

The bark of the Pacific yew contains paclitaxel, a substance that has proven effective in treating certain types of ovarian and breast cancer (see Chemical Connections 1A).

1.1 What Is Organic Chemistry?

Organic chemistry is the chemistry of the compounds of carbon. As you study Chapters 1–11 (organic chemistry) and 12–23 (biochemistry), you will see that organic compounds are everywhere around us. They are in our foods, flavors, and fragrances; in our medicines, toiletries, and cosmetics; in our plastics, films, fibers, and resins; in our paints, varnishes, and glues; and, of course, in our bodies and the bodies of all other living organisms.

OWL Online homework for this chapter may be assigned in **GOB OWL**.

FIGURE 1.1 Abundance of the elements in the Earth's crust.

Perhaps the most remarkable feature of organic compounds is that they involve the chemistry of carbon and only a few other elements—chiefly, hydrogen, oxygen, and nitrogen. While the majority of organic compounds contain carbon and just these three elements, many also contain sulfur, a halogen (fluorine, chlorine, bromine, or iodine), and phosphorus.

As of the writing of this text, there are 116 known elements. Organic chemistry concentrates on carbon, just one of the 116. The chemistry of the other 115 elements comes under the field of inorganic chemistry. As we see in Figure 1.1, carbon is far from being among the most abundant elements in the Earth's crust. In terms of elemental abundance, approximately 75% of the Earth's crust is composed of just two elements: oxygen and silicon. These two elements are the components of silicate minerals, clays, and sand. In fact, carbon is not even among the ten most abundant elements. Instead, it is merely one of the elements making up the remaining 0.9% of the Earth's crust. Why, then, do we pay such special attention to just one element from among 116?

The first reason is largely historical. In the early days of chemistry, scientists thought organic compounds were those produced by living organisms, and that inorganic compounds were those found in rocks and other nonliving matter. At that time, they believed that a "vital force," possessed only by living organisms, was necessary to produce organic compounds. In other words, chemists believed that they could not synthesize any organic compound by starting with only inorganic compounds. This theory was very easy to disprove if, indeed, it was wrong. It required only one experiment in which an organic compound was made from inorganic compounds. In 1828, Friedrich Wöhler (1800–1882) carried out just such an experiment. He heated an aqueous solution of ammonium chloride and silver cyanate, both inorganic compounds and—to his surprise—obtained urea, an "organic" compound found in urine.

Although this single experiment of Wöhler's was sufficient to disprove the "doctrine of vital force," it took several years and a number of additional experiments for the entire scientific community to accept the fact that organic compounds could be synthesized in the laboratory. This discovery meant that the terms "organic" and "inorganic" no longer had their original meanings because, as Wöhler demonstrated, organic compounds could be obtained from inorganic materials. A few years later, August Kekulé (1829–1896) put forth a new definition—organic compounds are those containing carbon—and his definition has been accepted ever since.

A second reason for the study of carbon compounds as a separate discipline is the sheer number of organic compounds. Chemists have discovered or synthesized more than 10 million of them, and an estimated 10,000 new ones are reported each year. By comparison, chemists have discovered or synthesized an estimated 1.7 million inorganic compounds. Thus, approximately 85% of all known compounds are organic compounds.

A third reason—and one particularly important for those of you going on to study biochemistry—is that biochemicals, including carbohydrates, lipids,

TABLE 1.1 A Comparison of Properties of Organic and Inorganic Compounds

Organic Compounds	Inorganic Compounds
Bonding is almost entirely covalent.	Most have ionic bonds.
Many are gases, liquids, or solids with low melting points (less than 360°C).	Most are solids with high melting points.
Most are insoluble in water.	Many are soluble in water.
Most are soluble in organic solvents such as diethyl ether, toluene, and dichloromethane.	Almost all are insoluble in organic solvents.
Aqueous solutions do not conduct electricity.	Aqueous solutions form ions that conduct electricity.
Almost all burn and decompose.	Very few burn.
Reactions are usually slow.	Reactions are often very fast.

Organic and inorganic compounds differ in their properties because they differ in their structure and composition—not because they obey different natural laws. One set of natural laws applies to all compounds.

proteins, enzymes, nucleic acids (DNA and RNA), hormones, vitamins, and almost all other important chemicals in living systems are organic compounds. Furthermore, their reactions are often strikingly similar to those occurring in test tubes. For this reason, knowledge of organic chemistry is essential for an understanding of biochemistry.

One final point about organic compounds. They generally differ from inorganic compounds in many of their properties, some of which are shown in Table 1.1. Most of these differences stem from the fact that the bonding in organic compounds is almost entirely covalent, while most inorganic compounds have ionic bonds.

Of course, the entries in Table 1.1 are generalizations, but they are largely true for the vast majority of compounds of both types.

1.2 Where Do We Obtain Organic Compounds?

Chemists obtain organic compounds in two principal ways: isolation from nature and synthesis in the laboratory.

A. Isolation from Nature

Living organisms are "chemical factories." Each terrestrial, marine, and freshwater plant (flora) and animal (fauna)—even microorganisms such as bacteria—make thousands of organic compounds by a process called biosynthesis. One way, then, to get organic compounds is to extract, isolate, and purify them from biological sources. In this book, we will encounter many compounds that are or have been isolated in this way. Some important examples include vitamin E, the penicillins, table sugar, insulin, quinine, and the anticancer drug paclitaxel (Taxol, see Chemical Connections 1A). Nature also supplies us with three other important sources of organic compounds: natural gas, petroleum, and coal. We will discuss them in Section 2.4.

B. Synthesis in the Laboratory

Ever since Wöhler synthesized urea, organic chemists have sought to develop more ways to synthesize the same compounds or design derivatives

Chemical Connections 1A

Taxol: A Story of Search and Discovery

In the early 1960s, the National Cancer Institute under-took a program to analyze samples of native plant materi-als in the hope of discovering substances that would prove effective in the fight against cancer. Among the materials tested was an extract of the bark of the Pacific yew, *Taxus brevifolia,* a slow-growing tree found in the old-growth for-ests of the Pacific Northwest. This biologically active ex-tract proved to be remarkably effective in treating certain types of ovarian and breast cancer, even in cases where other forms of chemotherapy failed. The structure of the cancer-fighting component of yew bark was determined in 1962, and the compound was named paclitaxel (Taxol).

Paclitaxel
(Taxol)

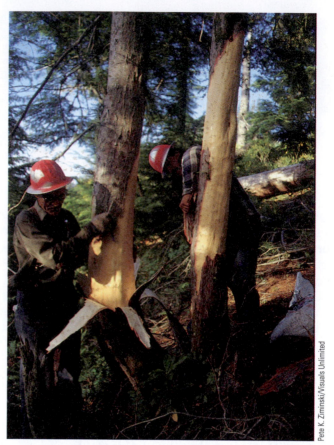

Pete K. Ziminski/Visuals Unlimited

Pacific yew bark being stripped for Taxol extraction

Unfortunately, the bark of a single 100-year-old yew tree yields only about 1 g of Taxol, not enough for effective treatment of even one cancer patient. Furthermore, iso-lating Taxol means stripping the bark from trees, which kills them. In 1994, chemists succeeded in synthesizing Taxol in the laboratory, but the cost of the synthetic drug was far too high to be economical. Fortunately, an alter-native natural source of the drug was found. Researchers in France discovered that the needles of a related plant, *Taxus baccata,* contain a compound that can be converted in the laboratory to Taxol. Because the needles can be gathered without harming the plant, it is not necessary to kill trees to obtain the drug.

Taxol inhibits cell division by acting on microtubules, a key component of the scaffolding of cells. Before cell division can take place, the cell must disassemble these microtubule units, and Taxol prevents this disassembly. Because cancer cells divide faster than normal cells, the drug effectively controls their spread.

The remarkable success of Taxol in the treatment of breast and ovarian cancer has stimulated research efforts to isolate and/or synthesize other substances that treat human diseases the same way in the body and that may be even more effective anticancer agents than Taxol.

of those found in nature. In recent years, the methods for doing so have become so sophisticated that there are few natural organic compounds, no matter how complicated, that chemists cannot synthesize in the laboratory.

Compounds made in the laboratory are identical in both chemical and physical properties to those found in nature—assuming, of course, that each is 100% pure. There is no way that anyone can tell whether a sample of any particular compound was made by chemists or obtained from nature. As a consequence, pure ethanol made by chemists has exactly the same physical and chemical properties as pure ethanol prepared by distilling wine. The same is true for ascorbic acid (vitamin C). There is no advantage, therefore, in paying more money for vitamin C obtained from a natural source than for synthetic vitamin C, because the two are identical in every way.

Organic chemists, however, have not been satisfied with merely duplicating nature's compounds. They have also designed and synthesized compounds not found in nature. In fact, the majority of the more than 10 million known organic compounds are purely synthetic and do not exist in living organisms. For example, many modern drugs—Valium, albuterol, Prozac, Zantac, Zoloft, Lasix, Viagra, and Enovid—are synthetic organic compounds not found in nature. Even the over-the-counter drugs aspirin and ibuprofen are synthetic organic compounds not found in nature.

The vitamin C in an orange is identical to its synthetic tablet form.

1.3 How Do We Write Structural Formulas of Organic Compounds?

A structural formula shows all the atoms present in a molecule as well as the bonds that connect the atoms to each other. The structural formula for ethanol, whose molecular formula is C_2H_6O, for example, shows all nine atoms and the eight bonds that connect them:

Ethanol

The Lewis model of bonding enables us to see how carbon forms four covalent bonds that may be various combinations of single, double, and triple bonds. Furthermore, the valence-shell electron-pair repulsion (VSEPR) model tells us that the most common bond angles about carbon atoms in covalent compounds are approximately 109.5°, 120°, and 180°, for tetrahedral, trigonal planar, and linear geometries, respectively.

Table 1.2 shows several covalent compounds containing carbon bonded to hydrogen, oxygen, nitrogen, and chlorine. From these examples, we see the following:

TABLE 1.2 Single, Double, and Triple Bonds in Compounds of Carbon. Bond angles and geometries for carbon are predicted using the VSEPR model.

- Carbon normally forms four covalent bonds and has no unshared pairs of electrons.
- Nitrogen normally forms three covalent bonds and has one unshared pair of electrons.
- Oxygen normally forms two covalent bonds and has two unshared pairs of electrons.
- Hydrogen forms one covalent bond and has no unshared pairs of electrons.
- A halogen (fluorine, chlorine, bromine, and iodine) normally forms one covalent bond and has three unshared pairs of electrons.

Example 1.1 Writing Structural Formulas

Following are structural formulas for acetic acid, CH_3COOH, and ethylamine, $CH_3CH_2NH_2$.

Acetic acid   Ethylamine

(a) Complete the Lewis structure for each molecule by adding unshared pairs of electrons so that each atom of carbon, oxygen, and nitrogen has a complete octet.
(b) Using the VSEPR model, predict all bond angles in each molecule.

Strategy and Solution

(a) Each carbon atom must be surrounded by eight valence electrons to have a complete octet. Each oxygen must have two bonds and two unshared pairs of electrons to have a complete octet. Each nitrogen must have three bonds and one unshared pair of electrons to have a complete octet.

(b) To predict bond angles about a carbon, nitrogen, or oxygen atom, count the number of regions of electron density (lone pairs and bonding pairs of electrons about it). If four regions of electron density surround the atom, the predicted bond angles are 109.5°. If three regions surround it, the predicted bond angles are 120°. If two regions surround it, the predicted bond angle is 180°.

Acetic acid

Ethylamine

Problem 1.1

The structural formulas for ethanol, CH_3CH_2OH, and propene, $CH_3CH=CH_2$, are

Ethanol   Propene

(a) Complete the Lewis structure for each molecule showing all valence electrons.
(b) Using the VSEPR model, predict all bond angles in each molecule.

1.4 What Is a Functional Group?

As noted earlier in this chapter, more than 10 million organic compounds have been discovered and synthesized by organic chemists. It might seem an almost impossible task to learn the physical and chemical properties of so many compounds. Fortunately, the study of organic compounds is not as formidable a task as you might think. While organic compounds can undergo a wide variety of chemical reactions, only certain portions of their structures undergo chemical transformations. We call the atoms or groups of atoms of an organic molecule that undergo predictable chemical reactions a **functional group.** As we will see, the same functional group, in whatever organic molecule it occurs, undergoes the same types of chemical reactions. Therefore, we do not have to study the chemical reactions of even a fraction of the 10 million known organic compounds. Instead, we need to identify only a few characteristic functional groups and then study the chemical reactions that each undergoes.

Functional groups are also important because they are the units by which we divide organic compounds into families of compounds. For example, we group those compounds that contain an —OH (hydroxyl) group bonded to a tetrahedral carbon into a family called alcohols; compounds containing a —COOH (carboxyl group) belong to a family called carboxylic acids. Table 1.3

Functional group An atom or group of atoms within a molecule that shows a characteristic set of predictable physical and chemical behaviors

TABLE 1.3 Six Common Functional Groups

Family	Functional Group	Example	Name
Alcohol	—OH	CH_3CH_2OH	Ethanol
Amine	—NH₂	$CH_3CH_2NH_2$	Ethanamine
Aldehyde	$-\overset{\overset{\text{O}}{\|\|}}{C}-H$	$CH_3\overset{\overset{\text{O}}{\|\|}}{C}H$	Ethanal
Ketone	$-\overset{\overset{\text{O}}{\|\|}}{C}-$	$CH_3\overset{\overset{\text{O}}{\|\|}}{C}CH_3$	Acetone
Carboxylic acid	$-\overset{\overset{\text{O}}{\|\|}}{C}-OH$	$CH_3\overset{\overset{\text{O}}{\|\|}}{C}OH$	Acetic acid
Ester	$-\overset{\overset{\text{O}}{\|\|}}{C}-OR$	$CH_3\overset{\overset{\text{O}}{\|\|}}{C}OCH_2CH_3$	Ethyl acetate

introduces six of the most common functional groups. A complete list of all functional groups that we will study appears on the inside back cover of the text.

At this point, our concern is simply pattern recognition—that is, how to recognize and identify one of these six common functional groups when you see it, and how to draw structural formulas of molecules containing them. We will have more to say about the physical and chemical properties of these and several other functional groups in Chapters 2–11.

Functional groups also serve as the basis for naming organic compounds. Ideally, each of the 10 million or more organic compounds must have a unique name different from the name of every other organic compound. We will show how these names are derived in Chapters 2–11 as we study individual functional groups in detail.

To summarize, functional groups

- Are sites of predictable chemical behavior—a particular functional group, in whatever compound it is found, undergoes the same types of chemical reactions.
- Determine in large measure the physical properties of a compound.
- Serve as the units by which we classify organic compounds into families.
- Serve as a basis for naming organic compounds.

A. Alcohols

Alcohol A compound containing an —OH (hydroxyl) group bonded to a tetrahedral carbon atom

Hydroxyl group An —OH group bonded to a tetrahedral carbon atom

As previously mentioned, the functional group of an **alcohol** is an **—OH (hydroxyl) group** bonded to a tetrahedral carbon atom (a carbon having bonds to four atoms). In the general formula of an alcohol (shown below on the left), the symbol R— indicates either a hydrogen or another carbon group. The important point of the general structure is the —OH group bonded to a tetrahedral carbon atom.

Functional group   Structural formula   Condensed structural formula

R = H or carbon group   An alcohol (Ethanol)

Here we represent the alcohol as a **condensed structural formula**, CH_3CH_2OH. In a condensed structural formula, CH_3 indicates a carbon bonded to three hydrogens, CH_2 indicates a carbon bonded to two hydrogens, and CH indicates a carbon bonded to one hydrogen. Unshared pairs of electrons are generally not shown in condensed structural formulas.

Alcohols are classified as **primary (1°), secondary (2°),** or **tertiary (3°)**, depending on the number of carbon atoms bonded to the carbon bearing the —OH group.

A 1° alcohol   A 2° alcohol   A 3° alcohol

Example 1.2 Drawing Structural Formulas of Alcohols

Draw Lewis structures and condensed structural formulas for the two alcohols with molecular formula C_3H_8O. Classify each as primary, secondary, or tertiary.

Strategy and Solution

Begin by drawing the three carbon atoms in a chain. The oxygen atom of the hydroxyl group may be bonded to the carbon chain at two different positions on the chain: either to an end carbon or to the middle carbon.

C—C—C   C—C—C—OH   C—C—C with OH on middle carbon
Carbon chain   The two locations for the —OH group

Finally, add seven more hydrogens, giving a total of eight as shown in the molecular formula. Show unshared electron pairs on the Lewis structures but not on the condensed structural formulas.

Lewis structures   Condensed structural   Ball-and-stick models
                        formulas

$CH_3CH_2CH_2OH$

A primary alcohol
(1-Propanol)

CH_3CHCH_3 with OH

A secondary alcohol
(2-Propanol)

The secondary alcohol 2-propanol, whose common name is isopropyl alcohol, is the cooling, soothing component in rubbing alcohol.

Problem 1.2

Draw Lewis structures and condensed structural formulas for the four alcohols with the molecular formula $C_4H_{10}O$. Classify each alcohol as primary, secondary, or tertiary. (*Hint*: First consider the connectivity of the four carbon atoms; they can be bonded either four in a chain or three in a chain with the fourth carbon as a branch on the middle carbon. Then consider the points at which the —OH group can be bonded to each carbon chain.)

2-Propanol (isopropyl alcohol) is used to disinfect cuts and scrapes.

Charles D. Winters/Cengage Learning

B. Amines

The functional group of an amine is an **amino group**—a nitrogen atom bonded to one, two, or three carbon atoms. In a **primary (1°) amine,** nitrogen is bonded to two hydrogens and one carbon group. In a **secondary (2°) amine,** it is bonded to one hydrogen and two carbon groups. In a **tertiary (3°) amine,** it is bonded to three carbon groups. The second and third structural formulas can be written in a more abbreviated form by collecting the CH_3

Amine An organic compound in which one, two, or three hydrogens of ammonia are replaced by carbon groups; RNH_2, R_2NH, or R_3N

Amino group A —NH_2, RNH_2, R_2NH, or R_3N group groups and writing them as $(CH_3)_2NH$ and $(CH_3)_3N$ respectively. The latter are known as condensed structural formulas.

CH_3NH_2

Methylamine
(a 1° amine)

CH_3NH or $(CH_3)_2NH$
|
CH_3

Dimethylamine
(a 2° amine)

CH_3NCH_3 or $(CH_3)_3N$
|
CH_3

Trimethylamine
(a 3° amine)

Example 1.3 Drawing Structural Formulas of Amines

Draw condensed structural formulas for the two primary amines with the molecular formula C_3H_9N.

Strategy and Solution

For a primary amine, draw a nitrogen atom bonded to two hydrogens and one carbon.

$$C-C-C-NH_2 \qquad \overset{\overset{\displaystyle NH_2}{|}}{C-C-C}$$

The three carbons may be
bonded to nitrogen in two ways

$$CH_3CH_2CH_2NH_2 \qquad \overset{\overset{\displaystyle NH_2}{|}}{CH_3CHCH_3}$$

Add seven hydrogens to give each carbon four
bonds and give the correct molecular formula

Problem 1.3

Draw structural formulas for the three secondary amines with the molecular formula $C_4H_{11}N$.

C. Aldehydes and Ketones

Carbonyl group A C=O group

Aldehyde A compound containing a carbonyl group bonded to a hydrogen; a —CHO group

Ketone A compound containing a carbonyl group bonded to two carbon groups

Both aldehydes and ketones contain a **C=O (carbonyl) group.** The **aldehyde** functional group contains a carbonyl group bonded to a hydrogen. Formaldehyde, CH_2O, the simplest aldehyde, has two hydrogens bonded to its carbonyl carbon. In a condensed structural formula, the aldehyde group may be written showing the carbon-oxygen double bond as CH=O or, alternatively, it may be written —CHO. The functional group of a **ketone** is a carbonyl group bonded to two carbon atoms. In the general structural formula of each functional group, we use the symbol R to represent other groups bonded to carbon to complete the tetravalence of carbon.

R—C—C—H   CH₃—C—H   Functional group   R—C—C—C—   CH₃—C—CH₃
 | | Acetaldehyde | Acetone
 R   (an aldehyde)   R   (a ketone)

Functional   Acetaldehyde   Functional   Acetone
group   (an aldehyde)   group   (a ketone)

Example 1.4 Drawing Structural Formulas of Aldehydes

Draw condensed structural formulas for the two aldehydes with the molecular formula C_4H_8O.

Strategy and Solution

First draw the functional group of an aldehyde, and then add the remaining carbons. These may be bonded in two ways. Then add seven hydrogens to complete the tetravalence of each carbon.

$$CH_3CHCH$$
(with O double-bonded to the last C, and CH₃ below)

or

$$CH_3CHCHO$$
(with CH₃ below)

$$CH_3CH_2CH_2CH$$
(with O double-bonded to last C)

or

$$CH_3CH_2CH_2CHO$$

Problem 1.4

Draw condensed structural formulas for the three ketones with the molecular formula $C_5H_{10}O$.

D. Carboxylic Acids

The functional group of a carboxylic acid is a —**COOH** (**carboxyl:** carbonyl + hydroxyl) **group.** In a condensed structural formula, a carboxyl group may also be written —CO_2H.

Carboxyl group A —COOH group

Carboxylic acid A compound containing a —COOH group

:O:
 ‖ ..
RCOH
 ..

Functional group

O
‖
CH₃COH

Acetic acid
(a carboxylic acid)

Example 1.5 Drawing Structural Formulas of Carboxylic Acids

Draw a condensed structural formula for the single carboxylic acid with the molecular formula $C_3H_6O_2$.

Srategy and Solution

The only way the carbon atoms can be written is three in a chain, and the —COOH group must be on an end carbon of the chain.

$$CH_3CH_2\overset{O}{\overset{\|}{C}}OH \quad \text{or} \quad CH_3CH_2CO_2H$$

Problem 1.5

Draw condensed structural formulas for the two carboxylic acids with the molecular formula $C_4H_8O_2$.

E. Carboxylic Esters

Carboxylic ester A derivative of a carboxylic acid in which the H of the carboxyl group is replaced by a carbon group

A **carboxylic ester,** commonly referred to as simply an **ester,** is a derivative of a carboxylic acid, in which the hydrogen of the carboxyl group is replaced by a carbon group. The ester group is written —COOR or —CO₂R in this text.

Functional group

$$CH_3-\overset{O}{\overset{\|}{C}}-O-CH_3 \quad \text{or} \quad CH_3COOCH_3$$

Methyl acetate
(an ester)

Example 1.6 Drawing Structural Formulas of Esters

The molecular formula of methyl acetate is $C_3H_6O_2$. Draw the structural formula of another ester with the same molecular formula.

Strategy and Solution

There is only one other ester with this molecular formula. Its structural formula is

$$H-\overset{O}{\overset{\|}{C}}-O-CH_2-CH_3$$

Ethyl formate

Problem 1.6

Draw structural formulas for the four esters with the molecular formula $C_4H_8O_2$.

Summary

⊍WL End-of-chapter problems identified in blue are assignable in **GOB OWL.**

Section 1.1 What Is Organic Chemistry?
Problem 1.7

- Organic Chemistry is the study of compounds containing carbon.

Section 1.2 Where Do We Obtain Organic Compounds?

- Chemists obtain organic compounds either by isolation from plant and animal sources or by synthesis in the laboratory.

Section 1.3 How Do We Write Structural Formulas of Organic Compounds?

- Carbon normally forms four bonds and has no unshared pairs of electrons. Its four bonds may be four single bonds, two single bonds and one double bond, or one single bond and one triple bond.

- Nitrogen normally forms three bonds and has one unshared pair of electrons. Its bonds may be three single bonds, one single bond and one double bond, or one triple bond.
- Oxygen normally forms two bonds and has two unshared pairs of electrons. Its bonds may be two single bonds or one double bond.

Section 1.4 What Is a Functional Group?
Problem 1.32

- **A Functional group** is a site of chemical reactivity; a particular functional group, in whatever compound it is found, always undergoes the same types of chemical reactions.
- In addition, functional groups are the characteristic structural units by which we both classify and name organic compounds. Important functional groups include the **hydroxyl group** of 1°, 2°, and 3° alcohols; the **amino group** of 1°, 2°, and 3° amines; the **carbonyl group** of aldehydes and ketones; the **carboxyl group** of carboxylic acids; and the **ester group.**

Problems

■ Indicates problems that are assignable in **GOB OWL.**

Blue numbered problems are applied.

Go to this book's companion website at **www.cengage. com/chemistry/bettelheim** for interactive versions of the How To tutorials and Active Figures, and to quiz yourself on this chapter.

Section 1.1 What Is Organic Chemistry?

1.7 ■ Answer true or false.
   (a) All organic compounds contain one or more atoms of carbon.
   (b) The majority of organic compounds are built from carbon, hydrogen, oxygen, and nitrogen.
   (c) By number of atoms, carbon is the most abundant element in the Earth's crust.
   (d) Most organic compounds are soluble in water.

Section 1.2 Where Do We Obtain Organic Compounds?

1.8 Answer true or false.
   (a) Organic compounds can only be synthesized in living organisms.
   (b) Organic compounds synthesized in the laboratory have the same chemical and physical properties as those synthesized in living organisms.
   (c) Chemists have synthesized many organic compounds that are not found in nature.

1.9 Is there any difference between vanillin made synthetically and vanillin extracted from vanilla beans?

1.10 Suppose that you are told that only organic substances are produced by living organisms. How would you rebut this assertion?

1.11 What important experiment did Wöhler carry out in 1828?

Section 1.3 How Do We Write Structural Formulas of Organic Compounds?

1.12 Answer true or false.
   (a) In organic compounds, carbon normally has four bonds and no unshared pairs of electrons.
   (b) When found in organic compounds, nitrogen normally has three bonds and one unshared pair of electrons.
   (c) The most common bond angles about carbon in organic compounds are approximately 109.5° and 180°.

1.13 List the four principal elements that make up organic compounds and give the number of bonds each typically forms.

1.14 Think about the types of substances in your immediate environment, and make a list of those that are organic—for example, textile fibers. We will ask you to return to this list later in the course and to refine, correct, and possibly expand it.

1.15 How many electrons are in the valence shell of each of the following atoms? Write a Lewis dot structure for an atom of each element. (*Hint*: Use the Periodic Table.)
   (a) Carbon   (b) Oxygen
   (c) Nitrogen   (d) Fluorine

■ **Problems assignable in GOB OWL**

1.16 What is the relationship between the number of electrons in the valence shell of each of the following atoms and the number of covalent bonds it forms?

(a) Carbon   (b) Oxygen

(c) Nitrogen   (d) Hydrogen

1.17 Write Lewis structures for these compounds. Show all valence electrons. None of them contains a ring of atoms. (*Hint*: Remember that carbon has four bonds, nitrogen has three bonds and one unshared pair of electrons, oxygen has two bonds and two unshared pairs of electrons, and each halogen has one bond and three unshared pairs of electrons.)

(a) H_2O_2 (b) N_2H_4

      Hydrogen peroxide   Hydrazine

(c) CH_3OH (d) CH_3SH

      Methanol   Methanethiol

(e) CH_3NH_2 (f) CH_3Cl

      Methylamine   Chloromethane

1.18 Write Lewis structures for these compounds. Show all valence electrons. None of them contains a ring of atoms.

(a) CH_3OCH_3 (b) C_2H_6

      Dimethyl ether   Ethane

(c) C_2H_4 (d) C_2H_2

      Ethylene   Acetylene

(e) CO_2 (f) CH_2O

      Carbon dioxide   Formaldehyde

(g) H_2CO_3 (h) CH_3COOH

      Carbonic acid   Acetic acid

1.19 Write Lewis structures for these ions.

(a) HCO_3^- (b) CO_3^{2-}

      Bicarbonate ion   Carbonate ion

(c) CH_3COO^- (d) Cl^-

      Acetate ion   Chloride ion

1.20 Why are the following molecular formulas impossible?

(a) CH_5 (b) C_2H_7

Review of the VSEPR Model

1.21 Explain how to use the valence-shell electron-pair repulsion (VSEPR) model to predict bond angles and geometry about atoms of carbon, oxygen, and nitrogen.

1.22 Suppose you forget to take into account the presence of the unshared pair of electrons on nitrogen in the molecule NH_3. What would you then predict for the H—N—H bond angles and the geometry (bond angles and shape) of ammonia?

1.23 Suppose you forget to take into account the presence of the two unshared pairs of electrons on the oxygen atom of ethanol, CH_3CH_2OH. What would you then predict for the C—O—H bond angle and the geometry of ethanol?

1.24 Use the VSEPR model to predict the bond angles and geometry about each highlighted atom. (*Hint*: Remember to take into account the presence of unshared pairs of electrons.)

1.25 Use the VSEPR model to predict the bond angles about each highlighted atom.

Section 1.4 What Is a Functional Group?

1.26 Answer true or false.

(a) A functional group is a group of atoms in an organic molecule that undergoes a predictable set of chemical reactions.

(b) The functional group of an alcohol, an aldehyde, and a ketone have in common the fact that each contains a single oxygen atom.

(c) A primary alcohol has one —OH group, a secondary alcohol has two —OH groups, and a tertiary alcohol has three —OH groups.

(d) There are two alcohols with the molecular formula C_3H_8O.

(e) There are three amines with the molecular formula C_3H_9N.

(f) Aldehydes, ketones, carboxylic acids, and esters all contain a carbonyl group.

(g) A compound with the molecular formula of C_3H_6O may be either an aldehyde, a ketone, or a carboxylic acid.

(h) Bond angles about the carbonyl carbon of an aldehyde, a ketone, a carboxylic acid, and an ester are all approximately 109.5°.

(i) The molecular formula of the smallest aldehyde is C_3H_6O, and that of the smallest ketone is also C_3H_6O.

(j) The molecular formula of the smallest carboxylic acid is $C_2H_4O_2$.

1.27 What is meant by the term *functional group*?

1.28 List three reasons why functional groups are important in organic chemistry.

1.29 Draw Lewis structures for each of the following functional groups. Show all valence electrons in each functional group.

(a) A carbonyl group

(b) A carboxyl group

(c) A hydroxyl group

(d) A primary amino group

(e) An ester group

1.30 Complete the following structural formulas by adding enough hydrogens to complete the tetravalence of each carbon. Then write the molecular formula of each compound.

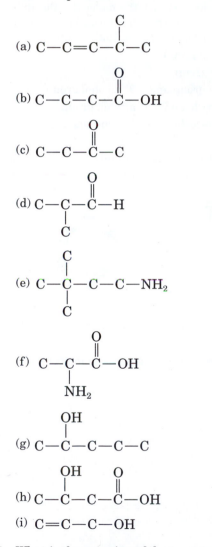

1.31 What is the meaning of the term *tertiary* (3°) when it is used to classify alcohols?

1.32 ■ Draw a structural formula for the one tertiary (3°) alcohol with the molecular formula $C_4H_{10}O$.

1.33 What is the meaning of the term *tertiary* (3°) when it is used to classify amines?

1.34 Draw condensed structural formulas for all compounds with the molecular formula C_4H_8O that contain a carbonyl group (there are two aldehydes and one ketone).

1.35 Draw structural formulas for each of the following:

(a) The four primary (1°) alcohols with the molecular formula $C_5H_{12}O$

(b) The three secondary (2°) alcohols with the molecular formula $C_5H_{12}O$

(c) The one tertiary (3°) alcohol with the molecular formula $C_5H_{12}O$

1.36 Draw structural formulas for the six ketones with the molecular formula $C_6H_{12}O$.

1.37 Draw structural formulas for the eight carboxylic acids with the molecular formula $C_6H_{12}O_2$.

1.38 Draw structural formulas for each of the following:

(a) The four primary (1°) amines with the molecular formula $C_4H_{11}N$

(b) The three secondary (2°) amines with the molecular formula $C_4H_{11}N$

(c) The one tertiary (3°) amine with the molecular formula $C_4H_{11}N$

Chemical Connections

1.39 (Chemical Connections 1A) How was Taxol discovered?

1.40 (Chemical Connections 1A) In what way does Taxol interfere with cell division?

Additional Problems

1.41 Use the VSEPR model to predict the bond angles about each atom of carbon, nitrogen, and oxygen in these molecules. (*Hint:* First add unshared pairs of electrons as necessary to complete the valence shell of each atom and then predict the bond angles.)

(a) $CH_3CH_2CH_2OH$

(b) $CH_3CH_2\overset{\displaystyle O}{\overset{\displaystyle \|}{C}}H$

(c) $CH_3CH{=}CH_2$

(d) $CH_3C{\equiv}CCH_3$

(e) $CH_3\overset{\displaystyle O}{\overset{\displaystyle \|}{C}}OCH_3$

(f) $CH_3\overset{\displaystyle CH_3}{\overset{\displaystyle |}{N}}CH_3$

1.42 Silicon is immediately below carbon in Group 4A of the Periodic Table. Predict the C—Si—C bond angles in tetramethylsilane, $(CH_3)_4Si$.

1.43 ■ Phosphorus is immediately below nitrogen in Group 5A of the Periodic Table. Predict the C—P—C bond angles in trimethylphosphine, $(CH_3)_3P$.

1.44 Draw the structure for a compound with the molecular formula

   (a) C_2H_6O that is an alcohol

   (b) C_3H_6O that is an aldehyde

   (c) C_3H_6O that is a ketone

   (d) $C_3H_6O_2$ that is a carboxylic acid

1.45 Draw structural formulas for the eight aldehydes with the molecular formula $C_6H_{12}O$.

1.46 Draw structural formulas for the three tertiary (3°) amines with the molecular formula $C_5H_{13}N$.

1.47 ■ Which of these covalent bonds are polar and which are nonpolar?

   (a) C—C   (b) C=C   (c) C—H   (d) C—O

   (e) O—H   (f) C—N   (g) N—H   (h) N—O

1.48 Of the bonds in Problem 1.50, which is the most polar? Which is the least polar?

1.49 Using the symbol $\delta+$ to indicate a partial positive charge and $\delta-$ to indicate a partial negative charge, indicate the polarity of the most polar bond (or bonds, if two or more have the same polarity) in each of the following molecules.

   (a) CH_3OH (b) CH_3NH_2

   (c) $HSCH_2CH_2NH_2$ (d) $CH_3\overset{\overset{\textstyle O}{\|}}{C}CH_3$

   (e) $H\overset{\overset{\textstyle O}{\|}}{C}H$ (f) $CH_3\overset{\overset{\textstyle O}{\|}}{C}OH$

Looking Ahead

1.50 ■ Identify the functional group(s) in each compound.

   (a) $CH_3CH_2\overset{\overset{\textstyle O}{\|}}{C}CH_3$

       2-Butanone
      (a solvent for
    paints and lacquers)

   (b) $HO\overset{\overset{\textstyle O}{\|}}{C}CH_2CH_2CH_2CH_2\overset{\overset{\textstyle O}{\|}}{C}OH$

      Hexanedioic acid
    (the second component
      of nylon-66)

   (c) $H_2NCH_2CH_2CH_2CH_2\overset{\overset{\textstyle O}{\|}}{C}\underset{\underset{\textstyle NH_2}{|}}{C}HOH$

         Lysine
    (one of the 20 amino acid
    building blocks of proteins)

   (d) $HOCH_2\overset{\overset{\textstyle O}{\|}}{C}CH_2OH$

      Dihydroxyacetone
    (a component of several
    artificial tanning lotions)

1.51 Consider molecules with the molecular formula $C_4H_8O_2$. Write the structural formula for a molecule with this molecular formula that contains

   (a) A carboxyl group

   (b) An ester group

   (c) A ketone group and a 2° alcohol group

   (d) An aldehyde and a 3° alcohol group

   (e) A carbon–carbon double bond and a 1° alcohol group.

1.52 Following is a structural formula and a ball-and-stick model of benzene, C_6H_6.

   (a) Predict each H—C—C and C—C—C bond angle in benzene.

   (b) Predict the shape of a benzene molecule.

Alkanes

A petroleum refinery. Petroleum, along with natural gas, provides nearly 90% of the organic materials for the synthesis and manufacture of synthetic fibers, plastics, drugs, dyes, adhesives, paints, and a multitude of other products.

J. L. Bohin / Photo Researchers, Inc.

2.1 How Do We Write Structural Formulas of Alkanes?

In this chapter, we examine the physical and chemical properties of **alkanes,** the simplest type of organic compounds. Actually, alkanes are members of a larger class of organic compounds called hydrocarbons. A **hydrocarbon** is a

Alkane A saturated hydrocarbon whose carbon atoms are arranged in a chain

Hydrocarbon A compound that contains only carbon and hydrogen atoms

OWL Online homework for this chapter may be assigned in **GOB OWL.**

FIGURE 2.1 The four classes of hydrocarbons.

compound composed of only carbon and hydrogen. Figure 2.1 shows the four classes of hydrocarbons, along with the characteristic type of bonding between carbon atoms in each class. Alkanes are **saturated hydrocarbons;** that is, they contain only carbon–carbon single bonds. Saturated in this context means that each carbon in the hydrocarbon has the maximum number of hydrogens bonded to it. A hydrocarbon that contains one or more carbon–carbon double bonds, triple bonds, or benzene rings is classified as an **unsaturated hydrocarbon.** We study alkanes (saturated hydrocarbons) in this chapter and alkenes, alkynes, and arenes (unsaturated hydrocarbons) in Chapters 3 and 4.

We often refer to alkanes as **aliphatic hydrocarbons** because the physical properties of the higher members of this class resemble those of the long carbon-chain molecules we find in animal fats and plant oils (Greek: *aleiphar*, fat or oil).

Methane, CH_4, and ethane, C_2H_6, are the first two members of the alkane family. Figure 2.2 shows Lewis structures and ball-and-stick models for these molecules. The shape of methane is tetrahedral, and all H—C—H bond angles are 109.5°. Each carbon atom in ethane is also tetrahedral, and the bond angles in it are all approximately 109.5° as well. Although the three-dimensional shapes of larger alkanes are more complex than those of methane and ethane, the four bonds about each carbon atom are still arranged in a tetrahedral manner, and all bond angles are still approximately 109.5°.

The next members of the alkane family are propane, butane, and pentane. In the following representations, these hydrocarbons are drawn as condensed structural formulas, which show all carbons and hydrogens. They can also be drawn in a more abbreviated form called a **line-angle formula.** In this type of representation, a line represents a carbon–carbon

Saturated hydrocarbon A hydrocarbon that contains only carbon–carbon single bonds

Aliphatic hydrocarbon An alkane

Line-angle formula An abbreviated way to draw structural formulas in which each vertex and line terminus represents a carbon atom and each line represents a bond

FIGURE 2.2 Methane and ethane.

Methane   Ethane

Ball-and-stick model			
Line-angle formula			
Condensed structural formula	$CH_3CH_2CH_3$	$CH_3CH_2CH_2CH_3$	$CH_3CH_2CH_2CH_2CH_3$
	Propane	Butane	Pentane

bond and a vertex represents a carbon atom. A line ending in space represents a —CH_3 group. To count hydrogens from a line-angle formula, you simply add enough hydrogens in your mind to give each carbon its required four bonds. Chemists use line-angle formulas because they are easier and faster to draw than condensed structural formulas.

Structural formulas for alkanes can be written in yet another condensed form. For example, the structural formula of pentane contains three CH_2 **(methylene)** groups in the middle of the chain. We can group them together and write the structural formula $CH_3(CH_2)_3CH_3$. Table 2.1 gives names and molecular formulas for the first ten alkanes with unbranched chains. Note that the names of all these alkanes end in "-ane." We will have more to say about naming alkanes in Section 2.3.

Butane, $CH_3CH_2CH_2CH_3$, is the fuel in this lighter. Butane molecules are present in both the liquid and gaseous states in the lighter.

Example 2.1 Drawing Line-Angle Formulas

Table 2.1 gives the condensed structural formula for hexane. Draw a line-angle formula for this alkane, and number the carbons on the chain beginning at one end and proceeding to the other end.

Strategy and Solution

Hexane contains six carbons in a chain. Its line-angle formula is

Problem 2.1

Following is a line-angle formula for an alkane. What are the name and the molecular formula of this alkane?

TABLE 2.1 The First Ten Alkanes with Unbranched Chains

Name	Molecular Formula	Condensed Structural Formula	Name	Molecular Formula	Condensed Structural Formula
methane	CH_4	CH_4	hexane	C_6H_{14}	$CH_3(CH_2)_4CH_3$
ethane	C_2H_6	CH_3CH_3	heptane	C_7H_{16}	$CH_3(CH_2)_5CH_3$
propane	C_3H_8	$CH_3CH_2CH_3$	octane	C_8H_{18}	$CH_3(CH_2)_6CH_3$
butane	C_4H_{10}	$CH_3(CH_2)_2CH_3$	nonane	C_9H_{20}	$CH_3(CH_2)_7CH_3$
pentane	C_5H_{12}	$CH_3(CH_2)_3CH_3$	decane	$C_{10}H_{22}$	$CH_3(CH_2)_8CH_3$

2.2 What Are Constitutional Isomers?

Constitutional isomers
Compounds with the same molecular formula but a different connectivity of their atoms

Constitutional isomers have also been called structural isomers, an older term that is still in use.

Constitutional isomers are compounds that have the same molecular formula but different structural formulas. By "different structural formulas" we mean that they differ in the kinds of bonds (single, double, or triple) and/or in the connectivity of their atoms. For the molecular formulas CH_4, C_2H_6, and C_3H_8, only one connectivity of their atoms is possible, so there are no constitutional isomers for these molecular formulas. For the molecular formula C_4H_{10}, however, two structural formulas are possible: In butane, the four carbons are bonded in a chain; in 2-methylpropane, three carbons are bonded in a chain with the fourth carbon is a branch on the chain. The two constitutional isomers with the molecular formula C_4H_{10} are drawn here both as condensed structural formulas and as line-angle formulas. Also shown are ball-and-stick models of each.

$CH_3CH_2CH_2CH_3$

Butane
(bp −0.5°C)

$$CH_3$$
$$CH_3CHCH_3$$

2-Methylpropane
(bp −11.6°C)

Butane and 2-methylpropane are different compounds and have different physical and chemical properties. Their boiling points, for example, differ by approximately 11°C.

In Section 1.4, we encountered several examples of constitutional isomers, although we did not call them that at the time. We saw, for example, that there are two alcohols with the molecular formula C_3H_8O, two primary amines with the molecular formula C_3H_9N, two aldehydes with the molecular formula C_4H_8O, and two carboxylic acids with the molecular formula $C_4H_8O_2$.

To determine whether two or more structural formulas represent constitutional isomers, write the molecular formula of each and then compare them. All compounds that have the same molecular formula but different structural formulas are constitutional isomers.

Example 2.2 Constitutional Isomerism

Do the structural formulas in each of the following sets represent the same compound or constitutional isomers? (*Hint*: You will find it helpful to redraw each molecule as a line-angle formula, which will make it easier for you to see similarities and differences in molecular structure.)

(a) $CH_3CH_2CH_2CH_2CH_2CH_3$ and $CH_3CH_2CH_2$ (Each is C_6H_{14})
$$\qquad\qquad\qquad\qquad\qquad\qquad\qquad\qquad CH_2CH_2CH_3$$

(b)
$$\qquad\qquad CH_3 \qquad\qquad\qquad CH_3$$
CH_3CHCH_2CH and $CH_3CH_2CHCHCH_3$ (Each is C_7H_{16})
$$\qquad\; CH_3 \quad CH_3 \qquad\qquad\qquad CH_3$$

Strategy

First, find the longest chain of carbon atoms. It makes no difference whether the chain is drawn as straight or bent; as structural formulas are drawn in this problem, there is no attempt to show three-dimensional shapes. Second, number the longest chain from the end nearest the first branch. Third, compare the lengths of the two chains and the sizes and locations of any branches. Structural formulas that have the same molecular formula and the same connectivity of their atoms represent the same compound; those that have the same molecular formula but different connectivities of their atoms represent constitutional isomers.

Solution

(a) Each structural formula has an unbranched chain of six carbons; they are identical and represent the same compound.

$$\overset{1}{}\overset{2}{}\overset{3}{}\overset{4}{}\overset{5}{}\overset{6}{}$$

$$\overset{1}{C}H_3\overset{2}{C}H_2\overset{3}{C}H_2\overset{4}{C}H_2\overset{5}{C}H_2\overset{6}{C}H_3 \quad \text{and} \quad \overset{1}{C}H_3\overset{2}{C}H_2\overset{3}{C}H_2$$

$$\overset{4}{C}H_2\overset{5}{C}H_2\overset{6}{C}H_3$$

(b) Each structural formula has the same molecular formula, C_7H_{16}. In addition, each has a chain of five carbons with two CH_3 branches. Although the branches are identical, they are at different locations on the chains. Therefore, these structural formulas represent constitutional isomers.

$$\overset{5}{C}H_3 \quad \overset{}{C}H_3$$
$$\overset{1}{C}H_3\overset{2}{C}H\overset{3}{C}H_2\overset{4}{C}H$$
$$\overset{}{C}H_3$$

and

$$\overset{}{C}H_3$$
$$\overset{5}{C}H_3\overset{4}{C}H_2\overset{3}{C}H\overset{2}{C}H\overset{1}{C}H_3$$
$$\overset{}{C}H_3$$

Problem 2.2

Do the line-angle formulas in each of the following sets represent the same compound or constitutional isomers?

(a)   and

(b)   and

Example 2.3 Constitutional Isomerism

Draw line-angle formulas for the five constitutional isomers with the molecular formula C_6H_{14}.

Strategy

In solving problems of this type, you should devise a strategy and then follow it. Here is one possible strategy. First, draw a line-angle formula for the constitutional isomer with all six carbons in an unbranched chain. Then, draw line-angle formulas for all constitutional isomers with five carbons in a chain and one carbon as a branch on the chain. Finally, draw line-angle formulas for all constitutional isomers with four carbons in a chain and two carbons as branches.

Solution

Here are line-angle formulas for all constitutional isomers with six, five, and four carbons in the longest chain. No constitutional isomers for C_6H_{14} having only three carbons in the longest chain are possible.

Six carbons in an
unbranched chain

Five carbons in a chain;
one carbon as a branch

Four carbons in a chain;
two carbons as branches

Problem 2.3

Draw structural formulas for the three constitutional isomers with the molecular formula C_5H_{12}.

Molecular Formula	Number of Constitutional Isomers
CH_4	1
C_5H_{12}	3
$C_{10}H_{22}$	75
$C_{15}H_{32}$	4347
$C_{25}H_{52}$	36,797,588
$C_{30}H_{62}$	4,111,846,763

The ability of carbon atoms to form strong, stable bonds with other carbon atoms results in a staggering number of constitutional isomers, as the following table shows:

Thus, for even a small number of carbon and hydrogen atoms, a very large number of constitutional isomers is possible. In fact, the potential for structural and functional group individuality among organic molecules made from just the basic building blocks of carbon, hydrogen, nitrogen, and oxygen is practically limitless.

2.3 How Do We Name Alkanes?

A. The IUPAC System

Ideally, every organic compound should have a name from which its structural formula can be drawn. For this purpose, chemists have adopted a set of rules established by the **International Union of Pure and Applied Chemistry (IUPAC).**

The IUPAC name for an alkane with an unbranched chain of carbon atoms consists of two parts: (1) a prefix that shows the number of carbon atoms in the chain and (2) the suffix **-ane,** which shows that the compound is a saturated hydrocarbon. Table 2.2 gives the prefixes used to show the presence of 1 to 20 carbon atoms.

The IUPAC chose the first four prefixes listed in Table 2.2 because they were well established long before the nomenclature was systematized. For example, the prefix *but-* appears in the name butyric acid, a compound of four carbon atoms formed by the air oxidation of butter fat (Latin: *butyrum,*

TABLE 2.2 Prefixes Used in the IUPAC System to Show the Presence of 1 to 20 Carbons in an Unbranched Chain

Prefix	Number of Carbon Atoms	Prefix	Number of Carbon Atoms	Prefix	Number of Carbon Atoms	Prefix	Number of Carbon Atoms
meth-	1	hex-	6	undec-	11	hexadec-	16
eth-	2	hept-	7	dodec-	12	heptadec-	17
prop-	3	oct-	8	tridec-	13	octadec-	18
but-	4	non-	9	tetradec-	14	nonadec-	19
pent-	5	dec-	10	pentadec-	15	eicos-	20

butter). The prefixes to show five or more carbons are derived from Latin numbers. Table 2.1 gives the names, molecular formulas, and condensed structural formulas for the first ten alkanes with unbranched chains.

IUPAC names of alkanes with branched chains consist of a parent name that shows the longest chain of carbon atoms and substituent names that indicate the groups bonded to the parent chain. For example,

4-Methyloctane

A substituent group derived from an alkane by removal of a hydrogen atom is called an **alkyl group** and is commonly represented by the symbol **R—**. Alkyl groups are named by dropping the -*ane* from the name of the parent alkane and adding the suffix -*yl*. Table 2.3 gives the names and condensed structural formulas for eight of the most common alkyl groups. The prefix *sec*- is an abbreviation for "secondary," meaning a carbon bonded to two other carbons. The prefix *tert*- is an abbreviation for "tertiary," meaning a carbon bonded to three other carbons.

The rules of the IUPAC system for naming alkanes are as follows:

1. The name for an alkane with an unbranched chain of carbon atoms consists of a prefix showing the number of carbon atoms in the parent chain and the suffix -*ane*.

Alkyl group A group derived by removing a hydrogen from an alkane; given the symbol R—

R— A symbol used to represent an alkyl group

TABLE 2.3 Names of the Eight Most Common Alkyl Groups

Name	Condensed Structural Formula	Name	Condensed Structural Formula
methyl	—CH₃	butyl	—CH₂CH₂CH₂CH₃
ethyl	—CH₂CH₃	isobutyl	—CH₂CHCH₃ | CH₃
propyl	—CH₂CH₂CH₃	*sec*-butyl	—CHCH₂CH₃ | CH₃
isopropyl	—CHCH₃ | CH₃	*tert*-butyl	CH₃ | —CCH₃ | CH₃

2. For branched-chain alkanes, take the longest chain of carbon atoms as the parent chain and its name becomes the root name.

3. Give each substituent on the parent chain a name and a number. The number shows the carbon atom of the parent chain to which the substituent is bonded. Use a hyphen to connect the number to the name.

2-Methylpropane

4. If there is one substituent, number the parent chain from the end that gives the substituent the lower number.

2-Methylpentane
(not 4-methylpentane)

5. If the same substituent occurs more than once, number the parent chain from the end that gives the lower number to the substituent encountered first. Indicate the number of times the substituent occurs by a prefix *di-*, *tri-*, *tetra-*, *penta-*, *hexa-*, and so on. Use a comma to separate position numbers.

2,4-Dimethylhexane
(not 3,5-dimethylhexane)

6. If there are two or more different substituents, list them in alphabetical order and number the chain from the end that gives the lower number to the substituent encountered first. If there are different substituents in equivalent positions on opposite ends of the parent chain, give the substituent of lower alphabetical order the lower number.

3-Ethyl-5-methylheptane
(not 3-methyl-5-ethylheptane)

7. Do not include the prefixes *di-*, *tri-*, *tetra-*, and so on or the hyphenated prefixes *sec-* and *tert-* in alphabetizing. Alphabetize the names of substituents first, and then insert these prefixes. In the following example, the alphabetizing parts are **ethyl** and **methyl**, not *ethyl* and *dimethyl*.

4-Ethyl-2,2-dimethylhexane
(not 2,2-dimethyl-4-ethylhexane)

Example 2.4 IUPAC Names of Alkanes

Write the molecular formula and IUPAC name for each alkane.

(a)   (b)

Strategy

If there is only one substituent on the parent chain, as in (a), number the parent chain from the end that gives the substituent the lowest possible number. If there are two or more substituents on the parent chain, as in (b), number the parent chain from the end that gives the substituent of lowest alphabetical order the lowest possible number.

Solution

The molecular formula of (a) is C_5H_{12}, and that of (b) is $C_{11}H_{24}$. In (a), number the longest chain from the end that gives the methyl substituent the lower number (rule 4). In (b), list isopropyl and methyl substituents in alphabetical order (rule 6).

(a)   2-Methylbutane   (b)   4-Isopropyl-2-methylheptane

Problem 2.4

Write the molecular formula and IUPAC name for each alkane.

(a)   (b)

B. Common Names

In the older system of **common nomenclature,** the total number of carbon atoms in an alkane, regardless of their arrangement, determines the name.

The first three alkanes are methane, ethane, and propane. All alkanes with the molecular formula C_4H_{10} are called butanes, all those with the molecular formula C_5H_{12} are called pentanes, and all those with the molecular formula C_6H_{14} are called hexanes. For alkanes beyond propane, **iso** indicates that one end of an otherwise unbranched chain terminates in a $(CH_3)_2CH$—group. Following are examples of common names:

CH₃CH₂CH₂CH₃   CH₃CHCH₃   CH₃CH₂CH₂CH₂CH₃   CH₃CH₂CHCH₃
Butane   Isobutane   Pentane   Isopentane

This system of common names has no way of handling other branching patterns and, therefore, for more complex alkanes, we must use the more flexible IUPAC system.

In this book, we concentrate on IUPAC names. From time to time, however, we also use common names, especially when chemists and biochemists use them almost exclusively in everyday discussions. When the text gives both IUPAC and common names for a compound, we will always give the

IUPAC name first, followed by the common name in parentheses. In this way, you should have no doubt about which name is which.

2.4 Where Do We Obtain Alkanes?

The two major sources of alkanes are natural gas and petroleum. **Natural gas** consists of approximately 90 to 95% methane, 5 to 10% ethane, and a mixture of other relatively low-boiling alkanes—chiefly propane, butane, and 2-methylpropane.

Petroleum is a thick, viscous liquid mixture of thousands of compounds, most of them hydrocarbons, formed from the decomposition of marine plants and animals. Petroleum and petroleum-derived products fuel automobiles, aircraft, and trains. They provide most of the greases and lubricants required for the machinery utilized by our highly industrialized society. Furthermore, petroleum, along with natural gas, provides nearly 90% of the organic raw materials for the synthesis and manufacture of synthetic fibers, plastics, detergents, drugs, dyes, and a multitude of other products.

The fundamental separation process in refining petroleum is fractional distillation (Figure 2.3). Practically all crude petroleum that enters a refinery goes to distillation units, where it is heated to temperatures as high as 370 to 425°C and separated into fractions. Each fraction contains a mixture of hydrocarbons that boils within a particular range.

2.5 What Are Cycloalkanes?

A hydrocarbon that contains carbon atoms joined to form a ring is called a **cyclic hydrocarbon.** When all carbons of the ring are saturated (only carbon–carbon single bonds are present), the hydrocarbon is called a **cycloalkane.** Cycloalkanes of ring sizes ranging from 3 to more than 30 carbon atoms are found in nature, and in principle there is no limit to ring

A petroleum fractional distillation tower.

Cycloalkane A saturated hydrocarbon that contains carbon atoms bonded to form a ring

Gases
Boiling point range below 20°C
(C_1–C_4 hydrocarbons; used as
fuels and reactants to make plastics)

Gasoline (naphthas) 20–200°C
(C_5–C_{12} hydrocarbons; used as
motor fuels and industrial solvents)

Kerosene 175–275°C (C_{12}–C_{16} hydrocarbons;
used for lamp oil, diesel fuel, starting
material for catalytic cracking)

Fuel oil 250–400°C (C_{15}–C_{18} hydrocarbons;
used for catalytic cracking, heating oil,
diesel fuel)

Lubricating oil above 350°C
(C_{16}–C_{20} hydrocarbons; used as lubricants)

Residue (asphalt)
(>C_{20} hydrocarbons)

Pipe still

Crude oil
and vapor
are preheated

FIGURE 2.3 Fractional distillation of petroleum. The lighter, more volatile fractions are removed from higher up the column; the heavier, less volatile fractions are removed from lower down.

Cyclobutane   Cyclopentane   Cyclohexane

FIGURE 2.4 Examples of cycloalkanes.

size. Five-membered (cyclopentane) and six-membered (cyclohexane) rings are especially abundant in nature; for this reason, we concentrate on them in this book.

Organic chemists rarely show all carbons and hydrogens when writing structural formulas for cycloalkanes. Rather, we use line-angle formulas to represent cycloalkane rings and represent each ring by a regular polygon having the same number of sides as there are carbon atoms in the ring. For example, we represent cyclobutane by a square, cyclopentane by a pentagon, and cyclohexane by a hexagon (Figure 2.4).

To name a cycloalkane, prefix the name of the corresponding open-chain alkane with *cyclo-*, and name each substituent on the ring. If there is only one substituent on the ring, there is no need to give it a location number. If there are two substituents, number the ring beginning with the substituent of lower alphabetical order.

Example 2.5 IUPAC Names of Cycloalkanes

Write the molecular formula and IUPAC name for each cycloalkane.

(a)   (b)

Strategy

For cycloalkanes, the parent name of the ring is the prefix *cyclo-*, plus the name of the alkane with the same number of carbon atoms as are in the ring. If there is only one substituent on the ring, as in (a), there is no need to give it a number. If there are two or more substituents on the ring, as in (b), number the carbon atoms of the ring beginning at the carbon with the substituent of lowest alphabetical order. If there are three or more substituents, number the atoms of the ring so as to give the substituents the lowest set of numbers and then list them in alphabetical order.

Solution

(a) The molecular formula of this compound is C_8H_{16}. Because there is only one substituent, there is no need to number the atoms of the ring. The IUPAC name of this cycloalkane is isopropylcyclopentane.
(b) The molecular formula is $C_{11}H_{22}$. To name this compound, first number the atoms of the cyclohexane ring beginning with *tert*-butyl, the substituent of lower alphabetical order (remember, alphabetical order here is determined by the *b* of butyl, and not by the *t* of *tert*-). The name of this cycloalkane is 1-*tert*-butyl-4-methylcyclohexane.

Problem 2.5

Write the molecular formula and IUPAC name for each cycloalkane.

(a)   (b)   (c)

2.6 What Are the Shapes of Alkanes and Cycloalkanes?

In this section, we concentrate on ways to visualize molecules as three-dimensional objects and to visualize bond angles and relative distances between various atoms and functional groups within a molecule. We urge you to build molecular models of these compounds and to study and manipulate those models. Organic molecules are three-dimensional objects, and it is essential that you become comfortable in dealing with them as such.

A. Alkanes

Although the **VSEPR** model gives us a way to predict the geometry about each carbon atom in an alkane, it provides us with no information about the three-dimensional shape of an entire molecule. There is, in fact, free rotation about each carbon–carbon bond in an alkane. As a result, even a molecule as simple as ethane has an infinite number of possible three-dimensional shapes, or **conformations.**

Figure 2.5 shows three conformations for a butane molecule. Conformation (a) is the most stable because the methyl groups at the ends of the four-carbon chain are farthest apart. Conformation (b) is formed by a rotation of 120° about the single bond joining carbons 2 and 3. In this conformation, some crowding of groups occurs because the two methyl groups are closer together than they are in conformation (a). Rotation about the C_2—C_3 single bond by another 60° gives conformation (c), which is the most crowded because the two methyl groups face each other.

Figure 2.5 shows only three of the possible conformations for a butane molecule. In fact, there are an infinite number of possible conformations that differ only in the angles of rotation about the various C—C bonds within the molecule. In an actual sample of butane, the conformation of each molecule constantly changes as a result of the molecule's collisions with other butane molecules and with the walls of the container. Even so, at any given time, a majority of butane molecules are in the most stable, fully extended conformation. There are the fewest butane molecules in the most crowded conformation.

To summarize, for any alkane (except, of course, for methane), there are an infinite number of conformations. The majority of molecules in any sample will be in the least crowded conformation; the fewest will be in the most crowded conformation.

At this point, you should review the use of the valence-shell electron-pair repulsion (VSEPR) model to predict bond angles and shapes of molecules.

Conformation Any three-dimensional arrangement of atoms in a molecule that results from rotation about a single bond

(a) Least crowded conformation; methyl groups are farthest apart.

(b) Intermediate crowding; methyl groups are closer to each other.

(c) Most crowded conformation; methyl groups are closest to each other.

ACTIVE FIGURE 2.5 Three conformations of a butane molecule. **Go to this book's companion website at www.cengage.com/chemistry/ bettelheim to explore an interactive version of this figure.**

B. Cycloalkanes

We limit our discussion to the conformations of cyclopentanes and cyclo-hexanes because they are the carbon rings most commonly found in nature. Nonplanar or puckered conformations are favored in all cycloalkanes larger than cyclopropane.

Cyclopentane

The most stable conformation of cyclopentane is the **envelope conforma-tion** shown in Figure 2.6. In it, four carbon atoms are in a plane, and the fifth carbon atom is bent out of the plane, like an envelope with its flap bent upward. All bond angles in cyclopentane are approximately 109.5°.

Cyclohexane

The most stable conformation of cyclohexane is the **chair conformation** (Figure 2.7), in which all bond angles are approximately 109.5°.

In a chair conformation, the 12 C—H bonds are arranged in two different orientations. Six of them are **axial bonds,** and the other six are **equatorial bonds.** One way to visualize the difference between these two types of bonds is to imagine an axis running through the center of the chair (Figure 2.8). Axial bonds are oriented parallel to this axis. Three of the axial bonds point up; the other three point down. Notice also that axial bonds alternate, first up and then down, as you move from one carbon to the next.

FIGURE 2.6 The most stable conformation of cyclopentane.

Equatorial position A position on a chair conformation of a cyclohexane ring that extends from the ring roughly perpendicular to the imaginary axis of the ring

Axial position A position on a chair conformation of a cyclohexane ring that extends from the ring parallel to the imaginary axis of the ring

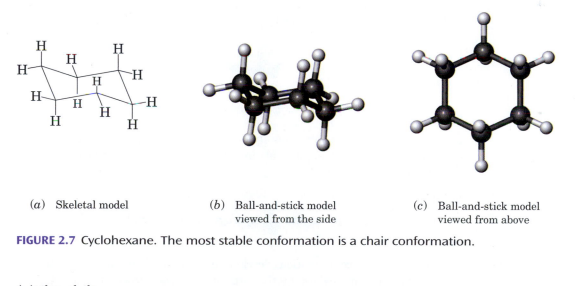

(a) Skeletal model

(b) Ball-and-stick model viewed from the side

(c) Ball-and-stick model viewed from above

FIGURE 2.7 Cyclohexane. The most stable conformation is a chair conformation.

Axis through the center of the ring

(a) Ball-and-stick model showing all 12 hydrogens

(b) The six equatorial C—H bonds shown in red

(c) The six axial C—H bonds shown in blue

FIGURE 2.8 Chair conformation of cyclohexane showing equatorial and axial C—H bonds.

Equatorial bonds are oriented approximately perpendicular to the imaginary axis of the ring and also alternate first slightly up and then slightly down as you move from one carbon to the next. Notice also that if the axial bond on a carbon points upward, the equatorial bond on that carbon points slightly downward. Conversely, if the axial bond on a particular carbon points downward, the equatorial bond on that carbon points slightly upward.

Finally, notice that each equatorial bond is oriented parallel to two ring bonds on opposite sides of the ring. A different pair of parallel C—H bonds is shown in each of the following structural formulas, along with the two ring bonds to which each pair is parallel.

Example 2.6 Chair Conformations of Cyclohexanes

Following is a chair conformation of methylcyclohexane showing a methyl group and one hydrogen. Indicate by a label whether each is equatorial or axial.

Strategy

Equatorial bonds are approximately perpendicular to the imaginary axis of the ring and form an equator about the ring. Axial bonds are parallel to the imaginary axis of the ring.

Solution

The methyl group is axial, and the hydrogen is equatorial.

Problem 2.6

Following is a chair conformation of cyclohexane with carbon atoms numbered 1 through 6. Draw methyl groups that are equatorial on carbons 1, 2, and 4.

Suppose that —CH_3 or another group on a cyclohexane ring may occupy either an equatorial or an axial position. Chemists have discovered that a six-membered ring is more stable when the maximum number of substituent groups are equatorial. Perhaps the simplest way to confirm this relationship is to examine molecular models. Figure 2.9(a) shows a space-filling model of methylcyclohexane with the methyl group in an equatorial position. In this position, the methyl group is as far away as possible from other atoms of the ring. When methyl is axial [Figure 2.9(b)], it quite literally bangs into two

The Poisonous Puffer Fish

Nature is by no means limited to carbon in six-membered rings. Tetrodotoxin, one of the most potent toxins known, is composed of a set of interconnected six-membered rings, each in a chair conformation. All but one of these rings contains atoms other than carbon.

Tetrodotoxin is produced in the liver and ovaries of many species of *Tetraodontidae*, one of which is the puffer fish, so called because it inflates itself to an almost spherical spiny ball when alarmed. It is evidently a species highly preoccupied with defense, but the Japanese are not put off by its prickly appearance. They regard the puffer, called *fugu* in Japanese, as a delicacy. To serve it in a public restaurant, a chef must be registered as sufficiently skilled in removing the toxic organs so as to make the flesh safe to eat.

Tetrodotoxin blocks the sodium ion channels, which are essential for neurotransmission (Section 16.3). This

A puffer fish with its body inflated.

blockage prevents communication between neurons and muscle cells and results in weakness, paralysis, and eventual death.

Tetrodotoxin

(a) Equatorial methylcyclohexane   (b) Axial methylcyclohexane

FIGURE 2.9 Methylcyclohexane. The three hydrogens of the methyl group are shown in green to make them stand out more clearly.

hydrogen atoms on the top side of the ring. Thus the more stable conformation of a substituted cyclohexane ring has the substituent group(s) as equatorial.

2.7 What Is *Cis-Trans* Isomerism in Cycloalkanes?

Cycloalkanes with substituents on two or more carbons of the ring show a type of isomerism called **cis-trans** isomerism. Cycloalkane *cis-trans* isomers have (1) the same molecular formula and (2) the same connectivity of

Cis-trans isomers Isomers that have the same connectivity of their atoms but a different arrangement of their atoms in space due to the presence of either a ring or a carbon–carbon double bond their atoms, but (3) a different arrangement of their atoms in space because of restricted rotation around the carbon–carbon single bonds of the ring. We study *cis-trans* isomerism in cycloalkanes in this chapter and that of alkenes in Chapter 3.

We can illustrate *cis-trans* isomerism in cycloalkanes by using 1,2-dimethyl-cyclopentane as an example. In the following structural formulas, the cyclopentane ring is drawn as a planar pentagon viewed through the plane of the ring. (In determining the number of *cis-trans* isomers in a substituted cycloalkane, it is adequate to draw the cycloalkane ring as a planar polygon.) Carbon–carbon bonds of the ring projecting toward you are shown as heavy lines. When viewed from this perspective, substituents bonded to the cyclopentane ring project above and below the plane of the ring. In one isomer of 1,2-dimethylcyclo-pentane, the methyl groups are on the same side of the ring (either both above or both below the plane of the ring); in the other isomer, they are on opposite sides of the ring (one above and one below the plane of the ring).

The prefix **cis** (Latin: on the same side) indicates that the substituents are on the same side of the ring; the prefix **trans** (Latin: across) indicates that they are on opposite sides of the ring.

> *Cis-trans* isomers have also been called geometric isomers.

> Planar representations of five- and six-membered rings are not spatially accurate because these rings normally exist as envelope and chair conformations. Planar representations are, however, adequate for showing *cis-trans* isomerism.

cis-1,2-Dimethyl-
cyclopentane

trans-1,2-Dimethyl-
cyclopentane

Alternatively, we can view the cyclopentane ring as a regular pentagon seen from above, with the ring in the plane of the page. Substituents on the ring then either project toward you (that is, they project up above the page) and are shown by solid wedges, or they project away from you (project down below the page) and are shown by broken wedges. In the following structural formulas, we show only the two methyl groups; we do not show hydrogen atoms of the ring.

> Occasionally hydrogen atoms are written before the carbon, H_3C— to avoid crowding or to emphasize the C—C bond, as in H_3C—CH_3.

cis-1,2-Dimethyl-
cyclopentane

trans-1,2-Dimethyl-
cyclopentane

> **Stereocenter** A tetrahedral atom, most commonly carbon, at which exchange of two groups produces a stereoisomer

We say that 1,2-dimethylcyclopentane has two stereocenters. A **stereocenter** is a tetrahedral atom, most commonly carbon, at which exchange of two groups produces a stereoisomer. Both carbons 1 and 2 of 1,2-dimethylcyclopentane, for example, are stereocenters; in this molecule, exchange of H and CH_3 groups at either stereocenter converts a *trans* isomer to a *cis* isomer, or vice versa.

> **Configuration** Refers to the arrangement of atoms about a stereocenter; that is, to the relative arrangement of parts of a molecule in space

Alternatively, we can refer to the stereoisomers of 1,2-dimethylcyclo-butane as having either a *cis* or a *trans* configuration. **Configuration** refers to the arrangement of atoms about a stereocenter. We say, for example, that exchange of groups at either stereocenter in the *cis* configuration gives the isomer with the *trans* configuration.

Cis and *trans* isomers are also possible for 1,2-dimethylcyclohexane. We can draw a cyclohexane ring as a planar hexagon and view it through the plane of the ring. Alternatively, we can view it as a regular hexagon viewed from above with substituent groups pointing toward us, shown by solid wedges, or pointing away from us, shown by broken wedges.

trans-1,4-Dimethylcyclohexane   *cis*-1,4-Dimethylcyclohexane

Because *cis-trans* isomers differ in the orientation of their atoms in space, they are **stereoisomers.** *Cis-trans* isomerism is one type of stereoisomerism. We will study another type of stereoisomerism, called enantiomerism, in Chapter 6.

Stereoisomers Isomers that have the same connectivity of their atoms but a different orientation of their atoms in space

Example 2.7 *Cis-Trans* Isomerism in Cycloalkanes

Which of the following cycloalkanes show *cis-trans* isomerism? For each that does, draw both isomers.

(a) Methylcyclopentane   (b) 1,1-Dimethylcyclopentane

(c) 1,3-Dimethylcyclobutane

Strategy

For a cycloalkane to show *cis-trans* isomerism, it must have at least two substituents, each on a different carbon of the ring.

Solution

(a) Methylcyclopentane does not show *cis-trans* isomerism; it has only one substituent on the ring.

(b) 1,1-Dimethylcyclobutane does not show *cis-trans* isomerism because only one arrangement is possible for the two methyl groups. Because both methyl groups are bonded to the same carbon, they must be *trans* to each other—one above the ring, the other below it.

(c) 1,3-Dimethylcyclobutane shows *cis-trans* isomerism. The two methyl groups may be *cis* or they may be *trans*.

cis-1,3-Dimethylcyclobutane   *trans*-1,3-Dimethylcyclobutane

Problem 2.7

Which of the following cycloalkanes show *cis-trans* isomerism? For each that does, draw both isomers.

(a) 1,3-Dimethylcyclopentane

(b) Ethylcyclopentane

(c) 1,3-Dimethylcyclohexane

2.8 What Are the Physical Properties of Alkanes?

The most important property of alkanes and cycloalkanes is their almost complete lack of polarity. The electronegativity difference between carbon and hydrogen is $2.5 - 2.1 = 0.4$ on the Pauling scale. Given this small difference, we classify a C—H bond as nonpolar covalent. Therefore, alkanes are nonpolar compounds and the only interactions between their molecules are the very weak London dispersion forces.

A. Melting and Boiling Points

The boiling points of alkanes are lower than those of almost any other type of compound with the same molecular weight. In general, both boiling and melting points of alkanes increase with increasing molecular weight (Table 2.4).

Alkanes containing 1 to 4 carbons are gases at room temperature. Alkanes containing 5 to 17 carbons are colorless liquids. High-molecular-weight alkanes (those containing 18 or more carbons) are white, waxy solids. Several plant waxes are high-molecular-weight alkanes. The wax found in apple skins, for example, is an unbranched alkane with the molecular formula $C_{27}H_{56}$. Paraffin wax, a mixture of high-molecular-weight alkanes, is used for wax candles, in lubricants, and to seal home-canned jams, jellies, and other preserves. Petrolatum, so named because it is derived from petroleum refining, is a liquid mixture of high-molecular-weight alkanes. It is sold as mineral oil and Vaseline, and is used as an ointment base in pharmaceuticals and cosmetics, and as a lubricant and rust preventive.

Alkanes that are constitutional isomers are different compounds and have different physical and chemical properties. Table 2.5 lists the boiling points of the five constitutional isomers with the molecular formula of C_6H_{14}. The boiling point of each branched-chain isomer is lower than that of hexane itself; the more branching, the lower the boiling point. These differences in boiling points are related to molecular shape in the following way. As branching increases, the alkane molecule becomes more compact and its surface area decreases. As surface area decreases, London dispersion forces act over a smaller surface area. Hence the attraction between molecules decreases and boiling point decreases. Thus, for any group of alkane constitutional isomers, the least-branched isomer generally has the highest boiling point and the most-branched isomer generally has the lowest boiling point.

Paraffin wax and mineral oil are mixtures of alkanes.

Charles D. Winters/Cengage Learning

TABLE 2.4 Physical Properties of Some Unbranched Alkanes

Name	Condensed Structural Formula	Molecular weight (amu)	Melting Point (°C)	Boiling Point (°C)	Density of Liquid (g/mL at 0°C)*
methane	CH_4	16.0	-182	-164	(a gas)
ethane	CH_3CH_3	30.1	-183	-88	(a gas)
propane	$CH_3CH_2CH_3$	44.1	-190	-42	(a gas)
butane	$CH_3(CH_2)_2CH_3$	58.1	-138	0	(a gas)
pentane	$CH_3(CH_2)_3CH_3$	72.2	-130	36	0.626
hexane	$CH_3(CH_2)_4CH_3$	86.2	-95	69	0.659
heptane	$CH_3(CH_2)_5CH_3$	100.2	-90	98	0.684
octane	$CH_3(CH_2)_6CH_3$	114.2	-57	126	0.703
nonane	$CH_3(CH_2)_7CH_3$	128.3	-51	151	0.718
decane	$CH_3(CH_2)_8CH_3$	142.3	-30	174	0.730

*For comparison, the density of H_2O is 1.000 g/mL at 4°C.

TABLE 2.5 Boiling Points of the Five Isomeric Alkanes with the Molecular Formula C_6H_{14}

Name	bp (°C)
hexane	68.7
3-methylpentane	63.3
2-methylpentane	60.3
2,3-dimethylbutane	58.0
2,2-dimethylbutane	49.7

Larger surface area, an increase in London dispersion forces, and a higher boiling point

Hexane (bp 68.7°)

Smaller surface area, a decrease in London dispersion forces, and a lower boiling point

2,2-Dimethylbutane (bp 49.7°)

B. Solubility: A Case of "Like Dissolves Like"

Because alkanes are nonpolar compounds, they are not soluble in water, which dissolves only ionic and polar compounds. Recall that water is a polar substance, and that its molecules associate with one another through hydrogen bonding. Alkanes do not dissolve in water because they cannot form hydrogen bonds with water. Alkanes, however, are soluble in each other, an example of "like dissolves like." Alkanes are also soluble in other nonpolar organic compounds, such as toluene and diethyl ether.

C. Density

The average density of the liquid alkanes listed in Table 2.4 is about 0.7 g/mL; that of higher-molecular-weight alkanes is about 0.8 g/mL. All liquid and solid alkanes are less dense than water (1.000 g/mL) and, because they are insoluble in water, they float on water.

Example 2.8 Physical Properties of Alkanes

Arrange the alkanes in each set in order of increasing boiling point.
(a) Butane, decane, and hexane
(b) 2-Methylheptane, octane, and 2,2,4-trimethylpentane

Strategy

The compounds in each set are alkanes, and the only forces of attraction between alkane molecules are very weak London dispersion forces. As the number of carbons in a hydrocarbon chain increases, London dispersion forces between chains increase and, therefore, boiling point also increases. For alkanes that are constitutional isomers, the strength of London dispersion forces between molecules depends on shape. The more compact the shape, the weaker the intermolecular forces of attraction and the lower the boiling point.

Solution

(a) All three compounds are unbranched alkanes. Decane has the longest carbon chain, the strongest London forces between its molecules, and the highest boiling point. Butane has the shortest carbon chain and the lowest boiling point.

Butane	Hexane	Decane
bp −0.5°C	bp 69°C	bp 174°C

(b) These three alkanes are constitutional isomers with the molecular formula C_8H_{18}. 2,2,4-Trimethylpentane is the most highly branched isomer and, therefore, has the smallest surface area and the lowest boiling point. Octane, the unbranched isomer, has the largest surface area and the highest boiling point.

2,2,4-Trimethylpentane	2-Methylheptane	Octane
(bp 99°C)	(bp 118°C)	(bp 126°C)

Problem 2.8

Arrange the alkanes in each set in order of increasing boiling point.
(a) 2-Methylbutane, pentane, and 2,2-dimethylpropane
(b) 3,3-Dimethylheptane, nonane, and 2,2,4-trimethylhexane

2.9 What Are the Characteristic Reactions of Alkanes?

The most important chemical property of alkanes and cycloalkanes is their inertness. They are quite unreactive toward normal ionic reaction conditions. Under certain conditions, however, alkanes react with oxygen, O_2. By far their most important reaction with oxygen is oxidation (combustion) to form carbon dioxide and water. They also react with bromine and chlorine to form halogenated hydrocarbons.

A. Reaction with Oxygen: Combustion

Oxidation of hydrocarbons, including alkanes and cycloalkanes, is the basis for their use as energy sources for heat [natural gas, liquefied petroleum gas (LPG), and fuel oil] and power (gasoline, diesel, and aviation fuel). Following are balanced equations for the complete combustion of methane, the major component of natural gas, and propane, the major component of LPG or bottled gas. The heat liberated when an alkane is oxidized to carbon dioxide and water is called its heat of combustion.

Chemical Connections 2B

Octane Rating: What Those Numbers at the Pump Mean

Gasoline is a complex mixture of C_6 to C_{12} hydrocarbons. The quality of gasoline as a fuel for internal combustion engines is expressed in terms of an octane rating. Engine knocking occurs when a portion of the air–fuel mixture explodes prior to the piston reaching the top of its stroke (usually as a result of heat developed during the compression) and independent of ignition by the spark plug. The resulting shockwave of the piston against the cylinder wall reverberates, creating a characteristic metallic "pinging" sound.

Two compounds were selected as reference fuels for rating gasoline quality. One of these, 2,2,4-trimethylpentane (isooctane) has very good antiknock properties and was assigned an octane rating of 100. Heptane, the other reference compound, has poor antiknock properties and was assigned an octane rating of 0.

The **octane rating** of a particular gasoline is the percent of 2,2,4-trimethylpentane in a mixture of 2,2,4-trimethylpentane and heptane that has antiknock properties equivalent to that of the test gasoline. For example, the antiknock properties of 2-methylhexane are the same as those of a mixture of 42% 2,2,4-trimethylpentane and 58% heptane; therefore, the octane rating of 2-methylhexane is 42. Ethanol, which is added to gasoline to produce gasohol, has an octane rating of 105. Octane itself has an octane rating of −20.

2,2,4-Trimethylpentane
(octane rating 100)

Heptane
(octane rating 0)

Typical octane ratings of commonly available gasolines.

$$CH_4 + 2O_2 \longrightarrow CO_2 + 2H_2O + 212 \text{ kcal/mol}$$
Methane

$$CH_3CH_2CH_3 + 5O_2 \longrightarrow 3CO_2 + 4H_2O + 530 \text{ kcal/mol}$$
Propane

B. Reaction with Halogens: Halogenation

If we mix methane with chlorine or bromine in the dark at room temperature, nothing happens. If, however, we heat the mixture to 100°C or higher or expose it to light, a reaction begins at once. The products of the reaction between methane and chlorine are chloromethane and hydrogen chloride. What occurs is a substitution reaction—in this case, the substitution of chlorine for hydrogen in methane.

$$CH_4 + Cl_2 \xrightarrow{\text{heat or light}} CH_3Cl + HCl$$
Methane   Chloromethane
(Methyl chloride)

If chloromethane is allowed to react with more chlorine, further chlorination produces a mixture of dichloromethane, trichloromethane, and tetrachloromethane.

$$CH_3Cl + Cl_2 \xrightarrow{\text{heat}} CH_2Cl_2 + HCl$$
Chloromethane   Dichloromethane
(Methyl chloride)   (Methylene chloride)

$$CH_2Cl_2 \xrightarrow[\text{heat}]{Cl_2} CHCl_3 \xrightarrow[\text{heat}]{Cl_2} CCl_4$$

Dichloromethane   Trichloromethane   Tetrachloromethane
(Methylene chloride)   (Chloroform)   (Carbon tetrachloride)

In the last equation, the reagent Cl_2 is placed over the reaction arrow and the equivalent amount of HCl formed is not shown. Placing reagents over reaction arrows and omitting by-products is commonly done to save space.

We derive IUPAC names of haloalkanes by naming the halogen atom as a substituent (*fluoro-, chloro-, bromo-,* and *iodo-*) and alphabetizing it along with other substituents. Common names consist of the common name of the alkyl group followed by the name of the halogen (chloride, bromide, and so forth) as a separate word. Dichloromethane (methylene chloride) is a widely used solvent for organic compounds.

Example 2.9 Halogenation of Alkanes

Write a balanced equation for the reaction of ethane with chlorine to form chloroethane, C_2H_5Cl.

Strategy

The reaction of ethane with chlorine results in the substitution of one of the hydrogen atoms of ethane by a chlorine atom.

Solution

$$CH_3CH_3 + Cl_2 \xrightarrow{\text{heat or light}} CH_3CH_2Cl + HCl$$

Ethane   Chloroethane
               (Ethyl chloride)

Problem 2.9

Reaction of propane with chlorine gives two products, each with the molecular formula C_3H_7Cl. Draw structural formulas for these two compounds, and give each an IUPAC name and a common name.

2.10 What Are Some Important Haloalkanes?

One of the major uses of haloalkanes is as intermediates in the synthesis of other organic compounds. Just as we can replace a hydrogen atom of an alkane, we can, in turn, replace the halogen atom by a number of other functional groups. In this way, we can construct more complex molecules. In contrast, alkanes that contain several halogens are often quite unreactive, a fact that has proved especially useful in the design of several classes of consumer products.

A. Chlorofluorocarbons

Of all the haloalkanes, the **chlorofluorocarbons (CFCs)** manufactured under the trade name Freons are the most widely known. CFCs are nontoxic, nonflammable, odorless, and noncorrosive. Originally, they

The Environmental Impact of Freons

Concern about the environmental impact of CFCs arose in the 1970s, when researchers found that more than 4.5×10^5 kg/yr of these compounds were being emitted into the atmosphere. In 1974, Sherwood Rowland and Mario Molina, both of the United States, announced their theory, which has since been amply confirmed, that these compounds destroy the stratospheric ozone layer. When released into the air, CFCs escape to the lower atmosphere. Because of their inertness, however, they do not decompose there. Slowly they find their way to the stratosphere, where they absorb ultraviolet radiation from the Sun and then decompose. As they do so, they set up chemical reactions that lead to the destruction of the stratospheric ozone layer, which shields the Earth against short-wavelength ultraviolet radiation from the Sun. An increase in short-wavelength ultraviolet radiation reaching the Earth is believed to promote the destruction of certain crops and agricultural species, and to increase the incidence of skin cancer in light-skinned individuals.

This concern prompted two conventions, one in Vienna in 1985 and one in Montreal in 1987, held by the United Nations Environmental Program. The 1987 meeting produced the Montreal Protocol, which set limits on the production and use of ozone-depleting CFCs and urged the complete phase-out of their production by 1996. This phase-out resulted in enormous costs for manufacturers and is not yet complete in developing countries.

Rowland, Molina, and Paul Crutzen (a Dutch chemist at the Max Planck Institute for Chemistry in Germany) were awarded the 1995 Nobel Prize for chemistry. As noted in the award citation by the Royal Swedish Academy of Sciences, "By explaining the chemical mechanisms that affect the thickness of the ozone layer, these three researchers have contributed to our salvation from a global environmental problem that could have catastrophic consequences."

The chemical industry has responded to this crisis by developing replacement refrigerants that have much lower ozone-depleting potential. The most prominent of these replacements are the hydrofluorocarbons (HFCs) and hydrochlorofluorocarbons (HCFCs).

HFC-134a   HCFC-141b

These compounds are much more chemically reactive in the atmosphere than the Freons and are destroyed before they reach the stratosphere. However, they cannot be used in air conditioners in 1994- and earlier-model cars.

seemed to be ideal replacements for the hazardous compounds such as ammonia and sulfur dioxide formerly used as heat-transfer agents in refrigeration systems. Among the CFCs most widely used for this purpose were trichlorofluoromethane (CCl_3F, Freon-11) and dichlorodifluoromethane (CCl_2F_2, Freon-12). The CFCs also found wide use as industrial cleaning solvents to prepare surfaces for coatings, to remove cutting oils and waxes from millings, and to remove protective coatings. In addition, they were employed as propellants for aerosol sprays.

B. Solvents

Several low-molecular-weight haloalkanes are excellent solvents in which to carry out organic reactions and to use as cleaners and degreasers. Carbon tetrachloride ("carbon tet") was the first of these compounds to find wide application, but its use for this purpose has since been discontinued because it is now known that carbon tet is both toxic and a carcinogen. Today, the most widely used haloalkane solvent is dichloromethane, CH_2Cl_2.

Summary of Key Questions

OWL End-of-chapter problems identified in blue are assignable in **GOB OWL**.

Section 2.1 How Do We Write Structural Formulas of Alkanes?

- A **hydrocarbon** is a compound that contains only carbon and hydrogen.
- A **saturated hydrocarbon** contains only single bonds. An **alkane** is a saturated hydrocarbon whose carbon atoms are arranged in an open chain.

Section 2.2 What Are Constitutional Isomers?

- **Constitutional isomers** have the same molecular formula but a different connectivity of their atoms.

Section 2.3 How Do We Name Alkanes?

Problem 2.22

- Alkanes are named according to a set of rules developed by the **International Union of Pure and Applied Chemistry (IUPAC).**
- The IUPAC name of an alkane consists of two parts: a prefix that tells the number of carbon atoms in the parent chain, and the ending **-ane.** Substituents derived from alkanes by removal of a hydrogen atom are called **alkyl groups** and denoted by the symbol **R—.**

Section 2.4 Where Do We Obtain Alkanes?

- **Natural gas** consists of 90 to 95% methane with lesser amounts of ethane and other lower-molecular-weight hydrocarbons.
- **Petroleum** is a liquid mixture of thousands of different hydrocarbons.

Section 2.5 What Are Cycloalkanes?

Problem 2.26

- A **cycloalkane** is an alkane that contains carbon atoms bonded to form a ring.
- To name a cycloalkane, prefix the name of the open-chain alkane with **cyclo-.**

Section 2.6 What Are the Shapes of Alkanes and Cycloalkanes?

- A **conformation** is any three-dimensional arrangement of the atoms of a molecule that results from rotation about a single bond.
- The lowest-energy conformation of cyclopentane is an **envelope conformation.**
- The lowest-energy conformation of cyclohexane is a **chair conformation.** In a chair conformation, six C—H bonds are **axial** and six C—H bonds are **equatorial.** A substituent on a six-membered ring is more stable when it is equatorial than when it is axial.

Section 2.7 What Is *Cis-Trans* Isomerism in Cycloalkanes?

- *Cis-trans* **isomers** of cycloalkanes have (1) the same molecular formula and (2) the same connectivity of their atoms, but (3) a different orientation of their atoms in space because of the restricted rotation around the C—C bonds of the ring.
- For *cis-trans* isomers of cycloalkanes, *cis* means that substituents are on the same side of the ring; *trans* means that they are on opposite sides of the ring.

Section 2.8 What Are the Physical Properties of Alkanes?

- Alkanes are nonpolar compounds, and the only forces of attraction between their molecules are London dispersion forces.
- At room temperature, low-molecular-weight alkanes are gases, higher-molecular-weight alkanes are liquids, and very-high-molecular-weight alkanes are waxy solids.
- For any group of alkane constitutional isomers, the least-branched isomer generally has the highest boiling point, and the most-branched isomer generally has the lowest boiling point.
- Alkanes are insoluble in water but soluble in each other and in other nonpolar organic solvents such as toluene. All liquid and solid alkanes are less dense than water.

Summary of Key Reactions

1. **Oxidation of Alkanes (Section 2.9A)** Oxidation of alkanes to carbon dioxide and water, an exothermic reaction, is the basis for our use of them as sources of heat and power.

$$CH_3CH_2CH_3 + 5O_2 \longrightarrow 3CO_2 + 4H_2O + 530 \text{ kcal/mol}$$
Propane

2. **Halogenation of Alkanes (Section 2.9B)** Reaction of an alkane with chlorine or bromine results in the substitution of a halogen atom for a hydrogen. **Problem 2.48**

$$CH_3CH_3 + Cl_2 \xrightarrow{\text{heat or light}} CH_3CH_2Cl + HCl$$
Ethane   Chloroethane
                                          (Ethyl chloride)

Problems

■ Indicates problems that are assignable in **GOB OWL**.

Blue numbered problems are applied.

Go to this book's companion website at **www.cengage. com/chemistry/bettelheim** for interactive versions of the How To tutorials and Active Figures, and to quiz yourself on this chapter.

Section 2.1 How Do We Write Structural Formulas of Alkanes?

2.10 Answer true or false.

(a) A hydrocarbon is composed only of the elements carbon and hydrogen.

(b) Alkanes are saturated hydrocarbons.

(c) The general formula of an alkane is C_nH_{2n+2}, where n is the number of carbons in the alkane.

(d) Alkenes and alkynes are unsaturated hydrocarbons.

2.11 Define:

(a) Hydrocarbon

(b) Alkane

(c) Saturated hydrocarbon

2.12 Why is it not accurate to describe an unbranched alkane as a "straight-chain" hydrocarbon?

2.13 What is meant by the term *line-angle formula* as applied to alkanes and cycloalkanes?

2.14 For each condensed structural formula, write a line-angle formula.

(a) $\underset{\underset{\displaystyle CH(CH_3)_2}{|}}{CH_3CH_2CHCHCH_2CHCH_3}$ with $\underset{CH_2CH_3}{|}$ and $\underset{CH_3}{|}$ on the respective carbons

(b) $\underset{\underset{\displaystyle CH_3}{|}}{\overset{\overset{\displaystyle CH_3}{|}}{CH_3CCH_3}}$

(c) $(CH_3)_2CHCH(CH_3)_2$

(d) $\underset{\underset{\displaystyle CH_2CH_3}{|}}{\overset{\overset{\displaystyle CH_2CH_3}{|}}{CH_3CH_2CCH_2CH_3}}$

(e) $(CH_3)_3CH$

(f) $CH_3(CH_2)_3CH(CH_3)_2$

2.15 Write the molecular formula for each alkane.

(a)　　　　　　　　(b)

(c)

Section 2.2 What Are Constitutional Isomers?

2.16 Answer true or false.

(a) Constitutional isomers have the same molecular formulas and the same connectivity of their atoms.

(b) There are two constitutional isomers with the molecular formula C_3H_8.

(c) There are four constitutional isomers with the molecular formula C_4H_{10}

(d) There are five constitutional isomers with the molecular formula C_5H_{12}.

2.17 Which statements are true about constitutional isomers?

(a) They have the same molecular formula.

(b) They have the same molecular weight.

(c) They have the same connectivity of their atoms.

(d) They have the same physical properties.

2.18 Each member of the following set of compounds is an alcohol; that is, each contains an —OH (hydroxyl group; see Section 1.4A). Which structural formulas represent the same compound, and which represent constitutional isomers?

(a)　　　　　　　　(b)

(c)　　　　　　　　(d)

(e) HO　　　　　　(f)

(g)　　　　　　　　(h)

2.19 Each member of the following set of compounds is either an aldehyde or ketone (Section 1.4A). Which structural formulas represent the same compound, and which represent constitutional isomers?

(a) (b) (c) (d) (e) (f) (g) (h)

2.20 Draw line-angle formulas for the nine constitutional isomers with the molecular formula C_7H_{16}.

Section 2.3 How Do We Name Alkanes?

2.21 Answer true or false.

(a) The parent name of an alkane is the name of the longest chain of carbon atoms in the alkane.

(b) Propyl and isopropyl groups are constitutional isomers.

(c) There are four alkyl groups with molecular formula C_4H_9.

2.22 ■ Name these alkyl groups:

(a) CH_3CH_2-

(b) CH_3CH- with CH_3 above

(c) CH_3CHCH_2- with CH_3 above

(d) CH_3C- with CH_3 above and CH_3 below

2.23 Write the IUPAC names for isobutane and isopentane.

Section 2.4 Where Do We Obtain Alkanes?

2.24 Answer true or false.

(a) The two major sources of alkanes the world over are petroleum and natural gas.

(b) The octane number of a particular gasoline is the number of grams of octane per liter of the fuel.

(c) Octane and 2,2,4-trimethylpentane are constitutional isomers and have the same octane number.

Section 2.5 What Are Cycloalkanes?

2.25 Answer true or false.

(a) Cycloalkanes are saturated hydrocarbons.

(b) Hexane and cyclohexane are constitutional isomers.

(c) The parent name of a cycloalkane is the name of the unbranched alkane with the same number of carbon atoms as are in the cycloalkane ring.

2.26 ■ Write the IUPAC names for these alkanes and cycloalkanes.

(a) $CH_3CHCH_2CH_2CH_3$ with CH_3 below

(b) $CH_3CHCH_2CH_2CHCH_3$ with CH_3 and CH_3 below

(c) $CH_3(CH_2)_4CHCH_2CH_3$ with CH_2CH_3 below

(d) (e) (f)

2.27 Write line-angle formulas for these alkanes and cycloalkanes.

(a) 2,2,4-Trimethylhexane

(b) 2,2-Dimethylpropane

(c) 3-Ethyl-2,4,5-trimethyloctane

(d) 5-Butyl-2,2-dimethylnonane

(e) 4-Isopropyloctane

(f) 3,3-Dimethylpentane

(g) *trans*-1,3-Dimethylcyclopentane

(h) *cis*-1,2-Diethylcyclobutane

Section 2.6 What Are the Shapes of Alkanes and Cycloalkanes?

2.28 Answer true or false.

(a) Conformations have the same molecular formula and the same connectivity, but differ in the three-dimensional arrangement of their atoms in space.

(b) In all conformations of ethane, propane, butane, and higher alkanes, all C—C—C and —C—H bond angles are approximately 109.5°.

(c) In a cyclohexane ring, if an axial bond is above the plane of the ring on a particular carbon, axial bonds on the two adjacent carbons are below the plane of the ring.

(d) In a cyclohexane ring, if an equatorial bond is above the plane of the ring on a particular carbon, equatorial bonds on the two adjacent carbons are below the plane of the ring.

(e) The more stable chair conformation of a cyclo-hexane ring has more substituent groups in equatorial positions.

2.29 The condensed structural formula of butane is $CH_3CH_2CH_2CH_3$. Explain why this formula does not show the geometry of the real molecule.

2.30 Draw a conformation of ethane in which hydrogen atoms on adjacent carbons are as far apart as possible. Also draw a conformation in which they are as close together as possible. In a sample of ethane molecules at room temperature, which conformation is the more likely?

Section 2.7 What Is *Cis-Trans* Isomerism in Cycloalkanes?

2.31 Answer true or false.

(a) *Cis* and *trans*-cycloalkanes have the same molecular formula but a different connectivity of their atoms.

(b) A *cis* isomer of a cycloalkane can be converted to its *trans* isomer by rotation about an appropriate carbon–carbon single bond.

(c) A *cis* isomer of a cycloalkane can be converted to its *trans* isomer by exchange of two groups at a stereocenter in the *cis*-cycloalkane

(d) Configuration refers to the arrangement in space of the atoms or groups of atoms at a stereocenter.

(e) *Cis*-1,4-dimethylcyclohexane and *trans*-1,4-dimethylcyclohexane are classified as conformations.

2.32 What structural feature of cycloalkanes makes *cis-trans* isomerism in them possible?

2.33 Is *cis-trans* isomerism possible in alkanes?

2.34 Name and draw structural formulas for the *cis* and *trans* isomers of 1,2-dimethylcyclopropane.

2.35 Name and draw structural formulas for the six cycloalkanes with the molecular formula C_5H_{10}. Include *cis-trans* isomers as well as constitutional isomers.

2.36 Why is equatorial methylcyclohexane more stable than axial methylcyclohexane?

Section 2.8 What Are the Physical Properties of Alkanes?

2.37 Answer true or false.

(a) Boiling points among alkanes with unbranched chains increase as the number of carbons in the chain increases.

(b) Alkanes that are liquid at room temperature are more dense than water.

(c) *Cis* and *trans* isomers have the same molecular formula, the same connectivity, and the same physical properties.

(d) Among alkane constitutional isomers, the least branched isomer generally has the lowest boiling point.

(e) Alkanes and cycloalkanes are insoluble in water.

(f) Liquid alkanes are soluble in each other.

2.38 In Problem 2.22, you drew structural formulas for the nine constitutional isomers with molecular formula C_7H_{16}. Predict which isomer has the lowest boiling point and which has the highest boiling point.

2.39 Which unbranched alkane (Table 2.4) has about the same boiling point as water? Calculate the molecular weight of this alkane, and compare it with the molecular weight of water.

2.40 What generalizations can you make about the densities of alkanes relative to the density of water?

2.41 What generalization can you make about the solubility of alkanes in water?

2.42 Suppose that you have samples of hexane and octane. Could you tell the difference by looking at them? What color would each be? How could you tell which is which?

2.43 As you can see from Table 2.4, each CH_2 group added to the carbon chain of an alkane increases its boiling point. This increase is greater going from CH_4 to C_2H_6 and from C_2H_6 to C_3H_8 than it is going from C_8H_{18} to C_9H_{20} or from C_9H_{20} to $C_{10}H_{22}$. What do you think is the reason for this difference?

2.44 How are the boiling points of hydrocarbons during petroleum refining related to their molecular weight?

Section 2.9 What Are the Characteristic Reactions of Alkanes?

2.45 Answer true or false.

(a) Combustion of alkanes is an endothermic reaction.

(b) The products of complete combustion of an alkane are carbon dioxide and water.

(c) Halogenation of an alkane converts it to a haloalkane.

2.46 Write balanced equations for the combustion of each of the following hydrocarbons. Assume that each is converted completely to carbon dioxide and water.

(a) Hexane

(b) Cyclohexane

(c) 2-Methylpentane

2.47 The heat of combustion of methane, a component of natural gas, is 212 kcal/mol. That of propane, a component of LP gas, is 530 kcal/mol. On a gram-for-gram basis, which hydrocarbon is the better source of heat energy?

2.48 ■ Draw structural formulas for these haloalkanes.

(a) Bromomethane

(b) Chlorocyclohexane

(c) 1,2-Dibromoethane

(d) 2-Chloro-2-methylpropane

(e) Dichlorodifluoromethane (Freon-12)

2.49 The reaction of chlorine with pentane gives a mixture of three chloroalkanes, each with the molecular formula $C_5H_{11}Cl$. Write a line-angle formula and the IUPAC name for each chloroalkane.

Section 2.10 What Are Some Important Haloalkanes?

2.50 Answer true or false.

(a) The Freons are members of a class of organic compounds called chlorofluorocarbons (CFCs)

(b) An advantage of Freons as heat-transfer agents in refrigeration systems, propellants in aerosol sprays, and solvents for industrial cleaning is that they are nontoxic, nonflammable, odorless, and noncorrosive.

(d) Freons in the stratosphere interact with ultraviolet radiation, and thereby set up chemical reactions that lead to the destruction of the stratospheric ozone layer.

(e) Alternative names for the important laboratory and industrial solvent CH_2Cl_2 are dichloromethane, methylene chloride, and chloroform.

Chemical Connections

2.51 (Chemical Connections 2A) How many rings in tetrodotoxin contain only carbon atoms? How many contain nitrogen atoms? How many contain two oxygen atoms?

2.52 (Chemical Connections 2B) What is an "octane rating"? What two reference hydrocarbons are used for setting the scale of octane ratings?

2.53 (Chemical Connections 2B) Octane has an octane rating of −20. Will it produce more or less engine knocking than heptane does?

2.54 (Chemical Connections 2B) Ethanol is added to gasoline to produce E-15 and E-85. It promotes more complete combustion of the gasoline and is an octane booster. Compare the heats of combustion of 2,2,4-trimethylpentane (1304 kcal/mol) and ethanol (327 kcal/mol). Which has the higher heat of combustion in kcal/mol? In kcal/g?

2.55 (Chemical Connections 2C) What are Freons? Why were they considered ideal compounds to use as heat-transfer agents in refrigeration systems? Give structural formulas of two Freons used for this purpose.

2.56 (Chemical Connections 2C) In what way do Freons negatively affect the environment?

2.57 (Chemical Connections 2C) What are HFCs and HCFCs? How does their use in refrigeration systems avoid the environmental problems associated with the use of Freons?

Additional Problems

2.58 ■ Tell whether the compounds in each set are constitutional isomers.

(a) CH_3CH_2OH and CH_3OCH_3

(b) $CH_3\overset{\text{O}}{\underset{\|}{C}}CH_3$ and $CH_3CH_2\overset{\text{O}}{\underset{\|}{C}}H$

(c) $CH_3\overset{\text{O}}{\underset{\|}{C}}OCH_3$ and $CH_3CH_2\overset{\text{O}}{\underset{\|}{C}}OH$

(d) $CH_3\overset{\text{OH}}{\underset{|}{C}HCH_2CH_3}$ and $CH_3\overset{\text{O}}{\underset{\|}{C}}CH_2CH_3$

(e) ⬠ and $CH_3CH_2CH_2CH_2CH_2CH_3$

(f) ⬠ and $CH_2{=}CHCH_2CH_2CH_3$

2.59 Explain why each of the following is an incorrect IUPAC name. Write the correct IUPAC name for the compound.

(a) 1,3-Dimethylbutane

(b) 4-Methylpentane

(c) 2,2-Diethylbutane

(d) 2-Ethyl-3-methylpentane

(e) 2-Propylpentane

(f) 2,2-Diethylheptane

(g) 2,2-Dimethylcyclopropane

(h) 1-Ethyl-5-methylcyclohexane

2.60 Which of the following compounds can exist as *cis-trans* isomers? For each that can, draw both isomers using solid and dashed wedges to show the orientation in space of the —OH and —CH₃ groups.

2.61 Tetradecane, $C_{14}H_{30}$, is an unbranched alkane with a melting point of 5.9°C and a boiling point of 254°C. Is tetradecane a solid, liquid, or gas at room temperature?

2.62 ■ Dodecane, $C_{12}H_{26}$, is an unbranched alkane. Predict the following:

(a) Will it dissolve in water?

(b) Will it dissolve in hexane?

(c) Will it burn when ignited?

(d) Is it a liquid, solid, or gas at room temperature and atmospheric pressure?

(e) Is it more or less dense than water?

Looking Ahead

2.63 Following is a structural formula for 2-isopropyl-5-methylcyclohexanol:

2-Isopropyl-5-methylcyclohexanol

Using a planar hexagon representation for the cyclohexane ring, draw a structural formula for the *cis-trans* isomer with isopropyl *trans* to —OH and methyl *cis* to —OH. If you answered this part correctly, you have drawn the isomer found in nature and given the name menthol.

2.64 On the left is a representation of the glucose molecule. Convert this representation to the alternative representations using the rings on the right. (We discuss the structure and chemistry of glucose in Chapter 12.)

Planar hexagon
representation

Chair conformation

2.65 On the left is a representation for 2-deoxy-D-ribose. This molecule is the "D" of DNA. Convert this representation to the alternative representation using the ring on the right. (We discuss the structure and chemistry of this compound in more detail in Chapter 12.)

2-Deoxy-D-ribose

2.66 As stated in Section 2.8, the wax found in apple skins is an unbranched alkane with the molecular formula $C_{27}H_{56}$. Explain how the presence of this alkane in apple skins prevents the loss of moisture from within the apple.

Alkenes and Alkynes

Charles D. Winters/Cengage Learning

Carotene is a naturally occurring polyene in carrots and tomatoes (Problems 3.61 and 3.62).

3.1 What Are Alkenes and Alkynes?

In this chapter, we begin our study of unsaturated hydrocarbons. Recall from Section 2.1 that these unsaturated compounds contain one or more carbon–carbon double bonds, triple bonds, or benzene-like rings. In this chapter we study **alkenes** and **alkynes. Alkynes** are unsaturated hydrocarbons that contain one or more carbon–carbon triple bonds. The simplest alkyne is acetylene.

Alkene An unsaturated hydrocarbon that contains a carbon–carbon double bond

Alkyne An unsaturated hydrocarbon that contains a carbon–carbon triple bond

OWL Online homework for this chapter may be assigned in **GOB OWL.**

Chemical Connections 3A

Ethylene: A Plant Growth Regulator

As we noted below, ethylene occurs only in trace amounts in nature. True enough, but scientists have discovered that this small molecule is a natural ripening agent for fruits. Thanks to this knowledge, fruit growers can now pick fruit while it is still green and less susceptible to bruising. Then, when they are ready to pack the fruit for shipment, the grower can treat it with ethylene gas. Alternatively, the fruit can be treated with ethephon (Ethrel), which slowly releases ethylene and initiates fruit ripening.

Ethephon

The next time you see ripe bananas in the market, you might wonder when they were picked and whether their ripening was artificially induced.

Ethylene
(an alkene)

Acetylene
(an alkyne)

Alkynes are unsaturated hydrocarbons that contain one or more carbon–carbon triple bonds. The simplest alkyne is acetylene. Because alkynes are not widespread in nature and have little importance in biochemistry, we will not study their chemistry in depth.

Compounds containing carbon–carbon double bonds are especially widespread in nature. Furthermore, several low-molecular-weight alkenes, including ethylene and propene, have enormous commercial importance in our modern, industrialized society. The organic chemical industry worldwide produces more pounds of ethylene than any other organic chemical. Annual production in the United States alone exceeds 55 billion pounds.

What is unusual about ethylene is that it occurs only in trace amounts in nature. The enormous amounts of it required to meet the needs of the chemical industry are derived the world over by thermal cracking of hydrocarbons. In the United States and other areas of the world with vast reserves of natural gas, the major process for the production of ethylene is thermal cracking of the small quantities of ethane extracted from natural gas. In **thermal cracking,** a saturated hydrocarbon is converted to an unsaturated hydrocarbon plus H_2. Ethane is thermally cracked by heating it in a furnace to 800–900°C for a fraction of a second.

$$CH_3CH_3 \xrightarrow[\text{(thermal cracking)}]{800-900°C} CH_2{=}CH_2 + H_2$$

Ethane   Ethylene

Europe, Japan, and other parts of the world with limited supplies of natural gas depend almost entirely on thermal cracking of petroleum for their ethylene.

From the perspective of the chemical industry, the single most important reaction of ethylene and other low-molecular-weight alkenes is polymerization, which we discuss in Section 3.7. The crucial point to recognize is that ethylene and all of the commercial and industrial products synthesized from it are derived from either natural gas or petroleum—both nonrenewable natural resources!

3.2 What Are the Structures of Alkenes and Alkynes?

A. Alkenes

Using the VSEPR model, we predict bond angles of 120° about each carbon in a double bond. The observed H—C—C bond angle in ethylene, for example, is 121.7°, close to the predicted value. In other alkenes, deviations from the predicted angle of 120° may be somewhat larger because of interactions between alkyl groups bonded to the doubly bonded carbons. The C—C—C bond angle in propene, for example, is 124.7°.

Ethylene　　　　Propene

If we look at a molecular model of ethylene, we see that the two carbons of the double bond and the four hydrogens bonded to them all lie in the same plane—that is, ethylene is a flat or planar molecule. Furthermore, chemists have discovered that, under normal conditions, no rotation is possible about the carbon–carbon double bond of ethylene or, for that matter, of any other alkene. Whereas free rotation occurs about each carbon–carbon single bond in an alkane (Section 2.6A), rotation about the carbon–carbon double bond in an alkene does not normally take place. For an important exception to this generalization about carbon–carbon double bonds, see Chemical Connections 3C on *cis-trans* isomerism in vision.

B. *Cis-Trans* Stereoisomerism in Alkenes

Cis-trans isomers Isomers that have the same connectivity of their atoms but a different arrangement of their atoms in space. Specifically, *cis* and *trans* stereoisomers result from the presence of either a ring or a carbon–carbon double bond

Because of the restricted rotation about a carbon–carbon double bond, an alkene in which each carbon of the double bond has two different groups bonded to it shows *cis-trans* isomerism (a type of stereoisomerism). For example, 2-butene has two *cis-trans* isomers. In *cis*-2-butene, the two methyl groups are located on the same side of the double bond, and the two hydrogens are on the other side. In *trans*-2-butene, the two methyl groups are located on opposite sides of the double bond. *Cis*-2-butene and *trans*-2-butene are different compounds and have different physical and chemical properties.

cis-2-Butene
mp −139°C, bp 4°C

trans-2-Butene
mp −106°C, bp 1°C

3.3 How Do We Name Alkenes and Alkynes?

Alkenes and alkynes are named using the IUPAC system of nomenclature. As we will see, some are still referred to by their common names.

A. IUPAC Names

The key to the IUPAC system of naming alkenes is the ending **-ene.** Just as the ending *-ane* tells us that a hydrocarbon chain contains only carbon–carbon single bonds, so the ending *-ene* tells us that it contains a carbon–carbon double bond. To name an alkene:

1. Find the longest carbon chain that includes the double bond. Indicate the length of the parent chain by using a prefix that tells the number of carbon atoms in it (see Table 2.2) and the suffix *-ene* to show that it is an alkene.

2. Number the chain from the end that gives the lower set of numbers to the carbon atoms of the double bond. Designate the position of the double bond by the number of its first carbon.

3. Branched alkenes are named in a manner similar to alkanes; substituent groups are located and named.

1-Hexene   4-Methyl-1-hexene   2,3-Diethyl-1-pentene

Note that, although 2,3-diethyl-1-pentene has a six-carbon chain, the longest chain that contains the double bond has only five carbons. The parent alkene is, therefore, a pentene rather than a hexene, and the molecule is named as a disubstituted 1-pentene.

The key to the IUPAC name of an alkyne is the ending **-yne,** which shows the presence of a carbon–carbon triple bond. Thus, $HC{\equiv}CH$ is ethyne (or acetylene) and $CH_3C{\equiv}CH$ is propyne. In higher alkynes, number the longest carbon chain that contains the triple bond from the end that gives the lower set of numbers to the triply bonded carbons. Indicate the location of the triple bond by the number of its first carbon atom.

3-Methyl-1-butyne   6,6-Dimethyl-3-heptyne

Example 3.1 IUPAC Names of Alkenes and Alkynes

Write the IUPAC name of each unsaturated hydrocarbon.

(a) $CH_2=CH(CH_2)_5CH_3$ (b)

$$\underset{H_3C}{\overset{H_3C}{>}}C=C\underset{H}{\overset{CH_3}{<}}$$

(c) $CH_3(CH_2)_2C\equiv CCH_3$

Strategy

Step 1: Locate the parent chain—the longest chain of carbon atoms that contains the carbon–carbon double or triple bonds.

Step 2: Number the parent chain from the direction that gives the carbons of the double or triple bond the lowest set of numbers. Show the presence of the multiple bond by the suffix *-ene* (for a double bond) or *-yne* (for a triple bond). Indicate the presence of the multiple bond by its first number.

Step 3: Name and locate all substituents on the parent chain. List them in alphabetical order.

Solution

(a) The parent chain contains eight carbons; thus the parent alkene is octene. To show the presence of the carbon–carbon double bond, use the suffix *-ene*. Number the chain beginning with the first carbon of the double bond. This alkene is 1-octene.

(b) Because there are four carbon atoms in the chain containing the carbon–carbon double bond, the parent alkene is butene. The double bond is between carbons 2 and 3 of the chain, and there is a methyl group on carbon 2. This alkene is 2-methyl-2-butene.

(c) There are six carbons in the parent chain, with the triple bond between carbons 2 and 3. This alkyne is 2-hexyne.

Problem 3.1

Write the IUPAC name of each unsaturated hydrocarbon.

B. Common Names

Despite the precision and universal acceptance of IUPAC nomenclature, some alkenes and alkynes—particularly those of low molecular weight—are known almost exclusively by their common names. Three examples follow:

	$CH_2=CH_2$	$CH_3CH=CH_2$	$\overset{\overset{\displaystyle CH_3}{\displaystyle |}}{CH_3C}=CH_2$
IUPAC name:	Ethene	Propene	2-Methylpropene
Common name:	Ethylene	Propylene	Isobutylene

We derive common names for alkynes by prefixing the names of the substituents on the carbon–carbon triple bond to the name *acetylene*:

	$HC\equiv CH$	$CH_3C\equiv CH$	$CH_3C\equiv CCH_3$
IUPAC name:	Ethyne	Propyne	2-Butyne
Common name:	Acetylene	Methylacetylene	Dimethylacetylene

C. *Cis* and *Trans* Configurations of Alkenes

The orientation of the carbon atoms of the parent chain determines whether an alkene is *cis* or *trans*. If the carbons of the parent chain are on the same side of the double bond, the alkene is *cis*; if they are on opposite sides, it is a *trans* alkene. In the first example below, they are on opposite sides and the compound is a *trans* alkene. In the second example, they are on the same side and the compound is a *cis* alkene.

trans-3-Hexene   *cis*-3,4-Dimethyl-2-pentene

Example 3.2 Naming Alkene *Cis* and *Trans* Isomers

Name each alkene and specify its configuration by indicating *cis* or *trans* where appropriate.

(a) $CH_3CH_2CH_2$ H
         C=C
     H CH_2CH_3

(b) $CH_3CH_2CH_2$ CH_2CH_3
         C=C
     CH_3 H

Strategy

For alkenes that show *cis-trans* isomerism, use the designator *cis* to show that the carbon atoms of the of the parent chain are on the same side of the double bond, and *trans* to show that they are on opposite sides of the double bond.

Solution

(a) The chain contains seven carbon atoms and is numbered from the right to give the lower number to the first carbon of the double bond. The carbon atoms of the parent chain are on opposite sides of the double bond. This alkene is *trans*-3-heptene.

(b) The longest chain contains seven carbon atoms and is numbered from the right so that the first carbon of the double bond is carbon 3 of the chain. The carbon atoms of the parent chain are on the same side of the double bond. This alkene is *cis*-4-methyl-3-heptene.

Problem 3.2

Name each alkene and specify its configuration.

(a)   (b)

Chemical Connections 3B

The Case of the Iowa and New York Strains of the European Corn Borer

Although humans communicate largely by sight and sound, the vast majority of other species in the animal world communicate by chemical signals. Often, communication within a species is specific for one of two or more stereoisomers. For example, a member of a given species might respond to the *cis* isomer of a chemical but not to the *trans* isomer. Alternatively, it might respond to a precise ratio of *cis* and *trans* isomers but not to other ratios of these same stereoisomers.

Several groups of scientists have studied the components of the sex **pheromones** of both the Iowa and New York strains of the European corn borer. Females of these closely related species secrete the sex attractant 11-tetradecenyl acetate. Males of the Iowa strain show their maximum response to a mixture containing 96% of the *cis* isomer and 4% of the *trans* isomer. When the pure *cis* isomer is used alone, males are only weakly attracted. Males of the New York strain show an entirely different response

pattern: They respond maximally to a mixture containing 3% of the *cis* isomer and 97% of the *trans* isomer.

trans-11-Tetradecenyl acetate

cis-11-Tetradecenyl acetate

Evidence suggests that an optimal response to a narrow range of stereoisomers, as we see here, is widespread in nature and that most insects maintain species isolation for mating and reproduction by the stereochemical mixtures of their pheromones.

D. Cycloalkenes

In naming **cycloalkenes,** number the carbon atoms of the ring double bond 1 and 2 in the direction that gives the substituent encountered first the lower number. It is not necessary to use a location number for the carbons of the double bond because, according to the IUPAC system of nomenclature, they will always be 1 and 2. Number substituents and list them in alphabetical order.

3-Methylcyclopentene
(not 5-methylcyclopentene)

4-Ethyl-1-methylcyclohexene
(not 5-ethyl-2-methylcyclohexene)

Example 3.3 Naming Cycloalkenes

Write the IUPAC name for each cycloalkene.

(a)　　　　(b)　　　　(c)

Strategy

In naming cycloalkenes, the carbon atoms of the double bond are always numbered 1 and 2 in the direction that gives the substituent encountered first, the lowest possible number. If there are multiple substituents, list them in alphabetical order.

Solution

(a) 3,3-Dimethylcyclohexene
(b) 1,2-Dimethylcyclopentene
(c) 4-Isopropyl-1-methylcyclohexene

Problem 3.3

Write the IUPAC name for each cycloalkene.

(a)   (b)   (c)

E. Dienes, Trienes, and Polyenes

We name alkenes that contain more than one double bond as alkadienes, alkatrienes, and so on. We often refer to those that contain several double bonds more generally as polyenes (Greek: *poly*, many). Following are three dienes:

$$CH_2{=}CHCH_2CH{=}CH_2$$

1,4-Pentadiene

$$CH_2{=}CCH{=}CH_2$$
$$\phantom{CH_2{=}}CH_3$$

2-Methyl-1,3-butadiene
(Isoprene)

1,3-Cyclopentadiene

We saw earlier that for an alkene with one carbon–carbon double bond that can show *cis-trans* isomerism, two stereoisomers are possible. For an alkene with n carbon–carbon double bonds, each of which can show *cis-trans* isomerism, 2^n stereoisomers are possible.

Example 3.4 *Cis-Trans* Isomerism

How many stereoisomers are possible for 2,4-heptadiene?

$$CH_3{-}CH{=}CH{-}CH{=}CH{-}CH_2{-}CH_3$$

2,4-Heptadiene

Strategy

To show *cis-trans* isomerism, each carbon of the double bond must have two different groups bonded to it.

Solution

This molecule has two carbon–carbon double bonds, each of which shows *cis-trans* isomerism. As shown in the following table, $2^2 = 4$ stereoisomers are possible. Line-angle formulas for two of these dienes are drawn here.

	Double-Bond	
	$C_2{-}C_3$	$C_4{-}C_5$
(1)	*trans*	*trans*
(2)	*trans*	*cis*
(3)	*cis*	*trans*
(4)	*cis*	*cis*

trans,trans-2,4-Heptadiene   *trans,cis*-2,4-Heptadiene

Problem 3.4

Draw structural formulas for the other two stereoisomers of 2,4-heptadiene.

Example 3.5 Drawing Alkene *Cis-Trans* Isomers

Draw all stereoisomers that are possible for the following unsaturated alcohol.

Strategy

To show *cis-trans* isomerism, each carbon of the double bond must have two different groups bonded to it. If a molecule has n double bonds about which *cis-trans* isomerism is possible, then 2^n isomers are possible, where n is the number of double bonds that show *cis-trans* isomerism.

Solution

Cis-trans isomerism is possible only about the double bond between carbons 2 and 3 of the chain. It is not possible for the other double bond because carbon 7 has two identical groups on it (review Section 3.2B). Thus, $2^1 = 2$ stereoisomers (one *cis-trans* pair) are possible. The *trans* isomer of this alcohol, named geraniol, is a major component of the oils of rose, citronella, and lemon grass.

The *trans* isomer   The *cis* isomer

Problem 3.5

How many stereoisomers are possible for the following unsaturated alcohol?

$$CH_3C{=}CHCH_2CH_2C{=}CHCH_2CH_2C{=}CHCH_2OH$$

with CH₃ groups: $$\overset{CH_3}{|} \quad \overset{CH_3}{|} \quad \overset{CH_3}{|}$$

An example of a biologically important polyunsaturated alcohol for which a number of *cis-trans* stereoisomers are possible is vitamin A. Each of the four carbon–carbon double bonds (shown in red) in the chain of carbon atoms bonded to the substituted cyclohexene ring has the potential for *cis-trans* isomerism. There are, therefore, $2^4 = 16$ stereoisomers possible for this structural formula. Vitamin A, the stereoisomer shown here, is the all-*trans* isomer.

Vitamin A (retinol)

Lemon grass from Shelton Herb Farm, North Carolina.

Chemical Connections 3C

Cis-Trans Isomerism in Vision

The retina, the light-detecting layer in the back of our eyes, contains reddish compounds called visual pigments. Their name, rhodopsin, is derived from the Greek word meaning "rose-colored." Each rhodopsin molecule is a combination of one molecule of a protein called opsin and one molecule of 11-*cis*-retinal, a derivative of vitamin A in which the CH_2OH group of carbon 15 is converted to an aldehyde group, —CH=O.

When rhodopsin absorbs light energy, the less stable 11-*cis* double bond is converted to the more stable 11-*trans* double bond. This isomerization changes the shape of the rhodopsin molecule, which in turn causes the neurons of the optic nerve to fire and produce a visual image.

The retinas of vertebrates contain two kinds of rhodopsin-containing cells: rods and cones. Cones function in bright light and are used for color vision; they are concentrated in the central portion of the retina, called the macula, and are responsible for the greatest visual acuity. The remaining area of the retina consists mostly of rods, which are used for peripheral and night vision. 11-*cis*-Retinal is present in both cones and rods. Rods have one kind of opsin, whereas cones have three kinds: one for blue, one for green, and one for red color vision.

3.4 What Are the Physical Properties of Alkenes and Alkynes?

Alkenes and alkynes are nonpolar compounds, and the only attractive forces between their molecules are very weak London dispersion forces. Their physical properties, therefore, are similar to those of alkanes with the same carbon skeletons. Alkenes and alkynes that are liquid at room temperature have densities less than 1.0 g/mL (they float on water). They are insoluble in water but soluble in one another and in other nonpolar organic liquids.

3.5 What Are Terpenes?

Terpene A compound whose carbon skeleton can be divided into two or more units identical to the five-carbon skeleton of isoprene

Among the compounds found in the essential oils of plants are a group of substances called **terpenes,** all of which have in common the fact that their carbon skeletons can be divided into two or more carbon units that are identical with the five-carbon skeleton of isoprene. Carbon 1 of an **isoprene unit** is called the head, and carbon 4 is called the tail. A terpene is a compound in which the tail of one isoprene unit becomes bonded to the head of another isoprene unit.

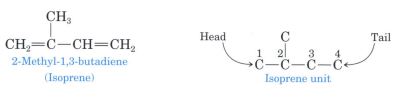

Terpenes are among the most widely distributed compounds in the biological world, and a study of their structure provides a glimpse into the wondrous diversity that nature can generate from a simple carbon skeleton. Terpenes also illustrate an important principle of the molecular logic of living systems: In building large molecules, small subunits are bonded together by a series of enzyme-catalyzed reactions and then chemically modified by additional enzyme-catalyzed reactions. Chemists use the same principles in the laboratory, but their methods cannot match the precision and selectivity of the enzyme-catalyzed reactions of cellular systems.

Probably the terpenes most familiar to you—at least by odor—are components of the so-called essential oils extracted from various parts of plants. Essential oils contain the relatively low-molecular-weight substances that are largely responsible for characteristic plant fragrances. Many essential oils, but particularly those from flowers, are used in perfumes.

One example of a terpene obtained from an essential oil is myrcene (Figure 3.1), a component of bayberry wax and oils of bay and verbena. Myrcene is a triene with a parent chain of eight carbon atoms and two one-carbon branches. The two isoprene units in myrcene are joined by bonding the tail of one unit (carbon 4) to the head of the other (carbon 1). Figure 3.1 also shows three more terpenes, each consisting of ten carbon atoms. In limonene and menthol, nature has formed an additional bond between two carbons to create a six-membered ring.

Farnesol, a terpene with a molecular formula of $C_{15}H_{26}O$, includes three isoprene units. Derivatives of both farnesol and geraniol are intermediates in the biosynthesis of cholesterol (Section 21.3).

FIGURE 3.1 Four terpenes, each derived from two isoprene units bonded from the tail of the first unit to the head of the second unit. In limonene and menthol, formation of an additional carbon–carbon bond creates a six-membered ring.

Farnesol
(Lily-of-the-valley)

Vitamin A (Section 3.3E), a terpene with a molecular formula of $C_{20}H_{30}O$, consists of four isoprene units bonded head-to-tail and cross-linked at one point to create a six-membered ring.

3.6 What Are the Characteristic Reactions of Alkenes?

California laurel, *Umbelluria californica*, one source of myrcene.

© Don Suzio.

The most characteristic reaction of alkenes is addition to the carbon–carbon double bond: The double bond is broken and in its place single bonds form to two new atoms or groups of atoms. Table 3.1 shows several examples of alkene addition reactions along with the descriptive name(s) associated with each reaction.

A. Addition of Hydrogen Halides (Hydrohalogenation)

The hydrogen halides HCl, HBr, and HI add to alkenes to give haloalkanes (alkyl halides). Addition of HCl to ethylene, for example, gives chloroethane (ethyl chloride):

$$CH_2=CH_2 + HCl \longrightarrow \overset{\overset{\displaystyle H}{|}}{CH_2}-\overset{\overset{\displaystyle Cl}{|}}{CH_2}$$

Ethylene   Chloroethane
(Ethyl chloride)

TABLE 3.1 Characteristic Addition Reactions of Alkenes

Reaction	Descriptive Name(s)
$\overset{\diagdown}{\diagup}C=C\overset{\diagup}{\diagdown} + HCl \longrightarrow -\overset{\overset{\displaystyle H}{\mid}}{C}-\overset{\overset{\displaystyle Cl}{\mid}}{C}-$	hydrochlorination
$\overset{\diagdown}{\diagup}C=C\overset{\diagup}{\diagdown} + H_2O \longrightarrow -\overset{\overset{\displaystyle H}{\mid}}{C}-\overset{\overset{\displaystyle OH}{\mid}}{C}-$	hydration
$\overset{\diagdown}{\diagup}C=C\overset{\diagup}{\diagdown} + Br_2 \longrightarrow -\overset{\overset{\displaystyle Br}{\mid}}{C}-\overset{\overset{\displaystyle Br}{\mid}}{C}-$	bromination
$\overset{\diagdown}{\diagup}C=C\overset{\diagup}{\diagdown} + H_2 \longrightarrow -\overset{\overset{\displaystyle H}{\mid}}{C}-\overset{\overset{\displaystyle H}{\mid}}{C}-$	hydrogenation (reduction)

Addition of HCl to propene gives 2-chloropropane (isopropyl chloride); hydrogen adds to carbon 1 of propene and chlorine adds to carbon 2. If the orientation of addition were reversed, 1-chloropropane (propyl chloride) would form. The observed result is that almost no 1-chloropropane forms. Because 2-chloropropane is the observed product, we say that addition of HCl to propene is **regioselective.**

Regioselective reaction A reaction in which one direction of bond forming or bond breaking occurs in preference to all other directions

This regioselectivity was noted by Vladimir Markovnikov (1838–1904), who made the following generalization, known as **Markovnikov's rule:** In the addition of HX (where X = halogen) to an alkene, hydrogen adds to the doubly bonded carbon that has the greater number of hydrogens bonded to it; halogen adds to the other carbon.

Markovnikov's rule In the addition of HX to an alkene, hydrogen adds to the carbon of the double bond having the greater number of hydrogens

Markovnikov's rule is often paraphrased as "the rich get richer."

Example 3.6 Addition of HX to an Alkene

Draw a structural formula for the product of each alkene addition reaction.

(a)
$$\underset{\overset{|}{CH_3}}{CH_3C} = CH_2 + HI \longrightarrow$$

(b) [cyclopentene with CH₃] + HCl ⟶

Strategy

Use Markovnikov's rule to predict the structural formula the product of each reaction. In the addition of HI and HCl, H adds to the carbon of the double bond that already has the greater number of H atoms bonded to it.

Solution

(a) Markovnikov's rule predicts that the hydrogen of HI adds to carbon 1 and iodine adds to carbon 2 to give 2-iodo-2-methylpropane.

(b) H adds to carbon 2 of the ring and Cl adds to carbon 1 to give 1-chloro-1-methylcyclopentane.

(a)
$$\underset{\overset{|}{I}}{\overset{\overset{CH_3}{|}}{CH_3CCH_3}}$$

2-Iodo-2-methylpropane

(b)

1-Chloro-1-methylcyclopentane

Problem 3.6

Draw a structural formula for the product of each alkene addition reaction.

(a) $CH_3CH = CH_2 + HBr \longrightarrow$ (b) [cyclohexane] $= CH_2 + HBr \longrightarrow$

Markovnikov's rule tells us what happens when we add HCl, HBr, or HI to a carbon–carbon double bond. We know that in the addition of HCl or other halogen acid, one bond of the double bond and the H—Cl bond are broken, and that new C—H and C—Cl bonds form. But chemists also want to know how this conversion happens. Are the C=C and H—X bonds broken and both new covalent bonds formed all at the same time? Or does this reaction take place in a series of steps? If the latter, what are these steps and in what order do they take place?

Chemists account for the addition of HX to an alkene by defining a two-step **reaction mechanism,** which we illustrate for the reaction of 2-butene with hydrogen chloride to give 2-chlorobutane. Step 1 is the addition of H^+ to 2-butene. To show this addition, we use a **curved arrow** that shows the repositioning of an electron pair from its origin (the tail of the arrow) to its new location (the head of the arrow). Recall that we used curved arrows in Section 7.1 to show bond breaking and bond formation in proton-transfer reactions. We now use curved arrows in the same way to show bond breaking and bond formation in a reaction mechanism.

Step 1 results in the formation of an organic cation. One carbon atom in this cation has only six electrons in its valence shell, so it carries a charge of $+1$. A species containing a positively charged carbon atom is called a **carbocation** (carbon + cation). Carbocations are classified as primary (1°), secondary (2°), or tertiary (3°), depending on the number of carbon groups bonded to the carbon bearing the positive charge.

Reaction mechanism A step-by-step description of how a chemical reaction occurs

Carbocation A species containing a carbon atom with only three bonds to it and bearing a positive charge

Mechanism: Addition of HCl to 2-Butene

Step 1: Reaction of the carbon–carbon double bond of the alkene with H^+ forms a 2° carbocation intermediate. In forming this intermediate, one bond of the double bond breaks and its pair of electrons forms a new covalent bond with H^+. One carbon of the double bond is left with only six electrons in its valence shell and, therefore, has a positive charge.

$$CH_3CH{=}CHCH_3 + H^+ \longrightarrow CH_3\overset{+}{C}H{-}\overset{\overset{\textstyle H}{|}}{C}HCH_3$$

A 2° carbocation intermediate

Step 2: Reaction of the 2° carbocation intermediate with chloride ion completes the valence shell of carbon and gives 2-chlorobutane.

$$:\!\overset{..}{\underset{..}{Cl}}\!:^- + CH_3\overset{+}{C}HCH_2CH_3 \longrightarrow CH_3\overset{\overset{\textstyle :\overset{..}{Cl}:}{|}}{C}HCH_2CH_3$$

Chloride ion   A 2° carbocation   2-chlorobutane
intermediate   (*sec*-Butyl chloride)

Example 3.7 Mechanism of Addition of HX to an Alkene

Propose a two-step mechanism for the addition of HI to methylenecyclohexane to give 1-iodo-1-methylcyclohexane.

Methylenecyclohexane   1-Iodo-1-methylcyclohexane

Strategy

The mechanism for the addition of HI to an alkene is similar to the two-step mechanism proposed for the addition of HCl to 2-butene.

Solution

Step 1: Reaction of H⁺ with the carbon–carbon double bond forms a new C—H bond to the carbon bearing the greater number of hydrogens and gives a 3° carbocation intermediate.

A 3° carbocation
intermediate

Step 2: Reaction of the 3° carbocation intermediate with iodide ion completes the valence shell of carbon and gives the product.

Problem 3.7

Propose a two-step mechanism for the addition of HBr to 1-methylcyclohexene to give 1-bromo-1-methylcyclohexane.

B. Addition of Water: Acid-Catalyzed Hydration

Hydration Addition of water

Most industrial ethanol is made by the acid-catalyzed hydration of ethylene.

In the presence of an acid catalyst, most commonly concentrated sulfuric acid, water adds to the carbon–carbon double bond of an alkene to give an alcohol. Addition of water is called **hydration.** In the case of simple alkenes, hydration follows Markovnikov's rule: H of H_2O adds to the carbon of the double bond with the greater number of hydrogens and OH of H_2O adds to the carbon with the smaller number of hydrogens.

$$CH_2{=\!\!=}CH_2 + H_2O \xrightarrow{H_2SO_4} \overset{H \quad OH}{CH_2{-}CH_2}$$

Ethylene   Ethanol

$$CH_3CH{=\!\!=}CH_2 + H_2O \xrightarrow{H_2SO_4} \overset{OH \quad H}{CH_3CH{-}CH_2}$$

Propene   2-Propanol

$$\overset{CH_3}{CH_3C}{=\!\!=}CH_2 + H_2O \xrightarrow{H_2SO_4} \overset{CH_3}{\underset{HO \quad H}{CH_3C{-}CH_2}}$$

2-Methylpropene   2-Methyl-2-propanol

Example 3.8 Acid-Catalyzed Hydration of an Alkene

Draw a structural formula for the alcohol formed by the acid-catalyzed hydration of 1-methylcyclohexene.

Strategy

Markovnikov's rule predicts that H adds to the carbon with the greater number of hydrogens.

Solution

H adds to carbon 2 of the cyclohexene ring and OH then adds to carbon 1.

1-Methylcyclohexene   1-Methylcyclohexanol

Problem 3.8

Draw a structural formula for the alcohol formed by acid-catalyzed hydration of each alkene:
(a) 2-Methyl-2-butene   (b) 2-Methyl-1-butene

The mechanism for the acid-catalyzed hydration of an alkene is similar to what we proposed for the addition of HCl, HBr, and HI to an alkene and is illustrated by the hydration of propene. This mechanism is consistent with the fact that acid is a catalyst. One H^+ is consumed in Step 1, but another is generated in Step 3.

Mechanism: Acid-Catalyzed Hydration of Propene

Step 1: Addition of H^+ to the carbon of the double bond with the greater number of hydrogens gives a 2° carbocation intermediate.

$$CH_3CH{=}CH_2 + H^+ \longrightarrow CH_3\overset{+}{C}H\overset{H}{\underset{}{CH_2}}$$

A 2° carbocation
intermediate

Step 2: The carbocation intermediate completes its valence shell by forming a new covalent bond with an unshared pair of electrons of the oxygen atom of H_2O to give an **oxonium ion**.

$$CH_3\overset{+}{C}HCH_3 + \overset{..}{:}\!\overset{}{O}{-}H \longrightarrow CH_3CHCH_3$$

An oxonium ion

Oxonium ion An ion in which oxygen is bonded to three other atoms and bears a positive charge

Step 3: Loss of H^+ from the oxonium ion gives the alcohol and generates a new H^+ catalyst.

Example 3.9 Acid-Catalyzed Hydration of an Alkene

Propose a three-step reaction mechanism for the acid-catalyzed hydration of methylenecyclohexane to give 1-methylcyclohexanol.

Strategy

The reaction mechanism for the acid-catalyzed hydration of methylene-cyclohexane is similar to the three-step mechanism proposed for the acid-catalyzed hydration of propene.

Solution

Step 1: Reaction of the carbon–carbon double bond with H^+ gives a 3° carbocation intermediate.

A 3° carbocation intermediate

Step 2: Reaction of the carbocation intermediate with water completes the valence shell of carbon and gives an oxonium ion.

An oxonium ion

Step 3: Loss of H^+ from the oxonium ion completes the reaction and generates a new H^+ catalyst.

Problem 3.9

Propose a three-step reaction mechanism for the acid-catalyzed hydration of 1-methylcyclohexene to give 1-methylcyclohexanol.

C. Addition of Bromine and Chlorine (Halogenation)

Chlorine, Cl_2, and bromine, Br_2, react with alkenes at room temperature by addition of halogen atoms to the carbon atoms of the double bond. This reaction is generally carried out either by using the pure reagents or by mixing them in an inert solvent, such as dichloromethane, CH_2Cl_2.

$$CH_3CH = CHCH_3 + Br_2 \xrightarrow{CH_2Cl_2} CH_3\overset{\underset{|}{Br}}{C}H - \overset{\underset{|}{Br}}{C}HCH_3$$

2-Butene   2,3-Dibromobutane

Cyclohexene   1,2-Dibromocyclohexane

Addition of bromine is a useful qualitative test for the presence of an alkene. If we dissolve bromine in carbon tetrachloride, the solution is red. In contrast, alkenes and dibromoalkanes are colorless. If we mix a few drops of the red bromine solution with an unknown sample suspected of being an alkene, disappearance of the red color as bromine adds to the double bond tells us that an alkene is, indeed, present.

Example 3.10 Addition of Halogens to an Alkene

Complete these reactions.

(a) $\bigcirc$ + Br$_2$ $\xrightarrow{CH_2Cl_2}$ (b) $\bigcirc$CH$_3$ + Cl$_2$ $\xrightarrow{CH_2Cl_2}$

Strategy

In addition of Br$_2$ or Cl$_2$ to a cycloalkene, one halogen adds to each carbon of the double bond.

Solution

(a) $\bigcirc$ + Br$_2$ $\xrightarrow{CH_2Cl_2}$ Br—$\bigcirc$—Br   (b) $\bigcirc$CH$_3$ + Cl$_2$ $\xrightarrow{CH_2Cl_2}$ CH$_3$, Cl, Cl

Problem 3.10

Complete these reactions.

(a) $\underset{\underset{CH_3}{|}}{\overset{\overset{CH_3}{|}}{CH_3C}}CH=CH_2$ + Br$_2$ $\xrightarrow{CH_2Cl_2}$ (b) $\bigcirc$=CH$_2$ + Cl$_2$ $\xrightarrow{CH_2Cl_2}$

D. Addition of Hydrogen: Reduction (Hydrogenation)

Virtually all alkenes react quantitatively with molecular hydrogen, H$_2$, in the presence of a transition metal catalyst to give alkanes. Commonly used transition metal catalysts include platinum, palladium, ruthenium, and nickel. Because the conversion of an alkene to an alkane involves reduction by hydrogen in the presence of a catalyst, the process is called **catalytic reduction** or, alternatively, **catalytic hydrogenation.**

In Section 13.3 we will see how catalytic hydrogenation is used to solidify liquid vegetable oils to margarines and semisolid cooking fats.

$$\underset{trans\text{-2-Butene}}{\overset{H_3C}{\underset{H}{}}C=C\overset{H}{\underset{CH_3}{}}} + H_2 \xrightarrow[25°C, 3\ atm]{Pd} \underset{Butane}{CH_3CH_2CH_2CH_3} \xleftarrow[25°C, 3\ atm]{Pd} H_2 + \underset{cis\text{-2-Butene}}{\overset{H}{\underset{CH_3}{}}C=C\overset{H}{\underset{CH_3}{}}}$$

$$\underset{Cyclohexene}{\bigcirc} + H_2 \xrightarrow[25°C, 3\ atm]{Pd} \underset{Cyclohexane}{\bigcirc}$$

Metal surface

ACTIVE FIGURE 3.2 The addition of hydrogen to an alkene involving a transition metal catalyst. (*a*) Hydrogen and the alkene are adsorbed on the metal surface and (*b*) one hydrogen atom is transferred to the alkene, forming one new C—H bond. The other carbon remains adsorbed on the metal surface. (*c*) A second C—H bond is formed and the alkene is desorbed. **Go to this book's companion website at www.cengage.com/chemistry/bettelheim to explore an interactive version of this figure.**

The metal catalyst is used in the form of a finely powdered solid. The reaction is carried out by dissolving the alkene in ethanol or another nonreacting organic solvent, adding the solid catalyst, and exposing the mixture to hydrogen gas at pressures ranging from 1 to 150 atm.

Mechanism: Catalytic Reduction

The transition metal catalysts used in catalytic hydrogenation are able to absorb large quantities of hydrogen onto their surfaces, probably by forming metal–hydrogen bonds. Similarly, alkenes are adsorbed on metal surfaces with the formation of carbon–metal bonds. Addition of hydrogen atoms to an alkene occurs in two steps (Figure 3.2).

3.7 What Are the Important Polymerization Reactions of Ethylene and Substituted Ethylenes?

A. Structure of Polyethylenes

Polymer From the Greek *poly*, many, and *meros*, part; any long-chain molecule synthesized by bonding together many single parts called monomers

Monomer From the Greek *mono*, single, and *meros*, part; the simplest nonredundant unit from which a polymer is synthesized

From the perspective of the chemical industry, the single most important reaction of alkenes is the formation of **chain-growth polymers** (Greek: *poly*, many, and *meros*, part). In the presence of certain compounds called initiators, many alkenes form polymers made by the stepwise addition of **monomers** (Greek: *mono*, one, and *meros*, part) to a growing polymer chain, as illustrated by the formation of polyethylene from ethylene. In alkene polymers of industrial and commercial importance, *n* is a large number, typically several thousand.

$$n\,CH_2{=}CH_2 \xrightarrow[\text{(polymerization)}]{\text{initiator}} {\left(CH_2CH_2\right)}_n$$

Ethylene   Polyethylene

To show the structure of a polymer, we place parentheses around the repeating monomer unit. The structure of an entire polymer chain can be reproduced by repeating this enclosed structure in both directions. A subscript

n is placed outside the parentheses to indicate that this unit is repeated n times, as illustrated for the conversion of propylene to polypropylene.

Monomer units shown in red

Propene

Part of an extended polymer chain of polypropylene

The repeating unit

Polyethylene films are produced by extruding molten plastic through a ring-like gap and inflating the film into a balloon.

The Stock Market

The most common method of naming a polymer is to attach the prefix **poly-** to the name of the monomer from which the polymer is synthesized—for example, polyethylene and polystyrene. When the name of the monomer consists of two words (for example, the monomer vinyl chloride), its name is enclosed in parentheses.

Vinyl chloride

Poly(vinyl chloride) (PVC)

Table 3.2 lists several important polymers derived from ethylene and substituted ethylene, along with their common names and most important uses.

TABLE 3.2 Polymers Derived from Ethylene and Substituted Ethylenes, Along with their Common Names and Most Important Uses.

Monomer Formula	Common Name	Polymer Name(s) and Common Uses
$CH_2{=}CH_2$	ethylene	polyethylene, Polythene; break-resistant containers and packaging materials
$CH_2{=}CHCH_3$	propylene	polypropylene, Herculon; textile and carpet fibers
$CH_2{=}CHCl$	vinyl chloride	poly(vinyl chloride), PVC; construction tubing
$CH_2{=}CCl_2$	1,1-dichloroethylene	poly(1,1-dichloroethylene); Saran Wrap is a copolymer with vinyl chloride
$CH_2{=}CHCN$	acrylonitrile	polyacrylonitrile, Orlon; acrylics and acrylates
$CF_2{=}CF_2$	tetrafluoroethylene	polytetrafluoroethylene, PTFE; Teflon, nonstick coatings
$CH_2{=}CHC_6H_5$	styrene	polystyrene, Styrofoam; insulating materials
$CH_2{=}CHCOOCH_2CH_3$	ethyl acrylate	poly(ethyl acrylate), latex paint
$CH_2{=}CCOOCH_3$ $\quad\ \vert$ $\quad\ CH_3$	methyl methacrylate	poly(methyl methacrylate), Lucite; Plexiglas; glass substitutes

Peroxide Any compound that contains an —O—O— bond as, for example, hydrogen peroxide, H—O—O—H

B. Low-Density Polyethylene (LDPE)

The first commercial process for ethylene polymerization used **peroxide** initiators at 500°C and 1000 atm and yielded a tough, transparent polymer known as **low-density polyethylene (LDPE).** At the molecular level, LDPE chains are highly branched, with the result that they do not pack well together and the London dispersion forces between them are weak.

Today approximately 65% of all LDPE is used for the manufacture of films by the blow-molding technique illustrated in Figure 3.3. LDPE film is inexpensive, which makes it ideal for packaging such consumer items as baked goods and vegetables and for the manufacture of trash bags.

C. High-Density Polyethylene (HDPE)

In the 1950s, Karl Ziegler of Germany and Giulio Natta of Italy developed an alternative method for the polymerization of alkenes, which does not rely on peroxide·initiators. Polyethylene from Ziegler-Natta systems,

(*b*)

(*a*)

(*c*) (*d*)

a, c, Beverly March; b, d, Charles D. Winters/Cengage Learning

Some articles made from chain-growth polymers. (*a*) Saran Wrap, a copolymer of vinyl chloride and 1,1-dichloroethylene. (*b*) Plastic containers for various supermarket products, made mostly from polyethylene and polypropylene. (*c*) Teflon-coated kitchenware. (*d*) Articles made from polystyrene.

termed **high-density polyethylene (HDPE),** has little chain branching. Consequently, its chains pack together more closely than those of LDPE, with the result that the London dispersion forces between chains of HDPE are stronger than those in LDPE.

Linear polyethylene
(high density)

HDPE has a higher melting point than LDPE and is three to ten times stronger.

Approximately 45 % of all HDPE products are made by the blow-molding process shown in Figure 3.4. HDPE is used for consumer items such as milk and water jugs, grocery bags, and squeezable bottles.

Rolled thin LDPE sheeting

Blown tube

Heater

Melted low-density polyethylene (LDPE)

Compressed air

FIGURE 3.3 Fabrication of LDPE film. A tube of melted LDPE along with a jet of compressed air is forced through an opening and blown into a gigantic, thin-walled bubble. The film is then cooled and taken up onto a roller. This double-walled film can be slit down the side to give LDPE film or sealed at points along its length to make LDPE bags.

(a)

Air tube

High-density polyethylene tube

Open die

(b)

Compressed air

(c)

Finished product

FIGURE 3.4 Blow molding an HDPE container. (a) A short length of HDPE tubing is placed in an open die and the die is closed, sealing the bottom of the tube. (b) Compressed air is forced into the hot polyethylene/die assembly, and the tubing is literally blown up to take the shape of the mold. (c) After the assembly cools, the die is opened, and there is the container!

Chemical Connections 3D

Recycling Plastics

Plastics are polymers that can be molded when hot and that retain their shape when cooled. Because they are durable and lightweight, plastics are probably the most versatile synthetic materials in existence. In fact, the current production of plastics in the United States exceeds the U.S. production of steel. Plastics have come under criticism, however, for their role in the solid waste crisis. They

and other lightweight materials. After any remaining labels and adhesives are eliminated with a detergent wash, the PET chips are dried. The PET produced by this method is 99.9% free of contaminants and sells for about half the price of the virgin material. The biggest market for recycled PET in 2005 was fibers. The carpet maker Mohawk Industries, for example, starts with about 250 million lb of recycled

Code	Polymer	Common Uses
1 PET	poly(ethylene terephthalate)	soft drink bottles, household chemical bottles, films, textile fibers
2 HDPE	high-density polyethylene	milk and water jugs, grocery bags, squeezable bottles
3 V	poly(vinyl chloride), PVC	shampoo bottles, pipes, shower curtains, vinyl siding, wire insulation, floor tiles
4 LDPE	low-density polyethylene	shrink wrap, trash and grocery bags, sandwich bags, squeeze bottles
5 PP	polypropylene	plastic lids, clothing fibers, bottle caps, toys, diaper linings
6 PS	polystyrene	Styrofoam cups, egg cartons, disposable utensils, packaging materials, appliances
7	all other plastics	various

account for approximately 21% of the volume and 8% of the weight of solid wastes, with most plastic waste consisting of disposable packaging and wrapping.

Six types of plastics are commonly used for packaging applications. In 1988, manufacturers adopted recycling code letters developed by the Society of the Plastics Industry as a means of identifying them.

Currently, only poly(ethylene terephthalate) (PET) and high-density polyethylene (HDPE) are recycled in large quantities. In fact, bottles made of these plastics account for more than 99% of the plastics recycled in the United States.

The synthesis and structure of PET, a polyester, is described in Section 11.6B.

The process for recycling most plastics is simple, with separation of the plastic from other contaminants being the most labor-intensive step. For example, PET soft drink bottles usually have a paper label and adhesive that must be removed before the PET can be reused. Recycling begins with hand or machine sorting, after which the bottles are chopped into small chips. Any ferrous metals are removed by magnets. Any nonferrous metal contaminants are removed by electric eddy currents that cause them to jump like fleas into a bin as they move down a conveyor belt during the soarting process. Air cyclone then removes paper

PET bottles per year and ends up with 80 to 100 million sq yards of carpet. The largest domestic use of recycled HDPE resins in 2005 was bottles.

These students are wearing jackets made from recycled PET soda bottles.

Summary of Key Questions

OWL End-of-chapter problems identified in blue are assignable in GOB OWL.

Section 3.1 What Are Alkenes and Alkynes?

- An **alkene** is an unsaturated hydrocarbon that contains a carbon–carbon double bond.
- An **alkyne** is an unsaturated hydrocarbon that contains a carbon–carbon triple bond.

Section 3.2 What Are the Structures of Alkenes and Alkynes? Problem 3.15

- The structural feature that makes **cis-trans stereo-isomerism** possible in alkenes is restricted rotation about the two carbons of the double bond. The *cis* or *trans* configuration of an alkene is determined by the orientation of the atoms of the parent chain about the double bond. If atoms of the parent chain are located on the same side of the double bond, the configuration of the alkene is *cis*; if they are located on opposite sides, the configuration is *trans*.

Section 3.3 How Do We Name Alkenes and Alkynes? Problem 3.17

- In IUPAC names, the presence of a carbon–carbon double bond is indicated by a prefix showing the number of carbons in the parent chain and the ending **-ene.** Substituents are numbered and named in alphabetical order.
- The presence of a carbon–carbon triple bond is indicated by a prefix that shows the number of carbons in the parent chain and the ending **-yne.**
- The carbon atoms of the double bond of a cycloalkene are numbered 1 and 2 in the direction that gives the smaller number to the first substituent.
- Compounds containing two double bonds are called **dienes,** those with three double bonds are called

trienes, and those containing four or more double bonds are called **polyenes.**

Section 3.4 What Are the Physical Properties of Alkenes and Alkynes?

- Because alkenes and alkynes are nonpolar compounds and the only interactions between their molecules are London dispersion forces, their physical properties are similar to those of alkanes with similar carbon skeletons.

Section 3.5 What Are Terpenes? Problem 3.32

- The characteristic structural feature of a **terpene** is a carbon skeleton that can be divided into two or more **isoprene units.** The most common pattern is the head of one unit bonded to the tail of the next unit.

Section 3.6 What Are the Characteristic Reactions of Alkenes? Problems 3.41, 3.46

- A characteristic reaction of alkenes is addition to the double bond.
- In addition, the double bond breaks and bonds to two new atoms, or groups of atoms form in its place.
- A **reaction mechanism** is a step-by-step description of how a chemical reaction occurs, including the role of the catalyst (if one is present).
- A **carbocation** contains a carbon with only six electrons in its valence shell and bears a positive charge.

Section 3.7 What Are the Important Polymerization Reactions of Ethylene and Substituted Ethylenes?

- **Polymerization** is the process of bonding together many small **monomers** into large, high-molecular-weight **polymers.**

Summary of Key Reactions

1. **Addition of HX (Hydrohalogenation) (Section 3.6A)** Addition of HX to the carbon–carbon double bond of an alkene follows Markovnikov's rule. The reaction occurs in two steps and involves formation of a carbocation intermediate.

2. **Acid-Catalyzed Hydration (Section 3.6B)** Addition of H_2O to the carbon–carbon double bond of an alkene follows Markovnikov's rule. Reaction occurs in three steps and involves formation of carbocation and oxonium ion intermediates.

3. **Addition of Bromine and Chlorine (Halogenation) (Section 3.6C)** Addition to a cycloalkene gives a 1,2-dihalocycloalkane.

Cyclohexene   1,2-Dibromocyclohexane

4. Reduction: Formation of Alkanes (Hydrogenation) (Section 3.6D) Catalytic reduction involves addition of hydrogen to form two new C—H bonds.

5. Polymerization of Ethylene and Substituted Ethylenes (Section 3.7A) In polymerization of alkenes, monomer units bond together without the loss of any atoms.

$$n CH_2{=}CH_2 \xrightarrow{\text{initiator}} -\!\!\left(CH_2CH_2\right)_n\!\!-$$

Problems

■ Indicates problems that are assignable in GOB OWL.

Blue numbered problems are applied.

Go to this book's companion website at **www.cengage. com/chemistry/bettelheim** for interactive versions of the How To tutorials and Active Figures, and to quiz yourself on this chapter.

Section 3.1 What Are Alkenes and Alkynes?

3.11 Answer true or false.
   (a) There are two classes of unsaturated hydrocarbons, alkenes and alkynes.
   (b) The bulk of the ethylene used by the chemical industry worldwide is obtained from renewable resources.
   (c) Ethylene and acetylene are constitutional isomers.
   (d) Cyclohexane and 1-hexene are constitutional isomers.

Section 3.2 What Are the Structures of Alkenes and Alkynes?

3.12 Answer true or false.
   (a) Both ethylene and acetylene are planar molecules.
   (b) An alkene in which each carbon of the double bond has two different groups bonded to it will show *cis-trans* isomerism.
   (c) *Cis-trans* isomers have the same molecular formula but a different connectivity of their atoms.
   (d) *Cis*-2-butene and *trans*-2-butene can be interconverted by rotation about the carbon–carbon double bond.
   (e) *Cis-trans* isomerism is possible only among appropriately substituted alkenes.
   (f) Both 2-hexene and 3-hexene can exist as pairs of *cis-trans* isomers.
   (g) Cyclohexene can exist as a pair of *cis-trans* isomers.
   (h) 1-Chloropropene can exist as a pair of *cis-trans* isomers.

3.13 What is the difference in structure between a saturated hydrocarbon and an unsaturated hydrocarbon?

3.14 Each carbon atom in ethane and in ethylene is surrounded by eight valence electrons and has four bonds to it. Explain how the VSEPR model predicts a bond angle of 109.5° about each carbon in ethane but an angle of 120° about each carbon in ethylene.

3.15 ■ Predict all bond angles about each highlighted carbon atom.

(c) $HC{\equiv}C{-}CH{=}CH_2$ (d)

Section 3.3 How Do We Name Alkenes and Alkynes?

3.16 Answer true or false.
   (a) The IUPAC name of an alkene is derived from the name of the longest carbon chain that contains the carbon–carbon double bond.
   (b) The IUPAC name of $CH_3CH{=}CHCH_3$ is 1,2-dimethylethene.
   (c) 2-Methyl-2-butene shows *cis-trans* isomerism.
   (d) 1,2-dimethylcyclohexene shows *cis-trans* isomerism.
   (e) The IUPAC name of $CH_2{=}CHCH{=}CHCH_3$ is 1,3-pentadiene.
   (f) 1,3-Butadiene has two carbon–carbon double bonds and $2^2 = 4$ stereoisomers are possible for it.

3.17 ■ Draw a structural formula for each compound.
   (a) *trans*-2-Methyl-3-hexene
   (b) 2-Methyl-3-hexyne
   (c) 2-Methyl-1-butene
   (d) 3-Ethyl-3-methyl-1-pentyne
   (e) 2,3-Dimethyl-2-pentene

3.18 Write the IUPAC name for each unsaturated hydrocarbon.

(a) $CH_2=CH(CH_2)_4CH_3$

(b)

(c)

(d) $(CH_3)_2CHCH=C(CH_3)_2$

(e) $CH_3(CH_2)_5C\equiv CH$

(f) $CH_3CH_2C\equiv CC(CH_3)_3$

3.19 Write the IUPAC name for each unsaturated hydrocarbon.

(a)   (b)

(c) $\begin{matrix} CH_2 \\ \| \\ CH_3CH_2CCH_3 \end{matrix}$

(d) $\begin{matrix} CH_3CH_2CH_2 \\ \quad\quad C=CH_2 \\ CH_3CH_2CH_2 \end{matrix}$

3.20 Explain why each name is incorrect, and then write a correct name.

(a) 1-Methylpropene   (b) 3-Pentene

(c) 2-Methylcyclohexene (d) 3,3-Dimethylpentene

(e) 4-Hexyne   (f) 2-Isopropyl-2-butene

3.21 Explain why each name is incorrect, and then write a correct name.

(a) 2-Ethyl-1-propene

(b) 5-Isopropylcyclohexene

(c) 4-Methyl-4-hexene

(d) 2-*sec*-Butyl-1-butene

(e) 6,6-Dimethylcyclohexene

(f) 2-Ethyl-2-hexene

3.22 What structural feature in alkenes makes *cis-trans* isomerism in them possible? What structural feature in cycloalkanes makes *cis-trans* isomerism in them possible? What do these two structural features have in common?

3.23 Which of these alkenes show *cis-trans* isomerism? For each that does, draw structural formulas for both isomers.

(a) 1-Hexene   (b) 2-Hexene

(c) 3-Hexene   (d) 2-Methyl-2-hexene

(e) 3-Methyl-2-hexene   (f) 2,3-Dimethyl-2-hexene

3.24 Name and draw structural formulas for all alkenes with the molecular formula C_5H_{10}. As you draw these alkenes, remember that *cis* and *trans* isomers are different compounds and must be counted separately.

3.25 Arachidonic acid is a naturally occurring polyunsaturated fatty acid. Draw a line-angle formula for arachidonic acid showing the *cis* configuration about each double bond.

$$CH_3(CH_2)_4(CH=CHCH_2)_4CH_2CH_2COOH$$
Arachidonic acid

3.26 Following is the structural formula of a naturally occurring unsaturated fatty acid.

$$CH_3(CH_2)_7CH=CH(CH_2)_7COOH$$

The *cis* stereoisomer is named oleic acid and the *trans* isomer is named elaidic acid. Draw a line-angle formula of each acid, showing clearly the configuration of the carbon–carbon double bond in each.

3.27 For each molecule that shows *cis-trans* isomerism, draw the *cis* isomer.

3.28 Draw structural formulas for all compounds with the molecular formula C_5H_{10} that are

(a) Alkenes that do not show *cis-trans* isomerism

(b) Alkenes that show *cis-trans* isomerism

(c) Cycloalkanes that do not show *cis-trans* isomerism

(d) Cycloalkanes that show *cis-trans* isomerism

3.29 β-Ocimene, a triene found in the fragrance of cotton blossoms and several essential oils, has the IUPAC name *cis*-3,7-dimethyl-1,3,6-octatriene. (*Cis* refers to the configuration of the double bond between carbons 3 and 4, the only double bond in this molecule about which *cis-trans* isomerism is possible.) Draw a structural formula for β-ocimene.

Section 3.4 What Are the Physical Properties of Alkenes and Alkynes?

3.30 Answer true or false.

(a) Alkenes and alkynes are nonpolar molecules.

(b) The physical properties of alkenes are similar to those of alkanes of the same carbon skeletons.

(c) Alkenes that are liquid at room temperature are insoluble in water and when added to water, will float on water.

Section 3.5 What Are Terpenes?

3.31 Answer true or false.

(a) Terpenes are identified by their carbon skeleton, namely one that can be divided into five-carbon units, each identical with the five-carbon skeleton of isoprene.

(b) Isoprene is a common name for 2-methyl-1,3-butadiene.

(c) Both geraniol and menthol (Figure 3.1) show *cis-trans* isomerism.

(d) *Cis-trans* isomerism is not possible in myrcene.

3.32 ■ Which of these terpenes show *cis-trans* isomerism?

(a) Myrcene (b) Geraniol

(c) Limonene (d) Farnesol

3.33 Show that the structural formula of vitamin A (Section 3.3E) can be divided into four isoprene units joined by head-to-tail linkages and cross-linked at one point to form the six-membered ring.

Section 3.6 What Are the Characteristic Reactions of Alkenes?

3.34 Answer true or false.

(a) Complete combustion of an alkene gives carbon dioxide and water.

(b) Addition reactions of alkenes involve breaking one of the bonds of the carbon–carbon double bond and formation of two new single bonds in its place.

(c) Markovnikov's rule refers to the regioselectivity of addition reactions of carbon–carbon double bonds.

(d) According to Markovnikov's rule, in the addition of HCl, HBr, or HI to an alkene, hydrogen adds to the carbon of the double bond that already has the greater number of hydrogen atoms bonded to it, and the halogen adds to the carbon that has the lesser number of hydrogens bonded to it.

(e) A carbocation is carbon with four bonds to it and that bears a positive charge.

(f) The carbocation derived from ethylene is $CH_3CH_2^+$.

(g) The reaction mechanism for the addition of a halogen acid (HX) to an alkene is divided into two steps, (1) formation of a carbocation and (2) reaction of the carbocation with halide ion, which complete the reaction.

(h) Acid-catalyzed addition of H_2O to an alkene is called *hydration*.

(i) If a compound fails to react with Br_2, it is unlikely that the compound contains a carbon–carbon double bond.

(j) Addition of H_2 to a double bond is a reduction reaction.

(k) Catalytic reduction of cyclohexene gives hexane.

(l) According to the mechanism presented in the text for acid-catalyzed hydration of an alkene, the H and —OH groups added to the carbon–carbon double bond both arise from the same molecule of H_2O.

(m) The conversion of ethylene, $CH_2=CH_2$, to ethanol, CH_3CH_2OH is an oxidation reaction.

(n) Acid-catalyzed hydration of 1-butene gives 1-butanol. Acid-catalyzed hydration of 2-butene gives 2-butanol.

3.35 Define *alkene addition reaction*. Write an equation for an addition reaction of propene.

3.36 What reagent and/or catalysts are necessary to bring about each conversion?

(a) $CH_3CH=CHCH_3 \longrightarrow CH_3CH_2\overset{\overset{\displaystyle Br}{|}}{C}HCH_3$

(b) $CH_3\overset{\overset{\displaystyle CH_3}{|}}{C}=CH_2 \longrightarrow CH_3\overset{\overset{\displaystyle CH_3}{|}}{\underset{\underset{\displaystyle OH}{|}}{C}}CH_3$

(c)

(d) $CH_3\overset{\overset{\displaystyle CH_3}{|}}{C}=CH_2 \longrightarrow CH_3\overset{\overset{\displaystyle CH_3}{|}}{\underset{\underset{\displaystyle Br}{|}}{C}}-\underset{\underset{\displaystyle Br}{|}}{C}H_2$

3.37 Complete these equations.

(a) —CH_2CH_3 + HCl ⟶

(b) (cyclopentene)—$CH_2CH_3 + H_2O \xrightarrow{H_2SO_4}$

(c) $CH_3(CH_2)_5CH=CH_2 + HI \longrightarrow$

(d) (cyclohexane)—$\overset{\overset{\displaystyle CH_2}{\|}}{\underset{\underset{\displaystyle CH_3}{}}{C}} + HCl \longrightarrow$

(e) $CH_3CH=CHCH_2CH_3 + H_2O \xrightarrow{H_2SO_4}$

(f) $CH_2=CHCH_2CH_2CH_3 + H_2O \xrightarrow{H_2SO_4}$

3.38 Draw structural formulas for all possible carbocations formed by the reaction of each alkene with HCl. Label each carbocation as primary, secondary, or tertiary.

(a) $CH_3CH_2\overset{\overset{\displaystyle CH_3}{|}}{C}=CHCH_3$

(b) $CH_3CH_2CH=CHCH_3$

(c) A cyclopentene ring with a CH_3 group

(d) A cyclohexane ring with $=CH_2$

3.39 Draw a structural formula for the product formed by treatment of 2-methyl-2-pentene with each reagent.

(a) HCl (b) H_2O in the presence of H_2SO_4

3.40 Draw a structural formula for the product of each reaction.

(a) 1-Methylcyclohexene + Br_2

(b) 1,2-Dimethylcyclopentene + Cl_2

3.41 ■ Draw a structural formula for an alkene with the indicated molecular formula that gives the compound shown as the major product. Note that more than one alkene may give the same compound as the major product.

(a) $C_5H_{10} + H_2O \xrightarrow{H_2SO_4} CH_3\overset{\overset{\displaystyle CH_3}{|}}{\underset{\underset{\displaystyle OH}{|}}{C}}CH_2CH_3$

(b) $C_5H_{10} + Br_2 \longrightarrow CH_3\overset{\overset{\displaystyle CH_3}{|}}{C}H\underset{\underset{\displaystyle Br}{|}}{C}H\underset{\underset{\displaystyle Br}{|}}{C}H_2$

(c) $C_7H_{12} + HCl \longrightarrow$ cyclohexane ring with CH_3 and Cl

3.42 Draw a structural formula for an alkene with the molecular formula C_5H_{10} that reacts with Br_2 to give each product.

(a) $CH_3\overset{\overset{\displaystyle CH_3}{|}}{\underset{\underset{\displaystyle Br}{|}}{C}}-\underset{\underset{\displaystyle Br}{|}}{C}HCH_3$

(b) $CH_2\overset{\overset{\displaystyle CH_3}{|}}{\underset{\underset{\displaystyle Br}{|}}{C}}\underset{\underset{\displaystyle Br}{|}}{C}H_2CH_3$

(c) $CH_2\underset{\underset{\displaystyle Br}{|}}{C}H\underset{\underset{\displaystyle Br}{|}}{C}HCH_2CH_2CH_3$

3.43 Draw a structural formula for an alkene with the molecular formula C_5H_{10} that reacts with HCl to

give the indicated chloroalkane as the major product. More than one alkene may give the same compound as the major product.

(a) $CH_3\overset{\overset{\displaystyle CH_3}{|}}{\underset{\underset{\displaystyle Cl}{|}}{C}}CH_2CH_3$

(b) $CH_3\overset{\overset{\displaystyle CH_3}{|}}{C}H\underset{\underset{\displaystyle Cl}{|}}{C}HCH_3$

(c) $CH_3\underset{\underset{\displaystyle Cl}{|}}{C}HCH_2CH_2CH_3$

3.44 Draw the structural formula of an alkene that undergoes acid-catalyzed hydration to give the indicated alcohol as the major product. More than one alkene may give each alcohol as the major product.

(a) 3-Hexanol

(b) 1-Methylcyclobutanol

(c) 2-Methyl-2-butanol

(d) 2-Propanol

3.45 Terpin, $C_{10}H_{20}O_2$, is prepared commercially by the acid-catalyzed hydration of limonene (Figure 3.1).

(a) Propose a structural formula for terpin.

(b) How many *cis-trans* isomers are possible for the structural formula you propose?

(c) Terpin hydrate, the isomer of terpin in which the methyl and isopropyl groups are *trans* to each other, is used as an expectorant in cough medicines. Draw a structural formula for terpin hydrate showing the *trans* orientation of these groups.

3.46 ■ Draw the product formed by treatment of each alkene with H_2/Ni.

3.47 Hydrocarbon A, C_5H_8, reacts with 2 moles of Br_2 to give 1,2,3,4-tetrabromo-2-methylbutane. What is the structure of hydrocarbon A?

3.48 Show how to convert ethylene to these compounds.

(a) Ethane (b) Ethanol

(c) Bromoethane (d) 1,2-Dibromoethane

(e) Chloroethane

3.49 Show how to convert 1-butene to these compounds.

(a) Butane (b) 2-Butanol

(c) 2-Bromobutane (d) 1,2-Dibromobutane

Section 3.7 What Are the Important Polymerizations Reactions of Ethylene and Substituted Ethylenes?

3.50 Answer true or false.

(a) Ethylene contains one carbon–carbon double bond and polyethylene contains many carbon–carbon double bonds.

(b) All C—C—C bond angles in both LDPE and HDPE are approximately 120°.

(c) Low-density polyethylene (LDPE) is a highly branched polymer.

(d) High-density polyethylene (HDPE) consists of carbon chains with little branching.

(e) The density of polyethylene polymers is directly related to the degree of chain branching; the greater the branching, the lower the density of the polymer.

(f) PS and PVC are currently recycled.

Chemical Connections

3.51 (Chemical Connections 3A) What is one function of ethylene as a plant growth regulator?

3.52 (Chemical Connections 3B) What is the meaning of the term *pheromone*?

3.53 (Chemical Connections 3B) What is the molecular formula of 11-tetradecenyl acetate? What is its molecular weight?

3.54 (Chemical Connections 3B) Assume that 1×10^{-12} g of 11-tetradecenyl acetate is secreted by a single corn borer. How many molecules is this?

3.55 (Chemical Connections 3C) What different functions are performed by the rods and cones in the eye?

3.56 (Chemical Connections 3C) In which isomer of retinal is the end-to-end distance longer, the all-*trans* isomer or the 11-*cis* isomer?

3.57 (Chemical Connections 3D) What types of consumer products are made of high-density polyethylene? What types of products are made of low-density polyethylene? One type of polyethylene is currently recyclable and the other is not. Which is which?

3.58 (Chemical Connections 3D) In recycling codes, what do these abbreviations stand for?

(a) V   (b) PP   (c) PS

Additional Problems

3.59 Write line-angle formulas for all compounds with the molecular formula C_4H_8. Which are sets of constitutional isomers? Which are sets of *cis-trans* isomers?

3.60 Name and draw structural formulas for all alkenes with the molecular formula C_6H_{12} that have these carbon skeletons. Remember to consider *cis* and *trans* isomers.

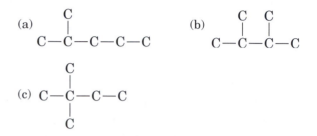

3.61 Following is the structural formula of lycopene, $C_{40}H_{56}$, a deep-red compound that is partially responsible for the red color of ripe fruits, especially tomatoes. Approximately 20 mg of lycopene can be isolated from 1 kg of fresh, ripe tomatoes.

(a) Show that lycopene is a terpene; that is, its carbon skeleton can be divided into two sets of four isoprene units with the units in each set joined head-to-tail.

(b) How many of the carbon–carbon double bonds in lycopene have the possibility for *cis-trans* isomerism? Lycopene is the all-*trans* isomer.

3.62 As you might suspect, β-carotene, $C_{40}H_{56}$, a precursor to vitamin A, was first isolated from carrots. Dilute solutions of β-carotene are yellow—hence its use as a food coloring. In plants, this compound is almost always present in combination with chlorophyll to assist in the harvesting of the energy of sunlight. As tree leaves die in the fall, the green of their chlorophyll molecules is replaced by the yellows and reds of carotene and carotene-related molecules (see β-carotene skeleton below.

Lycopene

β-Carotene

Compare the carbon skeletons of β-carotene and lycopene. What are the similarities? What are the differences?

3.63 Draw the structural formula for a cycloalkene with the molecular formula C_6H_{10} that reacts with Cl_2 to give each compound.

3.64 Propose a structural formula for the product(s) when each of the following alkenes is treated with H_2O/H_2SO_4. Why are two products formed in part (b), but only one in parts (a) and (c)?

(a) 1-Hexene gives one alcohol with a molecular formula of $C_6H_{14}O$.

(b) 2-Hexene gives two alcohols, each with a molecular formula of $C_6H_{14}O$.

(c) 3-Hexene gives one alcohol with a molecular formula of $C_6H_{14}O$.

3.65 *cis*-3-Hexene and *trans*-3-hexene are different compounds and have different physical and chemical properties. Yet, when treated with H_2O/H_2SO_4, each gives the same alcohol. What is this alcohol, and how do you account for the fact that each alkene gives the same one?

3.66 Draw the structural formula of an alkene that undergoes acid-catalyzed hydration to give each of the following alcohols as the major product. More than one alkene may give each compound as the major product.

3.67 Show how to convert cyclopentene into these compounds.

(a) 1,2-Dibromocyclopentane

(b) Cyclopentanol

(c) Iodocyclopentane

(d) Cyclopentane

Looking Ahead

3.68 Knowing what you do about the meaning of the terms "saturated" and "unsaturated" as applied to alkanes and alkenes, what do you think these same terms mean when they are used to describe animal fats such as those found in butter and animal meats? What might the term "polyunsaturated" mean in this same context?

3.69 In Chapter 21 on the biochemistry of lipids, we will study the three long-chain unsaturated carboxylic acids shown below. Each has 18 carbons and is a component of animal fats, vegetable oils, and biological membranes. Because of their presence in animal fats, they are called fatty acids. How many stereoisomers are possible for each fatty acid?

Oleic acid $CH_3(CH_2)_7CH{=}CH(CH_2)_7COOH$

Linoleic acid $CH_3(CH_2)_4(CH{=}CHCH_2)_2(CH_2)_6COOH$

Linolenic acid $CH_3CH_2(CH{=}CHCH_2)_3(CH_2)_6COOH$

3.70 The fatty acids in Problem 3.72 occur in animal fats, vegetable oils, and biological membranes almost exclusively as the all-*cis* isomers. Draw line-angle formulas for each fatty acid showing the *cis* configuration about each carbon–carbon double bond.

Benzene and Its Derivatives

Peppers of the Capsicum family (see Chemical Connections 4F).

Key Questions

4.1 What Is the Structure of Benzene?

4.2 How Do We Name Aromatic Compounds?

4.3 What Are the Characteristic Reactions of Benzene and Its Derivatives?

4.4 What Are Phenols?

4.1 What Is the Structure of Benzene?

So far we have described three classes of hydrocarbons—alkanes, alkenes, and alkynes—called aliphatic hydrocarbons. More than 150 years ago, organic chemists realized that yet another class of hydrocarbons exists, one whose properties are quite different from those of aliphatic hydrocarbons. Because some of these new hydrocarbons have pleasant odors, they were called **aromatic compounds.** Today we know that not all aromatic compounds share this characteristic. Some do have pleasant odors, but some

Aromatic compound
Benzene or one of its derivatives

OWL Online homework for this chapter may be assigned in **GOB OWL.**

have no odor at all, and others have downright unpleasant odors. A more appropriate definition of an aromatic compound is any compound that has one or more benzene-like rings.

We use the term **arene** to describe aromatic hydrocarbons. Just as a group derived by removal of an H from an alkane is called an alkyl group and given the symbol R—, a group derived by removal of an H from an arene is called an **aryl group** and given the symbol **Ar—**.

Benzene, the simplest aromatic hydrocarbon, was discovered in 1825 by Michael Faraday (1791–1867). Its structure presented an immediate problem to chemists of the day. Benzene has the molecular formula C_6H_6, and a compound with so few hydrogens for its six carbons (compare hexane, C_6H_{14}, and cyclohexane C_6H_{12}), chemists argued, should be unsaturated. But benzene does not behave like an alkene (the only class of unsaturated hydrocarbons known at that time). Whereas 1-hexene, for example, reacts instantly with Br_2 (Section 3.6C), benzene does not react at all with this reagent. Nor does benzene react with HBr, H_2O/H_2SO_4, or H_2/Pd—all reagents that normally add to carbon–carbon double bonds.

Arene A compound containing one or more benzene-like rings

Aryl group A group derived from an arene by removal of a H atom from an arene and given the symbol Ar—

Ar— The symbol used for an aryl group

Benzene is an important compound in both the chemical industry and the laboratory, but it must be handled carefully. Not only is it poisonous if ingested in liquid form, but its vapor is also toxic and can be absorbed either by breathing or through the skin. Long-term inhalation can cause liver damage and cancer.

A. Kekulé's Structure of Benzene

The first structure for benzene was proposed by Friedrich August Kekulé in 1872 and consisted of a six-membered ring with alternating single and double bonds, with one hydrogen bonded to each carbon.

A Kekulé structure
showing all atoms

A Kekulé structure
as a line-angle drawing

Although Kekulé's proposal was consistent with many of the chemical properties of benzene, it was contested for years. The major objection was its failure to account for the unusual chemical behavior of benzene. If benzene contains three double bonds, Kekulé's critics asked, why doesn't it undergo reactions typical of alkenes?

B. Resonance Structure of Benzene

The concept of resonance, developed by Linus Pauling in the 1930s, provided the first adequate description of the structure of benzene. According to the theory of resonance, certain molecules and ions are best described by writing two or more Lewis structures and considering the real molecule or ion to be a **resonance hybrid** of these structures. Each individual Lewis structure is called a **contributing structure.** To show that the real molecule is a resonance hybrid of the two Lewis structures, we position a double-headed arrow between them.

Resonance hybrid A molecule best described as a composite of two or more Lewis structures

Alternative Lewis contributing structures for benzene

A note about resonance hybrids. Do not confuse resonance contributing structures with equilibration among different chemical species. A molecule described as a resonance hybrid is not equilibrating among the electron configurations of the various contributing structures. Rather, the molecule has only one structure, which is best described as a hybrid of its various contributing structures. The colors of the color wheel provide a good analogy. Purple is not a primary color; the primary colors of blue and red are mixed to make purple. You can think of a molecule represented by a resonance hybrid as being purple. Purple is not sometimes blue and sometimes red. Purple is purple. In the analogous way, a molecule described as a resonance hybrid is not sometimes one contributing structure and sometimes another; it is a single structure all the time.

The resonance hybrid has some of the characteristics of each Lewis contributing structure. For example, the carbon–carbon bonds are neither single nor double but rather something intermediate between the two extremes. It has been determined experimentally that the length of the carbon–carbon bond in benzene is not as long as a carbon–carbon single bond nor as short as a carbon–carbon double bond, but rather is midway between the two. The closed loop of six electrons (two from the second bond of each double bond) characteristic of a benzene ring is sometimes called an **aromatic sextet.**

Wherever we find resonance, we find stability. The real structure is generally more stable than any of the hypothetical Lewis contributing structures. The benzene ring is greatly stabilized by resonance, which explains why it does not undergo the addition reactions typical of alkenes.

> The two contributing structures for benzene are often called Kekulé structures.

4.2 How Do We Name Aromatic Compounds?

A. One Substituent

Monosubstituted alkylbenzenes are named as derivatives of benzene—for example, ethylbenzene. The IUPAC system retains certain common names for several of the simpler monosubstituted alkylbenzenes, including **toluene** and **styrene.**

Ethylbenzene   Toluene   Styrene

The IUPAC system also retains common names for the following compounds:

Phenol   Anisole   Aniline   Benzaldehyde   Benzoic acid

> **Phenyl group** C_6H_5— the aryl group derived by removing a hydrogen atom from benzene

The substituent group derived by loss of an H from benzene is called a **phenyl group,** C_6H_5—, the common symbol for which is **Ph—.** In molecules containing other functional groups, phenyl groups are often named as substituents.

Phenyl group   1-Phenylcyclohexene   4-Phenyl-1-butene
(C_6H_5—; Ph—)

B. Two Substituents

When two substituents occur on a benzene ring, three isomers are possible. We locate the substituents either by numbering the atoms of the ring or by using the locators *ortho* (*o*), *meta* (*m*), and *para* (*p*). The numbers 1,2- are equivalent to *ortho* (Greek: straight); 1,3- to *meta* (Greek: after); and 1,4- to *para* (Greek: beyond).

1,2- or *ortho-*   1,3- or *meta*   1,4- or *para*

When one of the two substituents on the ring imparts a special name to the compound (for example, —CH_3, —OH, —NH_2, or —COOH), we name the compound as a derivative of that parent molecule and assume that the substituent occupies ring position number 1. The IUPAC system retains the common name **xylene** for the three isomeric dimethylbenzenes. Where neither substituent imparts a special name, we locate the two substituents and list them in alphabetical order before the ending "benzene." The carbon of the benzene ring with the substituent of lower alphabetical ranking is numbered C—1.

4-Bromobenzoic acid   3-Chloroaniline   1,3-Dimethylbenzene   1-Chloro-4-ethylbenzene
(*p*-Bromobenzoic acid)   (*m*-Chloroaniline)   (*m*-Xylene)   (*p*-Chloroethylbenzene)

C. Three or More Substituents

When three or more substituents are present on a benzene ring, specify their locations by numbers. If one of the substituents imparts a special name, then name the molecule as a derivative of that parent molecule. If none of the substituents imparts a special name, then locate the substituents, number them to give the smallest set of numbers, and list them in alphabetical order before the ending "benzene." In the following examples, the first compound is a derivative of toluene and the second is a derivative of phenol. Because no substituent in the third compound imparts a special name, list its three substituents in alphabetical order followed by the word "benzene."

p-Xylene is a starting material for the synthesis of poly(ethylene terephthalate). Consumer products derived from this polymer include Dacron polyester fibers and Mylar films (Section 11.6B).

4-Chloro-2-nitrotoluene   2,4,6-Tribromophenol   2-Bromo-1-ethyl-4-nitrobenzene

Example 4.1 Naming Aromatic Compounds

Write names for these compounds.

Strategy

First check to see if one of the substituents on the benzene ring imparts a special name. If one of them does, then name the compound as a derivative of that parent molecule.

Solution

(a) The parent is toluene, and the compound is 3-iodotoluene or *m*-iodotoluene.
(b) The parent is benzoic acid, and the compound is 3,5-dibromobenzoic acid.
(c) The parent is aniline, and the compound is 4-chloroaniline or *p*-chloroaniline.

Problem 4.1

Write names for these compounds.

D. Polynuclear Aromatic Hydrocarbons

Polynuclear aromatic hydrocarbon A hydrocarbon containing two or more benzene rings, each of which shares two carbon atoms with another benzene ring

Polynuclear aromatic hydrocarbons (PAHs) contain two or more benzene rings, with each pair of rings sharing two adjacent carbon atoms. Naphthalene, anthracene, and phenanthrene, the most common PAHs, and substances derived from them are found in coal tar and high-boiling petroleum residues.

Naphthalene   Anthracene   Phenanthrene

Chemical Connections 4A

Carcinogenic Polynuclear Aromatics and Smoking

A **carcinogen** is a compound that causes cancer. The first carcinogens to be identified were a group of polynuclear aromatic hydrocarbons, all of which have at least four aromatic rings. Among them is benzo[a]pyrene, one of the most carcinogenic of the aromatic hydrocarbons. It forms whenever incomplete combustion of organic compounds occurs. Benzo[a]pyrene is found, for example, in cigarette smoke, automobile exhaust, and charcoal-broiled meats.

Benzo[a]pyrene causes cancer in the following way. Once it is absorbed or ingested, the body attempts to convert it into a more water-soluble compound that can be excreted easily. By a series of enzyme-catalyzed reactions, benzo[a]pyrene is transformed into a **diol** (two —OH groups) **epoxide** (a three-membered ring, one atom of which is oxygen). This compound can bind to DNA by reacting with one of its amino groups, thereby altering the structure of DNA and producing a cancer-causing mutation.

Benzo[a]pyrene   A diol epoxide

At one time, naphthalene was used as mothballs and an insecticide in preserving woolens and furs, but its use decreased after the introduction of chlorinated hydrocarbons such as *p*-dichlorobenzene.

4.3 What Are the Characteristic Reactions of Benzene and Its Derivatives?

By far the most characteristic reaction of aromatic compounds is substitution at a ring carbon, which we give the name **aromatic substitution.** Groups we can introduce directly on the ring include the halogens, the nitro (—NO$_2$) group, and the sulfonic acid (—SO$_3$H) group.

A. Halogenation

As noted in Section 4.1, chlorine and bromine do not react with benzene, in contrast to their instantaneous reaction with cyclohexene and other alkenes (Section 3.6C). In the presence of an iron catalyst, however, chlorine reacts rapidly with benzene to give chlorobenzene and HCl:

Benzene   Chlorobenzene

Treatment of benzene with bromine in the presence of FeCl$_3$ results in formation of bromobenzene and HBr.

B. Nitration

When we heat benzene or one of its derivatives with a mixture of concentrated nitric and sulfuric acids, a nitro (—NO$_2$) group replaces one of the hydrogen atoms bonded to the ring.

Chemical Connections 4B

Iodide Ion and Goiter

One hundred years ago, goiter, an enlargement of the thyroid gland caused by iodine deficiency, was common in the central United States and central Canada. This disease results from underproduction of thyroxine, a hormone synthesized in the thyroid gland. Young mammals require this hormone for normal growth and development. A deficiency of thyroxine during fetal development results in mental retardation. Low levels of thyroxine in adults result in hypothyroidism, commonly called goiter, the symptoms of which are lethargy, obesity, and dry skin.

Iodine is an element that comes primarily from the sea. Rich sources of it, therefore, are fish and other seafoods. The iodine in our diets that doesn't come from the sea most commonly is derived from food additives. Most of the iodide ion in the North American diet comes from table salt fortified with sodium iodide, commonly referred to as iodized salt. Another source is dairy products, which accumulate iodide because of the iodine-containing additives used in cattle feeds and the iodine-containing disinfectants used on milking machines and milk storage tanks.

Thyroxine

Nitrobenzene

A particular value of nitration is that we can reduce the resulting —NO_2 group to a primary amino group, —NH_2, by catalytic reduction using hydrogen in the presence of a transition-metal catalyst. In the following example, neither the benzene ring nor the carboxyl group is affected by these experimental conditions:

4-Nitrobenzoic acid
(*p*-Nitrobenzoic acid)

4-Aminobenzoic acid
(*p*-Aminobenzoic acid, PABA)

Bacteria require *p*-aminobenzoic acid for the synthesis of folic acid (Section 22.4), which is in turn required for the synthesis of the heterocyclic aromatic amine bases of nucleic acids (Section 17.2). Whereas bacteria can synthesize folic acid from *p*-aminobenzoic acid, folic acid is a vitamin for humans and must be obtained through the diet.

C. Sulfonation

Heating an aromatic compound with concentrated sulfuric acid results in formation of an arenesulfonic acid, all of which are strong acids, comparable in strength to sulfuric acid.

Benzenesulfonic acid

Chemical Connections 4C

The Nitro Group in Explosives

Treatment of toluene with three moles of nitric acid in the presence of sulfuric acid as a catalyst results in nitration of toluene three times to form the explosive 2, 4, 6-trinitrotoluene, TNT. The presence of these three nitro groups confers the explosive properties to TNT. Similarly, the presence of three nitro groups leads to the explosive properties of nitroglycerin.

In recent years several new explosives have been discovered, all of which contain multiple nitro groups. Among them are RDX and PETN. The plastic explosive Semtex, for example, is a mixture of RDX and PETN. It was used in the destruction of Pan Am flight 103 over Lockerbie, Scotland, in December 1988.

A major use of sulfonation is in the preparation of synthetic detergents, an important example of which is sodium 4-dodecybenzenesulfonate. To prepare this type of detergent, a linear alkylbenzene such as dodecylbenzene is treated with concentrated sulfuric acid to give an alkylbenzenesulfonic acid. The sulfonic acid is then neutralized with sodium hydroxide.

$$CH_3(CH_2)_{10}CH_2 \text{—} \bigcirc \xrightarrow[\text{2. NaOH}]{\text{1. } H_2SO_4} CH_3(CH_2)_{10}CH_2 \text{—} \bigcirc \text{—} SO_3^-Na^+$$

Dodecylbenzene   Sodium 4-dodecylbenzenesulfonate, SDS
                                        (an anionic detergent)

Alkylbenzenesulfonate detergents were introduced in the late 1950s, and today they claim nearly 90% of the market once held by natural soaps. Section 10.4 discusses the chemistry and cleansing action of soaps and detergents.

4.4 What Are Phenols?

A. Structure and Nomenclature

The functional group of a **phenol** is a hydroxyl group bonded to a benzene ring. Substituted phenols are named either as derivatives of phenol or by common names.

Phenol A compound that contains an —OH group bonded to a benzene ring

Phenol 3-Methylphenol 1,2-Benzenediol 1,3-Benzenediol 1,4-Benzenediol
              (*m*-Cresol)   (Catechol)   (Resorcinol)   (Hydroquinone)

Phenol in crystalline form.

Charles D. Winters/Cengage Learning

Phenols are widely distributed in nature. Phenol itself and the isomeric cresols (*o-*, *m-*, and *p-*cresol) are found in coal tar. Thymol and vanillin are important constituents of thyme and vanilla beans, respectively. Urushiol is the main component of the irritating oil of poison ivy. It can cause severe contact dermatitis in sensitive individuals.

2-Isopropyl-5-methylphenol (Thymol)

4-Hydroxy-3-methoxy-benzaldehyde (Vanillin)

Urushiol

Poison ivy.

Charles D. Winters/Cengage Learning

B. Acidity of Phenols

Phenols are weak acids, with pK_a values of approximately 10 (Table 7.3). Most phenols are insoluble in water, but they react with strong bases, such as NaOH and KOH, to form water-soluble salts.

Phenol
pK_a = 9.95
(stronger acid)

Sodium hydroxide
(stonger base)

Sodium phenoxide
(weaker base)

Water
pK_a = 15.7
(weaker acid)

Most phenols are such weak acids that they do not react with weak bases such as sodium bicarbonate; that is, they do not dissolve in aqueous sodium bicarbonate.

C. Phenols as Antioxidants

An important reaction for living systems, foods, and other materials that contain carbon–carbon double bonds is **autoxidation**—that is, oxidation requiring oxygen and no other reactant. If you open a bottle of cooking oil that has stood for a long time, you may notice a hiss of air entering the bottle. This sound occurs because the consumption of oxygen by autoxidation of the oil creates a negative pressure inside the bottle.

Cooking oils contain esters of polyunsaturated fatty acids. We will discuss the structure and chemistry of esters in Chapter 11. The important point here is that all vegetable oils contain fatty acids with long hydrocarbon chains, many of which have one or more carbon–carbon double bonds (see Problems 3.29 and 3.72 for the structures of four of these fatty acids). Autoxidation takes place adjacent to one or more of their double bonds.

$$ \overset{\overset{\displaystyle H}{\displaystyle |}}{-CH_2CH=CH-CH-} + O_2 \xrightarrow[\text{or heat}]{\text{Light}} \overset{\overset{\displaystyle O-O-H}{\displaystyle |}}{-CH_2CH=CH-CH-} $$

Section of a fatty acid hydrocarbon chain

A hydroperoxide

Autoxidation is a radical-chain process that converts an R—H group to an R—O—O—H group, called a hydroperoxide. This process begins when a hydrogen atom with one of its electrons (H·) is removed from a carbon adjacent to one of the double bonds in a hydrocarbon chain. The carbon losing the H· has only seven electrons in its valence shell, one of which is unpaired. An atom or molecule with an unpaired electron is called a **radical.**

Chemical Connections 4D

FD & C No. 6 (a.k.a. Sunset Yellow)

Did you ever wonder what gives gelatin desserts their red, green, orange, or yellow color? Or what gives margarines their yellow color? Or what gives maraschino cherries their red color? If you read the content labels, you will see code names such as FD & C Yellow No. 6 and FD & C Red No. 40.

At one time, the only colorants for foods were compounds obtained from plant or animal materials. Beginning as early as the 1890s, however, chemists discovered a series of synthetic food dyes that offered several advantages over natural dyes, such as greater brightness, better stability, and lower cost. Opinion remains divided on the safety of their use. No synthetic food colorings are allowed, for example, in Norway and Sweden. In the United States, the Food and Drug Administration (FDA) has certified seven synthetic dyes for use in foods, drugs, and cosmetics

(FD & C)—two yellows, two reds, two blues, and one green. When these dyes are used alone or in combinations, they can approximate the color of almost any natural food.

Following are structural formulas for Allura Red (Red No. 40) and Sunset Yellow (Yellow No. 6). These and the other five food dyes certified in the United States have in common three or more benzene rings and two or more ionic groups, either the sodium salt of a carboxylic acid group, $-COO^-Na^+$, or the sodium salt of sulfonic acid group, $-SO_3^-Na^+$. These ionic groups make the dyes soluble in water.

To return to our original questions, maraschino cherries are colored with FD & C Red No. 40, and margarines are colored with FD & C Yellow No 6. Gelatin desserts use either one or a combination of the seven certified food dyes to create their color.

Allura Red
(FD & C Red No. 40)

Sunset Yellow
(FD & C Yellow No. 6)

Mechanism of Autoxidation

Step 1: Chain Initiation—Formation of a Radical from a Nonradical Compound
Removal of a hydrogen atom (H·) may be initiated by light or heat. The product formed is a carbon radical; that is, it contains a carbon atom with one unpaired electron.

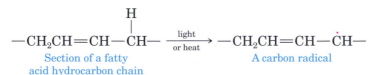

Section of a fatty
acid hydrocarbon chain

A carbon radical

Step 2a: Chain Propagation—Reaction of a Radical to Form a New Radical   The carbon radical reacts with oxygen, itself a diradical, to form a hydroperoxy radical. The new covalent bond of the hydroperoxy radical forms by the combination of one electron from the carbon radical and one electron from the oxygen diradical.

These two unpaired electrons combine to form a C—O single bond

Oxygen is a
diradical

A hydroperoxy radical

Chemical Connections 4E

Capsaicin, for Those Who Like It Hot

Capsaicin, the pungent principle from the fruit of various species of peppers (*Capsicum* and *Solanaceae*), was isolated in 1876, and its structure was determined in 1919. Capsaicin contains both a phenol and a phenol ether.

Capsaicin
(from various types of peppers)

The inflammatory properties of capsaicin are well known; the human tongue can detect as little as one drop in 5 L of water. We all know of the burning sensation in the mouth and sudden tearing in the eyes caused by a good dose of hot chili peppers. For this reason, capsaicin-containing extracts from these flaming foods are used in sprays to ward off dogs or other animals that might nip at your heels while you are running or cycling.

Paradoxically, capsaicin is able to both cause and relieve pain. Currently, two capsaicin-containing creams, Mioton and Zostrix, are prescribed to treat the burning pain associated with postherpetic neuralgia, a complication of the disease known as shingles. They are also prescribed for diabetics to relieve persistent foot and leg pain.

Red chilies being dried in the sun.

Step 2b: Chain Propagation—Reaction of a Radical to Form a New Radical The hydroperoxy radical removes a hydrogen atom (H·) from a new fatty acid hydrocarbon chain to complete the formation of a hydroperoxide and at the same time produce a new carbon radical.

The most important point about the pair of chain propagation steps (Steps 2a and 2b) is that they form a continuous cycle of reactions, in the following way. The new radical formed in Step 2b reacts with another molecule of O_2 by Step 2a to give a new hydroperoxy radical. This new hydroperoxy radical then reacts with a new hydrocarbon chain to repeat Step 2b, and so forth. This cycle of propagation steps repeats over and over in a chain reaction. Thus, once a radical is generated in Step 1, the cycle of propagation steps repeats many thousands of times and, in so doing, generates thousands of hydroperoxide molecules. The number of times the cycle of chain propagation steps repeats is called the **chain length.**

Hydroperoxides themselves are unstable and, under biological conditions, degrade to short-chain aldehydes and carboxylic acids with unpleasant "rancid" smells. These odors may be familiar to you if you have ever smelled old cooking oil or aged foods that contain polyunsaturated fats or oils. Similar formation of hydroperoxides in the low-density lipoproteins (Section 19.4) deposited on the walls of arteries leads to cardiovascular disease in humans. In addition, many effects of aging are thought to result from the formation and subsequent degradation of hydroperoxides.

Fortunately, nature has developed a series of defenses against the formation of these and other destructive hydroperoxides, including the phenol vitamin E (Section 22.6). This compound is one of "nature's scavengers." It inserts itself into either Step 2a or 2b, donates an H· from its —OH group to the carbon radical, and converts the carbon radical back to its original hydrocarbon chain. Because the vitamin E radical is stable, it breaks the cycle of chain propagation steps, thereby preventing further formation of destructive hydroperoxides. While some hydroperoxides may form, their numbers are very small and they are easily decomposed to harmless materials by one of several possible enzyme-catalyzed reactions.

Vitamin E

Butylated *hydroxy-toluene*
(BHT)

Butylated *hydroxy-anisole*
(BHA)

Unfortunately, vitamin E is removed in the processing of many foods and food products. To make up for this loss, phenols such as BHT and BHA are added to foods to "retard spoilage" (as they say on the packages) by autoxidation. Likewise, similar compounds are added to other materials, such as plastics and rubber, to protect them against autoxidation.

Summary of Key Questions

OWL End-of-chapter problems identified in blue are assignable in **GOB OWL.**

Section 4.1 What Is the Structure of Benzene? Problem 4.8

- **Benzene** and its alkyl derivatives are classified as **aromatic hydrocarbons** or **arenes**.
- The first structure for benzene was proposed by Friederich August Kekulé in 1872.
- The theory of **resonance**, developed by Linus Pauling in the 1930s, provided the first adequate structure for benzene.

Section 4.2 How Do We Name Aromatic Compounds? Problem 4.15

- Aromatic compounds are named according to the IUPAC system.

- The C_6H_5— group is named **phenyl.**
- Two substituents on a benzene ring may be located by either numbering the atoms of the ring or by using the locators *ortho* (*o*), *meta* (*m*), and *para* (*p*).
- **Polynuclear aromatic hydrocarbons** contain two or more benzene rings, each sharing two adjacent carbon atoms with another ring.

Section 4.3 What Are the Characteristic Reactions of Benzene and Its Derivatives? Problem 4.20

- A characteristic reaction of aromatic compounds is **aromatic substitution,** in which another atom or group of atoms is substituted for a hydrogen atom of the aromatic ring.
- Typical aromatic substitution reactions are halogenation, nitration, and sulfonation.

Section 4.4 What Are Phenols?

- The functional group of a phenol is an —OH group bonded to a benzene ring.
- Phenol and its derivatives are weak acids, with pK_a equal to approximately 10.0.

- Vitamin E, a phenolic compound, is a natural antioxidant.
- Phenolic compounds such as BHT and BHA are synthetic antioxidants.

Summary of Key Reactions

1. **Halogenation (Section 4.3A)** Treatment of an aromatic compound with Cl_2 or Br_2 in the presence of an $FeCl_3$ catalyst substitutes a halogen for an H.

2. **Nitration (Section 4.3B)** Heating an aromatic compound with a mixture of concentrated nitric and sulfuric acids substitutes a nitro group for an H.

3. **Sulfonation (Section 4.3C)** Heating an aromatic compound with concentrated sulfuric acid substitutes a sulfonic acid group for an H.

4. **Reaction of Phenols with Strong Bases (Section 4.4B)** Phenols are weak acids and react with strong bases to form water-soluble salts.

Problems

■ Indicates problems that are assignable in **GOB OWL**.

Blue numbered problems are applied.

Go to this book's companion website at **www.cengage. com/chemistry/bettelheim** for interactive versions of the How To tutorials and Active Figures, and to quiz yourself on this chapter.

Section 4.1 What Is the Structure of Benzene?

4.2 Answer true or false.
   (a) Alkenes, alkynes, and arenes are unsaturated hydrocarbons.
   (b) Aromatic compounds were so named because many of them have pleasant odors.
   (c) According to the resonance model of bonding, benzene is best described as a hybrid of two equivalent contributing structures.
   (d) Benzene is a planar molecule.

4.3 What is the difference in structure between a saturated and an unsaturated compound?

4.4 Define *aromatic compound*.

4.5 Why are alkenes, alkynes, and aromatic compounds said to be unsaturated?

4.6 Do aromatic rings have double bonds? Are they unsaturated? Explain.

4.7 Can an aromatic compound be a saturated compound?

4.8 ■ Draw at least two structural formulas for each of the following. (Several constitutional isomers are possible for each part.)
   (a) An alkene of six carbons
   (b) A cycloalkene of six carbons
   (c) An alkyne of six carbons
   (d) An aromatic hydrocarbon of eight carbons

4.9 Write a structural formula and the name for the simplest (a) alkane, (b) alkene, (c) alkyne, and (d) aromatic hydrocarbon.

4.10 Account for the fact that the six-membered ring in benzene is planar but the six-membered ring in cyclohexane is not.

4.11 The compound 1,4-dichlorobenzene (*p*-dichlorobenzene) has a rigid geometry that does not allow free rotation. Yet no *cis-trans* isomers exist for this structure. Explain why it does not show *cis-trans* isomerism.

4.12 One analogy often used to explain the concept of a resonance hybrid is to relate a rhinoceros to a unicorn and a dragon. Explain the reasoning in this analogy and how it might relate to a resonance hybrid.

Section 4.2 How Do We Name Aromatic Compounds?

4.13 Answer true or false.

(a) A phenyl group has the molecular formula C_6H_5, and is represented by the symbol Ph—.

(b) Para substituents occupy adjacent carbons on a benzene ring.

(c) 4-Bromobenzoic acid can be separated into *cis* and *trans* isomers.

(d) Naphthalene is a planar molecule.

(e) Benzene, naphthalene, and anthracene are polynuclear aromatic hydrocarbons (PAHs).

(f) Benzo[a]pyrene causes cancer by binding to DNA and producing a cancer-causing mutation.

4.14 Name these compounds.

(c) $C_6H_5CH_2CH_2CH_2Cl$

4.15 ■ Draw structural formulas for these compounds.

(a) 1-Bromo-2-chloro-4-ethylbenzene

(b) 4-Bromo-1,2-dimethylbenzene

(c) 2,4,6-Trinitrotoluene

(d) 4-Phenyl-1-pentene

(e) *p*-Cresol

(f) 2,4-Dichlorophenol

4.16 We say that naphthalene, anthracene, phenanthrene, and benzo[a]pyrene are polynuclear aromatic hydrocarbons. In this context, what does "polynuclear" mean? What does "aromatic" mean? What does "hydrocarbon" mean?

Section 4.3 What Are the Characteristic Reactions of Benzene and Its Derivatives?

4.17 Suppose you have unlabeled bottles of benzene and cyclohexene. What chemical reaction could you use

to tell which bottle contains which chemical? Explain what you would do, what you would expect to see, and how you would explain your observations.

4.18 Three products with the molecular formula C_6H_4BrCl form when bromobenzene is treated with chlorine, Cl_2, in the presence of $FeCl_3$ as a catalyst. Name and draw a structural formula for each product.

4.19 The reaction of bromine with toluene in the presence of $FeCl_3$ gives a mixture of three products, all with the molecular formula C_7H_7Br. Name and draw a structural formula for each product.

4.20 ■ What reagents and/or catalysts are necessary to carry out each conversion?

(a) Benzene to nitrobenzene

(b) 1,4-dichlorobenzene to 2-bromo-1,4-dichlorobenzene

(c) Benzene to aniline

4.21 What reagents and/or catalysts are necessary to carry out each conversion? Each conversion requires two steps.

(a) Benzene to 3-nitrobenzenesulfonic acid

(b) Benzene to 1-bromo-4-chlorobenzene

4.22 Aromatic substitution can be done on naphthalene. Treatment of naphthalene with concentrated H_2SO_4 gives two (and only two) different sulfonic acids. Draw a structural formula for each.

Section 4.4 What Are Phenols?

4.23 Answer true or false.

(a) Phenols and alcohols have in common the presence of an —OH group.

(b) Phenols are weak acids and react with strong bases to give water-soluble salts.

(c) The pK_a of phenol is smaller than that of acetic acid.

(d) Autoxidation converts an R—H group to an R—OH (hydroxyl) group.

(e) A carbon radical has only seven electrons in the valence shell of one of its carbons and this carbon bears a positive charge.

(f) A characteristic of a chain initiation step is conversion of a nonradical to a radical.

(g) Autoxidation is a radical-chain process.

(h) A characteristic of the chain propagation step is reaction of a radical and a molecule to form a new radical and a new molecule.

(i) Vitamin E and other natural antioxidants function by interrupting the cycle of chain propagation steps that occurs in autoxidation.

4.24 Both phenol and cyclohexanol are only slightly soluble in water. Account for the fact that phenol dissolves in aqueous sodium hydroxide but cyclohexanol does not.

4.25 Define *autoxidation.*

4.26 Autoxidation is described as a *radical-chain reaction.* What is meant by the term "radical" in this context? By the term "chain"? By the term "chain length"?

4.27 How does vitamin E function as an antioxidant?

4.28 What structural features are common to vitamin E, BHT, and BHA (the three antioxidants presented in Section 4.4C)?

Chemical Connections

4.29 (Chemical Connections 4A) What is a carcinogen? What kind of carcinogen is found in cigarette smoke?

4.30 (Chemical Connections 4B) In the absence of iodine in the diet, goiter develops. Explain why goiter is a regional disease.

4.31 (Chemical Connections 4C) Calculate the molecular weight of each explosive in this Chemical Connection. In which explosive do the nitro groups contribute the largest percentage of molecular weight?

4.32 (Chemical Connections 4D) What are the differences in structure between Allura Red and Sunset Yellow?

4.33 (Chemical Connections 4D) Which features of Allura Red and Sunset Yellow make them water-soluble?

4.34 (Chemical Connections 4D) What color would you get if you mixed Allura Red and Sunset Yellow? (*Hint*: Remember the color wheel.)

4.35 (Chemical Connections 4E) From what types of plants is capsaicin isolated?

4.36 (Chemical Connections 4E) How many *cis-trans* isomers are possible for capsaicin? Is the structural formula shown in this Chemical Connection the *cis* isomer or the *trans* isomer?

Additional Problems

4.37 The structure for naphthalene given in Section 4.2D is only one of three possible resonance structures. Draw the other two.

4.38 ■ Draw structural formulas for these compounds.
(a) 1-Phenylcyclopropanol   (b) Styrene
(c) *m*-Bromophenol   (d) 4-Nitrobenzoic acid
(e) Isobutylbenzene   (f) *m*-Xylene

4.39 2,6-Di-*tert*-butyl-4-methylphenol (BHT, Section 4.4C) is an antioxidant added to processed foods to "retard spoilage." How does BHT accomplish this goal?

4.40 Write the structural formula for the product of each reaction.

4.41 ■ Styrene reacts with bromine to give a compound with the molecular formula $C_8H_8Br_2$. Draw a structural formula for this compound.

Alcohols, Ethers, and Thiols

Fermentation vats of wine grapes at the Beaulieu Vineyards, California.

Earl Robber, Photo Researchers, Inc.

Key Questions

5.1 What Are the Structures, Names, and Physical Properties of Alcohols?

5.2 What Are the Characteristic Reactions of Alcohols?

5.3 What Are the Structures, Names, and Properties of Ethers?

5.4 What Are the Structures, Names, and Properties of Thiols?

5.5 What Are the Most Commercially Important Alcohols?

In this chapter, we study the physical and chemical properties of alcohols and ethers, two classes of oxygen-containing organic compounds. We also study thiols, a class of sulfur-containing organic compounds. Thiols are like alcohols in structure, except that they contain an —SH group rather than an —OH group.

$$CH_3CH_2OH \qquad CH_3CH_2OCH_2CH_3 \qquad CH_3CH_2SH$$

Ethanol   Diethyl ether   Ethanethiol

(An alcohol)   (An ether)   (A thiol)

Solutions in which ethanol is the solvent are called tinctures.

OWL Online homework for this chapter may be assigned in **GOB OWL**.

Problem 5.2

Classify each alcohol as primary, secondary, or tertiary.

(a)

(b)

(c)

(d)

In the IUPAC system, a compound containing two hydroxyl groups is named as a **diol,** one containing three hydroxyl groups as a **triol,** and so on. In IUPAC names for diols, triols, and so on, the final -*e* in the name of the parent alkane is retained—for example, 1,2-ethanediol.

Diol A compound containing two —OH (hydroxyl) groups

Glycol A compound with hydroxyl (—OH) groups on adjacent carbons

Ethylene glycol is colorless; the color of most antifreezes comes from additives.

As with many other organic compounds, common names for certain diols and triols have persisted. Compounds containing two hydroxyl groups on adjacent carbons are often referred to as **glycols.** Ethylene glycol and propylene glycol are synthesized from ethylene and propylene, respectively—hence their common names.

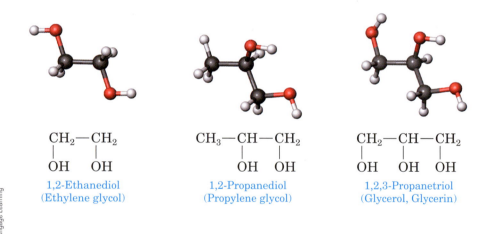

CH₂—CH₂
| |
OH   OH
1,2-Ethanediol
(Ethylene glycol)

CH₃—CH—CH₂
| |
OH   OH
1,2-Propanediol
(Propylene glycol)

CH₂—CH—CH₂
| | |
OH OH OH
1,2,3-Propanetriol
(Glycerol, Glycerin)

Ethylene glycol is a polar molecule and dissolves readily in the polar solvent water.

C. Physical Properties of Alcohols

The most important physical property of alcohols is the polarity of their —OH groups. Because of the large difference in electronegativity between oxygen and carbon ($3.5 - 2.5 = 1.0$) and between oxygen and hydrogen ($3.5 - 2.1 = 1.4$), both the C—O and O—H bonds of alcohols are polar covalent, and alcohols are polar molecules, as illustrated in Figure 5.2 for methanol.

FIGURE 5.2 Polarity of the C—O—H bonds in methanol. (*a*) There are partial positive charges on carbon and hydrogen and a partial negative charge on oxygen. (*b*) An electron density map showing the partial negative charge (in red) around oxygen and a partial positive charge (in blue) around hydrogen of the OH group.

(a)

(b)

Chemical Connections 5A

Nitroglycerin: An Explosive and a Drug

In 1847, Ascanio Sobrero (1812–1888) discovered that 1,2,3-propanetriol, more commonly called glycerin, reacts with nitric acid in the presence of sulfuric acid to give a pale yellow, oily liquid called nitroglycerin. Sobrero also discovered the explosive properties of this compound; when he heated a small quantity of it, it exploded!

$$\begin{array}{ccc}
\text{CH}_2\!-\!\text{OH} & & \text{CH}_2\!-\!\text{ONO}_2 \\
| & & | \\
\text{CH}\!-\!\text{OH} + 3\text{HNO}_3 \xrightarrow{\;\text{H}_2\text{SO}_4\;} & \text{CH}\!-\!\text{ONO}_2 + 3\text{H}_2\text{O} \\
| & & | \\
\text{CH}_2\!-\!\text{OH} & & \text{CH}_2\!-\!\text{ONO}_2 \\
\text{1,2,3-Propanetriol} & & \text{1,2,3-Propanetriol trinitrate} \\
\text{(Glycerol, Glycerin)} & & \text{(Nitroglycerin)}
\end{array}$$

Nitroglycerin very soon became widely used for blasting in the construction of canals, tunnels, roads, and mines and, of course, for warfare.

One problem with the use of nitroglycerin was soon recognized: It was difficult to handle safely, and accidental explosions occurred all too frequently. This problem was solved by the Swedish chemist Alfred Nobel (1833–1896),

who discovered that a clay-like substance called diatomaceous earth absorbs nitroglycerin so that it will not explode without a fuse. Nobel gave the name *dynamite* to this mixture of nitroglycerin, diatomaceous earth, and sodium carbonate.

Surprising as it may seem, nitroglycerin is used in medicine to treat angina pectoris, the symptoms of which are sharp chest pains caused by reduced flow of blood in the coronary artery. Nitroglycerin, which is available in liquid form (diluted with alcohol to render it nonexplosive), tablet form, and paste form, relaxes the smooth muscles of blood vessels, causing dilation of the coronary artery. This dilation, in turn, allows more blood to reach the heart.

When Nobel became ill with heart disease, his physicians advised him to take nitroglycerin to relieve his chest pains. He refused, saying he could not understand how the explosive could relieve chest pains. It took science more than 100 years to find the answer. We now know that nitric oxide, NO, derived from the nitro groups of nitroglycerin, relieves the pain (see Chemical Connections 16E).

(a)

(b)

(a) Nitroglycerin is more stable when absorbed onto an inert solid, a combination called dynamite. (b) The fortune of Alfred Nobel (1833–1896), built on the manufacture of dynamite, now funds the Nobel Prizes.

(a) Charles D. Winters (b) Bettmann Archives/Corbis

Alcohols have higher boiling points than do alkanes, alkenes, and alkynes of similar molecular weight (Table 5.1), because alcohol molecules associate with one another in the liquid state by **hydrogen bonding.** The strength of hydrogen bonding between alcohol molecules is approximately 2 to 5 kcal/mol, which means that extra energy is required to separate hydrogen-bonded alcohols from their neighbors (Figure 5.3).

Because of increased London dispersion forces between larger molecules, the boiling points of all types of compounds, including alcohols, increase with increasing molecular weight.

Alcohols are much more soluble in water than are hydrocarbons of similar molecular weight (Table 5.1), because alcohol molecules interact by hydrogen bonding with water molecules. Methanol, ethanol, and 1-propanol are

TABLE 5.1 Boiling Points and Solubilities in Water of Four
Sets of Alcohols and Alkanes of Similar Molecular Weight

Structural Formula	Name	Molecular Weight (amu)	Boiling Point (°C)	Solubility in Water
CH_3OH	methanol	32	65	infinite
CH_3CH_3	ethane	30	−89	insoluble
CH_3CH_2OH	ethanol	46	78	infinite
$CH_3CH_2CH_3$	propane	44	−42	insoluble
$CH_3CH_2CH_2OH$	1-propanol	60	97	infinite
$CH_3CH_2CH_2CH_3$	butane	58	0	insoluble
$CH_3CH_2CH_2CH_2OH$	1-butanol	74	117	8 g/100 g
$CH_3CH_2CH_2CH_2CH_3$	pentane	72	36	insoluble

FIGURE 5.3 The association of ethanol molecules in the liquid state. Each O—H can participate in up to three hydrogen bonds (one through hydrogen and two through oxygen). Only two of these three possible hydrogen bonds per molecule are shown.

Hydrogen bonding

soluble in water in all proportions. As molecular weight increases, the water solubility of alcohols becomes more like that of hydrocarbons of similar molecular weight. Higher-molecular-weight alcohols are much less soluble in water because the size of the hydrocarbon portion of their molecules (which decreases water solubility) becomes so large relative to the size of the —OH group (which increases water solubility).

5.2 What Are the Characteristic Reactions of Alcohols?

In this section, we study the acidity of alcohols, their dehydration to alkenes, and their oxidation to aldehydes, ketones, and carboxylic acids.

A. Acidity of Alcohols

Alcohols have about the same pK_a values as water (Table 7.3), which means that aqueous solutions of alcohols have approximately the same pH as that of pure water. In Section 4.4B, we studied the acidity of phenols, another class of compounds that contains an —OH group. Phenols are weak acids and react with aqueous sodium hydroxide to form water-soluble salts.

Phenol

Sodium phenoxide
(a water-soluble salt)

Alcohols are considerably weaker acids than phenols and do not react in this manner.

B. Acid-Catalyzed Dehydration of Alcohols

We can convert an alcohol to an alkene by eliminating a molecule of water from adjacent carbon atoms in a reaction called **dehydration.** In the laboratory, dehydration of an alcohol is most often brought about by heating it with either 85% phosphoric acid or concentrated sulfuric acid. Primary alcohols—the most difficult to dehydrate—generally require heating in concentrated sulfuric acid at temperatures as high as 180°C. Secondary alcohols undergo acid-catalyzed dehydration at somewhat lower temperatures. Tertiary alcohols generally undergo acid-catalyzed dehydration at temperatures only slightly above room temperature.

Dehydration Elimination of a molecule of water from an alcohol. In the dehydration of an alcohol, OH is removed from one carbon and an H is removed from an adjacent carbon.

Thus the ease of acid-catalyzed dehydration of alcohols follows this order:

1° alcohols   2° alcohols   3° alcohols

Ease of dehydration of alcohols

When the acid-catalyzed dehydration of an alcohol yields isomeric alkenes, the alkene having the greater number of alkyl groups on the double bond generally predominates. In the acid-catalyzed dehydration of 2-butanol, for example, the major product is 2-butene, which has two alkyl groups (two methyl groups) on its double bond. The minor product is 1-butene, which has only one alkyl group (an ethyl group) on its double bond.

$$\underset{\text{2-Butanol}}{CH_3CH_2\overset{\displaystyle OH}{\overset{|}{C}}HCH_3} \xrightarrow[\text{heat}]{H_3PO_4} \underset{\substack{\text{2-Butene}\\(80\%)}}{CH_3CH{=}CHCH_3} + \underset{\substack{\text{1-Butene}\\(20\%)}}{CH_3CH_2CH{=}CH_2} + H_2O$$

Example 5.3 Acid-Catalyzed Dehydration of Alcohols

Draw structural formulas for the alkenes formed by the acid-catalyzed dehydration of each alcohol. For each part, predict which alkene will be the major product.

(a) 3-Methyl-2-butanol   (b) 2-Methylcyclopentanol

Strategy

In acid-catalyzed dehydration of an alcohol, H and OH are removed from adjacent carbon atoms. When dehydration yields isomeric alkenes, the alkene with the greater number of alkyl groups on the carbon atoms of the double bond generally predominates.

Solution

(a) Elimination of H_2O from carbons 2–3 gives 2-methyl-2-butene; elimination of H_2O from carbons 1–2 gives 3-methyl-1-butene. 2-Methyl-2-butene has three alkyl groups (three methyl groups) on its double bond and is the major product. 3-Methyl-1-butene has only one alkyl group (an isopropyl group) on its double bond and is the minor product.

3-Methyl-2-butanol   2-Methyl-2-butene (major product)   3-Methyl-1-butene

(b) The major product, 1-methylcyclopentene, has three alkyl groups on its double bond. The minor product, 3-methylcyclopentene, has only two alkyl groups on its double bond.

2-Methylcyclopentanol   1-Methylcyclopentene (major product)   3-Methylcyclopentene

Problem 5.3

Draw structural formulas for the alkenes formed by the acid-catalyzed dehydration of each alcohol. For each part, predict which alkene will be the major product.

(a) 2-Methyl-2-butanol   (b) 1-Methylcyclopentanol

In Section 3.6B, we studied the acid-catalyzed hydration of alkenes to give alcohols. In this section, we study the acid-catalyzed dehydration of alcohols to give alkenes. In fact, hydration–dehydration reactions are reversible. Alkene hydration and alcohol dehydration are competing reactions, and the following equilibrium exists:

An alkene   An alcohol

In accordance with Le Chatelier's principle, large amounts of water (in other words, using dilute aqueous acid) favor alcohol formation, whereas scarcity of water (using concentrated acid) or experimental conditions where water is removed (heating the reaction mixture above 100°C) favor alkene formation. Thus, depending on the experimental conditions, we can use the hydration–dehydration equilibrium to prepare both alcohols and alkenes, each in high yields.

Example 5.4 Acid-Catalyzed Dehydration of Alcohols and Hydration of Alkenes

In part (a), acid-catalyzed dehydration of 2-methyl-3-pentanol gives predominantly Compound A. Treatment of Compound A with water in the presence of sulfuric acid in part (b) gives Compound B. Propose structural formulas for Compounds A and B.

(a)
$$\underset{\underset{\text{OH}}{|}}{\underset{\overset{\overset{\text{CH}_3}{|}}{}}{\text{CH}_3\text{CHCHCH}_2\text{CH}_3}} \xrightarrow[\substack{\text{acid-catalyzed} \\ \text{dehydration}}]{\text{H}_2\text{SO}_4} \text{Compound A}(\text{C}_6\text{H}_{12}) + \text{H}_2\text{O}$$

(b) $\text{Compound A}(\text{C}_6\text{H}_{12}) + \text{H}_2\text{O} \xrightarrow{\text{H}_2\text{SO}_4} \text{Compound B}(\text{C}_6\text{H}_{14}\text{O})$

Strategy

The key to part (a) is that when acid-catalyzed dehydration of an alcohol can yield isomeric alkenes, the alkene with the greater number of alkyl groups on the carbon atoms of the double bond generally predominates. After the structural formula of A is determined, use Markovnikov's Rule to predict the structural formula of compound B.

Solution

(a) Acid-catalyzed dehydration of 2-methyl-3-pentanol gives predominantly 2-methyl-2-pentene, an alkene with three substituents on its double bond: two methyl groups and one ethyl group.

$$\underset{\text{2-Methyl-3-butanol}}{\underset{\underset{\text{OH}}{|}}{\underset{\overset{\overset{\text{CH}_3}{|}}{}}{\text{CH}_3\text{CHCHCH}_2\text{CH}_3}}} \xrightarrow[\substack{\text{acid-catalyzed} \\ \text{dehydration}}]{\text{H}_2\text{SO}_4} \underset{\substack{\text{2-Methyl-2-pentene} \\ \text{(major product)}}}{\underset{\overset{\overset{\text{CH}_3}{|}}{}}{\text{CH}_3\text{C}=\text{CHCH}_2\text{CH}_3}} + \text{H}_2\text{O}$$

(b) Acid-catalyzed addition of water to this alkene gives 2-methyl-2-pentanol in accord with Markovnikov's rule (Section 3.6B).

$$\underset{\text{Compound A}(\text{C}_6\text{H}_{12})}{\underset{\overset{\overset{\text{CH}_3}{|}}{}}{\text{CH}_3\text{C}=\text{CHCH}_2\text{CH}_3}} + \text{H}_2\text{O} \xrightarrow[\substack{\text{acid-catalyzed} \\ \text{hydration}}]{\text{H}_2\text{SO}_4} \underset{\text{Compound B}(\text{C}_6\text{H}_{14}\text{O})}{\underset{\underset{\text{OH}}{|}}{\underset{\overset{\overset{\text{CH}_3}{|}}{}}{\text{CH}_3\text{CCH}_2\text{CH}_2\text{CH}_3}}}$$

Problem 5.4

Acid-catalyzed dehydration of 2-methylcyclohexanol gives predominantly Compound C (C_7H_{12}). Treatment of Compound C with water in the presence of sulfuric acid gives Compound D ($\text{C}_7\text{H}_{14}\text{O}$). Propose structural formulas for Compounds C and D.

C. Oxidation of Primary and Secondary Alcohols

A primary alcohol can be oxidized to an aldehyde or to a carboxylic acid, depending on the experimental conditions. Following is a series of transformations in which a primary alcohol is oxidized first to an aldehyde and then to a carboxylic acid. The letter O in brackets over the reaction arrow indicates that each transformation involves oxidation.

A primary alcohol → An aldehyde → A carboxylic acid

According to one definition, oxidation is either the loss of hydrogens or the gain of oxygens. Using this definition, conversion of a primary alcohol to an aldehyde is an oxidation reaction because the alcohol loses hydrogens. Conversion of an aldehyde to a carboxylic acid is also an oxidation reaction because the aldehyde gains an oxygen.

The reagent most commonly used in the laboratory for the oxidation of a primary alcohol to a carboxylic acid is potassium dichromate, $K_2Cr_2O_7$, dissolved in aqueous sulfuric acid. Using this reagent, oxidation of 1-octanol, for example, gives octanoic acid. This experimental condition is more than sufficient to oxidize the intermediate aldehyde to a carboxylic acid.

$$CH_3(CH_2)_6CH_2OH \xrightarrow[H_2SO_4]{K_2Cr_2O_7} CH_3(CH_2)_6\overset{\displaystyle O}{\overset{\|}{C}}H \xrightarrow[H_2SO_4]{K_2Cr_2O_7} CH_3(CH_2)_6\overset{\displaystyle O}{\overset{\|}{C}}OH$$

1-Octanol — Octanal — Octanoic acid

Although the usual product of oxidation of a primary alcohol is a carboxylic acid, it is often possible to stop the oxidation at the aldehyde stage by distilling the mixture. That is, the aldehyde, which usually has a lower boiling point than either the primary alcohol or the carboxylic acid, is removed from the reaction mixture before it can be oxidized further.

Secondary alcohols may be oxidized to ketones by using potassium dichromate as the oxidizing agent. Menthol, a secondary alcohol present in peppermint and other mint oils, is used in liqueurs, cigarettes, cough drops, perfumery, and nasal inhalers. Its oxidation product, menthone, is also used in perfumes and artificial flavors.

2-Isopropyl-5-methyl-cyclohexanol (Menthol) → 2-Isopropyl-5-methyl-cyclohexanone (Menthone)

Tertiary alcohols resist oxidation because the carbon bearing the —OH is bonded to three carbon atoms and, therefore, cannot form a carbon–oxygen double bond.

Example 5.5 Oxidation of Alcohols

Draw a structural formula for the product formed by oxidation of each alcohol with potassium dichromate.

(a) 1-Hexanol   (b) 2-Hexanol

Strategy

Oxidation of a 1-hexanol, a primary alcohol, gives either an aldehyde or a carboxylic acid, depending on the experimental conditions. Oxidation of 2-hexanol, a secondary alcohol, gives a ketone.

Solution

(a) Hexanal   or   Hexanoic acid   (b) 2-Hexanone

Problem 5.5

Draw the product formed by oxidation of each alcohol with potassium dichromate.

(a) Cyclohexanol   (b) 2-Pentanol

Chemical Connections 5B

Breath-Alcohol Screening

Potassium dichromate oxidation of ethanol to acetic acid is the basis for the original breath-alcohol screening test used by law enforcement agencies to determine a person's blood alcohol content (BAC). The test is based on the difference in color between the dichromate ion (reddish orange) in the reagent and the chromium(III) ion (green) in the product.

$$CH_3CH_2OH \ + \ Cr_2O_7{}^{2-}$$

Ethanol   Dichromate ion
               (reddish orange)

$$\xrightarrow[\text{H}_2\text{O}]{\text{H}_2\text{SO}_4} \ CH_3\overset{\overset{\displaystyle O}{\|}}{C}OH \ + \ Cr^{3+}$$

Acetic acid   Chromium(III) ion
              (green)

In its simplest form, breath-alcohol screening uses a sealed glass tube that contains a potassium dichromate–sulfuric acid reagent impregnated on silica gel. To administer the test, the ends of the tube are broken off, a mouthpiece is fitted to one end, and the other end is inserted into the neck of a plastic bag. The person being tested then blows into the mouthpiece to inflate the plastic bag.

As breath containing ethanol vapor passes through the tube, reddish orange dichromate ion is reduced to green chromium(III) ion. To estimate the concentration of ethanol in the breath, one measures how far the green color extends along the length of the tube. When it reaches beyond the halfway point, the person is judged as having a sufficiently high blood alcohol content to warrant further, more precise testing.

The test described here measures the alcohol content of the breath. The legal definition of being under the influence of alcohol, however, is based on alcohol content in the blood, not in the breath. The correlation between these two measurements is based on the fact that air deep within the lungs is in equilibrium with blood passing through the pulmonary arteries, and thus an equilibrium is established between blood alcohol and breath alcohol. Based on tests in persons drinking alcohol, researchers have determined that 2100 mL of breath contains the same amount of ethanol as 1.00 mL of blood.

Person forces breath through mouthpiece into the tube.

Glass tube containing potassium dichromate–sulfuric acid coated on silica gel particles

As the person blows into the tube, the plastic bag becomes inflated.

5.3 What Are the Structures, Names, and Properties of Ethers?

A. Structure

Ether A compound containing an oxygen atom bonded to two carbon atoms

The functional group of an **ether** is an atom of oxygen bonded to two carbon atoms. Figure 5.4 shows a Lewis structure and a ball-and-stick model of dimethyl ether, CH_3OCH_3, the simplest ether.

B. Nomenclature

Although the IUPAC system can be used to name ethers, chemists almost invariably use common names for low-molecular-weight ethers. Common names are derived by listing the alkyl groups bonded to oxygen in alphabetical order and adding the word *ether*. Alternatively, one of the groups on oxygen is named as an alkoxy group. The $-OCH_3$ group, for example, is named "methoxy" to indicate a methyl group bonded to oxygen.

(a)

(b)

FIGURE 5.4 Dimethyl ether, CH_3OCH_3. (*a*) Lewis structure and (*b*) ball-and-stick model. The C—O—C bond angle is 110.3°, close to the tetrahedral angle of 109.5°.

$$CH_3CH_2OCH_2CH_3$$

Diethyl ether

Cyclohexyl methyl ether
(Methoxycyclohexane)

Example 5.6 Common Names for Ethers

Write the common name for each ether.

(a)

(b)

Strategy

To derive the common name of an ether, list the groups bonded to oxygen in alphabetical order.

Solution

(a) The groups bonded to the ether oxygen are *tert*-butyl and ethyl. The compound's common name is *tert*-butyl ethyl ether.
(b) Two cyclohexyl groups are bonded to the ether oxygen. The compound's common name is dicyclohexyl ether.

Problem 5.6

Write the common name for each ether.

(a)

(b) —OCH₃

Chemical Connections 5C

Ethylene Oxide: A Chemical Sterilant

Ethylene oxide is a colorless, flammable gas with a boiling point of 11°C. Because it is such a highly strained molecule (the normal tetrahedral bond angles of both C and O are compressed from the normal tetrahedral angle of 109.5° to approximately 60°), ethylene oxide reacts with the amino ($-NH_2$) and sulfhydryl ($-SH$) groups present in biological materials.

At sufficiently high concentrations, it reacts with enough molecules in cells to cause the death of microorganisms. This toxic property is the basis for ethylene oxide's use as a fumigant in foodstuffs and textiles and its use in hospitals to sterilize surgical instruments.

In **cyclic ethers,** one of the atoms in a ring is oxygen. These ethers are also known by their common names. Ethylene oxide is an important building block for the organic chemical industry (Section 5.5). Tetrahydrofuran is a useful laboratory and industrial solvent.

Cyclic ether An ether in which the ether oxygen is one of the atoms of a ring

Ethylene oxide   Tetrahydrofuran
(THF)

C. Physical Properties

Ethers are polar compounds in which oxygen bears a partial negative charge and each attached carbon bears a partial positive charge (Figure 5.5). However, only weak forces of attraction exist between ether molecules in the pure liquid. Consequently, the boiling points of ethers are close to those of hydrocarbons of similar molecular weight.

The effect of hydrogen bonding on physical properties is illustrated dramatically by comparing the boiling points of ethanol (78°C) and its constitutional isomer dimethyl ether (−24°C). The difference in boiling point

FIGURE 5.5 Ethers are polar molecules, but only weak attractive interactions exist between ether molecules in the liquid state. Shown on the right is an electron density map of diethyl ether.

Chemical Connections 5D

Ethers and Anesthesia

Before the mid-1800s, surgery was performed only when absolutely necessary, because no truly effective general anesthetic was available. More often than not, patients were drugged, hypnotized, or simply tied down.

In 1772, Joseph Priestley isolated nitrous oxide, N_2O, a colorless gas. In 1799, Sir Humphry Davy demonstrated this compound's anesthetic effect, naming it "laughing gas." In 1844, an American dentist, Horace Wells, introduced nitrous oxide into general dental practice. One patient

This photo shows the first use of ether as an anesthetic in 1848. Dr. Robert John Collins was removing a tumor from the patient's neck and dentist W. T. G. Morton—who discovered the anesthetic properties—administered the ether.

Courtesy of Library of Congress

awakened prematurely, however, screaming with pain; another died during the procedure. Wells was forced to withdraw from practice, became embittered and depressed, and committed suicide at age 33. In the same period, a Boston chemist, Charles Jackson, anesthetized himself with diethyl ether; he also persuaded a dentist, William Morton, to use it. Subsequently, they persuaded a surgeon, John Warren, to give a public demonstration of surgery under anesthesia. The operation was completely successful, and soon general anesthesia by diethyl ether became routine practice for general surgery.

Diethyl ether was easy to use and caused excellent muscle relaxation. Blood pressure, pulse rate, and respiration were usually only slightly affected. Diethyl ether's chief drawbacks are its irritating effect on the respiratory passages and its aftereffect of nausea.

Among the inhalation anesthetics used today are several halogenated ethers, the most important of which are enflurane and isoflurane.

Enflurane
(Ethrane)

Isoflurane
(Forane)

between these two compounds is due to the presence in ethanol of a polar O—H group, which is capable of forming hydrogen bonds. This hydrogen bonding increases intermolecular associations, thereby giving ethanol a higher boiling point than dimethyl ether.

CH_3CH_2OH
Ethanol
bp 78°C

CH_3OCH_3
Dimethyl ether
bp −24°C

Ethers are more soluble in water than hydrocarbons of similar molecular weight and shape, but far less soluble than isomeric alcohols. Their greater solubility reflects the fact that the oxygen atom of an ether carries a partial negative charge and forms hydrogen bonds with water.

D. Reactions of Ethers

Ethers resemble hydrocarbons in their resistance to chemical reaction. For example, they do not react with oxidizing agents, such as potassium dichromate. Likewise, they do not react with reducing agents such as H_2 in the presence of a metal catalyst (Section 3.6D). Furthermore, most acids and bases at moderate temperatures do not affect them. Because of their general inertness to chemical reaction and their good solvent properties, ethers are excellent solvents in which to carry out many organic reactions. The most important ether solvents are diethyl ether and tetrahydrofuran.

5.4 What Are the Structures, Names, and Properties of Thiols?

A. Structure

The functional group of a **thiol** is an —SH **(sulfhydryl) group** bonded to a tetrahedral carbon atom. Figure 5.6 shows a Lewis structure and a ball-and-stick model of methanethiol, CH_3SH, the simplest thiol.

Thiol A compound that contains an —SH (sulfhydryl) group bonded to a tetrahedral carbon atom

(a)

(b)

FIGURE 5.6 Methanethiol, CH_3SH. (a) Lewis structure and (b) ball-and-stick model. The H—S—C bond angle is 100.3°, somewhat smaller than the tetrahedral angle of 109.5°.

B. Nomenclature

The sulfur analog of an alcohol is called a thiol (*thi-* from the Greek: *theion*, sulfur) or, in the older literature, a **mercaptan,** which literally means "mercury capturing." Thiols react with Hg^{2+} in aqueous solution to give sulfide salts as insoluble precipitates. Thiophenol, C_6H_5SH, for example, gives $(C_6H_5S)_2Hg$.

In the IUPAC system, thiols are named by selecting the longest carbon chain that contains the —SH group as the parent alkane. To show that the compound is a thiol, we add the suffix *-thiol* to the name of the parent alkane. The parent chain is numbered in the direction that gives the —SH group the lower number.

Common names for simple thiols are derived by naming the alkyl group bonded to —SH and adding the word *mercaptan*.

Mercaptan A common name for any molecule containing an —SH group

CH$_3$CH$_2$SH
Ethanethiol
(Ethyl mercaptan)

$\overset{\displaystyle CH_3}{\underset{\displaystyle }{|}}$
CH$_3$CHCH$_2$SH
2-Methyl-1-propanethiol
(Isobutyl mercaptan)

Example 5.7 Systematic Names of Thiols

Write the IUPAC name for each thiol.

(a)   (b) ⎯⎯⎯SH (with SH above)

Strategy

To derive the IUPAC name of a thiol, select as the parent alkane the longest carbon chain that contains the —SH group. Show that the compound is a thiol by adding the suffix *-thiol* to the name of the parent alkane. Number the parent chain in the direction that gives the —SH group the lower number.

Solution

(a) The parent alkane is pentane. Show the presence of the —SH group by adding "thiol" to the name of the parent alkane. The IUPAC name of this thiol is 1-pentanethiol. Its common name is pentyl mercaptan.

(b) The parent alkane is butane. The IUPAC name of this thiol is 2-butanethiol. Its common name is *sec*-butyl mercaptan.

Problem 5.7

Write the IUPAC name for each thiol.

(a)   (b)

The most notable property of low-molecular-weight **thiols** is their stench. They are responsible for unpleasant odors such as those from skunks, rotten eggs, and sewage. The scent of skunks is due primarily to two thiols:

$$CH_3CH=CHCH_2SH$$

2-Butene-1-thiol

$$CH_3CHCH_2CH_2SH$$
$$\overset{\displaystyle CH_3}{|}$$

3-Methyl-1-butanethiol

The scent of the spotted skunk, native to the Sonoran Desert, is a mixture of two thiols, 2-buten-1-thiol and 3-methyl-1-butanethiol.

C. Physical Properties

Because of the small difference in electronegativity between sulfur and hydrogen ($2.5 - 2.1 = 0.4$), we classify the S—H bond as nonpolar covalent. Because of this lack of polarity, thiols show little association by hydrogen bonding. Consequently, they have lower boiling points and are less soluble in water and other polar solvents than are alcohols of similar molecular weight. Table 5.2 gives boiling points for three low-molecular-weight thiols. Shown for comparison are boiling points of alcohols with the same number of carbon atoms.

Earlier we illustrated the importance of hydrogen bonding in alcohols by comparing the boiling points of ethanol (78°C) and its constitutional isomer dimethyl ether (−24°C). By contrast, the boiling point of ethanethiol is 35°C and that of its constitutional isomer dimethyl sulfide is 37°C. Because the boiling points of these constitutional isomers are almost identical, we know that little or no association by hydrogen bonding occurs between thiol molecules.

$$CH_3CH_2SH \qquad CH_3SCH_3$$

Ethanethiol   Dimethyl sulfide
bp 35°C   bp 37°C

D. Reactions of Thiols

Thiols are weak acids ($pK_a = 10$) that are comparable in strength to phenols (Section 4.4B). Thiols react with strong bases such as NaOH to form thiolate salts.

$$CH_3CH_2SH + NaOH \xrightarrow{\ H_2O\ } CH_3CH_2S^-Na^+ + H_2O$$

Ethanethiol   Sodium
(pK_a 10)   ethanethiolate

TABLE 5.2 Boiling Points of Three Thiols and Alcohols with the Same Number of Carbon Atoms

Thiol	Boiling Point (°C)	Alcohol	Boiling Point (°C)
methanethiol	6	methanol	65
ethanethiol	35	ethanol	78
1-butanethiol	98	1-butanol	117

The most common reaction of thiols in biological systems is their oxidation to disulfides, the functional group of which is a **disulfide** (—S—S—) bond. Thiols are readily oxidized to disulfides by molecular oxygen. In fact, they are so susceptible to oxidation that they must be protected from contact with air during their storage. Disulfides, in turn, are easily reduced to thiols by several reducing agents. This easy interconversion between thiols and disulfides is very important in protein chemistry, as we will see in Chapters 14 and 15.

Disulfide A compound containing an (—S—S—) group

$$2\text{HOCH}_2\text{CH}_2\text{SH} \underset{\text{reduction}}{\overset{\text{oxidation}}{\rightleftharpoons}} \text{HOCH}_2\text{CH}_2\text{S} - \text{SCH}_2\text{CH}_2\text{OH}$$

A thiol   A disulfide

To derive the common name of a disulfide, list the names of the groups bonded to sulfur and add the word *disulfide*.

5.5 What Are the Most Commercially Important Alcohols?

As you study the alcohols described in this section, you should pay particular attention to two key points. First, they are derived almost entirely from petroleum, natural gas, or coal—all nonrenewable resources. Second, many are themselves starting materials for the synthesis of valuable commercial products, without which our modern industrial society could not exist.

At one time **methanol** was derived by heating hard woods in a limited supply of air—hence the name "wood alcohol." Today methanol is obtained entirely from the catalytic reduction of carbon monoxide. Methanol, in turn, is the starting material for the preparation of several important industrial and commercial chemicals, including acetic acid and formaldehyde. Treatment of methanol with carbon monoxide in the presence of a rhodium catalyst gives acetic acid. Partial oxidation of methanol gives formaldehyde. An important use of this one-carbon aldehyde is in the preparation of phenol-formaldehyde and urea-formaldehyde glues and resins, which are used as molding materials and as adhesives for plywood and particle board for the construction industry.

$$\text{Coal or methane} \xrightarrow{[O]} \underset{\substack{\text{Carbon} \\ \text{monoxide}}}{\text{CO}} \xrightarrow{2\text{H}_2} \underset{\text{Methanol}}{\text{CH}_3\text{OH}} \Big\langle \begin{array}{l} \xrightarrow[\text{catalyst}]{\text{CO}} \underset{\text{Acetic acid}}{\text{CH}_3\text{COOH}} \\ \xrightarrow[\text{oxidation}]{\text{O}_2} \underset{\text{Formaldehyde}}{\text{CH}_2\text{O}} \end{array}$$

The bulk of the **ethanol** produced worldwide is prepared by acid-catalyzed hydration of ethylene, itself derived from the cracking of the ethane separated from natural gas (Section 2.4). Ethanol is also produced by fermentation of the carbohydrates in plant materials, particularly corn and molasses. The majority of the fermentation-derived ethanol is used as an "oxygenate" additive to produce E85, which is a blend of up to 85% ethanol in gasoline. Combustion of E85 produces less air pollution than combustion of gasoline itself.

$$\underset{\text{Ethylene}}{\text{CH}_2{=}\text{CH}_2} \Big\langle \begin{array}{l} \xrightarrow{\text{H}_2\text{O, H}_2\text{SO}_4} \underset{\text{Ethanol}}{\text{CH}_3\text{CH}_2\text{OH}} \xrightarrow[180°\text{C}]{\text{H}_2\text{SO}_4} \underset{\text{Diethyl ether}}{\text{CH}_3\text{CH}_2\text{OCH}_2\text{CH}_3} + \text{H}_2\text{O} \\ \\ \xrightarrow[\text{catalyst}]{\text{O}_2} \underset{\substack{\text{Ethylene} \\ \text{oxide}}}{\overset{O}{\text{H}_2\text{C}{-}\text{CH}_2}} \xrightarrow{\text{H}_2\text{O, H}_2\text{SO}_4} \underset{\text{Ethylene glycol}}{\text{HOCH}_2\text{CH}_2\text{OH}} \end{array}$$

Acid-catalyzed dehydration of ethanol gives **diethyl ether,** an important laboratory and industrial solvent. Ethylene is also the starting material for the preparation of **ethylene oxide.** This compound itself has few direct uses. Rather, ethylene oxide's importance derives from its role as an intermediate in the production of **ethylene glycol,** a major component of automobile antifreeze. Ethylene glycol freezes at $-12°C$ and boils at $199°C$, which makes it ideal for this purpose. In addition, reaction of ethylene glycol with the methyl ester of terephthalic acid gives the polymer poly(ethylene terephthalate), abbreviated PET or PETE (Section 11.6B). Ethylene glycol is also used as a solvent in the paint and plastics industry, and in the formulation of printer's inks, inkpads and in inks for ballpoint pens.

Isopropyl alcohol, the alcohol in rubbing alcohol, is made by acid-catalyzed hydration of propene. It is also used in hand lotions, after-shave lotions, and similar cosmetics. A multistep process also converts propene to epichlorohydrin, one of the key components in the production of epoxy glues and resins.

Glycerin is a by-product of the manufacture of soaps by saponification of animal fats and tropical oils (Section 13.3). The bulk of the glycerin used for industrial and commercial purposes, however, is prepared from propene. Perhaps the best-known use of glycerin is for the manufacture of nitroglycerin. Glycerin is also used as an emollient in skin care products and cosmetics, in liquid soaps, and printing inks.

Summary of Key Questions

Section 5.1 What Are the Structures, Names, and Physical Properties of Alcohols?

Problems 5.11, 5.22

- The functional group of an **alcohol** is an —OH **(hydroxyl)** group bonded to a tetrahedral carbon atom.
- The functional group of an **ether** is an atom of oxygen bonded to two carbon atoms.
- The IUPAC name of an alcohol is derived by changing the *-e* of the parent alkane to *-ol*. The parent chain is numbered from the end that gives the carbon bearing the —OH group the lower number.
- The common name for an alcohol is derived by naming the alkyl group bonded to the —OH group and adding the word "alcohol."

- Alcohols are classified as **1°, 2°,** or **3°,** depending on the number of carbon groups bonded to the carbon bearing the —OH group.
- Compounds containing hydroxyl groups on adjacent carbons are called **glycols.**
- Alcohols are polar compounds in which oxygen bears a partial negative charge and both the carbon and hydrogen bonded to it bear partial positive charges. Alcohols associate in the liquid state by **hydrogen bonding.** As a consequence, their boiling points are higher than those of hydrocarbons of similar molecular weight.
- Because of increased London dispersion forces, the boiling points of alcohols increase with their increasing molecular weight.
- Alcohols interact with water by hydrogen bonding and are more soluble in water than are hydrocarbons of similar molecular weight.

- Alcohols have about the same pK_a values as pure water. For this reason, aqueous solutions of alcohols have the same pH as that of pure water.

Section 5.3 What Are the Structures, Names, and Properties of Ethers? Problem 5.37

- Common names for ethers are derived by naming the two groups bonded to oxygen followed by the word "ether."
- In a **cyclic ether,** oxygen is one of the atoms in a ring.
- Ethers are weakly polar compounds. Their boiling points are close to those of hydrocarbons of similar molecular weight.
- Because ethers form hydrogen bonds with water, they are more soluble in water than are hydrocarbons of similar molecular weight.

Section 5.4 What Are the Structures, Names, and Properties of Thiols? Problem 5.39

- A **thiol** contains an —SH **(sulfhydryl)** group.
- Thiols are named in the same manner as alcohols, but the suffix *-e* of the parent alkane is retained and *-thiol* is added.
- Common names for thiols are derived by naming the alkyl group bonded to —SH and adding the word **"mercaptan."**
- The S—H bond is nonpolar, and the physical properties of thiols resemble those of hydrocarbons of similar molecular weight.

Summary of Key Reactions

1. Acid-Catalyzed Dehydration of an Alcohol (Section 5.2B)
When isomeric alkenes are possible, the major product is generally the more substituted alkene.

$$CH_3CH_2CHCH_3 \text{ (OH)} \xrightarrow[\text{heat}]{H_3PO_4} CH_3CH{=}CHCH_3 + CH_3CH_2CH{=}CH_2 + H_2O$$

Major product

2. Oxidation of a Primary Alcohol (Section 5.2C)
Problem 5.31
Oxidation of a primary alcohol by potassium dichromate gives either an aldehyde or a carboxylic acid, depending on the experimental conditions.

$$CH_3(CH_2)_6CH_2OH \xrightarrow[H_2SO_4]{K_2Cr_2O_7} CH_3(CH_2)_6\overset{O}{\overset{\|}{C}}H$$

$$\xrightarrow[H_2SO_4]{K_2Cr_2O_7} CH_3(CH_2)_6\overset{O}{\overset{\|}{C}}OH$$

3. Oxidation of a Secondary Alcohol (Section 5.2C)
Oxidization of a secondary alcohol by potassium dichromate gives a ketone.

$$CH_3(CH_2)_4\underset{\text{OH}}{CH}CH_3 \xrightarrow[H_2SO_4]{K_2Cr_2O_7} CH_3(CH_2)_4\overset{O}{\overset{\|}{C}}CH_3$$

4. Acidity of Thiols (Section 5.4D) Thiols are weak acids, with pK_a values of approximately 10. They react with strong bases to form water-soluble thiolate salts.

$$CH_3CH_2SH + NaOH \xrightarrow{H_2O} CH_3CH_2S^-\ Na^+ + H_2O$$

Ethanethiol   Sodium
(pK_a 10)   ethanethiolate

5. Oxidation of a Thiol to a Disulfide (Section 5.4D)
Oxidation of a thiol gives a disulfide. Reduction of a disulfide gives two thiols.

$$2HOCH_2CH_2SH \xrightleftharpoons[\text{reduction}]{\text{oxidation}} HOCH_2CH_2S{-}SCH_2CH_2OH$$

A thiol   A disulfide

Problems

■ Indicates problems that are assignable in **GOB OWL**.

Blue numbered problems are applied.

Go to this book's companion website at **www.cengage. com/chemistry/bettelheim** for interactive versions of the How To tutorials and Active Figures, and to quiz yourself on this chapter.

Section 5.1 What Are the Structures, Names, and Physical Properties of Alcohols?

5.8   Answer true or false.
  (a) The functional group of an alcohol is the —OH (hydroxyl) group.

(b) The parent name of an alcohol is the name of the longest carbon chain that contains the —OH group.

(c) A primary alcohol contains one —OH group, and a tertiary alcohol contains three —OH groups.

(d) In the IUPAC system, the presence of three —OH groups is shown by the ending *–triol*.

(e) A glycol is a compound that contains two —OH groups. The simplest glycol is ethylene glycol, $HOCH_2CH_2$—OH

(f) Because of the presence of an —OH group, all alcohols are polar compounds.

(g) The boiling points of alcohols increase with increasing molecular weight.

(h) The solubility of alcohols in water increases with increasing molecular weight.

5.9 What is the difference in structure between a primary, a secondary, and a tertiary alcohol?

5.10 Write the IUPAC name of each compound.

5.11 ■ Draw a structural formula for each alcohol.
(a) Isopropyl alcohol
(b) Propylene glycol
(c) 5-Methyl-2-hexanol
(d) 2-Methyl-2-propyl-1,3-propanediol
(e) 1-Octanol
(f) 3,3-Dimethylcyclohexanol

5.12 Both alcohols and phenols contain an —OH group. What structural feature distinguishes these two classes of compounds? Illustrate your answer by drawing the structural formulas of a phenol with six carbon atoms and an alcohol with six carbon atoms.

5.13 Name the functional groups in each compound.

(a)

Prednisone
(a synthetic anti-
inflammatory steroid)

(b)

Estradiol
(a female sex hormone;
Section 13.10)

5.14 Explain in terms of noncovalent interactions why the low-molecular-weight alcohols are soluble in water but the low-molecular-weight alkanes and alkynes are not.

5.15 Explain in terms of noncovalent interactions why the low-molecular-weight alcohols are more soluble in water than the low-molecular-weight ethers.

5.16 Why does the water solubility of low-molecular-weight alcohols decrease as molecular weight increases?

5.17 Show hydrogen bonding between methanol and water in the following ways.
(a) Between the oxygen of methanol and a hydrogen of water
(b) Between the hydrogen of methanol's OH group and the oxygen of water

5.18 Show hydrogen bonding between the oxygen of diethyl ether and a hydrogen of water.

5.19 Arrange these compounds in order of increasing boiling point. Values in °C are −42, 78, 117, and 198.
(a) $CH_3CH_2CH_2CH_2OH$
(b) CH_3CH_2OH
(c) $HOCH_2CH_2OH$
(d) $CH_3CH_2CH_3$

5.20 Arrange these compounds in order of increasing boiling point. Values in °C are 0, 35, and 97.
(a) $CH_3CH_2CH_2OH$
(b) $CH_3CH_2OCH_2CH_3$
(c) $CH_3CH_2CH_2CH_3$

5.21 Explain why glycerol is much thicker (more viscous) than ethylene glycol, which in turn is much thicker than ethanol.

5.22 ■ From each pair, select the compound that is more soluble in water.

(a) CH_3OH or CH_3OCH_3

(b)
$$\underset{\substack{|\\ OH}}{CH_3CHCH_3} \quad \text{or} \quad \underset{\substack{||\\ CH_2}}{CH_3CCH_3}$$

(c) $CH_3CH_2CH_2SH$ or $CH_3CH_2CH_2OH$

5.23 Arrange the compounds in each set in order of decreasing solubility in water.

(a) Ethanol, butane and diethyl ether

(b) 1-Hexanol, 1,2-hexanediol and hexane

Synthesis of Alcohols (Review Chapter 3)

5.24 Give the structural formula of an alkene or alkenes from which each alcohol can be prepared.

(a) 2-Butanol

(b) 1-Methylcyclohexanol

(c) 3-Hexanol

(d) 2-Methyl-2-pentanol

(e) Cyclopentanol

Section 5.2 What Are the Characteristic Reactions of Alcohols?

5.25 Answer true or false.

(a) The two most important reactions of alcohols are their acid-catalyzed dehydration to give alkenes and their oxidation to aldehydes, ketones, and carboxylic acids.

(b) The acidity of alcohols is comparable to that of water.

(c) Water-insoluble alcohols and water-insoluble phenols react with strong bases to give water-soluble salts.

(d) Acid-catalyzed dehydration of cyclohexanol gives cyclohexane.

(e) When the acid-catalyzed dehydration of an alkene can yield isomeric alkenes, the alkene with the greater number of hydrogens on the carbons of the double bond generally predominates.

(f) The acid-catalyzed dehydration of 2-butanol gives predominantly 1-butene.

(g) The oxidation of a primary alcohol gives either an aldehyde or carboxylic acid depending on experimental conditions.

(h) The oxidation of a secondary alcohol gives a carboxylic acid.

(i) Acetic acid, CH_3COOH, can be prepared from ethylene, $CH_2=CH_2$, by treatment of ethylene with H_2O/H_2SO_4, followed by treatment with $K_2Cr_2O_7/H_2SO_4$.

(j) Treatment of propene, $CH_3CH=CH_2$, with H_2O/H_2SO_4, followed by treatment with $K_2Cr_2O_7/H_2SO_4$ gives propanoic acid, CH_3CH_2COOH.

5.26 Show how to distinguish between cyclohexanol and cyclohexene by a simple chemical test. Tell what you would do, what you would expect to see, and how you would interpret your observation.

5.27 Compare the acidity of alcohols and phenols, which are both classes of organic compounds that contain an —OH group.

5.28 Both 2,6-diisopropylcyclohexanol and the intravenous anesthetic Propofol are insoluble in water. Show how these two compounds can be

distinguished by their reaction with aqueous sodium hydroxide.

2,6-Diisopropylcyclohexanol   2,6-Diisopropylphenol
(Propofol)

5.29 Write equations for the reaction of 1-butanol, a primary alcohol, with these reagents.

(a) H_2SO_4, heat

(b) $K_2Cr_2O_7$, H_2SO_4

5.30 Write equations for the reaction of 2-butanol with these reagents.

(a) H_2SO_4, heat

(b) $K_2Cr_2O_7$, H_2SO_4

5.31 ■ Write equations for reaction of each of the following compounds with $K_2Cr_2O_7/H_2SO_4$.

(a) 1-Octanol

(b) 1,4-Butanediol

5.32 Show how to convert cyclohexanol to these compounds.

(a) Cyclohexene

(b) Cyclohexane

(c) Cyclohexanone

(d) Bromocyclohexane

5.33 Show reagents and experimental conditions to synthesize each compound from 1-propanol.

5.34 Name two important alcohols derived from ethylene and give two important uses of each.

5.35 Name two important alcohols derived from propene and give two important uses of each.

Section 5.3 What Are the Structures, Names, and Properties of Ethers?

5.36 Answer true or false.
(a) Ethanol and dimethyl ether are constitutional isomers.
(b) The solubility of low-molecular-weight ethers in water is comparable to that of low-molecular-weight alcohols in water.
(c) Ethers undergo many of the same reactions that alcohols do.

5.37 ■ Write the common name for each ether.

(a)   (b) $[CH_3(CH_2)_4]_2O$

(c)
$$CH_3\underset{\underset{CH_3}{|}}{C}HO\underset{\underset{CH_3}{|}}{C}HCH_3$$

Section 5.4 What Are the Structures, Names, and Properties of Thiols?

5.38 Answer true or false.
(a) The functional group of a thiol is the —SH (sulfhydryl) group.
(b) The parent name of a thiol is the name of the longest carbon chain that contains the —SH group.
(c) The S—H bond is nonpolar covalent.
(d) The acidity of ethanethiol is comparable to that of phenol.
(e) Both phenols and thiols are classified as weak acids.
(f) The most common biological reaction of thiols is their oxidation to disulfides.
(g) The functional group of a disulfide is the —S—S— group.
(h) Conversion of a thiol to a disulfide is a reduction reaction.

5.39 ■ Write the IUPAC name for each thiol.

(a)
$$CH_3CH_2\underset{\underset{SH}{|}}{C}HCH_3$$
   (b) $CH_3CH_2CH_2CH_2SH$

(c)

5.40 Write the common name for each thiol in Problem 5.45.

5.41 Following are structural formulas for 1-butanol and 1-butanethiol. One of these compounds has a boiling point of 98°C and the other has a boiling point of 117°C. Which compound has which boiling point?

$CH_3CH_2CH_2CH_2OH$ $CH_3CH_2CH_2CH_2SH$
1-Butanol   1-Butanethiol

5.42 Explain why methanethiol, CH_3SH, has a lower boiling point (6°C) than methanol, CH_3OH (65°C), even though methanethiol has a higher molecular weight.

Section 5.5 What Are the Most Commercially Important Alcohols?

5.43 Answer true or false.
(a) Today, the major carbon sources for the synthesis of methanol are coal and methane (natural gas), both nonrenewable resources.
(b) Today the major carbon sources for the synthesis of ethanol are petroleum and natural gas, both nonrenewable resources.
(c) Intermolecular acid-catalyzed dehydration of ethanol gives diethyl ether.
(d) Conversion of ethylene to ethylene glycol involves oxidation to ethylene oxide, followed by acid-catalyzed hydration (addition of water) to ethylene oxide.
(e) Ethylene glycol is soluble in water in all proportions.
(f) A major use of ethylene glycol is as automobile antifreeze.

Chemical Connections

5.44 (Chemical Connections 5A) When was nitroglycerin discovered? Is this substance a solid, a liquid, or a gas?

5.45 (Chemical Connections 5A) What was Alfred Nobel's discovery that made nitroglycerin safer to handle?

5.46 (Chemical Connections 5A) What is the relationship between the medical use of nitroglycerin to relieve the sharp chest pains (angina) associated with heart disease and the gas nitric oxide, NO?

5.47 (Chemical Connections 5B) What is the color of dichromate ion, $Cr_2O_7^{2-}$? What is the color of chromium(III) ion, Cr^{3+}? Explain how the conversion of one to the other is used in breath-alcohol screening.

5.48 (Chemical Connections 5B) The legal definition of being under the influence of alcohol is based on blood alcohol content. What is the relationship between breath alcohol content and blood alcohol content?

5.49 (Chemical Connections 5C) What does it mean to say that ethylene oxide is a highly strained molecule?

5.50 (Chemical Connections 5D) What are the advantages and disadvantages of using diethyl ether as an anesthetic?

5.51 (Chemical Connections 5D) Show that enflurane and isoflurane are constitutional isomers.

5.52 (Chemical Connections 5D) Would you expect enflurane and isoflurane to be soluble in water? Would you expect them to be soluble in organic solvents such as hexane?

Additional Problems

5.53 Write a balanced equation for the complete combustion of ethanol, the alcohol added to gasoline to produce E85.

5.54 Knowing what you do about electronegativity, the polarity of covalent bonds, and hydrogen bonding, would you expect an N—H----N hydrogen bond to be stronger than, the same strength as, or weaker than an O—H----O hydrogen bond?

5.55 Draw structural formulas and write IUPAC names for the eight isomeric alcohols with the molecular formula $C_5H_{12}O$.

5.56 Draw structural formulas and write common names for the six isomeric ethers with the molecular formula $C_5H_{12}O$.

5.57 Explain why the boiling point of ethylene glycol (198°C) is so much higher than that of 1-propanol (97°C), even though their molecular weights are about the same.

5.58 1,4-Butanediol, hexane and 1-pentanol have similar molecular weights. Their boiling points, arranged from lowest to highest, are 69°C, 138°C, and 230°C. Which compound has which boiling point?

5.59 Of the three compounds given in Problem 5.65, one is insoluble in water, another has a solubility of 2.3 g/100 g water, and one is infinitely soluble in water. Which compound has which solubility?

5.60 Each of the following compounds is a common organic solvent. From each pair of compounds, select the solvent with the greater solubility in water.
(a) CH_2Cl_2 or CH_3CH_2OH
(b) $CH_3CH_2OCH_2CH_3$ or CH_3CH_2OH

5.61 Show how to prepare each compound from 2-methyl-1-propanol.
(a) 2-Methylpropene
(b) 2-Methyl-2-propanol
(c) 2-Methylpropanoic acid, $(CH_3)_2CHCOOH$

5.62 Show how to prepare each compound from 2-methylcyclohexanol.

Looking Ahead

5.63 Following is a structural formula for the amino acid cysteine:

$$HS-CH_2-CH-\overset{\overset{\displaystyle O}{\|}}{C}-OH$$
$$|$$
$$NH_2$$

(a) Name the three functional groups in cysteine.
(b) In the human body, cysteine is oxidized to cystine, a disulfide. Draw a structural formula for cystine.

Chirality: The Handedness of Molecules

6

Charles D. Winters/Cengage Learning

Median cross section through the shell of a chambered nautilus found in the deep waters of the Pacific Ocean. The shell shows handedness; this cross section is a left-handed spiral.

Key Questions

6.1 What Is Enantiomerism?

How To . . . Draw Enantiomers

6.2 How Do We Specify the Configuration of a Stereocenter?

6.3 How Many Stereoisomers Are Possible for Molecules with Two or More Stereocenters?

6.4 What Is Optical Activity, and How Is Chirality Detected in the Laboratory?

6.5 What Is the Significance of Chirality in the Biological World?

6.1 What Is Enantiomerism?

In Chapters 2 through 5, we studied two types of stereoisomers, namely the *cis-trans* isomers of certain disubstituted cycloalkanes and appropriately substituted alkenes. Recall that stereoisomers have the same connectivity of their atoms but a different orientation of their atoms in space.

OWL Online homework for this chapter may be assigned in **GOB OWL**.

cis-1,4-Dimethyl-cyclohexane trans-1,4-Dimethyl-cyclohexane cis-2-Butene trans-2-Butene

FIGURE 6.1 Relationships among isomers. In this chapter, we study enantiomers and diastereomers.

In this chapter, we study the relationship between objects and their **mirror images;** that is, we study stereoisomers called enantiomers and diastereomers. Figure 6.1 summarizes the relationship among these isomers and those we studied in Chapters 2 through 5.

The significance of enantiomers is that, except for inorganic compounds and a few simple organic compounds, the vast majority of molecules in the biological world show this type of isomerism. Furthermore, approximately one half of all medications used to treat humans display this property. As an example of enantiomerism, let us consider 2-butanol. As we discuss this molecule, we will focus on carbon 2, the carbon bearing the —OH group. What makes this carbon of interest is the fact that it has four different groups bonded to it: CH_3, H, OH, and CH_2CH_3.

$$\overset{\displaystyle OH}{\underset{\displaystyle \text{2-Butanol}}{CH_3CHCH_2CH_3}}$$

This structural formula does not show the three-dimensional shape of 2-butanol or the orientation of its atoms in space. To do so, we must consider the molecule as a three-dimensional object. On the left is what we will call the "original molecule" and a ball-and-stick model of it. In this drawing, the —OH and —CH_3 groups are in the plane of the paper, —H is behind the plane (shown as a broken wedge), and —CH_2CH_3 is in front of it (shown as a solid wedge). In the middle is a mirror. On the right is a **mirror image**

Original molecule

Mirror image

of the original molecule along with a ball-and-stick model of the mirror image. Every molecule—and, in fact, every object in the world around us—has a mirror image.

The question we now need to ask is "What is the relationship between the original molecule of 2-butanol and its mirror image?" To answer this question, we need to imagine that we can pick up the mirror image and move it in 3-D space in any way we wish. If we can move the mirror image in space and find that it fits over the original molecule so that every bond, atom, and detail of the mirror image matches exactly the bonds, atoms, and details of the original, then the two are **superposable.** In other words, the mirror image and the original represent the same molecule; they are merely oriented differently in space. If, however, no matter how we turn the mirror image in space, it will not fit exactly on the original with every detail matching, then the two are **nonsuperposable;** they are different molecules.

The terms "superposable" and "superimposable" mean the same thing and are both used currently.

One way to see that the mirror image of 2-butanol is not superposable on the original molecule is illustrated in the following drawings. Imagine that we hold the mirror image by the C—OH bond and rotate the bottom part of molecule by 180° about this bond. The —OH group retains its position in space, but the —CH₃ group, which was to the right and in the plane of the paper, remains in the plane of the paper but is now to the left. Similarly, the—CH₂CH₃ group, which was in front of the plane of the paper and to the left, is now behind the plane and to the right.

Original molecule   Mirror image of original molecule   Mirror image rotated by 180°

Now move the mirror image in space and try to fit it on the original molecule so that all bonds and atoms match.

The threads of a drill or screw twist along the axis of the helix, and some plants climb by sending out tendrils that twist helically. The drill bit shown here has a left-handed twist and the tendril has a right-handed twist.

By turning the mirror image as we did, its —OH and —CH₃ groups now fit exactly on top of the —OH and —CH₃ groups of the original molecule. However, the —H and the —CH₂CH₃ groups of the two do not match. The —H is away from us in the original but toward us in the mirror image; the —CH₂CH₃ group is toward us in the original but away from us in the mirror image. We conclude that the original molecule of 2-butanol and its mirror image are nonsuperposable and, therefore, they are different compounds.

(a)   (b)

Charles D. Winters/Cengage Learning

FIGURE 6.2 Mirror images. (a) Two woodcarvings. The mirror images cannot be superposed on the actual model. The man's right arm rests on the camera in the mirror image, but in the actual statue, the man's left arm rests on the camera. (b) Left- and right-handed sea shells. If you cup a right-handed shell in your right hand with your thumb pointing from the narrow end to the wide end, the opening will be on your right.

To summarize, we can turn and rotate the mirror image of 2-butanol in any direction in space, but as long as no bonds are broken and rearranged, we can make only two of the four groups bonded to carbon 2 of the mirror image coincide with those on its original molecule. Because 2-butanol and its mirror image are not superposable, they are isomers. Isomers such as these are called **enantiomers.** Enantiomers, like gloves, always occur in pairs.

Objects that are not superposable on their mirror images are said to be **chiral** (pronounced "ki-ral," rhymes with spiral; from the Greek: *cheir*, "hand"); that is, they show handedness. We encounter chirality in three-dimensional objects of all sorts. Our left hand is chiral, and so is our right hand. Thus our hands have an enantiomeric relationship. A spiral binding on a notebook is chiral. A machine screw with a right-handed twist is chiral. A ship's propeller is chiral. As you examine the objects in the world around you, you will undoubtedly conclude that the vast majority of them are chiral.

The most common cause of enantiomerism in organic molecules is the presence of a carbon with four different groups bonded to it. Let us examine this statement further by considering 2-propanol, which has no such carbon atom. In this molecule, carbon 2 is bonded to three different groups, but no carbon is bonded to four different groups.

On the left is a three-dimensional representation of 2-propanol; on the right is its mirror image. Also shown are ball-and-stick models of each molecule.

Enantiomers Stereoisomers that are nonsuperposable mirror images; refers to a relationship between pairs of objects

Chiral From the Greek *cheir*, "hand"; an object that is not superposable on its mirror image

Original molecule   Mirror image

The question we now ask is "What is the relationship of the mirror image to the original?" This time, let us rotate the mirror image by 120° about the C—OH bond, and then compare it to the original. After performing this rotation, we see that all atoms and bonds of the mirror image fit exactly on the original. Thus the structures we first drew for the original molecule and its mirror image are, in fact, the same molecule—just viewed from different perspectives (Figure 6.3).

ACTIVE FIGURE 6.3   Rotation of the mirror image about the C—OH bond by 120° does not change the configuration of the stereocenter, but makes it easier to see that the mirror image is superposable on the original molecule. **Go to this book's companion website at www.cengage.com/chemistry/ bettelheim to explore an interactive version of this figure.**

Achiral An object that lacks chirality; an object that is superposable on its mirror image

If an object and its mirror image are superposable, then they are identical and enantiomerism is not possible. We say that such an object is **achiral** (without chirality); that is, it has no handedness. Examples of achiral objects include an undecorated cup, an unmarked baseball bat, a regular tetrahedron, a cube, and a sphere.

To repeat, the most common cause of chirality in organic molecules is a tetrahedral carbon atom with four different groups bonded to it. We call such a chiral carbon atom a **stereocenter.** 2-Butanol has one stereocenter; 2-propanol has none. As another example of a molecule with a stereocenter, let us consider 2-hydroxypropanoic acid, more commonly named lactic acid. Lactic acid is a product of anaerobic glycolysis. (See Section 20.2 and Chemical Connections 20A.) It is also what gives sour cream its sour taste.

Stereocenter A tetrahedral carbon atom that has four different groups bonded to it

Figure 6.4 shows three-dimensional representations of lactic acid and its mirror image. In these representations, all bond angles about the central carbon atom are approximately 109.5°, and the four bonds from it are directed toward the corners of a regular tetrahedron. Lactic acid displays enantiomerism or chirality; that is, the original molecule and its mirror image are not superposable but rather are different compounds.

FIGURE 6.4   Three-dimensional representations of lactic acid and its mirror image.

Racemic mixture A mixture of equal amounts of two enantiomers

An equimolar mixture of two enantiomers is called a **racemic mixture,** a term derived from the name racemic acid (Latin: *racemus*, "a cluster of grapes"). Racemic acid is the name originally given to the equimolar mixture of the enantiomers of tartaric acid that forms as a by-product during the fermentation of grape juice to produce wine.

> ## How To . . .
> ## Draw Enantiomers
>
> Now that we know what enantiomers are, we can think about how to represent their three-dimensional structures on a two-dimensional surface. Let us take one of the enantiomers of 2-butanol as an example. Following are a

molecular model of one enantiomer and four different three-dimensional representations of it:

(1)   (2)   (3)   (4)

In our initial discussions of 2-butanol, we used representation (1) to show the tetrahedral geometry of the stereocenter. In this representation, two groups (OH and CH$_3$) are in the plane of the paper, one (CH$_2$CH$_3$) is coming out of the plane toward us, and one (H) is behind the plane and going away from us. We can turn representation (1) slightly in space and tip it a bit to place the carbon framework in the plane of the paper. Doing so gives us representation (2), in which there are still two groups in the plane of the paper, one coming toward us and one going away from us.

For an even more abbreviated representation of this enantiomer of 2-butanol, we can change representation (2) into the line-angle formula (3). Although we do not normally show hydrogens in a line-angle formula, we do so here just to remind ourselves that the fourth group on the stereocenter is really there and that it is H. Finally, we can carry the abbreviation a step further and write 2-butanol as the line-angle formula (4). Here, we omit the H on the stereocenter, but we know that it must be there (carbon needs four bonds), and we know that it must be behind the plane of the paper. Clearly, the abbreviated formulas (3) and (4) are the easiest to write, and we will rely on this type of representation throughout the remainder of the text.

When you have to write three-dimensional representations of stereocenters, try to keep the carbon framework in the plane of the paper and the other two atoms or groups of atoms on the stereocenter toward and away from you, respectively. Using representation (4) as a model, we get the following alternative representations of its mirror image:

One enantiomer   Alternative representations of
of 2-butanol   its mirror image

Example 6.1 Drawing Mirror Images

Each of the following molecules has one stereocenter marked by an asterisk. Draw three-dimensional representations for the enantiomers of each molecule.

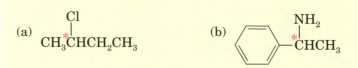

(a) $\overset{\text{Cl}}{\underset{\displaystyle \text{CH}_3\overset{*}{\text{C}}\text{HCH}_2\text{CH}_3}{|}}$

(b) phenyl–$\overset{\text{NH}_2}{\underset{\displaystyle \overset{*}{\text{C}}\text{HCH}_3}{|}}$

Strategy

First, draw the carbon stereocenter showing the tetrahedral orientation of its four bonds. One way to do this is to draw two bonds in the plane of the paper, a third bond toward you in front of the plane, and the fourth bond away from you behind the plane. Next, place the four groups bonded to the stereocenter on these positions. This completes the stereodrawing of one enantiomer. To draw the other enantiomer, interchange any two of the groups on the original stereodrawing.

Solution

To draw an original of (a), for example, place the CH_3 and CH_2CH_3 groups in the plane of the paper. Place H away from you and the Cl toward you; this orientation gives the enantiomer of (a) on the left. Its mirror image is on the right.

Problem 6.1

Each of the following molecules has one stereocenter marked by an asterisk. Draw three-dimensional representations for the enantiomers of each molecule.

(a) cyclopentyl–$\overset{\text{COOH}}{\underset{\displaystyle \overset{*}{\text{C}}\text{HCH}_3}{|}}$

(b) $\overset{\text{OH}}{\underset{\displaystyle \underset{\displaystyle \text{CH}_3}{\overset{\displaystyle |}{\text{CH}_3\text{CH}\overset{*}{\text{C}}\text{HCH}_3}}}{|}}$

6.2 How Do We Specify the Configuration of a Stereocenter?

Because enantiomers are different compounds, each must have a different name. The over-the-counter drug ibuprofen, for example, displays enantiomerism and can exist as the pair of enantiomers shown here:

The inactive enantiomer
of ibuprofen

The active enantiomer
of ibuprofen

Only one enantiomer of ibuprofen is biologically active. It reaches therapeutic concentrations in the human body in approximately 12 minutes, whereas the racemic mixture takes approximately 30 minutes to achieve this feat. However, in this case, the inactive enantiomer is not wasted. The body converts it to the active enantiomer, but this process takes time.

What we need is a system to designate which enantiomer of ibuprofen (or one of any other pair of enantiomers, for that matter) is which without having to draw and point to one or the other of the enantiomers. To do so, chemists have developed the **R,S system.** The first step in assigning an R or S configuration to a stereocenter is to arrange the groups bonded to it in order of priority. Priority is based on atomic number: The higher the atomic number, the higher the priority. If a priority cannot be assigned on the basis of the atoms bonded directly to the stereocenter, look at the next atom or set of atoms and continue to the first point of difference; that is, continue until you can assign a priority.

Table 6.1 shows the priorities of the most common groups we encounter in organic and biochemistry. In the R,S system, a C=O is treated as if carbon were bonded to two oxygens by single bonds; thus CH=O has a higher priority than —CH_2OH, in which carbon is bonded to only one oxygen.

R,S system A set of rules for specifying the configuration about a stereocenter

TABLE 6.1 *R,S* Priorities of Some Common Groups

Atom or Group	Reason for Priority: First Point of Difference (Atomic Number)
— I	iodine (53)
— Br	bromine (35)
— Cl	chlorine (17)
— SH	sulfur (16)
— OH	oxygen (8)
— NH_2	nitrogen (7)
$\overset{\text{O}}{\overset{\|}{—\text{COH}}}$	carbon to oxygen, oxygen, then oxygen (6 $\longrightarrow$ 8, 8, 8)
$\overset{\text{O}}{\overset{\|}{—\text{CNH2}}}$	carbon to oxygen, oxygen, then nitrogen (6 $\longrightarrow$ 8, 8, 7)
$\overset{\text{O}}{\overset{\|}{—\text{CH}}}$	carbon to oxygen, oxygen, then hydrogen (6 $\longrightarrow$ 8, 8, 1)
— CH2OH	carbon to oxygen (6 $\longrightarrow$ 8)
— CH2NH2	carbon to nitrogen (6 $\longrightarrow$ 7)
— CH2CH3	carbon to carbon (6 $\longrightarrow$ 6)
— CH2H	carbon to hydrogen (6 $\longrightarrow$ 1)
— H	hydrogen (1)

Increasing priority

Example 6.2 Using the R,S System

Assign priorities to the groups in each set.

(a) —CH₂OH and —CH₂CH₂OH
(b) —CH₂CH₂OH and —CH₂NH₂

Strategy and Solution

(a) The first point of difference is O of the —OH group compared to C of the —CH₂OH group.

First point of difference

—CH₂OH   —CH₂CH₂OH
Higher priority   Lower priority

(b) The first point of difference is C of the CH₂OH group compared to N of the NH₂ group.

First point of difference

—CH₂CH₂OH   —CH₂NH₂
Lower priority   **Higher priority**

Problem 6.2

Assign priorities to the groups in each set.

$$\text{(a)} \quad -CH_2OH \quad \text{and} \quad -CH_2CH_2\overset{\displaystyle O}{\overset{\displaystyle \|}{C}}H$$

$$\text{(b)} \quad -CH_2NH_2 \quad \text{and} \quad -CH_2\overset{\displaystyle O}{\overset{\displaystyle \|}{C}}H$$

R Used in the R,S system to show that, when the lowest-priority group is away from you, the order of priority of groups on a stereocenter is clockwise

S Used in the R,S system to show that, when the lowest-priority group is away from you, the order of priority of groups on a stereocenter is counterclockwise

Group of lowest priority points away from you

To assign an R or S configuration to a stereocenter:

1. Assign a priority from 1 (highest) to 4 (lowest) to each group bonded to the stereocenter.

2. Orient the molecule in space so that the lowest-priority group (4) is directed away from you, as would be, for instance, the steering column of a car. The three higher-priority groups (1–3) then project toward you, as would the spokes of a steering wheel.

3. Read the three groups projecting toward you in order from highest (1) to lowest (4) priority.

4. If reading the groups 1-2-3 proceeds in a clockwise direction (to the right), the configuration is designated as R (Latin: *rectus*, "straight"); if reading the groups 1-2-3 proceeds in a counterclockwise direction (to the left), the configuration is S (Latin: *sinister*, "left"). You can also visualize this system as follows: Turning the steering wheel to the right equals R and turning it to the left equals S.

Example 6.3 Adding an *R* or *S* Configuration

Assign an *R* or *S* configuration to each stereocenter.

(a) 2-Butanol

(b) Alanine

Strategy and Solution

View each molecule through the stereocenter and along the bond from the stereocenter to the group of lowest priority.

(a) The order of decreasing priority about the stereocenter in this enantiomer of 2-butanol is —OH > —CH₂CH₃ > —CH₃ > —H. Therefore, view the molecule along the C—H bond with the H pointing away from you. Reading the other three groups in the order 1-2-3 follows in the clockwise direction. Therefore, the configuration is *R* and this enantiomer is (*R*)-2-butanol.

(*R*)-2-Butanol

With —H, the lowest-priority group, pointing away from you, this is what you see

(b) The order of decreasing priority in this enantiomer of alanine is —NH₂ > —COOH > —CH₃ > —H. View the molecule along the C—H bond with H pointing away from you. Reading the groups in the order 1-2-3 follows in a clockwise direction; therefore, the configuration is *R* and this enantiomer is (*R*)-alanine.

(*R*)-Alanine

With —H, the lowest-priority group, pointing away from you, this is what you see

Problem 6.3

Assign an *R* or *S* configuration to the single stereocenter in glyceraldehyde, the simplest carbohydrate (Chapter 12).

Glyceraldehyde

Now let us return to our three-dimensional drawing of the enantiomers of ibuprofen and assign each an R or S configuration. In order of decreasing priority, the groups bonded to the stereocenter are —COOH (1) > —C_6H_5 (2) > —CH_3 (3) > H (4). In the enantiomer on the left, reading the groups on the stereocenter in order of priority is clockwise and, therefore, this enantiomer is (R)-ibuprofen. Its mirror image is (S)-ibuprofen.

(R)-Ibuprofen
(the inactive enantiomer)

(S)-Ibuprofen
(the active enantiomer)

The R,S system can be used to specify the configuration of any stereocenter in any molecule. It is not, however, the only system used for this purpose. There is also a D,L system, which is used primarily to specify the configuration of carbohydrates (Chapter 12) and amino acids (Chapter 14).

In closing, note that the purpose of Section 6.2 is to show you how chemists assign a configuration to a stereocenter that specifies the relative orientation of the four groups on the stereocenter. What is important is that when you see a name such as (S)-Naproxen or (R)-Plavix, you realize that the compound is chiral and that the compound is not a racemic mixture. Rather, it is a pure enantiomer. We use the symbol (R,S) to show that a compound is a racemic mixture, as for example, (R,S)-Naproxen.

6.3 How Many Stereoisomers Are Possible for Molecules with Two or More Stereocenters?

For a molecule with n stereocenters, the maximum number of stereoisomers possible is 2^n. We have already verified that, for a molecule with one stereocenter, $2^1 = 2$ stereoisomers (one pair of enantiomers) are possible. For a molecule with two stereocenters, a maximum of $2^2 = 4$ stereoisomers (two pairs of enantiomers) is possible; for a molecule with three stereocenters, a maximum of $2^3 = 8$ stereoisomers (four pairs of enantiomers) is possible; and so forth.

A. Molecules with Two Stereocenters

We begin our study of molecules with two stereocenters by considering 2,3,4-trihydroxybutanal, a molecule with two stereocenters.

2,3,4-Trihydroxybutanal
(2 stereocenters;
4 stereoisomers
are possible)

The maximum number of stereoisomers possible for this molecule is $2^2 = 4$, each of which is drawn in Figure 6.5.

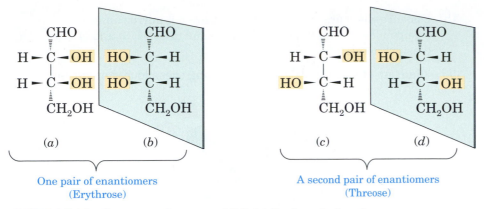

One pair of enantiomers
(Erythrose)

A second pair of enantiomers
(Threose)

FIGURE 6.5 The four stereoisomers of 2,3,4-trihydroxybutanal.

Stereoisomers (a) and (b) are nonsuperposable mirror images and are, therefore, a pair of enantiomers. Stereoisomers (c) and (d) are also nonsuperposable mirror images and are a second pair of enantiomers. We describe the four stereoisomers of 2,3,4-trihydroxybutanal by saying that they consist of two pairs of enantiomers. Enantiomers (a) and (b) are named **erythrose.** Erythrose is synthesized in erythrocytes (red blood cells); hence the derivation of its name. Enantiomers (c) and (d) are named **threose.** Erythrose and threose belong to the class of compounds called carbohydrates, which we discuss in Chapter 12.

We have specified the relationship between (a) and (b) and that between (c) and (d). What is the relationship between (a) and (c), between (a) and (d), between (b) and (c), and between (b) and (d)? The answer is that they are **diastereomers**—stereoisomers that are not mirror images.

Diastereomers Stereoisomers that are not mirror images

Example 6.4 Enantiomers and Diastereomers

1,2,3-Butanetriol has two stereocenters (carbons 2 and 3); thus $2^2 = 4$ stereoisomers are possible for it. Following are three-dimensional representations for each.

(a) Which stereoisomers are pairs of enantiomers?
(b) Which stereoisomers are diastereomers?

Strategy

First, identify those structures that are mirror images. These, then, are the pairs of enantiomers. All other pairs of structures are diastereomers.

Solution

(a) Enantiomers are stereoisomers that are nonsuperposable mirror images. Compounds (1) and (4) are one pair of enantiomers and compounds (2) and (3) are a second pair of enantiomers.

(b) Diastereomers are stereoisomers that are not mirror images. Compounds (1) and (2), (1) and (3), (2) and (4), and (3) and (4) are diastereomers.

The diagram shows the relationship among these four stereoisomers.

Problem 6.4

3-Amino-2-butanol has two stereocenters (carbons 2 and 3); thus $2^2 = 4$ stereoisomers are possible for it.

(a) Which stereoisomers are pairs of enantiomers?
(b) Which sets of stereoisomers are diastereomers?

We can analyze chirality in cyclic molecules with two stereocenters in the same way we analyzed it in acyclic compounds.

Example 6.5 Enentriomerism in Cyclic Compounds

How many stereoisomers are possible for 3-methylcyclopentanol?

Strategy and Solution

Carbons 1 and 3 of this compound are stereocenters. Therefore, $2^2 = 4$ stereoisomers are possible for this molecule. The *cis* isomer exists as one pair of enantiomers, the *trans* isomer exists as a second pair of enantiomers.

cis-2-Methylcyclopentanol
(a pair of enantiomers)

trans-2-Methylcyclopentanol
(a second pair of enantiomers)

Problem 6.5

How many stereoisomers are possible for 3-methylcyclohexanol?

Example 6.6 Locating Stereocenters

Mark the stereocenters in each compound with an asterisk. How many stereoisomers are possible for each?

Strategy

A stereocenter is a carbon atom that has four different groups bonded to it. Therefore, you are being asked to identify each carbon bonded to four different groups.

Solution

Each stereocenter is marked with an asterisk, and the number of stereoisomers possible for it appears under each compound. In (a), the carbon bearing the two methyl groups is not a stereocenter; this carbon has only three different groups bonded to it.

Problem 6.6

Mark all stereocenters in each compound with an asterisk. How many stereoisomers are possible for each?

B. Molecules with Three or More Stereocenters

The 2^n rule applies equally well to molecules with three or more stereocenters. The following disubstituted cyclohexanol has three stereocenters, each marked with an asterisk. A maximum of $2^3 = 8$ stereoisomers is possible for this molecule. Menthol, one of the eight, has the configuration shown in the middle and on the right. Menthol is present in peppermint and other mint oils.

2-Isopropyl-5-methyl-cyclohexanol
(three stereocenters; eight stereoisomers are possible)

Menthol
(one of the eight possible stereoisomers)

Menthol drawn as a chair conformation
(note that the three groups on the cyclohexane ring are all equatorial)

Chemical Connections 6A

Chiral Drugs

Some common drugs used in human medicine—for example, aspirin—are achiral. Others, such as the penicillin and erythromycin classes of antibiotics and the drug Captopril, are chiral and are sold as single enantiomers. Captopril is very effective for the treatment of high blood pressure and congestive heart failure (Chemical Connections 14F). It is manufactured and sold as the (S,S)-stereoisomer.

Captopril

A large number of chiral drugs, however, are sold as racemic mixtures. The popular analgesic ibuprofen (the active ingredient in Motrin, Advil, and many other non-aspirin analgesics) is an example.

Recently, the U.S. Food and Drug Administration established new guidelines for the testing and marketing of chiral drugs. After reviewing these guidelines, many drug companies have decided to develop only single enantiomers of new chiral drugs.

In addition to regulatory pressure, pharmaceutical developers must deal with patent considerations. If a company has a patent on a racemic mixture of a drug, a new patent can often be taken out on one of its enantiomers.

Cholesterol, a more complicated molecule, has eight stereocenters. To identify them, remember to add an appropriate number of hydrogens to complete the tetravalence of each carbon you think might be a stereocenter.

Cholesterol has eight stereocenters; 256 stereoisomers are possible

This is the stereoisomer found in human metabolism

6.4 What Is Optical Activity, and How Is Chirality Detected in the Laboratory?

A. Plane-Polarized Light

As we have already established, the two members of a pair of enantiomers are different compounds, and we must expect, therefore, that some of their properties differ. One such property relates to their effect on the plane of polarized light. Each member of a pair of enantiomers rotates the plane of polarized light; for this reason, each enantiomer is said to be **optically active.** To understand how optical activity is detected in the laboratory, we must first understand what plane-polarized light is and how a polarimeter, the instrument used to detect optical activity, works.

Ordinary light consists of waves vibrating in all planes perpendicular to its direction of propagation. Certain materials, such as a Polaroid sheet (a plastic film like that used in polarized sunglasses), selectively transmit

Optically active Showing that a compound rotates the plane of polarized light

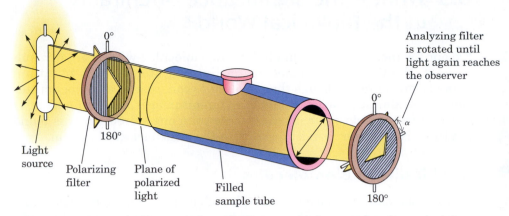

FIGURE 6.6 Schematic diagram of a polarimeter with its sample tube containing a solution of an optically active compound. The analyzer has been turned clockwise by α degrees to restore the light field.

light waves vibrating only in parallel planes. Electromagnetic radiation vibrating in only parallel planes is said to be **plane polarized.**

B. A Polarimeter

A **polarimeter** consists of a light source emitting unpolarized light, a polarizer, an analyzer, and a sample tube (Figure 6.6). If the sample tube is empty, the intensity of light reaching the detector (in this case, your eye) is at its maximum when the axes of the polarizer and analyzer are parallel to each other. If the analyzer is turned either clockwise or counterclockwise, less light is transmitted. When the axis of the analyzer is at right angles to the axis of the polarizer, the field of view is dark (no light passes through).

When a solution of an optically active compound is placed in the sample tube, it rotates the plane of the polarized light. If it rotates the plane clockwise, we say it is **dextrorotatory;** if it rotates the plane counterclockwise, we say it is **levorotatory.** Each member of a pair of enantiomers rotates the plane of polarized light by the same number of degrees, but in opposite directions. If one enantiomer is dextrorotatory, the other is levorotatory. Thus racemic mixtures (as well as achiral compounds) do not display optical activity.

The number of degrees by which an optically active compound rotates the plane of polarized light is called its **specific rotation** and is given the symbol $[\alpha]$. Specific rotation is defined as the observed rotation of an optically active substance at a concentration of 1 g/mL in a sample tube that is 10 cm long. A dextrorotatory compound is indicated by a plus sign in parentheses, (+), and a levorotatory compound is indicated by a minus sign in parentheses, (−). It is common practice to report the temperature (in °C) at which the measurement is made and the wavelength of light used. The most common wavelength of light used in polarimetry is the sodium D line, the same wavelength responsible for the yellow color of sodium-vapor lamps.

Following are specific rotations for the enantiomers of lactic acid measured at 21°C and using the D line of a sodium-vapor lamp as the light source:

Plane-polarized light Light with waves vibrating in only parallel planes

Dextrorotatory The clockwise (to the right) rotation of the plane of polarized light in a polarimeter

Levorotatory The counterclockwise (to the left) rotation of the plane of polarized light in a polarimeter

The (+) enantiomer of lactic acid is produced by muscle tissue in humans. The (−) enantiomer is found in sour cream and sour milk.

(S)-(+)-Lactic acid
$[\alpha]_D^{21} = +2.6°$

(R)-(−)-Lactic acid
$[\alpha]_D^{21} = -2.6°$

6.5 What Is the Significance of Chirality in the Biological World?

Except for inorganic salts and a few low-molecular-weight organic substances, the majority of molecules in living systems—both plant and animal—are chiral. Although these molecules can exist as a number of stereoisomers, almost invariably only one stereoisomer is found in nature. Of course, instances do occur in which more than one stereoisomer is found, but these isomers rarely exist together in the same biological system.

The horns of this African Gazelle show chirality; one horn is the mirror image of the other.

A. Chirality in Biomolecules

Perhaps the most conspicuous examples of chirality among biological molecules are the enzymes, all of which have many stereocenters. Consider chymotrypsin, an enzyme in the intestines of animals that catalyzes the digestion of proteins (Chapter 15). Chymotrypsin has 251 stereocenters. The maximum number of stereoisomers possible is 2^{251}—a staggeringly large number, almost beyond comprehension. Fortunately, nature does not squander its precious energy and resources unnecessarily; any given organism produces and uses only one of these stereoisomers.

B. How Does an Enzyme Distinguish Between a Molecule and Its Enantiomer?

An enzyme catalyzes a biological reaction of a molecule by first positioning the molecule at a **binding site** on the enzyme surface. An enzyme with specific binding sites for three of the four groups on a stereocenter can distinguish between a chiral molecule and its enantiomer or one of its diastereomers. Assume, for example, that an enzyme involved in catalyzing a reaction of glyceraldehyde has three binding sites: one specific for —H, a second specific for —OH, and a third specific for —CHO. Assume further that the three sites are arranged on the enzyme surface as shown in Figure 6.7. The enzyme can distinguish (R)-glyceraldehyde (the natural or biologically active form) from its enantiomer because the natural enantiomer can be adsorbed with three groups interacting with their appropriate binding sites; for the S enantiomer, at best only two groups can interact with these three binding sites.

Because interactions between molecules in living systems take place in a chiral environment, it should come as no surprise that a molecule and its enantiomer or one of its diastereomers elicit different physiological responses. As we have already seen, (S)-ibuprofen is active as a pain and fever reliever,

FIGURE 6.7 A schematic diagram of an enzyme surface that can interact with (R)-glyceraldehyde at three binding sites, but with (S)-glyceraldehyde at only two of these sites.

(R)-Glyceraldehyde
fits the three binding
sites on surface

(S)-Glyceraldehyde
fits only two of the
three binding sites

while its *R* enantiomer is inactive. The *S* enantiomer of the closely related analgesic naproxen is also the active pain reliever of this compound, but its *R* enantiomer is a liver toxin!

(*S*)-Ibuprofen (*S*)-Naproxen

Summary

Section 6.1 What Is Enantiomerism?

- A **mirror image** is the reflection of an object in a mirror.
- **Enantiomers** are a pair of stereoisomers that are non-superposable mirror images.
- A **racemic mixture** contains equal amounts of two enantiomers and does not rotate the plane of polarized light.
- **Diastereomers** are stereoisomers that are not mirror images.
- An object that is not superposable on its mirror image is said to be **chiral;** it has handedness. An **achiral** object lacks chirality (handedness); that is, it has a superposable mirror image.
- The most common cause of chirality in organic molecules is the presence of a tetrahedral carbon atom with four different groups bonded to it. Such a carbon is called a **stereocenter.**

Section 6.2 How Do We Specify the Configuration of a Stereocenter?

- We use the **R,S system** to specify the configuration of a stereocenter.

Section 6.3 How Many Stereoisomers Are Possible for Molecules with Two or More Stereocenters?
Problem 6.26

- For a molecule with n stereocenters, the maximum number of stereoisomers possible is 2^n.

Section 6.4 What Is Optical Activity, and How Is Chirality Detected in the Laboratory?

- Light with waves that vibrate in only parallel planes is said to be **plane polarized.**
- We use a **polarimeter** to measure optical activity. A compound is said to be **optically active** if it rotates the plane of polarized light.
- If a compound rotates the plane clockwise, it is **dextrorotatory;** if it rotates the plane counterclockwise, it is **levorotatory.**
- Each member of a pair of enantiomers rotates the plane of polarized light an equal number of degrees, but in opposite directions.

Section 6.5 What Is the Significance of Chirality in the Biological World?

- An enzyme catalyzes biological reactions of molecules by first positioning them at binding sites on its surface. An enzyme with binding sites specific for three of the four groups on a stereocenter can distinguish between a molecule and its enantiomer or one of its diastereomers.

Problems

■ Indicates problems that are assignable in **GOB OWL**.

Blue numbered problems are applied.

Go to this book's companion website at **www.cengage.com/chemistry/bettelheim** for interactive versions of the How To tutorials and Active Figures, and to quiz yourself on this chapter.

Section 6.1 What Is Enantiomerism?

6.7 Answer true or false.

(a) The *cis* and *trans* stereoisomers of 2-butene are achiral.

(b) The carbonyl carbon of an aldehyde, ketone, carboxylic acid, or ester cannot be a stereocenter.

(c) Stereoisomers have the same connectivity of their atoms.

(d) Constitutional isomers have the same connectivity of their atoms.

(e) An unmarked cube is achiral

(f) A human foot is chiral.

(g) Every object in nature has a mirror image.

(h) The most common cause of chirality in organic molecules is the presence of a tetrahedral carbon atom with four different groups bonded to it.

(i) If a molecule is not superposable on its mirror image, the molecule is chiral.

6.8 What does the term "chiral" mean? Give an example of a chiral molecule.

6.9 What does the term "achiral" mean? Give an example of an achiral molecule.

6.10 Define the term "stereoisomer." Name three types of stereoisomers.

6.11 In what way are constitutional isomers different from stereoisomers? In what way are they the same?

6.12 Which of the following objects are chiral (assume that there is no label or other identifying mark)?
(a) Pair of scissors   (b) Tennis ball
(c) Paper clip   (d) Beaker
(e) The swirl created in water as it drains out of a sink or bathtub

6.13 2-Pentanol is chiral but 3-pentanol is not. Explain.

6.14 2-Butene exists as a pair of *cis-trans* isomers. Is *cis*-2-butene chiral? Is *trans*-2-butene chiral? Explain.

6.15 Explain why the carbon of a carbonyl group cannot be a stereocenter.

6.16 Which of the following compounds contain stereocenters?
(a) 2-Chloropentane   (b) 3-Chloropentane
(c) 3-Chloro-1-butene   (d) 1,2-Dichloropropane

6.17 Which of the following compounds contain stereocenters?
(a) Cyclopentanol
(b) 1-Chloro-2-propanol
(c) 2-Methylcyclopentanol
(d) 1-Phenyl-1-propanol

6.18 Using only C, H, and O, write structural formulas for the lowest-molecular-weight chiral molecule of each class.
(a) Alkane   (b) Alkene
(c) Alcohol   (d) Aldehyde
(e) Ketone   (f) Carboxylic acid

6.19 Draw the mirror image for each molecule:

6.20 Draw the mirror image for each molecule:

Section 6.3 How Many Stereoisomers Are Possible for Molecules with Two or More Stereocenters?

6.21 Answer true or false.
(a) For a molecule with two stereocenters, $2^2 = 4$ stereoisomers are possible.
(b) For a molecule with three stereocenters, $3^2 = 9$ stereoisomers are possible.
(c) Enantiomers, like gloves, occur in pairs.
(d) 2-Pentanol and 3-pentanol are both chiral and show enantiomerism.
(e) 1-Methylcyclohexanol is achiral and does not show enantiomerism.
(f) Diastereomers are stereoisomers that are not mirror images.

6.22 Mark each stereocenter in these molecules with an asterisk. Note that not all contain stereocenters.

6.23 Mark each stereocenter in these molecules with an asterisk. Note that not all contain stereocenters.

6.24 Label all stereocenters in each molecule with an asterisk. How many stereoisomers are possible for each molecule?

(a)
$$CH_3CHCHCOOH$$
with OH (top) and OH (bottom)

(b)
$$CH_2COOH$$
$$CHCOOH$$
$$HO-CHCOOH$$

(c) (d)

6.25 Label all stereocenters in each molecule with an asterisk. How many stereoisomers are possible for each molecule?

(a)   (b)

(c)   (d)

6.26 ■ For centuries, Chinese herbal medicine has used extracts of *Ephedra sinica* to treat asthma. The asthma-relieving component of this plant is ephedrine, a very potent dilator of the air passages of the lungs. The naturally occurring stereoisomer is levorotatory and has the following structure.

Ephedrine $[\alpha]_D^{21} = -41°$

(a) Mark each stereocenter in epinephrine with an asterisk.

(b) How many stereoisomers are possible for this compound?

6.27 The specific rotation of naturally occurring ephedrine, shown in Problem 6.26, is −41°. What is the specific rotation of its enantiomer?

6.28 What is a racemic mixture? Is a racemic mixture optically active? That is, will it rotate the plane of polarized light?

Section 6.4 What Is Optical Activity, and How Is Chirality Detected in the Laboratory?

6.29 Answer true or false.

(a) If a chiral compound is dextrorotatory, its enantiomer is levorotatory by the same number of degrees.

(b) A racemic mixture is optically inactive.

(c) All stereoisomers are optically active.

(d) Plane-polarized light consists of light waves vibrating in parallel planes.

Chemical Connections

6.30 (Chemical Connections 6A) What does it mean to say that a drug is *chiral*? If a drug is chiral, will it be optically active? That is, will it rotate the plane of polarized light?

Additional Problems

6.31 Which of the eight alcohols with a molecular formula of $C_5H_{12}O$ are chiral?

6.32 ■ Write the structural formula of an alcohol with the molecular formula $C_6H_{14}O$ that contains two stereocenters.

6.33 Which carboxylic acids with a molecular formula of $C_6H_{12}O_2$ are chiral?

6.34 Following are structural formulas for three of the drugs most widely prescribed to treat depression. Label all stereocenters in each, and state the number of stereoisomers possible for each.

(a)

Fluoxetine
(Prozac)

(b)

Sertraline
(Zoloft)

(c)

Paroxetine
(Paxil)

6.35 ■ Label the four stereocenters in amoxicillin, which belongs to the family of semisynthetic penicillins.

Amoxicillin

6.36 ■ Consider a cyclohexane ring substituted with one hydroxyl group and one methyl group. Draw a structural formula for a compound of this composition that
(a) Does not show *cis-trans* isomerism and has no stereocenters.
(b) Shows *cis-trans* isomerism but has no stereocenters.
(c) Shows *cis-trans* isomerism and has two stereocenters.

6.37 The next time you have an opportunity to examine any of the seemingly endless varieties of blond spiral pasta (rotini, fusilli, radiatori, tortiglione, and so forth), examine their twists. Do the twists of any one kind all have a right-handed twist or a left-handed twist, or are they a racemic mixture?

6.38 Think about the helical coil of a telephone cord or the spiral binding on a notebook. Suppose that you view the spiral from one end and find that it has a left-handed twist. If you view the same spiral from the other end, does it have a left-handed twist from that end as well or does it have a right-handed twist?

Looking Ahead

6.39 ■ Triamcinolone acetonide, the active ingredient in Azmacort Inhalation Aerosol, is a steroid used to treat bronchial asthma.

Triamcinolone acetonide

(a) Label the eight stereocenters in this molecule.
(b) How many stereoisomers are possible for it? (Of these, the stereoisomer with the configuration shown here is the active ingredient in Azmacort.)

Acids and Bases

7

Charles D. Winters/Cengage Learning

Some foods and household products are very acidic while others are basic. From your prior experiences, can you tell which ones belong to which category?

7.1 What Are Acids and Bases?

We frequently encounter acids and bases in our daily lives. Oranges, lemons, and vinegar are examples of acidic foods, and sulfuric acid is in our automobile batteries. As for bases, we take antacid tablets for heartburn and use household ammonia as a cleaning agent. What do these substances have in common? Why are acids and bases usually discussed together?

ᴕWL Online howmework for this chapter may be assigned in **GOB OWL**.

In 1884, a young Swedish chemist named Svante Arrhenius (1859–1927) answered the first question by proposing what was then a new definition of acids and bases. According to the Arrhenius definition, an **acid** is a substance that produces H_3O^+ ions in aqueous solution, and a **base** is a substance that produces OH^- ions in aqueous solution.

This definition of acid is a slight modification of the original Arrhenius definition, which stated that an acid produces hydrogen ions, H^+. Today we know that H^+ ions cannot exist in water. An H^+ ion is a bare proton, and a charge of $+1$ is too concentrated to exist on such a tiny particle. Therefore, an H^+ ion in water immediately combines with an H_2O molecule to give a **hydronium ion,** H_3O^+.

Hydronium ion The H_3O^+ ion

$$H^+(aq) + H_2O(\ell) \longrightarrow H_3O^+(aq)$$

Hydronium ion

Apart from this modification, the Arrhenius definitions of acid and base are still valid and useful today, as long as we are talking about aqueous solutions. Although we know that acidic aqueous solutions do not contain H^+ ions, we frequently use the terms "H^+" and "proton" when we really mean "H_3O^+." The three terms are generally used interchangeably.

When an acid dissolves in water, it reacts with the water to produce H_3O^+. For example, hydrogen chloride, HCl, in its pure state is a poisonous gas. When HCl dissolves in water, it reacts with a water molecule to give hydronium ion and chloride ion:

$$H_2O(\ell) + HCl(aq) \longrightarrow H_3O^+(aq) + Cl^-(aq)$$

Thus a bottle labeled aqueous "HCl" is actually not HCl at all, but rather an aqueous solution of H_3O^+ and Cl^- ions in water.

We can show the transfer of a proton from an acid to a base by using a curved arrow. First we write the Lewis structure of each reactant and product. Then we use curved arrows to show the change in position of electron pairs during the reaction. The tail of the curved arrow is located at the electron pair. The head of the curved arrow shows the new position of the electron pair.

In this equation, the curved arrow on the left shows that an unshared pair of electrons on oxygen forms a new covalent bond with hydrogen. The curved arrow on the right shows that the pair of electrons of the H—Cl bond is given entirely to chlorine to form a chloride ion. Thus, in the reaction of HCl with H_2O, a proton is transferred from HCl to H_2O and, in the process, an O—H bond forms and an H—Cl bond is broken.

With bases, the situation is slightly different. Many bases are metal hydroxides, such as KOH, NaOH, $Mg(OH)_2$, and $Ca(OH)_2$. When these ionic solids dissolve in water, their ions merely separate, and each ion is solvated by water molecules. For example,

$$NaOH(s) \xrightarrow{H_2O} Na^+(aq) + OH^-(aq)$$

Other bases are not hydroxides. Instead, they produce OH^- ions in water by reacting with water molecules. The most important example of this kind of base is ammonia, NH_3, a poisonous gas. When ammonia dissolves in water, it reacts with water to produce ammonium ions and hydroxide ions.

$$NH_3(aq) + H_2O(\ell) \rightleftharpoons NH_4^+(aq) + OH^-(aq)$$

As we will see in Section 7.2, ammonia is a weak base, and the position of the equilibrium for its reaction with water lies considerably toward the left. In a 1.0 M solution of NH_3 in water, for example, only about 4 molecules of NH_3 out of every 1000 react with water to form NH_4^+ and OH^-. Thus, when ammonia is dissolved in water, it exists primarily as NH_3 molecules. Nevertheless, some OH^- ions are produced and, therefore, NH_3 is a base.

Bottles of NH_3 in water are sometimes labeled "ammonium hydroxide" or "NH_4OH," but this gives a false impression of what is really in the bottle. Most of the NH_3 molecules have not reacted with the water, so the bottle contains mostly NH_3 and H_2O and only a little NH_4^+ and OH^-.

We indicate how the reaction of ammonia with water takes place by using curved arrows to show the transfer of a proton from a water molecule to an ammonia molecule. Here, the curved arrow on the left shows that the unshared pair of electrons on nitrogen forms a new covalent bond with a hydrogen of a water molecule. At the same time as the new N—H bond forms, an O—H bond of a water molecule breaks and the pair of electrons forming the H—O bond moves entirely to oxygen, forming OH^-.

Thus ammonia produces an OH^- ion by taking H^+ from a water molecule and leaving OH^- behind.

7.2 How Do We Define the Strength of Acids and Bases?

All acids are not equally strong. According to the Arrhenius definition, a **strong acid** is one that reacts completely or almost completely with water to form H_3O^+ ions. Table 7.1 gives the names and molecular formulas for six of the most common strong acids. They are strong acids because, when they dissolve in water, they dissociate completely to give H_3O^+ ions.

Weak acids produce a much smaller concentration of H_3O^+ ions. Acetic acid, for example, is a weak acid. In water it exists primarily as acetic acid molecules; only a few acetic acid molecules (4 out of every 1000) are converted to acetate ions.

$$CH_3COOH(aq) + H_2O(\ell) \rightleftharpoons CH_3COO^-(aq) + H_3O^+(aq)$$
$$\text{Acetic acid} \qquad\qquad\qquad \text{Acetate ion}$$

There are four common **strong bases** (Table 7.1), all of which are metal hydroxides. They are strong bases because, when they dissolve in water, they ionize completely to give OH^- ions. Another base, $Mg(OH)_2$, dissociates almost completely once dissolved, but it is also very insoluble in water to

Strong acid An acid that ionizes completely in aqueous solution

Weak acid An acid that is only partially ionized in aqueous solution

Strong base A base that ionizes completely in aqueous solution

TABLE 7.1 Strong Acids and Bases

Acid Formula	Name	Base Formula	Name
HCl	Hydrochloric acid	LiOH	Lithium hydroxide
HBr	Hydrobromic acid	NaOH	Sodium hydroxide
HI	Hydroiodic acid	KOH	Potassium hydroxide
HNO_3	Nitric acid	$Ba(OH)_2$	Barium hydroxide
H_2SO_4	Sulfuric acid		
$HClO_4$	Perchloric acid		

Chemical Connections 7A

Some Important Acids and Bases

STRONG ACIDS Sulfuric acid, H_2SO_4, is used in many industrial processes, such as manufacturing fertilizer, dyes and pigments, and rayon. In fact, sulfuric acid is one of the most widely produced single chemicals in the United States.

Hydrochloric acid, HCl, is an important acid in chemistry laboratories. Pure HCl is a gas, and the HCl in laboratories is an aqueous solution. HCl is the acid in the gastric fluid in your stomach, where it is secreted at a strength of about 5% w/v.

Nitric acid, HNO_3, is a strong oxidizing agent. A drop of it causes the skin to turn yellow because the acid reacts with skin proteins. A yellow color upon contact with nitric acid has long been a test for proteins.

Weak acids are found in many common materials. In the foreground are strips of litmus paper that have been dipped into solutions of these materials. Acids turn litmus paper red.

Charles D. Winters/Cengage Learning

Weak bases are also common in many household products. These cleaning agents all contain weak bases. Bases turn litmus paper blue.

Charles D. Winters/Cengage Learning

WEAK ACIDS Acetic acid, CH_3COOH, is present in vinegar (about 5%). Pure acetic acid is called glacial acetic acid because of its melting point of 17°C, which means that it freezes on a moderately cold day.

Boric acid, H_3BO_3, is a solid. Solutions of boric acid in water were once used as antiseptics, especially for eyes. Boric acid is toxic when swallowed.

Phosphoric acid, H_3PO_4, is one of the strongest of the weak acids. The ions produced from it—$H_2PO_4^-$, HPO_4^{2-}, and PO_4^{3-}—are important in biochemistry (see also Section 19.3).

STRONG BASES Sodium hydroxide, NaOH, also called lye, is the most important of the strong bases. It is a solid whose aqueous solutions are used in many industrial processes, including the manufacture of glass and soap. Potassium hydroxide, KOH, also a solid, is used for many of the same purposes as NaOH.

WEAK BASES Ammonia, NH_3, the most important weak base, is a gas with many industrial uses. One of its chief uses is for fertilizers. A 5% solution is sold in supermarkets as a cleaning agent, and weaker solutions are used as "spirits of ammonia" to revive people who have fainted.

Magnesium hydroxide, $Mg(OH)_2$, is a solid that is insoluble in water. A suspension of about 8% $Mg(OH)_2$ in water is called milk of magnesia and is used as a laxative. $Mg(OH)_2$ is also used to treat wastewater in metal-processing plants and as a flame retardant in plastics.

Weak base A base that is only partially ionized in aqueous solution

begin with. As we saw in Section 7.1, ammonia is a **weak base** because the equilibrium for its reaction with water lies far to the left.

It is important to understand that the strength of an acid or a base is not related to its concentration. HCl is a strong acid, whether it is concentrated or dilute, because it dissociates completely in water to chloride ions and hydronium ions. Acetic acid is a weak acid, whether it is concentrated or dilute, because the equilibrium for its reaction with water lies far to the left. When acetic acid dissolves in water, most of it is present as undissociated CH_3COOH molecules.

$$HCl(aq) + H_2O(\ell) \longrightarrow Cl^-(aq) + H_3O^+(aq)$$

$$\underset{\text{Acetic acid}}{CH_3COOH(aq)} + H_2O(\ell) \rightleftharpoons \underset{\text{Acetate ion}}{CH_3COO^-(aq)} + H_3O^+(aq)$$

We saw that electrolytes (substances that produce ions in aqueous solution) can be strong or weak. The strong acids and bases in Table 7.1 are strong electrolytes. Almost all other acids and bases are weak electrolytes.

7.3 What Are Conjugate Acid–Base Pairs?

The Arrhenius definitions of acid and base are very useful in aqueous solutions. But what if water is not involved? In 1923, the Danish chemist Johannes Brønsted and the English chemist Thomas Lowry independently proposed the following definitions: An **acid** is a proton donor, a **base** is a proton acceptor, and an **acid–base reaction** is a proton-transfer reaction. Furthermore, according to the Brønsted-Lowry definitions, any pair of molecules or ions that can be interconverted by transfer of a proton is called a **conjugate acid–base pair.** When an acid transfers a proton to a base, the acid is converted to its **conjugate base.** When a base accepts a proton, it is converted to its **conjugate acid.**

We can illustrate these relationships by examining the reaction between acetic acid and ammonia:

$$CH_3COOH + NH_3 \rightleftharpoons CH_3COO^- + NH_4^+$$

Acetic acid — Ammonia — Acetate ion — Ammonium ion

(Acid) — (Base) — (Conjugate base of acetic acid) — (Conjugate acid of ammonia)

Conjugate acid–base pair A pair of molecules or ions that are related to one another by the gain or loss of a proton

Conjugate base In the Brønsted-Lowry theory, a substance formed when an acid donates a proton to another molecule or ion

Conjugate acid In the Brønsted-Lowry theory, a substance formed when a base accepts a proton

We can use curved arrows to show how this reaction takes place. The curved arrow on the right shows that the unshared pair of electrons on nitrogen becomes shared to form a new H—N bond. At the same time that the H—N bond forms, the O—H bond breaks and the electron pair of the O—H bond moves entirely to oxygen to form —O⁻ of the acetate ion. The result of these two electron-pair shifts is the transfer of a proton from an acetic acid molecule to an ammonia molecule:

Acetic acid (Proton donor) — Ammonia (Proton acceptor) — Acetate ion — Ammonium ion

Table 7.2 gives examples of common acids and their conjugate bases. As you study the examples of conjugate acid–base pairs in Table 7.2, note the following points:

1. An acid can be positively charged, neutral, or negatively charged. Examples of these charge types are H_3O^+, H_2CO_3, and $H_2PO_4^-$, respectively.

2. A base can be negatively charged or neutral. Examples of these charge types are PO_4^{3-} and NH_3, respectively.

3. Acids are classified as monoprotic, diprotic, or triprotic depending on the number of protons each may give up. Examples of **monoprotic acids** include HCl, HNO_3, and CH_3COOH. Examples of **diprotic acids** include H_2SO_4 and H_2CO_3. An example of a **triprotic acid** is H_3PO_4.

A box of Arm & Hammer baking soda (sodium bicarbonate). Sodium bicarbonate is composed of Na^+ and HCO_3^-, the amphiprotic bicarbonate ion.

Monoprotic acid An acid that can give up only one proton

Diprotic acid An acid that can give up two protons

Triprotic acid An acid that can give up three protons

7.4 How Can We Tell the Position of Equilibrium in an Acid–Base Reaction?

We know that HCl reacts with H_2O according to the following equilibrium:

$$HCl + H_2O \rightleftharpoons Cl^- + H_3O^+$$

We also know that HCl is a strong acid, which means the position of this equilibrium lies very far to the right. In fact, this equilibrium lies so far to the right that out of every 10,000 HCl molecules dissolved in water, all but one react with water molecules to give Cl^- and H_3O^+.

For this reason, we usually write the acid reaction of HCl with a unidirectional arrow, as follows:

$$HCl + H_2O \longrightarrow Cl^- + H_3O^+$$

As we have also seen, acetic acid reacts with H_2O according to the following equilibrium:

$$\underset{\text{Acetic acid}}{CH_3COOH} + H_2O \rightleftharpoons \underset{\text{Acetate ion}}{CH_3COO^-} + H_3O^+$$

Acetic acid is a weak acid. Only a few acetic acid molecules react with water to give acetate ions and hydronium ions, and the major species present in equilibrium in aqueous solution are CH_3COOH and H_2O. The position of this equilibrium, therefore, lies very far to the left.

In these two acid–base reactions, water is the base. But what if we have a base other than water as the proton acceptor? How can we determine which are the major species present at equilibrium? That is, how can we determine if the position of equilibrium lies toward the left or toward the right?

As an example, let us examine the acid–base reaction between acetic acid and ammonia to form acetate ion and ammonium ion. As indicated by the question mark over the equilibrium arrow, we want to determine whether the position of this equilibrium lies toward the left or toward the right.

$$\underset{\substack{\text{Acetic acid}\\(\text{Acid})}}{CH_3COOH} + \underset{\substack{\text{Ammonia}\\(\text{Base})}}{NH_3} \overset{?}{\rightleftharpoons} \underset{\substack{\text{Acetate ion}\\(\text{Conjugate base}\\\text{of }CH_3COOH)}}{CH_3COO^-} + \underset{\substack{\text{Ammonium ion}\\(\text{Conjugate acid}\\\text{of }NH_3)}}{NH_4^+}$$

In this equilibrium there are two acids present: acetic acid and ammonium ion. There are also two bases present: ammonia and acetate ion. One way to analyze this equilibrium is to view it as a competition of the two bases, ammonia and acetate ion, for a proton. Which is the stronger base? The information we need to answer this question is found in Table 7.2. We first determine which conjugate acid is the stronger acid and then use this information along with the fact that the stronger the acid, the weaker its conjugate base. From Table 7.2, we see that CH_3COOH is the stronger acid, which means that CH_3COO^- is the weaker base. Conversely, NH_4^+ is the weaker acid, which means that NH_3 is the stronger base. We can now label the relative strengths of each acid and base in this equilibrium:

$$\underset{\substack{\text{Acetic acid}\\(\text{Stronger acid})}}{CH_3COOH} + \underset{\substack{\text{Ammonia}\\(\text{Stronger base})}}{NH_3} \overset{?}{\rightleftharpoons} \underset{\substack{\text{Acetate ion}\\(\text{Weaker base})}}{CH_3COO^-} + \underset{\substack{\text{Ammonium ion}\\(\text{Weaker acid})}}{NH_4^+}$$

In an acid–base reaction, the equilibrium position always favors reaction of the stronger acid and stronger base to form the weaker acid and weaker base. Thus, at equilibrium, the major species present are the weaker acid and the weaker base. In the reaction between acetic acid and ammonia, therefore, the equilibrium lies to the right and the major species present are acetate ion and ammonium ion:

$$CH_3COOH + NH_3 \rightleftharpoons CH_3COO^- + NH_4^+$$

Acetic acid   Ammonia   Acetate ion   Ammonium ion
(Stronger acid) (Stronger base) (Weaker base)   (Weaker acid)

To summarize, we use the following four steps to determine the position of an acid–base equilibrium:

1. Identify the two acids in the equilibrium; one is on the left side of the equilibrium, and the other on the right side.
2. Using the information in Table 7.2, determine which acid is the stronger acid and which acid is the weaker acid.
3. Identify the stronger base and the weaker base. Remember that the stronger acid gives the weaker conjugate base and the weaker acid gives the stronger conjugate base.
4. The stronger acid and stronger base react to give the weaker acid and weaker base. The position of equilibrium, therefore, lies on the side of the weaker acid and weaker base.

Example 7.2 Acid/Base Pairs

For each acid–base equilibrium, label the stronger acid, the stronger base, the weaker acid, and the weaker base. Then predict whether the position of equilibrium lies toward the right or toward the left.

(a) $H_2CO_3 + OH^- \rightleftharpoons HCO_3^- + H_2O$
(b) $HPO_4^{2-} + NH_3 \rightleftharpoons PO_4^{3-} + NH_4^+$

Strategy

Use Table 7.2 to identify the stronger acid from the weaker acid and the stronger base from the weaker base. Once you have done that, determine in which direction the equilibrium lies. It always lies in the direction of the stronger components moving towards the weaker components.

Solution

Arrows connect the conjugate acid–base pairs, with the red arrows showing the stronger acid. The position of equilibrium in (a) lies toward the right. In (b) it lies toward the left.

(a)

$$H_2CO_3 + OH^- \rightleftharpoons HCO_3^- + H_2O$$

Stronger acid   Stronger base   Weaker base   Weaker acid

(b)

$$HPO_4^{2-} + NH_3 \rightleftharpoons PO_4^{3-} + NH_4^+$$

Weaker acid   Weaker base   Stronger base   Stronger acid

pK_a is $-\log K_a$

7.5 How Do We Use Acid Ionization Constants?

In Section 7.2, we learned that acids vary in the extent to which they produce H_3O^+ when added to water. Because the ionizations of weak acids in water are all equilibria, we can use equilibrium constants to tell us quantitatively just how strong any weak acid is. The reaction that takes place when a weak acid, HA, is added to water is

$$HA + H_2O \rightleftharpoons A^- + H_3O^+$$

The equilibrium constant expression for this ionization is

$$K = \frac{[A^-][H_3O^+]}{[HA][H_2O]}$$

Notice that this expression contains the concentration of water. Because water is the solvent and its concentration changes very little when we add HA to it, we can treat the concentration of water, $[H_2O]$, as a constant equal to 1000 g/L or approximately 55.49 mol/L. We can then combine these two constants (K and $[H_2O]$) to define a new constant called an **acid ionization constant, K_a.**

Acid ionization constant (K_a) An equilibrium constant for the ionization of an acid in aqueous solution to H_3O^+ and its conjugate base; also called an acid dissociation constant

$$K_a = K[H_2O] = \frac{[A^-][H_3O^+]}{[HA]}$$

The value of the acid ionization constant for acetic acid, for example, is 1.8×10^{-5}. Because acid ionization constants for weak acids are numbers with negative exponents, we often use an algebraic trick to turn them into numbers that are easier to use. To do so, we take the negative logarithm of the number. Acid strengths are therefore expressed as $-\log K_a$, which we call the pK_a. The "p" of anything is just the negative logarithm of that thing. The pK_a of acetic acid is 4.75. Table 7.3 gives names, molecular formulas, and values of K_a and pK_a for some weak acids. As you study the entries in

TABLE 7.3 K_a and pK_a Values for Some Weak Acids

Formula	Name	K_a	pK_a
H_3PO_4	Phosphoric acid	7.5×10^{-3}	2.12
HCOOH	Formic acid	1.8×10^{-4}	3.75
$CH_3CH(OH)COOH$	Lactic acid	1.4×10^{-4}	3.86
CH_3COOH	Acetic acid	1.8×10^{-5}	4.75
H_2CO_3	Carbonic acid	4.3×10^{-7}	6.37
$H_2PO_4^-$	Dihydrogen phosphate ion	6.2×10^{-8}	7.21
H_3BO_3	Boric acid	7.3×10^{-10}	9.14
NH_4^+	Ammonium ion	5.6×10^{-10}	9.25
HCN	Hydrocyanic acid	4.9×10^{-10}	9.31
C_6H_5OH	Phenol	1.3×10^{-10}	9.89
HCO_3^-	Bicarbonate ion	5.6×10^{-11}	10.25
HPO_4^{2-}	Hydrogen phosphate ion	2.2×10^{-13}	12.66

Increasing acid strength

this table, note the inverse relationship between the values of K_a and pK_a. The weaker the acid, the smaller its K_a, but the larger its pK_a.

One reason for the importance of K_a is that it immediately tells us how strong an acid is. For example, Table 7.3 shows us that although acetic acid, formic acid, and phenol are all weak acids, their strengths as acids are not the same. Formic acid, with a K_a of 1.8×10^{-4}, is stronger than acetic acid, whereas phenol, with a K_a of 1.3×10^{-10}, is much weaker than acetic acid. Phosphoric acid is the strongest of the weak acids. We can tell that an acid is classified as a weak acid by the fact that we list a pK_a for it, and the pK_a is a positive number. If we tried to take the negative logarithm of the K_a for a strong acid, we would get a negative number.

Example 7.3 pK_as

K_a for benzoic acid is 6.5×10^{-5}. What is the pK_a of this acid?

Strategy

The pK_a is $-\log K_a$. Thus, use your calculator to find the log of the K_a and then take the negative of it.

Solution

Take the logarithm of 6.5×10^{-5} on your scientific calculator. The answer is -4.19. Because pK_a is equal to $-\log K_a$, you must multiply this value by -1 to get pK_a. The pK_a of benzoic acid is 4.19.

Problem 7.3

K_a for hydrocyanic acid, HCN, is 4.9×10^{-10}. What is its pK_a?

Example 7.4 Acid Strength

Which is the stronger acid:

(a) Benzoic acid with a K_a of 6.5×10^{-5} or hydrocyanic acid with a K_a of 4.9×10^{-10}?

(b) Boric acid with a pK_a of 9.14 or carbonic acid with a pK_a of 6.37?

Strategy

Relative acid strength is determined by comparing the K_a values or the pK_a values. If using K_a values, the stronger acid has the larger K_a. If using pK_a values, the stronger acid has the smaller pK_a.

Solution

(a) Benzoic acid is the stronger acid; it has the larger K_a value.

(b) Carbonic acid is the stronger acid; it has the smaller pK_a.

Problem 7.4

Which is the stronger acid:
(a) Carbonic acid, $pK_a = 6.37$, or ascorbic acid (vitamin C), $pK_a = 4.1$?
(b) Aspirin, $pK_a = 3.49$, or acetic acid, $pK_a = 4.75$?

All of these fruits and fruit drinks contain organic acids.

Charles D. Winters/Cengage Learning

7.6 What Are the Properties of Acids and Bases?

Today's chemists do not taste the substances they work with, but 200 years ago they routinely did so. That is how we know that acids taste sour and

Charles D. Winters/Cengage Learning

ACTIVE FIGURE 7.1 Acids react with metals. A ribbon of magnesium metal reacts with aqueous HCl to give H$_2$ gas and aqueous MgCl$_2$. **Go to this book's companion website at www.cengage.com/chemistry/bettelheim to explore an interactive version of this figure.**

bases taste bitter. The sour taste of lemons, vinegar, and many other foods, for example, is due to the acids they contain.

A. Neutralization

The most important reaction of acids and bases is that they react with each other in a process called neutralization. This name is appropriate because, when a strong corrosive acid such as hydrochloric acid reacts with a strong corrosive base such as sodium hydroxide, the product (a solution of ordinary table salt in water) has neither acidic nor basic properties. We call such a solution neutral. Section 7.9 discusses neutralization reactions in detail.

B. Reaction with Metals

Strong acids react with certain metals (called active metals) to produce hydrogen gas, H$_2$, and a salt. Hydrochloric acid, for example, reacts with magnesium metal to give the salt magnesium chloride and hydrogen gas (Figure 7.1).

$$\underset{\substack{\text{Magnesium}}}{Mg(s)} + \underset{\substack{\text{Hydrochloric}\\\text{acid}}}{2HCl(aq)} \longrightarrow \underset{\substack{\text{Magnesium}\\\text{chloride}}}{MgCl_2(aq)} + \underset{\substack{\text{Hydrogen}}}{H_2(g)}$$

The reaction of an acid with an active metal to give a salt and hydrogen gas is a redox reaction. The metal is oxidized to a metal ion and H$^+$ is reduced to H$_2$.

C. Reaction with Metal Hydroxides

Acids react with metal hydroxides to give a salt and water.

$$\underset{\substack{\text{Hydrochloric}\\\text{acid}}}{HCl(aq)} + \underset{\substack{\text{Potassium}\\\text{hydroxide}}}{KOH(aq)} \longrightarrow \underset{\substack{\text{Water}}}{H_2O(\ell)} + \underset{\substack{\text{Potassium}\\\text{chloride}}}{KCl(aq)}$$

Both the acid and the metal hydroxide are ionized in aqueous solution. Furthermore, the salt formed is an ionic compound that is present in aqueous solution as anions and cations. Therefore, the actual equation for the reaction of HCl and KOH could be written showing all of the ions present:

$$H_3O^+ + Cl^- + K^+ + OH^- \longrightarrow 2H_2O + Cl^- + K^+$$

We usually simplify this equation by omitting the spectator ions, which gives the following equation for the net ionic reaction of any strong acid and strong base to give a salt and water:

$$H_3O^+ + OH^- \longrightarrow 2H_2O$$

D. Reaction with Metal Oxides

Strong acids react with metal oxides to give water and a salt, as shown in the following net ionic equation:

$$2H_3O^+(aq) + \underset{\substack{\text{Calcium}\\\text{oxide}}}{CaO(s)} \longrightarrow 3H_2O(\ell) + Ca^{2+}(aq)$$

E. Reaction with Carbonates and Bicarbonates

When a strong acid is added to a carbonate such as sodium carbonate, bubbles of carbon dioxide gas are rapidly given off. The overall reaction is a summation of two reactions. In the first reaction, carbonate ion reacts with

H_3O^+ to give carbonic acid. Almost immediately, in the second reaction, carbonic acid decomposes to carbon dioxide and water. The following equations show the individual reactions and then the overall reaction:

$$2H_3O^+(aq) + CO_3^{2-}(aq) \longrightarrow H_2CO_3(aq) + 2H_2O(\ell)$$
$$H_2CO_3(aq) \longrightarrow CO_2(g) + H_2O(\ell)$$
$$\overline{2H_3O^+(aq) + CO_3^{2-}(aq) \longrightarrow CO_2(g) + 3H_2O(\ell)}$$

Strong acids also react with bicarbonates such as potassium bicarbonate to give carbon dioxide and water:

$$H_3O^+(aq) + HCO_3^-(aq) \longrightarrow H_2CO_3(aq) + H_2O(\ell)$$
$$H_2CO_3(aq) \longrightarrow CO_2(g) + H_2O(\ell)$$
$$\overline{H_3O^+(aq) + HCO_3^-(aq) \longrightarrow CO_2(g) + 2H_2O(\ell)}$$

To generalize, any acid stronger than carbonic acid will react with carbonate or bicarbonate ion to give CO_2 gas.

The production of CO_2 is what makes bread doughs and cake batters rise. The earliest method used to generate CO_2 for this purpose involved the addition of yeast, which catalyzes the fermentation of carbohydrates to produce carbon dioxide and ethanol (Chapter 20):

$$\underset{\text{Glucose}}{C_6H_{12}O_6} \xrightarrow{\text{Yeast}} 2CO_2 + \underset{\text{Ethanol}}{2C_2H_5OH}$$

The production of CO_2 by fermentation, however, is slow. Sometimes it is desirable to have its production take place more rapidly, in which case bakers use the reaction of $NaHCO_3$ (sodium bicarbonate, also called **baking soda**) and a weak acid. But which weak acid? Vinegar (a 5% solution of acetic acid in water) would work, but it has a potential disadvantage—it imparts a particular flavor to foods. For a weak acid that imparts little or no flavor, bakers use either sodium dihydrogen phosphate, NaH_2PO_4, or potassium dihydrogen phosphate, KH_2PO_4. The two salts do not react when they are dry but, when mixed with water in a dough or batter, they react quite rapidly to produce CO_2. The production of CO_2 is even more rapid in an oven!

$$H_2PO_4^-(aq) + H_2O(\ell) \rightleftharpoons HPO_4^{2-}(aq) + H_3O^+(aq)$$
$$HCO_3^-(aq) + H_3O^+(aq) \longrightarrow CO_2(g) + 2H_2O(\ell)$$
$$\overline{H_2PO_4^-(aq) + HCO_3^-(aq) \longrightarrow HPO_4^{2-}(aq) + CO_2(g) + H_2O(\ell)}$$

Baking powder contains a weak acid, either sodium or potassium dihydrogen phosphate, and sodium or potassium bicarbonate. When they are mixed with water, they react to produce the bubbles of CO_2 seen in this picture.

F. Reaction with Ammonia and Amines

Any acid stronger than NH_4^+ (Table 7.2) is strong enough to react with NH_3 to form a salt. In the following reaction, the salt formed is ammonium chloride, NH_4Cl, which is shown as it would be ionized in aqueous solution:

$$HCl(aq) + NH_3(aq) \longrightarrow NH_4^+(aq) + Cl^-(aq)$$

In Chapter 8 we will meet a family of compounds called amines, which are similar to ammonia except that one or more of the three hydrogen atoms of ammonia are replaced by carbon groups. A typical amine is methylamine, CH_3NH_2. The base strength of most amines is similar to that of NH_3, which means that amines also react with acids to form salts. The salt formed in the reaction of methylamine with HCl is methylammonium chloride, shown here as it would be ionized in aqueous solution:

How To . . .

Use Logs and Antilogs

When dealing with acids, bases, and buffers, we often have to use common or base 10 logarithms (logs). To most people, a logarithm is just a button they push on a calculator. Here we describe briefly how to handle logs and antilogs.

1. What is a logarithm and how is it calculated?

A common logarithm is the power to which you raise 10 to get another number. For example, the log of 100 is 2, because you must raise 10 to the second power to get 100.

$$\log 100 = 2 \quad \text{since} \quad 10^2 = 100$$

Other examples are

$$\log 1000 = 3 \quad \text{since} \quad 10^3 = 1000$$
$$\log 10 = 1 \quad \text{since} \quad 10^1 = 10$$
$$\log 1 = 0 \quad \text{since} \quad 10^0 = 1$$
$$\log 0.1 = -1 \quad \text{since} \quad 10^{-1} = 0.1$$

The common logarithm of a number other than a simple power is usually obtained from a calculator by entering the number and then pressing log. For example,

$$\log 52 = 1.72$$
$$\log 4.5 = 0.653$$
$$\log 0.25 = -0.602$$

Try it now. Enter 100 and then press log. Did you get 2? If so you did it right. Try again with 52. Enter 52 and press log. Did you get 1.72 (rounded to two decimal places)? Some calculators may have you press log first and then the number. Try it both ways to make sure you know how your calculator works.

2. What are antilogarithms (antilogs)?

An antilog is the reverse of a log. It is also called the inverse log. If you take 10 and raise it to a power, you are taking an antilog. For example,

$$\text{antilog } 5 = 100,000$$

because taking the antilog of 5 means raising 10 to the power of 5 or

$$10^5 = 100,000$$

Try it now on your calculator. What is the antilog of 3? Enter 3 on your calculator. Press INV (inverse) or 2nd (second function), and then press log. The answer should be 1000. Your calculator may be different, but the INV or 2nd function keys are the most common.

3. What is the difference between antilog and −log?

There is a huge and very important difference. Antilog 3 means that we take 10 and raise it to the power of 3, so we get 1000. In contrast, −log 3 means that we take the log of 3, which equals 0.477, and take the negative of it. Thus −log 3 equals −0.48. For example,

$$\text{antilog } 2 = 100$$
$$-\log 2 = -0.30$$

In Section 7.8, we will use negative logs to calculate pH. The pH equals $-\log[H^+]$. Thus, if we know that $[H^+]$ is 0.01 M, to find the pH we enter 0.01 into our calculator and press log. That gives an answer of -2. Then we take the negative of that value to give a pH of 2.

In the last example, what answer would we have gotten if we had taken the antilog instead of the negative log? We would have gotten what we started with: 0.01. Why? Because all we calculated was antilog log 0.01. If we take the antilog of the log, we have not done anything at all.

ⓌWL Online homework for this How To tutorial may be assigned in GOB OWL. Go to this book's companion website at **www.cengage.com/chemistry/bettelheim** to view an interactive version of this tutorial.

7.8 What Are pH and pOH?

Because hydronium ion concentrations for most solutions are numbers with negative exponents, these concentrations are more conveniently expressed as pH, where

$$pH = -\log[H_3O^+]$$

similarly to how we expressed pK_a values in Section 7.5.

In Section 7.7, we saw that a solution is acidic if its $[H_3O^+]$ is greater than $1.0 \times 10^{-7}\,M$, and that it is basic if its $[H_3O^+]$ is less than $1.0 \times 10^{-7}\,M$. We can now state the definitions of acidic and basic solutions in terms of pH.

> A solution is acidic if its pH is less than 7.0
>
> A solution is basic if its pH is greater than 7.0
>
> A solution is neutral if its pH is equal to 7.0

The pH of this soft drink is 3.12. Soft drinks are often quite acidic.

Charles D. Winters/Cengage Learning

Example 7.6 Calculating pH

(a) The $[H_3O^+]$ of a certain liquid detergent is $1.4 \times 10^{-9}\,M$. What is its pH? Is this solution acidic, basic, or neutral?

(b) The pH of black coffee is 5.3. What is its $[H_3O^+]$? Is it acidic, basic, or neutral?

Strategy

To determine the pH when given the concentration of a hydrogen ion, just take the negative of the log. If it is less than 7, the solution is acidic. If it is greater than 7, it is basic.

If given the pH, you can immediately determine if it is acidic, basic, or neutral according to how the number relates to 7. To convert the pH to the $[H_3O^+]$, take the inverse log of $-pH$.

Solution

(a) On your calculator, take the log of 1.4×10^{-9}. The answer is -8.85. Multiply this value by -1 to give the pH of 8.85. This solution is basic.

(b) Enter 5.3 into your calculator and then press the $+/-$ key to change the sign to minus and give -5.3. Then take the antilog of this number. The $[H_3O^+]$ of black coffee is 5×10^{-6}. This solution is acidic.

Problem 7.6

(a) The $[H_3O^+]$ of an acidic solution is 3.5×10^{-3} M. What is its pH?

(b) The pH of tomato juice is 4.1. What is its $[H_3O^+]$? Is this solution acidic, basic, or neutral?

Just as pH is a convenient way to designate the concentration of H_3O^+, pOH is a convenient way to designate the concentration of OH^-.

$$pOH = -\log [OH^-]$$

As we saw in the previous section, in aqueous solutions, the ion product of water, K_w, is 1×10^{-14}, which is equal to the product of the concentration of H^+ and OH^-:

$$K_w = 1 \times 10^{-14} = [H^+][OH^-]$$

By taking the logarithm of both sides, and the fact that $-\log (1 \times 10^{-14}) = 14$, we can rewrite this equation as shown below:

$$14 = pH + pOH$$

Thus, once we know the pH of a solution, we can easily calculate the pOH.

Example 7.7 Calculating pOH

The $[OH^-]$ of a strongly basic solution is 1.0×10^{-2}. What are the pOH and pH of this solution?

Strategy

When given the $[OH^-]$, determine the pOH by taking the negative logarithm. To calculate the pH, subtract the pOH from 14.

Solution

The pOH is $-\log 1.0 \times 10^{-2}$ or 2, and the pH is $14 - 2 = 12$.

Problem 7.7

The $[OH^-]$ of a solution is 1.0×10^{-4} M. What are the pOH and pH of this solution?

The pH of three household substances. The colors of the acid–base indicators in the flasks show that vinegar is more acidic than club soda, and the cleaner is basic.

Charles D. Winters/Cengage Learning

All fluids in the human body are aqueous; that is, the only solvent present is water. Consequently, all body fluids have a pH value. Some of them have a narrow pH range; others have a wide pH range. The pH of blood, for example, must be between 7.35 and 7.45 (slightly basic). If it goes outside these limits, illness and even death may result (Chemical Connections 7C). In contrast, the pH of urine can vary from 5.5 to 7.5. Table 7.4 gives pH values for some common materials.

One thing you must remember when you see a pH value is that, because pH is a logarithmic scale, an increase (or decrease) of one pH unit means a tenfold decrease (or increase) in the $[H_3O^+]$. For example, a pH of 3 does not sound very different from a pH of 4. The first, however, means a $[H_3O^+]$ of

TABLE 7.4 pH Values of Some Common Materials

Material	pH	Material	pH
Battery acid	0.5	Saliva	6.5–7.5
Gastric juice	1.0–3.0	Pure water	7.0
Lemon juice	2.2–2.4	Blood	7.35–7.45
Vinegar	2.4–3.4	Bile	6.8–7.0
Tomato juice	4.0–4.4	Pancreatic fluid	7.8–8.0
Carbonated beverages	4.0–5.0	Sea water	8.0–9.0
Black coffee	5.0–5.1	Soap	8.0–10.0
Urine	5.5–7.5	Milk of magnesia	10.5
Rain (unpolluted)	6.2	Household ammonia	11.7
Milk	6.3–6.6	Lye (1.0 M NaOH)	14.0

Strips of paper impregnated with indicator are used to find an approximate pH.

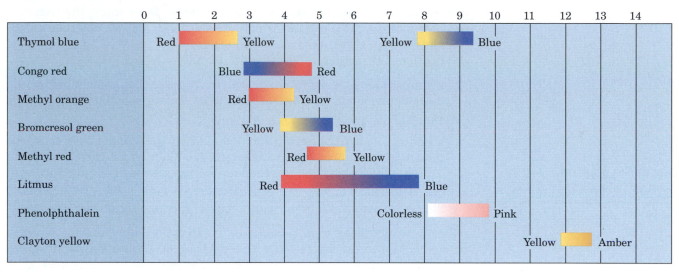

FIGURE 7.2 Some acid–base indicators. Note that some indicators have two color changes.

10^{-3} M, whereas the second means a $[H_3O^+]$ of 10^{-4} M. The $[H_3O^+]$ of the pH 3 solution is ten times the $[H_3O^+]$ of the pH 4 solution.

There are two ways to measure the pH of an aqueous solution. One way is to use pH paper, which is made by soaking plain paper with a mixture of pH indicators. A pH **indicator** is a substance that changes color at a certain pH. When we place a drop of solution on this paper, the paper turns a certain color. To determine the pH, we compare the color of the paper with the colors on a chart supplied with the paper.

One example of an acid–base indicator is the compound methyl orange. When a drop of methyl orange is added to an aqueous solution with a pH of 3.2 or lower, this indicator turns red and the entire solution becomes red. When added to an aqueous solution with a pH of 4.4 or higher, this indicator turns yellow. These particular limits and colors apply only to methyl orange. Other indicators have other limits and colors (Figure 7.2). With pH indicators, the chemical form of the indicator determines its color. The lower pH color is due to the acid form of the indicator, while the higher pH color is associated with the conjugate base form of the indicator.

The second way of determining pH is more accurate and more precise. In this method, we use a pH meter (Figure 7.3). We dip the electrode of the pH meter into the solution whose pH is to be measured, and then read the pH on a display. The most commonly used pH meters read pH to the nearest hundredth of a unit. It should be mentioned that the accuracy of a pH meter, like that of any instrument, depends on correct calibration.

FIGURE 7.3 A pH meter can rapidly and accurately measure the pH of an aqueous solution.

7.9 How Do We Use Titrations to Calculate Concentration?

Laboratories, whether medical, academic, or industrial, are frequently asked to determine the exact concentration of a particular substance in solution, such as the concentration of acetic acid in a given sample of vinegar, or the concentrations of iron, calcium, and magnesium ions in a sample of "hard" water. Determinations of solution concentrations can be made using an analytical technique called a **titration.**

Titration An analytical procedure whereby we react a known volume of a solution of known concentration with a known volume of a solution of unknown concentration

In a titration, we react a known volume of a solution of known concentration with a known volume of a solution of unknown concentration. The solution of unknown concentration may contain an acid (such as stomach acid), a base (such as ammonia), an ion (such as Fe^{2+} ion), or any other substance whose concentration we are asked to determine. If we know the titration volumes and the mole ratio in which the solutes react, we can then calculate the concentration of the second solution.

Titrations must meet several requirements:

1. We must know the equation for the reaction so that we can determine the stoichiometric ratio of reactants to use in our calculations.

2. The reaction must be rapid and complete.

3. When the reactants have combined exactly, there must be a clear-cut change in some measurable property of the reaction mixture. We call the point at which the reactants combine exactly the **equivalence point** of the titration.

Equivalence point The point at which there is an equal amount of acid and base in a neutralization reaction

4. We must have accurate measurements of the amount of each reactant.

Let us apply these requirements to the titration of a solution of sulfuric acid of known concentration with a solution of sodium hydroxide of unknown concentration. We know the balanced equation for this acid–base reaction, so requirement 1 is met.

$$2NaOH(aq) + H_2SO_4(aq) \longrightarrow Na_2SO_4(aq) + 2H_2O(\ell)$$
(Concentration not known)   (Concentration known)

Sodium hydroxide ionizes in water to form sodium ions and hydroxide ions; sulfuric acid ionizes to form hydronium ions and sulfate ions. The reaction between hydroxide and hydronium ions is rapid and complete, so requirement 2 is met.

To meet requirement 3, we must be able to observe a clear-cut change in some measurable property of the reaction mixture at the equivalence point. For acid–base titrations, we use the sudden pH change that occurs at this point. Suppose we add the sodium hydroxide solution slowly. As it is added, it reacts with hydronium ions to form water. As long as any unreacted hydronium ions are present, the solution is acidic. When the number of hydroxide ions added exactly equals the original number of hydronium ions, the solution becomes neutral. Then, as soon as any extra hydroxide ions are added, the solution becomes basic. We can observe this sudden change in pH by reading a pH meter.

Another way to observe the change in pH at the equivalence point is to use an acid–base indicator (Section 7.8). Such an indicator changes color when the solution changes pH. Phenolphthalein, for example, is colorless in acid solution and pink in basic solution. If this indicator is added to the original sulfuric acid solution, the solution remains colorless as long as excess hydronium ions are present. After enough sodium hydroxide solution has been added to react with all of the hydronium ions, the next drop of base

FIGURE 7.4 An acid–base titration. (*a*) An acid of known concentration is in the Erlenmeyer flask. (*b*) When a base is added from the buret, the acid is neutralized. (*c*) The end point is reached when the color of the indicator changes from colorless to pink.

provides excess hydroxide ions, and the solution turns pink (Figure 7.4). Thus we have a clear-cut indication of the equivalence point. The point at which an indicator changes color is called the **end point** of the titration. It is convenient if the **end point** and the equivalence point are the same, but there are many pH indicators whose end points are not at pH 7.

To meet requirement 4, which is that the volume of each solution used must be known, we use volumetric glassware such as volumetric flasks, burets, and pipets.

Data for a typical acid–base titration are given in Example 7.8. Note that the experiment is run in triplicate, a standard procedure for checking the precision of a titration.

Example 7.8 Titrations

Following are data for the titration of 0.108 *M* H_2SO_4 with a solution of NaOH of unknown concentration. What is the concentration of the NaOH solution?

	Volume of 0.108 *M* H_2SO_4	Volume of NaOH
Trial I	25.0 mL	33.48 mL
Trial II	25.0 mL	33.46 mL
Trial III	25.0 mL	33.50 mL

Strategy

Use the volume of the acid and its concentration to calculate how many moles of hydrogen ions are available to be titrated. At the equivalence point, the moles of base used will equal the moles of H^+ available. Divide the moles of H^+ by the volume of base used in liters to calculate the concentration of the base.

Solution

From the balanced equation for this acid–base reaction, we know the stoichiometry: Two moles of NaOH react with one mole of H_2SO_4. From the three trials, we calculate that the average volume of the NaOH required for complete reaction is 33.48 mL. Because the units of molarity are moles/liter, we must convert volumes of reactants from milliliters to liters. We can then use the factor-label method to calculate the molarity of the NaOH solution. What we wish to calculate is the number of moles of NaOH per liter of NaOH.

$$\frac{\text{mol NaOH}}{\text{L NaOH}} = \frac{0.108 \text{ mol } H_2SO_4}{1 \text{ L } H_2SO_4} \times \frac{0.0250 \text{ L } H_2SO_4}{0.03348 \text{ L NaOH}} \times \frac{2 \text{ mol NaOH}}{1 \text{ mol } H_2SO_4}$$

$$= \frac{0.161 \text{ mol NaOH}}{\text{L NaOH}} = 0.161 \, M$$

Problem 7.8

Calculate the concentration of an acetic acid solution using the following data. Three 25.0-mL samples of acetic acid were titrated to a phenolphthalein end point with 0.121 M NaOH. The volumes of NaOH were 19.96 mL, 19.73 mL, and 19.79 mL.

It is important to understand that a titration is not a method for determining the acidity (or basicity) of a solution. If we want to do that, we must measure the sample's pH, which is the only measurement of solution acidity or basicity. Rather, titration is a method for determining the total acid or base concentration of a solution, which is not the same as the acidity. For example, a 0.1 M solution of HCl in water has a pH of 1, but a 0.1 M solution of acetic acid has a pH of 2.9. These two solutions have the same concentration of acid and each neutralizes the same volume of NaOH solution, but they have very different acidities.

7.10 What Are Buffers?

Buffer A solution that resists change in pH when limited amounts of an acid or a base are added to it; an aqueous solution containing a weak acid and its conjugate base

As noted earlier, the body must keep the pH of blood between 7.35 and 7.45. Yet we frequently eat acidic foods such as oranges, lemons, sauerkraut, and tomatoes, and doing so eventually adds considerable quantities of H_3O^+ to the blood. Despite these additions of acidic or basic substances, the body manages to keep the pH of blood remarkably constant. The body manages this feat by using buffers. A **buffer** is a solution whose pH changes very little when small amounts of H_3O^+ or OH^- ions are added to it. In a sense, a pH buffer is an acid or base "shock absorber."

The most common buffers consist of approximately equal molar amounts of a weak acid and a salt of the weak acid. Put another way, they consist of approximately equal amounts of a weak acid and its conjugate base. For example, if we dissolve 1.0 mol of acetic acid (a weak acid) and 1.0 mol of its conjugate base (in the form of CH_3COONa, sodium acetate) in 1.0 L of water, we have a good buffer solution. The equilibrium present in this buffer solution is

Added as CH_3COOH Added as $CH_3COO^-Na^+$

$$CH_3COOH + H_2O \rightleftharpoons CH_3COO^- + H_3O^+$$

Acetic acid   Acetate ion
(A weak acid)   (Conjugate base of a weak acid)

A. How Do Buffers Work?

A buffer resists any change in pH upon the addition of small quantities of acid or base. To see how, we will use an acetic acid–sodium acetate buffer as an example. If a strong acid such as HCl is added to this buffer solution, the added H_3O^+ ions react with CH_3COO^- ions and are removed from solution.

$$CH_3COO^- + H_3O^+ \longrightarrow CH_3COOH + H_2O$$

Acetate ion · Acetic acid
(Conjugate base · (A weak acid)
of a weak acid)

There is a slight increase in the concentration of CH_3COOH as well as a slight decrease in the concentration of CH_3COO^-, but there is no appreciable change in pH. We say that this solution is buffered because it resists a change in pH upon the addition of small quantities of a strong acid.

If NaOH or another strong base is added to the buffer solution, the added OH^- ions react with CH_3COOH molecules and are removed from solution:

$$CH_3COOH + OH^- \longrightarrow CH_3COO^- + H_2O$$

Acetic acid · Acetate ion
(A weak acid) · (Conjugate base
of a weak acid)

Here there is a slight decrease in the concentration of CH_3COOH as well as a slight increase in the concentration of CH_3COO^-, but, again, there is no appreciable change in pH.

The important point about this or any other buffer solution is that when the conjugate base of the weak acid removes H_3O^+, it is converted to the undissociated weak acid. Because a substantial amount of weak acid is already present, there is no appreciable change in its concentration and, because H_3O^+ ions are removed from solution, there is no appreciable change in pH. By the same token, when the weak acid removes OH^- ions from solution, it is converted to its conjugate base. Because OH^- ions are removed from solution, there is no appreciable change in pH.

The effect of a buffer can be quite powerful. Addition of either dilute HCl or NaOH to pure water, for example, causes a dramatic change in pH (Figure 7.5).

(a) pH 7.00   (b) pH 2.00   (c) pH 12.00

FIGURE 7.5 The addition of HCl and NaOH to pure water. (a) The pH of pure water is 7.0. (b) The addition of 0.01 mol of HCl to 1 L of pure water causes the pH to decrease to 2. (c) The addition of 0.010 mol of NaOH to 1 L of pure water causes the pH to increase to 12.

When HCl or NaOH is added to a phosphate buffer, the results are quite different. Suppose we have a phosphate buffer solution of pH 7.21 prepared by dissolving 0.10 mol NaH_2PO_4 (a weak acid) and 0.10 mol Na_2HPO_4 (its conjugate base) in enough water to make 1.00 L of solution. If we add 0.010 mol of HCl to 1.0 L of this solution, the pH decreases to only 7.12. If we add 0.01 mol of NaOH, the pH increases to only 7.30.

Phosphate buffer (pH 7.21) + 0.010 mol HCl   pH 7.21 $\longrightarrow$ 7.12

Phosphate buffer (pH 7.21) + 0.010 mol NaOH   pH 7.21 $\longrightarrow$ 7.30

Had the same amount of acid or base been added to 1 liter of pure water, the resulting pH values would have been 2 and 12, respectively.

Figure 7.6 shows the effect of adding acid to a buffer solution.

B. Buffer pH

In the previous example, the pH of the buffer containing equal molar amounts of $H_2PO_4^-$ and HPO_4^{2-} is 7.21. From Table 7.3, we see that 7.21 is the pK_a of the acid $H_2PO_4^-$. This is not a coincidence. If we make a buffer solution by mixing equimolar concentrations of any weak acid and its conjugate base, the pH of the solution will equal the pK_a of the weak acid.

This fact allows us to prepare buffer solutions to maintain almost any pH. For example, if we want to maintain a pH of 9.14, we could make a buffer solution from boric acid, H_3BO_3, and sodium dihydrogen borate, NaH_2BO_3, the sodium salt of its conjugate base (see Table 7.3).

(a)

(b)

FIGURE 7.6 Buffer solutions. The solution in the Erlenmeyer flask on the right in both (a) and (b) is a buffer of pH 7.40, the same pH as human blood. The buffer solution also contains bromcresol green, an acid–base indicator that is blue at pH 7.40 (see Figure 7.2). (a) The beaker contains some of the pH 7.40 buffer and the bromcresol green indicator to which has been added 5 mL of 0.1 *M* HCl. After the addition of the HCl, the pH of the buffer solution drops only 0.65 unit to 6.75. (b) The beaker contains pure water and bromcresol green indicator to which has been added 5 mL of 0.10 *M* HCl. After the addition of the HCl, the pH of the unbuffered solution drops to 3.02.

Example 7.9 Buffers

What is the pH of a buffer solution containing equimolar quantities of

(a) H_3PO_4 and NaH_2PO_4?   (b) H_2CO_3 and $NaHCO_3$?

Strategy

When there are equimolar quantities of a weak acid and its conjugate base in a buffer solution, the pH is always the same as the pK_a of the weak acid. Look up the pK_a of the weak acid in Table 7.3.

Solution

Because we are adding equimolar quantities of a weak acid and its conjugate base, the pH is equal to the pK_a of the weak acid, which we find in Table 7.3:

(a) pH = 2.12   (b) pH = 6.37

Problem 7.9

What is the pH of a buffer solution containing equimolar quantities of

(a) NH_4Cl and NH_3?   (b) CH_3COOH and CH_3COONa?

C. Buffer Capacity

Buffer capacity is the amount of hydronium or hydroxide ions that a buffer can absorb without a significant change in its pH. We have already mentioned that a pH buffer is an acid–base "shock absorber." We now ask what makes one solution a better acid–base shock absorber than another solution. The nature of the buffer capacity of a pH buffer depends on both its pH relative to its pK_a and its concentration.

Buffer capacity The extent to which a buffer solution can prevent a significant change in pH of a solution upon addition of a strong acid or a strong base

pH:	The closer the pH of the buffer is to the pK_a of the weak acid, the more symmetric the buffer capacity, meaning the buffer can resist a pH change with added acid or added base.
Concentration:	The greater the concentration of the weak acid and its conjugate base, the greater the buffer capacity.

An effective buffer has a pH equal to the pK_a of the weak acid ± 1. For acetic acid, for example, the pK_a is 4.75. Therefore, a solution of acetic acid and sodium acetate functions as an effective buffer within the pH range of approximately 3.75–5.75. When the pH of the buffer solution is equal to the pK_a of the conjugate acid, the solution will have equal capacity with respect to additions of either acid or base. If the pH of the buffer is below the pK_a, the capacity will be greater toward addition of base. When the pH is above the pK_a, the acid buffer capacity will be greater than the base buffer capacity.

Buffer capacity also depends on concentration. The greater the concentration of the weak acid and its conjugate base, the greater the buffer capacity. We could make a buffer solution by dissolving 1.0 mol each of CH_3COONa and CH_3COOH in 1 L of H_2O, or we could use only 0.10 mol of each. Both solutions have the same pH of 4.75. However, the former has a buffer capacity ten times that of the latter. If we add 0.2 mol of HCl to the former solution, it performs the way we expect—the pH drops to 4.57. If we add 0.2 mol of HCl to the latter solution, however, the pH drops to 1.0 because the buffer has been swamped out. That is, the amount of H_3O^+ added has exceeded the buffer capacity. The first 0.10 mol of HCl completely neutralizes essentially all the CH_3COO^- present. After that, the solution contains only CH_3COOH and is no longer a buffer, so the second 0.10 mol of HCl decreases the pH to 1.0.

D. Blood Buffers

The average pH of human blood is 7.4. Any change larger than 0.10 pH unit in either direction may cause illness. If the pH goes below 6.8 or above 7.8, death may result. To hold the pH of the blood close to 7.4, the body uses three buffer systems: carbonate, phosphate, and proteins (proteins are discussed in Chapter 14).

The most important of these systems is the carbonate buffer. The weak acid of this buffer is carbonic acid, H_2CO_3; the conjugate base is the bicarbonate ion, HCO_3^-. The pK_a of H_2CO_3 is 6.37 (from Table 7.3). Because the pH of an equal mixture of a weak acid and its salt is equal to the pK_a of the weak acid, a buffer with equal concentrations of H_2CO_3 and HCO_3^- has a pH of 6.37.

Blood, however, has a pH of 7.4. The carbonate buffer can maintain this pH only if $[H_2CO_3]$ and $[HCO_3^-]$ are not equal. In fact, the necessary $[HCO_3^-]/[H_2CO_3]$ ratio is about $10:1$. The normal concentrations of these species in blood are about $0.025\ M$ HCO_3^- and $0.0025\ M$ H_2CO_3. This buffer works because any added H_3O^+ is neutralized by the HCO_3^- and any added OH^- is neutralized by the H_2CO_3.

The fact that the $[HCO_3^-]/[H_2CO_3]$ ratio is $10:1$ means that this system is a better buffer for acids, which lower the ratio and thus improve buffer efficiency, than for bases, which raise the ratio and decrease buffer capacity. This is in harmony with the actual functioning of the body because, under normal conditions, larger amounts of acidic than basic substances enter the blood. The $10:1$ ratio is easily maintained under normal conditions, because the body can very quickly increase or decrease the amount of CO_2 entering the blood.

The second most important buffering system of the blood is a phosphate buffer made up of hydrogen phosphate ion, HPO_4^{2-}, and dihydrogen phosphate ion, $H_2PO_4^-$. In this case, a $1.6:1$ $[HPO_4^{2-}]/[H_2PO_4^-]$ ratio is necessary to maintain a pH of 7.4. This ratio is well within the limits of good buffering action.

7.11 How Do We Calculate the pH of a Buffer?

Suppose we want to make a phosphate buffer solution of pH 7.00. The weak acid with a pK_a closest to this desired pH is $H_2PO_4^-$; it has a pK_a of 7.21. If we use equal concentrations of NaH_2PO_4 and Na_2HPO_4, however, we will have a buffer of pH 7.21. We want a phosphate buffer that is slightly more acidic than 7.21, so it would seem reasonable to use more of the weak acid, $H_2PO_4^-$, and less of its conjugate base, HPO_4^{2-}. But what proportions of these two salts do we use? Fortunately, we can calculate these proportions using the **Henderson-Hasselbalch** equation.

The Henderson-Hasselbalch equation is a mathematical relationship between pH, the pK_a of a weak acid, and the concentrations of the weak acid and its conjugate base. The equation is derived in the following way. Assume that we are dealing with weak acid, HA, and its conjugate base, A^-.

$$HA + H_2O \rightleftharpoons A^- + H_3O^+$$

$$K_a = \frac{[A^-][H_3O^+]}{[HA]}$$

Taking the logarithm of this equation gives

$$\log K_a = \log [H_3O^+] + \log \frac{[A^-]}{[HA]}$$

Rearranging terms gives us a new expression, in which $-\log K_a$ is, by definition, pK_a, and $-\log [H_3O^+]$ is, by definition, pH. Making these substitutions gives the Henderson-Hasselbalch equation.

$$-\log [H_3O^+] = -\log K_a + \log \frac{[A^-]}{[HA]}$$

$$pH = pK_a + \log \frac{[A^-]}{[HA]} \quad \text{Henderson-Hasselbalch equation}$$

The Henderson-Hasselbalch equation gives us a convenient way to calculate the pH of a buffer when the concentrations of the weak acid and its conjugate base are not equal.

Example 7.10 Buffer pH Calculation

What is the pH of a phosphate buffer solution containing 1.0 mol/L of sodium dihydrogen phosphate, NaH_2PO_4, and 0.50 mol/L of sodium hydrogen phosphate, Na_2HPO_4?

Strategy

Use the Henderson-Hasselbalch equation to determine the pH. You must know either the number of moles of both the conjugate acid and base or the concentrations of the conjugate acid or base. Divide the conjugate base by the conjugate acid, take the log of that ratio and add it to the pK_a of the conjugate acid.

Solution

The weak acid in this problem is $H_2PO_4^-$; its ionization produces HPO_4^{2-}. The pK_a of this acid is 7.21 (from Table 7.3). Under the weak acid and its conjugate base are shown their concentrations.

$$H_2PO_4^- + H_2O \rightleftharpoons HPO_4^{2-} + H_3O^+ \quad pK_a = 7.21$$
1.0 mol/L   0.50 mol/L

Substituting these values in the Henderson-Hasselbalch equation gives a pH of 6.91.

$$pH = 7.21 + \log \frac{0.50}{1.0}$$

$$= 7.21 - 0.30 = 6.91$$

Problem 7.10

What is the pH of a boric acid buffer solution containing 0.25 mol/L of boric acid, H_3BO_3, and 0.50 mol/L of its conjugate base? See Table 7.3 for the pK_a of boric acid.

Returning to the problem posed at the beginning of this section, how do we calculate the proportions of NaH_2PO_4 and Na_2HPO_4 needed to make up a phosphate buffer of pH 7.00? We know that the pK_a of $H_2PO_4^-$ is 7.21 and that the buffer we wish to prepare has a pH of 7.00. We can substitute these two values in the Henderson-Hasselbalch equation as follows:

$$7.00 = 7.21 + \log \frac{[HPO_4^{2-}]}{[H_2PO_4^-]}$$

Rearranging and solving gives

$$\log \frac{[HPO_4^{2-}]}{[H_2PO_4^-]} = 7.00 - 7.21 = -0.21$$

$$\frac{[HPO_4^{2-}]}{[H_2PO_4^-]} = 10^{-0.21} = \frac{0.62}{1.0}$$

Thus, to prepare a phosphate buffer of pH 7.00, we can use 0.62 mol of Na_2HPO_4 and 1.0 mol of NaH_2PO_4. Alternatively, we can use any other amounts of these two salts, as long as their mole ratio is 0.62 : 1.0.

7.12 What Are TRIS, HEPES, and These Buffers with the Strange Names?

The original buffers used in the lab were made from simple weak acids and bases, such as acetic acid, phosphoric acid, and citric acid. It was eventually discovered that many of these buffers had limitations. For example, they often changed their pH too much if the solution was diluted or if the temperature changed. They often permeated cells in solution, thereby changing the chemistry of the interior of the cell. To overcome these shortcomings, a scientist named N. E. Good developed a series of buffers that consist of zwitterions, molecules with both positive and negative charges. Zwitterions do not readily permeate cell membranes. Zwitterionic buffers also are more resistant to concentration and temperature changes.

Most of the common synthetic buffers used today have complicated formulas, such as 3-[N-morpholino]propanesulfonic acid, which we abbreviate MOPS. Table 7.5 gives a few examples.

The important thing to remember is that you don't really need to know the structure of these odd-sounding buffers to use them correctly. The important considerations are the pK_a of the buffer and the concentration you want to have. The Henderson-Hasselbalch equation works just fine whether or not you know the structure of the compound in question.

Example 7.11 Buffer pH Calculation

What is the pH of a solution if you mix 100 mL of 0.2 M HEPES in the acid form with 200 mL of 0.2 M HEPES in the basic form?

Strategy

To use the Henderson-Hasselbalch equation, you need the ratio of the conjugate base to weak acid forms of the buffer. Since the HEPES solutions have equal concentrations, the ratio of the volumes will give you the ratio of the moles used. Divide the volume of the conjugate base form by the volume of the weak acid form. Take the log of the ratio and add it to the pK_a for HEPES.

Solution

First we must find the pK_a, which we see from Table 7.5 is 7.55. Then we must calculate the ratio of the conjugate base to the acid. The formula calls for the concentration, but in this situation, the ratio of the concentrations will be the same as the ratio of the moles, which will be the same as the ratio of the volumes, because both solutions had the same starting concentration of 0.2 M. Thus, we can see that the ratio of base to acid is 2 : 1 because we added twice the volume of base.

$$pH = pK_a + \log ([A^-]/[HA]) = 7.55 + \log (2) = 7.85$$

Notice that we did not have to know anything about the structure of HEPES to work out this example.

Problem 7.11

What is the pH of a solution made by mixing 0.2 mol of TRIS acid and 0.05 mol of TRIS base in 500 mL of water?

Chemical Connections 7C

Respiratory and Metabolic Acidosis

The pH of blood is normally between 7.35 and 7.45. If the pH goes lower than that level, the condition is called **acidosis.** Acidosis leads to depression of the nervous system. Mild acidosis can result in dizziness, disorientation, or fainting; a more severe case can cause coma. If the acidosis persists for a sufficient period of time, or if the pH gets too far away from 7.35 to 7.45, death may result.

Acidosis has several causes. One type, called **respiratory acidosis,** results from difficulty in breathing (hypoventilation). An obstruction in the windpipe or diseases such as pneumonia, emphysema, asthma, or congestive heart failure may diminish the amount of oxygen that reaches the tissues and the amount of CO_2 that leaves the body through the lungs. You can even produce mild acidosis by holding your breath. If you have ever tried to see how long you could swim underwater in a pool without surfacing, you will have noticed a deep burning sensation in all your muscles when you finally came up for air. The pH of the blood decreases because the CO_2, unable to escape fast enough, remains in the blood, where it lowers the $[HCO_3^-]/[H_2CO_3]$ ratio. Rapid breathing as a result of physical exertion is more about getting rid of CO_2 than it is about breathing in O_2.

Acidosis caused by other factors is called **metabolic acidosis.** Two causes of this condition are starvation (or fasting) and heavy exercise. When the body doesn't get enough food, it burns its own fat, and the products of this reaction are acidic compounds that enter the blood. This problem sometimes happens to people on fad diets. Heavy exercise causes the muscles to produce excessive amounts of lactic acid, which makes muscles feel tired and sore. The lowering of the blood pH due to lactic acid is also what leads to the rapid breathing, dizziness, and nausea that athletes feel at the end of a sprint. In addition, metabolic acidosis is caused by a number of metabolic irregularities. For example, the disease diabetes mellitus produces acidic compounds called ketone bodies (Section 20.6).

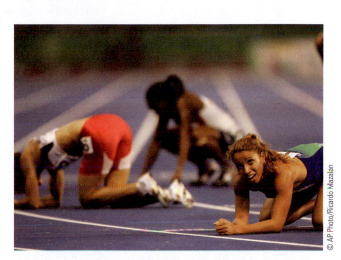

These runners have just competed for the gold medal in the 4 × 400 m relay race at the 1996 Olympic Games. The buildup of lactic acid and lowered blood pH has caused severe muscle pain and breathlessness.

© AP Photo/Ricardo Mazalan

Both types of acidosis can be related. When cells are deprived of oxygen, respiratory acidosis results. These cells are unable to produce the energy they need through aerobic (*oxygen-requiring*) pathways that we will learn about in Chapters 19 and 20. To survive, the cells must use the anaerobic (*without oxygen*) pathway called glycolysis. This pathway has lactic acid as an end product, leading to metabolic acidosis. The lactic acid is the body's way of buying time and keeping the cells alive and functioning a little longer. Eventually the lack of oxygen, called an oxygen debt, must be repaid, and the lactic acid must be cleared out. In extreme cases, the oxygen debt is too great, and the individual can die. This was the case of a famous cyclist, Tom Simpson, who died on the slopes of Mont Ventoux during the 1967 Tour de France. Under the influence of amphetamines, he rode so hard that he built up a fatal oxygen debt.

TABLE 7.5 Acid and Base Forms of Some Useful Biochemical Buffers

Acid Form		Base Form	pKa
TRIS—H$^+$ (protonated form) $(HOCH_2)_3CNH_3^+$	N—*tris*[hydroxymethyl]aminomethane (TRIS) $\rightleftharpoons$	TRIS* (free amine) $(HOCH_2)_3CNH_2$	8.3
$^-$TES—H$^+$ (zwitterionic form) $(HOCH_2)_3\overset{+}{C}NH2CH_2CH_2SO_3^-$	N—*tris*[hydroxymethyl]methyl-2-aminoethane sulfonate (TES) $\rightleftharpoons$	$^-$TES (anionic form) $(HOCH_2)_3CNHCH_2CH_2SO_3^-$	7.55
$^-$HEPES—H$^+$ (zwitterionic form) $HOCH_2CH_2N^+\!\!\bigcirc\!\!NCH_2CH_2SO_3^-$ H	N—2—hydroxyethylpiperazine-N'-2-ethane sulfonate (HEPES) $\rightleftharpoons$	$^-$HEPES (anionic form) $HOCH_2CH_2N\!\!\bigcirc\!\!NCH_2CH_2SO_3^-$	7.55
$^-$MOPS—H$^+$ (zwitterionic form) $O\!\!\bigcirc\!\!^+NCH_2CH_2CH_2SO_3^-$ H	3—[N—morpholino]propane-sulfonic acid (MOPS) $\rightleftharpoons$	$^-$MOPS (anionic form) $O\!\!\bigcirc\!\!NCH_2CH_2CH_2SO_3^-$	7.2
$^{2-}$PIPES—H$^+$ (protonated dianion) $^-O_3SCH_2CH_2N\!\!\bigcirc\!\!^+NCH_2CH_2SO_3^-$ H	Piperazine—N,N'-*bis*[2-ethanesulfonic acid] (PIPES) $\rightleftharpoons$	$^{2-}$PIPES (dianion) $^-O_3SCH_2CH_2N\!\!\bigcirc\!\!NCH_2CH_2SO_3^-$	6.8

*Note that TRIS is not a zwitterion.

Chemical Connections 7D

Alkalosis and the Sprinter's Trick

Reduced pH is not the only irregularity that can occur in the blood. The pH may also be elevated, a condition called **alkalosis** (blood pH higher than 7.45). It leads to overstimulation of the nervous system, muscle cramps, dizziness, and convulsions. It arises from rapid or heavy breathing, called hyperventilation, which may be caused by fever, infection, the action of certain drugs, or even hysteria. In this case, the excessive loss of CO_2 raises both the ratio of $[HCO_3^-]/[H_2CO_3]$ and the pH.

Athletes who compete in short-distance races that take about a minute to finish have learned how to use hyperventilation to their advantage. By hyperventilating right before the start, they force extra CO_2 out of their lungs. This causes more H_2CO_3 to dissociate into CO_2 and H_2O to replace the lost CO_2. In turn, the loss of the HA form of the bicarbonate blood buffer raises the pH of the blood. When an athlete starts an event with a slightly higher blood pH, he or she can absorb more lactic acid before the blood pH drops to the point where performance is impaired. Of course, the timing of this hyperventilation must be perfect. If the athlete artificially raises blood pH and then the race does not start quickly, the same effects of dizziness will occur.

© AP Photo/Peter DeJong

Athletes often hyperventilate before the start of a short distance event. This raises the pH of the blood allowing it to absorb more H$^+$ before their performance declines.

Summary

OWL End-of-chapter problems identified in blue are assignable in **GOB OWL**.

Section 7.1 What Are Acids and Bases?

- By the **Arrhenius definitions,** acids are substances that produce H_3O^+ ions in aqueous solution.
- Bases are substances that produce OH^- ions in aqueous solution.

Section 7.2 How Do We Define the Strength of Acids and Bases?

- A strong acid reacts completely or almost completely with water to form H_3O^+ ions.
- A strong base reacts completely or almost completely with water to form OH^- ions.

Section 7.3 What Are Conjugate Acid–Base Pairs? Problem 7.20

- The **Brønsted-Lowry definitions** expand the definitions of acid and base to beyond water.
- An acid is a proton donor; a base is a proton acceptor.
- Every acid has a **conjugate base,** and every base has a **conjugate acid.** The stronger the acid, the weaker its conjugate base. Conversely, the stronger the base, the weaker its conjugate acid.
- An **amphiprotic substance,** such as water, can act as either an acid or a base.

Section 7.4 How Can We Tell the Position of Equilibrium in an Acid–Base Reaction?

- In an acid–base reaction, the position of equilibrium favors the reaction of the stronger acid and the stronger base to form the weaker acid and the weaker base.

Section 7.5 How Do We Use Acid Ionization Constants?

- The strength of a weak acid is expressed by its **ionization constant, K_a.**
- The larger the value of K_a, the stronger the acid. $pK_a = -\log [K_a]$.

Section 7.6 What Are the Properties of Acids and Bases?

- Acids react with metals, metal hydroxides, and metal oxides to give **salts,** which are ionic compounds made up of cations from the base and anions from the acid.
- Acids also react with carbonates, bicarbonates, ammonia, and amines to give salts.

Section 7.7 What Are the Acidic and Basic Properties of Pure Water? Problem 7.34

- In pure water, a small percentage of molecules undergo self-ionization:

$$H_2O + H_2O \rightleftharpoons H_3O^+ + OH^-$$

- As a result, pure water has a concentration of 10^{-7} M for H_3O^+ and 10^{-7} M for OH^-.
- The **ion product of water, K_w,** is equal to 1.0×10^{-14}. $pK_w = 14$.

Section 7.8 What Are pH and pOH?

- Hydronium ion concentrations are generally expressed in **pH** units, with $pH = -\log [H_3O^+]$.
- $pOH = -\log [OH^-]$.
- Solutions with pH less than 7 are acidic; those with pH greater than 7 are basic. A **neutral solution** has a pH of 7.
- The pH of an aqueous solution is measured with an acid–base indicator or with a pH meter.

Section 7.9 How Do We Use Titrations to Calculate Concentration? Problem 7.46

- We can measure the concentration of aqueous solutions of acids and bases using titration. In an acid–base titration, a base of known concentration is added to an acid of unknown concentration (or vice versa) until an equivalence point is reached, at which point the acid or base being titrated is completely neutralized.

Section 7.10 What Are Buffers?

- A **buffer** does not change its pH very much when either hydronium ions or hydroxide ions are added to it.
- Buffer solutions consist of approximately equal concentrations of a weak acid and its conjugate base.
- The **buffer capacity** depends on both its pH relative to its pK_a and its concentration. The most effective buffer solutions have a pH equal to the pK_a of the weak acid. The greater the concentration of the weak acid and its conjugate base, the greater the buffer capacity.
- The most important buffers for blood are bicarbonate and phosphate.

Section 7.11 How Do We Calculate the pH of a Buffer? Problem 7.64

- The **Henderson-Hasselbalch** equation is a mathematical relationship between pH, the pK_a of a weak acid, and the concentrations of the weak acid and its conjugate base:

$$pH = pK_a + \log \frac{[A^-]}{[HA]}$$

Section 7.12 What Are TRIS, HEPES, and These Buffers with the Strange Names?

- Many modern buffers have been designed, and their names are often abbreviated.
- These buffers have qualities useful to scientists, such as not crossing membranes and resisting pH change with dilution or temperature change.
- You do not have to understand the structure of these buffers to use them. The important things to know are the molecular weight and the pK_a of the weak acid form of the buffer.

Problems

■ Indicates problems that are assignable in **GOB OWL**.

Blue numbered problems are applied.

Go to this book's companion website at **www.cengage. com/chemistry/bettelheim** for interactive versions of the How To tutorials and Active Figures, and to quiz yourself on this chapter.

Section 7.1 What Are Acids and Bases?

7.12 Define (a) an Arrhenius acid and (b) an Arrhenius base.

7.13 Write an equation for the reaction that takes place when each acid is added to water. For a diprotic or triprotic acid, consider only its first ionization.
(a) HNO_3 (b) HBr   (c) H_2SO_3
(d) H_2SO_4 (e) HCO_3^- (f) NH_4^+

7.14 Write an equation for the reaction that takes place when each base is added to water.
(a) LiOH   (b) $(CH_3)_2NH$

Section 7.2 How Do We Define the Strength of Acids and Bases?

7.15 For each of the following, tell whether the acid is strong or weak.
(a) Acetic acid   (b) HCl
(c) H_3PO_4 (d) H_2SO_4
(e) HCN   (f) H_2CO_3

7.16 For each of the following, tell whether the base is strong or weak.
(a) NaOH   (b) Sodium acetate
(c) KOH   (d) Ammonia
(e) Water

7.17 Answer True or False
(a) If an acid has a pK_a of 2.1, it is a strong acid
(b) The pH of 0.1 M HCl is the same as the pH of 0.1 M acetic acid
(c) HCl and HNO_3 are both strong acids
(d) The concentration of $[H^+]$ is always higher in a solution of strong acid than weak acid
(e) If two monoprotic acids have the same concentration, the hydrogen ion concentration will be higher in the stronger acid.
(f) If two strong acids have the same concentration, the hydrogen ion will be higher in a polyprotic acid than a monoprotic one.
(g) Ammonia is a strong base.
(h) Carbonic acid is a strong acid.

Section 7.3 What Are Conjugate Acid–Base Pairs?

7.18 Which of these acids are monoprotic, which are diprotic, and which are triprotic? Which are amphiprotic?

(a) $H_2PO_4^-$ (b) HBO_3^{2-} (c) $HClO_4$ (d) C_2H_5OH
(e) HSO_3^- (f) HS^- (g) H_2CO_3

7.19 Define (a) a Brønsted-Lowry acid and (b) a Brønsted-Lowry base.

7.20 ■ Write the formula for the conjugate base of each acid.
(a) H_2SO_4 (b) H_3BO_3 (c) HI
(d) H_3O^+ (e) NH_4^+ (f) HPO_4^{2-}

7.21 Write the formula for the conjugate base of each acid.
(a) $H_2PO_4^-$ (b) H_2S
(c) HCO_3^- (d) CH_3CH_2OH
(e) H_2O

7.22 Write the formula for the conjugate acid of each base.
(a) OH^- (b) HS^- (c) NH_3
(d) $C_6H_5O^-$ (e) CO_3^{2-} (f) HCO_3^-

7.23 Write the formula for the conjugate acid of each base.
(a) H_2O (b) HPO_4^{2-} (c) CH_3NH_2
(d) PO_4^{3-} (e) NH_3

Section 7.4 How Can We Tell the Position of Equilibrium in an Acid–Base Reaction?

7.24 For each equilibrium, label the stronger acid, stronger base, weaker acid, and weaker base. For which reaction(s) does the position of equilibrium lie toward the right? For which does it lie toward the left?
(a) $H_3PO_4 + OH^- \rightleftharpoons H_2PO_4^- + H_2O$
(b) $H_2O + Cl^- \rightleftharpoons HCl + OH^-$
(c) $HCO_3^- + OH^- \rightleftharpoons CO_3^{2-} + H_2O$

7.25 For each equilibrium, label the stronger acid, stronger base, weaker acid, and weaker base. For which reaction(s) does the position of equilibrium lie toward the right? For which does it lie toward the left?
(a) $C_6H_5OH + C_2H_5O^- \rightleftharpoons C_6H_5O^- + C_2H_5OH$
(b) $HCO_3^- + H_2O \rightleftharpoons H_2CO_3 + OH^-$
(c) $CH_3COOH + H_2PO_4^- \rightleftharpoons CH_3COO^- + H_3PO_4$

7.26 Will carbon dioxide be evolved as a gas when sodium bicarbonate is added to an aqueous solution of each compound? Explain.
(a) Sulfuric acid
(b) Ethanol, C_2H_5OH
(c) Ammonium chloride, NH_4Cl

Section 7.5 How Do We Use Acid Ionization Constants?

7.27 Which has the larger numerical value?
(a) The pK_a of a strong acid or the pK_a of a weak acid
(b) The K_a of a strong acid or the K_a of a weak acid

7.28 In each pair, select the stronger acid.
(a) Pyruvic acid ($pK_a = 2.49$) or lactic acid ($pK_a = 3.08$)

(b) Citric acid ($pK_a = 3.08$) or phosphoric acid ($pK_a = 2.10$)

(c) Benzoic acid ($K_a = 6.5 \times 10^{-5}$) or lactic acid ($K_a = 8.4 \times 10^{-4}$)

(d) Carbonic acid ($K_a = 4.3 \times 10^{-7}$) or boric acid ($K_a = 7.3 \times 10^{-10}$)

7.29 Which solution will be more acidic; that is, which will have a lower pH?

(a) 0.10 M CH_3COOH or 0.10 M HCl

(b) 0.10 M CH_3COOH or 0.10 M H_3PO_4

(c) 0.010 M H_2CO_3 or 0.010 M $NaHCO_3$

(d) 0.10 M NaH_2PO_4 or 0.10 M Na_2HPO_4

(e) 0.10 M aspirin ($pK_a = 3.47$) or 0.10 M acetic acid

7.30 Which solution will be more acidic; that is, which will have a lower pH?

(a) 0.10 M C_6H_5OH (phenol) or 0.10 M C_2H_5OH (ethanol)

(b) 0.10 M NH_3 or 0.10 M NH_4Cl

(c) 0.10 M NaCl or 0.10 M NH_4Cl

(d) 0.10 M $CH_3CH(OH)COOH$ (lactic acid) or 0.10 M CH_3COOH

(e) 0.10 M ascorbic acid (vitamin C, $pK_a = 4.1$) or 0.10 M acetic acid

Section 7.6 What Are the Properties of Acids and Bases?

7.31 Write an equation for the reaction of HCl with each compound. Which are acid–base reactions? Which are redox reactions?

(a) Na_2CO_3 (b) Mg (c) NaOH (d) Fe_2O_3 (e) NH_3 (f) CH_3NH_2 (g) $NaHCO_3$

7.32 When a solution of sodium hydroxide is added to a solution of ammonium carbonate and is heated, ammonia gas, NH_3, is released. Write a net ionic equation for this reaction. Both NaOH and $(NH_4)_2CO_3$ exist as dissociated ions in aqueous solution.

Section 7.7 What Are the Acidic and Basic Properties of Pure Water?

7.33 Given the following values of $[H_3O^+]$, calculate the corresponding value of $[OH^-]$ for each solution.

(a) $10^{-11}\,M$ (b) $10^{-4}\,M$ (c) $10^{-7}\,M$ (d) $10\,M$

7.34 ■ Given the following values of $[OH^-]$, calculate the corresponding value of $[H_3O^+]$ for each solution.

(a) $10^{-10}\,M$ (b) $10^{-2}\,M$ (c) $10^{-7}\,M$ (d) $10\,M$

Section 7.8 What Are pH and pOH?

7.35 What is the pH of each solution, given the following values of $[H_3O^+]$? Which solutions are acidic, which are basic, and which are neutral?

(a) $10^{-8}\,M$ (b) $10^{-10}\,M$ (c) $10^{-2}\,M$ (d) $10^0\,M$ (e) $10^{-7}\,M$

7.36 What is the pH and pOH of each solution given the following values of $[OH^-]$? Which solutions are acidic, which are basic, and which are neutral?

(a) $10^{-3}\,M$ (b) $10^{-1}\,M$ (c) $10^{-5}\,M$ (d) $10^{-7}\,M$

7.37 What is the pH of each solution, given the following values of $[H_3O^+]$? Which solutions are acidic, which are basic, and which are neutral?

(a) $3.0 \times 10^{-9}\,M$ (b) $6.0 \times 10^{-2}\,M$

(c) $8.0 \times 10^{-12}\,M$ (d) $5.0 \times 10^{-7}\,M$

7.38 Which is more acidic, a beer with $[H_3O^+] = 3.16 \times 10^{-5}$ or a wine with $[H_3O^+] = 5.01 \times 10^{-4}$?

7.39 What is the $[OH^-]$ and pOH of each solution?

(a) 0.10 M KOH, pH = 13.0

(b) 0.10 M Na_2CO_3, pH = 11.6

(c) 0.10 M Na_3PO_4, pH = 12.0

(d) 0.10 M $NaHCO_3$, pH = 8.4

Section 7.9 How Do We Use Titrations to Calculate Concentration?

7.40 What is the purpose of an acid–base titration?

7.41 What is the molarity of a solution made by dissolving 12.7 g of HCl in enough water to make 1.00 L of solution?

7.42 What is the molarity of a solution made by dissolving 3.4 g of $Ba(OH)_2$ in enough water to make 450 mL of solution? Assume that $Ba(OH)_2$ ionizes completely in water to Ba^{2+} and OH^- ions. What is the pH of the solution?

7.43 Describe how you would prepare each of the following solutions (in each case assume that you have the solid bases).

(a) 400.0 mL of 0.75 M NaOH

(b) 1.0 L of 0.071 M $Ba(OH)_2$

(c) 500.0 mL of 0.1 M KOH

(d) 2.0 L of 0.3 M sodium acetate

7.44 If 25.0 mL of an aqueous solution of H_2SO_4 requires 19.7 mL of 0.72 M NaOH to reach the end point, what is the molarity of the H_2SO_4 solution?

7.45 A sample of 27.0 mL of 0.310 M NaOH is titrated with 0.740 M H_2SO_4. How many milliliters of the H_2SO_4 solution are required to reach the end point?

7.46 ■ A 0.300 M solution of H_2SO_4 was used to titrate 10.00 mL of NaOH; 15.00 mL of acid was required to neutralize the basic solution. What was the molarity of the base?

7.47 A solution of NaOH base was titrated with 0.150 M HCl, and 22.0 mL of acid was needed to reach the end point of the titration. How many moles of the unknown base were in the solution?

7.48 The usual concentration of HCO_3^- ions in blood plasma is approximately 24 millimoles per liter (mmol/L). How would you make up 1.00 L of a solution containing this concentration of HCO_3^- ions?

7.49 What is the end point of a titration?

7.50 Why does a titration not tell us the acidity or basicity of a solution?

Section 7.10 What Are Buffers?

7.51 Write equations to show what happens when, to a buffer solution containing equimolar amounts of CH_3COOH and CH_3COO^-, we add
(a) H_3O^+ (b) OH^-

7.52 Write equations to show what happens when, to a buffer solution containing equimolar amounts of HPO_4^{2-} and $H_2PO_4^-$, we add
(a) H_3O^+ (b) OH^-

7.53 We commonly refer to a buffer as consisting of approximately equal molar amounts of a weak acid and its conjugate base—for example, CH_3COOH and CH_3COO^-. Is it also possible to have a buffer consisting of approximately equal molar amounts of a weak base and its conjugate acid? Explain.

7.54 What is meant by buffer capacity?

7.55 How can you change the pH of a buffer? How can you change the capacity of a buffer?

7.56 What is the connection between buffer action and Le Chatelier's principle?

7.57 Give two examples of a situation where you would want a buffer to have unequal amounts of the conjugate acid and the conjugate base.

7.58 How is the buffer capacity affected by the ratio of the conjugate base to the conjugate acid?

7.59 Can 100 mL of 0.1 M phosphate buffer at pH 7.2 act as an effective buffer against 20 mL of 1 M NaOH?

Section 7.11 How Do We Calculate the pH of a Buffer?

7.60 What is the pH of a buffer solution made by dissolving 0.10 mol of formic acid, HCOOH, and 0.10 mol of sodium formate, HCOONa, in 1 L of water?

7.61 The pH of a solution made by dissolving 1.0 mol of propanoic acid and 1.0 mol of sodium propanoate in 1.0 L of water is 4.85.
(a) What would the pH be if we used 0.10 mol of each (in 1 L of water) instead of 1.0 mol?
(b) With respect to buffer capacity, how would the two solutions differ?

7.62 Show that when the concentration of the weak acid, [HA], in an acid–base buffer equals that of the conjugate base of the weak acid, [A$^-$], the pH of the buffer solution is equal to the pK_a of the weak acid.

7.63 Show that the pH of a buffer is 1 unit higher than its pK_a when the ratio of A$^-$ to HA is ten to 1.

7.64 ■ Calculate the pH of an aqueous solution containing the following:
(a) 0.80 M lactic acid and 0.40 M lactate ion
(b) 0.30 M NH$_3$ and 1.50 M NH$_4^+$

7.65 The pH of 0.10 M HCl is 1.0. When 0.10 mol of sodium acetate, CH_3COONa, is added to this solution, its pH changes to 2.9. Explain why the pH changes, and why it changes to this particular value.

7.66 If you have 100 mL of a 0.1 M buffer made of NaH_2PO_4 and Na_2HPO_4 that is at pH 6.8, and you add 10 mL of 1 M HCl, will you still have a usable buffer? Why or why not?

Section 7.12 What Are TRIS, HEPES, and These Buffers with the Strange Names?

7.67 Write an equation showing the reaction of TRIS in the acid form with sodium hydroxide (do not write out the chemical formula for TRIS).

7.68 What is the pH of a solution that is 0.1 M in TRIS in the acid form and 0.05 M in TRIS in the basic form?

7.69 Explain why you do not need to know the chemical formula of a buffer compound to use it.

7.70 If you have a HEPES buffer at pH 4.75, will it be a usable buffer? Why or why not?

7.71 Which of the compounds listed in Table 7.5 would be the most effective for making a buffer at pH 8.15? Why?

7.72 Which of the compounds listed in Table 7.5 would be the most effective for making a buffer at pH 7.0?

Chemical Connections

7.73 (Chemical Connections 7A) Which weak base is used as a flame retardant in plastics?

7.74 (Chemical Connections 7B) Name the most common bases used in over-the-counter antacids.

7.75 (Chemical Connections 7C) What causes (a) respiratory acidosis and (b) metabolic acidosis?

7.76 (Chemical Connections 7D) Explain how the sprinter's trick works. Why would an athlete want to raise the pH of his or her blood?

7.77 (Chemical Connections 7D) Another form of the sprinter's trick is to drink a sodium bicarbonate shake before the event. What would be the purpose of doing so? Give the relevant equations.

Additional Problems

7.78 4-Methylphenol, $CH_3C_6H_4OH$ (pK_a = 10.26), is only slightly soluble in water, but its sodium salt, $CH_3C_6H_4O^-Na^+$, is quite soluble in water. In which of the following solutions will 4-methylphenol dissolve more readily than in pure water?
(a) Aqueous NaOH   (b) Aqueous NaHCO$_3$
(c) Aqueous NH$_3$

7.79 Benzoic acid, C_6H_5COOH (pK_a = 4.19), is only slightly soluble in water, but its sodium salt, $C_6H_5COO^-Na^+$, is quite soluble in water. In which of the following solutions will benzoic acid dissolve more readily than in pure water?
(a) Aqueous NaOH   (b) Aqueous NaHCO$_3$
(c) Aqueous Na$_2$CO$_3$

7.80 Assume that you have a dilute solution of HCl (0.10 M) and a concentrated solution of acetic acid (5.0 M). Which solution is more acidic? Explain.

7.81 Which of the two solutions from Problem 7.80 would take a greater amount of NaOH to hit a phenolphthalein end point, assuming you had equal volumes of the two? Explain.

7.82 If the $[OH^-]$ of a solution is 1×10^{-14},
(a) What is the pH of the solution?
(b) What is the $[H_3O^+]$?

7.83 What is the molarity of a solution made by dissolving 0.583 g of the diprotic acid oxalic acid, $H_2C_2O_4$, in enough water to make 1.75 L of solution?

7.84 Following are three organic acids and the pK_a of each: butanoic acid, 4.82; barbituric acid, 5.00; and lactic acid, 3.85.
(a) What is the K_a of each acid?
(b) Which of the three is the strongest acid and which is the weakest?
(c) What information would you need to predict which one of the three acids would require the most NaOH to reach a phenolphthalein end point?

7.85 The pK_a value of barbituric acid is 5.0. If the H_3O^+ and barbiturate ion concentrations are each 0.0030 M, what is the concentration of the undissociated barbituric acid?

7.86 If pure water self-ionizes to give H_3O^+ and OH^- ions, why doesn't pure water conduct an electric current?

7.87 Can an aqueous solution have a pH of zero? Explain your answer using aqueous HCl as your example.

7.88 If an acid, HA, dissolves in water such that the K_a is 1000, what is the pK_a of that acid? Is this scenario possible?

7.89 A scale of K_b values for bases could be set up in a manner similar to that for the K_a scale for acids. However, this setup is generally considered unnecessary. Explain.

7.90 Do a 1.0 M CH_3COOH solution and a 1.0 M HCl solution have the same pH? Explain.

7.91 Do a 1.0 M CH_3COOH solution and a 1.0 M HCl solution require the same amount of 1.0 M NaOH to hit a titration end point? Explain.

7.92 ■ Suppose you wish to make a buffer whose pH is 8.21. You have available 1 L of 0.100 M NaH_2PO_4 and solid Na_2HPO_4. How many grams of the solid Na_2HPO_4 must be added to the stock solution to accomplish this task? (Assume that the volume remains 1 L.)

7.93 In the past, boric acid was used to rinse an inflamed eye. What is the $H_3BO_3/H_2BO_3^-$ ratio in a borate buffer solution that has a pH of 8.40?

7.94 Suppose you want to make a CH_3COOH/CH_3COO^- buffer solution with a pH of 5.60. The acetic acid concentration is to be 0.10 M. What should the acetate ion concentration be?

7.95 For an acid–base reaction, one way to determine the position of equilibrium is to say that the larger of the equilibrium arrow pair points to the acid with the higher value of pK_a. For example,

$$CH_3COOH + HCO_3^- \rightleftharpoons CH_3COO^- + H_2CO_3$$
$$pK_a = 4.75 \qquad\qquad\qquad\qquad pK_a = 6.37$$

Explain why this rule works.

7.96 When a solution prepared by dissolving 4.00 g of an unknown monoprotic acid in 1.00 L of water is titrated with 0.600 M NaOH, 38.7 mL of the NaOH solution is needed to neutralize the acid. What was the molarity of the acid solution? What is the molecular weight of the unknown acid?

7.97 Write equations to show what happens when, to a buffer solution containing equal amounts of HCOOH and $HCOO^-$, we add
(a) H_3O^+ (b) OH^-

7.98 If we add 0.10 mol of NH_3 to 0.50 mol of HCl dissolved in enough water to make 1.0 L of solution, what happens to the NH_3? Will any NH_3 remain? Explain.

7.99 Suppose you have an aqueous solution prepared by dissolving 0.050 mol of NaH_2PO_4 in 1 L of water. This solution is not a buffer, but suppose you want to make it into one. How many moles of solid Na_2HPO_4 must you add to this aqueous solution to make it into
(a) A buffer of pH 7.21
(b) A buffer of pH 6.21
(c) A buffer of pH 8.21

7.100 The pH of a 0.10 M solution of acetic acid is 2.93. When 0.10 mol of sodium acetate, CH_3COONa, is added to this solution, its pH changes to 4.74. Explain why the pH changes, and why it changes to this particular value.

7.101 Suppose you have a phosphate buffer ($H_2PO_4^-/HPO_4^{2-}$) of pH 7.21. If you add more solid NaH_2PO_4 to this buffer, would you expect the pH of the buffer to increase, decrease, or remain unchanged? Explain.

7.102 Suppose you have a bicarbonate buffer containing carbonic acid, H_2CO_3, and sodium bicarbonate, $NaHCO_3$, and that the pH of the buffer is 6.37. If you add more solid $NaHCO_3$ to this buffer solution, would you expect its pH to increase, decrease, or remain unchanged? Explain.

7.103 A student pulls a bottle of TRIS off of a shelf and notes that the bottle says, "TRIS (basic form), $pK_a = 8.3$." The student tells you that if you add 0.1 mol of this compound to 100 mL of water, the pH will be 8.3. Is the student correct? Explain.

Looking Ahead

7.104 Unless under pressure, carbonic acid in aqueous solution breaks down into carbon dioxide and water, and carbon dioxide is evolved as bubbles of gas. Write an equation for the conversion of carbonic acid to carbon dioxide and water.

7.105 Following are pH ranges for several human biological materials. From the pH at the midpoint of each range, calculate the corresponding $[H_3O^+]$. Which materials are acidic, which are basic, and which are neutral?

(a) Milk, pH 6.6–7.6

(b) Gastric contents, pH 1.0–3.0

(c) Spinal fluid, pH 7.3–7.5

(d) Saliva, pH 6.5–7.5

(e) Urine, pH 4.8–8.4

(f) Blood plasma, pH 7.35–7.45

(g) Feces, pH 4.6–8.4

(h) Bile, pH 6.8–7.0

7.106 What is the ratio of $HPO_4^{2-}/H_2PO_4^-$ in a phosphate buffer of pH 7.40 (the average pH of human blood plasma)?

7.107 What is the ratio of $HPO_4^{2-}/H_2PO_4^-$ in a phosphate buffer of pH 7.9 (the pH of human pancreatic fluid)?

Amines

Key Questions

8.1 What Are amines?

8.2 How Do We Name Amines?

8.3 What Are the Physical Properties of Amines?

8.4 How Do We Describe the Basicity of Amines?

8.5 What Are the Characteristic Reactions of Amines?

This inhaler delivers a puff of albuterol (Proventil), a potent synthetic bronchodilator whose structure is patterned after that of epinephrine. See Chemical Connections 8E, "Epinephrine: A Prototype for the Development of New Bronchodilators."

8.1 What Are Amines?

Carbon, hydrogen, and oxygen are the three most common elements in organic compounds. Because of the wide distribution of amines in the biological world, nitrogen is the fourth most common element of organic compounds. The most important chemical property of amines is their basicity.

Amines (Section 1.4B) are classified as **primary (1°)**, **secondary (2°)**, or **tertiary (3°)**, depending on the number of carbon groups bonded to nitrogen.

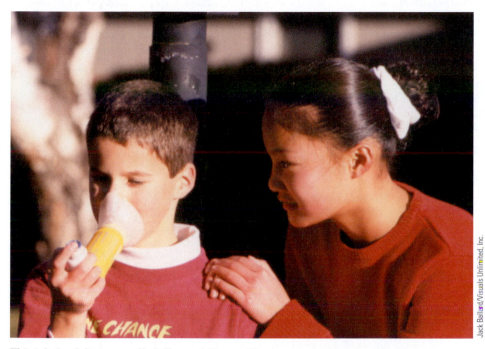

$$CH_3{-}NH_2$$
Methylamine
(a 1° amine)

$$CH_3{-}\overset{\overset{\textstyle H}{|}}{N}{-}CH_3$$
Dimethylamine
(a 2° amine)

$$CH_3{-}\overset{\overset{\textstyle CH_3}{|}}{N}{-}CH_3$$
Trimethylamine
(a 3° amine)

OWL Online homework for this chapter may be assigned in **GOB OWL.**

Chemical Connections 8A

Amphetamines (Pep Pills)

Amphetamine, methamphetamine, and phentermine—all synthetic amines—are powerful stimulants of the central nervous system. Like most other amines, they are stored and administered as their salts. The sulfate salt of amphetamine is named Benzedrine, the hydrochloride salt of the *S* enantiomer of methamphetamine is named Methedrine, and the hydrochloride salt of phentermine is named Fastin.

These three amines have similar physiological effects and are referred to by the general name **amphetamines.** Structurally, they have in common a benzene ring with a three-carbon side chain and an amine nitrogen on the second carbon of the side chain. Physiologically, they share an ability to reduce fatigue and diminish hunger by raising the glucose level of the blood. Because of these properties, amphetamines are widely prescribed to counter mild depression, reduce hyperactivity in children, and suppress appetite in people who are trying to lose weight. These drugs are also used illegally to reduce fatigue and elevate mood.

Abuse of amphetamines can have severe effects on both body and mind. They are addictive, concentrate in the brain and nervous system, and can lead to long periods of sleeplessness, loss of weight, and paranoia.

The action of amphetamines is similar to that of epinephrine (Chemical Connections 8E), the hydrochloride salt of which is named adrenaline.

Amphetamine
(Benzedrine)

(*s*)-Methamphetamine
(Methedrine)

Phentermine
(Fastin)

Aliphatic amine An amine in which nitrogen is bonded only to alkyl groups or hydrogens

Aromatic amine An amine in which nitrogen is bonded to one or more aromatic rings

Amines are further classified as aliphatic or aromatic. An **aliphatic amine** is one in which all the carbons bonded to nitrogen are derived from alkyl groups. An **aromatic amine** is one in which one or more of the groups bonded to nitrogen are aryl groups.

Aniline
(a 1° aromatic amine)

N-Methylaniline
(a 2° aromatic amine)

Benzyldimethylamine
(a 3° aliphatic amine)

Heterocyclic amine An amine in which nitrogen is one of the atoms of a ring

Heterocyclic aromatic amine An amine in which nitrogen is one of the atoms of an aromatic ring

An amine in which the nitrogen atom is part of a ring is classified as a **heterocyclic amine.** When the ring is saturated, the amine is classified as a **heterocyclic aliphatic amine.** When the nitrogen is part of an aromatic ring (Section 4.1), the amine is classified as a **heterocyclic aromatic amine.** Two of the most important heterocyclic aromatic amines are pyridine and pyrimidine, in which nitrogen atoms replace first one and then two CH groups of a benzene ring. Pyrimidine and purine serve as the building blocks for the amine bases of DNA and RNA (Chapter 17).

Pyrrolidine   Piperidine
(heterocyclic aliphatic amines)

Pyridine   Pyrimidine   Imidazole   Purine   Pyrrole
(heterocyclic aromatic amines)

Example 8.1 Structure of Amines

How many hydrogen atoms does piperidine have? How many hydrogen atoms does pyridine have? Write the molecular formula of each amine.

Strategy

Remember that hydrogen atoms bonded to carbon are not shown in line-angle formulas. To determine the number of hydrogens present, add a sufficient number to give four bonds to each carbon and three bonds to each nitrogen.

Solution

Piperidine has 11 hydrogen atoms, and its molecular formula is $C_5H_{11}N$. Pyridine has 5 hydrogen atoms, and its molecular formula is C_5H_5N.

Problem 8.1

How many hydrogen atoms does pyrrolidine have? How many does purine have? Write the molecular formula of each amine.

Chemical Connections 8B

Alkaloids

Alkaloids are basic nitrogen-containing compounds found in the roots, bark, leaves, berries, or fruits of plants. In almost all alkaloids, the nitrogen atom is part of a ring. The name "alkaloid" was chosen because these compounds are alkali-like (*alkali* is an older term for a basic substance) and react with strong acids to give water-soluble salts. Thousands of different alkaloids, many of which are used in modern medicine, have been extracted from plant sources.

When administered to animals, including humans, alkaloids have pronounced physiological effects. Whatever their individual effects, most alkaloids are toxic in large enough doses. For some, the toxic dose is very small!

(S)-Coniine is the toxic principle of water hemlock (a member of the carrot family). Its ingestion can cause weakness, labored respiration, paralysis, and eventually death. It was the toxic substance in the "poison hemlock" used in the death of Socrates. Water hemlock is easily confused with Queen Anne's lace, a type of wild carrot— a mistake that has killed numerous people.

(S)-Nicotine occurs in the tobacco plant. In small doses, it is an addictive stimulant. In larger doses, this substance causes depression, nausea, and vomiting. In still larger doses, it is a deadly poison. Solutions of nicotine in water are used as insecticides.

(S)-Coniine   (S)-Nicotine

Cocaine

Tobacco plants.

Inga Spencer/Visuals Unlimited

Cocaine is a central nervous system stimulant obtained from the leaves of the coca plant. In small doses, it decreases fatigue and gives a sense of well-being. Prolonged use of cocaine leads to physical addiction and depression.

TABLE 8.1 Approximate Base Strengths of Amines

Class	pK_b	Example	Name	
Aliphatic	3.0–4.0	$CH_3CH_2NH_2$	Ethanamine	Stronger base
Ammonia	4.74			
Aromatic	8.5–9.5	⬡—NH_2	Aniline	Weaker base

Given the basicities of amines, we can determine which form of an amine exists in body fluids—say, blood. In a normal, healthy person, the pH of blood is approximately 7.40, which is slightly basic. If an aliphatic amine is dissolved in blood, it is present predominantly as its protonated or conjugate acid form.

Dopamine

Conjugate acid of dopamine
(the major form present
in blood plasma)

We can show that an aliphatic amine such as dopamine dissolved in blood is present largely as its protonated or conjugate acid form in the following way. Assume that the amine, RNH_2, has a pK_b of 3.50 and that it is dissolved in blood, pH 7.40. We first write the base dissociation constant for the amine and then solve for the ratio of RNH_3^+ to RNH_2.

$$RNH_2 + H_2O \rightleftharpoons RNH_3^+ + OH^-$$

$$K_b = \frac{[RNH_3^+][OH^-]}{[RNH_2]}$$

$$\frac{K_b}{[OH^-]} = \frac{[RNH_3^+]}{[RNH_2]}$$

We now substitute the appropriate values for K_b and $[OH^-]$ in this equation. Taking the antilog of 3.50 gives a K_b of 3.2×10^{-4}. Calculating the concentration of hydroxide requires two steps. First recall from Section 7.8 that pH + pOH = 14. If the pH of blood is 7.40, then its pOH is 6.60 and its $[OH^-]$ is 2.5×10^{-7}. Substituting these values in the appropriate equation gives a ratio of 1300 parts RNH_3^+ to 1 part RNH_2.

$$\frac{3.2 \times 10^{-4}}{2.5 \times 10^{-7}} = \frac{[RNH_3^+]}{[RNH_2]} = 1300$$

As this calculation demonstrates, an aliphatic amine present in blood is more than 99.9% in the protonated form. Thus, even though we may write the structural formula for dopamine as the free amine, it is present in blood as the protonated form. It is important to realize, however, that the amine and ammonium ion forms are always in equilibrium, so some of the unprotonated form is nevertheless present in solution.

Aromatic amines, by contrast, are considerably weaker bases than aliphatic amines and are present in blood largely in the unprotonated form. Performing the same type of calculation for an aromatic amine, $ArNH_2$, with pK_b of approximately 10, we find that the aromatic amine is more than 99.0% in its unprotonated ($ArNH_2$) form.

Example 8.4 Basicity of Amines

Select the stronger base in each pair of amines.

Strategy

Determine whether the amine is an aromatic or an aliphatic amine. Aliphatic amines are stronger bases than aromatic amines.

Solution

(a) Morpholine (B), a 2° aliphatic amine, is the stronger base. Pyridine (A), a heterocyclic aromatic amine, is the weaker base.

(b) Benzylamine (D), a 1° aliphatic amine, is the stronger base. Even though it contains an aromatic ring, it is not an aromatic amine because the amine nitrogen is not bonded to the aromatic ring. o-Toluidine (C), a 1° aromatic amine, is the weaker base.

Problem 8.4

Select the stronger base from each pair of amines.

8.5 What Are the Characteristic Reactions of Amines?

The most important chemical property of amines is their basicity. Amines, whether soluble or insoluble in water, react quantitatively with strong acids to form water-soluble salts, as illustrated by the reaction of (R)-norepinephrine (noradrenaline) with aqueous HCl to form a hydrochloride salt.

(R)-Norepinephrine   (R)-Norepinephrine hydrochloride
(only slightly soluble in water)   (a water-soluble salt)

Chemical Connections 8D

The Solubility of Drugs in Body Fluids

Many drugs have "•HCl" or some other acid as part of their chemical formula and occasionally as part of their generic name. Invariably these drugs are amines that are insoluble in aqueous body fluids such as blood plasma and cerebrospinal fluid. For the administered drug to be absorbed and carried by body fluids, it must be treated with an acid to form a water-soluble ammonium salt. Methadone, a narcotic analgesic, is marketed as its water-soluble hydrochloride salt. Novocain, one of the first local anesthetics, is the hydrochloride salt of procaine.

Methadone ·HCl

Procaine ·HCl
(Novocain, a local anesthetic)

These two drugs are amino salts and are labeled as hydrochlorides.

Beverly March

There is another reason besides increased water solubility for preparing these and other amine drugs as salts. Amines are very susceptible to oxidation and decomposition by atmospheric oxygen, with a corresponding loss of biological activity. By comparison, their amine salts are far less susceptible to oxidation; they retain their effectiveness for a much longer time.

Example 8.5 Basicity of Amines

Complete the equation for each acid–base reaction, and name the salt formed.

(a) $(CH_3CH_2)_2NH + HCl \longrightarrow$

(b) [pyridine] $+ CH_3COOH \longrightarrow$

Strategy

Each acid–base reaction involves a proton transfer from the acid to the amino group (a base). The product is named as an ammonium salt.

Solution

(a) $(CH_3CH_2)_2NH_2{}^+Cl^-$
Diethylammonium chloride

(b) [pyridinium] CH_3COO^-

Pyridinium acetate

Problem 8.5

Complete the equation for each acid–base reaction and name the salt formed.

(a) $(CH_3CH_2)_3N + HCl \longrightarrow$

(b)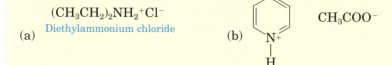

Chemical Connections 8E

Epinephrine: A Prototype for the Development of New Bronchodilators

Epinephrine was first isolated in pure form in 1897 and its structure determined in 1901. It occurs in the adrenal gland (hence the common name adrenalin) as a single enantiomer with the *R* configuration at its stereocenter. Epinephrine is commonly referred to as a catecholamine: the common name of 1,2-dihydroxybenzene is catechol (Section 4.4A), and amines containing a benzene ring with *ortho*-hydroxyl groups are called catecholamines.

Early on, it was recognized that epinephrine is a vasoconstrictor, a bronchodilator, and a cardiac stimulant. The fact that it has these three major effects stimulated much research, one line of which sought to develop compounds that are even more effective bronchodilators than epinephrine but, at the same time, are free from epinephrine's cardiac-stimulating and vasoconstricting effects.

Soon after epinephrine became commercially available, it emerged as an important treatment of asthma and hay fever. It has been marketed for the relief of bronchospasms under several trade names, including Bronkaid Mist and Primatine Mist.

enzyme-catalyzed reaction that converts one of the two —OH groups on the catechol unit to an OCH_3 group. A strategy to circumvent this enzyme-catalyzed inactivation was to replace the catechol unit with one that would allow the drug to bind to the catecholamine receptors in the bronchi but would not be inactivated by this enzyme.

In terbutaline (Brethaire), inactivation is prevented by placing the —OH groups *meta* to each other on the aromatic ring. In addition, the isopropyl group of isoproterenol is replaced by a *tert*-butyl group. In albuterol (Proventil), the commercially most successful of the anti-asthma medications, one —OH group of the catechol unit is replaced by a —CH_2OH group and the isopropyl group is replaced by a *tert*-butyl group. When terbutaline and albuterol were introduced into clinical medicine in the 1960s, they almost immediately replaced isoproterenol as the drugs of choice for the treatment of asthmatic attacks. The *R* enantiomer of albuterol is 68 times more effective in the treatment of asthma than the *S* enantiomer.

Epinephrine

Terbutaline

(R)-Isoproterenol

(R)-Albuterol

One of the most important of the first synthetic catecholamines was isoproterenol, the levorotatory enantiomer of which retains the bronchodilating effects of epinephrine but is free from its cardiac-stimulating effects. (*R*)-Isoproterenol was introduced into clinical medicine in 1951; for the next two decades, it was the drug of choice for the treatment of asthmatic attacks. Interestingly, the hydrochloride salt of (*S*)-isoproterenol is a nasal decongestant and was marketed under several trade names, including Sudafed.

A problem with the first synthetic catecholamines (and with epinephrine itself) is that they are inactivated by an

In their search for a longer-acting bronchodilator, scientists reasoned that extending the side chain on nitrogen might strengthen the binding of the drug to the adrenoreceptors in the lungs, thereby increasing the duration of the drug's action. This line of reasoning led to the synthesis and introduction of salmeterol (Serevent), a bronchodilator that is approximately ten times more potent than albuterol and much longer acting.

Salmeterol

The basicity of amines and the solubility of amine salts in water gives us a way to separate water-insoluble amines from water-insoluble nonbasic compounds. Figure 8.2 is a flowchart for the separation of aniline from cyclohexanol, a neutral compound.

FIGURE 8.2 Separation and purification of an amine and a neutral compound.

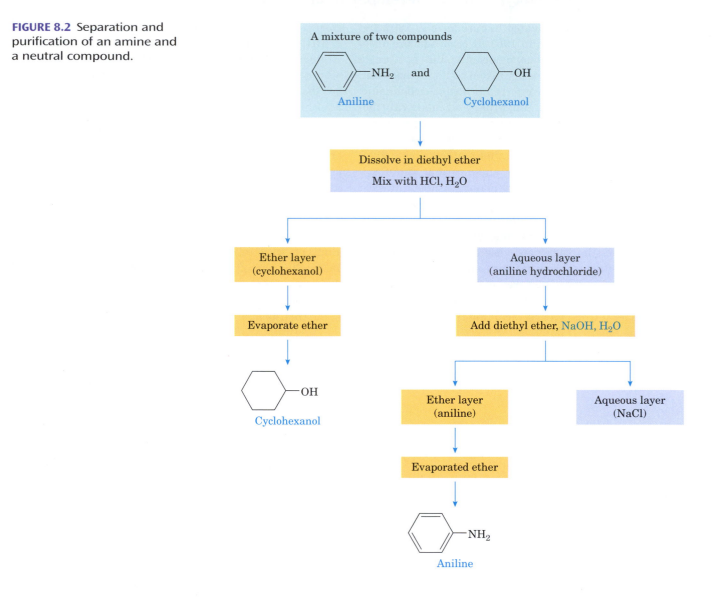

Summary of Key Questions

🦉**WL** End-of-chapter problems identified in blue are assignable in **GOB OWL.**

Section 8.1 What Are Amines?

- Amines are classified as **primary, secondary** or **tertiary,** depending on the number of carbon atoms bonded to nitrogen.
- In an **aliphatic amine,** all carbon atoms bonded to nitrogen are derived from alkyl groups.
- In an **aromatic amine,** one or more of the groups bonded to nitrogen are aryl groups.
- In a **heterocyclic amine,** the nitrogen atom is part of a ring.

Section 8.2 How Do We Name Amines?

- In IUPAC nomenclature, aliphatic amines are named by changing the final **-e** of the parent alkane to -**amine** and using a number to locate the amino group on the parent chain.
- In the common system of nomenclature, aliphatic amines are named by listing the carbon groups bonded to nitrogen in alphabetical order in one word ending in the suffix -**amine**.

Section 8.3 What Are the Physical Properties of Amines?

- Amines are polar compounds, and primary and secondary amines associate by intermolecular hydrogen bonding.
- All classes of amines form hydrogen bonds with water and are more soluble in water than are hydrocarbons of comparable molecular weight.

Section 8.4 How Do We Describe the Basicity of Amines? **Problems 8.20, 8.21**

- Amines are weak bases, and aqueous solutions of amines are basic.

- The base ionization constant for an amine in water is denoted by the symbol K_b.
- Aliphatic amines are stronger bases than aromatic amines.

Section 8.5 What Are the Characteristic Reactions of Amines? **Problem 8.26**

- All amines, whether soluble or insoluble in water, react with strong acids to form water-soluble salts.
- We can use this property to separate water-insoluble amines from water-insoluble nonbasic compounds.

Summary of Key Reactions

1. **Basicity of Aliphatic Amines (Section 8.4)** Most aliphatic amines have about the same basicity (pK_b 3.0 − 4.0) and are slightly stronger bases than ammonia (pK_b 4.74).

$$CH_3NH_2 + H_2O \rightleftharpoons CH_3NH_3^+ + OH^- \quad pK_b = 3.36$$

2. **Basicity of Aromatic Amines (Section 8.4)** Most aromatic amines (pK_b 9.0 − 10.0) are considerably weaker bases than ammonia and aliphatic amines.

$$\text{—}NH_2 + H_2O \rightleftharpoons \text{—}NH_3^+ + OH^- \quad pK_b = 9.36$$

3. **Reaction with Acids (Section 8.5)** All amines, whether water-soluble or water-insoluble, react quantitatively with strong acids to form water-soluble salts.

Insoluble in water   A water-soluble salt

Problems

■ Indicates problems that are assignable in **GOB OWL**.

Blue numbered problems are applied.

Go to this book's companion website at **www. cengage.com/chemistry/bettelheim** for interactive versions of the How To tutorials and Active Figures, and to quiz yourself on this chapter.

Section 8.1 What Are Amines?

8.6 What is the difference in structure between an aliphatic amine and an aromatic amine?

8.7 In what way are pyridine and pyrimidine related to benzene?

Section 8.2 How Do We Name Amines?

8.8 Answer true or false.
- (a) In the IUPAC system, primary aliphatic amines are named as alkanamines.
- (b) The IUPAC name of $CH_3CH_2CH_2CH_2CH_2NH_2$ is 1-pentylamine.

- (c) 2-Butanamine is chiral and shows enantiomerism.
- (d) N,N-Dimethylaniline is a 3° aromatic amine.

8.9 Draw a structural formula for each amine.
- (a) 2-Butanamine
- (b) 1-Octanamine
- (c) 2,2-Dimethyl-1-propanamine
- (d) 1,5-Pentanediamine
- (e) 2-Bromoaniline
- (f) Tributylamine

8.10 Classify each amino group as primary, secondary, or tertiary, and as aliphatic or aromatic.

(a)

Serotonin
(a neurotransmitter)

(b)

Benzocaine
(a topical anesthetic)

(c)

Diphenhydramine
(the hydrochloride salt is
the antihistamine Benadryl)

(d)

Chloroquine
(an antimalaria drug)

8.11 There are eight constitutional isomers with the molecular formula $C_4H_{11}N$.

  (a) Name and draw a structural formula for each amine.

  (b) Classify each amine as primary, secondary, or tertiary.

  (c) Which are chiral?

8.12 There are eight primary amines with the molecular formula $C_5H_{13}N$.

  (a) Name and draw a structural formula for each amine.

  (b) Which are chiral?

Section 8.3 What Are the Physical Properties of Amines?

8.13 Answer true or false.

  (a) Hydrogen bonding between 2° amines is stronger than that between 2° alcohols.

  (b) Primary and secondary amines generally have higher boiling points than hydrocarbons with comparable carbon skeletons.

  (c) The boiling points of amines increase as the molecular weight of the amine increases.

8.14 Propylamine (bp 48°C), ethylmethylamine (bp 37°C), and trimethylamine (bp 3°C) are constitutional isomers with the molecular formula C_3H_9N. Account for the fact that trimethylamine has the lowest boiling point of the three and propylamine has the highest boiling point.

8.15 Account for the fact that 1-butanamine (bp 78°C) has a lower boiling point than 1-butanol (bp 117°C).

8.16 2-Methylpropane (bp −12°C), 2-propanol (bp 82°C), and 2-propanamine (bp 32°C) all have approximately the same molecular weight, yet their boiling points are quite different. Explain the reason for these differences.

8.17 Account for the fact that most low-molecular-weight amines are very soluble in water whereas low-molecular-weight hydrocarbons are not.

Section 8.4 How Do We Describe the Basicity of Amines?

8.18 Answer true or false.

  (a) Aqueous solutions of amines are basic.

  (b) Aromatic amines, such as aniline, in general are weaker bases than aliphatic amines, such as cyclohexanamine.

  (c) Aliphatic amines are stronger bases than inorganic bases, such as NaOH and KOH.

  (d) Water-insoluble amines react with strong aqueous acids, such as HCl, to form water-soluble salts.

  (e) If the pH of an aqueous solution of a 1° aliphatic amine, RNH_2, is adjusted to pH 2.0 by the addition of concentrated HCl, the amine will be present in solution almost entirely as its conjugate acid, RNH_3^+.

  (f) If the pH of an aqueous solution of a 1° aliphatic amine, RNH_2, is adjusted to pH 10.0 by the addition of NaOH, the amine will be present in solution almost entirely as the free base, RNH_2.

  (g) For a 1° aliphatic amine, the concentrations of RNH_3^+ and RNH_2 will be equal when the pH of the solution is equal to the pK_b of the amine.

8.19 Compare the basicities of amines with those of alcohols.

8.20 ■ Write a structural formula for each amine salt.

  (a) Ethyltrimethylammonium hydroxide

  (b) Dimethylammonium iodide

  (c) Tetramethylammonium chloride

  (d) Anilinium bromide

8.21 ■ Name these amine salts.

  (a) $CH_3CH_2NH_3^+Cl^-$

  (b) $(CH_3CH_2)_2NH_2^+Cl^-$

  (c) $-NH_3^+HSO_4^-$

8.22 From each pair of compounds, select the stronger base.

(a)

(b)

or

(c)

or

8.23 The pK_b of amphetamine is approximately 3.2.

Amphetamine

(a) Which form of amphetamine (the base or its conjugate acid) would you expect to be present at pH 1.0, the pH of stomach acid?

(b) Which form of amphetamine would you expect to be present at pH 7.40, the pH of blood plasma?

Section 8.5 What Are the Characteristic Reactions of Amines?

8.24 Suppose you have two test tubes, one containing 2-methylcyclohexanol and the other containing 2-methylcyclohexanamine (both of which are insoluble in water) and that you do not know which test tube contains which compound. Describe a simple chemical test by which you could tell which compound is the alcohol and which is the amine.

8.25 Complete the equations for the following acid–base reactions.

(a)

Acetic acid   Pyridine

(b)

1-Phenyl-2-propanamine
(Amphetamine)

(c)

Methamphetamine

8.26 ■ Pyridoxamine is one form of vitamin B$_6$.

Pyridoxamine
(Vitamin B$_6$)

(a) Which nitrogen atom of pyridoxamine is the stronger base?

(b) Draw a structural formula for the salt formed when pyridoxamine is treated with one mole of HCl.

8.27 Many tumors of the breast are correlated with estrogen levels in the body. Drugs that interfere with estrogen binding have antitumor activity and may even help prevent tumor occurrence. A widely used antiestrogen drug is tamoxifen.

Tamoxifen

(a) Name the functional groups in tamoxifen.

(b) Classify the amino group in tamoxifen as primary, secondary, or tertiary.

(c) How many stereoisomers are possible for tamoxifen?

(d) Would you expect tamoxifen to be soluble or insoluble in water? In blood?

Chemical Connections

8.28 (Chemical Connections 8A) What are the differences in structure between the natural hormone epinephrine (Chemical Connections 8E) and the synthetic pep pill amphetamine? Between amphetamine and methamphetamine?

8.29 (Chemical Connections 8A) What are the possible negative effects of illegal use of amphetamines such as methamphetamine?

8.30 (Chemical Connections 8B) What is an alkaloid? Are all alkaloids basic to litmus?

8.31 (Chemical Connections 8B) Identify all stereocenters in coniine and nicotine. How many stereoisomers are possible for each?

8.32 (Chemical Connections 8B) Which of the two nitrogen atoms in nicotine is converted to its salt by reaction with one mole of HCl? Draw a structural formula for this salt.

8.33 (Chemical Connections 8B) Cocaine has four stereocenters. Identify each. Draw a structural formula for the salt formed by treatment of cocaine with one mole of HCl.

8.34 (Chemical Connections 8C) What structural feature is common to all benzodiazepines?

8.35 (Chemical Connections 8C) Is Librium chiral? Is Valium chiral?

8.36 (Chemical Connections 8C) Benzodiazepines affect neural pathways in the central nervous system that are mediated by GABA, whose IUPAC name is 4-aminobutanoic acid. Draw a structural formula for GABA.

8.37 (Chemical Connections 8D) Suppose you saw this label on a decongestant: phenylephrine · HCl. Should you worry about being exposed to a strong acid such as HCl? Explain.

8.38 (Chemical Connections 8D) Give two reasons why amine-containing drugs are most commonly administered as their salts.

8.39 (Chemical Connections 8E) Classify each amino group in epinephrine and albuterol as primary, secondary, or tertiary. In addition, list the similarities and differences between the structural formulas of these two compounds.

Additional Problems

8.40 Draw a structural formula for a compound with the given molecular formula that is:
   (a) A 2° aromatic amine, C_7H_9N
   (b) A 3° aromatic amine, $C_8H_{11}N$
   (c) A 1° aliphatic amine, C_7H_9N
   (d) A chiral 1° amine, $C_4H_{11}N$
   (e) A 3° heterocyclic amine, $C_5H_{11}N$
   (f) A trisubstituted 1° aromatic amine, $C_9H_{13}N$
   (g) A chiral quaternary ammonium salt, $C_9H_{22}NCl$

8.41 Arrange these three compounds in order of decreasing ability to form intermolecular hydrogen bonds: CH_3OH, CH_3SH, and $(CH_3)_2NH$.

8.42 ■ Consider these three compounds: CH_3OH, CH_3SH, and $(CH_3)_2NH$.
   (a) Which is the strongest acid?
   (b) Which is the strongest base?
   (c) Which has the highest boiling point?
   (d) Which forms the strongest intermolecular hydrogen bonds in the pure state?

8.43 Arrange these compounds in order of increasing boiling point: $CH_3CH_2CH_2CH_3$, $CH_3CH_2CH_2OH$, and $CH_3CH_2CH_2NH_2$. Boiling point values from lowest to highest are −0.5°C, 7.2°C, and 77.8°C.

8.44 Account for the fact that amines have about the same solubility in water as alcohols of similar molecular weight.

8.45 The compound phenylpropanolamine hydrochloride is used as both a decongestant and an anorexic. The IUPAC name of this compound is 1-phenyl-2-amino-1-propanol.
   (a) Draw a structural formula for 1-phenyl-2-amino-1-propanol.
   (b) How many stereocenters are present in this molecule? How many stereoisomers are possible for it?

8.46 Several poisonous plants, including *Atropa belladonna*, contain the alkaloid atropine. The name "belladonna" (which means "beautiful lady") probably comes from the fact that Roman women

used extracts from this plant to make themselves more attractive. Atropine is widely used by ophthalmologists and optometrists to dilate the pupils for eye examination.

Atropine

(a) Classify the amino group in atropine as primary, secondary, or tertiary.
(b) Locate all stereocenters in atropine.
(c) Account for the fact that atropine is almost insoluble in water (1 g in 455 mL of cold water), but atropine hydrogen sulfate is very soluble (1 g in 5 mL of cold water).
(d) Account for the fact that a dilute aqueous solution of atropine is basic (pH approximately 10.0).

8.47 ■ Epibatadine, a colorless oil isolated from the skin of the Equadorian poison arrow frog *Epipedobates tricolor*, has several times the analgesic potency of morphine. It is the first chlorine-containing, non-opioid (nonmorphine-like in structure) analgesic ever isolated from a natural source.
   (a) Which of the two nitrogen atoms in epibatadine is the stronger base?
   (b) Mark the three stereocenters in this molecule.

Epibatadine

8.48 Following are two structural formulas for 4-aminobutanoic acid, a neurotransmitter. Is this compound better represented by structural formula (A) or (B)? Explain.

(A)   or   (B)

8.49 Alanine, $C_3H_7O_2N$, is one of the 20 amino acid building blocks of proteins (Chapter 14). Alanine contains a primary amino group ($-NH_2$) and a carboxyl group ($-COOH$), and has one stereocenter. Given this information, draw a structural formula for alanine.

Aldehydes and Ketones

Benzaldehyde is found in the kernels of bitter almonds, and cinnamaldehyde is found in Ceylonese and Chinese cinnamon oils.

9.1 What Are Aldehydes and Ketones?

In this and the three following chapters, we study the physical and chemical properties of compounds containing the **carbonyl group,** C=O. Because the carbonyl group is present in aldehydes, ketones, and carboxylic acids and their derivatives, as well as in carbohydrates, it is one of the most

⌁WL Online homework for this chapter may be assigned in **GOB OWL.**

important functional groups in organic chemistry. Its chemical properties are straightforward, and an understanding of its characteristic reaction patterns leads very quickly to an understanding of a wide variety of organic and biochemical reactions.

The functional group of an **aldehyde** is a carbonyl group bonded to a hydrogen atom (Section 1.4C). In methanal, the simplest aldehyde, the carbonyl group is bonded to two hydrogen atoms. In other aldehydes, it is bonded to one hydrogen atom and one carbon atom. The functional group of a **ketone** is a carbonyl group bonded to two carbon atoms (Section 1.4C). Acetone is the simplest ketone.

$$\begin{array}{c} O \\ \parallel \\ HCH \end{array} \qquad \begin{array}{c} O \\ \parallel \\ CH_3CH \end{array} \qquad \begin{array}{c} O \\ \parallel \\ CH_3CCH_3 \end{array}$$

Methanal
(Formaldehyde)

Ethanal
(Acetaldehyde)

Propanone
(Acetone)

Because aldehydes always contain at least one hydrogen bonded to the C=O group, they are often written RCH=O or RCHO. Similarly, ketones are often written RCOR'.

9.2 How Do We Name Aldehydes and Ketones?

A. IUPAC Names

The IUPAC names for aldehydes and ketones follow the familiar pattern of selecting as the parent alkane the longest chain of carbon atoms that contains the functional group (Section 2.3A). To name an aldehyde, we change the suffix -e of the parent alkane to -al. Because the carbonyl group of an aldehyde can appear only at the end of a parent chain and numbering must start with it as carbon 1, there is no need to use a number to locate the aldehyde group.

For **unsaturated aldehydes,** we show the presence of the carbon–carbon double bond and the aldehyde by changing the ending of the parent alkane from -ane to -enal: "-en-" to show the carbon–carbon double bond, and "-al" to show the aldehyde. We show the location of the carbon–carbon double bond by the number of its first carbon.

Hexanal

3-Methylbutanal

2-Propenal
(Acrolein)

In the IUPAC system, we name ketones by selecting as the parent alkane the longest chain that contains the carbonyl group and then indicating the presence of this group by changing the -e of the parent alkane to -one. The parent chain is numbered from the direction that gives the smaller number to the carbonyl carbon. While the systematic name of the simplest ketone is 2-propanone, the IUPAC retains its common name, acetone.

Acetone

5-Methyl-3-hexanone

2-Methylcyclohexanone

Example 9.1 IUPAC Names for Aldehydes and Ketones

Write the IUPAC name for each compound:

(a)   (b)   (c)

Strategy and Solution

(a) The longest chain has six carbons, but the longest chain that contains the carbonyl carbon has only five carbons. Its IUPAC name is 2-ethyl-3-methylpentanal.

(a) ι)

2-Ethyl-3-methylpentanal

(b) Number the six-membered ring beginning with the carbonyl carbon. Its IUPAC name is 3,3-dimethylcyclohexanone.

(c) This molecule is derived from benzaldehyde. Its IUPAC name is 2-ethylbenzaldehyde.

Problem 9.1

Write the IUPAC name for each compound.

(a)   (b)   (c)

Example 9.2 Structural Formulas for Ketones

Write structural formulas for all ketones with the molecular formula $C_6H_{12}O$ and give the IUPAC name of each. Which of these ketones are chiral?

Strategy and Solution

There are six ketones with this molecular formula: two with a six-carbon chain, three with a five-carbon chain and a methyl branch, and one with a four-carbon chain and two methyl branches. Only 3-methyl-2-pentanone has a stereocenter and is chiral.

2-Hexanone   3-Hexanone   4-Methyl-2-pentanone

3-Methyl-2-pentanone   2-Methyl-3-pentanone   3,3-Dimethyl-2-butanone

Problem 9.2

Write structural formulas for all aldehydes with the molecular formula $C_6H_{12}O$ and give the IUPAC name of each. Which of these aldehydes are chiral?

In naming aldehydes or ketones that also contain an —OH or —NH_2 group elsewhere in the molecule, the parent chain is numbered to give the carbonyl group the lower number. An —OH substituent is indicated by *hydroxy*, and an —NH_2 substituent is indicated by *amino-*. Hydroxy and amino substituents are numbered and alphabetized along with any other substituents that might be present.

Example 9.3 Naming Difunctional Aldehydes and Ketones

Write the IUPAC name for each compound.

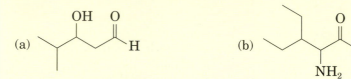

(a)   (b)

Strategy and Solution

(a) We number the parent chain beginning with CHO as carbon 1. There is a hydroxyl group on carbon 3 and a methyl group on carbon 4. The IUPAC name of this compound is 3-hydroxy-4-methylpentanal. Note that this hydroxyaldehyde is chiral and can exist as a pair of enantiomers.

(b) The longest chain that contains the carbonyl is six carbons; the carbonyl group is on carbon 2 and the amino group on carbon 3. The IUPAC name of this compound is 3-amino-4-ethyl-2-hexanone. Note that this ketoamine is also chiral and can exist as a pair of enantiomers.

Problem 9.3

Write the IUPAC name for each compound.

(a)   (b)   (c)

B. Common Names

We derive the common name for an aldehyde from the common name of the corresponding carboxylic acid. The word "acid" is dropped and the suffix *-ic* or *-oic* is changed to *-aldehyde*. Because we have not yet studied common names for carboxylic acids, we are not in a position to discuss common names for aldehydes. We can, however, illustrate how they are derived by reference to two common names with which you are familiar. The name formaldehyde is derived from formic acid, and the name acetaldehyde is derived from acetic acid.

Chemical Connections 9A

Some Naturally Occurring Aldehydes and Ketones

Benzaldehyde
(oil of almonds)

Cinnamaldehyde
(oil of cinnamon)

Citronellal
(citronella oils; also in
lemon and lemon grass oils)

Muscone
(from the musk deer;
used in perfumes)

Vanillin
(vanilla bean)

β-Ionone
(from violets)

We derive common names for ketones by naming each alkyl or aryl group bonded to the carbonyl group as a separate word, followed by the word "ketone." The alkyl or aryl groups are generally listed in order of increasing molecular weight.

Ethyl isopropyl ketone   Methyl ethyl ketone   Dicyclohexyl ketone

2-Butanone, more commonly called methyl ethyl ketone (MEK), is used as a solvent for paints and varnishes.

9.3 What Are the Physical Properties of Aldehydes and Ketones?

Oxygen is more electronegative than carbon. Therefore a carbon–oxygen double bond is polar, with oxygen bearing a partial negative charge and carbon bearing a partial positive charge (Figure 9.1).

In liquid aldehydes and ketones, intermolecular attractions occur between the partial positive charge on the carbonyl carbon of one molecule and the partial negative charge on the carbonyl oxygen of another molecule. There is no possibility for hydrogen bonding between aldehyde or ketone molecules, which explains why these compounds have lower boiling points than alcohols (Section 5.1C) and carboxylic acids (Section 10.3D), compounds in which hydrogen bonding between molecules does occur.

Table 9.1 lists structural formulas and boiling points of six compounds of similar molecular weight. Of the six, pentane and diethyl ether have the lowest boiling points. The boiling point of 1-butanol, which can associate by intermolecular hydrogen bonding, is higher than that of either butanal or 2-butanone. Propanoic acid, in which intermolecular association by hydrogen bonding is the strongest, has the highest boiling point.

Because the oxygen atom of each carbonyl group is a hydrogen bond acceptor, the low-molecular-weight aldehydes and ketones are more soluble in

Polarity of a
carbonyl group

FIGURE 9.1 The polarity of a carbonyl group. The carbonyl oxygen bears a partial negative charge and the carbonyl carbon bears a partial positive charge.

TABLE 9.1 Boiling Points of Six Compounds of Comparable Molecular Weight

Name	Structural Formula	Molecular Weight	Boiling Point (°C)
diethyl ether	$CH_3CH_2OCH_2CH_3$	74	34
pentane	$CH_3CH_2CH_2CH_2CH_3$	72	36
butanal	$CH_3CH_2CH_2CHO$	72	76
2-butanone	$CH_3CH_2COCH_3$	72	80
1-butanol	$CH_3CH_2CH_2CH_2OH$	74	117
propanoic acid	CH_3CH_2COOH	74	141

water than are nonpolar compounds of comparable molecular weight. Formaldehyde, acetaldehyde, and acetone are infinitely soluble in water. As the hydrocarbon portion of the molecule increases in size, aldehydes and ketones become less soluble in water.

Most aldehydes and ketones have strong odors. The odors of ketones are generally pleasant, and many are used in perfumes and as flavoring agents. The odors of aldehydes vary. You may be familiar with the smell of formaldehyde; if so, you know that it is not pleasant. Many higher aldehydes, however, have pleasant odors and are used in perfumes.

9.4 What Are the Characteristic Reactions of Aldehydes and Ketones?

A. Oxidation

Aldehydes are oxidized to carboxylic acids by a variety of oxidizing agents, including potassium dichromate (Section 5.2C).

The body uses nicotinamide adenine dinucleotide, NAD^+, for this type of oxidation (Section 19.3).

Hexanal → Hexanoic acid

Aldehydes are also oxidized to carboxylic acids by the oxygen in the air. In fact, aldehydes that are liquid at room temperature are so sensitive to oxidation that they must be protected from contact with air during storage. Often this is done by sealing the aldehyde in a container under an atmosphere of nitrogen.

Benzaldehyde → Benzoic acid

Ketones, in contrast, resist oxidation by most oxidizing agents, including potassium dichromate and molecular oxygen.

The fact that aldehydes are so easy to oxidize and ketones are not allows us to use simple chemical tests to distinguish between these types of compounds. Suppose that we have a compound we know is either an aldehyde or a ketone. To determine which it is, we can treat the compound with a mild oxidizing agent. If it can be oxidized, it is an aldehyde; otherwise, it is a ketone. One reagent that has been used for this purpose is Tollens' reagent.

Tollens' reagent contains silver nitrate and ammonia in water. When these two compounds are mixed, silver ion combines with NH_3 to form the complex ion $Ag(NH_3)_2^+$. When this solution is added to an aldehyde, the aldehyde acts as a reducing agent and reduces the complexed silver ion to silver metal. If this reaction is carried out properly, the silver metal precipitates as a smooth, mirror-like deposit on the inner surface of the reaction vessel, leading to the name **silver-mirror test.** If the remaining solution is then acidified with HCl, the carboxylic anion, $RCOO^-$, formed during the aldehyde's oxidation is converted to the carboxylic acid, RCOOH.

Today, silver(I) is rarely used for the oxidation of aldehydes because of its high cost and because of the availability of other, more convenient methods for this oxidation. This reaction, however, is still used for making (silvering) mirrors.

A silver mirror has been deposited on the inside of this flask by the reaction between an aldehyde and Tollens' reagent.

Charles D. Winters/Cengage Learning

Example 9.4 Oxidation of Aldehydes and Ketones

Draw a structural formula for the product formed by treating each compound with Tollens' reagent followed by acidification with aqueous HCl.

(a) Pentanal　　　　　　　　(b) 4-Hydroxybenzaldehyde

Strategy and Solution

The aldehyde group in each compound is oxidized to a carboxylic anion, $-COO^-$. Acidification with HCl converts the anion to a carboxylic acid, $-COOH$.

(a) Pentanoic acid

(b) 4-Hydroxybenzoic acid

Problem 9.4

Complete equations for these oxidations.
(a) Hexanedial + $O_2 \longrightarrow$
(b) 3-Phenylpropanal + $Ag(NH_3)_2^+ \longrightarrow$

B. Reduction

In Section 3.3D, we saw that the C=C double bond of an alkene can be reduced by hydrogen in the presence of a transition metal catalyst to a C—C single bond. The same is true of the C=O double bond of an aldehyde

or ketone. Aldehydes are reduced to primary alcohols and ketones are reduced to secondary alcohols.

The reduction of a C=O double bond under these conditions is slower than
the reduction of a C=C double bond. Thus, if the same molecule contains
both C=O and C=C double bonds, the C=C double bond is reduced first.

The reagent most commonly used in the laboratory for the reduction of
an aldehyde or ketone is sodium borohydride, NaBH$_4$. This reagent behaves
as if it were a source of hydride ions, H:$^-$. In the hydride ion, hydrogen
has two valence electrons and bears a negative charge. In a reduction
by sodium borohydride, hydride ion is attracted to and then adds to the
partially positive carbonyl carbon, which leaves a negative charge on the
carbonyl oxygen. Reaction of this alkoxide intermediate with aqueous acid
gives the alcohol.

Of the two hydrogens added to the carbonyl group in this reduction,
one comes from the reducing agent and the other comes from aqueous
acid. Reduction of cyclohexanone, for example, with this reagent gives
cyclohexanol:

An advantage of using NaBH$_4$ over the H$_2$/metal reduction is that NaBH$_4$
does not reduce carbon–carbon double bonds. The reason for this selectivity
is quite straightforward. There is no polarity (no partial positive or negative
charges) on a carbon–carbon double bond. Therefore, a C=C double bond
has no partially positive site to attract the negatively charged hydride ion.
In the following example, NaBH$_4$ selectively reduces the aldehyde to a primary alcohol:

In biological systems, the agent for the reduction of aldehydes and ketones
is the reduced form of the coenzyme nicotinamide adenine dinucleotide, abbreviated NADH (Section 19.3). This reducing agent, like NaBH$_4$, delivers

a hydride ion to the carbonyl carbon of the aldehyde or ketone. Reduction of pyruvate, for example, by NADH gives lactate:

Pyruvate is the end product of glycolysis, a series of enzyme-catalyzed reactions that converts glucose to two molecules of this ketoacid (Section 20.1). Under anaerobic conditions, NADH reduces pyruvate to lactate. The build-up of lactate in the bloodstream leads to acidosis and in muscle tissue is associated with muscle fatigue. When blood lactate reaches a concentration of about 0.4 mg/100 mL, muscle tissue becomes almost completely exhausted.

Example 9.5 Reduction of Aldehydes and Ketones

Complete the equations for these reductions.

Strategy and Solution

The carbonyl group of the aldehyde in (a) is reduced to a primary alcohol and that of the ketone in (b) is reduced to a secondary alcohol.

Problem 9.5

Which aldehyde or ketone gives these alcohols upon reduction with H_2/metal catalyst?

C. Addition of Alcohols

Addition of a molecule of alcohol to the carbonyl group of an aldehyde or ketone forms a **hemiacetal** (a half-acetal). The functional group of a

Hemiacetal A molecule containing a carbon bonded to one —OH group and one —OR group; the product of adding one molecule of alcohol to the carbonyl group of an aldehyde or ketone hemiacetal is a carbon bonded to one —OH group and one —OR group. In forming a hemiacetal, the H of the alcohol adds to the carbonyl oxygen and the OR group of the alcohol adds to the carbonyl carbon. Shown here are the hemiacetals formed by addition of one molecule of ethanol to benzaldehyde and to cyclohexanone:

Hemiacetals are generally unstable and are only minor components of an equilibrium mixture, except in one very important type of molecule. When a hydroxyl group is part of the same molecule that contains the carbonyl group and a five- or six-membered ring can form, the compound exists almost entirely in a cyclic hemiacetal form. In this case, the —OH group adds to the C=O group of the same molecule. We will have much more to say about cyclic hemiacetals when we consider the chemistry of carbohydrates in Chapter 12.

Acetal A molecule containing two —OR groups bonded to the same carbon

Hemiacetals can react further with alcohols to form **acetals** plus water. This reaction is acid-catalyzed. The functional group of an acetal is a carbon bonded to two —OR groups.

All steps in hemiacetal and acetal formation are reversible. As with any other equilibrium, we can make this one go in either direction by using Le Chatelier's principle. If we want to drive it to the right (formation of the acetal), we either use a large excess of alcohol or remove water from

the equilibrium mixture. If we want to drive it to the left (hydrolysis of the acetal to the original aldehyde or ketone and alcohol), we use a large excess of water.

Example 9.6 Formation of Hemiacetals and Acetals

Show the reaction of 2-butanone with one molecule of ethanol to form a hemiacetal and then with a second molecule of ethanol to form an acetal.

Strategy and Solution

Given are structural formulas for the hemiacetal and then the acetal.

2-Butanone   A hemiacetal   An acetal

Problem 9.6

Show the reaction of benzaldehyde with one molecule of methanol to form a hemiacetal and then with a second molecule of methanol to form an acetal.

Example 9.7 Recognizing the Presence of a Hemiacetal and an Acetal

Identify all hemiacetals and acetals in the following structures, and tell whether each is formed from an aldehyde or a ketone.

(a) $CH_3CH_2\overset{\overset{\displaystyle OCH_3}{|}}{\underset{\underset{\displaystyle OCH_3}{|}}{C}}CH_2CH_3$ (b) $CH_3CH_2OCH_2CH_2OH$ (c)

Strategy

An acetal contains a carbon atom bonded to two OR groups; a hemiacetal contains a carbon atom bonded to one —OH group and one —OR group.

Solution

Compound (a) is an acetal derived from a ketone. Compound (b) is neither a hemiacetal nor an acetal because it does not have a carbon bonded to two oxygens; its functional groups are an ether and a primary alcohol. Compound (c) is a hemiacetal derived from an aldehyde.

(a)   An acetal   2-Pentanone

(c)   A hemiacetal   5-Hydroxypentanal

Problem 9.7

Identify all hemiacetals and acetals in the following structures, and tell whether each is formed from an aldehyde or a ketone.

$$\text{(a)} \quad CH_3CH_2\overset{\overset{\displaystyle OH}{|}}{\underset{\underset{\displaystyle OCH_2CH_3}{|}}{C}}CH_2CH_3 \qquad \text{(b)} \quad CH_3OCH_2CH_2OCH_3 \qquad \text{(c)}$$

9.5 What Is Keto-Enol Tautomerism?

A carbon atom adjacent to a carbonyl group is called an α-carbon, and a hydrogen atom bonded to it is called an α-hydrogen.

Enol A molecule containing an —OH group bonded to a carbon of a carbon–carbon double bond

A carbonyl compound that has a hydrogen on an α-carbon is in equilibrium with a constitutional isomer called an **enol.** The name "enol" is derived from the IUPAC designation of it as both an alkene (-*en*-) and an alcohol (-*ol*).

$$CH_3-\overset{\overset{\displaystyle O}{||}}{C}-CH_3 \rightleftharpoons CH_3-\overset{\overset{\displaystyle OH}{|}}{C}=CH_2$$

Acetone (keto form)   Acetone (enol form)

Tautomers Constitutional isomers that differ in the location of a hydrogen atom and a double bond

Keto and enol forms are examples of **tautomers,** constitutional isomers in equilibrium with each other that differ in the location of a hydrogen atom and a double bond. This type of isomerism is called **keto-enol tautomerism.** For any pair of keto-enol tautomers, the keto form generally predominates at equilibrium.

Example 9.8 Keto-Enol Tautomerism

Draw structural formulas for the two enol forms for each ketone.

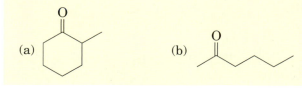

(a)   (b)

Strategy and Solution

Any aldehyde or ketone with one hydrogen on its α-carbon can show keto-enol tautomerism.

Cis & trans isomers

Problem 9.8

Draw a structural formula for the keto form of each enol.

(a) ⬡—OH   (b) ⬡(—OH)(—OH)   (c) ⬡=CHOH

Summary of Key Questions

Section 9.1 What Are Aldehydes and Ketones?
Problem 9.15

- An **aldehyde** contains a carbonyl group bonded to at least one hydrogen atom.
- A **ketone** contains a carbonyl group bonded to two carbon atoms.

Section 9.2 How Do We Name Aldehydes and Ketones?

- We derive the IUPAC name of an aldehyde by changing the -e of the parent alkane to -al.
- We derive the IUPAC name of a ketone by changing the -e of the parent alkane to -one and using a number to locate the carbonyl carbon.

Section 9.3 What Are the Physical Properties of Aldehydes and Ketones?

- Aldehydes and ketones are polar compounds. They have higher boiling points and are more soluble in water than nonpolar compounds of comparable molecular weight.

Section 9.4 What Are the Characteristic Reactions of Aldehydes and Ketones?
Problems 9.28, 9.35

- Aldehydes are oxidized to carboxylic acids but ketones resist oxidation.
- **Tollens' reagent** is used to test for the presence of aldehydes.
- Aldehydes can be reduced to primary alcohols and ketones to secondary alcohols.
- Addition of a molecule of alcohol to an aldehyde or ketone produces a **hemiacetal.** A hemiacetal can react with another molecule of alcohol to produce an **acetal.**

Section 9.5 What Is Keto-Enol Tautomerization?

- A molecule containing an —OH group bonded to a carbon of a carbon–carbon double bond is called an **enol.**
- Constitutional isomers that differ in the location of a hydrogen atom and a double bond are called **tautomers.**

Summary of Key Reactions

1. **Oxidation of an Aldehyde to a Carboxylic Acid (Section 9.4A)**
   The aldehyde group is among the most easily oxidized organic functional groups. Oxidizing agents include $K_2Cr_2O_7$, Tollens' reagent, and O_2.

2. Reduction (Section 9.4B)

Aldehydes are reduced to primary alcohols and ketones to secondary alcohols by H_2 in the presence of a transition metal catalyst such as Pt or Ni. They are also reduced to alcohols by sodium borohydride, $NaBH_4$, followed by protonation.

3. Addition of Alcohols to Form Hemiacetals (Section 9.4C)

Hemiacetals are only minor components of an equilibrium mixture of an aldehyde or ketone and an alcohol, except where the —OH and C=O groups are parts of the same molecule and a five- or six-membered ring can form.

4. Addition of Alcohols to Form Acetals (Section 9.4C)

Formation of acetals is catalyzed by acid. Acetals are hydrolyzed in aqueous acid to an aldehyde or ketone and two molecules of an alcohol.

Cyclohexanone   Methanol   An acetal

5. Keto-Enol Tautomerism (Section 9.5)

The keto form generally predominates at equilibrium.

$$CH_3CCH_3 \rightleftharpoons CH_3C=CH_2$$

Keto form   Enol form
(approximately 99.9%)

Problems

■ Indicates problems that are assignable in **GOB OWL**.

Blue numbered problems are applied.

Go to this book's companion website at **www. cengage.com/chemistry/bettelheim** for interactive versions of the How To tutorials and Active Figures, and to quiz yourself on this chapter.

Section 9.1 What Are Aldehydes and Ketones?

9.9 Answer true or false.

(a) The one aldehyde and the one ketone with a molecular formula of C_3H_6O are constitutional isomers.

(b) Aldehydes and ketones both contain a carbonyl group.

(c) The VSEPR model predicts bond angles of 120° about the carbonyl carbon of aldehydes and ketones.

(d) The carbonyl carbon of a ketone is a stereocenter.

9.10 What is the difference in structure between an aldehyde and a ketone?

9.11 What is the difference in structure between an aromatic aldehyde and an aliphatic aldehyde?

9.12 Is it possible for the carbon atom of a carbonyl group to be a stereocenter? Explain.

9.13 Which compounds contain carbonyl groups?

9.14 Following are structural formulas for two steroid hormones.

Cortisone

Aldosterone

(a) Name the functional groups in each.

(b) Mark all stereocenters in each hormone and state how many stereoisomers are possible for each.

9.15 ■ Draw structural formulas for the four aldehydes with the molecular formula $C_5H_{10}O$. Which of these aldehydes are chiral?

Section 9.2 How Do We Name Aldehydes and Ketones?

9.16 Answer true or false.

(a) An aldehyde is named as an alkanal and a ketone is named as an alkanone.

(b) The names for aldehydes and ketones are derived from the name of the longest carbon chain that contains the carbonyl group.

(c) In an aromatic aldehyde, the carbonyl carbon is bonded to an aromatic ring.

9.17 Draw structural formulas for these aldehydes.

(a) Formaldehyde   (b) Propanal

(c) 3,7-Dimethyloctanal   (d) Decanal

(e) 4-Hydroxybenzaldehyde

(f) 2,3-Dihydroxypropanal

9.18 Draw structural formulas for these ketones.

(a) Ethyl isopropyl ketone

(b) 2-Chlorocyclohexanone

(c) 2,4-Dimethyl-3-pentanone

(d) Diisopropyl ketone

(e) Acetone

(f) 2,5-Dimethylcyclohexanone

9.19 Write IUPAC names for these compounds.

9.20 Write IUPAC names for these compounds.

Section 9.3 What Are the Physical Properties of Aldehydes and Ketones?

9.21 Answer true or false.

(a) Aldehydes and ketones are polar compounds.

(b) Aldehydes have lower boiling points than alcohols with comparable carbon skeletons.

(c) Low-molecular-weight aldehydes and ketones are very soluble in water.

(d) There is no possibility for hydrogen bonding between molecules of aldehydes and ketones.

9.22 In each pair of compounds, select the one with the higher boiling point.

(a) Acetaldehyde or ethanol

(b) Acetone or 3-pentanone

(c) Butanal or butane

(d) Butanone or 2-butanol

9.23 Acetone is completely soluble in water, but 4-heptanone is completely insoluble in water. Explain.

9.24 Account for the fact that acetone has a higher boiling point (56°C) than ethyl methyl ether (11°C), even though their molecular weights are almost the same.

9.25 Pentane, 1-butanol, and butanal all have approximately the same molecular weights but different boiling points. Arrange them in order of increasing boiling point. Explain the basis for your ranking.

9.26 Show how acetaldehyde can form hydrogen bonds with water.

9.27 Why can't two molecules of acetone form a hydrogen bond with each other?

Section 9.4 What Are the Characteristic Reactions of Aldehydes and Ketones?

9.28 ■ Draw a structural formula for the principal organic product formed when each compound is treated with $K_2Cr_2O_7/H_2SO_4$. If there is no reaction, say so.

(a) Butanal   (b) Benzaldehyde

(c) Cyclohexanone   (d) Cyclohexanol

9.29 Draw a structural formula for the principal organic product formed when each compound in Problem 9.28 is treated with Tollens' reagent. If there is no reaction, say so.

9.30 What simple chemical test could you use to distinguish between the members of each pair of compounds? Tell what you would do, what you would expect to observe, and how you would interpret your experimental observation.

(a) Pentanal and 2-pentanone

(b) 2-Pentanone and 2-pentanol

9.31 Explain why liquid aldehydes are often stored under an atmosphere of nitrogen rather than in air.

9.32 Suppose that you take a bottle of benzaldehyde (a liquid, bp 179°C) from a shelf and find a white solid in the bottom of the bottle. The solid turns litmus red; that is, it is acidic. Yet aldehydes are neutral compounds. How can you explain these observations?

9.33 Write a structural formula for the principal organic product formed by treating each compound with H₂/transition metal catalyst. Which products are chiral?

9.34 Write a structural formula for the principal organic product formed by treating each compound in Problem 9.33 with NaBH₄ followed by H₂O.

9.35 ■ 1,3-Dihydroxy-2-propanone, more commonly known as dihydroxyacetone, is the active ingredient in artificial tanning agents, such as Man-Tan and Magic Tan.

(a) Write a structural formula for this compound.

(b) Would you expect it to be soluble or insoluble in water?

(c) Write a structural formula for the product formed by its reduction with NaBH₄.

9.36 Draw a structural formula for the product formed by treatment of butanal with each set of reagents.

(a) H₂/metal catalyst

(b) NaBH₄, then H₂O

(c) Ag(NH₃)₂⁺ (Tollens' reagent)

(d) K₂Cr₂O₇/H₂SO₄

9.37 Draw a structural formula for the product formed by treatment of acetophenone, C₆H₅COCH₃, with each set of reagents given in Problem 9.36.

Section 9.5 What Is Keto-Enol Tautomerism?

9.38 Mark each statement true or false.

(a) Keto and enol tautomers are constitutional isomers.

(b) For a pair of keto-enol tautomers, the keto form generally predominates.

9.39 Which of these compounds undergo keto-enol tautomerism?

9.40 Draw all enol forms of each aldehyde and ketone.

9.41 Draw a structural formula for the keto form of each enol.

Addition of Alcohols

9.42 What is the characteristic structural feature of a hemiacetal? Of an acetal?

9.43 ■ Which compounds are hemiacetals, which are acetals, and which are neither?

(a)

(b) CH₃CH₂CHOCH₃ with OH above

(c) CH₃OCH₂OCH₃

(d)

(e)

(f)

9.44 Draw the hemiacetal and then the acetal formed in each reaction. In each case, assume an excess of the alcohol.

(a) Propanal + methanol →

(b) Cyclopentanone + methanol →

9.45 Draw the structures of the aldehydes or ketones and alcohols formed when these acetals are treated with aqueous acid and hydrolyzed.

9.46 The following compound is a component of the fragrance of jasmine:

From which carbonyl-containing compound and alcohol is this compound derived?

9.47 What is the difference in meaning between the terms "hydration" and "hydrolysis"? Give an example of each.

9.48 What is the difference in meaning between the terms "hydration" and "dehydration"? Give an example of each.

9.49 Show reagents and experimental conditions to convert cyclohexanone to each of the following compounds.

9.50 Draw a structural formula for an aldehyde or ketone that can be reduced to produce each alcohol. If none exists, say so.

(b)

9.51 Draw a structural formula for an aldehyde or ketone that can be reduced to produce each alcohol. If none exists, say so.

(c) CH₃OH

(d)

9.52 1-Propanol can be prepared by the reduction of an aldehyde, but it cannot be prepared by the acid-catalyzed hydration of an alkene. Explain why it cannot be prepared from an alkene.

9.53 Show how to bring about these conversions. In addition to the given starting material, use any other organic or inorganic reagents as necessary.

(a) $C_6H_5\overset{O}{\overset{\|}{C}}CH_2CH_3 \longrightarrow C_6H_5\overset{OH}{\overset{|}{C}}HCH_2CH_3$

$\longrightarrow C_6H_5CH=CHCH_3$

(b) [cyclopentanone] $=O \longrightarrow$ [cyclopentane]$-OH \longrightarrow$ [cyclopentene]

$\longrightarrow$ [cyclopentane]$-Cl$

9.54 Show how to bring about these conversions. In addition to the given starting material, use any other organic or inorganic reagents as necessary.
(a) 1-Pentene to 2-pentanone
(b) Cyclohexene to cyclohexanone

9.55 Describe a simple chemical test by which you could distinguish between the members of each pair of compounds.
(a) Cyclohexanone and aniline
(b) Cyclohexene and cyclohexanol
(c) Benzaldehyde and cinnamaldehyde

Additional Problems

9.56 Indicate the aldehyde or ketone group in these compounds.

(c) HOCH₂CHCH

(d)

(e)

(f) HO

9.57 Draw a structural formula for the product formed by treating each compound in Problem 9.56 with sodium borohydride, NaBH₄.

9.58 Draw structural formulas for the (a) one ketone and (b) two aldehydes with the molecular formula C_4H_8O.

9.59 Draw structural formulas for these compounds.
  (a) 1-Chloro-2-propanone
  (b) 3-Hydroxybutanal
  (c) 4-Hydroxy-4-methyl-2-pentanone
  (d) 3-Methyl-3-phenylbutanal
  (e) 1,3-Cyclohexanedione
  (f) 5-Hydroxyhexanal

9.60 Why does acetone have a lower boiling point (56°C) than 2-propanol (82°C), even though their molecular weights are almost the same?

9.61 Propanal (bp 49°C) and 1-propanol (bp 97°C) have about the same molecular weight, yet their boiling points differ by almost 50°C. Explain this fact.

9.62 What simple chemical test could you use to distinguish between the members of each pair of compounds? Tell what you would do, what you would expect to observe, and how you would interpret your experimental observation.
  (a) Benzaldehyde and cyclohexanone
  (b) Acetaldehyde and acetone

9.63 ■ 5-Hydroxyhexanal forms a six-membered cyclic hemiacetal, which predominates at equilibrium in aqueous solution.
  (a) Draw a structural formula for this cyclic hemiacetal.
  (b) How many stereoisomers are possible for 5-hydroxyhexanal?
  (c) How many stereoisomers are possible for this cyclic hemiacetal?

9.64 The following molecule is an enediol; each carbon of the double bond carries an —OH group. Draw structural formulas for the α-hydroxyketone and the α-hydroxyaldehyde with which this enediol is in equilibrium.

$$\alpha\text{-hydroxyaldehyde} \rightleftharpoons \underset{\underset{\displaystyle CH_3}{|}}{\overset{\overset{\displaystyle HC-OH}{\|}}{C-OH}} \rightleftharpoons \alpha\text{-hydroxyketone}$$

An enediol

9.65 Alcohols can be prepared by the acid-catalyzed hydration of alkenes (Section 3.6B) and by the reduction of aldehydes and ketones (Section 9.4B). Show how you might prepare each of the following alcohols by (1) acid-catalyzed hydration of an alkene and (2) reduction of an aldehyde or ketone.
  (a) Ethanol   (b) Cyclohexanol
  (c) 2-Propanol   (d) 1-Phenylethanol

Looking Ahead

9.66 Glucose, $C_6H_{12}O_6$, contains an aldehyde group but exists predominantly in the form of the cyclic hemiacetal shown here. We discuss this cyclic form of glucose in Chapter 12.

β-D-Glucose

(a) A cyclic hemiacetal is formed when the —OH group of one carbon bonds to the carbonyl group of another carbon. Which carbon in glucose provides the —OH group and which provides the CHO group?

9.67 Ribose, $C_5H_{10}O_5$, contains an aldehyde group but exists predominantly in the form of the cyclic hemiacetal shown here. We discuss this cyclic form of ribose in Chapter 12.

β-D-Ribose

(a) Which carbon of ribose provides the —OH group and which provides the CHO group for formation of this cyclic hemiacetal?

9.68 Sodium borohydride is a laboratory reducing agent. NADH is a biological reducing agent. In what way is the chemistry by which they reduce aldehydes and ketones similar?

9.69 Write an equation for each conversion.
  (a) 1-Pentanol to pentanal
  (b) 1-Pentanol to pentanoic acid
  (c) 2-Pentanol to 2-pentanone
  (d) 2-Propanol to acetone
  (e) Cyclohexanol to cyclohexanone

Carboxylic Acids

Citrus fruits are sources of citric acid, a tricarboxylic acid.

10.1 What Are Carboxylic Acids?

In this chapter, we study carboxylic acids, another class of organic compounds containing the carbonyl group. The functional group of a **carboxylic acid** is a **carboxyl group** (Section 1.4D), which can be represented in any one of three ways:

$$\overset{\displaystyle O}{\underset{}{-\overset{\|}{C}-OH}} \qquad -COOH \qquad -CO_2H$$

10.2 How Do We Name Carboxylic Acids?

A. IUPAC Names

We derive the IUPAC name of an acyclic carboxylic acid from the name of the longest carbon chain that contains the carboxyl group. Drop the

OWL Online homework for this chapter may be assigned in **GOB OWL.**

final *-e* from the name of the parent alkane and replace it by *-oic acid*. Number the chain beginning with the carbon of the carboxyl group. Because the carboxyl carbon is understood to be carbon 1, there is no need to give it a number. In the following examples, the common name is given in parentheses.

Hexanoic acid
(Caproic acid)

3-Methylbutanoic acid
(Isovaleric acid)

When a carboxylic acid also contains an —OH (hydroxyl) group, we indicate its presence by adding the prefix *hydroxy-*. When it contains a primary (1°) amine, we indicate the presence of the —NH_2 group by *amino-*.

5-Hydroxyhexanoic acid

4-Aminobenzoic acid
(*p*-Aminobenzoic acid)

To name dicarboxylic acids, we add the suffix *-dioic acid* to the name of the parent alkane that contains both carboxyl groups. The numbers of the carboxyl carbons are not indicated because they can be only at the ends of the parent chain.

Formic acid was first obtained in 1670 from the destructive distillation of ants, whose Latin genus is *Formica*. It is one of the components of the venom injected by stinging ants.

Ethanedioic acid
(Oxalic acid)

Propanedioic acid
(Malonic acid)

Butanedioic acid
(Succinic acid)

Pentanedioic acid
(Glutaric acid)

Hexanedioic acid
(Adipic acid)

The name *oxalic acid* is derived from one of its sources in the biological world—plants of the genus *Oxalis*, one of which is rhubarb. Oxalic acid also occurs in human and animal urine, and calcium oxalate is a major component of kidney stones. Succinic acid is an intermediate in the citric acid cycle (Section 19.4). Adipic acid is one of the two monomers required for the synthesis of the polymer nylon-66 (Section 11.6B).

B. Common Names

Common names for aliphatic carboxylic acids, many of which were known long before the development of IUPAC nomenclature, are often derived from the name of a natural substance from which the acid could be isolated. Table 10.1 lists several of the unbranched aliphatic carboxylic acids found in the biological world along with the common name of each. Those with

TABLE 10.1 Several Aliphatic Carboxylic Acids and Their Common Names

Structure	IUPAC Name	Common Name	Derivation
HCOOH	methanoic acid	formic acid	Latin: *formica,* ant
CH_3COOH	ethanoic acid	acetic acid	Latin: *acetum,* vinegar
CH_3CH_2COOH	propanoic acid	propionic acid	Greek: *propion,* first fat
$CH_3(CH_2)_2COOH$	butanoic acid	butyric acid	Latin: *butyrum,* butter
$CH_3(CH_2)_3COOH$	pentanoic acid	valeric acid	Latin: *valere,* to be strong
$CH_3(CH_2)_4COOH$	hexanoic acid	caproic acid	Latin: *caper,* goat
$CH_3(CH_2)_6COOH$	octanoic acid	caprylic acid	Latin: *caper,* goat
$CH_3(CH_2)_8COOH$	decanoic acid	capric acid	Latin: *caper,* goat
$CH_3(CH_2)_{10}COOH$	dodecanoic acid	lauric acid	Latin: *laurus,* laurel
$CH_3(CH_2)_{12}COOH$	tetradecanoic acid	myristic acid	Greek: *myristikos,* fragrant
$CH_3(CH_2)_{14}COOH$	hexadecanoic acid	palmitic acid	Latin: *palma,* palm tree
$CH_3(CH_2)_{16}COOH$	octadecanoic acid	stearic acid	Greek: *stear,* solid fat
$CH_3(CH_2)_{18}COOH$	eicosanoic acid	arachidic acid	Greek: *arachis,* peanut

The unbranched carboxylic acids having between 12 and 20 carbon atoms are known as fatty acids. We study them further in Chapter 13.

16, 18, and 20 carbon atoms are particularly abundant in animal fats and vegetable oils (Section 13.2), and the phospholipid components of biological membranes (Section 13.5).

When common names are used, the Greek letters alpha (α), beta (β), gamma (γ), and so forth are often added as a prefix to locate substituents.

4-Aminobutanoic acid
(γ-Aminobutyric acid; GABA)

GABA is a neurotransmitter in the central nervous system.

Example 10.1 IUPAC Names of Carboxylic Acids

Write the IUPAC name for each carboxylic acid:

Strategy and Solution

(a) The longest carbon chain that contains the carboxyl group has five carbons and, therefore, the parent alkane is pentane. The IUPAC name is 2-ethylpentanoic acid.

(b) 4-Hydroxybenzoic acid

(c) *trans*-3-Phenyl-2-propenoic acid (cinnamic acid)

Problem 10.1

Each of the following compounds has a well-recognized and widely used common name. A derivative of glyceric acid is an intermediate in glycolysis (Section 20.2). β-Alanine is a building block of pantothenic acid

(Section 19.5). Mevalonic acid is an intermediate in the biosynthesis of steroids (Section 19.4). Write the IUPAC name for each compound.

(a)
$$\begin{array}{c} \text{COOH} \\ | \\ \text{CHOH} \\ | \\ \text{CH}_2\text{OH} \end{array}$$

Glyceric acid

(b) $H_2NCH_2CH_2COOH$

β-Alanine

(c)

Mevalonic acid

10.3 What Are the Physical Properties of Carboxylic Acids?

A major feature of carboxylic acids is the polarity of the carboxyl group (Figure 10.1). This group contains three polar covalent bonds: C=O, C—O, and O—H. The polarity of these bonds determines the major physical properties of carboxylic acids.

Carboxylic acids have significantly higher boiling points than other types of organic compounds of comparable molecular weight (Table 10.2). Their higher boiling points result from their polarity and the fact that hydrogen bonding between two carboxyl groups creates a dimer that behaves as a higher-molecular-weight compound.

A hydrogen-bonded dimer of acetic acid

Carboxylic acids are more soluble in water than are alcohols, ethers, aldehydes, and ketones of comparable molecular weight. This increased solubility is due to their strong association with water molecules by hydrogen bonding through both their carbonyl and hydroxyl groups. The first four aliphatic carboxylic acids (formic, acetic, propanoic, and butanoic) are infinitely soluble in water. As the size of the hydrocarbon chain increases relative to that of the carboxyl group, however, water solubility decreases. The solubility of hexanoic acid (six carbons) in water is 1.0 g/100 mL water.

FIGURE 10.1 Polarity of a carboxyl group.

$$\begin{array}{c} \text{O} \\ \| \\ \text{CH}_3\text{COH} \end{array}$$

Acetic acid

TABLE 10.2 Boiling Points and Solubilities in Water of Two Groups of Compounds of Comparable Molecular Weight

Structure	Name	Molecular Weight	Boiling Point (°C)	Solubility (g/100 mL H_2O)
CH_3COOH	acetic acid	60.1	118	infinite
$CH_3CH_2CH_2OH$	1-propanol	60.1	97	infinite
CH_3CH_2CHO	propanal	58.1	48	16
$CH_3(CH_2)_2COOH$	butanoic acid	88.1	163	infinite
$CH_3(CH_2)_3CH_2OH$	1-pentanol	88.1	137	2.3
$CH_3(CH_2)_3CHO$	pentanal	86.1	103	slight

We must mention two other properties of carboxylic acids. First, the liquid carboxylic acids from propanoic acid to decanoic acid have sharp, often disagreeable odors. Butanoic acid is found in stale perspiration and is a major component of "locker room odor." Pentanoic acid smells even worse, and goats, which secrete C_6, C_8, and C_{10} carboxylic acids (Table 10.1), are not famous for their pleasant odors. Second, carboxylic acids have a characteristic sour taste. The sour taste of pickles and sauerkraut, for example, is due to the presence of lactic acid. The sour tastes of limes (pH 1.9), lemons (pH 2.3), and grapefruit (pH 3.2) are due to the presence of citric and other acids.

10.4 What Are Soaps and Detergents?

A. Fatty Acids

More than 500 different **fatty acids** have been isolated from various cells and tissues. Given in Table 10.3 are common names and structural formulas for the most abundant fatty acids. The number of carbons in a fatty acid and the number of carbon–carbon double bonds in its hydrocarbon chain are shown by two numbers separated by a colon. In this notation, linoleic acid, for

Fatty acids are long, unbranched-chain carboxylic acids, most commonly consisting of 12 to 20 carbons. They are derived from the hydrolysis of animal fats, vegetable oils, and the phospholipids of biological membranes (Chapter 13).

TABLE 10.3 The Most Abundant Fatty Acids in Animal Fats, Vegetable Oils, and Biological Membranes

Carbon Atoms: Double Bonds*	Structure	Common Name	Melting Point (°C)
Saturated Fatty Acids			
12:0	$CH_3(CH_2)_{10}COOH$	lauric acid	44
14:0	$CH_3(CH_2)_{12}COOH$	myristic acid	58
16:0	$CH_3(CH_2)_{14}COOH$	palmitic acid	63
18:0	$CH_3(CH_2)_{16}COOH$	stearic acid	70
20:0	$CH_3(CH_2)_{18}COOH$	arachidic acid	77
Unsaturated Fatty Acids			
16:1	$CH_3(CH_2)_5CH{=}CH(CH_2)_7COOH$	palmitoleic acid	1
18:1	$CH_3(CH_2)_7CH{=}CH(CH_2)_7COOH$	oleic acid	16
18:2	$CH_3(CH_2)_4(CH{=}CHCH_2)_2(CH_2)_6COOH$	linoleic acid	−5
18:3	$CH_3CH_2(CH{=}CHCH_2)_3(CH_2)_6COOH$	linolenic acid	−11
20:4	$CH_3(CH_2)_4(CH{=}CHCH_2)_4(CH_2)_2COOH$	arachidonic acid	−49

*The first number is the number of carbons in the fatty acid; the second number is the number of carbon–carbon double bonds in its hydrocarbon chain.

example, is designated as an 18:2 fatty acid; its 18-carbon chain contains two carbon–carbon double bonds.

Following are several characteristics of the most abundant fatty acids in higher plants and animals:

1. Nearly all fatty acids have an even number of carbon atoms, most between 12 and 20, in an unbranched chain.

2. The three most abundant fatty acids in nature are palmitic acid (16:0), stearic acid (18:0), and oleic acid (18:1).

3. In most unsaturated fatty acids, the *cis* isomer predominates; the *trans* isomer is rare.

4. Unsaturated fatty acids have lower melting points than their saturated counterparts. The greater the degree of unsaturation, the lower the melting point. Compare, for example, the melting points of the following 18-carbon fatty acids: Linolenic acid, with three carbon–carbon double bonds, has the lowest melting point of the four fatty acids.

Stearic acid (18:0) (mp 70°C)

Oleic acid (18:1) (mp 16°C)

Linoleic acid (18:2) (mp –5°C)

Linolenic acid (18:3) (mp –11°C)

Fatty acids can be divided into two groups: saturated and unsaturated. Saturated fatty acids have only carbon–carbon single bonds in their hydrocarbon chains. Unsaturated fatty acids have at least one C=C double bond in the chain. All unsaturated fatty acids listed in Table 10.3 are the *cis* isomer.

Saturated fatty acids are solids at room temperature, because the regular nature of their hydrocarbon chains allows their molecules to pack together in close parallel alignment. When packed in this manner, the attractive interactions between adjacent hydrocarbon chains (London dispersion forces) are maximized. Although London dispersion forces are weak interactions, the regular packing of hydrocarbon chains allows these forces to operate over a large portion of their chains, ensuring that a considerable amount of energy is needed to separate and melt them.

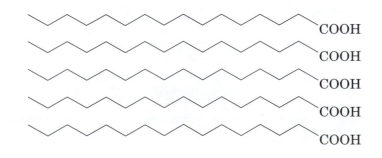

Chemical Connections 10A

Trans Fatty Acids: What Are They and How Do You Avoid Them?

Animal fats are rich in saturated fatty acids, whereas plant oils (for example, corn, soybean, canola, olive, and palm oils) are rich in unsaturated fatty acids. Fats are added to processed foods to provide a desirable firmness along with a moist texture and pleasant taste. To supply the demand for dietary fats of the appropriate consistency, the *cis* double bonds of vegetable oils are partially hydrogenated. The greater the extent of hydrogenation, the higher the melting point of the triglyceride. The extent of hydrogenation is carefully controlled, usually by employing a Ni catalyst and a calculated amount of H_2 as a limiting reagent. Under these conditions, the H_2 is used up before all double bonds are reduced, so that only partial hydrogenation and the desired overall consistency is achieved. For example, by controlling the degree of hydrogenation, an oil with a melting point below room temperature can be converted to a semisolid or even a solid product.

The mechanism of catalytic hydrogenation of alkenes was discussed in Section 3.6D. Recall that a key step in this mechanism involves interaction of the carbon–carbon double bond of the alkene with the metal catalyst to form a carbon–metal bond. Because the interaction of a carbon–carbon double bond with the Ni catalyst is reversible, many of the double bonds remaining in the oil may be isomerized from the less stable *cis* configuration to the more stable *trans* configuration. Thus the equilibration between the *cis* and *trans* configurations may occur when H_2 is the limiting reagent. For example, elaidic acid is the *trans* C_{18} fatty acid analog of oleic acid, a common C_{18} *cis* fatty acid.

The oils used for frying in fast-food restaurants are usually partially hydrogenated plant oils and, therefore, they contain substantial amounts of *trans* fatty acids that are transferred to the foods cooked in them. Other major sources of *trans* fatty acids in the diet include stick margarine, certain commercial bakery products, creme-filled cookies, potato and corn chips, frozen breakfast foods, and cake mixes.

Recent studies have shown that consuming significant amounts of *trans* fatty acids can lead to serious health problems related to serum cholesterol levels. Low overall serum cholesterol and a decreased ratio of low-density lipoprotein (LDL) cholesterol to high-density lipoprotein (HDL) cholesterol are associated with good cardiovascular health. High serum cholesterol levels and an elevated ratio of LDL cholesterol to HDL cholesterol are linked to a high incidence of cardiovascular disease, especially atherosclerosis. Research has indicated that consuming a diet high in either saturated fatty acids or *trans* fatty acids substantially increases the risk of cardiovascular disease.

The FDA has announced that processed foods must list the amount of *trans* fatty acids they contain, so that consumers can make better choices about the foods they eat. A diet low in saturated and *trans* fatty acids is recommended, along with consumption of more fish, whole grains, fruits, and vegetables, and especially daily exercise, which is tremendously beneficial regardless of diet.

Monounsaturated and polyunsaturated fatty acids have not produced similar health risks in most studies, although too much fat of any kind in the diet can lead to obesity, a major health problem that is associated with several diseases, one of which is diabetes. Some polyunsaturated (*cis*) fatty acids, such as those found in certain types of fish, have even been shown to have beneficial effects in some studies. These are the so-called omega-3 fatty acids.

In omega-3 fatty acids, the last carbon of the last double bond of the hydrocarbon chain ends three carbons in from the methyl terminal end of the chain. The last carbon of the hydrocarbon chain is called the omega (the last letter of the Greek alphabet) carbon—hence the designation of omega-3. The two most commonly found in health food supplements are eicosapentaenoic acid and docosahexaenoic acid.

Eicosapentaenoic acid, $C_{20}H_{30}O_2$, is an important fatty acid in the marine food chain and serves as a precursor in humans of several members of the prostacyclin and thromboxane families (Chapter 13). Note how the name of this fatty acid is derived. *Eicosa-* is the prefix indicating 20 carbons in the chain: *-pentaene-* indicates five carbon–carbon double bonds, and *–oic acid* shows the carboxyl functional group.

Docosahexaenoic acid, $C_{22}H_{32}O_2$, is found in fish oils and many phospholipids. It is a major structural component of excitable membranes in the retina and brain, and is synthesized in the liver from linoleic acid.

Elaidic acid
mp 46°C
(a *trans* C_{18} fatty acid)

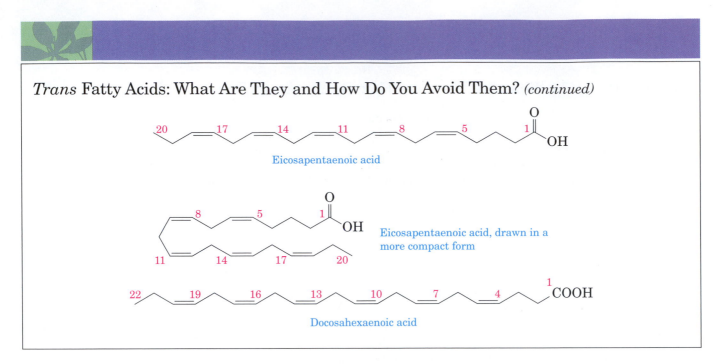

Trans Fatty Acids: What Are They and How Do You Avoid Them? *(continued)*

Eicosapentaenoic acid

Eicosapentaenoic acid, drawn in a more compact form

Docosahexaenoic acid

All common *cis* unsaturated fatty acids are liquids at room temperature because the *cis* double bonds interrupt the regular packing of the chains and the London dispersion forces act over only shorter segments of the chains, so that less energy is required to melt them. The greater the degree of unsaturation, the lower the melting point, because each double bond introduces more disorder into the packing of the fatty acid molecules.

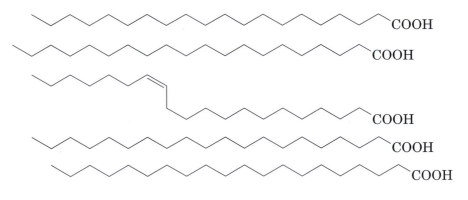

B. Structure and Preparation of Soaps

Natural **soaps** are most commonly prepared from a blend of tallow and coconut oils. In the preparation of tallow, the solid fats of cattle are melted with steam, and the tallow layer that forms on the top is removed. The preparation of soaps begins by boiling these triglycerides with sodium hydroxide. The reaction that takes place is called **saponification** (Latin: *saponem*, "soap"):

A triglyceride   1,2,3-Propanetriol   Sodium soaps
(Glycerol; Glycerin)

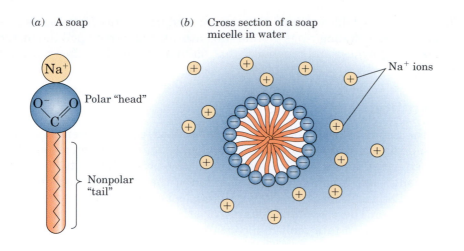

(a) A soap

(b) Cross section of a soap micelle in water

Na⁺ ions

Polar "head"

Nonpolar "tail"

FIGURE 10.2 Soap micelles. Nonpolar (hydrophobic) hydrocarbon chains cluster in the interior of the micelle and polar (hydrophilic) carboxylate groups are on the surface of the micelle. Soap micelles repel each other because of their negative surface charges.

At the molecular level, saponification corresponds to base-promoted hydrolysis (Section 11.4A) of the ester groups in triglycerides. A triglyceride is a triester of glycerol. The resulting soaps contain mainly the sodium salts of palmitic, stearic, and oleic acids from tallow and the sodium salts of lauric and myristic acids from coconut oil.

After hydrolysis is complete, sodium chloride is added to precipitate the sodium salts as thick curds of soap. The water layer is then drawn off, and glycerol is recovered by vacuum distillation. The crude soap contains sodium chloride, sodium hydroxide, and other impurities that are removed by boiling the curd in water and reprecipitating with more sodium chloride. After several purifications, the soap can be used as an inexpensive industrial soap without further processing. Other treatments transform the crude soap into pH-controlled cosmetic soaps, medicated soaps, and the like.

C. How Soap Cleans

Soap owes its remarkable cleansing properties to its ability to act as an emulsifying agent. Because the long hydrocarbon chains of natural soaps are insoluble in water, soap molecules tend to cluster in such a way as to minimize contact of their hydrocarbon chains with surrounding water molecules. The polar carboxylate groups, by contrast, tend to remain in contact with the surrounding water molecules. Thus, in water, soap molecules spontaneously cluster into **micelles** (Figure 10.2).

Many of the things we commonly think of as dirt (such as grease, oil, and fat stains) are nonpolar and insoluble in water. When soap and this type of dirt are mixed together, as in a washing machine, the nonpolar hydrocarbon inner parts of the soap micelles "dissolve" the nonpolar substances. In effect, new soap micelles form, with the nonpolar dirt molecules in the center (Figure 10.3). In this way, nonpolar organic grease, oil, and fat are "dissolved" and washed away in the polar wash water.

Soaps, however, have their disadvantages, foremost among which is the fact that they form water-insoluble salts when used in water containing Ca(II), Mg(II), or Fe(III) ions (hard water):

Micelle A spherical arrangement of molecules in aqueous solution such that their hydrophobic (water-hating) parts are shielded from the aqueous environment and their hydrophilic (water-loving) parts are on the surface of the sphere and in contact with the aqueous environment.

Soap micelle with "dissolved" grease

Grease

Soap

FIGURE 10.3 A soap micelle with a "dissolved" oil or grease droplet.

$$2CH_3(CH_2)_{14}COO^-Na^+ + Ca^{2+} \longrightarrow [CH_3(CH_2)_{14}COO^-]_2Ca^{2+} + 2Na^+$$

A sodium soap (soluble in water as micelles)

Calcium salt of a fatty acid (insoluble in water)

These water-insoluble calcium, magnesium, and iron salts of fatty acids create problems, including rings around the bathtub, films that spoil the luster of hair, and grayness and roughness that build up on textiles after repeated washings.

D. Synthetic Detergents

After the cleansing action of soaps was understood, chemists were in a position to design a synthetic detergent. Molecules of a good detergent, they reasoned, must have a long hydrocarbon chain—preferably 12 to 20 carbon atoms long—and a polar group at one end of the molecule that does not form insoluble salts with the Ca(II), Mg(II), or Fe(III) ions that are present in hard water. These essential characteristics of a soap, they recognized, could be produced in a molecule containing a sulfonate ($-SO_3^-$) group instead of a carboxylate ($-COO^-$) group. Calcium, magnesium, and iron salts of alkylsulfonic acids ($R-SO_3H$) are much more soluble in water than comparable salts of fatty acids.

The most widely used synthetic detergents today are the linear alkylbenzenesulfonates (LAS). One of the most common of these is sodium 4-dodecylbenzenesulfonate. To prepare this type of detergent, a linear alkylbenzene is treated with sulfuric acid to form an alkylbenzenesulfonic acid (Section 4.3C), followed by neutralization of the sulfonic acid with sodium hydroxide:

Dodecylbenzene

Sodium 4-dodecylbenzenesulfonate
(an anionic detergent)

The product is mixed with builders and then spray-dried to give a smooth-flowing powder. The most common builder is sodium silicate. Alkylbenzenesulfonate detergents were introduced in the late 1950s, and today they account for close to 90% of the market that was once held by natural soaps.

Among the most common additives to detergent preparations are foam stabilizers, bleaches, and optical brighteners. A common foam stabilizer added to liquid soaps, but not to laundry detergents (for obvious reasons: think of a top-loading washing machine with foam spewing out of the lid!), is the amide prepared from dodecanoic acid (lauric acid) and 2-aminoethanol (ethanolamine). The most common bleach is sodium perborate tetrahydrate, which decomposes at temperatures higher than 50°C to give hydrogen peroxide, the actual bleaching agent.

N-(2-Hydroxyethyl)dodecanamide
(a foam stabilizer)

$O=B-O-O^-$ Na^+ • $4H_2O$

Sodium perborate tetrahydrate
(a bleach)

Also added to laundry detergents are optical brighteners (optical bleaches). These substances are absorbed into fabrics and, after absorbing ambient light, fluoresce with a blue color, offsetting the yellow color that develops in fabric as it ages. Optical brighteners produce a "whiter-than-white" appearance. You most certainly have observed their effects if you have seen the glow of white T-shirts or blouses when they are exposed to black light (UV radiation).

10.5 What Are the Characteristic Reactions of Carboxylic Acids?

A. Acidity

Carboxylic acids are weak acids. Values of K_a for most unsubstituted aliphatic and aromatic carboxylic acids fall within the range of 10^{-4} to 10^{-5} ($pK_a = 4.0 - 5.0$). The value of K_a for acetic acid, for example, is 1.74×10^{-5}; its pK_a is 4.76 (Section 7.5).

$$CH_3\overset{\displaystyle O}{\overset{\|}{C}}OH + H_2O \rightleftharpoons CH_3\overset{\displaystyle O}{\overset{\|}{C}}O^- + H_3O^+ \qquad K_a = \frac{[CH_3COO^-][H_3O^+]}{[CH_3COOH]} = 1.74 \times 10^{-5}$$

$$pK_a = 4.76$$

Substituents of high electronegativity (especially —OH, —Cl, and —NH$_3^+$) near the carboxyl group increase the acidity of carboxylic acids, often by several orders of magnitude. Compare, for example, the acidities of acetic acid and the chlorine-substituted acetic acids. Both dichloroacetic acid and trichloroacetic acid are stronger acids than acetic acid (pK_a 4.75) and H_3PO_4 (pK_a 2.12).

Dichloroacetic acid is used as a topical astringent and as a treatment for genital warts in males.

Dentists use a 50% aqueous solution of trichloroacetic acid to cauterize gums. This strong acid stops the bleeding, kills diseased tissue, and allows the growth of healthy gum tissue.

Formula:	CH_3COOH	$ClCH_2COOH$	$Cl_2CHCOOH$	Cl_3CCOOH
Name:	Acetic acid	Chloroacetic acid	Dichloroacetic acid	Trichloroacetic acid
pK_a:	4.76	2.86	1.48	0.70

Increasing acid strength →

Electronegative atoms on the carbon adjacent to a carboxyl group increase acidity because they pull electron density away from the O—H bond, thereby facilitating ionization of the carboxyl group and making it a stronger acid.

One final point about carboxylic acids: When a carboxylic acid is dissolved in an aqueous solution, the form of the carboxylic acid present depends on the pH of the solution in which it is dissolved. Consider typical carboxylic acids, which have pK_a values in the range of 4.0 to 5.0. When the pH of the solution is equal to the pK_a of the carboxylic acid (that is, when the pH of the solution is in the range $4.0 - 5.0$), the acid, RCOOH, and its conjugate base, RCOO$^-$, are present in equal concentrations, which we can demonstrate by using the Henderson-Hasselbalch equation (Section 7.11).

$$pH = pK_a + \log\frac{[A^-]}{[HA]} \qquad \text{Henderson-Hasselbalch Equation}$$

Consider the ionization of a weak acid, HA in aqueous solution. When the pH of the solution is equal to the pK_a of the carboxylic acid, the Henderson-Hasselbalch equation reduces to

$$\log\frac{[A^-]}{[HA]} = 0$$

Taking the antilog gives us the ratio of [A$^-$] to [HA], and tells us that the concentrations the two are equal.

$$\frac{[A^-]}{[HA]} = 1$$

If the pH of the solution is adjusted to 2.0 or lower by the addition of a strong acid, the carboxylic acid then is present in solution almost entirely as RCOOH. If the pH of the solution is adjusted to 7.0 or higher, the carboxylic acid is present almost entirely as its anion. Thus, even in a neutral solution (pH 7.0), a carboxylic acid is present predominantly as its anion.

Predominant species when pH of the solution is ≤2.0   Present in equal concentrations when pH of the solution = pK_a of the acid   Predominant species when pH of the solution is ≥7.0

B. Reaction with Bases

All carboxylic acids, whether soluble or insoluble in water, react with NaOH, KOH, and other strong bases to form water-soluble salts.

Benzoic acid (slightly soluble in water)   Sodium benzoate (60 g/100 mL water)

Sodium benzoate, a fungal growth inhibitor, is often added to baked goods "to retard spoilage." Calcium propanoate is used for the same purpose. Carboxylic acids also form water-soluble salts with ammonia and amines.

Benzoic acid (slightly soluble in water)   Ammonium benzoate (20 g/100 mL water)

Carboxylic acids react with sodium bicarbonate and sodium carbonate to form water-soluble sodium salts and carbonic acid (a weaker acid). Carbonic acid, in turn, decomposes to give water and carbon dioxide, which evolves as a gas (Section 7.6E).

$$CH_3COOH(aq) + NaHCO_3(aq) \longrightarrow CH_3COO^-Na^+(aq) + CO_2(g) + H_2O(l)$$

Salts of carboxylic acids are named in the same manner as the salts of inorganic acids: The cation is named first and then the anion. The name of the anion is derived from the name of the carboxylic acid by dropping the suffix *-ic acid* and adding the suffix *-ate*.

Sodium benzoate and calcium propanoate are fungal growth inhibitors, and are added to baked goods "to retard spoilage."

Example 10.2 Acidity of Carboxylic Acids

Complete each acid–base reaction and name the carboxylate salt formed.

(a) ＼／＼COOH + NaOH ⟶   (b) with OH ⟶ COOH + NaHCO₃ ⟶
(s)-Lactic acid

Strategy and Solution

Each carboxylic acid is converted to its sodium salt. In (b), carbonic acid is formed, and decomposes to carbon dioxide and water.

Problem 10.2

Write equations for the reaction of each acid in Example 10.2 with ammonia and name the carboxylate salt formed.

A consequence of the water solubility of carboxylic acid salts is that water-insoluble carboxylic acids can be converted to water-soluble ammonium or alkali metal salts and extracted into aqueous solution. The salt, in turn, can be transformed back to the free carboxylic acid by treatment with HCl, H_2SO_4, or another strong acid. These reactions allow an easy separation of water-insoluble carboxylic acids from water-insoluble nonacidic compounds.

Shown in Figure 10.4 is a flowchart for the separation of benzoic acid, a water-insoluble carboxylic acid, from benzyl alcohol, a nonacidic compound.

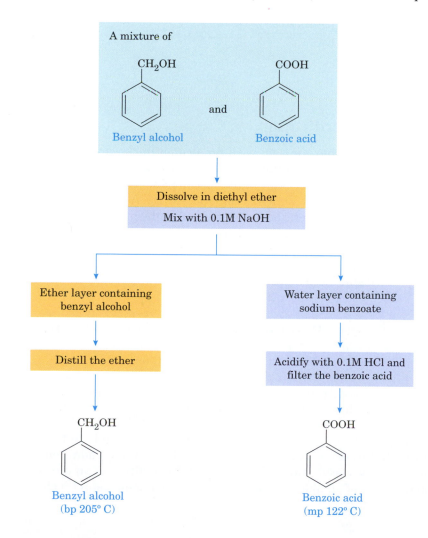

FIGURE 10.4 Flow chart for separation of benzoic acid from benzyl alcohol.

First, the mixture of benzoic acid and benzyl alcohol is dissolved in diethyl ether. When the ether solution is shaken with aqueous NaOH or another strong base, benzoic acid is converted to its water-soluble sodium salt. Then the ether and aqueous phases are separated. The ether solution is distilled, yielding first diethyl ether (bp 35°C) and then benzyl alcohol (bp 205°C). The aqueous solution is acidified with HCl, and benzoic acid precipitates as a white, crystalline solid (mp 122°C), which is recovered by filtration.

C. Reduction

The carboxyl group is one of the organic functional groups that is most resistant to reduction. It is not affected by catalytic reduction under conditions that readily reduce alkenes to alkanes (Section 3.6D) or by sodium borohydride, ($NaBH_4$), which readily reduces aldehydes to 1° alcohols and ketones to 2° alcohols (Section 9.4B).

The most common reagent for the reduction of a carboxylic acid to a 1° alcohol is the very powerful reducing agent lithium aluminum hydride, $LiAlH_4$. Reduction of a carboxyl group with this reagent is commonly carried out in diethyl ether. The initial product is an aluminum alkoxide, which is then treated with water to give the primary alcohol and lithium and aluminum hydroxides. These two hydroxides are insoluble in diethyl ether and are removed by filtration. Evaporation of the ether solvent yields the primary alcohol.

$$\underset{\substack{\text{3-Cyclopentene-}\\\text{carboxylic acid}}}{\ce{COH}} \xrightarrow[\text{2. H2O}]{\text{1. LiAlH4, ether}} \underset{\substack{\text{4-Hydroxymethyl-}\\\text{cyclopentene}}}{\ce{CH2OH}} + \text{LiOH} + \text{Al(OH)}_3$$

D. Fischer Esterification

Ester A compound in which the —OH of a carboxyl group, RCOOH, is replaced by an alkoxy or aryloxy group

Fischer esterification The process of forming an ester by refluxing a carboxylic acid and an alcohol in the presence of an acid catalyst, commonly sulfuric acid

Treatment of a carboxylic acid with an alcohol in the presence of an acid catalyst—most commonly, concentrated sulfuric acid—gives an **ester.** This method of forming an ester is given the special name **Fischer esterification,** after the German chemist Emil Fischer (1852–1919). As an example of Fischer esterification, treating acetic acid with ethanol in the presence of concentrated sulfuric acid gives ethyl acetate and water:

$$\underset{\substack{\text{Ethanoic acid}\\\text{(Acetic acid)}}}{\ce{CH3C-OH}} + \underset{\substack{\text{Ethanol}\\\text{(Ethyl alcohol)}}}{\ce{H-OCH2CH3}} \xrightleftharpoons{\ce{H2SO4}} \underset{\substack{\text{Ethyl ethanoate}\\\text{(Ethyl acetate)}}}{\ce{CH3COCH2CH3}} + \ce{H2O}$$

We study the structure, nomenclature, and reactions of esters in detail in Chapter 11. In this chapter, we discuss only their preparation from carboxylic acids.

In the process of Fischer esterification, the alcohol adds to the carbonyl group of the carboxylic acid to form a tetrahedral carbonyl addition intermediate. Note how closely this step resembles the addition of an alcohol to the carbonyl group of an aldehyde or ketone to form a hemiacetal (Section 9.4C).

In the case of Fischer esterification, the intermediate loses a molecule of water to give the ester.

A tetrahedral carbonyl addition intermediate

Acid-catalyzed esterification is reversible, and, at equilibrium, the quantities of remaining carboxylic acid and alcohol are generally appreciable. By controlling the experimental conditions, however, we can use Fischer esterification to prepare esters in high yields. If the alcohol is inexpensive compared with the carboxylic acid, we can use a large excess of the alcohol (one of the starting reagents) to drive the equilibrium to the right and achieve a high conversion of carboxylic acid to its ester. Alternatively, we can remove water (one of the products of the reaction) as it is formed and drive the equilibrium to the right.

Chemical Connections 10B

Esters as Flavoring Agents

Flavoring agents are the largest class of food additives. At present, more than 1000 synthetic and natural flavors are available. The majority of these are concentrates or extracts from the material whose flavor is desired; they are often complex mixtures of from tens to hundreds of compounds. A number of ester flavoring agents are synthesized industrially. Many have flavors very close to the target flavor, and so adding only one or a few of them is sufficient to make ice cream, soft drinks, or candies taste natural. The table shows the structures of a few of the esters used as flavoring agents.

Structure	Name	Flavor
	Ethyl formate	Rum
	Isopentyl acetate	Banana
	Octyl acetate	Orange
	Methyl butanoate	Apple
	Ethyl butanoate	Pineapple
	Methyl 2-aminobenzoate (Methyl anthranilate)	Grape

Example 10.3 Fischer Esterification

Complete these Fischer esterification reactions (assume an excess of the alcohol). The stoichiometry of each reaction is indicated in the problem.

(a) Benzoic acid

(b) Butanedioic acid (Succinic acid) (excess)

Strategy and Solution

Substitution of the —OR group of the alcohol for the —OH group of the carboxylic acid gives an ester. Here are the structural formulas and names for the ester produced in each reaction.

(a) Methyl benzoate

(b) Diethyl butanedioate (Diethyl succinate)

Problem 10.3

Complete these Fischer esterification reactions:

(b) (a cyclic ester)

E. Decarboxylation

Decarboxylation is the loss of CO_2 from a carboxyl group. Almost any carboxylic acid, when heated to a very high temperature, undergoes thermal decarboxylation:

Most carboxylic acids, however, are quite resistant to moderate heat and melt or even boil without decarboxylation. Exceptions are carboxylic acids that have a carbonyl group β to the carboxyl group. This type of carboxylic acid undergoes decarboxylation quite readily on mild heating. For example, when 3-oxobutanoic acid (acetoacetic acid) is heated moderately, it undergoes decarboxylation to give acetone and carbon dioxide:

3-Oxobutanoic acid
(Acetoacetic acid)
(a β-ketoacid)

Acetone

Decarboxylation on moderate heating is a unique property of β-ketoacids and is not observed with other classes of ketoacids.

Mechanism: Decarboxylation of a β-Ketocarboxylic Acid

Step 1: Redistribution of six electrons in a cyclic six-membered transition state gives carbon dioxide and an enol:

Cyclic six-membered
transition state

Step 2: Keto-enol tautomerism (Section 9.5) of the enol gives the more stable keto form of the product:

An important example of decarboxylation of a β-ketoacid in the biological world occurs during the oxidation of foodstuffs in the tricarboxylic acid (TCA) cycle (Chapter 20). Oxalosuccinic acid, one of the intermediates in this cycle, undergoes spontaneous decarboxylation to produce α-ketoglutaric acid. Only one of the three carboxyl groups of oxalosuccinic acid has a carbonyl group in the position β to it, and this carboxyl group is lost as CO_2:

Only this carboxyl
has a C=O beta to it

Oxalosuccinic acid

α-Ketoglutaric acid

Note that thermal decarboxylation is a reaction unique to β-ketoacids—it does not occur with α-ketoacids. In the biochemistry chapters that follow, however, we will see examples of decarboxylation of α-ketoacids—as for example, the decarboxylation of α-ketoglutarate. Because the decarboxylation of α-ketoacids requires an oxidizing agent (NAD^+), this reaction is called oxidative decarboxylation.

Chemical Connections 10C

Ketone Bodies and Diabetes

3-Oxobutanoic acid (acetoacetic acid) and its reduction product, 3-hydroxybutanoic acid, are synthesized in the liver from acetyl-CoA (Section 20.5), a product of the metabolism of fatty acids and certain amino acids.

3-Oxobutanoic acid
(Acetoacetic acid)

3-Hydroxybutanoic acid
(β-Hydroxybutyric acid)

3-Oxobutanoic acid and 3-hydroxybutanoic acid are known collectively as ketone bodies.

The concentration of ketone bodies in the blood of healthy, well-fed humans is approximately 0.01 mM/L. However, in persons suffering from starvation or diabetes mellitus, the concentration of ketone bodies may increase to as much as 500 times the normal level. Under these conditions, the concentration of acetoacetic acid increases to the point where it undergoes spontaneous decarboxylation to form acetone and carbon dioxide. Acetone is not metabolized by humans and is excreted through the kidneys and the lungs. The odor of acetone is responsible for the characteristic "sweet smell" on the breath of severely diabetic patients.

Summary of Key Questions

OWL End-of-chapter problems identified in blue are assignable in **GOB OWL**.

Section 10.1 What Are Carboxylic Acids?

• The functional group of a **carboxylic acid** is the **carboxyl group,** — COOH.

Section 10.2 How Do We Name Carboxylic Acids? Problem 10.8

• IUPAC names of carboxylic acids are derived from the name of the parent alkane by dropping the suffix -e and adding -oic acid.
• Dicarboxylic acids are named as -dioic acids.
• Common names for many carboxylic and dicarboxylic acids are still widely used.

Section 10.3 What Are the Physical Properties of Carboxylic Acids? Problem 10.19

• Carboxylic acids are polar compounds. Consequently, they have higher boiling points and are more soluble in water than alcohols, aldehydes, ketones, and ethers of comparable molecular weight.

Section 10.4 What Are Soaps and Detergents?

• Fatty acids are long, unbranched-chain carboxylic acids. They can be saturated or unsaturated.
• A **triglyceride** is a triester of glycerol.
• A **micelle** is a spherical arrangement of molecules in an aqueous environment in which the hydrocarbon parts are on the inside and the hydrophilic parts are on the surface.

Section 10.5 What Are the Characteristic Reactions of Carboxylic Acids? Problem 10.36

• Carboxylic acids are weak acids, which react with strong bases to form water-soluble salts.
• Treatment of a carboxylic acid with an alcohol in the presence of an acid catalyst gives an ester.
• When exposed to a very high temperature, carboxylic acids can undergo decarboxylation.

Summary of Key Reactions

1. **Acidity of Carboxylic Acids (Section 10.5A)** Values of pK_a for most unsubstituted carboxylic acids are within the range of 4 to 5.

$$CH_3COH + H_2O \rightleftharpoons CH_3CO^- + H_3O^+$$

$$K_a = \frac{[CH_3COO^-][H_3O^+]}{[CH_3COOH]} = 1.74 \times 10^{-5}$$

$$pK_a = 4.76$$

2. **Reaction of Carboxylic Acids with Bases (Section 10.5B)** Carboxylic acids, whether water-soluble or insoluble, react with alkali metal hydroxides, carbonates and bicarbonates, and ammonia and amines to form water-soluble salts.

$$\text{C}_6\text{H}_5{-}\text{COOH} + \text{NaOH} \xrightarrow{\text{H}_2\text{O}}$$

Benzoic acid
(slightly soluble in water)

$$\text{C}_6\text{H}_5{-}\text{COO}^-\text{Na}^+ + \text{H}_2\text{O}$$

Sodium benzoate
(60 g/100 mL water)

$$\text{CH}_3\text{COOH} + \text{NaHCO}_3 \longrightarrow$$

$$\text{CH}_3\text{COO}^-\text{Na}^+ + \text{CO}_2 + \text{H}_2\text{O}$$

3. **Reduction by Lithium Aluminum Hydride (Section 10.5C)** Lithium aluminum hydride reduces a carboxyl group to a primary alcohol. This reagent does not normally reduce carbon–carbon double bonds, but it does reduce aldehydes to 1° alcohols and ketones to 2° alcohols.

3-Cyclopentene-
carboxylic acid

$$\xrightarrow[\text{2. H}_2\text{O}]{\text{1. LiAlH}_4,\ \text{ether}}$$

4-Hydroxymethyl-
cyclopentene

$$\text{CH}_2\text{OH} + \text{LiOH} + \text{Al(OH)}_3$$

4. **Fischer Esterification (Section 10.5D)** Fischer esterification is reversible.

$$\text{CH}_3\overset{\text{O}}{\overset{\|}{\text{C}}}\text{OH} + \text{CH}_3\text{CH}_2\text{OH} \underset{}{\overset{\text{H}_2\text{SO}_4}{\rightleftharpoons}}$$

Ethanoic acid
(Acetic acid)   Ethanol
(Ethyl alcohol)

$$\text{CH}_3\overset{\text{O}}{\overset{\|}{\text{C}}}\text{OCH}_2\text{CH}_3 + \text{H}_2\text{O}$$

Ethyl ethanoate
(Ethyl acetate)

One way to force the equilibrium to the right is to use an excess of the alcohol. Alternatively, water can be removed from the reaction mixture as it is formed.

5. **Decarboxylation (Section 10.5E)** Thermal decarboxylation is a unique property of β-ketoacids. The immediate products of thermal decarboxylation of β-ketoacids are carbon dioxide and an enol. Loss of CO_2 is followed immediately by keto-enol tautomerism.

$$\xrightarrow{\text{warm}}$$

3-Oxobutanoic acid
(Acetoacetic acid)
(a β-ketoacid)   Acetone   + CO_2

Problems

■ Indicates problems that are assignable in **GOB OWL**.

Blue numbered problems are applied.

Go to this book's companion website at **www.cengage. com/chemistry/bettelheim** for interactive versions of the How To tutorials and Active Figures, and to quiz yourself on this chapter.

> For preparation of carboxylic acids, review Chapters 5, Alcohols, Esters, and Thiols, and 9, Aldehydes and Ketones

Section 10.2 How Do We Name Carboxylic Acids?

10.4 Name and draw structural formulas for the four carboxylic acids with molecular formula $C_5H_{10}O_2$. Which of these carboxylic acids are chiral?

10.5 ■ Write the IUPAC name for each carboxylic acid.

(a)   (b)

(c)

10.6 Write the IUPAC name for each carboxylic acid.

(a) HOOC   COOH (with OH group)   (b)   COOH / OH (benzene ring)

(c) CCl₃COOH

10.7 Draw a structural formula for each carboxylic acid.
(a) 4-Nitrophenylacetic acid
(b) 4-Aminobutanoic acid
(c) 4-Phenylbutanoic acid
(d) *cis*-3-Hexenedioic acid

10.8 ■ Draw a structural formula for each carboxylic acid.
(a) 2-Aminopropanoic acid
(b) 3,5-Dinitrobenzoic acid
(c) Dichloroacetic acid
(d) *o*-Aminobenzoic acid

10.9 Draw a structural formula for each salt.
(a) Sodium benzoate   (b) Lithium acetate
(c) Ammonium acetate   (d) Disodium adipate
(e) Sodium salicylate   (f) Calcium butanoate

10.10 Calcium oxalate is a major component of kidney stones. Draw a structural formula for this compound.

10.11 The monopotassium salt of oxalic acid is present in certain leafy vegetables, including rhubarb. Both oxalic acid and its salts are poisonous in high concentrations. Draw a structural formula for monopotassium oxalate.

Section 10.3 What Are the Physical Properties of Carboxylic Acids?

10.12 Answer true or false.
(a) Carboxylic acids are polar compounds.
(b) The most polar bond of a carboxyl group is the C—O single bond.
(c) Carboxylic acids have significantly higher boiling points than aldehydes, ketones, and alcohols of comparable molecular weight.
(d) The low-molecular-weight carboxylic acids (formic, acetic, propanoic, and butanoic acids) are infinitely soluble in water.

$$\underset{I}{HOCCH_2CH_2COH} < \underset{II}{HOCCH_2CH_2CH_3} <$$

$$\underset{III}{HOCCH_2CH_2CH_2CH_2CH_3} < \underset{IV}{HOCCH_2CH_2CH_2CH_2COH}$$

10.13 Draw a structural formula for the dimer formed when two molecules of formic acid interact by hydrogen bonding.

10.14 Propanedioic (malonic) acid forms an internal hydrogen bond in which the H of one COOH group forms a hydrogen bond with an O of the other COOH group. Draw a structural formula to show this internal hydrogen bonding. (There are two possible answers.)

10.15 Hexanoic (caproic) acid has a solubility in water of about 1 g/100 m L water. Which part of the molecule contributes to water solubility and which part prevents solubility?

10.16 Propanoic acid and methyl acetate are constitutional isomers, and both are liquids at room temperature. One of these compounds has a boiling point of 141°C; the other has a boiling point of 57°C. Which compound has which boiling point? Explain.

$$\underset{\text{Propanoic acid}}{CH_3—CH_2—\overset{O}{\overset{||}{C}}—OH} \qquad \underset{\text{Methyl acetate}}{CH_3—\overset{O}{\overset{||}{C}}—OCH_3}$$

10.17 The following compounds have approximately the same molecular weight: hexanoic acid, heptanal, and 1-heptanol. Arrange them in order of increasing boiling point.

10.18 The following compounds have approximately the same molecular weight: propanoic acid, 1-butanol, and diethyl ether. Arrange them in order of increasing boiling point.

10.19 ■ Arrange these compounds in order of increasing solubility in water: acetic acid, pentanoic acid, decanoic acid.

Section 10.4 What Are Soaps and Detergents?

10.20 Answer true or false.
(a) Fatty acids are long-chain carboxylic acids, with most consisting of between 12 to 20 carbons in an unbranched chain.
(b) An unsaturated fatty acid contains one or more carbon–carbon double bonds in its hydrocarbon chain.
(c) In most unsaturated fatty acids found in animal fats, vegetable oils, and biological membranes, the *cis* isomer predominates.
(d) In general, unsaturated fatty acids have lower melting points than saturated fatty acids with the same number of carbon atoms.
(e) Natural soaps are sodium or potassium salts of fatty acids.
(f) Soaps remove grease, oil, and fat stains by incorporating these substances into the nonpolar interior of soap micelles.

(g) "Hard water," by definition, is water that contains Ca^{2+}, Mg^{2+}, or Fe^{3+} ions, all of which react with soap molecules to form water-insoluble salts.

(h) The structure of synthetic detergents is patterned after that of natural soaps.

(i) The most widely used synthetic detergents are the linear alkylbenzenesulfonates (LAS).

(j) Present-day synthetic detergents do not form water-insoluble salts with hard water.

(k) Most detergent preparations contain foam stabilizers, a bleach, and optical brighteners (optical bleaches).

Section 10.5 What Are the Characteristic Reactions of Carboxylic Acids?

10.21 Answer true or false.

(a) Carboxylic acids are weak acids compared to mineral acids such as HCl, H_2SO_4, and HNO_3.

(b) Phenols, alcohols, and carboxylic acids have in common the presence of an —OH group.

(c) Carboxylic acids are stronger acids than alcohols but weaker acids than phenols.

(d) The order of acidity of the following carboxylic acids is:

(a)   >   (b)   >

(c)   >   (d)

(e) The reaction of benzoic acid with aqueous sodium hydroxide gives sodium benzoate.

(f) A mixture of the following compounds is extracted in order with (1) 1 M HCl, (2) 1 M NaOH, and (3) diethyl ether. Only compound II is extracted into the basic layer.

I   II   III

(g) The following ester can be prepared by treating benzoic acid with 1-butanol in the presence of a catalytic amount of H_2SO_4:

(h) Thermal decarboxylation of this β-ketoacid gives benzoic acid and carbon dioxide:

(i) Thermal decarboxylation of this β-ketoacid gives 2-pentanone and carbon dioxide.

$$CH_3CH_2CH_2\overset{O}{\overset{\|}{C}}CH_2\overset{O}{\overset{\|}{C}}OH$$

10.22 Alcohols, phenols, and carboxylic acids all contain an —OH group. Which are the strongest acids? Which are the weakest acids?

10.23 Arrange these compounds in order of increasing acidity: benzoic acid, benzyl alcohol, phenol.

10.24 Complete the equations for these acid–base reactions.

(a)   ⟨ring⟩—CH_2COOH + NaOH ⟶

(b)   ⟨chain⟩$COOH$ + $NaHCO_3$ ⟶

(c)   ⟨ring⟩ with $COOH$ and OCH_3 + $NaHCO_3$ ⟶

(d) $CH_3\overset{OH}{\overset{|}{C}H}COOH + H_2NCH_2CH_2OH$ ⟶

(e)   ⟨chain⟩COO^-Na^+ + HCl ⟶

10.25 Complete the equations for these acid–base reactions.

(a)   <image description: benzene ring with OH and CH_3> + NaOH ⟶

(b)

(c)

COOH + $H_2NCH_2CH_2OH$ ⟶

OCH$_3$

(d)

◯—COOH + $NaHCO_3$ ⟶

10.26 Formic acid is one of the components responsible for the sting of biting ants and is injected under the skin by bee and wasp stings. The pain can be relieved by rubbing the area of the sting with a paste of baking soda ($NaHCO_3$) and water that neutralizes the acid. Write an equation for this reaction.

10.27 Starting with the definition of K_a of a weak acid, HA, as

$$HA + H_2O \rightleftharpoons A^- + H_3O^+ \qquad K_a = \frac{[A^-][H_3O^+]}{[HA]}$$

show that

$$\frac{[A^-]}{[HA]} = \frac{K_a}{[H_3O^+]}$$

10.28 Using the equation from Problem 10.29 that shows the relationship between K_a, $[H_3O^+]$, $[A^-]$ and $[HA]$, calculate the ratio of $[A^-]$ to $[HA]$ in a solution whose pH is

(a) 2.0 (b) 5.0
(c) 7.0 (d) 9.0
(e) 11.0.

Assume that the pK_a of the weak acid is 5.0.

10.29 The normal pH range for blood plasma is 7.35 to 7.45. Under these conditions, would you expect the carboxyl group of lactic acid (pK_a 4.07) to exist primarily as a carboxyl group or as a carboxylate anion? Explain.

10.30 The pK_a of ascorbic acid (Chemical Connections 12B) is 4.10. Would you expect ascorbic acid dissolved in blood plasma, pH 7.35–7.45, to exist primarily as ascorbic acid or as ascorbate anion? Explain.

10.31 Complete the equations for the following acid–base reactions. Assume one mole of NaOH per mole of amino acid. (*Hint*: review Section 7.4.)

(a) $CH_3CHCOOH + NaOH \xrightarrow{H_2O}$
$\qquad\;\; |$
$\qquad\;\; NH_3{}^+$

(b) $CH_3CHCOO^-Na^+ + NaOH \xrightarrow{H_2O}$
$\qquad\;\; |$
$\qquad\;\; NH_3{}^+$

10.32 Which is the stronger base: $CH_3CH_2NH_2$ or $CH_3CH_2COO^-$? Explain.

10.33 Complete the equations for the following acid–base reactions. Assume one mole of HCl per mole of amino acid.

(a) $CH_3CHCOO^-Na^+ + HCl \xrightarrow{H_2O}$
$\qquad\;\; |$
$\qquad\;\; NH_2$

(b) $CH_3CHCOO^-Na^+ + HCl \xrightarrow{H_2O}$
$\qquad\;\; |$
$\qquad\;\; NH_3{}^+$

10.34 Define and give an example of Fischer esterification.

10.35 Complete these examples of Fischer esterification. In each case, assume an excess of the alcohol.

(a) $CH_3COOH + HO$‒ ⟍⟋ $\xrightarrow{H^+}$ ⇌

(b) $CH_3COOH + HO$‒◯ $\xrightarrow{H^+}$ ⇌

(c)

COOH

◯

COOH $+ CH_3CH_2OH \xrightarrow{H^+}$ ⇌

10.36 ■ From what carboxylic acid and alcohol is each ester derived?

(a)

$$CH_3\overset{\displaystyle O}{\overset{\|}{C}}O—◯—O\overset{\displaystyle O}{\overset{\|}{C}}CH_3$$

(b)

$$◯—\overset{\displaystyle O}{\overset{\|}{C}}OCH_3$$

(c)

$$CH_3O\overset{\displaystyle O}{\overset{\|}{C}}CH_2CH_2\overset{\displaystyle O}{\overset{\|}{C}}OCH_3$$

(d)

10.37 Methyl 2-hydroxybenzoate (methyl salicylate) has the odor of oil of wintergreen. This compound is prepared by Fischer esterification of 2-hydroxybenzoic acid (salicylic acid) with methanol. Draw a structural formula for methyl 2-hydroxybenzoate.

10.38 Show how you could convert cinnamic acid to each compound.

trans-3-Phenyl-2-propenoic acid
(Cinnamic acid)

Additional Problems

10.39 ■ Give the expected organic product formed when phenylacetic acid, $C_6H_5CH_2COOH$, is treated with each of the following reagents:

(a) $NaHCO_3$, H_2O (b) $NaOH$, H_2O

(c) NH_3, H_2O (d) $LiAlH_4$ then H_2O

(e) $NaBH_4$ then H_2O (f) $CH_3OH + H_2SO_4$ (catalyst)

(g) H_2/Ni

10.40 Methylparaben and propylparaben are used as preservatives in foods, beverages, and cosmetics.

4-Aminobenzoic acid
(*p*-Aminobenzoic acid)

Methyl 4-aminobenzoate
(Methylparaben)

Propyl 4-aminobenzoate
(Propylparaben)

Show how each of these preservatives can be prepared from 4-aminobenzoic acid.

10.41 ■ 4-Aminobenzoic acid is prepared from benzoic acid by the following two steps.

Benzoic acid

4-Nitrobenzoic acid

4-Aminobenzoic acid

Show reagents and experimental conditions to bring about each step.

Looking Ahead

10.42 When 5-hydroxypentanoic acid is treated with an acid catalyst, it forms a lactone (a cyclic ester). Draw a structural formula for this lactone.

10.43 We have seen that esters can be prepared by treatment of a carboxylic acid and an alcohol in the presence of an acid catalyst. Suppose you start instead with a dicarboxylic acid, such as 1,6-hexanedioic acid (adipic acid), and a diol, such as 1,2-ethanediol (ethylene glycol).

1,6-Hexanedioic acid
(Adipic acid)

1,2-Ethanediol
(Ethylene glycol)

A polyester

Show how Fischer esterification in this case can produce a polymer (a macromolecule with molecular weight several thousand times that of the starting materials).

10.44 Draw the structural formula of a compound of the given molecular formula that, on oxidation by potassium dichromate in aqueous sulfuric acid, gives the carboxylic acid or dicarboxylic acid shown.

(a) $C_6H_{14}O$ — oxidation →

(b) $C_6H_{12}O$ — oxidation →

(c) $C_6H_{14}O_2$ — oxidation →

Carboxylic Anhydrides, Esters, and Amides

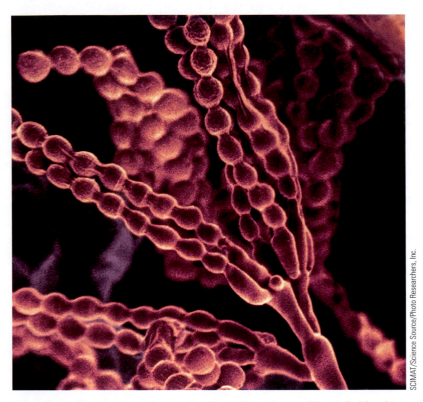

Colored scanning electron micrograph of *Penicillium* fungus. The stalk-like objects are condiophores to which are attached numerous round condia. The condia are the fruiting bodies of the fungus. See Chemical Connections 11B, "The Penicillins and Cephalosporins β-Lactam Antibiotics."

Key Questions

11.1 What Are Carboxylic Anhydrides, Esters, and Amides?

11.2 How Do We Prepare Esters?

11.3 How Do We Prepare Amides?

11.4 What Are the Characteristic Reactions of Anhydrides, Esters, and Amides?

11.5 What Are Phosphoric Anhydrides and Phosphoric Esters?

11.6 What Is Step-Growth Polymerization?

11.1 What Are Carboxylic Anhydrides, Esters, and Amides?

In Chapter 10, we studied the structure and preparation of esters, a class of organic compounds derived from carboxylic acids. In this chapter, we study anhydrides and amides, two more classes of carboxylic derivatives. Below, under the general formula of each carboxylic acid derivative, is a drawing to help you see how the functional group of each derivative is formally related to a carboxyl group. The loss of —OH from a carboxyl group and —H from

OWL Online homework for this chapter may be assigned in **GOB OWL**.

an alcohol, for example, gives an ester. Loss of —OH from a carboxyl group and —H from ammonia or an amine gives an amide.

$$\underset{\text{A carboxylic acid}}{\overset{\overset{\displaystyle O}{\parallel}}{RCOH}} \qquad \underset{\text{An anhydride}}{\overset{\overset{\displaystyle O\;\;\;O}{\parallel\;\;\parallel}}{RCOCR'}} \qquad \underset{\text{An ester}}{\overset{\overset{\displaystyle O}{\parallel}}{RCOR'}} \qquad \underset{\text{An amide}}{\overset{\overset{\displaystyle O}{\parallel}}{RCNH_2}}$$

Of these three carboxylic derivatives, anhydrides are so reactive that they are rarely found in nature. Esters and amides, however, are widespread in the biological world.

A. Anhydrides

The functional group of an **anhydride** consists of two carbonyl groups bonded to an oxygen atom. The anhydride may be symmetrical (from two identical **acyl groups**) or mixed (from two different acyl groups). To name anhydrides, we drop the word *acid* from the name of the carboxylic acid from which the anhydride is derived and add the word *anhydride*.

$$\underset{\text{Acetic anhydride}}{\overset{\overset{\displaystyle O\qquad\;\; O}{\parallel\qquad\;\;\parallel}}{CH_3C-O-CCH_3}}$$

B. Esters

The functional group of an ester is a carbonyl group bonded to an —OR group. Both IUPAC and common names of esters are derived from the names of the parent carboxylic acids (Chapter 10). The alkyl group bonded to oxygen is named first, followed by the name of the acid in which the suffix **-ic acid** is replaced by the suffix **-ate.**

$$\underset{\substack{\text{Ethyl ethanoate}\\ \text{(Ethyl acetate)}}}{\overset{\overset{\displaystyle O}{\parallel}}{CH_3COCH_2CH_3}} \qquad\qquad \underset{\substack{\text{Diethyl pentanedioate}\\ \text{(Diethyl glutarate)}}}{}$$

Recall that cyclic esters are called **lactones.**

C. Amides

The functional group of an **amide** is a carbonyl group bonded to a nitrogen atom. We name amides by dropping the suffix **-oic acid** from the IUPAC name of the parent acid, or **-ic acid** from its common name, and add **-amide.** If the nitrogen atom of the amide is bonded to an alkyl or aryl group, the group is named and its location on nitrogen is indicated by *N-.* Two alkyl groups are indicated by *N,N*-di-.

$$\underset{\substack{\text{Acetamide}\\ \text{(a 1° amide)}}}{\overset{\overset{\displaystyle O}{\parallel}}{CH_3CNH_2}} \qquad \underset{\substack{N\text{-Methylacetamide}\\ \text{(a 2° amide)}}}{\overset{\overset{\displaystyle O}{\parallel}}{CH_3CNHCH_3}} \qquad \underset{\substack{N,N\text{-Dimethylformamide}\\ \text{(a 3° amide)}}}{\overset{\overset{\displaystyle O}{\parallel}}{HCN(CH_3)_2}}$$

Chemical Connections 11A

The Pyrethrins; Natural Insecticides of Plant Origin

Pyrethrym is a natural insecticide obtained from the powdered flower heads of several species of *Chrysanthemum*, particularly *C. cinerariaefolium*. The active substances in pyrethrum, principally pyrethrins I and II, are contact poisons for insects and cold-blooded vertebrates. Because their concentrations in the pyrethrum powder used in *Chrysanthemum*-based insecticides are nontoxic to plants and higher animals, pyrethrum powder has found wide application in household and livestock sprays as well as in dusts for edible plants. Natural pyrethrins are esters of chrysanthemic acid.

While pyrethrum powders are effective insecticides, the active substances in them are destroyed rapidly in the environment. In an effort to develop synthetic compounds that are as effective as these natural insecticides but that offer greater biostability, chemists have prepared a series of esters related in structure to chrysanthemic acid. Permethrin is one of the most commonly used synthetic pyrethrin-like compounds in household and agricultural products.

Pyrethrin I

Permethrin

Cyclic amides are called **lactams.** following are structural formulas for a four-membered and a seven-membered lactam. A four-membered lactam is essential to the function of the penicillin and cephalosporin antibiotics (Chemical Connections 11B).

A four-membered lactam

(α β-lactam)

A seven-membered lactam

Example 11.1 IUPAC Names for Amides

Write the IUPAC name for each amide.

$$\text{(a)} \quad CH_3CH_2CH_2\overset{O}{\overset{\|}{C}}NH_2 \qquad \text{(b)}$$

Strategy and Solution

To name an amide, start with the systematic name of the corresponding carboxylic acid, drop the suffix **-oic acid** and replace it by **-amide.** Given is the IUPAC name and then, in parentheses, the common name.

(a) Butanamide (butyramide, from butyric acid)
(b) Hexanediamide (adipamide, from adipic acid)

Problem 11.1

Draw a structural formula for each amide.

(a) *N*-Cyclohexylacetamide
(b) Benzamide

Chemical Connections 11B

The Penicillins and Cephalosporins: β-Lactam Antibiotics

The **penicillins** were discovered in 1928 by the Scottish bacteriologist Sir Alexander Fleming. Thanks to the brilliant experimental work of Sir Howard Florey, an Australian pathologist, and Ernst Chain, a German chemist who fled Nazi Germany, penicillin G was introduced into the practice of medicine in 1943. For their pioneering work in developing one of the most effective antibiotics of all time, in 1945, Fleming, Florey, and Chain were awarded the Nobel Prize for physiology or medicine.

The mold from which Fleming discovered penicillin was *Penicillium notatum,* a strain that gives a relatively low yield of penicillin. It was replaced in commercial production of the antibiotic by *P. chrysogenum,* a strain cultured from a mold found growing on a grapefruit in a market in Peoria, Illinois. The structural feature common to all penicillins is the four-membered β-lactam ring, bonded to a five-membered, sulfur-containing ring. The penicillins owe their antibacterial activity to a common mechanism that inhibits the biosynthesis of a vital part of bacterial cell walls.

The penicillins differ in the group bonded to the carbonyl carbon

The β-lactam ring

Penicillin G

Amoxicillin

Soon after the penicillins were introduced into medical practice, however, penicillin-resistant strains of bacteria began to appear. They have since proliferated dramatically. One approach to combating resistant strains is to synthesize newer, more effective penicillins, such as ampicillin, methicillin, and amoxicillin.

Another approach is to search for newer, more effective β-lactam antibiotics. The most effective of these agents discovered so far are the **cephalosporins,** first of which was isolated from the fungus *Cephalosporium acremonium.* This class of β-lactam antibiotics has an even broader spectrum of antibacterial activity than the penicillins and is effective against many penicillin-resistant bacterial strains. Cephalexin (Keflex) is currently one of the most widely prescribed of the cephalosporin antibiotics.

β-Lactam ring

... and the group bonded to this carbon of the six-membered ring

The cephalosporins differ in the group bonded to the carbonyl carbon ...

Cephalexin
(a β-lactam antibiotic)

The commonly prescribed formulation Augmentin is a combination of amoxicillin trihydrate, a penicillin, and clavulanic acid, a β-lactamase inhibitor that is isolated from *Streptomyces clavuligerus.*

β-Lactam ring

Clavulanic acid

Clavulanic acid, which also contains a β-lactam ring, reacts with and inhibits the β-lactamase enzyme before the enzyme can catalyze the inactivation of the penicillin. Augmentin is used as a second line of defense against childhood ear infections when penicillin resistance is suspected. Most children know it as a white liquid with a banana taste.

11.2 How Do We Prepare Esters?

The most common method for the preparation of esters is Fischer esterification (Section 10.5D). As an example of Fischer esterification, treating acetic acid with ethanol in the presence of concentrated sulfuric acid gives ethyl acetate and water:

$$\underset{\substack{\text{Ethanoic acid} \\ \text{(Acetic acid)}}}{CH_3COH} + \underset{\substack{\text{Ethanol} \\ \text{(Ethyl alcohol)}}}{CH_3CH_2OH} \underset{H_2SO_4}{\rightleftharpoons} \underset{\substack{\text{Ethyl ethanoate} \\ \text{(Ethyl acetate)}}}{CH_3COCH_2CH_3} + H_2O$$

Chemical Connections 11C

From Willow Bark to Aspirin and Beyond

The story of the development of this modern pain reliever goes back more than 2000 years. In 400 BCE the Greek physician Hippocrates recommended chewing the bark of the willow tree to alleviate the pain of childbirth and to treat eye infections.

The active component of willow bark was found to be salicin, a compound composed of salicyl alcohol bonded to a unit of β-D-glucose (Section 12.4A). Hydrolysis of salicin in aqueous acid followed by oxidation gave salicylic acid. Salicylic acid proved to be an even more effective reliever of pain, fever, and inflammation than salicin, and without the latter's extremely bitter taste. Unfortunately, patients quickly recognized salicylic acid's major side effect: It causes severe irritation of the mucous membrane lining of the stomach.

Salicin · Salicylic acid

In the search for less irritating but still effective derivatives of salicylic acid, chemists at the Bayer division of I. G. Farben in Germany in 1883 treated salicylic acid with acetic anhydride and prepared acetylsalicylic acid. They gave this new compound the name aspirin.

Acetylsalicylic acid (Aspirin)

Aspirin proved to be less irritating to the stomach than salicylic acid as well as more effective in relieving the pain and inflammation of rheumatoid arthritis. Aspirin, however, remains irritating to the stomach and frequent use of it can cause duodenal ulcers in susceptible persons.

In the 1960s, in a search for even more effective and less irritating analgesics and nonsteroidal anti-inflammatory drugs (NSAIDs), chemists at the Boots Pure Drug Company in England, who were studying compounds structurally related to salicylic acid, discovered an even more potent compound, which they named ibuprofen. Soon thereafter, Syntex Corporation in the United States developed naproxen, the active ingredient in Aleve. Both ibuprofen and naproxen have one stereocenter and can exist as a pair of enantiomers. For each drug, the active form is the S enantiomer. Naproxen is administered as its water-soluble sodium salt.

(S)-Ibuprofen

(S)-Naproxen

In the 1960s, researchers discovered that aspirin acts by inhibiting cyclooxygenase (COX), a key enzyme in the conversion of arachidonic acid to prostaglandins (Chemical Connections 13H). With this discovery, it became clear why only one enantiomer of ibuprofen and naproxen is active: Only the S enantiomer of each has the correct handedness to bind to COX and inhibit its activity.

11.3 How Do We Prepare Amides?

In principle, we can form an amide by treating a carboxylic acid with an amine and removing —OH from the acid and an —H from the amine. In practice, mixing these two leads to an acid–base reaction that forms an ammonium salt. If this salt is heated to a high enough temperature, water splits out and an amide forms.

Removal of OH and one H gives the amide

Acetic acid · Ethanamine (Ethylamine) · An ammonium salt · Heat · An amide

It is much more common, however, to prepare amides by treating an anhydride with an amine (Section 11.4C).

$$CH_3\overset{\overset{\displaystyle O}{\|}}{C}-O-\overset{\overset{\displaystyle O}{\|}}{C}CH_3 + H_2NCH_2CH_3 \longrightarrow CH_3\overset{\overset{\displaystyle O}{\|}}{C}-NHCH_2CH_3 + CH_3\overset{\overset{\displaystyle O}{\|}}{C}OH$$

Acetic anhydride   An amide

11.4 What Are the Characteristic Reactions of Anhydrides, Esters, and Amides?

The most common reaction of each of these three functional groups is with compounds that contain either an —OH group, as in water (H—OH) or an alcohol (H—OR), or an H—N group, as in ammonia (H—NH$_2$), or in a primary or secondary amine (H—NR$_2$) or H—NHR). These reactions have in common the addition of the oxygen or nitrogen atom to the carboxyl carbon and the hydrogen atom to the carbonyl oxygen to give a tetrahedral carbonyl addition intermediate. This intermediate then collapses to regenerate the carbonyl group and give either a new carboxyl derivative or a carboxylic acid itself. We illustrate here with the reaction of an ester with water:

Compare the formation of this tetrahedral carbonyl addition intermediate with that formed by the addition of an alcohol to the carbonyl group of an aldehyde or ketone and formation of a hemiacetal (Section 9.4C), and that formed by the addition of an alcohol to the carbonyl group of a carboxylic acid during Fischer esterification (Section 10.5D).

A. Reaction with Water: Hydrolysis

Hydrolysis is a chemical decomposition involving breaking a bond and the addition of the elements of water.

Anhydrides

Carboxylic anhydrides, particularly the low-molecular-weight ones, react readily with water to give two carboxylic acids. In the hydrolysis of an anhydride, one of the C—O bonds breaks, and OH is added to carbon and H is added to oxygen of what was the C—O bond. Hydrolysis of acetic anhydride gives two molecules of acetic acid.

$$CH_3\overset{\overset{\displaystyle O}{\|}}{C}O\overset{\overset{\displaystyle O}{\|}}{C}CH_3 + H_2O \longrightarrow CH_3\overset{\overset{\displaystyle O}{\|}}{C}OH + HO\overset{\overset{\displaystyle O}{\|}}{C}CH_3$$

Acetic anhydride   Acetic acid   Acetic acid

Esters

Esters are hydrolyzed only very slowly, even in boiling water. Hydrolysis becomes considerably more rapid, however, when the ester is heated in aqueous acid or base. When we discussed acid-catalyzed Fischer esterification in Section 10.5D, we pointed out that it is an equilibrium reaction. Hydrolysis of esters in aqueous acid, also an equilibrium reaction, is the reverse of Fischer esterification. A large excess of water drives the equilibrium to the right to form the carboxylic acid and alcohol.

Hydrolysis of an ester can also be carried out using a hot aqueous base, such as aqueous NaOH. This reaction is often called **saponification,** a reference to its use in the manufacture of soaps (Section 10.4B). The carboxylic acid formed in the hydrolysis reacts with hydroxide ion to form a carboxylic acid anion. Thus each mole of ester hydrolyzed requires one mole of base, as shown in the following balanced equation:

$$CH_3\overset{\displaystyle O}{\overset{\|}{C}}OCH_2CH_3 + NaOH \xrightarrow{H_2O} CH_3\overset{\displaystyle O}{\overset{\|}{C}}O^-Na^+ + CH_3CH_2OH$$

Ethyl acetate   Sodium hydroxide   Sodium acetate   Ethanol

There are two major differences between hydrolysis of esters in aqueous acid and aqueous base.

1. For hydrolysis of an ester in aqueous acid, acid is required in only catalytic amounts. For hydrolysis in aqueous base, base is required in stoichiometric amounts (one mole of base per mole of ester), because base is a reactant, not merely a catalyst.

2. Hydrolysis of an ester in aqueous acid is reversible. Hydrolysis in aqueous base is irreversible because the carboxylate anion does not react with water or hydroxide ion.

Example 11.2 Hydrolysis of an Ester

Complete the equation for each hydrolysis reaction. Show the products as they are ionized under the given experimental conditions.

(b) $CH_3\overset{\displaystyle O}{\overset{\|}{C}}OCH_2CH_2\overset{\displaystyle O}{\overset{\|}{C}}CH_3 + 2NaOH \xrightarrow{H_2O}$

Strategy

The products of the hydrolysis of an ester are a carboxylic acid and an alcohol. If hydrolysis is carried out in aqueous NaOH, the carboxylic acid is converted to its sodium salt.

Solution

The products of hydrolysis of compound (a) are benzoic acid and 2-propanol. In aqueous NaOH, benzoic acid is converted to its sodium salt. In this reaction, one mole of NaOH is required for hydrolysis of each mole of this ester.

Compound (b) is a diester of ethylene glycol and requires two moles of NaOH for its complete hydrolysis.

(a)

| Sodium | 2-Propanol |
| benzoate | (Isopropyl alcohol) |

(b) $2CH_3CO^-Na^+ + HOCH_2CH_2OH$

Sodium acetate   1,2-Ethanediol
(Ethylene glycol)

Problem 11.2

Complete the equation for each hydrolysis reaction. Show all products as they are ionized under these experimental conditions.

(a) $+ \; 2NaOH \xrightarrow{H_2O}$

(b) $+ \; H_2O \xrightarrow{HCl}$

Amides

Amides require more vigorous conditions for hydrolysis in both acid and base than do esters. Hydrolysis in hot aqueous acid gives a carboxylic acid and an ammonium ion. This reaction is driven to completion by the acid–base reaction between ammonia or the amine and the acid to form an ammonium ion. Complete hydrolysis requires one mole of acid per mole of amide.

$$CH_3CH_2CH_2\overset{O}{\overset{\|}{C}}NH_2 + H_2O + HCl \xrightarrow[\text{heat}]{H_2O} CH_3CH_2CH_2\overset{O}{\overset{\|}{C}}OH + NH_4^+ \; Cl^-$$

Butanamide   Butanoic acid

The products of amide hydrolysis in aqueous base are a carboxylic acid salt and ammonia or an amine. The acid–base reaction between the carboxylic acid and base to form a carboxylic salt drives this hydrolysis to completion. Thus, complete hydrolysis of an amide requires one mole of base per mole of amide.

Acetanilide   Sodium acetate   Aniline

Chemical Connections 11D

Ultraviolet Sunscreens and Sunblocks

Ultraviolet (UV) radiation penetrating the Earth's ozone layer is arbitrarily divided by wavelength into two regions: UVB (290–320 nm) and UVA (320–400 nm). UVB, a more energetic form of radiation than UVA, creates more radicals and hence does more oxidative damage to tissue (Section 17.7). UVB radiation interacts directly with biomolecules of the skin and eyes, causing skin cancer, skin aging, eye damage leading to cataracts, and delayed sunburn that appears 12–24 hours after exposure. UVA radiation, by contrast, causes tanning. It also damages skin, albeit much less efficiently than UVB. Its role in promoting skin cancer is less well understood.

Commercial sunscreen products are rated according to their sun protection factor (SPF), which is defined as the minimum effective dose of UV radiation that produces a delayed sunburn on protected skin compared to unprotected skin. Two types of active ingredients are found in commercial sunblocks and sunscreens. The most common sunblock agent is zinc oxide, ZnO, which reflects and scatters UV radiation. Sunscreens, the second type of active ingredient, absorb UV radiation and then reradiate it as heat. These compounds are most effective in screening UVB radiation, but they do not screen UVA radiation. Thus they allow tanning but prevent the UVB-associated damage. Given here are structural formulas for three common esters used as UVB-screening agents, along with the name by which each is most commonly listed in the Active Ingredients labels on commercial products:

Octyl *p*-methoxycinnamate   Homosalate   Padimate A

Example 11.3 Hydrolysis of an Amide

Write a balanced equation for the hydrolysis of each amide in concentrated aqueous HCl. Show all products as they exist in aqueous HCl.

(a) $CH_3CN(CH_3)_2$ (with O double bonded to C)

(b)

Strategy

Hydrolysis of an amide gives a carboxylic acid and an amine. If the hydrolysis is carried out in aqueous acid, the amine is converted to its ammonium salt. Hydrolysis of an amide in aqueous acid requires one mole of acid per mole of amide.

Solution

(a) Hydrolysis of *N,N*-dimethylacetamide gives acetic acid and dimethylamine. Dimethylamine, a base, reacts with HCl to form dimethylammonium ion, shown here as dimethylammonium chloride.

$$CH_3CN(CH_3)_2 + H_2O + HCl \xrightarrow{\text{Heat}} CH_3COH + (CH_3)_2NH_2{}^+Cl^-$$

(b) Hydrolysis of this lactam gives the protonated form of 5-amino-pentanoic acid.

Problem 11.3

Write a balanced equation for the hydrolysis of each amide in Example 11.3 in concentrated aqueous NaOH. Show all products as they exist in aqueous NaOH.

B. Reaction with Alcohols

Anhydrides

Anhydrides react with alcohols and phenols to give one mole of ester and one mole of a carboxylic acid.

Thus, the reaction of an alcohol with an anhydride is a useful method for the synthesis of esters. Aspirin (Chemical Connections 11C) is synthesized on an industrial scale by the reaction of acetic anhydride with salicylic acid.

C. Reaction with Ammonia and Amines

Anhydrides

Anhydrides react with ammonia and with 1° and 2° amines to form amides. Two moles of amine are required: one to form the amide and one to neutralize the carboxylic acid by product. We show this reaction here in two steps: (1) formation of the amide and the carboxylic acid by-product, and (2) an acid-base reaction of the carboxylic acid by-product with the second mole of ammonia to give an ammonium salt.

Chemical Connections 11E

Barbiturates

In 1864, Adolph von Baeyer (1835–1917) discovered that heating the diethyl ester of malonic acid with urea in the presence of sodium ethoxide (like sodium hydroxide, a strong base) gives a cyclic compound that he named barbituric acid. Some say that Baeyer named it after a friend of his named Barbara. Others claim that he named it after St. Barbara, the patron saint of artillerymen.

Diethyl propanedioate   Urea
(Diethyl malonate)

Barbituric acid

A number of derivatives of barbituric acid have powerful sedative and hypnotic effects. One such derivative is pentobarbital. Like other derivatives of barbituric acid, pentobarbital is quite insoluble in water and body fluids. To increase its solubility in these fluids, pentobarbital is converted to its sodium salt, which is given the name Nembutal. Phenobarbital, also administered as its sodium salt, is an anticonvulsant, sedative, and hypnotic.

Technically speaking, only the sodium salts of these compounds should be called barbiturates. In practice, however, all derivatives of barbituric acid are called barbiturates,

whether they are the un-ionized form or the ionized, water-soluble salt form.

Pentobarbital

Sodium pentobarbital
(Nembutal)

Phenobarbital

Barbiturates have two principal effects. In small doses, they are sedatives (tranquilizers); in larger doses, they induce sleep. Barbituric acid, in contrast, has neither of these effects. Barbiturates are dangerous because they are addictive, which means that a regular user will suffer withdrawal symptoms when their use is stopped. They are especially dangerous when taken with alcohol because the combined effect (called a synergistic effect) is usually greater than the sum of the effects of either drug taken separately.

Esters

Esters react with ammonia and with 1° and 2° amines to form amides.

Ethyl 2-phenyl acetate   2-Phenylacetamide

Thus, as seen in this section, amides can be prepared readily from esters. Because carboxylic acids are easily converted to esters by Fischer esterification, we have a good way to convert a carboxylic acid to an amide. This method of amide formation is, in fact, much more useful and applicable than converting a carboxylic acid to an ammonium salt and then heating the salt to form an amide.

Amides

Amides do not react with ammonia or primary or secondary amines.

11.5 What Are Phosphoric Anhydrides and Phosphoric Esters?

A. Phosphoric Anhydrides

Because of the special importance of phosphoric anhydrides in biochemical systems, we discuss them here to show the similarity between them and the anhydrides of carboxylic acids. The functional group of a **phosphoric anhydride** consists of two phosphoryl ($P=O$) groups bonded to the same oxygen atom. Shown here are structural formulas for two anhydrides of phosphoric acid and the ions derived by ionization of the acidic hydrogens of each:

Diphosphoric acid
(Pyrophosphoric acid)

Diphosphate ion
(Pyrophosphate ion)

Triphosphoric acid

Triphosphate ion

B. Phosphoric Esters

Phosphoric acid has three —OH groups and forms mono-, di-, and triphosphoric esters, which we name by giving the name(s) of the alkyl group(s) bonded to oxygen followed by the word "phosphate" (for example, dimethyl phosphate). In more complex **phosphoric esters,** it is common practice to name the organic molecule and then indicate the presence of the phosphoric ester by including either the word "phosphate" or the prefix *phospho-*. Dihydroxyacetone phosphate, for example, is an intermediate in glycolysis (Section 20.2). Pyridoxal phosphate is one of the metabolically active forms of vitamin B_6. The last two phosphoric esters are shown here as they are ionized at pH 7.4, the pH of blood plasma.

Dimethyl phosphate

Dihydroxyacetone phosphate

Pyridoxal phosphate

11.6 What Is Step-Growth Polymerization?

Step-growth polymers form from the reaction of molecules containing two functional groups, with each new bond being created in a separate step. In this section, we discuss three types of step-growth polymers: polyamides, polyesters, and polycarbonates.

A. Polyamides

In the early 1930s, chemists at E. I. DuPont de Nemours & Company began fundamental research into the reactions between dicarboxylic acids and diamines to form **polyamides.** In 1934, they synthesized nylon-66, the first purely synthetic fiber. Nylon-66 is so named because it is synthesized from two different monomers, each containing six carbon atoms.

In the synthesis of nylon-66, hexanedioic acid and 1,6-hexanediamine are dissolved in aqueous ethanol and then heated in an autoclave to 250°C and an internal pressure of 15 atm. Under these conditions —COOH and —NH$_2$ groups react by loss of H$_2$O to form a polyamide, similar to the formation of amides we described in Section 11.3.

Hexanedioic acid
(Adipic acid)

1,6-Hexanediamine
(Hexamethylenediamine)

Nylon-66
(a polyamide)

Based on extensive research into the relationships between molecular structure and bulk physical properties, scientists at DuPont reasoned that a polyamide containing benzene rings would be even stronger than nylon-66. This line of reasoning eventually produced a polyamide that DuPont named Kevlar.

1,4-Benzenedicarboxylic acid
(Terephthalic acid)

1,4-Benzenediamine
(*p*-Phenylenediamine)

Kevlar
(a polyaromatic amide)

One remarkable feature of Kevlar is that it weighs less than other materials of similar strength. For example, a cable woven of Kevlar has a strength equal to that of a similarly woven steel cable. Yet the Kevlar cable has only 20% of the weight of the steel cable! Kevlar now finds use in such articles as anchor cables for offshore drilling rigs and reinforcement fibers for automobile tires. It is also woven into a fabric that is so tough that it can be used for bulletproof vests, jackets, and raincoats.

B. Polyesters

The first **polyester,** developed in the 1940s, involved polymerization of benzene 1,4-dicarboxylic acid with 1,2-ethanediol to give poly(ethylene terephthalate), abbreviated PET. Virtually all PET is now made from the dimethyl ester of terephthalic acid by the following reaction:

Dimethyl terephthalate   1,2-Ethanediol   Poly(ethylene terephthalate)
                           (Ethylene glycol)   (Dacron, Mylar)

The crude polyester can be melted, extruded, and then drawn to form the textile fiber called Dacron polyester. Dacron's outstanding features include its stiffness (about four times that of nylon-66), very high tensile strength, and remarkable resistance to creasing and wrinkling. Because the early Dacron polyester fibers were harsh to the touch due to their stiffness, they were usually blended with cotton or wool to make acceptable textile fibers. Newly developed fabrication techniques now produce less-harsh Dacron polyester textile fibers. PET is also fabricated into Mylar films and recyclable plastic beverage containers.

Mylar can be made into extremely strong films. Because the film has a very small pore size, it is used for balloons that can be inflated with helium; the helium atoms diffuse only slowly through the pores of the film.

C. Polycarbonates

A **polycarbonate,** the most familiar of which is Lexan, forms from the reaction between the disodium salt of bisphenol A and phosgene. Phosgene is a derivative of carbonic acid, H_2CO_3, in which both —OH groups have been replaced by chlorine atoms. An ester of carbonic acid is called a carbonate.

Carbonic acid   Phosgene   Diethyl carbonate
(H_2CO_3)   (a carbonate ester)

In forming a polycarbonate, each mole of phosgene reacts with two moles of the sodium salt of a phenol called bisphenol A (BPA).

A polycarbonate hockey mask.

Disodium salt of Bisphenol A   Phosgene   Lexan
                                              (a polycarbonate)

Lexan is a tough, transparent polymer that has high impact and tensile strength and retains its properties over a wide temperature range. It is used in sporting equipment (helmets and face masks); to make light, impact-resistant housings for household appliances; and in the manufacture of safety glass and unbreakable windows.

Chemical Connections 11F

Stitches That Dissolve

As the technological capabilities of medicine have expanded, the demand for synthetic materials that can be used inside the body has increased as well. Polymers already have many of the characteristics of an ideal biomaterial: They are lightweight and strong, are inert or biodegradable depending on their chemical structure, and have physical properties (softness, rigidity, elasticity) that are easily tailored to match those of natural tissues.

Even though most medical uses of polymeric materials require biostability, some applications require them to be biodegradable. An example is the polyester of glycolic acid

and lactic acid used in absorbable sutures, which are marketed under the trade name of Lactomer.

A health care specialist must remove traditional suture materials such as catgut after they have served their purpose. Stitches of Lactomer, however, are hydrolyzed slowly over a period of approximately two weeks. By the time the torn tissues have healed, the stitches have hydrolyzed, and no suture removal is necessary. The body metabolizes and excretes the glycolic and lactic acids formed during this hydrolysis.

Glycolic acid   Lactic acid   A polymer of glycolic acid and lactic acid

Summary of Key Questions

◯WL End-of-chapter problems identified in blue are assignable in **GOB OWL.**

Section 11.1 What Are Carboxylic Anhydrides, Esters, and Amides?

- A **carboxylic anhydride** contains two carbonyl groups bonded to the same oxygen.
- A carboxylic **ester** contains a carbonyl group bonded to an — OR group derived from an alcohol or a phenol.
- A carboxylic **amide** contains a carbonyl group bonded to a nitrogen atom derived from an amine.

Section 11.2 How Do We Prepare Esters?

- The most common laboratory method for the preparation of esters is Fischer esterification (Section 10.5D).

Section 11.3 How Do We Prepare Amides?

- Amides can be prepared by the reaction of an amine with a carboxylic anhydride.

Section 11.4 What Are the Characteristic Reactions of Carboxylic Anhydrides, Esters, and Amides? Problem 11.9

- Hydrolysis is a chemical process in which a bond is split and the elements of H_2O are added.

- Hydrolysis of a carboxylic anhydride gives two molecules of carboxylic acid.
- Hydrolysis of a carboxylic ester requires the presence of either concentrated aqueous acid or base. Acid is a catalyst and the reaction is the reverse of Fischer esterification. Base is a reactant and is required in stoichiometric amounts.
- Hydrolysis of a carboxylic amide requires the presence of either aqueous acid or base. Both acid and base are reactants and are required in stoichiometric amounts.

Section 11.5 What Are Phosphoric Anhydrides and Phosphoric Esters? Problems 11.16, 11.17

- Phosphoric anhydrides consist of two phosphoryl groups (P=O) bonded to the same oxygen atom.

Section 11.6 What Is Step-Growth Polymerization?

- Step-growth polymerization involves the stepwise reaction of difunctional monomers. Important commercial polymers synthesized through step-growth processes include polyamides, polyesters, and polycarbonates.

Summary of Key Reactions

1. **Fischer Esterification (Section 11.2)** Fischer esterification is reversible. To achieve high yields of ester, it is necessary to force the equilibrium to the right. One way to maximize the yield of ester is to use an excess of the alcohol. Another way is to remove the water as it is formed.

$$CH_3COH \ + \ CH_3CH_2CH_2OH$$

$$\overset{H_2SO_4}{\rightleftharpoons} \ CH_3COCH_2CH_2CH_3 \ + \ H_2O$$

2. **Preparation of an Amide (Section 11.3)** Reaction of an anhydride with ammonia or a 1° or 2° amine gives an amide.

$$CH_3C\!-\!O\!-\!CCH_3 \ + \ H_2NCH_2CH_3$$

$$\longrightarrow \ CH_3C\!-\!NHCH_2CH_3 \ + \ CH_3COH$$

3. **Hydrolysis of an Anhydride (Section 11.4A)** Anhydrides, particularly low-molecular-weight ones, react readily with water to give two carboxylic acids.

$$CH_3COCCH_3 \ + \ H_2O$$

$$\longrightarrow \ CH_3COH \ + \ HOCCH_3$$

4. **Hydrolysis of an Ester (Section 11.4A)** Esters are hydrolyzed rapidly only in the presence of acid or base. Acid-catalyzed hydrolysis is the reverse of Fischer esterification. Acid is a catalyst. Base is a reactant and therefore is required in an equimolar amount.

$$CH_3COCH_2CH_3 \ + \ H_2O$$

$$\overset{H^+}{\rightleftharpoons} \ CH_3COH \ + \ HOCH_2CH_3$$

$$CH_3COCH_2CH_3 \ + \ NaOH$$

$$\overset{H_2O}{\longrightarrow} \ CH_3CO^-Na^+ \ + \ CH_3CH_2OH$$

5. **Hydrolysis of an Amide (Section 11.4A)** Amides require more vigorous conditions for hydrolysis than

do esters. Either acid or base is required in an amount equivalent to that of the amide: acid to convert the resulting amine to an ammonium salt and base to convert the resulting carboxylic acid to a carboxylate salt.

$$CH_3CH_2CH_2CNH_2 \ + \ H_2O \ + \ HCl$$

$$\overset{H_2O}{\underset{heat}{\longrightarrow}} \ CH_3CH_2CH_2COH \ + \ NH_4^+Cl^-$$

$$CH_3CH_2CH_2CNH_2 \ + \ NaOH$$

$$\overset{H_2O}{\underset{heat}{\longrightarrow}} \ CH_3CH_2CH_2CO^-Na^+ \ + \ NH_3$$

6. **Reaction of Anhydrides with Alcohols (Section 11.4B)** Anhydrides react with alcohols to give one mole of ester and one mole of a carboxylic acid.

$$CH_3COCCH_3 \ + \ HOCH_2CH_3$$

$$\longrightarrow \ CH_3COCH_2CH_3 \ + \ HOCCH_3$$

7. **Reaction of Anhydrides with Ammonia and Amines (Section 11.4C)** Anhydrides react with ammonia and with 1° and 2° amines to give amides. Two moles of amine are required: one mole to give the amide and one mole to neutralize the carboxylic acid by-product.

$$CH_3C\!-\!O\!-\!CCH_3 \ + \ 2NH_3$$

$$\longrightarrow \ CH_3CNH_2 \ + \ CH_3CO^-NH_4^+$$

8. **Reaction of Esters with Ammonia and with 1° and 2° Amines (Section 11.4C)** Esters react with ammonia and with 1° and 2° amines to give an amide and an alcohol.

$$\longrightarrow \qquad\qquad\qquad NH_2 \ + \ CH_3CH_2OH$$

Problems

■ Indicates problems that are assignable in **GOB OWL**.

Blue numbered problems are applied.

Go to this book's companion website at **www.cengage. com/chemistry/bettelheim** for interactive versions of the How To tutorials and Active Figures, and to quiz yourself on this chapter.

Section 11.1 What Are Carboxylic Anhydrides, Esters, and Amides?

11.4 Draw a structural formula for each compound.
- (a) Dimethyl carbonate
- (b) *p*-Nitrobenzamide
- (c) Ethyl 3-hydroxybutanoate
- (d) Diethyl oxalate
- (e) Ethyl *trans*-2-pentenoate
- (f) Butanoic anhydride

11.5 Write the IUPAC name for each compound.

(a)

(b) $CH_3(CH_2)_8COCH_3$

(c) $CH_3(CH_2)_4CNHCH_3$

(d)

(e)

(f)

Section 11.4 What Are the Characteristic Reactions of Anhydrides, Esters, and Amides?

11.6 What product forms when ethyl benzoate is treated with each reagent?
- (a) H_2O, NaOH, heat   (b) H_2O, HCl, heat

11.7 What product forms when benzamide, $C_6H_5CONH_2$, is treated with each reagent?
- (a) H_2O, NaOH, heat   (b) H_2O, HCl, heat

11.8 Complete the equations for these reactions.

(a)

(b)

11.9 ■ The analgesic phenacetin is synthesized by treating 4-ethoxyaniline with acetic anhydride. Draw a structural formula for phenacetin.

4-Ethoxyaniline

11.10 Phenobarbital is a long-acting sedative, hypnotic, and anticonvulsant.
- (a) Name all functional groups in this compound.
- (b) Draw structural formulas for the products from complete hydrolysis of all amide groups in aqueous NaOH.

Phenobarbital

11.11 Following is a structural formula for aspartame, an artificial sweetener about 180 times as sweet as sucrose (table sugar).

Aspartame

- (a) Is aspartame chiral? If so, how many stereoisomers are possible for it?
- (b) Name each functional group in aspartame.
- (c) Estimate the net charge on an aspartame molecule in aqueous solution at pH 7.0.
- (d) Would you expect aspartame to be soluble in water? Explain.
- (e) Draw structural formulas for the products of complete hydrolysis of aspartame in aqueous HCl.

Show each product as it would be ionized in this solution.

(f) Draw structural formulas for the products of complete hydrolysis of aspartame in aqueous NaOH. Show each product as it would be ionized in this solution.

11.12 Why are nylon-66 and Kevlar referred to as polyamides?

11.13 Draw short sections of two parallel chains of nylon-66 (each chain running in the same direction) and show how it is possible to align them such that there is hydrogen bonding between the N—H groups of one chain and the C=O groups of the parallel chain.

11.14 Why are Dacron and Mylar referred to as polyesters?

Section 11.5 What Are Phosphoric Anhydrides and Phosphoric Esters?

11.15 What type of structural feature do the anhydrides of phosphoric acid and carboxylic acids have in common?

11.16 ■ Draw structural formulas for the mono-, di-, and triethyl esters of phosphoric acid.

11.17 ■ 1,3-Dihydroxy-2-propanone (dihydroxyacetone) and phosphoric acid form a monoester called dihydroxyacetone phosphate, which is an intermediate in glycolysis (Section 20.2). Draw a structural formula for this monophosphate ester.

11.18 Write an equation for the hydrolysis of trimethyl phosphate to dimethyl phosphate and methanol in aqueous base. Show each product as it would be ionized in this solution.

Chemical Connections

11.19 (Chemical Connections 11A) Locate the ester group in pyrethrin I and draw a structural formula for chrysanthemic acid, the carboxylic acid from which this ester is derived.

11.20 (Chemical Connections 11A) What structural features do pyrethrin I (a natural insecticide) and permethrin (a synthetic pyrethrenoid) have in common?

11.21 (Chemical Connections 11A) A commercial Clothing & Gear Insect Repellant gives the following information about permethrin, its active ingredient:

Cis/trans ratio: Minimum 35% (+/−) *cis* and maximum 65% (+/−) *trans*

(a) To what does the *cis/trans* ratio refer?
(b) To what does the designation "(+/−)" refer?

11.22 (Chemical Connections 11B) Identify the β-lactam portion of amoxicillin and cephalexin.

11.23 (Chemical Connections 11C) What is the compound in willow bark that is responsible for its ability to relieve pain? How is this compound related to salicylic acid?

11.24 (Chemical Connections 11C) Once it has been opened, and particularly if it has been left open to the air, a bottle of aspirin may develop a vinegar-like odor. Explain how this might happen.

11.25 (Chemical Connections 11C) What is the structural relationship between aspirin and ibuprofen? Between aspirin and naproxen?

11.26 (Chemical Connections 11D) What is the difference in meaning between *sunblock* and *sunscreen*?

11.27 (Chemical Connections 11D) How do sunscreens prevent UV radiation from reaching the skin?

11.28 (Chemical Connections 11D) What structural features do the three sunscreens given in this Chemical Connection have in common?

11.29 (Chemical Connections 11E) Barbiturates are derived from urea. Identify the portion of the structure of pentobarbital and phenobarbital that is derived from urea.

11.30 (Chemical Connections 11F) Why do Lactomer stitches dissolve within 2 to 3 weeks following surgery?

Additional Problems

11.31 ■ Benzocaine, a topical anesthetic, is prepared by treating 4-aminobenzoic acid with ethanol in the presence of an acid catalyst followed by neutralization. Draw a structural formula for benzocaine.

11.32 The analgesic acetaminophen is synthesized by treating 4-aminophenol with one equivalent of acetic anhydride. Write an equation for the formation of acetaminophen. (Hint: The —NH₂ group is more reactive with acetic anhydride than the —OH group.)

11.33 1,3-Diphosphoglycerate, an intermediate in glycolysis (Section 20.2), contains a mixed anhydride (an anhydride of a carboxylic acid and phosphoric acid) and a phosphoric ester. Draw structural formulas for the products formed by hydrolysis of the anhydride and ester bonds in this molecule. Show each product as it would exist in solution at pH 7.4.

1,3-Diphosphoglycerate

11.34 ■ *N,N*-Diethyl *m*-toluamide (DEET) is the active ingredient in several common insect repellents. From what acid and amine can DEET by synthesized?

N,N-Diethyl *m*-toluamide
(DEET)

11.35 Following are structural formulas for two local anesthetics used in dentistry. Lidocaine was introduced in 1948 and is now the most widely used local anesthetic for infiltration and regional anesthesia. Its hydrochloride salt is marketed under the name Xylocaine. Mepivacaine is faster acting and somewhat longer in duration than lidocaine. Its hydrochloride salt is marketed under the name Carbocaine.

Lidocaine
(Xylocaine)

Mepivacaine
(Carbocaine)

(a) Name the functional groups in each anesthetic.
(b) What similarities in structure do you find between these compounds?

Looking Ahead

11.36 We have seen that an amide can be formed from a carboxylic acid and an amine. Suppose that you start instead with an amino acid such as alanine. Show how amide formation in this case can lead to a macromolecule of molecular weight several thousands of times that of the starting materials. We will study these polyamides in Chapter 14 (proteins).

Alanine   Alanine

+ H$_2$N—CH—C—OH etc. ⟶ a polyamide

Alanine

11.37 We will encounter the following molecule in our discussion of glycolysis, the biochemical pathway that converts glucose to pyruvic acid (Section 20.2).

Phosphoenolpyruvate

(a) Draw structural formulas for the products of hydrolysis of the ester bond in phosphoenolpyruvate.
(b) Why are the letters *enol* a part of the name of this compound?

Carbohydrates

Breads, grains, and pasta are sources of carbohydrates.

12.1 Carbohydrates: What Are Monosaccharides?

Carbohydrates are the most abundant organic compounds in the plant world. They act as storehouses of chemical energy (glucose, starch, glycogen); are components of supportive structures in plants (cellulose), crustacean shells (chitin), and connective tissues in animals (acidic polysaccharides); and are essential components of nucleic acids (D-ribose and 2-deoxy-D-ribose). Carbohydrates account for approximately three-fourths of the dry weight of plants. Animals (including humans) get their carbohydrates by eating plants, but they do not store much of what they consume. In fact, less than 1% of the body weight of animals is made up of carbohydrates.

OWL Online homework for this chapter may be assigned in **GOB OWL**.

The word *carbohydrate* means "hydrate of carbon" and derives from the formula $C_n(H_2O)_m$. Two examples of carbohydrates with this general molecular formula that can be written alternatively as hydrates of carbon are

- Glucose (blood sugar): $C_6H_{12}O_6$, which can be written as $C_6(H_2O)_6$
- Sucrose (table sugar): $C_{12}H_{22}O_{11}$, which can be written as $C_{12}(H_2O)_{11}$

Not all carbohydrates, however, have this general formula. Some contain too few oxygen atoms to fit it; some contain too many oxygens. Some also contain nitrogen. The term *carbohydrate* has become so firmly rooted in the chemical nomenclature that, although not completely accurate, it persists as the name for this class of compounds.

At the molecular level, most **carbohydrates** are polyhydroxyaldehydes, polyhydroxyketones, or compounds that yield them after hydrolysis. The simpler members of the carbohydrate family are often referred to as **saccharides** because of their sweet taste (Latin: *saccharum,* "sugar"). Carbohydrates are classified as monosaccharides, oligosaccharides, or polysaccharides depending on the number of simple sugars they contain.

Carbohydrate A polyhydroxyaldehyde or polyhydroxyketone, or a substance that gives these compounds on hydrolysis

A. Structure and Nomenclature

Monosaccharide A carbohydrate that cannot be hydrolyzed to a simpler compound

Monosaccharides have the general formula $C_nH_{2n}O_n$, with one of the carbons being the carbonyl group of either an aldehyde or a ketone. The most common monosaccharides have three to nine carbon atoms. The suffix **-ose** indicates that a molecule is a carbohydrate, and the prefixes **tri-, tetr-, pent-,** and so forth indicate the number of carbon atoms in the chain. Monosaccharides containing an aldehyde group are classified as **aldoses;** those containing a ketone group are classified as **ketoses.**

Aldose A monosaccharide containing an aldehyde group

Ketose A monosaccharide containing a ketone group

There are only two trioses: the aldotriose glyceraldehyde and the ketotriose dihydroxyacetone.

$$
\begin{array}{cc}
\text{CHO} & \text{CH}_2\text{OH} \\
| & | \\
\text{CHOH} & \text{C}=\text{O} \\
| & | \\
\text{CH}_2\text{OH} & \text{CH}_2\text{OH} \\
\text{Glyceraldehyde} & \text{Dihydroxyacetone} \\
\text{(an aldotriose)} & \text{(a ketotriose)}
\end{array}
$$

Often the designations *aldo-* and *keto-* are omitted, and these molecules are referred to simply as trioses, tetroses, and the like.

B. Fischer Projection Formulas

Glyceraldehyde contains a stereocenter and therefore exists as a pair of enantiomers (Figure 12.1).

Fischer projection A two-dimensional representation showing the configuration of a stereocenter; horizontal lines represent bonds projecting forward from the stereocenter, and vertical lines represent bonds projecting toward the rear

Chemists commonly use two-dimensional representations called **Fischer projections** to show the configuration of carbohydrates. To draw a Fischer projection, draw a three-dimensional representation of the molecule oriented so that the vertical bonds from the stereocenter are directed away from you and the horizontal bonds from it are directed toward you (none of the bonds to stereocenter are in the plane of the paper). Then write the molecule as a cross, with the stereocenter indicated by the point at which the bonds cross.

Emil Fischer, who in 1902 became the second Nobel Prize winner in chemistry, made many fundamental discoveries in the chemistry of carbohydrates, proteins, and other areas of organic and biochemistry.

(R)-Glyceraldehyde (three-dimensional representation)   (R)-Glyceraldehyde (Fischer projection)

CHO
H▸C◂OH
CH₂OH

CHO
HO▸C◂H
CH₂OH

FIGURE 12.1 The enantiomers of glyceraldehyde.

The horizontal segments of this Fischer projection represent bonds directed toward you, and the vertical segments represent bonds directed away from you. The only atom in the plane of the paper is the stereocenter.

C. D- and L-Monosaccharides

Even though the R,S system is widely accepted today as a standard for designating configuration, the configuration of carbohydrates is commonly designated using the D,L system proposed by Emil Fischer in 1891. At that time, it was known that one enantiomer of glyceraldehyde has a specific rotation (Section 6.4B) of $+13.5°$; the other has a specific rotation of $-13.5°$. Fischer proposed that these enantiomers be designated D and L, but he had no experimental way to determine which enantiomer has which specific rotation. Fischer, therefore, did the only possible thing—he made an arbitrary assignment. He assigned the dextrorotatory enantiomer the following configuration and named it D-glyceraldehyde. He named its enantiomer L-glyceraldehyde. Fischer could have been wrong, but by a stroke of good fortune, he wasn't. In 1952, scientists proved that his assignment of the D,L-configuration to the enantiomers of glyceraldehyde is correct.

CHO
H————OH
CH₂OH
D-Glyceraldehyde
$[\alpha]_D^{25} = +13.5°$

CHO
HO————H
CH₂OH
L-Glyceraldehyde
$[\alpha]_D^{25} = -13.5°$

D-glyceraldehyde and L-glyceraldehyde serve as reference points for the assignment of relative configurations to all other aldoses and ketoses. The reference point is the penultimate carbon—that is, the next-to-the-last carbon on the chain. A **D-monosaccharide** has the same configuration at its penultimate carbon as D-glyceraldehyde (its —OH group is on the right) in a Fischer projection; an **L-monosaccharide** has the same configuration at its penultimate carbon as L-glyceraldehyde (its —OH group is on the left).

Tables 12.1 and 12.2 show names and Fischer projections for all D-aldo- and D-2-ketotetroses, pentoses, and hexoses. Each name consists of three parts. The D specifies the configuration at the stereocenter farthest from the carbonyl group. Prefixes such as *rib-*, *arabin-*, and *gluc-* specify the configuration of all other stereocenters in the monosaccharide relative to one another. The suffix *-ose* indicates that the compound is a carbohydrate.

The three most abundant hexoses in the biological world are D-glucose, D-galactose, and D-fructose. The first two are D-aldohexoses; the third is a D-2-ketohexose. Glucose, by far the most abundant of the three, is

D-Monosaccharide A monosaccharide that, when written as a Fischer projection, has the —OH group on its penultimate carbon to the right

L-Monosaccharide A monosaccharide that, when written as a Fischer projection, has the —OH group on its penultimate carbon to the left

TABLE 12.1 Configurational Relationships among the Isomeric D-Aldotetroses, D-Aldopentoses, and D-Aldohexoses *

*The configuration of the reference —OH group on the penultimate carbon is shown in color.

also known as dextrose because it is dextrorotatory. Other names for this monosaccharide include grape sugar and blood sugar. Human blood normally contains 65–110 mg of glucose/100 mL of blood.

Example 12.1 Drawing Fischer Projections

Draw Fischer projections for the four aldotetroses. Which are D-monosaccharides, which are L-monosaccharides, and which are enantiomers? Refer to Table 12.1 and write the name of each aldotetrose.

Strategy

Start with the Fischer projections of the two aldotrioses, D-glyceraldehyde and L-glyceraldehyde. Draw structures with four carbons, adding the fourth carbon between the one that determines the D,L desgination and the aldehyde carbon.

Solution

Following are Fischer projections for the four aldotetroses. The D- and L- refer to the configuration of the penultimate carbon, which, in the case of aldotetroses, is carbon 3. In the Fischer projection of a D-aldotetrose, the —OH group on carbon 3 is on the right; in an L-aldotetrose, it is on the left.

TABLE 12.2 Configurational Relationships among the D-2-Ketopentoses and D-2-Ketohexoses

Gloved hand holding an intravenous (IV) drip bag containing 0.15% potassium chloride (saline) and 5% glucose.

Problem 12.1

Draw Fischer projections for all 2-ketopentoses. Which are D-2-ketopentoses, which are L-2-ketopentoses, and which are enantiomers? Refer to Table 12.2 and write the name of each 2-ketopentose.

D. Amino Sugars

Amino sugars contain an —NH_2 group in place of an —OH group. Only three amino sugars are common in nature: D-glucosamine, D-mannosamine, and D-galactosamine.

Amino sugar A monosaccharide in which an —OH group is replaced by an —NH_2 group

N-Acetyl-D-glucosamine, a derivative of D-glucosamine, is a component of many polysaccharides, including connective tissue such as cartilage. It is

Chemical Connections 12A

Galactosemia

One out of every 18,000 infants is born with a genetic defect that renders the child unable to utilize the monosaccharide galactose. Galactose is part of lactose (milk sugar, Section 12.4B). When the body cannot absorb galactose, it accumulates in the blood and in the urine. This buildup in the blood is harmful because it can lead to mental retardation, failure to grow, cataract formation in the eye, and, in severe cases, death due to liver damage. When galactose accumulation results from a transient disorder in infants, known as galactosuria, it has only mild symptoms. When the enzyme galactose-1-phosphate uridinyltransferase is defective, however, the disorder is called galactosemia, and its symptoms are severe.

The deleterious effects of galactosemia can be avoided by giving the infant a milk formula in which sucrose is substituted for lactose. Because sucrose contains no galactose, the infant consumes a galactose-free diet. A galactose-free diet is critical only in infancy. With maturation, most children develop another enzyme capable of metabolizing galactose. As a consequence, they are able to tolerate galactose as they mature.

A number of products are available to meet calcium needs of those who are lactose intolerant.

also a component of chitin, the hard, shell-like exoskeleton of lobsters, crabs, shrimp, and other shellfish. Several other amino sugars are components of naturally occurring antibiotics.

E. Physical Properties of Monosaccharides

Monosaccharides are colorless, crystalline solids. Because hydrogen bonding is possible between their polar —OH groups and water, all monosaccharides are very soluble in water. They are only slightly soluble in ethanol and are insoluble in nonpolar solvents such as diethyl ether, dichloromethane, and benzene.

12.2 What Are the Cyclic Structures of Monosaccharides?

In Section 9.4C, we saw that aldehydes and ketones react with alcohols to form **hemiacetals.** We also saw that cyclic hemiacetals form very readily when hydroxyl and carbonyl groups are part of the same molecule and that their interaction produces a ring. For example, 4-hydroxypentanal forms a five-membered cyclic hemiacetal. Note that 4-hydroxypentanal contains one stereocenter and that hemiacetal formation generates a second stereocenter at carbon 1.

Monosaccharides have hydroxyl and carbonyl groups in the same molecule. As a result, they exist almost exclusively as five- and six-membered cyclic hemiacetals.

A. Haworth Projections

A common way of representing the cyclic structure of monosaccharides is the **Haworth projection,** named after the English chemist Sir Walter N. Haworth (Nobel Prize for chemistry, 1937). In a Haworth projection, a five- or six-membered cyclic hemiacetal is represented as a planar pentagon or hexagon, respectively, lying roughly perpendicular to the plane of the paper. Groups bonded to the carbons of the ring then lie either above or below the plane of the ring. The new carbon stereocenter created in forming the cyclic structure is called an **anomeric carbon.** Stereoisomers that differ in configuration only at the anomeric carbon are called **anomers.** The anomeric carbon of an aldose is carbon 1; that of the most common ketoses is carbon 2.

Typically, Haworth projections are most commonly drawn with the anomeric carbon to the right and the hemiacetal oxygen to the back (Figure 12.2).

In the terminology of carbohydrate chemistry, the designation β means that the —OH on the anomeric carbon of the cyclic hemiacetal lies on the same side of the ring as the terminal —CH_2OH. Conversely, the designation α means that the —OH on the anomeric carbon of the cyclic hemiacetal lies on the side of the ring opposite from the terminal —CH_2OH.

A six-membered hemiacetal ring is indicated by **-pyran-,** and a five-membered hemiacetal ring is indicated by **-furan-.** The terms **furanose** and **pyranose** are used because monosaccharide five- and six-membered rings correspond to the heterocyclic compounds furan and pyran.

Haworth projection A way to view furanose and pyranose forms of monosaccharides; the ring is drawn flat and viewed through its edge, with the anomeric carbon on the right and the oxygen atom to the rear

Anomeric carbon The hemiacetal carbon of the cyclic form of a monosaccharide

Anomers Monosaccharides that differ in configuration only at their anomeric carbons

Furanose A five-membered cyclic hemiacetal form of a monosaccharide

Pyranose A six-membered cyclic hemiacetal form of a monosaccharide

Furan   Pyran

Because the α and β forms of glucose are six-membered cyclic hemiacetals, they are named α-D-glucopyranose and β-D-glucopyranose, respectively. The

FIGURE 12.2 Haworth projections for α-D-glucopyranose and β-D-glucopyranose.

D-Glucose

Redraw to show the —OH on carbon 5 close to the aldehyde on carbon 1

Anomeric carbon

Anomeric carbon

β-D-Glucopyranose (β-D-Glucose)

α-D-Glucopyranose (α-D-Glucose)

designations *-furan-* and *-pyran-* are not always used in monosaccharide names, however. Thus, the glucopyranoses, for example, are often named simply α-D-glucose and β-D-glucose.

You would do well to remember the configurations of the groups on the Haworth projections of α-D-glucopyranose and β-D-glucopyranose as reference structures. Knowing how the open-chain configuration of any other aldohexose differs from that of D-glucose, you can construct Haworth projections for the aldohexose by referring to the Haworth projection of D-glucose.

Example 12.2 Drawing Haworth Projections

Draw Haworth projections for the α and β anomers of D-galactopyranose.

Strategy

A comparison of α and β anomers appears in Figure 12.2, with glucose as the example. The only modification needed is to change the structure of glucose to that of galactose.

Solution

One way to arrive at these projections is to use the α and β forms of D-glucopyranose as references and to remember (or discover by looking at Table 12.1) that D-galactose differs from D-glucose only in the configuration at carbon 4. Thus you can begin with the Haworth projection shown in Figure 12.2 and then invert the configuration at carbon 4.

α-D-Galactopyranose
(α-D-Galactose)

β-D-Galactopyranose
(β-D-Galactose)

Problem 12.2

D-Mannose exists in aqueous solution as a mixture of α-D-mannopyranose and β-D-mannopyranose. Draw Haworth projections for these molecules.

Aldopentoses also form cyclic hemiacetals. The most prevalent forms of D-ribose and other pentoses in the biological world are furanoses. Following are Haworth projections for α-D-ribofuranose (α-D-ribose) and β-2-deoxy-D-ribofuranose (β-2-Deoxy-D-ribose). The prefix *2-deoxy* indicates the absence of oxygen at carbon 2. Units of D-ribose and 2-deoxy-D-ribose in nucleic acids and most other biological molecules are found almost exclusively in the β configuration.

α-D-Ribofuranose
(α-D-Ribose)

β-2-Deoxy-D-ribofuranose
(β-2-Deoxy-D-ribose)

Fructose also forms five-membered cyclic hemiacetals. β-D-fructofuranose, for example, is found in the disaccharide sucrose (Section 12.4A).

α-D-Fructofuranose
(α-D-Fructose)

D-Fructose

β-D-Fructofuranose
(β-D-Fructose)

Anomeric carbon

B. Conformation Representations

A five-membered furanose ring is so close to being planar that Haworth projections provide adequate representations of furanoses. For pyranoses, however, the six-membered ring is more accurately represented as a **chair conformation** (Section 2.6B). Following are structural formulas for α-D-glucopyranose and β-D-glucopyranose, both drawn as chair conformations. Also shown is the open-chain or free aldehyde form with which the cyclic hemiacetal forms are in equilibrium in aqueous solution. Notice that each group, including the anomeric —OH group, on the chair conformation of β-D-glucopyranose is equatorial. Notice also that the —OH group on the anomeric carbon is axial in α-D-glucopyranose. Because the —OH on the anomeric carbon of β-D-glucopyranose is in the more stable equatorial position (Section 2.6B), the β anomer predominates in aqueous solution.

We do not show hydrogen atoms bonded to the ring in chair conformations. We often show them, however, in Haworth projections.

Anomeric carbon

β-D-Glucopyranose
$[\alpha]_D^{25} = +18.7°$

D-Glucose

α-D-Glucopyranose
$[\alpha]_D^{25} = 112°$

At this point, let us compare the relative orientations of groups on the D-glucopyranose ring in the Haworth projection and the chair conformation. The orientations of groups on carbons 1 through 5 of β-D-glucopyranose, for example, are up, down, up, down, and up in both representations. Note that in β-D-glucopyranose, all groups other than hydrogen atoms are in the stable equatorial position.

β-D-Glucopyranose
(Haworth projection)

β-D-Glucopyranose
(chair conformation)

Chemical Connections 12B

L-Ascorbic Acid (Vitamin C)

The structure of L-ascorbic acid (vitamin C) resembles that of a monosaccharide. In fact, this vitamin is synthesized both biochemically by plants and some animals and commercially from D-glucose. Humans do not have the enzymes required to carry out this synthesis. For this reason, we must obtain vitamin C in the food we eat or as a vitamin supplement. Approximately 66 million kg of vitamin C is synthesized annually in the United States.

L-Ascorbic acid
(Vitamin C)

oxidation ⇌ reduction

L-Dehydroascorbic acid

Charles D. Winters/Cengage Learning

Oranges are a major source of vitamin C.

Example 12.3 Chair Conformations

Draw chair conformations for α-D-galactopyranose and β-D-galactopyranose. Label the anomeric carbon in each.

Strategy and Solution

The configuration of D-galactose differs from that of D-glucose only at carbon 4. Therefore, draw the α and β forms of D-glucopyranose and then interchange the positions of the —OH and —H groups on carbon 4.

β-D-Galactopyranose
(β-D-Galactose)

D-Galactose

α-D-Galactopyranose
(α-D-Galactose)

Problem 12.3

Draw chair conformations for α-D-mannopyranose and β-D-mannopyranose. Label the anomeric carbon in each.

C. Mutarotation

Mutarotation is the change in specific rotation that accompanies the equilibration of α and β anomers in aqueous solution. For example, a solution prepared by dissolving crystalline α-D-glucopyranose in water has a specific rotation of +112°, which gradually decreases to an equilibrium value of +52.7° as α-D-glucopyranose reaches equilibrium with β-D-glucopyranose. A solution of β-D-glucopyranose also undergoes mutarotation, during which the specific rotation changes from +18.7° to the same equilibrium value of +52.7°. The equilibrium mixture consists of 64% β-D-glucopyranose and 36% α-D-glucopyranose, with only a trace (0.003%) of the open-chain form. Mutarotation is common to all carbohydrates that exist in hemiacetal forms.

Mutarotation The change in specific rotation that occurs when an α or β form of a carbohydrate is converted to an equilibrium mixture of the two forms

β-D-Glucopyranose
$[\alpha]_D^{25} = +18.7°$

Open-chain form

α-D-Glucopyranose
$[\alpha]_D^{25} = +112°$

12.3 What Are the Characteristic Reactions of Monosaccharides?

A. Formation of Glycosides (Acetals)

As we saw in Section 9.4C, treatment of an aldehyde or ketone with one molecule of alcohol yields a hemiacetal, and treatment of the hemiacetal with a molecule of alcohol yields an acetal. Treatment of a monosaccharide—all forms of which exist almost exclusively as cyclic hemiacetals—with an alcohol also yields an acetal, as illustrated by the reaction of β-D-glucopyranose with methanol.

β-D-Glucopyranose
(β-D-Glucose)

Methyl β-D-glucopyranoside
(Methyl β-D-glucoside)

Methyl α-D-glucopyranoside
(Methyl α-D-glucoside)

A cyclic acetal derived from a monosaccharide is called a **glycoside,** and the bond from the anomeric carbon to the —OR group is called a **glycosidic bond.** Mutarotation is not possible in a glycoside because an acetal—unlike a hemiacetal—is no longer in equilibrium with the open-chain carbonyl-containing compound. Glycosides are stable in water and aqueous base; like other acetals (Section 9.4C), however, they are hydrolyzed in aqueous acid to an alcohol and a monosaccharide.

Glycoside A carbohydrate in which the —OH group on its anomeric carbon is replaced by an —OR group

Glycosidic bond The bond from the anomeric carbon of a glycoside to an —OR group

We name glycosides by listing the alkyl or aryl group bonded to oxygen, followed by the name of the carbohydrate in which the ending **-e** is replaced by **-ide**. For example, the methyl glycoside derived from β-D-glucopyranose is named methyl β-D-glucopyranoside; that derived from β-D-ribofuranose is named methyl β-D-ribofuranoside.

Example 12.4 Finding the Anomeric Carbon and Glycosidic Bond

Draw a structural formula for methyl β-D-ribofuranoside (methyl β-D-riboside). Label the anomeric carbon and the glycosidic bond.

Strategy

Furanosides are five-membered rings. The anomeric carbon is carbon one, and the glycosidic bond is formed at the anomeric carbon.

Solution

Methyl β-D-ribofuranoside
(Methyl β-D-riboside)

Problem 12.4

Draw a Haworth projection and a chair conformation for methyl α-D-mannopyranoside (methyl α-D-mannoside). Label the anomeric carbon and the glycosidic bond.

B. Reduction to Alditols

Oxidations and reductions of monosaccharides in nature are catalyzed by specific enzymes classified as oxidases—for example, glucose oxidase.

Alditol The product formed when the CHO group of a monosaccharide is reduced to a CH_2OH group

The carbonyl group of a monosaccharide can be reduced to a hydroxyl group by a variety of reducing agents, including hydrogen in the presence of a transition metal catalyst and sodium borohydride (Section 9.4C). The reduction products are known as **alditols.** Reduction of D-glucose gives D-glucitol, more commonly known as D-sorbitol. Here, D-glucose is shown in the open-chain form. Only a small amount of this form is present in solution but, as it is reduced, the equilibrium between cyclic hemiacetal forms (only the β form is shown here) and the open-chain form shifts to replace it.

β-D-Glucopyranose   D-Glucose   D-Glucitol
(D-Sorbitol)

We name alditols by dropping the **-ose** from the name of the monosaccharide and adding **-itol.** Sorbitol is found in the plant world in many berries and in cherries, plums, pears, apples, seaweed, and algae. It is about 60% as sweet as sucrose (table sugar) and is used in the manufacture of candies and as a sugar substitute for diabetics. Other alditols commonly found in the biological world include erythritol, D-mannitol, and xylitol. Xylitol is used as a sweetening agent in "sugarless" gum, candy, and sweet cereals.

Many "sugar-free" products contain sugar alcohols, such as sorbitol and xylitol.

C. Oxidation to Aldonic Acids (Reducing Sugars)

As we saw in Section 9.4A, aldehydes (RCHO) are oxidized to carboxylic acids (RCOOH) by several agents, including oxygen, O_2. Similarly, the aldehyde group of an aldose can be oxidized, under basic conditions, to a carboxylate group. Under these conditions, the cyclic form of the aldose is in equilibrium with the open-chain form, which is then oxidized by the mild oxidizing agent. D-Glucose, for example, is oxidized to D-gluconate (the anion of D-gluconic acid).

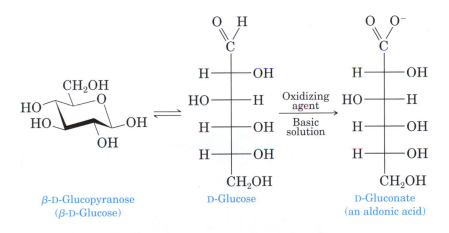

Any carbohydrate that reacts with an oxidizing agent to form an aldonic acid is classified as a **reducing sugar** (it reduces the oxidizing agent).

Surprisingly, 2-ketoses are also reducing sugars. Carbon 1 (a CH_2OH group) of a ketose is not oxidized directly. Instead, under the basic conditions of this oxidation, a 2-ketose exists in equilibrium with an aldose by way of an enediol intermediate. The aldose is then oxidized by the mild oxidizing agent.

Reducing sugar A carbohydrate that reacts with a mild oxidizing agent under basic conditions to give an aldonic acid; the carbohydrate reduces the oxidizing agent

Chemical Connections 12C

Testing for Glucose

The analytical procedure most often performed in a clinical chemistry laboratory is the determination of glucose in blood, urine, or other biological fluids. The high frequency with which this test is performed reflects the high incidence of diabetes mellitus. Nearly 20 million known diabetics live in the United States, and it is estimated that millions more remain undiagnosed.

Diabetes mellitus (Chemical Connections 16F) is characterized by insufficient blood levels of the hormone insulin. If the blood concentration of insulin is too low, muscle and liver cells do not absorb glucose from the blood; this problem, in turn, leads to increased levels of blood glucose (hyperglycemia), impaired metabolism of fats and proteins, ketosis, and possible diabetic coma. A rapid test for blood glucose levels is critical for early diagnosis and effective management of this disease. In addition to giving results rapidly, a test must be specific for D-glucose; that is, it must give a positive test for glucose but not react with any other substances normally present in biological fluids.

Today, blood glucose levels are measured by an enzyme-based procedure using the enzyme glucose oxidase. This enzyme catalyzes the oxidation of β-D-glucose to D-gluconic acid.

Glucose oxidase is specific for β-D-glucose. Therefore, complete oxidation of any sample containing both β-D-glucose and α-D-glucose requires conversion of the α form to the β form. Fortunately, this interconversion is rapid and complete in the short time required for the test.

Molecular oxygen, O_2, is the oxidizing agent in this reaction and is reduced to hydrogen peroxide, H_2O_2. In one procedure, hydrogen peroxide formed in the glucose oxidase–catalyzed reaction oxidizes colorless o-toluidine to a colored product in a reaction catalyzed by the enzyme peroxidase. The concentration of the colored oxidation product is determined spectrophotometrically and is proportional to the concentration of glucose in the test solution.

2-Methylaniline
(o-Toluidine)

Several commercially available test kits use the glucose oxidase reaction for qualitative determination of glucose in urine.

β-D-Glucopyranose
(β-D-Glucose)

D-Gluconic acid

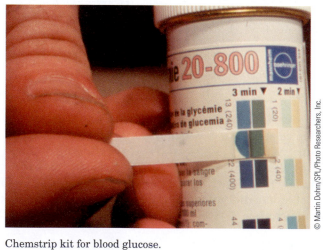

Chemstrip kit for blood glucose.

D. Oxidation to Uronic Acids

Enzyme-catalyzed oxidation of the primary alcohol at carbon 6 of a hexose yields a uronic acid. Enzyme-catalyzed oxidation of D-glucose, for example, yields D-glucuronic acid, shown here in both its open-chain and cyclic hemiacetal forms:

| Fischer projection | Chair conformation |

D-Glucuronic acid
(a uronic acid)

D-glucuronic acid is widely distributed in both the plant and animal worlds. In humans, it serves as an important component of the acidic polysaccharides of connective tissues (Section 12.6A). The body also uses D-glucuronic acid to detoxify foreign phenols and alcohols. In the liver, these compounds are converted to glycosides of glucuronic acid (glucuronides) and excreted in the urine. The intravenous anesthetic propofol, for example, is converted to the following glucuronide and then excreted in urine:

Propofol   A urine-soluble D-glucuronide

E. The Formation of Phosphoric Esters

Mono- and diphosphoric esters are important intermediates in the metabolism of monosaccharides. For example, the first step in glycolysis (Section 20.2) involves conversion of glucose to glucose 6-phosphate. Note that phosphoric acid is a strong enough acid so that at the pH of cellular and intercellular fluids, both acidic protons of a phosphoric ester are ionized, giving the ester a charge of −2.

D-Glucose   D-Glucose 6-phosphate   α-D-Glucose 6-phosphate

Chemical Connections 12D

A, B, AB, and O Blood Types

Membranes of animal plasma cells have large numbers of relatively small carbohydrates bound to them. In fact, the outsides of most plasma cell membranes are literally "sugar-coated." These membrane-bound carbohydrates are part of the mechanism by which cell types recognize one another and, in effect, act as biochemical markers. Typically, they contain from 4 to 17 monosaccharide units consisting primarily of relatively few monosaccharides, the most common of which are D-galactose, D-mannose, L-fucose, *N*-acetyl-D-glucosamine, and *N*-acetyl-D-galactosamine. L-Fucose is a 6-deoxyaldohexose.

An L-monosaccharide because this —OH group is on the left in the Fischer projection

Carbon 6 is —CH₃ rather than —CH₂OH

L-Fucose

To see the importance of these membrane-bound carbohydrates, consider the ABO blood group system, discovered in 1900 by Karl Landsteiner (1868–1943). Whether an individual belongs to type A, B, AB, or O is genetically determined and depends on the type of trisaccharide or tetrasaccharide bound to the surface of the red blood cells. These surface-bound carbohydrates, designated as A, B, and O, act as antigens. The type of glycosidic bond joining each monosaccharide is shown in the figure.

The blood carries antibodies against foreign substances. When a person receives a blood transfusion, the antibodies clump together (aggregate) the foreign blood cells. Type A blood, for example, has A antigens (*N*-acetyl-D-galactosamine) on the surfaces of its red blood cells and carries anti-B antibodies (against B antigen). B-type blood carries B antigen (D-galactose) and has anti-A antibodies (against A antigens). Transfusion of type A blood into a person with type B blood can be fatal, and vice versa. The relationships between blood type and donor–receiver interactions are summarized in the figure.

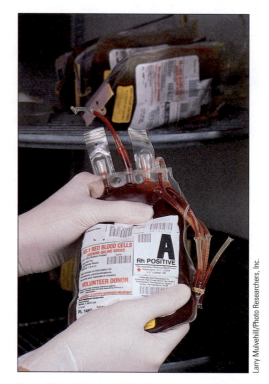

Bag of blood showing blood type.

Chemical Connections 12D

A, B, AB, and O Blood Types (continued)

Sugar on cell surface: O
Has antibodies against: A and B
Can receive blood from: O
Can donate blood to: O, A, B, and AB

Type O

Type A

Sugar on cell surface: A
Has antibodies against: B
Can receive blood from: A and O
Can donate blood to: A and AB

Type B

Sugar on cell surface: B
Has antibodies against: A
Can receive blood from: B and O
Can donate blood to: B and AB

Type AB

Sugars on cell surface: A and B
Has antibodies against: None
Can receive blood from: O, A, B, and AB
Can donate blood to: AB

People with type O blood are universal donors, and those with type AB blood are universal acceptors. People with type A blood can accept blood from type A or type O donors only. Those with type B blood can accept blood from type B or type O donors only. Type O persons can accept blood only from type O donors.

12.4 What Are Disaccharides and Oligosaccharides?

Most carbohydrates in nature contain more than one monosaccharide unit. Those that contain two units are called **disaccharides,** those that contain three units are called **trisaccharides,** and so forth. We use the general term **oligosaccharide** to describe carbohydrates that contain from six to ten monosaccharide units. Carbohydrates containing larger numbers of monosaccharide units are called **polysaccharides.**

In a disaccharide, two monosaccharide units are joined by a glycosidic bond between the anomeric carbon of one unit and an —OH group of the other unit. Three important disaccharides are sucrose, lactose, and maltose.

A. Sucrose

Sucrose (table sugar) is the most abundant disaccharide in the biological world. It is obtained principally from the juice of sugar cane and sugar beets. In sucrose, carbon 1 of α-D-glucopyranose bonds to carbon 2 of D-fructofuranose by an α-1,2-glycosidic bond. Because the anomeric carbons of both the glucopyranose and fructofuranose units are involved in formation of the glycosidic

Disaccharide A carbohydrate containing two monosaccharide units joined by a glycosidic bond

Oligosaccharide A carbohydrate containing from six to ten monosaccharide units, each joined to the next by a glycosidic bond

Polysaccharide A carbohydrate containing a large number of monosaccharide units, each joined to the next by one or more glycosidic bonds

In the production of sucrose, sugar cane or sugar beet is boiled with water, and the resulting solution is cooled. Sucrose crystals separate and are collected. Subsequent boiling to concentrate the solution followed by cooling yields a dark, thick syrup known as molasses.

bond, neither monosaccharide unit is in equilibrium with its open-chain form. Thus sucrose is a nonreducing sugar.

Sucrose

B. Lactose

Lactose is the principal sugar present in milk. It accounts for 5 to 8% of human milk and 4 to 6% of cow's milk. This disaccharide consists of D-galactopyranose bonded by a β-1,4-glycosidic bond to carbon 4 of D-glucopyranose. Lactose is a reducing sugar, because the cyclic hemiacetal of the D-glucopyranose unit is in equilibrium with its open-chain form and can be oxidized to a carboxyl group.

Lactose

C. Maltose

Maltose is an ingredient in most syrups.

Maltose derives its name from its presence in malt, the juice from sprouted barley and other cereal grains. It consists of two units of D-glucopyranose joined by a glycosidic bond between carbon 1 (the anomeric carbon) of one unit and carbon 4 of the other unit. Because the oxygen atom on the anomeric carbon of the first glucopyranose unit is alpha, the bond joining the two units is an α-1,4-glycosidic bond. Following are a Haworth projection and a chair conformation for β-maltose, so named because the —OH groups on the anomeric carbon of the glucose unit on the right are beta.

Maltose

TABLE 12.3 Relative Sweetness of Some Carbohydrate and Artificial Sweetening Agents

Carbohydrate	Sweetness Relative to Sucrose	Artificial Sweetener	Sweetness Relative to Sucrose
fructose	1.74	saccharine	450
sucrose (table sugar)	1.00	acesulfame-K	200
honey	0.97	aspartame	180
glucose	0.74	sucralose	600
maltose	0.33		
galactose	0.32		
lactose (milk sugar)	0.16		

Maltose is a reducing sugar; the hemiacetal group on the right unit of D-glucopyranose is in equilibrium with the free aldehyde and can be oxidized to a carboxylic acid.

D. Relative Sweetness

Among the disaccharide sweetening agents, D-fructose tastes the sweetest—even sweeter than sucrose (Table 12.3). The sweet taste of honey is due largely to D-fructose and D-glucose. Lactose has almost no sweetness and is sometimes added to foods as a filler. Some people cannot tolerate lactose well, however, and should avoid these foods.

We have no mechanical way to measure sweetness. Such testing is done by having a group of people taste and rate the sweetness of solutions of varying sweetening agents.

Example 12.5 Drawing Chair Conformations for a Disaccharide

Draw a chair conformation for the β anomer of a disaccharide in which two units of D-glucopyranose are joined by an α-1,6-glycosidic bond.

Strategy

Three points are needed here. The first is the chair conformation of α-D-glucopyranose. The second is the α-1,6-glycosidic bond between the two glucopyranose molecules. The third is the correct confomation of the anomeric carbon at the reducing end, β in this case.

Solution

First draw a chair conformation of α-D-glucopyranose. Then connect the anomeric carbon of this monosaccharide to carbon 6 of a second D-glucopyranose unit by an α-1,6-glycosidic bond. The resulting molecule is either α or β depending on the orientation of the —OH group on the reducing end of the disaccharide. The disaccharide shown here is the β form.

Cellulose molecules act much like stiff rods, a characteristic that enables them to align themselves side by side into well-organized, water-insoluble fibers in which the OH groups form numerous intermolecular hydrogen bonds. This arrangement of parallel chains in bundles gives cellulose fibers their high mechanical strength. It also explains why cellulose is insoluble in water. When a piece of cellulose-containing material is placed in water, there are not enough —OH groups on the surface of the fiber to pull individual cellulose molecules away from the strongly hydrogen-bonded fiber.

Humans and other animals cannot use cellulose as food because our digestive systems do not contain β-glucosidases, enzymes that catalyze the hydrolysis of β-glucosidic bonds. Instead, we have only α-glucosidases; hence, we use the polysaccharides starch and glycogen as sources of glucose. In contrast, many bacteria and microorganisms do contain β-glucosidases and so can digest cellulose. Termites (much to our regret) have such bacteria in their intestines and can use wood as their principal food. Ruminants (cud-chewing animals) and horses can also digest grasses and hay because β-glucosidase-containing microorganisms are present in their alimentary systems.

12.6 What Are Acidic Polysaccharides?

Acidic polysaccharides are a group of polysaccharides that contain carboxyl groups and/or sulfuric ester groups. Acidic polysaccharides play important roles in the structure and function of connective tissues. Because they contain amino sugars, a more current name for these substances is glycosaminoglycans. There is no single general type of connective tissue. Rather, a large number of highly specialized forms exist, such as cartilage, bone, synovial fluid, skin, tendons, blood vessels, intervertebral disks, and cornea. Most connective tissues consist of collagen, a structural protein, combined with a variety of acidic polysaccharides (glycosaminoglycans) that interact with collagen to form tight or loose networks.

In rheumatoid arthritis, inflammation of the synovial tissue results in swelling of the joints.

A. Hyaluronic Acid

Hyaluronic acid is the simplest acidic polysaccharide present in connective tissue. It has a molecular weight of between 10^5 and 10^7 g/mol and contains from 300 to 100,000 repeating units, depending on the organ in which it occurs. It is most abundant in embryonic tissues and in specialized connective tissues such as synovial fluid, the lubricant of joints in the body, and the vitreous of the eye, where it provides a clear, elastic gel that holds the retina in its proper position. Hyaluronic acid is also a common ingredient in lotions, moisturizers, and cosmetics.

Hyaluronic acid is composed of D-glucuronic acid joined by a β-1,3-glycosidic bond to N-acetyl-D-glucosamine, which is in turn linked to D-glucuronic acid by a β-1,4-glycosidic bond.

The repeating unit of hyaluronic acid

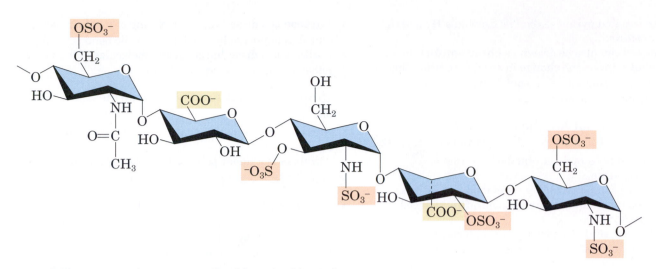

FIGURE 12.5 A repeating pentasaccharide unit of heparin.

B. Heparin

Heparin is a heterogeneous mixture of variably sulfonated polysaccharide chains, ranging in molecular weight from 6000 to 30,000 g/mol. This acidic polysaccharide is synthesized and stored in mast cells (cells that are part of the immune system and that occur in several types of tissues) of various tissues—particularly the liver, lungs, and gut. Heparin has many biological functions, the best known and fully understood of which is its anticoagulant activity. It binds strongly to antithrombin III, a plasma protein involved in terminating the clotting process. A heparin preparation with good anticoagulant activity contains a minimum of eight repeating units (Figure 12.5). The larger the molecule, the greater its anticoagulant activity. Because of this anticoagulant activity, it is widely used in medicine.

Summary of Key Questions

OWL End-of-chapter problems identified in blue are assignable in **GOB OWL.**

Section 12.1 Carbohydrates: What Are Monosaccharides? Problem 12.14

- **Monosaccharides** are polyhydroxyaldehydes or polyhydroxyketones.
- The most common have the general formula $C_nH_{2n}O_n$, where n varies from 3 to 8.
- Their names contain the suffix *-ose*, and the prefixes *tri-*, *tetr-*, and so on indicate the number of carbon atoms in the chain. The prefix *aldo-* indicates an aldehyde, and the prefix *keto-* indicates a ketone.

Section 12.2 What Are the Cyclic Structures of Monosaccharides?

- In a **Fischer projection** of a monosaccharide, we write the carbon chain vertically with the most highly oxidized carbon toward the top. Horizontal lines represent groups projecting above the plane of the page; vertical

lines represent groups projecting behind the plane of the page.
- The penultimate carbon of a monosaccharide is the next-to-last carbon of a Fischer projection.
- A monosaccharide that has the same configuration at the penultimate carbon as D-glyceraldehyde is called a **D-monosaccharide;** one that has the same configuration at the penultimate carbon as L-glyceraldehyde is called an **L-monosaccharide.**
- Monosaccharides exist primarily as cyclic hemiacetals.
- A six-membered cyclic hemiacetal is a **pyranose;** a five-membered cyclic hemiacetal is a **furanose.**
- The new stereocenter resulting from hemiacetal formation is called an **anomeric carbon,** and the stereoisomers formed in this way are called **anomers.**
- The symbol **β-** indicates that the —OH group on the anomeric carbon lies on the same side of the ring as the terminal —CH₂OH.
- The symbol **α-** indicates that the —OH group on the anomeric carbon lies on the opposite side from the terminal —CH₂OH.

- Furanoses and pyranoses can be drawn as **Haworth projections.**
- Pyranoses can also be drawn as **chair conformations.**
- **Mutarotation** is the change in specific rotation that accompanies formation of an equilibrium mixture of α and β anomers in aqueous solution.

Section 12.3 What Are the Characteristic Reactions of Monosaccharides? **Problems 12.34, 12.37**

- A **glycoside** is a cyclic acetal derived from a monosaccharide.
- An **alditol** is a polyhydroxy compound formed when the carbonyl group of a monosaccharide is reduced to a hydroxyl group.
- An **aldonic acid** is a carboxylic acid formed when the aldehyde group of an aldose is oxidized to a carboxyl group.
- Any carbohydrate that reacts with an oxidizing agent to form an aldonic acid is classified as a **reducing sugar** (it reduces the oxidizing agent).

Section 12.4 What Are Disaccharides and Oligosaccharides?

- A **disaccharide** contains two monosaccharide units joined by a glycosidic bond.
- Terms applied to carbohydrates containing larger numbers of monosaccharides are **trisaccharide, tetrasaccharide, oligosaccharide,** and **polysaccharide.**
- **Sucrose** is a disaccharide consisting of D-glucose joined to D-fructose by an α-1,2-glycosidic bond.

- **Lactose** is a disaccharide consisting of D-galactose joined to D-glucose by a β-1,4-glycosidic bond.
- **Maltose** is a disaccharide of two molecules of D-glucose joined by an α-1,4-glycosidic bond.

Section 12.5 What Are Polysaccharides?
Problem 12.44

- **Starch** can be separated into two fractions: amylose and amylopectin.
- **Amylose** is a linear polysaccharide of as many as 4000 units of D-glucopyranose joined by α-1,4-glycosidic bonds.
- **Amylopectin** is a highly branched polysaccharide of D-glucose joined by α-1,4-glycosidic bonds and, at branch points, by α-1,6-glycosidic bonds.
- **Glycogen,** the reserve carbohydrate of animals, is a highly branched polysaccharide of D-glucopyranose joined by α-1,4-glycosidic bonds and, at branch points, by α-1,6-glycosidic bonds.
- **Cellulose,** the skeletal polysaccharide of plants, is a linear polysaccharide of D-glucopyranose joined by β-1,4 glycosidic bonds.

Section 12.6 What Are Acidic Polysaccharides?

- The carboxyl and sulfate groups of **acidic polysaccharides** are ionized to —COO⁻ and —SO₃⁻ at the pH of body fluids, which gives these polysaccharides net negative charges.

Summary of Key Reactions

1. **Formation of Cyclic Hemiacetals (Section 12.2)** A monosaccharide existing as a five-membered ring is a furanose; one existing as a six-membered ring is a pyranose. A pyranose is most commonly drawn as either a Haworth projection or a chair conformation.

D-Glucose → β-D-Glucopyranose (β-D-Glucose)

β-D-Glucopyranose
$[\alpha]_D^{25} = +18.7°$

Open-chain form

α-D-Glucopyranose
$[\alpha]_D^{25} = 112°$

2. **Mutarotation (Section 12.2C)** Anomeric forms of a monosaccharide are in equilibrium in aqueous solution. Mutarotation is the change in specific rotation that accompanies this equilibration.

3. **Formation of Glycosides (Section 12.3A)** Treatment of a monosaccharide with an alcohol in the presence of an acid catalyst forms a cyclic acetal called

a glycoside. The bond to the new —OR group is called a glycosidic bond.

$\xrightarrow[-H_2O]{H^+}$

4. Reduction to Alditols (Section 12.3B) Reduction of the carbonyl group of an aldose or ketose to a hydroxyl group yields a polyhydroxy compound called an alditol.

β-D-Glucopyranose   D-Glucose   D-Glucitol (D-Sorbitol)

5. Oxidation to an Aldonic Acid (Section 12.3C) Oxidation of the aldehyde group of an aldose to a carboxyl group by a mild oxidizing agent gives a polyhydroxy-carboxylic acid called an aldonic acid.

D-Glucose   D-Gluconic acid

Problems

■ Indicates problems that are assignable in **GOB OWL**.

Blue numbered problems are applied.

Go to this book's companion website at **www. cengage.com/chemistry/bettelheim** for interactive versions of the How To tutorials and Active Figures, and to quiz yourself on this chapter.

Section 12.1 Carbohydrates: What Are Monosaccharides?

12.6 Define *carbohydrate*.

12.7 What is the difference in structure between an aldose and a ketose? Between an aldopentose and a ketopentose?

12.8 Of the eight D-aldohexoses, which is the most abundant in the biological world?

12.9 Name the three most abundant hexoses in the biological world. Which are aldohexoses, and which are ketohexoses?

12.10 Which hexose is also known as "dextrose"?

12.11 What does it mean to say that D- and L-glyceraldehyde are enantiomers?

12.12 Explain the meaning of the designations D and L as used to specify the configuration of a monosaccharide.

12.13 Which carbon of an aldopentose determines whether the pentose has a D or L configuration?

12.14 ■ How many stereocenters are present in D-glucose? In D-ribose? How many stereoisomers are possible for each monosaccharide?

12.15 Which of the following compounds are D-monosaccharides, and which are L-monosaccharides?

12.16 Draw Fischer projections for L-ribose and L-arabinose.

12.17 Draw a Fischer projection for a D-2-ketoheptose.

12.18 Explain why all mono- and disaccharides are soluble in water.

12.19 What is an amino sugar? Name the three amino sugars most commonly found in nature.

Section 12.2 What Are the Cyclic Structures of Monosaccharides?

12.20 Define the term *anomeric carbon*. Which carbon is the anomeric carbon in glucose? In fructose?

12.21 Define (a) pyranose and (b) furanose.

12.22 Explain the conventions for using α and β to designate the configurations of the cyclic forms of monosaccharides.

12.23 Are α-D-glucose and β-D-glucose anomers? Explain. Are they enantiomers? Explain.

12.24 Are the hydroxyl groups on carbons 1, 2, 3, and 4 of α-D-glucose all in equatorial positions?

12.25 In what way are chair conformations a more accurate representation of the molecular shapes of hexopyranoses than Haworth projections?

12.26 Convert each of the following Haworth projections to an open-chain form and to a Fischer projection. Name the monosaccharides you have drawn.

12.27 Convert each of the following chair conformations to an open-chain form and to a Fischer projection. Name the monosaccharides you have drawn.

12.28 Explain the phenomenon of mutarotation. How is it detected?

12.29 The specific rotation of α-D-glucose is +112.2°. What is the specific rotation of α-L-glucose?

12.30 When α-D-glucose is dissolved in water, the specific rotation of the solution changes from +112.2° to +52.7°. Does the specific rotation of α-L-glucose also change when it is dissolved in water? If so, to what value does it change?

12.31 Define the terms *glycoside* and *glycosidic bond*.

12.32 What is the difference in meaning between the terms *glycosidic bond* and *glucosidic bond*?

12.33 Do glycosides undergo mutarotation?

Section 12.3 What Are the Characteristic Reactions of Monosaccharides?

12.34 ■ Draw Fischer projections for the product formed by treating each of the following monosaccharides with sodium borohydride, $NaBH_4$, in water.

(a) D-Galactose (b) D-Ribose

12.35 Reduction of D-glucose by $NaBH_4$ gives D-sorbitol, a compound used in the manufacture of sugar-free gums and candies. Draw a structural formula for D-sorbitol.

12.36 Reduction of D-fructose by $NaBH_4$ gives two alditols, one of which is D-sorbitol. Name and draw a structural formula for the other alditol.

12.37 ■ Ribitol and β-D-ribose 1-phosphate are derivatives of D-ribose. Draw a structural formula for each compound.

Section 12.4 What Are Disaccharides and Oligosaccharides?

12.38 Name three important disaccharides. From which monosaccharides is each derived?

12.39 What does it mean to describe a glycosidic bond as β-1,4-? To describe it as α-1,6-?

12.40 Both maltose and lactose are reducing sugars, but sucrose is a nonreducing sugar. Explain why.

12.41 Following is a structural formula for a disaccharide.

(a) Name the two monosaccharide units in the disaccharide.

(b) Describe the glycosidic bond.

(c) Is this disaccharide a reducing sugar or a nonreducing sugar?

(d) Will this disaccharide undergo mutarotation?

12.42 The disaccharide trehalose is found in young mushrooms and is the chief carbohydrate in the blood of certain insects.

Trehalose

(a) Name the two monosaccharide units in trehalose.

(b) Describe the glycosidic bond in trehalose.

(c) Is trehalose a reducing sugar or a nonreducing sugar?

(d) Will trehalose undergo mutarotation?

Section 12.5 What Are Polysaccharides?

12.43 What is the difference in structure between oligosaccharides and polysaccharides?

12.44 ■ Name three polysaccharides that are composed of units of D-glucose. In which polysaccharide are the glucose units joined by α-glycosidic bonds? In which are they joined by β-glycosidic bonds?

12.45 Starch can be separated into two principal polysaccharides, amylose and amylopectin. What is the major difference in structure between the two?

12.46 Where is glycogen stored in the human body?

12.47 Why is cellulose insoluble in water?

12.48 How is it possible that cows can digest grass but humans cannot?

12.49 A Fischer projection of N-acetyl-D-glucosamine is given in Section 12.1D.

(a) Draw a Haworth projection and a chair conformation for the β-pyranose form of this monosaccharide.

(b) Draw a Haworth projection and a chair conformation for the disaccharide formed by joining two units of the pyranose form of N-acetyl-D-glucosamine by a β-1,4-glycosidic bond. If you draw them correctly, you will have a structural formula for the repeating dimer of chitin, the structural polysaccharide component of the shell of lobsters and other crustaceans.

12.50 Propose structural formulas for the repeating disaccharide unit in these polysaccharides.

(a) Alginic acid, isolated from seaweed, is used as a thickening agent in ice cream and other foods. Alginic acid is a polymer of D-mannuronic acid in the pyranose form joined by β-1,4-glycosidic bonds.

(b) Pectic acid is the main component of pectin, which is responsible for the formation of jellies from fruits and berries. Pectic acid is a polymer of D-galacturonic acid in the pyranose form joined by α-1,4-glycosidic bonds.

D-Mannuronic acid   D-Galacturonic acid

Section 12.6 What Are Acidic Polysaccharides?

12.51 Hyaluronic acid acts as a lubricant in the synovial fluid of joints. In rheumatoid arthritis, inflammation breaks hyaluronic acid down to smaller molecules. Under these conditions, what happens to the lubricating power of the synovial fluid?

12.52 The anticlotting property of heparin is partly due to the negative charges it carries.

(a) Identify the functional groups that provide the negative charges.

(b) Which type of heparin is a better anticoagulant, one with a high degree or a low degree of polymerization?

Chemical Connections

12.53 (Chemical Connections 12A) Why does congenital galactosemia appear only in infants? Why can galactosemia be relieved by feeding an affected infant a formula containing sucrose as the only carbohydrate?

12.54 (Chemical Connections 12B) What is the difference in structure between L-ascorbic acid and L-dehydroascorbic acid? What does the designation L indicate in these names?

12.55 (Chemical Connections 12B) When L-ascorbic acid participates in a redox reaction, it is converted to L-dehydroascorbic acid. In this reaction, is L-ascorbic acid oxidized or reduced? Is L-ascorbic acid a biological oxidizing agent or a biological reducing agent?

12.56 (Chemical Connections 12C) Why is the glucose assay one of the most common analytical tests performed in clinical chemistry laboratories?

12.57 (Chemical Connections 12D) What monosaccharides do types A, B, and O blood have in common? In which monosaccharides do they differ?

12.58 (Chemical Connections 12D) L-Fucose is a monosaccharide unit common to A, B, AB, and O blood types.
   (a) Is this monosaccharide an aldose or a ketose? A hexose or a pentose?
   (b) What is unusual about this monosaccharide?
   (c) If L-fucose were to undergo a reaction in which its terminal —CH₃ group was converted to a —CH₂OH group, which monosaccharide would result?

12.59 (Chemical Connections 12D) Why can't a person with type A blood donate to a person with type B blood?

12.60 (Chemical Connections 12E) How can chitosan dressings stop blood loss in wounds even though gauze dressings cannot do so?

Additional Problems

12.61 ■ 2,6-Dideoxy-D-altrose, also known as D-digitoxose, is a monosaccharide obtained from the hydrolysis of digitoxin, a natural product extracted from purple foxglove (*Digitalis purpurea*). Digitoxin has found wide use in cardiology because it reduces pulse rate, regularizes heart rhythm, and strengthens heart beat. Draw a structural formula for the open-chain form of 2,6-dideoxy-D-altrose.

12.62 In making candy or sugar syrups, sucrose is boiled in water with a little acid, such as lemon juice. Why does the product mixture taste sweeter than the initial sucrose solution?

12.63 Hot-water extracts of ground willow bark are an effective pain reliever (Chemical Connections 11C). Unfortunately, the liquid is so bitter that most people refuse it. The pain reliever in these infusions is salicin. Name the monosaccharide unit in salicin.

Salicin

12.64 Show how D-sorbitol, used in "sugarless" chewing gum is produced from D-glucose.

12.65 What is the source of chitosan, a polysaccharide used in newly developed dressings for wounds?

12.66 Does the planar Haworth representation give a reasonably good three-dimensional picture of the structure of furanoses, such as ribose?

12.67 In Section 12.4A, two structures are shown for sucrose. In one, both the glucose and fructose moieties are shown in the Haworth representation.

In the other, the chair form of the glucose moiety is shown with a glycosidic bond to the fructose moiety shown in the Haworth form. Why does the Haworth form for the fructose moiety appear in both structures?

12.68 Blood samples for research or medical tests sometimes have heparin added. Why is this done?

12.69 What is the difference in the glycosidic bonds in starch and cellulose? How does this difference affect their biological function?

12.70 What are the structural differences between vitamin C and sugars? Do these structural differences play a role in the susceptibility of this vitamin to air oxidation?

12.71 What is the role of amino sugars in the structure of carbohydrates?

Looking Ahead

12.72 One pathway for the metabolism of D-glucose 6-phosphate is its enzyme-catalyzed conversion to D-fructose 6-phosphate. Show that this transformation can be regarded as two enzyme-catalyzed keto-enol tautomerizations (Section 9.5).

D-Glucose 6-phosphate   D-Fructose 6-phosphate

12.73 One step in glycolysis, the pathway that converts glucose to pyruvate (Section 20.2), involves an enzyme-catalyzed conversion of dihydroxyacetone phosphate to D-glyceraldehyde 3-phosphate. Show that this transformation can be regarded as two enzyme-catalyzed keto-enol tautomerizations (Section 9.5).

Dihydroxyacetone phosphate   D-Glyceraldehyde 3-phosphate

12.74 Following is a Haworth projection and a chair conformation for the repeating disaccharide unit in chondroitin 6-sulfate. This biopolymer acts as the flexible connecting matrix between the tough protein filaments in cartilage. It is available as a dietary supplement, often combined with D-glucosamine sulfate.

Some believe this combination can strengthen and improve joint flexibility.

(a) From what two monosaccharide units is the repeating disaccharide unit of chondroitin 6-sulfate derived?

(b) Describe the glycosidic bond between the two units.

12.75 Following is the structural formula of coenzyme A, an important biomolecule.

(a) Is coenzyme A chiral?

(b) Name each functional group in coenzyme A.

(c) Would you expect coenzyme A to be soluble in water? Explain.

(d) Draw structural formulas for the products of complete hydrolysis of coenzyme A in aqueous HCl. Show each product as it would be ionized in this solution.

(e) Draw structural formulas for the products of complete hydrolysis of coenzyme A in aqueous NaOH. Show each product as it would be ionized in this solution.

Coenzyme A

Lipids

Sea lions are marine mammals that require a heavy layer of fat to survive in cold waters.

13.1 What Are Lipids?

Found in living organisms, **lipids** are a family of substances that are insoluble in water but soluble in nonpolar solvents and solvents of low polarity, such as diethyl ether. Unlike the case with carbohydrates, we define lipids in terms of a property and not in terms of their structure.

A. Classification by Function

Lipids play three major roles in human biochemistry: (1) They store energy within fat cells, (2) they are parts of membranes that separate compartments of aqueous solutions from each other, and (3) they serve as chemical messengers.

Storage

An important use for lipids, especially in animals, is the storage of energy. As we saw in Section 12.5, plants store energy in the form of starch. Animals (including humans) find it more economical to use fats instead. Although our bodies do store some carbohydrates in the form of glycogen for quick energy when we need it, energy stored in the form of fats has much greater importance for us. The reason is simple: The burning of fats produces more than twice as much energy (about 9 kcal/g) as the burning of an equal weight of carbohydrates (about 4 kcal/g).

Membrane Components

The lack of water solubility of lipids is an important property because our body chemistry is so heavily based on water. Most body constituents, including carbohydrates and proteins, are soluble in water. However, the body also needs insoluble compounds for the membranes that separate compartments containing aqueous solutions, whether they are cells or organelles within the cells. Lipids provide these membranes. Their water insolubility derives from the fact that the polar groups they contain are much smaller than their alkane-like (nonpolar) portions. These nonpolar portions provide the water-repellent, or *hydrophobic,* property.

Messengers

Lipids also serve as chemical messengers. Primary messengers, such as steroid hormones, deliver signals from one part of the body to another part. Secondary messengers, such as prostaglandins and thromboxanes, mediate the hormonal response.

B. Classification by Structure

For purposes of study, we can classify lipids into four groups: (1) simple lipids, such as fats and waxes; (2) complex lipids; (3) steroids; and (4) prostaglandins, thromboxanes, and leukotrienes.

13.2 What Are the Structures of Triglycerides?

Animal fats and plant oils are triglycerides. **Triglycerides** are triesters of glycerol and long-chain carboxylic acids called fatty acids. In Section 11.1, we saw that esters are made up of an alcohol part and an acid part. As the name indicates, the alcohol of triglycerides is always glycerol.

CH_2—OH
|
CH—OH
|
CH_2—OH
Glycerol

In contrast to the alcohol part, the acid component of triglycerides may be any number of fatty acids (Section 10.4A). These fatty acids do, however, have certain things in common:

1. Fatty acids are practically all unbranched carboxylic acids.

2. They range in size from about 10 to 20 carbons.

3. They contain an even number of carbon atoms.

4. Apart from the —COOH group, they have no functional groups, except that some do have double bonds.

5. In most fatty acids that have double bonds, the *cis* isomers predominate.

Only even-numbered acids are found in triglycerides because the body builds these acids entirely from acetate units and therefore puts the carbons in two at a time (Section 21.2).

In **triglycerides** (also called **triacylglycerols**), all three hydroxyl groups of glycerol are esterified. Thus a typical triglyceride molecule might be

A triglyceride

Triglycerides are the most common lipid materials, although **mono-** and **diglycerides** are not infrequent. In the latter two types, only one or two —OH groups of the glycerol are esterified by fatty acids.

Triglycerides are complex mixtures. Although some of the molecules contain three identical fatty acids, in most cases two or three different acids are present. The hydrophobic character of triglycerides is caused by the long hydrocarbon chains. The ester groups ($-\overset{\overset{\textstyle O}{\|}}{C}-O-$), although polar themselves, are buried in a nonpolar environment, which makes the triglycerides insoluble in water.

13.3 What Are Some Properties of Triglycerides?

A. Physical State

Fat A mixture of triglycerides containing a high proportion of long-chain, saturated fatty acids

Oil A mixture of triglycerides containing a high proportion of long-chain, unsaturated fatty acids or short-chain, saturated fatty acids

With some exceptions, **fats** that come from animals are generally solids at room temperature, and those from plants or fish are usually liquids. Liquid fats are often called **oils,** even though they are esters of glycerol just like solid fats and should not be confused with petroleum, which is mostly alkanes.

What is the structural difference between solid fats and liquid oils? In most cases, it is the degree of unsaturation. The physical properties of the fatty acids are carried over to the physical properties of the triglycerides. Solid animal fats contain mainly saturated fatty acids, whereas vegetable oils contain high amounts of unsaturated fatty acids. Table 13.1 shows the average fatty acid content of some common fats and oils. Note that even solid fats contain some unsaturated acids and that liquid fats contain some saturated acids. Some unsaturated fatty acids (linoleic and linolenic acids) are called *essential fatty acids* because the body cannot synthesize them from precursors; they must, therefore, be consumed as part of the diet.

Although most vegetable oils contain high amounts of unsaturated fatty acids, there are exceptions. Coconut oil, for example, has only a small amount of unsaturated fatty acids. This oil is a liquid not because it contains many double bonds, but rather because it is rich in low-molecular-weight fatty acids (chiefly lauric acid).

TABLE 13.1 Average Percentage of Fatty Acids of Some Common Fats and Oils

	Saturated				Unsaturated			
	Lauric	Myristic	Palmitic	Stearic	Oleic	Linoleic	Linolenic	Other
Animal Fats								
Beef tallow	—	6.3	27.4	14.1	49.6	2.5	—	0.1
Butter	2.5	11.1	29.0	9.2	26.7	3.6	—	17.9
Human	—	2.7	24.0	8.4	46.9	10.2	—	7.8
Lard	—	1.3	28.3	11.9	47.5	6.0	—	5.0
Vegetable Oils								
Coconut	45.4	18.0	10.5	2.3	7.5	—	—	16.3
Corn	—	1.4	10.2	3.0	49.6	34.3	—	1.5
Cottonseed	—	1.4	23.4	1.1	22.9	47.8	—	3.4
Linseed	—	—	6.3	2.5	19.0	24.1	47.4	0.7
Olive	—	—	6.9	2.3	84.4	4.6	—	1.8
Peanut	—	—	8.3	3.1	56.0	26.0	—	6.6
Safflower	—	—	6.8	—	18.6	70.1	3.4	1.1
Soybean	0.2	0.1	9.8	2.4	28.9	52.3	3.6	2.7
Sunflower	—	—	6.1	2.6	25.1	66.2	—	—

Oils with an average of more than one double bond per fatty acid chain are called *polyunsaturated*. Their role in the human diet is discussed in Section 22.4.

Pure fats and oils are colorless, odorless, and tasteless. This statement may seem surprising because we all know the tastes and colors of such fats and oils as butter and olive oil. The tastes, odors, and colors are caused by small amounts of other substances dissolved in the fat or oil.

B. Hydrogenation

In Section 3.6D, we learned that we can reduce carbon–carbon double bonds to single bonds by treating them with hydrogen and a catalyst. It is, therefore, not difficult to convert unsaturated liquid oils to solids. For example:

This hydrogenation is carried out on a large scale to produce the solid shortening sold in stores under such brand names as Crisco, Spry, and Dexo. In making such products, manufacturers must be careful not to hydrogenate all of the double bonds, because a fat with no double bonds would be *too* solid. Partial, but not complete, hydrogenation results in a product with just the right consistency for cooking. Margarine is also made by partial hydrogenation of vegetable oils. Because less hydrogen is used, margarine contains more unsaturation than fully hydrogenated shortenings. The hydrogenation process is the source of *trans* fatty acids, as we have

Chemical Connections 13A

Rancidity

The double bonds in fats and oils are subject to oxidation by the air (see Section 4.4C). When a fat or oil is allowed to stand out in the open, this reaction slowly turns some of the molecules into aldehydes and other compounds with foul tastes and odors. We then say that the fat or oil has become *rancid* and is no longer edible. Plant oils, which generally contain more double bonds, are more susceptible to this transformation than solid fats, but even fats contain some double bonds, so rancidity can be a problem here, too.

Another cause of unpleasant taste is hydrolysis. The hydrolysis of triglycerides may produce short-chain fatty acids, such as butanoic acid (butyric acid), which have unpleasant odors. To prevent rancidity, fats and oils should be kept refrigerated (these reactions occur more slowly at low temperatures) and in dark bottles (the oxidation is catalyzed by ultraviolet light). In addition, antioxidants are often added to fats and oils to prevent rancidity.

© Steven Lunetta Photography, 2008

Storing oils in dark bottles helps prevent the formation of compounds with unpleasant tastes and odors.

already seen (Chemical Connections 10A). The food-processing industry is taking steps to address this issue. Many food labels specifically call attention to the fact that there are "no trans fats" in the product.

C. Saponification

Glycerides, being esters, are subject to hydrolysis, which can be carried out with either acids or bases. As we saw in Section 11.4, the use of bases is more practical. An example of the saponification of a typical fat is

Chemical Connections 13B

Waxes

Plant and animal waxes are simple esters. They are solids because of their high molecular weights. As in fats, the acid portions of the esters consist of a mixture of fatty acids; the alcohol portions are not glycerol, however, but rather simple long-chain alcohols. For example, a major component of beeswax is 1-triacontyl palmitate:

Palmitic acid portion   1-Triacontanol portion

$$CH_3(CH_2)_{13}CH_2\overset{O}{\underset{\|}{C}}-OCH_2(CH_2)_{28}CH_3$$

1-Triacontyl palmitate

Waxes generally have higher melting points than fats (60 to 100°C) and are harder. Animals and plants often use them for protective coatings. For example, the leaves of most plants are coated with wax, which helps to prevent microorganisms from attacking them and allows the plants to conserve water. The feathers of birds and the fur of animals are also coated with wax.

Important waxes include carnauba wax (from a Brazilian palm tree), lanolin (from lamb's wool), beeswax, and spermaceti (from whales). These substances are used to make cosmetics, polishes, candles, and ointments. Paraffin waxes are not esters, but rather mixtures of high-molecular-weight alkanes. Neither is ear wax a simple ester. This gland secretion contains a mixture of fats (triglycerides), phospholipids, and esters of cholesterol.

A triglyceride → 1,2,3-Propanetriol (Glycerol; Glycerin) + Sodium soaps

Many common products contain hydrogenated vegetable oils.

Thus saponification is the base-promoted hydrolysis of fats and oils producing glycerol and a mixture of fatty acid salts called soaps. **Soap** has been used for thousands of years, and saponification is one of the oldest known chemical reactions.

13.4 What Are the Structures of Complex Lipids?

The triglycerides discussed in the previous sections are significant components of fat storage cells. Other kinds of lipids, called complex lipids, are important in a different way. They constitute the main components of membranes (Section 13.5). Complex lipids can be classified into two groups: phospholipids and glycolipids.

Phospholipids contain an alcohol, two fatty acids, and a phosphate group. There are two types: **glycerophospholipids** and **sphingolipids.** In glycerophospholipids, the alcohol is glycerol (Section 13.6). In sphingolipids, the alcohol is sphingosine (Section 13.7).

Glycolipids are complex lipids that contain carbohydrates (Section 13.8). Figure 13.1 shows schematic structures for all of these lipids.

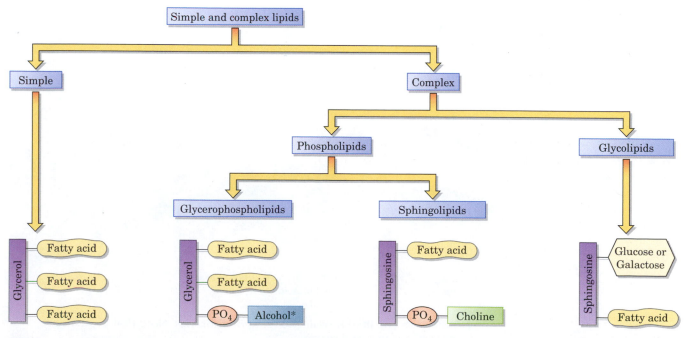

FIGURE 13.1 Schematic diagram of simple and complex lipids.*

*The alcohol can be choline, serine, ethanolamine, inositol, or certain others.

13.5 What Role Do Lipids Play in the Structure of Membranes?

The complex lipids mentioned in Section 13.4 form the **membranes** around body cells and around small structures inside the cells. (These small structures inside the cell are called *organelles*.) Unsaturated fatty acids are important components of these lipids. Most lipid molecules in the bilayer contain at least one unsaturated fatty acid. The cell membranes separate cells from the external environment and provide selective transport for nutrients and waste products into and out of cells.

These membranes are made of **lipid bilayers** (Figure 13.2). In a lipid bilayer, two rows (layers) of complex lipid molecules are arranged tail to tail. The hydrophobic tails point toward each other, which enables them to get as far away from the water as possible. This arrangement leaves the hydrophilic heads projecting to the inner and outer surfaces of the membrane. Cholesterol (Section 13.9), another membrane component, also positions the hydrophilic portion of the molecule on the surface of the membranes and the hydrophobic portion inside the bilayer.

The unsaturated fatty acids prevent the tight packing of the hydrophobic chains in the lipid bilayer, thereby providing a liquid-like character to the membranes. This effect is similar to the one that causes unsaturated fatty acids to have lower melting points than saturated fatty acids. This property of membrane fluidity is of extreme importance because many products of the body's biochemical processes must cross the membrane, and the liquid nature of the lipid bilayer allows such transport.

The lipid part of the membrane serves as a barrier against any movement of ions or polar compounds into and out of the cells. In the lipid bilayer, protein molecules are either suspended on the surface (peripheral proteins) or partly or fully embedded in the bilayer (integral proteins). These proteins stick out either on the inside or on the outside of the membrane. Others

FIGURE 13.2 The fluid mosaic model of membranes. Note that proteins are embedded in the lipid matrix.

are thoroughly embedded, going through the bilayer and projecting from both sides. The model shown in Figure 13.2, called the **fluid mosaic model** of membranes, allows the passage of nonpolar compounds by diffusion, as these compounds are soluble in the lipid membranes. The term *mosaic* refers to the topography of the bilayers: protein molecules dispersed in the lipid. The term *fluid* is used because the free lateral motion in the bilayers makes membranes liquid-like. In contrast, polar compounds are transported either via specific channels through the protein regions or by a mechanism called active transport (see Chemical Connections 13C). For any transport process, the membrane must behave like a nonrigid liquid so that the proteins can move sideways within the membrane.

13.6 What Are Glycerophospholipids?

The structure of glycerophospholipids (also called phosphoglycerides) is very similar to that of fats. Glycerophospholipids are membrane components of cells throughout the body. The alcohol is glycerol. Two of the three hydroxyl groups are esterified by fatty acids. As with the simple fats, these fatty acids may be any long-chain carboxylic acids, with or without double bonds. In all glycerophospholipids, lecithins, cephalins, and phosphatidylinositols, the fatty acid on carbon 2 of glycerol is always unsaturated. The third group is esterified not by a fatty acid, but rather by a phosphate group, which is also esterified to another alcohol. If the other alcohol is choline, a quaternary ammonium compound, the glycerophospholipids are called **phosphatidylcholines** (common name **lecithin**):

This typical lecithin molecule has stearic acid on one end and linoleic acid in the middle. Other lecithin molecules contain other fatty acids, but the one on the end is always saturated and the one in the middle is always unsaturated. Lecithin is a major component of egg yolk. Because it includes both polar and nonpolar portions within one molecule, it is an excellent emulsifier and is used in mayonnaise.

Note that lecithin has a negatively charged phosphate group and a positively charged quaternary nitrogen from the choline. These charged parts of the molecule provide a strongly hydrophilic head, whereas the rest of the molecule is hydrophobic. Thus, when a phospholipid such as lecithin is part of a lipid bilayer the hydrophobic tail points toward the middle of the bilayer, and the hydrophilic heads line both the inner and outer surfaces of the membranes (Figures 13.2 and 13.3).

Lecithins are just one example of glycerophospholipids. Another is the **cephalins,** which are similar to the lecithins in every way except that,

FIGURE 13.3 Space-filling molecular models of complex lipids in a bilayer.

From *Biochemistry* by Lubert Stryer © 1975, 1981, 1988, 1995 by Lubert Stryer; © 2002 by W. H. Freeman and Company. Used with the permission of W. H. Freeman and Company.

Chemical Connections 13C

Transport Across Cell Membranes

Membranes are not just random assemblies of complex lipids that provide a nondescript barrier. In human red blood cells, for example, the outer part of the bilayer is made largely of phosphatidylcholine and sphingomyelin, while the inner part consists mostly of phosphatidylethanolamine and phosphatidylserine (Sections 13.6 and 13.7). In another example, in the membrane called sarcoplasmic reticulum in the heart muscles, phosphatidylethanolamine is found in the outer part of the membrane, phosphatidylserine is found in the inner part, and phosphatidylcholine is equally distributed in the two layers of the membrane.

Membranes are not static structures, either. In many processes, they fuse with one another; in other processes, they disintegrate and their building blocks are used elsewhere. When membranes fuse in vacuole fusions inside of cells, for example, certain restrictions prevent incompatible membranes from intermixing.

The protein molecules are not dispersed randomly in the bilayer. Sometimes they cluster in patches; at other times they appear in regular geometric patterns. An example of the latter are **gap junctions,** channels made of six proteins that create a central pore. These channels allow neighboring cells to communicate. Gap junctions are an example of **passive transport.** Small polar molecules—which include such essential nutrients as inorganic ions, sugars, amino acids, and nucleotides—can readily pass through gap junctions. Large molecules, such as proteins, polysaccharides, and nucleic acids, cannot.

In **facilitated transport,** a specific interaction takes place between the transporter and the transported molecule. Consider the **anion transporter** of the red blood cells, through which chloride and bicarbonate ions are exchanged in a 1 : 1 ratio. The transporter is a protein with 14 helical structures that span the membrane. One side of the helices contains the hydrophobic parts of the protein, which can interact with the lipid membrane. The other side of the helices forms a channel. This channel contains the hydrophilic portions of the protein, which can interact with the hydrated ions. In this manner, the anion passes through the erythrocyte membrane.

Active transport involves the passage of ions through a concentration gradient. For example, a higher concentration of K^+ is found inside cells than outside the cells in the surrounding environment. Nevertheless, potassium ions can be transported from the outside into a cell, albeit at the expense of energy. The transporter, a membrane protein called Na^+, K^+, ATPase, uses the energy from the hydrolysis of ATP molecules to change the conformation of the transporter, which brings in K^+ and exports Na^+. Detailed studies of K^+ ion channels have revealed that K^+ ions enter the channel in pairs. Each of the hydrated ions carries eight water molecules in its solvation layers, with the negative pole of the water molecules (the oxygen atoms) surrounding the positive ion. Deeper in the channel, K^+ ions encounter a constriction, called a selectivity filter. To pass through, the K^+ ions must shed their water molecules. The pairing of the naked K^+ ions generates enough electrostatic repulsion to keep the ions moving through the channel. The channel itself is lined with oxygen atoms to provide an attractive environment similar to that offered by the stable hydrated form before entry.

Polar compounds, in general, are transported through specific **transmembrane channels.**

Gap junctions are made of six cylindrical protein subunits. They line up in two plasma membranes parallel to each other, forming a pore. The pores of gap junctions are closed by a sliding and twisting motion of the cylindrical subunits.

instead of choline, they contain other alcohols, such as ethanolamine or serine:

A phosphatidylethanolamine
(a cephalin)

A phosphatidylserine
(a cephalin)

R = hydrocarbon tail of fatty acid portion

Another important group of glycerophospholipids is the **phosphatidylino-sitols (PI).** In PI, the alcohol inositol is bonded to the rest of the molecule by a phosphate ester bond. Such compounds not only are integral structural parts of the biological membranes, but also, in their higher phosphorylated form, such as **phosphatidylinositol 4,5-bisphosphates (PIP2),** serve as signaling molecules in chemical communication (see Chapter 16).

Phosphatidylinositols, PI

13.7 What Are Sphingolipids?

Myelin, the coating of nerve axons, contains a different kind of complex lipid: **sphingolipids.** In sphingolipids, the alcohol portion is sphingosine:

Sphingosine

A long-chain fatty acid is connected to the —NH$_2$ group by an amide bond, and the —OH group at the end of the chain is esterified by phosphorylcholine:

A sphingomyelin
(a sphingolipid)

Sphingomyelin
(schematic diagram)

The combination of a fatty acid and sphingosine (shown in color) is called the **ceramide** portion of the molecule, because many of these compounds are also found in cerebrosides (Section 13.8). The ceramide part of complex lipids may contain different fatty acids. Stearic acid, for example, occurs mainly in sphingomyelin.

Sphingomyelins are the most important lipids in the myelin sheaths of nerve cells and are associated with diseases such as multiple sclerosis (Chemical Connections 13D). Sphingolipids are not randomly distributed in membranes. In viral membranes, for example, most of the sphingomyelin appears on the inside of the membrane. Johann Thudichum, who discovered sphingolipids in 1874, named these brain lipids after a monster of Greek mythology, the sphinx. Part woman and part winged lion, the sphinx devoured all who could not provide the correct answer to her riddles. Sphingolipids appeared to Thudichum as part of a dangerous riddle of the brain.

13.8 What Are Glycolipids?

Glycolipids are complex lipids that contain carbohydrates and ceramides. One group, the **cerebrosides,** consists of ceramide mono- or oligosaccharides. Other groups, such as the **gangliosides,** contain a more complex carbohydrate structure (see Chemical Connections 13E). In cerebrosides, the fatty acid of the ceramide part may contain either 18-carbon or 24-carbon chains; the latter form is found only in these complex lipids. A glucose or galactose carbohydrate unit forms a beta-glycosidic bond with the ceramide portion of the molecule. The cerebrosides occur primarily in the brain (accounting for 7% of the brain's dry weight) and at nerve synapses.

Glucocerebroside

Example 13.1 Lipid Structures

A lipid isolated from the membrane of red blood cells has the following structure:

$$
\begin{array}{l}
\quad\quad\quad\quad\quad O \\
\quad\quad\quad\quad\quad \| \\
CH_2-O-C-(CH_2)_{14}CH_3 \\
\quad\quad\quad\quad\quad O \\
\quad\quad\quad\quad\quad \| \\
CH-O-C-(CH_2)_7CH{=}CH(CH_2)_7CH_3 \\
\quad\quad\quad\quad\quad O \\
\quad\quad\quad\quad\quad \| \\
CH_2-O-P-O-CH_2CH_2NH_3{}^+ \\
\quad\quad\quad\quad\quad | \\
\quad\quad\quad\quad\quad O^-
\end{array}
$$

(a) To what group of complex lipids does this compound belong?
(b) What are the components?

Strategy

Part (b) of the question, about the component parts, is key to the whole answer. Once the parts are identified, they indicate the class of compound.

Solution

(a) The molecule is a triester of glycerol and contains a phosphate group; therefore, it is a glycerophospholipid.
(b) Besides glycerol and phosphate, it has palmitic acid and oleic acid components. The other alcohol is ethanolamine. Therefore, it belongs to the subgroup of cephalins.

Problem 13.1

A complex lipid has the following structure:

$$
\begin{array}{l}
\quad\quad\quad\quad\quad O \\
\quad\quad\quad\quad\quad \| \\
CH_2-O-C-(CH_2)_{12}CH_3 \\
\quad\quad\quad\quad\quad O \\
\quad\quad\quad\quad\quad \| \\
CH-O-C-(CH_2)_7CH{=}CHCH_2CH{=}CH(CH_2)_4CH_3 \\
\quad\quad\quad\quad\quad O \\
\quad\quad\quad\quad\quad \| \\
CH_2-O-P-O-CH_2CHCOO^- \\
\quad\quad\quad\quad\quad | \quad\quad\quad\quad | \\
\quad\quad\quad\quad\quad O^- \quad\quad\quad NH_3{}^+
\end{array}
$$

(a) To what group of complex lipids does this compound belong?
(b) What are the components?

Chemical Connections 13D

The Myelin Sheath and Multiple Sclerosis

The human brain and spinal cord can be divided into gray and white regions. Forty percent of the human brain is white matter. Microscopic examination reveals that the white matter consists of nerve axons wrapped in a white lipid coating, called the **myelin sheath,** which provides insulation and allows the rapid conduction of electrical signals. The myelin sheath consists of 70% lipids and 30% proteins in the usual lipid bilayer structure.

Specialized cells, called **Schwann cells,** wrap themselves around the peripheral nerve axons to form numerous concentric layers. In the brain, other cells perform the wrapping in a similar manner.

Multiple sclerosis affects 250,000 people in the United States. In this disease, the myelin sheath gradually deteriorates. Symptoms, which include muscle weariness, lack of coordination, and loss of vision, may vanish for a time but later return with greater severity. Autopsy of multiple sclerotic brains shows scar-like plaques of white matter, with bare axons not covered by myelin sheaths. The symptoms occur because the demyelinated axons cannot conduct nerve impulses. A secondary effect of the demyelination is damage to the axon itself.

Similar demyelination occurs in the Guillain-Barré syndrome that follows certain viral infections. In 1976, fears of a "swine flu" epidemic prompted a vaccination program that precipitated a number of cases of Guillain-Barré syndrome.

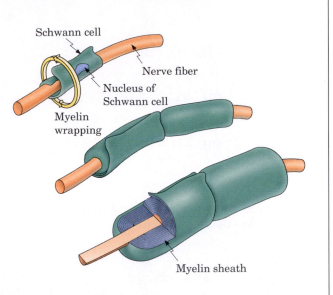

Myelination of a nerve axon outside the brain by a Schwann cell. The myelin sheath is produced by the Schwann cell and is rolled around the nerve axon for insulation.

This disease can lead to paralysis, which can cause death unless artificial breathing is supplied. In the 1976 cases, the U.S. government assumed legal responsibility for the few bad vaccines and paid compensation to the victims and their families.

13.9 What Are Steroids?

The third major class of lipids is the **steroids,** which are compounds containing the following ring system:

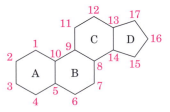

In this structure, three cyclohexane rings (A, B, and C) are connected in the same way as in phenanthrene (Section 4.2D); a fused cyclopentane ring (D) is also present. Steroids are thus completely different in structure from the lipids already discussed. Note that they are not necessarily esters, although some of them are.

Chemical Connections 13E

Lipid Storage Diseases

Complex lipids are constantly being synthesized and decomposed in the body. In several genetic diseases classified as lipid storage diseases, some of the enzymes needed to decompose the complex lipids are defective or missing. As a consequence, the complex lipids accumulate and cause an enlarged liver and spleen, mental retardation, blindness, and, in certain cases, early death. Table 13E summarizes some of these diseases and indicates the missing enzyme and the accumulating complex lipid in each.

At present, no treatment is available for these diseases. The best way to prevent them is by genetic counseling. Some of the diseases can be diagnosed during fetal development. For example, Tay-Sachs disease, which affects about 1 in every 30 Jewish Americans (versus 1 in 300 in the non-Jewish population), can be diagnosed from amniotic fluid obtained by amniocentesis.

© CNRI / Photo Researchers, Inc.

The accumulation of glucocerebrosides in the cell of a patient with Gaucher's disease. These Gaucher cells infiltrate the bone marrow.

TABLE 13E Lipid Storage Diseases

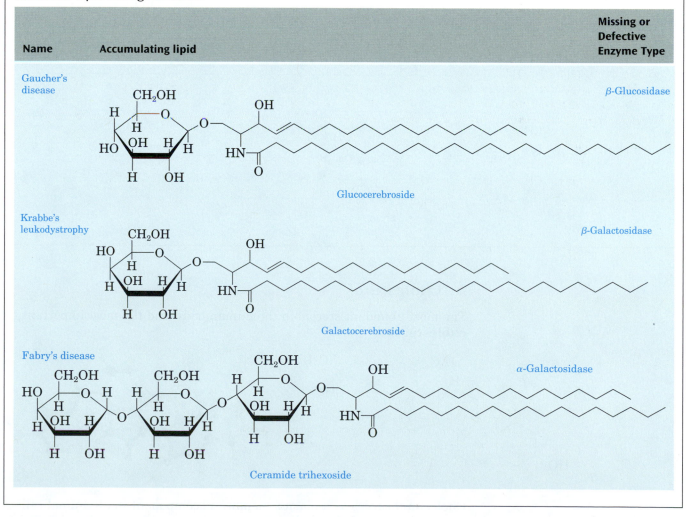

Name	Accumulating lipid	Missing or Defective Enzyme Type
Gaucher's disease	Glucocerebroside	β-Glucosidase
Krabbe's leukodystrophy	Galactocerebroside	β-Galactosidase
Fabry's disease	Ceramide trihexoside	α-Galactosidase

Chemical Connections 13E

Lipid Storage Diseases (continued)

TABLE 13E Lipid Storage Diseases

Name	Accumulating lipid	Missing or Defective Enzyme Type

Tay-Sachs disease

Ceramide oligosaccharide (a ganglioside)

Hexosaminidase A

Niemann-Pick disease

Sphingomyelinase

Sphingomyelin

A. Cholesterol

The most abundant steroid in the human body, and the most important, is **cholesterol**:

Cholesterol

Cholesterol serves as a plasma membrane component in all animal cells—for example, in red blood cells. Its second important function is to serve as a raw material for the synthesis of other steroids, such as the sex and adreno-corticoid hormones (Section 13.10) and bile salts (Section 13.11).

Cholesterol exists both in the free form and esterified with fatty acids. Gallstones contain free cholesterol (Figure 13.4).

Because the correlation between high serum cholesterol levels and such diseases as atherosclerosis has received so much publicity, many people are afraid of cholesterol and regard it as some kind of poison. Far from being poisonous, cholesterol is, in fact, necessary for human life. In essence, our livers manufacture cholesterol that satisfies our needs even without dietary intake. When the cholesterol level exceeds 150 mg/100 mL, cholesterol synthesis in the liver is reduced to half the normal rate of production. The amount of cholesterol is regulated, and it is an excess—rather than the presence—of cholesterol that is associated with disease.

Cholesterol in the body is in a dynamic state. It constantly circulates in the blood. Cholesterol and esters of cholesterol, being hydrophobic, need a water-soluble carrier to circulate in the aqueous medium of blood.

FIGURE 13.4 A human gallstone is almost pure cholesterol. This gallstone measures 5 mm in diameter.

B. Lipoproteins: Carriers of Cholesterol

Cholesterol, along with fat, is transported by **lipoproteins.** Most lipoproteins contain a core of hydrophobic lipid molecules surrounded by a shell of hydrophilic molecules such as proteins and phospholipids (Figure 13.5). As summarized in Table 13.2, there are four kinds of lipoproteins:

- **High-density lipoprotein (HDL) ("good cholesterol"),** which consists of about 33% protein and about 30% cholesterol
- **Low-density lipoprotein (LDL) ("bad cholesterol"),** which contains only 25% protein but 50% cholesterol
- **Very-low-density lipoprotein (VLDL),** which mostly carries triglycerides (fats) synthesized by the liver
- **Chylomicrons,** which carry dietary lipids synthesized in the intestines

Lipoproteins Spherically shaped clusters containing both lipid molecules and protein molecules

C. Transport of Cholesterol in LDL

The transport of cholesterol from the liver starts out as a large VLDL particle (55 nanometers in diameter). The core of VLDL contains triglycerides and cholesteryl esters, mainly cholesteryl linoleate. It is surrounded by a polar coat of phospholipids and proteins (Figure 13.5). The VLDL is carried in the serum. When the capillaries reach muscle or fat tissues, the triglycerides and all proteins except a protein called apoB-100 are removed from the VLDL. At this point, the diameter of the lipoprotein shrinks to 22 nanometers and its core contains only cholesteryl esters. Because of the removal of fat, its density increases and it becomes LDL. Low-density lipoprotein stays in the plasma for about 2.5 days.

The LDL carries cholesterol to the cells, where specific LDL-receptor molecules line the cell surface in certain concentrated areas called **coated pits.** The apoB-100 protein on the surface of the LDL binds specifically to the LDL-receptor molecules in the coated pits. After such binding, the LDL is taken inside the cell (endocytosis), where enzymes break down the lipoprotein. In the process, they liberate free cholesterol from cholesteryl esters. In this manner, the cell can, for example, use cholesterol as a component of a membrane. This is the normal fate of LDL and the normal course of cholesterol transport. Michael Brown and Joseph Goldstein of the University of Texas shared the Nobel Prize in medicine in 1986 for the discovery of the LDL-receptor–mediated pathway. If the LDL receptors are not sufficient in

FIGURE 13.5 Low-density lipoprotein.

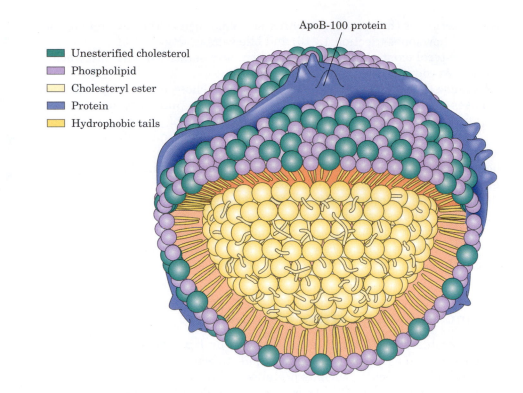

TABLE 13.2 Compositions and Properties of Human Lipoproteins

Property	HDL	LDL	VLDL	Chylomicrons
Core				
Cholesterol and cholesteryl esters (%)	30	50	22	8
Triglycerides (%)	8	4	50	84
Surface				
Phospholipids (%)	29	21	18	7
Proteins (%)	33	25	10	1–2
Density (g/mL)	1.05–1.21	1.02–1.06	0.95–1.00	<0.95
Diameter (nm)	5–15	18–28	30–80	100–500

Percentages are given as % dry weight.

number, cholesterol accumulates in the blood; this accumulation can happen even with low intake of dietary cholesterol. Both genetics and diet play a role in determining cholesterol levels in the blood.

D. Transport of Cholesterol in HDL

High-density lipoprotein transports cholesterol from peripheral tissues to the liver and transfers cholesterol to LDL. While in the serum, the free cholesterols in HDL are converted to cholesteryl esters. These esterified cholesterols are delivered to the liver for synthesis of bile acids and steroid hormones. The cholesterol uptake from HDL differs from that noted with LDL. This process does not involve endocytosis and degradation of the lipoprotein particle. Instead, in a selective lipid uptake, the HDL binds to the liver cell surface and transfers its cholesteryl ester to the cell. The HDL, depleted from its lipid content, then reenters the circulation. It is desirable to have a high level of HDL in the blood because of the way it removes cholesterol from the bloodstream.

E. Levels of LDL and HDL

Like all lipids, cholesterol is insoluble in water. If its level is elevated in the blood serum, plaque-like deposits may form on the inner surfaces of the arteries. The resulting decrease in the diameter of the blood vessels may, in turn, decrease the flow of blood. This **atherosclerosis,** along with accompanying high blood pressure, may lead to heart attack, stroke, or kidney dysfunction.

Atherosclerosis may exacerbate the blockage of some arteries by a clot at the point where the arteries are constricted by plaque. Furthermore, blockage may deprive cells of oxygen, causing them to cease to function. The death of heart muscles due to lack of oxygen is called *myocardial infarction*.

Most cholesterol is transported by low-density lipoproteins. If a sufficient number of LDL receptors are found on the surface of the cells, LDL is effectively removed from the circulation and its concentration in the blood plasma drops. The number of LDL receptors on the surface of cells is controlled by a feedback mechanism (see Section 15.6). That is, when the concentration of cholesterol molecules inside the cells is high, the synthesis of the LDL receptor is suppressed. As a consequence, less LDL is taken into the cells from the plasma and the LDL concentration in the plasma rises. Conversely, when the cholesterol level inside the cells is low, the synthesis of the LDL receptor increases. As a consequence, the LDL is taken up more rapidly and its level in the plasma falls.

In certain cases, however, there are not enough LDL receptors. In the disease called *familial hypercholesterolemia*, the cholesterol level in the plasma may be as high as 680 mg/100 mL, compared to 175 mg/100 mL in normal subjects. These high levels of cholesterol can lead to premature atherosclerosis and heart attack. The high plasma cholesterol levels in affected patients occur because the body lacks enough functional LDL receptors or, if enough are produced, they are not concentrated in the coated pits.

In general, high LDL content means high cholesterol content in the plasma because LDL cannot enter the cells and be metabolized. Therefore, a high LDL level combined with a low HDL level is a symptom of faulty cholesterol transport and a warning for possible atherosclerosis.

The serum cholesterol level controls the amount of cholesterol synthesized by the liver. When serum cholesterol is high, synthesis is at a low level. Conversely, when the serum cholesterol level is low, synthesis of cholesterol increases.

Diets low in cholesterol and saturated fatty acids usually reduce the serum cholesterol level, and a number of drugs can inhibit the synthesis of cholesterol in the liver. Commonly used statin drugs, such as atorvastatin (Lipitor) and simvastatin (Zocor), inhibit one of the key enzymes (biocatalysts) in cholesterol synthesis, HMG-CoA reductase (Section 21.4). In this way, they block the synthesis of cholesterol inside the cells and stimulate the synthesis of LDL-receptor proteins. More LDL then enters the cells, diminishing the amount of cholesterol that will be deposited on the inner walls of arteries.

It is generally considered desirable to have high levels of HDL and low levels of LDL in the bloodstream. High-density lipoproteins carry cholesterol from plaques deposited in the arteries to the liver, thereby reducing the risk of atherosclerosis. Premenopausal women have more HDL than men, which is why women have a lower risk of coronary heart disease. HDL levels can be increased by exercise and weight loss.

13.10 What Are Some of the Physiological Roles of Steroid Hormones?

Cholesterol is the starting material for the synthesis of steroid hormones. In this process, the aliphatic side chain on the D ring is shortened by the removal of a six-carbon unit, and the secondary alcohol group on carbon 3 is oxidized to a ketone. The resulting molecule, *progesterone*, serves as the starting compound for both the sex hormones and the adrenocorticoid hormones (Figure 13.6).

A. Adrenocorticoid Hormones

The adrenocorticoid hormones (Figure 13.6) are products of the adrenal glands. The term *adrenal* means "adjacent to the renal" (which refers to the kidney). We classify these hormones into two groups according to function: *Mineralocorticoids* regulate the concentrations of ions (mainly Na^+ and K^+), and *glucocorticoids* control carbohydrate metabolism. The term *corticoid* indicates that the site of the secretion is the cortex (outer part) of the gland.

Aldosterone is one of the most important mineralocorticoids. Increased secretion of aldosterone enhances the reabsorption of Na^+ and Cl^- ions in the kidney tubules and increases the loss of K^+. Because Na^+ concentration controls water retention in the tissues, aldosterone controls tissue swelling.

Cortisol is the major glucocorticoid. Its function is to increase the glucose and glycogen concentrations in the body. This accumulation occurs at the expense of other nutrients. Fatty acids from fat storage cells and amino

FIGURE 13.6 The biosynthesis of hormones from progesterone.

acids from body proteins are transported to the liver, which, under the influence of cortisol, manufactures glucose and glycogen from these sources.

Cortisol and its ketone derivative, *cortisone*, have remarkable anti-inflammatory effects. These or similar synthetic derivatives, such as prednisolone, are used to treat inflammatory diseases of many organs, rheumatoid arthritis, and bronchial asthma.

B. Sex Hormones

The most important male sex hormone is testosterone (Figure 13.6). This hormone, which promotes the normal growth of the male genital organs, is synthesized in the testes from cholesterol. During puberty, increased testosterone production leads to such secondary male sexual characteristics as deep voice and facial and body hair.

Female sex hormones, the most important of which is estradiol (Figure 13.6), are synthesized from the corresponding male hormone (testosterone) by aromatization of the A ring:

Testosterone
(partial structure)

Estradiol
(partial structure)

Estradiol, together with its precursor progesterone, regulates the cyclic changes occurring in the uterus and ovaries known as the *menstrual cycle*. As the cycle begins, the level of estradiol in the body rises, which in turn causes the lining of the uterus to thicken. Another hormone, called luteinizing hormone (LH), then triggers ovulation. If the ovum is fertilized, increased progesterone levels will inhibit any further ovulation. Both estradiol and progesterone promote further preparation of the uterine lining to receive the fertilized ovum. If no fertilization takes place, progesterone production stops altogether, and estradiol production decreases. This halt decreases the thickening of the uterine lining, which is sloughed off with accompanying bleeding during menstruation (Figure 13.7).

Because progesterone is essential for the implantation of the fertilized ovum, blocking its action leads to termination of pregnancy (see Chemical Connections 13G). Progesterone interacts with a receptor (a protein molecule) in the nucleus of cells. The receptor changes its shape when progesterone binds to it (see Section 16.7).

A drug, now widely used in France and China, called mifepristone or RU486 acts as a competitor to progesterone:

Mifepristone
(RU486)

Chemical Connections 13F

Anabolic Steroids

Testosterone, the principal male hormone, is responsible for the buildup of muscles in men. Recognizing this fact, many athletes have taken this drug in an effort to increase their muscular development. The practice is especially common among athletes in sports in which strength and muscle mass are important, including weight lifting, shot put, and hammer throw. Participants in other sports, such as running, swimming, and cycling, would also like larger and stronger muscles.

Although used by many athletes, testosterone has two disadvantages:

1. Besides its effect on muscles, it affects secondary sexual characteristics, and too much of it can result in undesired side effects.
2. It is not very effective when taken orally and must be injected to achieve the best results.

For these reasons, a large number of other anabolic steroids, all of them synthetic, have been developed. Examples include the following compounds:

Former home run champion Mark McGwire used androstenedione, a muscle-building dietary supplement that is allowed in baseball, but is banned in professional football, college athletics, and the Olympic Games.

Methandienone

Methenolone

Nandrolone decanoate

4-Androstene-3,17-dione

The use of anabolic steroids is forbidden in many sporting events, especially in international competition, largely for two reasons: (1) It gives some competitors an unfair advantage, and (2) these drugs can have many unwanted and even dangerous side effects, ranging from acne to liver tumors. Side effects can be especially disadvantageous for women; they can include growth of facial hair, baldness, deepening of the voice, and menstrual irregularities.

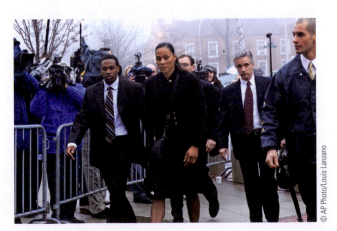

Olympic sprinter Marion Jones went to prison for lying under oath about steriod use.

Some women athletes use anabolic steroids, just as their male counterparts do. Because their bodies produce only small amounts of testosterone, women actually have much more to gain from anabolic steroids than men.

Another way to increase testosterone concentration is to use prohormones, which the body converts to testosterone. One such prohormone is 4-androstenedione, or "andro." Some athletes have used it to enhance performance.

Chemical Connections 13F

Anabolic Steroids (continued)

All athletes participating in the Olympic Games are required to pass a urine test for anabolic steroids. A number of medal-winning athletes have had their victories taken away because they tested positive for steroid use. For example, the Canadian Ben Johnson, a world-class sprinter, was stripped of both his world record and his gold medal in the 1988 Olympiad. A positive test for andro resulted in the U.S. shot put champion Randy Barnes being banned from competition in 1998. Prohormones such as andro are not listed under the Anabolic Steroid Act of 1990; hence, their nonmedical use is not a federal offense, as is the case with anabolic steroids. Mark McGwire hit his record-breaking home runs in 1998 while taking andro, because baseball rules did not prohibit its usage. Even so, the International Olympic Committee has banned the use of both prohormones and anabolic steroids.

The use of steroids in sports continues to cause controversy. In early 2008, a commission led by former Senator George Mitchell announced that a number of baseball players had used steroids. Congressional hearings followed, along with a number of suggestions about how to deal with the situation. Much of the controversy centered on whether prominent athletes had lied under oath during the congressional hearings, exposing them to accusations of perjury. In November, 2007, Barry Bonds was indicted on charges of perjury and obstruction of justice, The case is not resolved at this writing. As of May 2008, only one prominent athlete had been convicted of perjury in a high-profile case. In October, 2007, former Olympic sprinter Marion Jones, winner of five medals at the 2000 games in Sydney, admitted to steroid use for a two-year period including the Olympics. Before that, she had vehemently denied steroid use. She was sentenced to six months in prison for lying under oath and started her term in federal prison in March 2008.

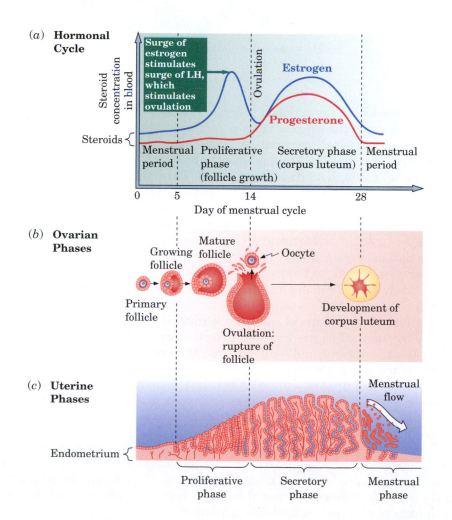

(a) **Hormonal Cycle**

(b) **Ovarian Phases**

(c) **Uterine Phases**

FIGURE 13.7 Events of the menstrual cycle. (a) Levels of sex hormones in the bloodstream during the phases of one menstrual cycle in which pregnancy does not occur. (b) Development of an ovarian follicle during the cycle. (c) Phases of development of the endometrium, the lining of the uterus. The endometrium thickens during the proliferative phase. In the secretory phase, which follows ovulation, the endometrium continues to thicken and the glands secrete a glycogen-rich nutritive material in preparation to receive an embryo. If no embryo is implanted, the new outer layers of the endometrium disintegrate and the blood vessels rupture, producing the menstrual flow.

Chemical Connections 13G

Oral Contraception

Because progesterone prevents ovulation during pregnancy, it occurred to investigators that progesterone-like compounds might be used for birth control. Synthetic analogs of progesterone proved to be more effective than the natural compound. In "the Pill," a synthetic progesterone-like compound is supplied together with an estradiol-like compound (the latter prevents irregular menstrual flow). Triple-bond derivatives of testosterone, such as norethindrone, norethynodrel, and ethynodiol diacetate, are used most often in birth-control pills:

The packaging of oral contraceptives is a reminder of their daily use.

Mifepristone blocks the action of progesterone by binding to the same receptor sites. Because the progesterone molecule cannot reach the receptor molecule, the uterus is not prepared for the implantation of the fertilized ovum, and the ovum is aborted. Once pregnancy has been established, RU486 can be taken up through 49 days of gestation. This chemical form of abortion has been approved by the U.S. Food and Drug Administration (FDA) and in recent years has found clinical application as a supplement to surgical abortion. RU486 binds to the receptors of glucocorticoid hormones as well. Its use as an antiglucocorticoid is also recommended to alleviate a disease known as Cushing syndrome, involving the overproduction of cortisone.

A completely different approach is the "morning after pill," which can be taken orally up to 72 hours after unprotected intercourse. The "morning after pill" is not an abortion pill, because it acts before pregnancy takes place. Actually, the components of the pill are regular contraceptives. Two kinds are on the market as prescription drugs: a progesterone-like compound, called levonorgesterel; and a combination of levonorgesterel and ethynil estradiol marketed as Preven.

Estradiol and progesterone also regulate secondary female sex characteristics, such as the growth of breasts. Thanks to this property, RU486, as an antiprogesterone, has been reported to be effective against certain types of breast cancer.

Testosterone and estradiol are not exclusive to either males or females. A small amount of estradiol production occurs in males, and a small amount of testosterone production is normal in females. Only when the proportion of these two hormones (hormonal balance) becomes upset can one observe symptoms of abnormal sexual differentiation.

13.11 What Are Bile Salts?

Bile salts are oxidation products of cholesterol. First the cholesterol is oxidized to the trihydroxy derivative, and the end of the aliphatic chain is oxidized to the carboxylic acid. The latter, in turn, forms an amide bond with an amino acid, either glycine or taurine:

Glycocholate   Taurocholate

Taurine has developed a certain amount of commercial importance in recent years as an ingredient in sports drinks. The drink marketed under the trade name Red Bull (*taurus* is the Latin word for "bull") contains various sugars (Chapter 12), caffeine, and B vitamins (Section 22.6) in addition to taurine.

Bile salts are powerful detergents. One end of the molecule is strongly hydrophilic because of the negative charge, and the rest of the molecule is largely hydrophobic. As a consequence, bile salts can disperse dietary lipids in the small intestine into fine emulsions, thereby facilitating digestion. The dispersion of dietary lipids by bile salts is similar to the action of soap on dirt.

Because they are eliminated in the feces, bile salts remove excess cholesterol in two ways: (1) They are themselves breakdown products of cholesterol (so cholesterol is eliminated via bile salts), and (2) they solubilize deposited cholesterol in the form of bile salt–cholesterol particles.

13.12 What Are Prostaglandins, Thromboxanes, and Leukotrienes?

Prostaglandins, a group of fatty-acid-like substances, were discovered by Kurzrok and Leib in the 1930s, when they demonstrated that seminal fluid caused a hysterectomized uterus to contract. Ulf von Euler of Sweden (winner of the Nobel Prize in physiology and medicine in 1970) isolated these compounds from human semen and, thinking that they had come from the prostate gland, named them **prostaglandin.** Even though the seminal gland secretes 0.1 mg of prostaglandin per day in mature males, small amounts of prostaglandins are present throughout the body in both sexes.

Prostaglandins are synthesized in the body from arachidonic acid by a ring closure at carbons 8 and 12. The enzyme catalyzing this reaction is called **cyclooxygenase** (COX, for short). The product, known as PGG_2, is the common precursor of other prostaglandins, including PGE and PGF. The prostaglandin E group (PGE) has a carbonyl group at carbon 9; the subscript indicates the number of double bonds in the hydrocarbon chain. The prostaglandin F group (PGF) has two hydroxyl groups on the ring

Arachidonic acid

Aspirin and other nonsteroidal anti-inflammatory drugs (NSAIDs) inhibit this enzyme

Cyclooxygenase (COX)

PGG$_2$

PGE$_2$

PGF$_{2\alpha}$

at carbons 9 and 11. Other prostaglandins (PGAs and PGBs) are derived from PGE.

The COX enzyme comes in two forms in the body: COX-1 and COX-2. COX-1 catalyzes the normal physiological production of prostaglandins, which are always present in the body. For example, PGE$_2$ and PGF$_{2\alpha}$ stimulate uterine contractions and induce labor. PGE$_2$ lowers blood pressure by relaxing the muscles around blood vessels. In aerosol form, this prostaglandin is used to treat asthma; it opens up the bronchial tubes by relaxing the surrounding muscles. PGE$_1$ is used as a decongestant; it opens up nasal passages by constricting blood vessels.

COX-2, by contrast, is responsible for the production of prostaglandins in inflammation. When a tissue is injured or damaged, special inflammatory cells invade the injured tissue and interact with resident cells—for example, smooth muscle cells. This interaction activates the COX-2 enzyme, and prostaglandins are synthesized. Such tissue injury may occur in a heart attack (myocardial infarction), rheumatic arthritis, and ulcerative colitis. Nonsteroidal anti-inflammatory drugs (NSAIDs), such as aspirin, inhibit both COX enzymes (see Chemical Connections 13H).

Another class of arachidonic acid derivatives is the **thromboxanes.** Their synthesis also includes a ring closure. These substances are derived from PGH$_2$, but their ring is a cyclic acetal. Thromboxane is known to induce platelet aggregation. When a blood vessel is ruptured, the first line of defense is the platelets circulating in the blood, which form an incipient clot. Thromboxane A$_2$ causes other platelets to clump, thereby increasing the size of the blood clot. Aspirin and similar anti-inflammatory agents inhibit the COX enzyme. Consequently, PGH$_2$ and thromboxane synthesis is inhibited, and blood clotting is impaired. This effect has prompted many physicians to recommend a daily dose of 81 mg aspirin for people at risk for heart attack or stroke. It also explains why physicians forbid patients to use aspirin and other anti-inflammatory agents for a week before a planned surgery—aspirin and other NSAIDs may cause excessive bleeding.

Chemical Connections 13H

Action of Anti-inflammatory Drugs

Anti-inflammatory steroids (such as cortisone; Section 13.10) exert their function by inhibiting phospholipase A_2, the enzyme that releases unsaturated fatty acids from complex lipids in the membranes. For example, arachidonic acid, one of the components of membranes, is made available to the cell through this process. Because arachidonic acid is the precursor of prostaglandins, thromboxanes, and leukotrienes, inhibiting its release stops the synthesis of these compounds and prevents inflammation.

Steroids such as cortisone are associated with many undesirable side effects (duodenal ulcer and cataract formation, among others). Therefore, their use must be controlled. A variety of nonsteroidal anti-inflammatory agents, including aspirin, ibuprofen, ketoprofen, and indomethacin, are available to serve this function.

Aspirin and other NSAIDs (see Chemical Connections 11C) inhibit the cyclooxygenase enzymes, which synthesize prostaglandins and thromboxanes. Aspirin (acetyl salicylic acid), for example, acetylates the enzymes, thereby blocking the entrance of arachidonic acid to the active site. This inhibition of both COX-1 and COX-2 explains why aspirin and the other anti-inflammatory agents have undesirable side effects. NSAIDs also interfere with the COX-1 isoform of the enzyme, which is needed for normal physiological function. Their side effects include stomach and duodenal ulceration and renal (kidney) toxicity.

Obviously, it would be desirable to have an anti-inflammatory agent without such side effects, and one that inhibits only the COX-2 isoform. To date, the FDA has approved two COX-2 inhibitor drugs: Celebrex, which quickly became the most frequently prescribed drug, and Vioxx, a more recent entrant. Despite their selective inhibition of COX-2, however, these drugs also have ulcer-causing side effects. Many other COX-2 inhibitors remain in the clinical trial stage of development.

The use of COX-2 inhibitors is not limited to rheumatoid arthritis and osteoarthritis. Celebrex has been approved by FDA to treat a type of colon cancer called familial adenomateous polyposis, in an approach called chemoprevention. All anti-inflammatory agents reduce pain and relieve fever and swelling by reducing the prostaglandin production, but they do not affect the leukotriene production. As a consequence, asthmatic patients must beware of using these anti-inflammatory agents. Even though they inhibit the prostaglandin synthesis,

© AP Photo/Mary Altaffer

Packages of the NSAID Celebrex contain a warning label.

these drugs may shift the available arachidonic acid to leukotriene production, which could precipitate a severe asthma reaction.

During the fall of 2004, studies indicated that high doses of Vioxx correlate with higher incidence of heart attacks and strokes; concerns were also raised about other COX-2 inhibitors, particularly Celebrex. The inhibition of prostaglandin synthesis allows formation of other lipids, including those that build up in atherosclerotic plaque. Vioxx was taken off the U.S. market, and some physicians became hesitant to prescribe Celebrex. These events caused consternation among patients who had come to depend on these drugs as well as among the pharmaceutical companies that produced them. In February 2005, a panel looked into the controversy for the FDA. This group concluded that COX-2 inhibitors should stay on the market, but their use should be strictly monitored. Warning labels must now appear on the packaging for these drugs.

A variety of NSAIDs inhibit COX enzymes. Ibuprofen and indometha-cin, both powerful painkillers, can block the inhibitory effect of aspirin and thus eliminate its anticlotting benefits. Therefore, the use of these NSAIDs together with aspirin is not recommended. Other painkillers, such as acet-aminophen and diclofenac, do not interfere with aspirin's anticlotting abil-ity and therefore can be taken together.

The **leukotrienes** are another group of substances that act to mediate hormonal responses. Like prostaglandins, they are derived from arachi-donic acid by an oxidative mechanism. However, in this case, there is no ring closure.

Leukotrienes occur mainly in white blood cells (leukocytes) but are also found in other tissues of the body. They produce long-lasting muscle con-tractions, especially in the lungs, and can cause asthma-like attacks. In fact, they are 100 times more potent than histamines. Both prostaglandins and leukotrienes cause inflammation and fever, so the inhibition of their produc-tion in the body is a major pharmacological concern. One way to counteract the effects of leukotrienes is to inhibit their uptake by leukotriene receptors (LTRs) in the body. A new antagonist of LTRs, zafirlukast (brand name Ac-colate), is used to treat and control chronic asthma. Another antiasthmatic drug, zileuton, inhibits 5-lipoxygenase, which is the initial enzyme in leu-kotriene biosynthesis from arachidonic acid.

Summary

Section 13.1 What Are Lipids?

- **Lipids** are water-insoluble substances.
- Lipids are classified into four groups: fats (triglycerides); complex lipids; steroids; and prostaglandins, throm-boxanes, and leukotrienes.

Section 13.2 What Are the Structures of Triglycerides?

- **Fats** consist of fatty acids and glycerol. In saturated fatty acids, the hydrocarbon chains have only single bonds; unsaturated fatty acids have hydrocarbon chains with one or more double bonds, all in the *cis* configuration.

Section 13.3 What Are Some Properties of Triglycerides? **Problems 13.11, 13.13**

- Solid fats contain mostly saturated fatty acids, whereas **oils** contain substantial amounts of unsaturated fatty acids.
- The alkali salts of fatty acids are called **soaps.**

Section 13.4 What Are the Structures of Complex Lipids?

- **Complex lipids** can be classified into two groups: phos-pholipids and glycolipids.
- **Phospholipids** are made of a central alcohol (glycerol or sphingosine), fatty acids, and a nitrogen-containing phosphate ester, such as phosphorylcholine or inositol phosphate.

- **Glycolipids** contain sphingosine and a fatty acid, collectively known as the ceramide portion of the molecule, and a carbohydrate portion.

Section 13.5 What Role Do Lipids Play in the Structure of Membranes?

- Many phospholipids and glycolipids are important components of cell **membranes.**
- Membranes are made of a **lipid bilayer,** in which the hydrophobic parts of phospholipids (fatty acid residues) point toward the middle of the bilayer, and the hydrophilic parts point toward the inner and outer surfaces of the membrane.

Section 13.6 What Are Glycerophospholipids?

- **Glycerophospholipids** are complex lipids that consist of a central glycerol moiety to which two fatty acids are esterified. The third alcohol group of the glycerol is esterified to a nitrogen-containing phosphate ester.

Section 13.7 What Are Sphingolipids?

- **Sphinogolipids** are complex lipids that consist of the long-chain alcohol sphingosine esterified to a fatty acid (the ceramide moiety). Nitrogen-containing phosphate esters are also bonded to the sphingosine moiety.

Section 13.8 What Are Glycolipids?

- **Glycolipids** are complex lipids that consist of two parts: a ceramide portion and a carbohydrate portion.

Section 13.9 What Are Steroids?

- The third major group of lipids comprises the **steroids.** The characteristic feature of the steroid structure is a fused four-ring nucleus.
- The most common steroid, **cholesterol,** serves as a starting material for the synthesis of other steroids, such as bile salts and sex and other hormones. Cholesterol is also an integral part of membranes,

occupying the hydrophobic region of the lipid bilayer. Because of its low solubility in water, cholesterol deposits are implicated in the formation of gallstones and the plaque-like deposits of atherosclerosis.

- Cholesterol is transported in the blood plasma mainly by two kinds of lipoproteins: **HDL** and **LDL.** LDL delivers cholesterol to the cells to be used mostly as a membrane component. HDL delivers cholesteryl esters mainly to the liver to be used in the synthesis of bile acids and steroid hormones.
- High levels of LDL and low levels of HDL are symptoms of faulty cholesterol transport, indicating greater risk of atherosclerosis.

Section 13.10 What Are Some of the Physiological Roles of Steroid Hormones?

- An oxidation product of cholesterol is progesterone, a **sex hormone.** It also gives rise to the synthesis of other sex hormones, such as testosterone and estradiol.
- Progesterone is also a precursor of the **adrenocorticoid hormones.** Within this group, cortisol and cortisone are best known for their anti-inflammatory action.

Section 13.11 What Are Bile Salts? Problem 13.44

- **Bile salts** are oxidation products of cholesterol. They emulsify all kinds of lipids, including cholesterol, and are essential in the digestion of fats.

Section 13.12 What Are Prostaglandins, Thromboxanes, and Leukotrienes?

- **Prostaglandins, thromboxanes,** and **leukotrienes** are derived from arachidonic acid. They have a wide variety of effects on body chemistry. Among other things, they can lower or raise blood pressure, cause inflammation and blood clotting, and induce labor. In general, they mediate hormone action.

Problems

■ Indicates problems that are assignable in **GOB OWL**.

Blue numbered problems are applied.

Go to this book's companion website at **www.cengage. com/chemistry/bettelheim** for interactive versions of the How To tutorials and Active Figures, and to quiz yourself on this chapter.

Section 13.1 What Are Lipids?

13.2 Why are fats a good source of energy for storage in the body?

13.3 What is the meaning of the term *hydrophobic?* Why is the hydrophobic nature of lipids important?

Section 13.2 What Are the Structures of Triglycerides?

13.4 Draw the structural formula of a fat molecule (triglyceride) made of myristic acid, oleic acid, palmitic acid, and glycerol.

13.5 Oleic acid has a melting point of 16°C. If you converted the *cis* double bond into a *trans* double bond, what would happen to the melting point? Explain.

13.6 Draw schematic formulas for all possible 1,3-diglycerides made up of glycerol, oleic acid, or stearic acid. How many are there? Draw the structure of one of the diglycerides.

Section 13.3 What Are Some Properties of Triglycerides?

13.7 For the diglycerides in Problem 13.6, predict which two will have the highest melting points and which two will have the lowest melting points.

13.8 Predict which acid in each pair has the higher melting point and explain why.

(a) Palmitic acid or stearic acid

(b) Arachidonic acid or arachidic acid

13.9 Which has the higher melting point: (a) a triglyceride containing only lauric acid and glycerol or (b) a triglyceride containing only stearic acid and glycerol?

13.10 Explain why the melting points of the saturated fatty acids increase as we move from lauric acid to stearic acid.

13.11 ■ Predict the order of the melting points of triglycerides containing fatty acids, as follows:

(a) Palmitic, palmitic, stearic

(b) Oleic, stearic, palmitic

(c) Oleic, linoleic, oleic

13.12 Look at Table 13.1. Which animal fat has the highest percentage of unsaturated fatty acids?

13.13 ■ Rank the following in order of increasing solubility in water (assuming that all are made with the same fatty acids): (a) triglycerides, (b) diglycerides, and (c) monoglycerides. Explain your answer.

13.14 How many moles of H_2 are used up in the catalytic hydrogenation of one mole of a triglyceride containing glycerol, palmitic acid, oleic acid, and linoleic acid?

13.15 Name the products of the saponification of this triglyceride:

$$
\begin{array}{l}
\quad\quad\quad\quad O \\
\quad\quad\quad\quad \| \\
CH_2-O-C-(CH_2)_{14}CH_3 \\
\quad\quad\quad\quad O \\
\quad\quad\quad\quad \| \\
CH-O-C-(CH_2)_{16}CH_3 \\
\quad\quad\quad\quad O \\
\quad\quad\quad\quad \| \\
CH_2-O-C-(CH_2)_7(CH=CHCH_2)_3CH_3
\end{array}
$$

13.16 Using the equation in Section 13.3C as a guideline for stoichiometry, calculate the number of moles of NaOH needed to saponify 5 mol of (a) triglycerides, (b) diglycerides, and (c) monoglycerides.

Section 13.4 What Are the Structures of Complex Lipids?

13.17 What are the main types of complex lipids, and what are the main characteristics of their structures?

Section 13.5 What Role Do Lipids Play in the Structure of Membranes?

13.18 Which portion of the phosphatidylinositol molecule contributes to (a) the fluidity of the bilayer and (b) the surface polarity of the bilayer?

13.19 How do the unsaturated fatty acids of the complex lipids contribute to the fluidity of a membrane?

13.20 Which type of lipid molecule is most likely to be present in membranes?

13.21 What is the difference between an integral and a peripheral membrane protein?

Section 13.6 What Are Glycerophospholipids?

13.22 Which glycerophospholipid has the most polar groups capable of forming hydrogen bonds with water?

13.23 Draw the structure of a phosphatidylinositol that contains oleic acid and arachidonic acid.

13.24 Among the glycerophospholipids containing palmitic acid and linolenic acid, which will have the greatest solubility in water: (a) phosphatidylcholine, (b) phosphatidylethanolamine, or (c) phosphatidylserine? Explain.

Section 13.7 What Are Sphingolipids?

13.25 Name all the groups of complex lipids that contain ceramides.

13.26 Are the various phospholipids randomly distributed in membranes? Give an example.

Section 13.8 What Are Glycolipids?

13.27 Enumerate the functional groups that contribute to the hydrophilic character of (a) glucocerebroside and (b) sphingomyelin.

Section 13.9 What Are Steroids?

13.28 Cholesterol has a fused four-ring steroid nucleus and is a part of body membranes. The —OH group on carbon 3 is the polar head, and the rest of the molecule provides the hydrophobic tail that does not fit into the zig-zag packing of the hydrocarbon portion of the saturated fatty acids. Considering this structure, tell whether small amounts of cholesterol that are well dispersed in the membrane contribute to the stiffening (rigidity) or the fluidity of a membrane. Explain.

13.29 Where can pure cholesterol crystals be found in the body?

13.30 (a) Find all of the carbon stereocenters in a cholesterol molecule.

(b) How many total stereoisomers are possible?

(c) How many of these stereoisomers do you think are found in nature?

13.31 Look at the structures of cholesterol and the hormones shown in Figure 13.6. Which ring of the steroid structure undergoes the most substitution?

13.32 What makes LDL soluble in blood plasma?

13.33 How does LDL deliver its cholesterol to the cells?

13.34 How does lovastatin reduce the severity of atherosclerosis?

13.35 How does VLDL become LDL?

13.36 How does HDL deliver its cholesteryl esters to liver cells?

13.37 How does the serum cholesterol level control both cholesterol synthesis in the liver and LDL uptake?

Section 13.10 What Are Some of the Physiological Roles of Steroid Hormones?

13.38 What physiological functions are associated with cortisol?

13.39 Estradiol in the body is synthesized starting from progesterone. What chemical modifications occur when estradiol is synthesized?

13.40 Describe the difference in structure between the male hormone testosterone and the female hormone estradiol.

13.41 Considering that RU486 can bind to the receptors of progesterone as well as to the receptors of cortisone and cortisol, what can you say regarding the importance of the functional group on carbon 11 of the steroid ring in drug and receptor binding?

13.42 (a) How does the structure of RU486 resemble that of progesterone?
(b) How do the two structures differ?

13.43 What are the structural features common to oral contraceptive pills, including mifepristone?

Section 13.11 What Are Bile Salts?

13.44 ■ List all of the functional groups that make taurocholate water-soluble.

13.45 Explain how the constant elimination of bile salts through the feces can reduce the danger of plaque formation in atherosclerosis.

Section 13.12 What Are Prostaglandins, Thromboxanes, and Leukotrienes?

13.46 What is the basic structural difference between:
(a) Arachidonic acid and prostaglandin PGE_2?
(b) PGE_2 and $PGF_{2\alpha}$?

13.47 Find and name all of the functional groups in (a) glycocholate, (b) cortisone, (c) prostaglandin PGE_2 and (d) leukotriene B4.

13.48 What are the chemical and physiological functions of the COX-2 enzyme?

13.49 How does aspirin, an anti-inflammatory drug, prevent strokes caused by blood clots in the brain?

Chemical Connections

13.50 (Chemical Connections 13A) What causes rancidity? How can it be prevented?

13.51 (Chemical Connections 13B) What makes waxes harder and more difficult to melt than fats?

13.52 (Chemical Connections 13C) How do the gap junctions prevent the passage of proteins from cell to cell?

13.53 (Chemical Connections 13C) How does the anion transporter provide a suitable environment for the passage of hydrated chloride ions?

13.54 (Chemical Connections 13C) In what sense is the active transport of K^+ selective? How does K^+ pass through the transporter?

13.55 (Chemical Connections 13D)
(a) What role does sphingomyelin play in the conductance of nerve signals?
(b) What happens to this process in multiple sclerosis?

13.56 (Chemical Connections 13E) Compare the complex lipid structures listed for the lipid storage diseases with the missing or defective enzymes. Explain why the missing enzyme in Fabry's disease is α-galactosidase and not β-galactosidase.

13.57 (Chemical Connections 13E) Identify the monosaccharides in the accumulating glycolipid of Fabry's disease.

13.58 (Chemical Connections 13F) How does the oral anabolic steroid methenolone differ structurally from testosterone?

13.59 (Chemical Connections 13G) What is the role of progesterone and similar compounds in contraceptive pills?

13.60 (Chemical Connections 13H) How does cortisone prevent inflammation?

13.61 (Chemical Connections 13H) How does indomethacin act in the body to reduce inflammation?

13.62 (Chemical Connections 13H) What kind of prostaglandins are synthesized by COX-1 and COX-2 enzymes?

13.63 (Chemical Connections 13H) Steroids prevent asthma-causing leukotriene synthesis as well as inflammation-causing prostaglandin synthesis. Nonsteroidal anti-inflammatory agents (NSAIDs) such as aspirin reduce only prostaglandin production. Why do NSAIDs not affect leukotriene production?

Additional Problems

13.64 What is the role of taurine in lipid digestion?

13.65 Draw a schematic diagram of a lipid bilayer. Show how the bilayer prevents the passage by diffusion of a polar molecule such as glucose. Show why nonpolar molecules, such as $CH_3CH_2\!-\!O\!-\!CH_2CH_3$, can diffuse through the membrane.

13.66 How many different triglycerides can you create using three different fatty acids (A, B, and C) in each case?

13.67 Prostaglandins have a five-membered ring closure; thromboxanes have a six-membered ring closure. The synthesis of both groups of compounds is prevented by COX inhibitors; the COX enzymes catalyze ring closure. How can these facts be correlated?

13.68 Which lipoprotein is instrumental in removing the cholesterol deposited in the plaques on arteries?

13.69 What are coated pits? What is their function?

13.70 What are the constituents of sphingomyelin?

13.71 (Chemical Connections 13C) What is the difference between a facilitated transporter and an active transporter?

13.72 Which part of LDL interacts with the LDL receptor?

13.73 What is the major difference between aldosterone and the other hormones listed in Figure 13.6?

13.74 (Chemical Connections 13H) The anti-inflammatory drug Celebrex does not have the usual side effect of stomach upset or ulceration commonly observed with the other NSAIDs. Why?

13.75 ■ How many grams of H_2 are needed to saturate 100.0 g of a triglyceride made of glycerol and one unit each of lauric, oleic, and linoleic acids?

13.76 Prednisolone is the synthetic glucocorticoid medicine most frequently prescribed to combat autoimmune diseases. Compare its structure to the natural glucocorticoid hormone, cortisone. What are the similarities and differences in structure?

13.77 You have just isolated a pure lipid that contains only sphingosine and a fatty acid. To what class of lipid does it belong?

13.78 Suggest a reason why the same protein system moves both sodium and potassium ions into and out of the cell.

13.79 Do all proteins associated with membranes span the membrane from one side to the other?

13.80 In the preparation of sauces that involve mixing water and melted butter, egg yolks are added to prevent separation. How do the egg yolks prevent separation? (Hint: Egg yolks are rich in phosphatidylcholine [lecithin].)

13.81 ■ Which of the following statements is (are) consistent with what is known about membranes?
(a) A membrane consists of a layer of proteins sandwiched between two layers of lipids.
(b) The compositions of the inner and outer lipid layers are the same in any individual membrane.
(c) Membranes contain glycolipids and glycoproteins.
(d) Lipid bilayers are an important component of membranes.
(e) Covalent bonding takes place between lipids and proteins in most membranes.

13.82 Suggest a reason why animals that live in cold climates tend to have higher proportions of polyunsaturated fatty acid residues in their lipids than do animals that live in warm climates.

13.83 Which statements are consistent with the fluid mosaic model of membranes?
(a) All membrane proteins are bound to the interior of the membrane.
(b) Both proteins and lipids undergo transverse (flip-flop) diffusion from the inside to the outside of the membrane.
(c) Some proteins and lipids undergo lateral diffusion along the inner or outer surface of the membrane.

13.84 Suggest a reason why the cell membranes of bacteria grown at 20°C tend to have a higher proportion of unsaturated fatty acids than the membranes of bacteria of the same species grown at 37°C. In other words, the bacteria grown at 37°C have a higher proportion of saturated fatty acids in their cell membranes.

Tying It Together

13.85 Lipids and carbohydrates are both vehicles for energy storage. How are they similar in terms of molecular structure, and how do they differ? What does the molecular structure of each class of substance imply about the polarity of typical molecules?

13.86 To what extent do lipids and carbohydrates play structural roles in living organisms? Do these roles differ in plants and in animals?

13.87 Which substances would you expect to consist primarily of carbohydrates and which primarily of lipids: olive oil, butter, cotton, cotton candy?

13.88 To what extent would you expect to find the following functional groups in lipids and in carbohydrates: aldehyde groups, carboxylic acid groups, ester bonds, hydroxyl groups?

Looking Ahead

13.89 Sports drinks tend to contain large amounts of sugars, and some contain taurine in small amounts. Would you expect more of the effect of these performance aids to come from dietary carbohydrates or from the role of taurine in breaking down fats?

13.90 Which of the following foods consist primarily of carbohydrates, and which of fats: soft drinks (not diet drinks), salad dressing, canned fruit, cream cheese?

13.91 The ester bonds in lipids do not give rise to macromolecules, but the amide bonds in proteins do. Comment on the underlying reason for this difference.

13.92 Given the structural differences between steroids and other kinds of lipids, would you expect the synthesis of steroids in living organisms to differ from the synthesis of other lipids?

Challenge Problems

13.93 Some of the lipid molecules that occur in membranes are bulkier than others. Are the bulkier molecules more or less likely to be found on the cytoplasmic side of the cell membrane or on the side facing the exterior of the cell?

13.94 What are the functions of a cell membrane? To what extent is a bilayer that consists entirely of lipids able to carry out these functions?

13.95 Glycerophospholipids tend to have both a positive charge and a negative charge in their hydrophilic portions. Does this fact help or hinder lipid packing in membranes, and why?

13.96 Leukotrienes differ from prostaglandins and thromboxanes in that they lack a ring closure. They also differ from prostaglandins and thromboxanes (and from all other lipids) in another feature of their structure. What is that structural feature? (*Hint:* It has to do with the position of their double bonds.)

Proteins

Spider silk is a fibrous protein that exhibits unmatched strength and toughness.

14.1 What Are the Many Functions of Proteins?

Proteins are by far the most important of all biological compounds. The very word "protein" is derived from the Greek *proteios,* meaning "of first importance," and the scientists who named these compounds more than 100 years ago chose an appropriate term. Many types of proteins exist, and they perform a variety of functions, including the following roles:

1. **Structure** In Section 12.5, we saw that the main structural material for plants is cellulose. For animals, it is structural proteins, which are the chief constituents of skin, bones, hair, and nails. Two important structural proteins are collagen and keratin.

Protein A large biological molecule made of numerous amino acids linked by amide bonds

OWL Online homework for this chapter may be assigned in **GOB OWL.**

2. **Catalysis** Virtually all the reactions that take place in living organisms are catalyzed by proteins called enzymes. Without enzymes, the reactions would take place so slowly as to be useless. We will discuss enzymes in depth in Chapter 15.

3. **Movement** Every time we crook a finger, climb stairs, or blink an eye, we use our muscles. Muscle expansion and contraction are involved in every movement we make. Muscles are made up of protein molecules called myosin and actin.

4. **Transport** A large number of proteins perform transportation duties. For example, hemoglobin, a protein in the blood, carries oxygen from the lungs to the cells in which it is used and carbon dioxide from the cells to the lungs. Other proteins transport molecules across cell membranes.

5. **Hormones** Many hormones are proteins, including insulin, erythropoietin, and human growth hormone.

6. **Protection** When a protein from an outside source or some other foreign substance (called an antigen) enters the body, the body makes its own proteins (called antibodies) to counteract the foreign protein. This antibody production is one of the major mechanisms that the body uses to fight disease. Blood clotting is another protective function carried out by a protein, this one called fibrinogen. Without blood clotting, we would bleed to death from any small wound.

7. **Storage** Some proteins store materials in the way that starch and glycogen store energy. For example, casein in milk and ovalbumin in eggs store nutrients for newborn mammals and birds. Ferritin, a protein in the liver, stores iron.

8. **Regulation** Some proteins not only control the expression of genes, thereby regulating the kind of proteins synthesized in a particular cell, but also dictate when such manufacture takes place.

These are not the only functions of proteins, but they are among the most important. Clearly, any individual needs a great many proteins to carry out these varied functions. A typical cell contains about 9000 different proteins; the entire human body has about 100,000 different proteins.

We can classify proteins into two major types: **fibrous proteins,** which are insoluble in water and are used mainly for structural purposes, and **globular proteins,** which are more or less soluble in water and are used mainly for nonstructural purposes.

14.2 What Are Amino Acids?

Although a wide variety of proteins exist, they all have basically the same structure: They are chains of amino acids. As its name implies, an **amino acid** is an organic compound containing an amino group and a carboxyl group. Organic chemists can synthesize many thousands of amino acids, but nature is much more restrictive and uses 20 common amino acids to make up proteins. Furthermore, all but one of the 20 fit the formula:

Alpha (α) amino acid An amino acid in which the amino group is linked to the carbon atom next to the —COOH carbon

α-Carbon

Side chain

Amino group   Carboxyl group   Ball-and-stick model   Amino acids are tetrahedral structures

TABLE 14.1 The 20 Amino Acids Commonly Found in Proteins

Name	3-Letter Abbreviation	1-Letter Abbreviation	Isoelectric Point
Alanine	Ala	A	6.01
Arginine	Arg	R	10.76
Asparagine	Asn	N	5.41
Aspartic acid	Asp	D	2.77
Cysteine	Cys	C	5.07
Glutamic acid	Glu	E	3.22
Glutamine	Gln	Q	5.65
Glycine	Gly	G	5.97
Histidine	His	H	7.59
Isoleucine	Ile	I	6.02
Leucine	Leu	L	5.98
Lysine	Lys	K	9.74
Methionine	Met	M	5.74
Phenylalanine	Phe	F	5.48
Proline	Pro	P	6.48
Serine	Ser	S	5.68
Threonine	Thr	T	5.87
Tryptophan	Trp	W	5.88
Tyrosine	Tyr	Y	5.66
Valine	Val	V	5.97

Even the one amino acid that doesn't fit this formula (proline) comes fairly close: It differs only in that it has a bond between the R and the N. The 20 amino acids commonly found in proteins are called **alpha amino acids.** They are listed in Table 14.1, which also shows the one- and three-letter abbreviations that chemists and biochemists use for them.

The most important aspect of the R groups is their polarity. On that basis we can classify amino acids into four groups, as shown in Figure 14.1: nonpolar, polar but neutral, acidic, and basic. Note that the nonpolar side chains are *hydrophobic* (they repel water), whereas polar but neutral, acidic, and basic side chains are *hydrophilic* (attracted to water). This aspect of the R groups is very important in determining both the structure and the function of each protein molecule.

When we look at the general formula for the 20 amino acids, we see at once that all of them (except glycine, in which R = H) are chiral with (carbon) stereocenters, since R, H, COOH, and NH_2 are four different groups. Thus each of the amino acids with one stereocenter exists as two enantiomers. As is the case for most examples of this kind, nature makes only one of the two possible enantiomers for each amino acid, and it is virtually always the L-isomer. Except for glycine, which is achiral, all the amino acids in all the proteins in your body are the L-isomer. D amino acids are extremely rare in nature; some are found, for example, in the cell walls of a few types of bacteria.

In Section 12.1C, we learned about the systematic use of the D,L system. There we used glyceraldehyde as a reference point for the assignment of relative configuration. Here again, we can use glyceraldehyde as a reference point with amino acids, as shown in Figure 14.2. The spatial relationship of the functional groups around the carbon stereocenter in L-amino acids, as in L-alanine, can be compared to that of L-glyceraldehyde. When we put the carbonyl groups of both compounds in the same position (top), the —OH of L-glyceraldehyde and the NH_3^+ of L-alanine lie to the left of the carbon stereocenter.

(a) Nonpolar (hydrophobic)

Leucine (Leu, L)

Proline (Pro, P)

Alanine (Ala, A)

Valine (Val, V)

(b) Polar, uncharged

Glycine (Gly, G)

Serine (Ser, S)

Asparagine (Asn, N)

Glutamine (Gln, Q)

(c) Acidic

Aspartic acid (Asp, D)

Glutamic acid (Glu, E)

FIGURE 14.1 The 20 amino acids that are the building blocks of proteins can be classified as (a) nonpolar (hydrophobic), (b) polar but neutral, (c) acidic, or (d) basic. Also shown here are the one-letter and three-letter codes used to denote amino acids. For each amino acid, the ball-and-stick (*left*) and space-filling (*right*) models show only the side chain. (*Irving Geis*)

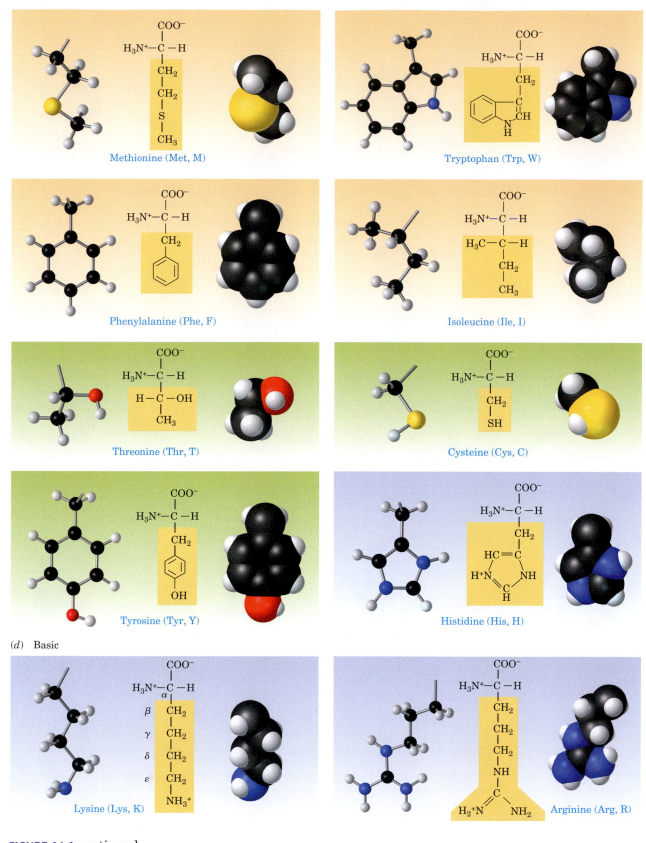

Methionine (Met, M)

Tryptophan (Trp, W)

Phenylalanine (Phe, F)

Isoleucine (Ile, I)

Threonine (Thr, T)

Cysteine (Cys, C)

Tyrosine (Tyr, Y)

Histidine (His, H)

(d) Basic

Lysine (Lys, K)

Arginine (Arg, R)

FIGURE 14.1 continued

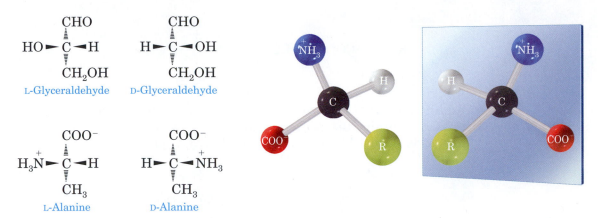

ACTIVE FIGURE 14.2 Stereochemistry of alanine and glyceraldehyde. The amino acids found in proteins have the same chirality as L-glyceraldehyde, which is opposite to that of D-glyceraldehyde. **Go to this book's companion website at www.cengage.com/chemistry/bettelheim to explore an interactive version of this figure.**

14.3 What Are Zwitterions?

In Section 10.5B we learned that carboxylic acids, RCOOH, cannot exist in the presence of a moderately weak base (such as NH_3). They donate a proton to become carboxylate ions, $RCOO^-$. Likewise, amines, RNH_2 (Section 8.5), cannot exist as such in the presence of a moderately weak acid (such as acetic acid). They gain a proton to become substituted ammonium ions, RNH_3^+.

An amino acid has —COOH and —NH_2 groups in the same molecule. Therefore, in water solution, the —COOH donates a proton to the —NH_2 so that an amino acid actually has the structure

$$R—\underset{\underset{\displaystyle NH_3^+}{|}}{\overset{\overset{\displaystyle H}{|}}{C}}—COO^-$$

Compounds that have a positive charge on one atom and a negative charge on another are called **zwitterions,** from the German word *zwitter,* meaning "hybrid." Amino acids are zwitterions, not only in water solution but also in the solid state. They are therefore ionic compounds—that is, internal salts. *Un-ionized RCH(NH$_2$)COOH molecules do not actually exist, in any form.*

The fact that amino acids are zwitterions explains their physical properties. All of them are solids with high melting points (for example, glycine melts at 262°C), just as we would expect for ionic compounds. The 20 common amino acids are also fairly soluble in water, as ionic compounds generally are. If they had no charges, we would expect only the smaller ones to be soluble.

If we add an amino acid to water, it dissolves and then has the same zwitterionic structure that it has in the solid state. Let us see what happens if we change the pH of the solution, as we can easily do by adding a source of H_3O^+, such as HCl solution (to lower the pH), or a strong base, such as NaOH (to raise the pH). Because H_3O^+ is a stronger acid than a typical carboxylic acid (Section 10.1), it donates a proton to the —COO^- group, turning the zwitterion into a positive ion. This happens to all amino acids if the pH is sufficiently lowered—say, to pH 0.

$$R-\overset{\overset{\displaystyle H}{|}}{\underset{\underset{\displaystyle NH_3^+}{|}}{C}}-COO^- + H_3O^+ \longrightarrow R-\overset{\overset{\displaystyle H}{|}}{\underset{\underset{\displaystyle NH_3^+}{|}}{C}}-COOH + H_2O$$

Addition of OH^- to the zwitterion causes the $-NH_3^+$ to donate its proton to OH^-, turning the zwitterion into a negative ion. This happens to all amino acids if the pH is sufficiently raised—say, to pH 14.

$$R-\overset{\overset{\displaystyle H}{|}}{\underset{\underset{\displaystyle NH_3^+}{|}}{C}}-COO^- + OH^- \longrightarrow R-\overset{\overset{\displaystyle H}{|}}{\underset{\underset{\displaystyle NH_2}{|}}{C}}-COO^- + H_2O$$

In both cases, the amino acid is still an ion so it is still soluble in water. There is no pH at which an amino acid has no ionic character at all. If the amino acid is a positive ion at low pH and a negative ion at high pH, there must be some pH at which all the molecules have equal positive and negative charges. This pH is called the **isoelectric point (pI).**

Isoelectric point (pI) A pH at which a sample of amino acids or protein has an equal number of positive and negative charges

Every amino acid has a different isoelectric point, although most of them are not very far apart (see the values in Table 14.1). Fifteen of the 20 amino acids have isoelectric points near 6. However, the three basic amino acids have higher isoelectric points, and the two acidic amino acids have lower values.

At or near the isoelectric point, amino acids exist in aqueous solution largely or entirely as zwitterions. As we have seen, they react with either a strong acid, by taking a proton (the $-COO^-$ becomes $-COOH$), or a strong base, by giving a proton (the $-NH_3^+$ becomes $-NH_2$). To summarize:

$$R-\overset{\overset{\displaystyle H}{|}}{\underset{\underset{\displaystyle NH_3^+}{|}}{C}}-COOH \underset{H_3O^+}{\overset{OH^-}{\rightleftharpoons}} R-\overset{\overset{\displaystyle H}{|}}{\underset{\underset{\displaystyle NH_3^+}{|}}{C}}-COO^- \underset{H_3O^+}{\overset{OH^-}{\rightleftharpoons}} R-\overset{\overset{\displaystyle H}{|}}{\underset{\underset{\displaystyle NH_2}{|}}{C}}-COO^-$$

In Section 7.3, we learned that a compound that is both an acid and a base is called amphiprotic. We also learned, in Section 7.10, that a solution that neutralizes both acid and base is a buffer solution. Amino acids are therefore *amphiprotic* compounds, and aqueous solutions of them are *buffers*.

14.4 What Determines the Characteristics of Amino Acids?

Since the side chains are the only differences between the amino acids, ultimately the functions of amino acids and their polymers, proteins, are determined by these side chains. For example, one of the 20 amino acids in Table 14.1 has a chemical property not shared by any of the others. This amino acid, cysteine, can easily be dimerized by many mild oxidizing agents:

A disulfide bond

$$2HS-CH_2-\underset{\underset{\displaystyle NH_3^+}{|}}{CH}-COO^- \underset{[H]}{\overset{[O]}{\rightleftharpoons}} {}^-OOC-\underset{\underset{\displaystyle NH_3^+}{|}}{CH}-CH_2-S-S-CH_2-\underset{\underset{\displaystyle NH_3^+}{|}}{CH}-COO^-$$

Cysteine   Cystine

The dimer of cysteine, which is called **cystine,** can in turn be fairly easily reduced to give two molecules of cysteine. As we shall see, the presence of cystine has important consequences for the chemical structure and shape of the protein molecules of which it is part. The bond (shown in color) is also called a **disulfide bond** (Section 5.4D).

Several of the amino acids have acidic or basic properties. Two amino acids, glutamic acid and aspartic acid, have carboxyl groups in their side chains in addition to the one present in all amino acids. A carboxyl group can lose a proton, forming the corresponding carboxylate anion—glutamate and aspartate, respectively, in the case of these two amino acids. Because of the presence of the carboxylate, the side chains of these two amino acids are negatively charged at neutral pH. Three amino acids—histidine, lysine, and arginine—have basic side chains. The side chains of lysine and arginine are positively charged at or near neutral pH. In lysine, the side-chain amino group is attached to an aliphatic hydrocarbon tail. In arginine, the side-chain basic group, the guanidino group, is more complex in structure than the amino group, but it is also bonded to an aliphatic hydrocarbon tail. In free histidine, the pK_a of the side-chain imidazole group is 6.0, which is not far from physiological pH. The pK_a values for amino acids depend on the environment and can change significantly within the confines of a protein. Histidine can be found in the protonated or unprotonated forms in proteins, and the properties of many proteins depend on whether individual histidine residues are charged or not. The charged amino acids are often found in the active sites of enzymes, which we will study in Chapter 15.

The amino acids phenylalanine, tryptophan, and tyrosine have aromatic rings in their side chains. They are important for a number of reasons. As a matter of practicality, these amino acids allow us to locate and measure proteins because the aromatic rings absorb strongly at 280 nm and can be detected using a spectrophotometer. These amino acids are also very important physiologically because they are both key precursors to neurotransmitters (substances involved in the transmission of nerve impulses). Tryptophan is converted to serotonin, more properly called 5-hydroxytryptamine, which has a calming effect. Very low levels of serotonin are associated with depression, whereas extremely high levels produce a manic state. Manic-depressive schizophrenia (also called bipolar disorder) can be managed by controlling the levels of serotonin and its further metabolites.

Tyrosine, itself normally derived from phenylalanine, is converted to the neurotransmitter class called catecholamines, which includes epinephrine, commonly known by its proprietary name, adrenalin.

L-Dihydroxyphenylalanine (L-dopa) is an intermediate in the conversion of tyrosine. Lower than normal levels of L-dopa are involved in Parkinson's disease. Tyrosine or phenylalanine supplements might increase the levels of dopamine, although L-dopa, the immediate precursor, is usually prescribed because it passes into the brain quickly through the blood–brain barrier.

Tyrosine and phenylalanine are precursors to norepinephrine and epinephrine, both of which are stimulatory. Epinephrine is commonly known as the "flight or fight" hormone. It causes the release of glucose and other nutrients into the blood and stimulates brain function.

It has been suggested that tyrosine and phenylalanine may have unexpected effects in some people. For example, a growing body of evidence indicates that some people get headaches from the phenylalanine in aspartame, an artificial sweetener commonly found in diet soft drinks. Some people insist that supplements of tyrosine give them a morning lift and that tryptophan helps them sleep at night. Milk proteins have high levels of tryptophan; a glass of warm milk before bed is widely believed to be an aid in inducing sleep.

14.5 What Are Uncommon Amino Acids?

Many other amino acids in addition to the ones listed in Table 14.1 are known to exist. They occur in some, but by no means all, proteins. Figure 14.3 shows some examples of the many possibilities. These uncommon amino acids are derived from the common amino acids and are produced by modification of the parent amino acid after the protein is synthesized by the organism in a process called post-translational modification (Chapter 18). Hydroxyproline and hydroxylysine differ from their parent amino acids in that they have hydroxyl groups on their side chains; they are found only in a few connective tissue proteins, such as collagen. Thyroxine differs from tyrosine in that it has an extra iodine-containing aromatic group on the side chain; it is found only in the thyroid gland, where it is formed by post-translational modification of tyrosine residues in the protein thyroglobulin. Thyroxine is then released as a hormone by proteolysis of thyroglobulin. Both animals and humans that exhibit sluggishness and slow metabolism are often given thyroxine to help ramp up their metabolism.

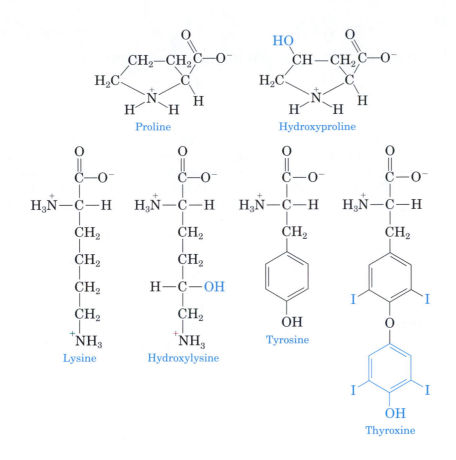

FIGURE 14.3 Structures of hydroxyproline, hydroxylysine, and thyroxine. The structures of the parent amino acids—proline for hydroxyproline, lysine for hydroxylysine, and tyrosine for thyroxine—are shown for comparison. All amino acids are shown in their predominant ionic forms at pH 7.

14.6 How Do Amino Acids Combine to Form Proteins?

Each amino acid has a carboxyl group and an amino group. In Chapter 11, we saw that a carboxylic acid and an amine could be combined to form an amide:

$$R-\overset{O}{\overset{\|}{C}}-O^- + R'-NH_3^+ \longrightarrow R-\overset{O}{\overset{\|}{C}}-\overset{}{\underset{H}{N}}-R' + H_2O$$

In the same way, the —COO⁻ group of one amino acid molecule—say, glycine—can combine with the —NH₃⁺ group of a second molecule—say, alanine:

Glycine   Alanine   Glycylalanine (Gly—Ala)

This reaction takes place in the cells by a mechanism that we will examine in Section 18.5. The product is an amide. The two amino acids are joined together by a **peptide bond** (also called a **peptide linkage**). The product is a **dipeptide.**

It is important to realize that glycine and alanine could also be linked the other way:

Peptide bond An amide bond that links two amino acids

Alanine   Glycine   Alanylglycine
                                  (Ala—Gly)

In this case, we get a *different* dipeptide. The two dipeptides are constitutional isomers, of course; they are different compounds in all respects, with different properties. The phrase "do much, talk little" has the same words as "do little, talk much," but the meaning of the two are quite different. In the same way, the order of amino acids in a peptide or protein is critical to both the structure and function.

Example 14.1 Peptide Formation

Show how to form the dipeptide aspartylserine (Asp—Ser).

Strategy

Start by drawing the two amino acids. Orient them so that both read (from left to right) amino group, alpha carbon, carboxyl group. Then draw the reaction between the first amino acid's carboxyl group and the second amino acid's amino group to give the peptide bond.

Solution

The name implies that this dipeptide is made of two amino acids, aspartic acid (Asp) and serine (Ser), with the amide being formed between the α-carboxyl group of aspartic acid and the α-amino group of serine. Therefore, we write the formula of aspartic acid with its amino group on the left side. Next, we place the formula of serine to the right, with its amino group facing the α-carboxyl group of aspartic acid. Finally, we eliminate a water molecule between the —COO⁻ and —NH₃⁺ groups that are next to each other, forming the peptide bond:

Problem 14.1
Show how to form the dipeptide valylphenylalanine (Val—Phe).

Any two amino acids, whether the same or different, can be linked together to form dipeptides in a similar manner. But the possibilities do not end there. Each dipeptide still contains a —COO⁻ and an —NH₃⁺ group.

Chemical Connections 14A

Aspartame, the Sweet Peptide

The dipeptide L-aspartyl-L-phenylalanine is of considerable commercial importance. The aspartyl residue has a free α-amino group, the N-terminal end of the molecule, and the phenylalanyl residue has a free carboxyl group, the C-terminal end. This dipeptide is about 200 times sweeter than sugar. A methyl ester derivative of this dipeptide is of even greater commercial importance than the dipeptide itself. The derivative has a methyl group at the C-terminal end in an ester linkage to the carboxyl group. The methyl ester derivative is called *aspartame* and is marketed as a sugar substitute under the trade name NutraSweet.

Common table sugar is consumed in the United States at about 100 pounds per person per year. Many people want to curtail their sugar intake in the interest of fight-ing obesity. Others must limit their sugar intake because of diabetes. One of the most common ways of doing so is by drinking diet soft drinks. The soft-drink industry is one of the largest markets for aspartame. The use of this sweet-ener was approved by the U.S. Food and Drug Adminis-tration in 1981 after extensive testing, although there is still considerable controversy about its safety. Diet soft drinks sweetened with aspartame carry warning labels about the presence of phenylalanine. This information is of vital importance to people who have phenylketonuria, a genetic disease of phenylalanine metabolism. Note that both amino acids have the L configuration. If a D-amino acid is substituted for either amino acid or for both of them, the resulting derivative is bitter rather than sweet.

L-Aspartyl-L-phenylalanine (methyl ester)

We can, therefore, add a third amino acid to alanylglycine—say, lysine: The product is a **tripeptide.** Because it also contains a —COO⁻ and an —NH₃⁺ group, we can continue the process to get a tetrapeptide, a pentapeptide, and so on, until we have a chain containing hundreds or even thousands of amino acids. These chains of amino acids are the proteins that serve so many important functions in living organisms.

Ala—Gly   Lys

Ala—Gly—Lys
A tripeptide

A word must be said about the terms used to describe these compounds. The shortest chains are often simply called **peptides,** longer ones are **polypeptides,** and still longer ones are **proteins,** but chemists differ about where to draw the line. Many chemists use the terms "polypeptide" and "protein" almost interchangeably. In this book, we will consider a protein to be a polypeptide chain that contains a minimum of 30 to 50 amino acids. The amino acids in a chain are often called **residues.** It is customary to use either the one-letter or the three-letter abbreviations shown in Table 14.1 to represent peptides and proteins. For example, the tripeptide shown on the previous page, alanylglycyllysine, is AGK or Ala—Gly—Lys. The **C-terminal amino acid** or **C-terminus** is the residue with the free α-COO$^-$ group (lysine in Ala—Gly—Lys), and the **N-terminal amino acid** or **N-terminus** is the residue with the free α-NH$_3^+$ group (alanine in Ala—Gly—Lys). It is the universal custom to write peptide and protein chains with the N-terminal residue on the left. This decision is not as arbitrary as it might seem. We read left to right, and proteins are synthesized from N-terminus to C-terminus, as we will see in Chapter 18.

C-terminus The amino acid at the end of a peptide that has a free α-carboxyl group

N-terminus The amino acid at the end of a peptide that has a free α-amino group

14.7 What Are the Properties of Proteins?

The properties of proteins are based on properties of the peptide backbone and properties of the side chains. The peptide backbone consists of the repeating structure shown by the horizontal line of atoms in Figure 14.4. The atoms along the backbone are linked N—C—C—N—C—C— and so on. By convention, peptides are shown with the N-terminus on the left. As it turns out, much of the structure of a protein is due to the interactions of the atoms in the backbone without taking into account the nature of the R groups on the side chains.

FIGURE 14.4 A small peptide showing the direction of the peptide chain (N-terminal to C-terminal)

Although the peptide bond is typically written as a carbonyl group bonded to an N—H group, as we saw in Section 9.5, such bonds can exhibit keto-enol tautomerism. The carbon–nitrogen bond actually has around 40% double bond character, as shown in Figure 14.5. As a result, the peptide group that forms the link between the two amino acids is actually planar. This

FIGURE 14.5 The resonance structures of the peptide bond lead to a planar group. (Illustration, Irving Geis. Rights owned by Howard Hughes Medical Institute. Not to be reproduced without permission.)

Chemical Connections 14B

AGE and Aging

A reaction can take place between a primary amine and an aldehyde or a ketone, linking the two molecules (shown here for an aldehyde):

$$R-\overset{\overset{\displaystyle O}{\|}}{C}-H + H_2N-R' \longrightarrow R-CH=N-R' + H_2O$$

An imine

Because proteins have NH_2 groups and carbohydrates have aldehyde or ketone groups, they can undergo this reaction, establishing a link between a sugar and a protein molecule. When this reaction is not catalyzed by enzymes, it is called *glycation* of proteins. The process, however, does not stop there. When these linked products are heated in a test tube, high-molecular-weight, water-insoluble, brownish complexes form. These complexes are called **advanced glycation end-products (AGE)**. In the body, they cannot be heated, but the same result happens over long periods of time.

The longer we live, and the higher the blood sugar concentration becomes, the more AGE products accumulate in the body. These AGEs can alter the function of proteins. Such AGE-dependent changes are thought to contribute to circulation, joint, and vision problems in people with diabetes. People with diabetes have high blood sugar due to a lack of transport of glucose out of the blood and into the cells. AGE products show up in all of the afflicted organs of diabetic patients: in the lens of the eye (cataracts), in the capillary blood vessels of the retina (diabetic retinopathy), and in the glomeruli of the kidneys (kidney failure). AGEs have been linked to atherosclerosis, as AGE-modified cells can bind to endothelial cells in blood vessels. AGE-modified collagen causes stiffening of arteries.

For people who do not have diabetes, these harmful protein modifications become disturbing only in an individual's advanced years. In a young person, metabolism functions properly and the AGE products decompose and are eliminated from the body. In an older person, metabolism slows and the AGE products accumulate. The AGE products themselves are thought to enhance oxidative damages.

Scientists are searching for ways to combat the harmful effects of AGEs. One approach is to use antioxidants, including the B vitamin, thiamine. A few other anti-AGE drugs have been developed, including aminoguanidine and metformin. Both of these drugs have been studied in animal models, but have not yet reached large-scale use in humans. Another approach that is being studied for several metabolic problems, including normal aging, is caloric restriction. A vast amount of evidence from animal models and human models alike indicates that life span can be extended by living a lean existence. This lifestyle has also been shown to reduce the level of AGEs.

National Eye Institute, National Institutes of Health

Cataracts in the eye are caused by Advanced Glycation End-products.

grouping is called the amide plane, and it has a tremendous influence on protein structure. There is freedom of rotation about the two bonds from the alpha carbon, but there is no rotation of the carbon–nitrogen bonds. A chain of amino acids linked via peptide bonds can be thought of as a series of playing cards linked by a swivel at their corners, as shown in Figure 14.6. The rigidity of the amide plane limits the possible orientation of the peptide.

The 20 different amino acid side chains supply variety and determine the rest of the physical and chemical properties of proteins. Among these properties, acid–base behavior is one of the most important. Like amino acids (Section 14.3), proteins behave as zwitterions. The side chains of glutamic and aspartic acids provide acidic groups, whereas lysine and arginine provide basic groups (histidine does as well, but this side chain is less basic than the other two). (See the structures of these amino acids in Figure 14.1.)

The isoelectric point of a protein occurs at the pH at which there are an equal number of positive and negative charges (the protein has no *net* charge). At any pH above the isoelectric point, the protein molecules have a net negative charge; at any pH below the isoelectric point, they have a net positive charge. Some proteins, such as hemoglobin, have an almost equal number of acidic and basic groups; the isoelectric point of hemoglobin is at pH 6.8. Others, such as serum albumin, have more acidic groups than basic groups; the isoelectric point of this protein is at pH 4.9. In each case, however, because proteins behave like zwitterions, they act as buffers—for example, in the blood (Figure 14.7).

FIGURE 14.6 Planar nature of peptide bond. The rigid planar peptide groups (called "playing cards" in the text) are shaded. (Illustration Irving Geis. Rights owned by Howard Hughes Medical Institute. Not to be reproduced without permission.)

FIGURE 14.7 Schematic diagram of a protein (a) at its isoelectric point and its buffering action when (b) H^+ or (c) OH^- ions are added.

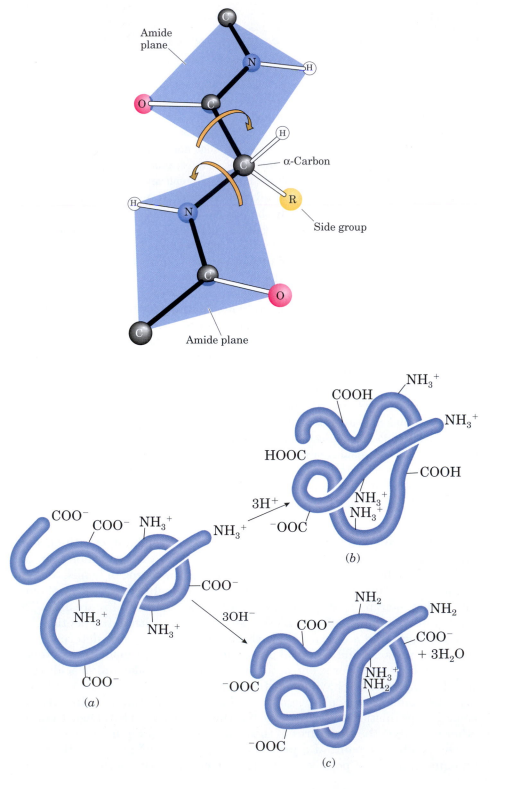

The water solubility of large molecules such as proteins often depends on the repulsive forces between like charges on their surfaces. When protein molecules are at a pH at which they have a net positive or negative charge, the presence of these like charges causes the protein molecules to repel one another. These repulsive forces are smallest at the isoelectric point, when the net charges are zero. When there are no repulsive forces, the protein molecules tend to clump together to form aggregates of two or more molecules, reducing their solubility. As a consequence, *proteins are least soluble in water at their isoelectric points and can be precipitated from their solutions.*

As we pointed out in Section 14.1, proteins have many functions. To understand these functions, we must look at four levels of organization in their structures. The *primary structure* describes the linear sequence of amino acids in the polypeptide chain. The *secondary structure* refers to certain repeating patterns, such as the α-helix conformation or the pleated sheet (Section 14.9), or the absence of a repeating pattern, as with the random coil (Section 14.9). The *tertiary structure* describes the overall conformation of the polypeptide chain (Section 14.10). The *quaternary structure* (Section 14.11) applies mainly to proteins containing more than one polypeptide chain (subunit) and deals with how the different chains are spatially related to one another.

14.8 What Is the Primary Structure of a Protein?

Very simply, the **primary structure** of a protein consists of the sequence of amino acids that makes up the chain. Each of the very large number of peptide and protein molecules in biological organisms has a different sequence of amino acids—and that sequence allows the protein to carry out its function, whatever it may be.

How can so many different proteins arise from different sequences of 20 amino acids? Let us look at a little arithmetic, starting with a dipeptide. How many different dipeptides can be made from 20 amino acids? There are 20 possibilities for the N-terminal amino acid, and for each of these 20 there are 20 possibilities for the C-terminal amino acid. This means that there are $20 \times 20 = 400$ different dipeptides possible from the 20 amino acids. What about tripeptides? We can form a tripeptide by taking any of the 400 dipeptides and adding any of the 20 amino acids. Thus, there are $20 \times 20 \times 20 = 8000$ tripeptides, all different. It is easy to see that we can calculate the total number of possible peptides or proteins for a chain of n amino acids simply by raising 20 to the nth power (20^n).

Taking a typical small protein to be one with 60 amino acid residues, the number of proteins that can be made from the 20 amino acids is $20^{60} = 10^{78}$. This is an enormous number, possibly greater than the total number of atoms in the universe. Clearly, only a tiny fraction of all possible protein molecules have ever been made by biological organisms.

Each peptide or protein in the body has its own unique sequence of amino acids. As with naming of peptides, the *assignment of positions of the amino acids in the sequence starts at the N-terminal end.* Thus, in Figure 14.8, glycine is in the number 1 position on the A chain and phenylalanine is number 1 on the B chain. We mentioned that proteins also have secondary, tertiary, and, in some cases, quaternary structures. We will deal with these in Sections 14.9, 14.10, and 14.11, but here we can say that *the primary structure of a protein determines to a large extent the native*

Primary structure The sequence of amino acids in a protein

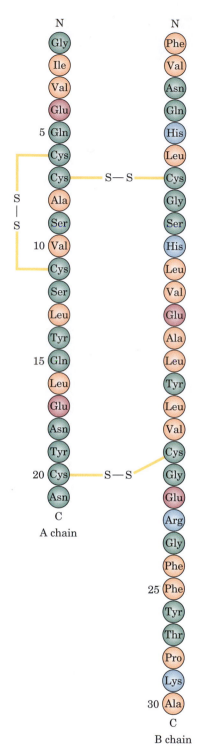

FIGURE 14.8 The hormone insulin consists of two polypeptide chains, A and B, held together by two disulfide cross-bridges (S—S). The sequence shown is for bovine insulin.

(most frequently occurring) *secondary and tertiary structures*. That is, the particular sequence of amino acids on the chain enables the whole chain to fold and curl in such a way as to assume its final shape. As we will see in Section 14.12, without its particular three-dimensional shape, a protein cannot function.

Just how important is the exact amino acid sequence to the function of a protein? Can a protein perform the same function if its sequence is a little different? The answer to this question is that a change in amino acid sequence may or may not matter, depending on what kind of a change it is. Consider, as an example, cytochrome c, which is a protein of terrestrial vertebrates. Its chain consists of 104 amino acid residues. It performs the same function (electron transport) in humans, chimpanzees, sheep, and other animals. While humans and chimpanzees have exactly the same amino acid sequence of this protein, sheep cytochrome c differs in 10 positions out of the 104. (You can find more about biochemical evolution in Chemical Connections 18F.)

Another example is the hormone insulin. Human insulin consists of two chains having a total of 51 amino acids. The two chains are connected by disulfide bonds. Figure 14.8 shows the sequence of amino acids. Insulin is necessary for proper utilization of carbohydrates (Section 20.1), and people with severe diabetes (Chemical Connections 14C) must take insulin injections. The amount of human insulin available is far too small to meet the need for it, so bovine insulin (from cattle) or insulin from hogs or sheep is used instead. Insulin from these sources is similar, but not identical, to human insulin. The differences are entirely in the 8, 9, and 10 positions of the A chain and the C-terminal position (30) of the B chain, as shown in Table 14.2. The remainder of the molecule is the same in all four varieties of insulin. Despite the slight differences in structure, all of these insulins perform the same function and even can be used by humans. However, none of the other three is quite as effective in humans as human insulin. This is one of the reasons that recombinant DNA techniques are now used to produce human insulin from bacteria (Section 18.8 and Chemical Connections 14C).

Another factor showing the effect of substituting one amino acid for another is that sometimes patients become allergic to, say, bovine insulin but can switch to hog or sheep insulin without experiencing an allergic reaction.

In contrast to the previous examples, some small changes in amino acid sequence make a great deal of difference. Consider two peptide hormones, oxytocin and vasopressin (Figure 14.9). These nonapeptides have identical structures, including a disulfide bond, except for different amino acids in positions 2 and 7. Yet their biological functions are quite different. Vasopressin is an antidiuretic hormone. It increases the amount of water reabsorbed by the kidneys and raises blood pressure. Oxytocin has no effect on water in the kidneys and slightly lowers blood pressure. It affects contractions of the uterus in childbirth and the muscles in the breast that aid in the secretion

TABLE 14.2 Amino Acid Sequence Differences for Human, Bovine, Hog, and Sheep Insulin

	A Chain			B Chain
	8	9	10	30
Human	—Thr—	Ser—	Ile—	—Thr
Bovine	—Ala—	Ser—	Val—	—Ala
Hog	—Thr—	Ser—	Ile—	—Ala
Sheep	—Ala—	Gly—	Val—	—Ala

Chemical Connections 14C

The Use of Human Insulin

Although human insulin, as manufactured by recombinant DNA techniques (see Section 18.8), is on the market, many diabetic patients continue to use hog or sheep insulin because it is less expensive. Changing from animal to human insulin creates an occasional problem for diabetics. All diabetics experience an insulin reaction (hypoglycemia) when the insulin level in the blood is too high relative to the blood sugar level. Hypoglycemia is preceded by symptoms of hunger, sweating, and poor coordination. These symptoms, called hypoglycemic awareness, signal the patient that hypoglycemia is coming and that it must be reversed, which the patient can do by eating sugar.

Some diabetics who changed from animal to human insulin reported that the hypoglycemic awareness from recombinant DNA human insulin is not as strong as that from animal insulin. This lack of recognition can create some hazards, and the effect is probably due to different rates of absorption in the body. The literature supplied with human insulin now incorporates a warning that hypoglycemic awareness may be altered.

Insulin medication can be made from animal sources, such as hog or sheep, or created by recombinant DNA techniques to give a human protein.

of milk. Vasopressin also stimulates uterine contractions, albeit to a much lesser extent than oxytocin.

Another instance where a minor change makes a major difference is in the blood protein hemoglobin. A change in only one amino acid in a chain of 146 is enough to cause a fatal disease—sickle cell anemia (Chemical Connections 14D).

In some cases, slight changes in amino acid sequence make little or no difference to the functioning of peptides and proteins, but most times, the sequence is highly important. The sequences of tens of thousands of protein and peptide molecules have now been determined. The methods for doing so are complicated and will not be discussed in this book.

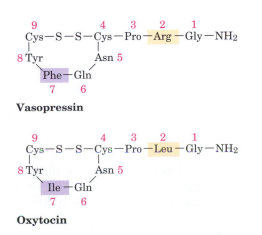

FIGURE 14.9 The structures of vasopressin and oxytocin. Differences are shown in color.

Chemical Connections 14D

Sickle Cell Anemia

Normal adult human hemoglobin has two alpha chains and two beta chains (see Figure 14.17). Some people, however, have a slightly different kind of hemoglobin in their blood. This hemoglobin (called HbS) differs from the normal type only in the beta chains and only in one position on these two chains: The glutamic acid in the sixth position of normal Hb is replaced by a valine residue in HbS.

	4	5	6	7	8	9
Normal Hb	—Thr—	Pro—	Glu—	Glu—	Lys—	Ala—
Sickle cell Hb	—Thr—	Pro—	Val—	Glu—	Lys—	Ala—

This change affects only two positions in a molecule containing 574 amino acid residues, yet it is enough to produce a very serious disease, **sickle cell anemia.**

Red blood cells carrying HbS behave normally when there is an ample oxygen supply. When the oxygen pressure decreases, however, the red blood cells become sickle-shaped, as shown in the figure. This malformation occurs in the capillaries. As a result of this change in shape, the cells may clog the capillaries. The body's defenses destroy the clogging cells, and the loss of the blood cells causes anemia.

This change at only a single position of a chain consisting of 146 amino acids is severe enough to cause a high death rate. A child who inherits two genes programmed to produce sickle cell hemoglobin (a homozygote) has an 80% smaller chance of surviving to adulthood than a child with only one such gene (a heterozygote) or a child with two normal genes. Despite the high mortality of homozygotes, the genetic trait survives. In central Africa, 40% of the population in malaria-ridden areas carry the sickle cell gene, and 4% are homozygotes. It seems that the sickle cell genes help to acquire immunity against malaria in early childhood so that in malaria-ridden areas the transmission of these genes is advantageous.

There is no known cure for sickle cell anemia. Recently the U.S. Food and Drug Administration approved hydroxyurea

Blood cells from a patient with sickle cell anemia. Both normal cells (round) and sickle cells (shriveled) are visible.

G. W. Willis/Visuals Unlimited

(sold under the name Droxia) to treat and control the symptoms of the disease.

Hydroxyurea

Hydroxyurea prompts the bone marrow to manufacture fetal hemoglobin (HbF), which does not have beta chains where the mutation occurs. Thus red blood cells containing HbF do not sickle and do not clog the capillaries. With hydroxyurea therapy, the bone marrow still manufactures mutated HbS, but the presence of cells with fetal hemoglobin dilutes the concentration of the sickling cells, thereby relieving the symptoms of the disease.

14.9 What Is the Secondary Structure of a Protein?

Secondary structure A repetitive conformation of the protein backbone

Proteins can fold or align themselves in such a manner that certain patterns repeat themselves. These repeating patterns are referred to as **secondary structures.** The two most common secondary structures encountered in proteins are the α-helix and the β-pleated sheet (Figure 14.10), which were proposed by Linus Pauling and Robert Corey in the 1940s. In contrast, those protein conformations that do not exhibit a repeated pattern are called random coils (Figure 14.11).

● Carbon atom ● Oxygen atom ● Nitrogen atom ● Hydrogen atom ● R group

FIGURE 14.10 (*a*) The α-helix. (*b*) The β-pleated sheet structure.

In the **α-helix** form, a single protein chain twists in such a manner that its shape resembles a right-handed coiled spring—that is, a helix. The shape of the helix is maintained by numerous **intramolecular hydrogen bonds** that exist between the backbone —C=O and H—N— groups. As shown in Figure 14.10, there is a hydrogen bond between the —C=O oxygen atom of each peptide linkage and the —N—H hydrogen atom of another peptide linkage four amino acid residues farther along the chain. These hydrogen bonds are in just the right position to cause the molecule (or a portion of it) to maintain a helical shape. Each —N—H points upward and each C=O points downward, roughly parallel to the axis of the helix. All the amino acid side chains point outward from the helix.

The other important orderly structure in proteins is the **β-pleated sheet.** In this case, the orderly alignment of protein chains is maintained by **intermolecular** or **intramolecular hydrogen bonds.** The β-sheet structure can occur between molecules when polypeptide chains run parallel (all N-terminal ends on one side) or antiparallel (neighboring N-terminal ends on opposite sides). β-Pleated sheets can also occur intramolecularly, when the polypeptide chain makes a U-turn, forming a hairpin structure, and the pleated sheet is antiparallel (Figure 14.10).

In all secondary structures, the hydrogen bonding is between backbone —C=O and H—N— groups, a characteristic that distinguishes between secondary and tertiary structures. In the latter, as we shall see, the hydrogen bonding can take place between R groups on the side chains.

Few proteins have predominantly α-helix or β-sheet structures. Most proteins, especially globular ones, have only certain portions of their molecules in these conformations. The rest of the molecule consists of **random coil.** Many globular proteins contain all three kinds of secondary structures in different parts of their molecules: α-helix, β-sheet, and random coil. Figure 14.12 shows a schematic representation of such a structure.

Keratin, a fibrous protein of hair, fingernails, horns, and wool, is one protein that does have a predominantly α-helix structure. Silk is made of fibroin, another fibrous protein, which exists mainly in the β-pleated sheet form. Silkworm silk and especially spider silk exhibit a combination of strength and toughness that is unmatched by high-performance synthetic fibers. In its primary structure, silk contains sections that consist of only alanine (25%) and glycine (42%). The formation of β-pleated sheets, largely

Alpha (α)-helix A secondary structure where the protein folds into a coil held together by hydrogen bonds parallel to the axis of the coil

Beta (β)-pleated sheet A secondary protein structure in which the backbone of two protein chains in the same or different molecules is held together by hydrogen bonds

FIGURE 14.11 A random coil.

FIGURE 14.12 Schematic structure of the enzyme carboxypeptidase. The β-pleated sheet portions are shown in blue, the green structures are the α-helix portions, and the orange strings are the random coil areas.

Tertiary structure The complete three-dimensional arrangement of the atoms in a protein

by the alanine sections, allows microcrystals to orient themselves along the fiber axis, which accounts for the material's superior tensile strength.

Another repeating pattern classified as a secondary structure is the **extended helix** of collagen (Figure 14.13). It is quite different from the α-helix. Collagen is the structural protein of connective tissues (bone, cartilage, tendon, blood vessels, skin), where it provides strength and elasticity. The most abundant protein in humans, it makes up about 30% by weight of all the body's protein. The extended helix structure is made possible by the primary structure of collagen. Each strand of collagen consists of repetitive units that can be symbolized as Gly—X—Y; that is, every third amino acid in the chain is glycine. Glycine, of course, has the shortest side chain (—H) of all amino acids. About one-third of the X amino acid is proline, and the Y is often hydroxyproline.

14.10 What Is the Tertiary Structure of a Protein?

The **tertiary structure** of a protein is the three-dimensional arrangement of every atom in the molecule. Unlike the secondary structure, it includes interactions of the side chains, and not just the peptide backbone. In general, tertiary structures are stabilized five ways:

1. **Covalent Bonds** The covalent bond most often involved in stabilization of the tertiary structure of proteins is the disulfide bond. In Section 14.4, we noted that the amino acid cysteine is easily converted to the dimer cystine. When a cysteine residue is in one chain and another cysteine residue is in another chain (or in another part of the same chain), formation of a disulfide bond provides a covalent linkage that binds together the two chains or the two parts of the same chain:

$$\text{—SH HS—} \xrightarrow{[O]} \text{—S—S—}$$

Examples of both types are found in the structure of insulin (Figure 14.8).

2. **Hydrogen Bonding** In Section 14.9, we saw that secondary structures are stabilized by hydrogen bonding between backbone —C=O and —N—H groups. Tertiary structures are stabilized by hydrogen bonding between polar groups on side chains or between side chains and the peptide backbone [Figure 14.14(a)].

3. **Salt Bridges** Salt bridges, also called electrostatic attractions, occur between two amino acids with ionized side chains—that is, between an acidic amino acid (—COO⁻) and a basic amino acid (—NH₃⁺ or =NH₂⁺) side chain. The two are held together by simple ion–ion attraction [Figure 14.14(b)].

4. **Hydrophobic Interactions** In aqueous solution, globular proteins usually turn their polar groups outward, toward the aqueous solvent, and their nonpolar groups inward, away from the water molecules. The nonpolar groups prefer to interact with each other, excluding water from these regions. The result is a series of hydrophobic interactions (see Section 13.1) [Figure 14.14(c)]. Although this type of interaction is weaker than hydrogen bonding or salt bridges, it usually acts over large surface areas, so that the interactions are collectively strong enough to stabilize a loop or some other tertiary structure formation.

5. **Metal Ion Coordination** Two side chains with the same charge would normally repel each other, but they can also be linked via a metal ion. For example, two glutamic acid side chains (—COO⁻) would both

FIGURE 14.13 The triple helix of collagen.

FIGURE 14.14 Noncovalent interactions that stabilize the tertiary and quaternary structures of proteins: (*a*) hydrogen bonding, (*b*) salt bridge (electrostatic interaction), (*c*) hydrophobic interaction, and (*d*) metal ion coordination.

be attracted to a magnesium ion (Mg^{2+}), forming a bridge. This is one reason that the human body requires certain trace minerals—they are necessary components of proteins [Figure 14.14(d)].

Example 14.2 Amino Acid Interactions

What kind of noncovalent interaction occurs between the side chains of serine and glutamine?

Strategy

Analyze the types of functional groups in the side chains and then look for potential interactions.

Solution

The side chain of serine ends in an —OH group; that of glutamine ends in an amide, the CO—NH$_2$ group. The two groups can form hydrogen bonds.

Problem 14.2

What kind of noncovalent interaction occurs between the side chains of arginine and glutamic acid?

In Section 14.8, we pointed out that the primary structure of a protein largely determines its secondary and tertiary structures. We can now see the reason for this relationship. When the particular R groups are in the proper positions, all of the hydrogen bonds, salt bridges, disulfide linkages, and hydrophobic interactions that stabilize the three-dimensional structure of that molecule can form. Figure 14.15 illustrates the possible combinations of forces that lead to tertiary structure.

The side chains of some proteins allow them to fold (form a tertiary structure) in only one way; other proteins, especially those with long polypeptide chains, can fold in a number of possible ways. Certain proteins in living

Chemical Connections 14E

Protein/Peptide Conformation–Dependent Diseases

In a number of diseases, a normal protein or peptide becomes pathological when its conformation changes. A common feature of these proteins is the property to self-assemble into β-sheet–forming amyloid (starch-like) plaques. These amyloid structures appear in several diseases.

One example of this process involves the prion protein, the discovery of which brought Stanley Prusiner of the University of California, San Francisco, the Nobel Prize in 1997. Prions are small proteins found in nerve tissue, although their exact function remains a mystery. When prions undergo conformational change, they can cause diseases such as mad cow disease and scrapie in sheep. During the conformational change, the α-helical content of the normal prion protein unfolds and reassembles in the β-sheet form. This new form has the potential to cause more normal prion proteins to undergo conformational change. In humans, it causes spongiform encephalitis; Creutzfeld-Jakob disease is one variant that mainly afflicts elderly people. Although the transmission of this infection from diseased cows to humans is rare, fear of it caused the wholesale slaughter of British cattle in 1998 and, for a while, an embargo was placed on the importation of such meat in most of Europe and America. β-Amyloid plaques also appear in the brains of patients with Alzheimer's disease (see Chemical Connections 16C).

The modus operandi of prion diseases stumped scientists for many years. On the one hand, human spongiform encephalopathies behave like inheritable diseases, in that they can be traced through families. On the other hand, they behave like infectious diseases which can be acquired from someone else. It is now believed that the mechanism of spread is a combination of the two. There is a genetic component in that a person could have a 100% wild-type prion protein that would not adopt the alternate form. Several mutations that lead to the abnormal prion form have been identified. However, there appears to be a need for a triggering event as well. This characteristic was seen in studies of sheep in New Zealand, where isolated groups were found to have the right mutations to get a prion disease, but none of the sheep did, generation after generation, because they were never infected with a mutant prion.

Schematic representation of a possible mechanism of amyloid fibril formation. After synthesis, the protein is assumed to fold in native (N) secondary structure aided by chaperones. Under certain conditions, the native structure can partially unfold (I) and form sheets of amyloid fibrils or even completely unfold (U) as a random coil.

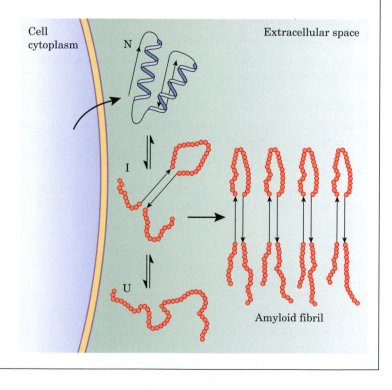

Chaperone A protein that helps other proteins to fold into the biologically active conformation and enables partially denatured proteins to regain their biologically active conformation

cells, called **chaperones,** help a newly synthesized polypeptide chain assume the proper secondary and tertiary structures that are necessary for the functioning of that molecule and prevent foldings that would yield biologically inactive molecules.

FIGURE 14.15 Forces that stabilize the tertiary structures of proteins. Note that the helical structure and the sheet structure are two kinds of backbone hydrogen bonding. Although the backbone hydrogen bonding is part of the secondary structure, the conformation of the backbone puts constraints on the possible arrangement of the side chains.

14.11 What Is the Quaternary Structure of a Protein?

The highest level of protein organization is the **quaternary structure,** which applies to proteins with more than one polypeptide chain. Figure 14.16 summarizes schematically the four levels of protein structure. Quaternary structure determines how the different subunits of the protein fit into an organized whole. The subunits are packed and held together by hydrogen bonds, salt bridges, and hydrophobic interactions—the same forces that operate within tertiary structures.

Quaternary structure The spatial relationship and interactions between subunits in a protein that has more than one polypeptide chain

1. **Hemoglobin** Hemoglobin in adult humans is made of four chains (called globins): two identical α chains of 141 amino acid residues each and two identical β chains of 146 residues each. Figure 14.17 shows how the four chains fit together.

   In hemoglobin, each globin chain surrounds an iron-containing heme unit, the structure of which is shown in Figure 14.18. Proteins that contain non-amino acid portions are called **conjugated proteins.** The non-amino acid portion of a conjugated protein is called a **prosthetic group.** In hemoglobin, the globins are the amino acid portions and the heme units are the prosthetic groups.

The designations α and β with respect to hemoglobin have nothing to do with the same designations for the α-helix and β-pleated sheet.

   Hemoglobin containing two alpha and two beta chains is not the only kind existing in the human body. In the early development stage of the fetus, the hemoglobin contains two alpha and two gamma chains. The fetal hemoglobin has greater affinity for oxygen than does the adult hemoglobin. In this way, the mother's red blood cells carrying oxygen can pass it to the fetus for its own use. Fetal hemoglobin also alleviates some of the symptoms of sickle cell anemia (see Chemical Connections 14D).

2. **Collagen** Another example of quaternary structure and higher organizations of subunits can be seen in collagen. The triple helix units, called *tropocollagen,* constitute the soluble form of collagen; they are stabilized by hydrogen bonding between the backbones of the three chains. Collagen

Chemical Connections 14F

Proteomics, Ahoy!

Proteins in the body are in a state of dynamic flux. Their multiple functions necessitate that they change constantly: Some are rapidly synthesized, others have their synthesis inhibited; some are degraded, others are modified. The complement of proteins expressed by a genome is called its **proteome.** Today, a concerted effort is being made to catalogue all the proteins in their various forms in a particular cell or a tissue. The name of this venture is *proteomics,* a term coined as an analogy to *genomics* (see Chemical Connections 17D), in which all the genes of an organism and their locations in the chromosomes are determined. The approximately 30,000 genes defined by the Human Genome Project translate into 300,000 to 1 million proteins when alternate splicing and post-translational modifications are considered (Chapter 18).

While a genome remains unchanged to a large extent, the proteins in any particular cell change dramatically as genes are turned on and off in response to its environment. In proteomics, all the proteins and peptides of a cell or tissue are separated and then studied by a variety of procedures, including some very new technologies. The first procedure is necessary to separate proteins from one another. The main way this separation has been achieved is two-dimensional polyacrylamide gel electrophoresis (2-D PAGE; see Chapter 17 for more on electrophoresis). 2-D PAGE can achieve the separation of several thousand different proteins in one gel. High-resolution 2-D PAGE can resolve as many as 10,000 protein spots per gel. In one dimension, the proteins are separated by charge (isoelectric point; Section 14.3); in the second dimension, they are separated by mass. In isoelectric focusing, the proteins migrate in a pH gradient to the pH at which they have no net charge (the isoelectric point). Most commonly, proteins are separated by size in the vertical direction and by isoelectric point in the horizontal direction.

Mass spectrometry is used for mass determination and can be adapted for protein identification. A mass spectrometer separates proteins according to their mass-to-charge (m/Z) ratio. The molecule is ionized by one of several techniques, and the ion is propelled into a mass analyzer by an electric field that resolves each ion according to its m/Z ratio. The detector passes the information to the computer for analysis.

A new technology that holds much promise for the future of protein analysis is **protein microarrays.** They can be used for protein purification, expression profiling, or protein interaction profiling. Many types of substances may be bound to protein arrays, including antibodies,

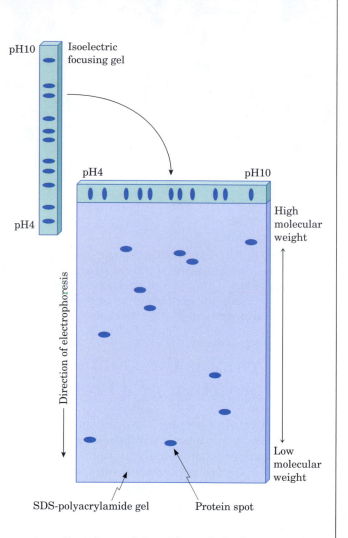

receptors, ligands, nucleic acids, carbohydrates, or chromatographic (cationic, anionic, hydrophobic, hydrophilic) surfaces. Some surfaces have broad specificity and bind entire classes of proteins; others are highly specific and bind only a few proteins from a complex sample. Some protein arrays contain antibodies (Chapter 23) that are covalently immobilized onto the array surface and that capture corresponding antigens from a complex mixture. Many analyses can follow from this binding. Other proteins of interest can also be immobilized on the array. Bound receptors can reveal ligands, and binding domains for protein–protein interactions can be detected.

The goal of such techniques is to obtain information about the dynamic states of a large number of proteins and the status of a cell or tissue, thereby ascertaining whether it is healthy or pathological.

FIGURE 14.16 Primary, secondary, tertiary, and quaternary structures of a protein.

consists of many tropocollagen units. Tropocollagen is found only in fetal or young connective tissues. With aging, the triple helixes (Figure 14.13) that organize themselves into fibrils cross-link and form insoluble collagen. In collagen, the **cross-linking** consists of covalent bonds that link together two lysine residues on adjacent chains of the helix. This cross-linking of collagen is an example of the tertiary structures that stabilize the three-dimensional conformations of protein molecules.

3. Integral Membrane Proteins Integral membrane proteins traverse partly or completely a membrane bilayer. (See Figure 13.2.) An estimated one-third of all proteins are integral membrane proteins. To keep the protein stable in the nonpolar environment of a lipid bilayer, it must form quaternary structures in which the outer surface is largely nonpolar and interacts with the lipid bilayer. Thus most of the polar groups of the protein must turn inward. Two such quaternary structures exist in integral membrane proteins: (1) often 6 to 10 α-helices crossing the membrane and (2) β-barrels made of 8, 12, 16, or 18 antiparallel β-sheets (Figures 14.19 and 14.20).

FIGURE 14.17 The quaternary structure of hemoglobin.

FIGURE 14.18 The structure of heme.

FIGURE 14.19 Integral membrane protein of rhodopsin, made of α-helices.

FIGURE 14.20 An integral membrane protein from the outer mitochondrial membrane forming a β-barrel from eight β-pleated sheets.

14.12 How Are Proteins Denatured?

Protein conformations are stabilized in their native states by secondary and tertiary structures and through the aggregation of subunits in quaternary structure. Any physical or chemical agent that destroys these stabilizing structures changes the conformation of the protein (Table 14.3). We call this process **denaturation.**

For example, heat cleaves hydrogen bonds, so boiling a protein solution destroys the α-helical and β-pleated sheet structure. In collagen, the triple helixes disappear upon boiling, and the molecules have a largely random-coil conformation in the denatured state, which is gelatin. In other proteins, especially globular proteins, heat causes the unfolding of the polypeptide chains; because of subsequent intermolecular protein–protein interactions, precipitation or coagulation then takes place. That is what happens when we boil an egg.

Denaturation The loss of the secondary, tertiary, and quaternary structures of a protein by a chemical or physical agent that leaves the primary structure intact

TABLE 14.3 Modes of Protein Denaturation (Destruction of Secondary and Higher Structures)

Denaturing Agent	Affected Regions
Heat	H bonds
6 M urea	H bonds
Detergents	Hydrophobic regions
Acids, bases	Salt bridges, H bonds
Salts	Salt bridges
Reducing agents	Disulfide bonds
Heavy metals	Disulfide bonds
Alcohol	Hydration layers

Chemical Connections 14G

Quaternary Structure and Allosteric Proteins

Quaternary structure is a property of proteins that consist of more than one polypeptide chain. Each chain is called a subunit. The number of chains can range from two to more than a dozen, and the chains may be identical or different. The chains interact with one another noncovalently via electrostatic attractions, hydrogen bonds, and hydrophobic interactions. As a result of these noncovalent interactions, subtle changes in structure at one site on a protein molecule may cause drastic changes in properties at a distant site. Proteins that exhibit this property are called **allosteric proteins.** Not all multisubunit proteins exhibit allosteric effects, but many do.

A classic illustration of the quaternary structure of proteins and its effect on properties is a comparison of hemoglobin, an allosteric protein, with myoglobin, which consists of a single polypeptide chain. Both hemoglobin and myoglobin bind to oxygen via a heme group (Figure 14.18). As we have seen, hemoglobin is a **tetramer,** a molecule consisting of four polypeptide chains: two α chains and two β chains. The two α chains of hemoglobin are identical, as are the two β chains. The overall structure of hemoglobin is $\alpha_2\beta_2$ in Greek-letter notation. Both the α and β chains of hemoglobin are very similar to the myoglobin chain.

The α chain is 141 residues long, and the β chain is 146 residues long. For comparison, the myoglobin chain is 153 residues long. Many of the amino acids of the α chain, the β chain, and myoglobin are *homologous;* that is, the same amino acid residues occupy the same positions.

The heme group is the same in both myoglobin and hemoglobin. One molecule of myoglobin binds to one oxygen molecule. Four molecules of oxygen can bind to one hemoglobin molecule. Both hemoglobin and myoglobin bind oxygen reversibly, but the binding of oxygen to hemoglobin exhibits **positive cooperativity,** whereas oxygen binding to myoglobin does not. Positive cooperativity means that

Quaternary Structure and Allosteric Proteins *(continued)*

when one oxygen molecule is bound, it becomes easier for the next molecule to bind. A graph of the oxygen-binding properties of hemoglobin and myoglobin is one of the best ways to illustrate this point.

When the degree of saturation of myoglobin with oxygen is plotted against oxygen pressure, a steady rise is observed until complete saturation is approached and the curve levels off. The oxygen-binding curve of myoglobin is thus said to be **hyperbolic.** In contrast, the oxygen-binding curve for hemoglobin is **sigmoidal.** This shape indicates that the binding of the first oxygen molecule facilitates the binding of the second oxygen, which facilitates the binding of the third oxygen, which in turn facilitates the binding of the fourth oxygen. This is precisely what is meant by the term "cooperative binding."

The two types of behavior are related to the functions of these proteins. Myoglobin has the function of oxygen *storage* in muscle. It must bind strongly to oxygen at very low pressures, and it is 50% saturated at a partial pressure of oxygen of 1 torr. The function of hemoglobin is oxygen *transport,* and it must be able both to bind strongly to oxygen and to release oxygen easily, depending upon conditions. In the alveoli of lungs (where hemoglobin must bind oxygen for transport to the tissues), the oxygen pressure is 100 torr. At this pressure, hemoglobin is 100% saturated with oxygen. In the capillaries running through active

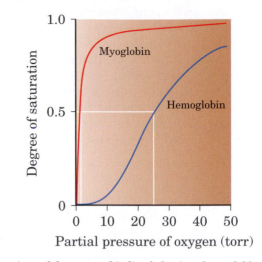

A comparison of the oxygen-binding behavior of myoglobin and hemoglobin. The oxygen-binding curve of myoglobin is hyperbolic, whereas that of hemoglobin is sigmoidal. Myoglobin is 50% saturated with oxygen at 1 torr partial pressure; hemoglobin does not reach 50% saturation until the partial pressure of oxygen reaches 26 torr.

muscles, the pressure of oxygen is 20 torr, corresponding to less than 50% saturation of hemoglobin, which occurs at 26 torr. In other words, hemoglobin gives up oxygen easily in capillaries, where the need for oxygen is great.

The structure of myoglobin.

The structure of hemoglobin.

Similar conformational changes can be brought about by the addition of denaturing chemicals. Solutions such as 6 *M* aqueous urea, $H_2N-CO-NH_2$, break hydrogen bonds and cause the unfolding of globular proteins. Surface-active agents (detergents) change protein conformation by opening up the

Chemical Connections 14H

Laser Surgery and Protein Denaturation

Proteins can be denatured by physical means, most notably by heat. For instance, bacteria are killed and surgical instruments are sterilized by heat. A special method of heat denaturation that is seeing increasing use in medicine relies on lasers. A laser beam (a highly coherent light beam of a single wavelength) is absorbed by tissues, and its energy is converted to heat energy. This process can be used to cauterize incisions so that a minimal amount of blood is lost during the operation.

Laser beams can be delivered by an instrument called a **fiberscope.** The laser beam is guided through tiny fibers, thousands of which are fitted into a tube only 1 mm in diameter. In this way, the laser delivers the energy for denaturation only where it is needed. It can, for example, seal wounds or join blood vessels without the necessity of cutting through healthy tissues. Fiberscopes have been used successfully to diagnose and treat many bleeding ulcers in the stomach, intestines, and colon.

A novel use of the laser fiberscope is in treating tumors that cannot be reached for surgical removal. A drug called Photofrin, which is activated by light, is given to patients intravenously. The drug in this form is inactive and harmless. The patient then waits 24 to 48 hours, during which time the drug accumulates in the tumor but is removed and excreted from healthy tissues. A laser fiberscope with 630 nm of red light is then directed toward the tumor. An exposure between 10 and 30 minutes is applied. The energy of the laser beam activates the Photofrin, which destroys the tumor.

This technique does not offer a complete cure, because the tumor may grow back or it may have spread before the treatment. The treatment has only one side effect: The patient remains sensitive to exposure to strong light for approximately 30 days (so sunlight must be avoided). Of course, this inconvenience is minor compared to the pain, nausea, hair loss, and other side effects that accompany radiation or chemotherapy of tumors.

Argon krypton laser surgery.

In the United States, Photofrin is approved only to treat esophageal cancer. In Europe, Japan, and Canada, it is also used to treat lung, bladder, gastric, and cervical cancers. The light that activates Photofrin penetrates only a few millimeters, but the new drugs under development may use radiation in the near-infrared spectrum that can penetrate tumors up to a few centimeters.

The most common use of laser technology in surgery is its application to correct near-sightedness and astigmatism. In a computer-assisted laser surgery process, the curvature of the cornea is changed. Using the energy of the laser beam, physicians remove part of the cornea. In the procedure called photorefractive keratectomy (PRK), the outer layers of the cornea are denatured—that is, burned off. In the LASIK (*la*ser *i*n *s*itu *k*eratomileusis) procedure, the surgeon creates a flap or a hinge of the outer layers of the cornea and then with the laser beam burns off a computer-programmed amount under the flap to change the shape of the cornea. After the 5- to 10-minute procedure is complete, the flap is put back, and it heals without stitches. In successful surgeries, patients regain good vision one day after the surgery and no longer need prescription lenses.

hydrophobic regions, whereas acids, bases, and salts affect both salt bridges and hydrogen bonds.

Reducing agents, such as 2-mercaptoethanol ($HOCH_2CH_2SH$), can break the —S—S— disulfide bonds, reducing them to —SH groups. The processes of permanent waving and straightening of curly hair are examples of the latter effect (Figure 14.21). The protein keratin, which makes up human hair, contains a high percentage of disulfide bonds. These bonds are primarily responsible for the shape of the hair, whether straight or curly. In either permanent waving or straightening, the hair is first treated with a reducing agent that cleaves some of the —S—S— bonds. This treatment allows the molecules to lose their rigid orientations and become more flexible. The hair

FIGURE 14.21 Permanent wave alters the shape of hair through reduction and oxidation of disulfides and thiols.

Raw egg whites are an antidote to heavy metal poisoning.

is then set into the desired shape, using curlers or rollers, and an oxidizing agent is applied. The oxidizing agent reverses the preceding reaction, forming new disulfide bonds, which now hold the molecules together in the desired positions.

Heavy metal ions (for example, Pb^{2+}, Hg^{2+}, and Cd^{2+}) also denature protein by attacking the —SH groups. They form salt bridges, as in —S^-Hg^{2+} ^-S—. This very feature is taken advantage of in the antidote for heavy metal poisoning: raw egg whites and milk. The egg and milk proteins are denatured by the metal ions, forming insoluble precipitates in the stomach. These must be pumped out or removed by inducing vomiting. In this way, the poisonous metal ions are removed from the body. If the antidote is not pumped out of the stomach, the digestive enzymes would degrade the proteins and release the poisonous heavy metal ions, which would then be absorbed into the bloodstream.

Other chemical agents such as alcohol also denature proteins, coagulating them. This process is used in sterilizing the skin before injections. At a concentration of 70%, ethanol penetrates bacteria and kills them by coagulating their proteins, whereas 95% alcohol denatures only surface proteins.

Denaturation changes secondary, tertiary, and quaternary structures. It does not affect primary structures (that is, the sequence of amino acids that make up the chain). If these changes occur to a small extent, denaturation can be reversed. For example, when we remove a denatured protein from a urea solution and put it back into water, it often reassumes its secondary and tertiary structures. This process is called reversible denaturation. In living cells, some denaturation caused by heat can be reversed by chaperones. These proteins help a partially heat-denatured protein to regain its native secondary, tertiary, and quaternary structures. Some denaturation, however, is irreversible. We cannot unboil a hard-boiled egg, for example.

Summary

OWL End-of-chapter problems identified in blue are assignable in **GOB OWL**.

Section 14.1 What Are the Many Functions of Proteins?

- **Proteins** are giant molecules made of amino acids linked together by **peptide bonds.**
- Proteins have many functions: structural (collagen), enzymatic, carrier (hemoglobin), storage (casein), protective (immunoglobulin), and hormonal (insulin).

Section 14.2 What Are Amino Acids?
Problems 14.14

- **Amino acids** are organic compounds containing an amino and a carboxy group.
- The 20 common amino acids found in proteins are classified by their side chains: nonpolar, polar but neutral, acidic, and basic.
- All amino acids in human tissues are L-amino acids.

Section 14.3 What Are Zwitterions?
Problems 14.20

- Amino acids in the solid state, as well as in water, carry both positive and negative charges; they are called **zwitterions.**
- The pH at which the number of positive charges equals the number of negative charges is the **isoelectric point** of an amino acid or protein.

Section 14.4 What Determines the Characteristics of Amino Acids?

- Amino acids are nearly identical in most ways except for their side chain (R—) groups.
- It is the unique nature of the side chain that gives an amino acid its particular properties.
- Some amino acids have charged side chains (Glu, Asp, Lys, Arg, His).
- Cysteine is a special amino acid because its side chain (—SH) can form disulfide bridges with another cysteine.

- The aromatic amino acids (Phe, Tyr, Trp) are important physiologically, because they are precursors of neurotransmitters. They also absorb ultraviolet light and allow us to easily measure and locate them.

Section 14.5 What Are Uncommon Amino Acids?

- Besides the 20 common amino acids found in proteins, other amino acids are known.
- These amino acids are normally produced after one of the standard amino acids has been incorporated into a protein.
- Examples include hydroxyproline (collagen), hydroxylysine, and thyroxine.

Section 14.6 How Do Amino Acids Combine to Form Proteins? Problems 14.36, 14.38

- When the amino group of one amino acid condenses with the carboxyl group of another amino acid, an amide (peptide) bond is formed, with the elimination of water.
- Two amino acids form a dipeptide. Three amino acids form a tripeptide.
- Many amino acids form a **polypeptide chain.** Proteins are made of one or more polypeptide chains.

Section 14.7 What Are the Properties of Proteins?

- The properties of proteins are based on properties of the peptide backbone and the properties of the side chains.
- Although the peptide bond is typically written as a carbonyl group bonded to an N—H group, such bonds can exhibit keto-enol tautomerism. As a result, the peptide bond that links two amino acids is planar.
- The planar nature of the peptide bond limits the possible orientations that peptides and proteins can take.
- The nature of the amino acid side chains determines most of the nature of a protein.
- Some amino acids have acid or basic side chains. The isoelectric point of a protein is the pH where all the negative charges match all the positive charges and the net charge on the protein is zero.

Section 14.8 What Is the Primary Structure of a Protein?

- The linear sequence of amino acids is the **primary structure** of the protein.

- The primary structure is largely responsible for the eventual higher-order structures of proteins.

Section 14.9 What Is the Secondary Structure of a Protein? Problem 14.49

- The repeating short-range conformations (**α-helix, β-pleated sheet, extended helix of collagen,** and **random coil**) are the **secondary structures** of proteins.
- Secondary structure refers to those repetitive structures that are held together via hydrogen bonds between groups on the peptide backbone only.

Section 14.10 What Is the Tertiary Structure of a Protein?

- The **tertiary structure** is the three-dimensional conformation of the protein molecule.
- Tertiary structures are maintained by covalent cross-links such as **disulfide bonds** and by **salt bridges, hydrogen bonds, metal ion coordination,** and **hydrophobic interactions** between the side chains.

Section 14.11 What Is the Quaternary Structure of a Protein?

- The precise fit of polypeptide subunits into an aggregated whole is called the **quaternary structure.**
- Not all proteins have a quaternary structure—only those proteins that have subunits.
- Hemoglobin is an example of a protein that exhibits a quaternary structure.

Section 14.12 How Are Proteins Denatured?

- Secondary and tertiary structures stabilize the native conformations of proteins.
- Physical and chemical agents, such as heat or urea, destroy these structures and **denature** proteins.
- Protein functions depend on native conformation; when a protein is denatured, it can no longer carry out its function.
- Some (but not all) denaturation is reversible; in some cases, **chaperone** molecules may reverse denaturation.

Problems

■ Indicates problems that are assignable in **GOB OWL**.

Blue-numbered problems are applied.

Go to this book's companion website at **www.cengage. com/chemistry/bettelheim** for interactive versions of the How To tutorials and Active Figures, and to quiz yourself on this chapter.

Section 14.1 What Are the Many Functions of Proteins?

14.3 What are the functions of (a) ovalbumin and (b) myosin?

14.4 The members of which class of proteins are insoluble in water and can serve as structural materials?

■ Problems assignable in GOB OWL

14.5 What is the function of an immunoglobulin?

14.6 What are the two basic types of proteins?

Section 14.2 What Are Amino Acids?

14.7 What is the difference in structure between tyrosine and phenylalanine?

14.8 Classify the following amino acids as nonpolar, polar but neutral, acidic, or basic.

(a) Arginine (b) Leucine

(c) Glutamic acid (d) Asparagine

(e) Tyrosine (f) Phenylalanine

(g) Glycine

14.9 Which amino acid has the highest percentage nitrogen (g N/100 g amino acid)?

14.10 Why does glycine have no D or L form?

14.11 Draw the structure of proline. To which class of heterocyclic compounds does this molecule belong? (See Section 8.1.)

14.12 Which amino acid is also a thiol?

14.13 Why is it necessary to have proteins in our diets?

14.14 ■ Which amino acids in Table 14.1 have more than one stereocenter?

14.15 What are the similarities and differences in the structures of alanine and phenylalanine?

14.16 Draw the structures of L- and D-valine.

Section 14.3 What Are Zwitterions?

14.17 Why are all amino acids solids at room temperature?

14.18 Show how alanine, in solution at its isoelectric point, acts as a buffer (write equations to show why the pH does not change much if we add an acid or a base).

14.19 Explain why an amino acid cannot exist in an unionized form at any pH.

14.20 ■ Draw the structure of valine at pH 1 and at pH 12.

14.21 Draw the most predominant form of aspartic acid at its isoelectric point.

14.22 Draw the most predominant form of histidine at its isoelectric point.

14.23 Draw the most predominant form of lysine at its isoelectric point.

14.24 Draw the sequential transition of glutamic acid as it passes from its fully protonated form to its fully deprotonated form as the pH rises.

Section 14.4 What Determines the Characteristics of Amino Acids?

14.25 Which of the three functional groups on histidine is the most unique?

14.26 How are aromatic amino acids related to neurotransmitters?

14.27 Why is histidine considered a basic amino acid when the pK_a of its side chain is 6.0?

14.28 Which are the acidic amino acids?

14.29 Which are the basic amino acids?

14.30 Why does proline not absorb light at 280 nm?

Section 14.5 What Are Uncommon Amino Acids?

14.31 Two of the 20 amino acids listed in Table 14.1 can be obtained by hydroxylation of other amino acids. What are those two, and what are their precursor amino acids?

14.32 When a protein contains hydroxyproline, at what point in the production of the protein is the proline hydroxylated?

14.33 What is the effect of thyroxine on metabolism?

14.34 How is thyroxine made?

Section 14.6 How Do Amino Acids Combine to Form Proteins?

14.35 Show by chemical equations how alanine and glutamine can be combined to give two different dipeptides.

14.36 ■ A tetrapeptide is abbreviated as DPKH. Which amino acid is at the N-terminal and which is at the C-terminal?

14.37 Draw the structure of a tripeptide made of threonine, arginine, and methionine.

14.38 ■ (a) Use the three-letter abbreviations to write the representation of the following tripeptide:

$$\overset{+}{H_3N}-CH-\overset{O}{\overset{\|}{C}}-N-CH-\overset{O}{\overset{\|}{C}}-N-CH-\overset{O}{\overset{\|}{C}}-O^-$$

with side chains: CH_2, H, CH_2, H, CH_2; CH_2, $CH-CH_3$, COOH; S, CH_3; CH_3

(b) Which amino acid is the C-terminal end, and which is the N-terminal end?

14.39 A polypeptide chain is made of alternating valine and phenylalanine. Which part of the polypeptide is polar (hydrophilic)?

Section 14.7 What Are the Properties of Proteins?

14.40 (a) How many atoms of the peptide bond lie in the same plane?

(b) Which atoms are they?

14.41 (a) Draw the structural formula of the tripeptide met—ser—cys.

(b) Draw the different ionic structures of this tripeptide at pH 2.0, 7.0, and 10.0.

14.42 How can a protein act as a buffer?

14.43 Proteins are least soluble at their isoelectric points. What would happen to a protein precipitated at its isoelectric point if a few drops of dilute HCl were added?

Section 14.8 What Is the Primary Structure of a Protein?

14.44 How many different tripeptides can be made (a) using one, two, or three residues each of leucine, threonine, and valine and (b) using all 20 amino acids?

14.45 How many different tetrapeptides can be made (a) if the peptides contain the residues of asparagine, proline, serine, and methionine and (b) if all 20 amino acids can be used?

14.46 How many amino acid residues in the A chain of insulin are the same in insulin from humans, cattle (bovine), hogs, and sheep?

14.48 Based on your knowledge of the chemical properties of amino acid side chains, suggest a substitution for leucine in the primary structure of a protein that would probably not change the character of the protein very much.

Section 14.9 What Is the Secondary Structure of a Protein?

14.48 Is a random coil a (a) primary, (b) secondary, (c) tertiary, or (d) quaternary structure? Explain.

14.49 ■ Decide whether the following structures that exist in collagen are primary, secondary, tertiary, or quaternary.

(a) Tropocollagen
(b) Collagen fibril
(c) Collagen fiber
(d) The proline—hydroxyproline—glycine repeating sequence

14.50 Proline is often called an α-helix terminator; that is, it is usually in the random-coil secondary structure following an α-helical portion of a protein chain. Why does proline not fit easily into an α-helix structure?

Section 14.10 What Is the Tertiary Structure of a Protein?

14.51 Polyglutamic acid (a polypeptide chain made only of glutamic acid residues) has an α-helix conformation below pH 6.0 and a random-coil conformation above pH 6.0. What is the reason for this conformational change?

14.52 Distinguish between intermolecular and intramolecular hydrogen bonding between backbone groups. Where in protein structures do you find one and where do you find the other?

14.53 Identify the primary, secondary, and tertiary structures in the numbered boxes:

Section 14.11 What Is the Quaternary Structure of a Protein?

14.54 If both cysteine residues on the B chain of insulin were changed to alanine residues, how would it affect the quaternary structure of insulin?

14.55 (a) What is the difference in the quaternary structure between fetal hemoglobin and adult hemoglobin?

(b) Which can carry more oxygen?

(c) What would the oxygen saturation curve of fetal hemoglobin look like compared to that of mygoglobin and regular adult hemoglobin?

14.56 Where are the nonpolar side chains of proteins located in an integral membrane protein?

14.57 The cytochrome c protein is important in producing energy from food. It contains a heme surrounded by a polypeptide chain. What kind of structure do these two entities form? To which group of proteins does cytochrome c belong?

14.58 Hemoglobin is an important protein for many reasons and has interesting physical characteristics. How would you classify hemoglobin?

Section 14.12 How Are Proteins Denatured?

14.59 In a 6 *M* urea solution, a protein that contained mostly antiparallel β-sheet became a random coil. Which groups and bonds were affected by urea?

14.60 What kind of changes are necessary to transform a protein having a predominantly α-helical structure into one having a β-pleated sheet structure?

14.61 Which amino acid side chain is most frequently involved in denaturation by reduction?

14.62 What does the reducing agent do in straightening curly hair?

14.63 Silver nitrate is sometimes put into the eyes of newborn infants as a preventive measure against gonorrhea. Silver is a heavy metal. Explain how this treatment may work against bacteria.

14.64 Why do nurses and physicians use 70% alcohol to wipe the skin before giving injections?

Chemical Connections

14.65 (Chemical Connections 14A) Why must some people avoid drinking diet sodas with Nutrasweet?

14.66 (Chemical Connections 14B) AGE products become disturbing only in elderly people, even though they also form in younger people. Why don't they harm younger people?

14.67 (Chemical Connections 14C) Define *hypoglycemic awareness*.

14.68 (Chemical Connections 14D) How does hydroxyurea therapy alleviate the symptoms of sickle cell anemia?

14.69 (Chemical Connections 14E) What is the difference in the conformation between normal prion protein and the amyloid prion that causes mad cow disease?

14.70 (Chemical Connections 14F) What is the aim of proteomics?

14.71 (Chemical Connections 14G) Explain the difference in oxygen-binding behavior of myoglobin and hemoglobin.

14.72 (Chemical Connections 14H) How does the fiberscope help to heal bleeding ulcers?

Additional Problems

14.73 Which diseases are associated with amyloid plaques?

14.74 How many different dipeptides can be made (a) using only alanine, tryptophan, glutamic acid, and arginine and (b) using all 20 amino acids?

14.75 Denaturation is usually associated with transitions from helical structures to random coils. If an imaginary process were to transform the keratin in your hair from an α-helix to a β-pleated sheet structure, would you call the process denaturation? Explain.

14.76 Draw the structure of lysine (a) above, (b) below, and (c) at its isoelectric point.

14.77 In collagen, some of the chains of the triple helices in tropocollagen are cross-linked by covalent bonds between two lysine residues. What kind of structure is formed by these cross-links? Explain.

14.78 Considering the vast number of animal and plant species on Earth (including those now extinct) and the large variety of protein molecules in each organism, have all possible protein molecules been used already by some species or other? Explain.

14.79 What kind of noncovalent interaction occurs between the following amino acids?

(a) Valine and isoleucine

(b) Glutamic acid and lysine

(c) Tyrosine and threonine

(d) Alanine and alanine

14.80 How many different decapeptides (peptides containing 10 amino acids each) can be made from the 20 amino acids?

14.81 Which amino acid does not rotate the plane of polarized light?

14.82 Write the expected products of the acid hydrolysis of the following tetrapeptide:

14.83 What charges are on aspartic acid at pH 2.0?

14.84 How many ways can you link the two amino acids, lysine and valine, in a dipeptide? Which of these peptide bonds will you find in proteins?

Looking Ahead

14.85 Enzymes are biological catalysts and usually proteins. They catalyze common organic reactions. Why are amino acids, such as histidine, aspartic acid, and serine, found more often near the reaction catalysis site than amino acids such as leucine and valine?

14.86 Hormones are molecules that are released from one tissue but have their effect in another tissue. Give an example of a hormone encountered in this chapter that would be ineffective if taken orally. Give an example of one that could be effective if taken orally.

14.87 Using what you know about protein denaturation, what is one reason that you must maintain a body temperature in a strict range?

14.88 What is the difference between genomics and proteomics?

14.89 Why does knowing the complete genome of an organism not necessarily tell you about the nature of all the proteins in the organism?

14.90 Why is collagen not a very good source of dietary protein?

14.91 A recent diet supplement advertised that it would repair your muscles while allowing you to burn fat because the product had collagen protein. Evaluate this claim.

Enzymes

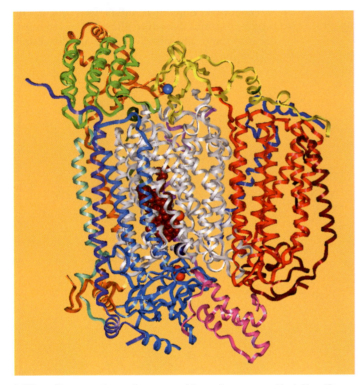

Ribbon diagram of cytochrome c oxidase, the enzyme that directly uses oxygen during respiration.

15.1 What Are Enzymes?

The cells in your body are chemical factories. Only a few of the thousands of compounds necessary for the operation of the human organism are obtained from the diet. Instead, most of these substances are synthesized within the cells, which means that hundreds of chemical reactions take place in your cells every second of your life.

Nearly all of these reactions are catalyzed by **enzymes**, which are large molecules that increase the rates of chemical reactions without themselves undergoing any change. Without enzymes to act as biological catalysts, life as we know it would not be possible.

The vast majority of all known enzymes are globular proteins, and we will devote most of our study to protein-based enzymes. However, proteins are not the only biological catalysts. **Ribozymes** are enzymes made of ribonucleic acids. They catalyze the self-cleavage of certain portions of their

OWL Online homework for this chapter may be assigned in **GOB OWL**.

own molecules and have been implicated in the reaction that generates peptide bonds (Chapter 14). Many biochemists believe that during evolution RNA catalysts emerged first, with protein enzymes arriving on the scene later. (We will learn more about RNA catalysts in Section 17.4.)

Like all catalysts, enzymes do not change the position of equilibrium. That is, enzymes cannot make a reaction take place that would not occur without them. Instead, they increase the reaction rate: They cause reactions to take place faster by lowering the activation energy. As catalysts, enzymes are remarkable in two respects:

1. They are extremely effective, increasing reaction rates by anywhere from 10^9 to 10^{20} times.

2. Most of them are extremely **specific**.

Substrate specificity The limitation of an enzyme to catalyze specific reactions with specific substrates

As an example of their effectiveness, consider the oxidation of glucose. A lump of glucose or even a glucose solution exposed to oxygen under sterile conditions would show no appreciable change for months. In the human body, however, the same glucose is oxidized within seconds.

Every organism has many enzymes—many more than 3000 in a single cell. Most enzymes are very specific, each of them speeding up only one particular reaction or class of reactions. For example, the enzyme urease catalyzes only the hydrolysis of urea and not that of other amides, even closely related ones.

$$(NH_2)_2C{=}O + H_2O \xrightarrow{\text{urease}} 2\,NH_3 + CO_2$$
Urea

Another type of specificity can be seen with trypsin, an enzyme that cleaves the peptide bonds of protein molecules—but not every peptide bond, only those on the carboxyl side of lysine and arginine residues (Figure 15.1).

The enzyme carboxypeptidase specifically catalyzes the hydrolysis on only the last amino acid on a protein chain—the one at the C-terminal end. Lipases are less specific: They catalyze the hydrolysis of any triglyceride, but they still do not affect carbohydrates or proteins.

The specificity of enzymes also extends to stereospecificity. The enzyme arginase hydrolyzes the amino acid L-arginine (the naturally occurring form) to a compound called L-ornithine and urea (Section 20.8) but has no effect on its mirror image, D-arginine.

Enzymes are distributed according to the body's need to catalyze specific reactions. A large number of protein-splitting enzymes are in the blood, ready to promote clotting. Digestive enzymes, which also catalyze the hydrolysis of proteins, are located in the secretions of the stomach and pancreas. Even within the cells themselves, some enzymes are localized according to the need for specific reactions. The enzymes that catalyze the oxidation of compounds that are part of the citric acid cycle (Section 19.4) are located in the mitochondria, for example, and special organelles such as lysosomes contain an enzyme (lysozyme) that catalyzes the dissolution of bacterial cell walls.

FIGURE 15.1 A typical amino acid sequence. The enzyme trypsin catalyzes the hydrolysis of this chain only at the points marked with an arrow (the carboxyl side of lysine and arginine).

Chemical Connections 15A

Muscle Relaxants and Enzyme Specificity

In the body, nerves transmit signals to the muscles. Acetylcholine is a neurotransmitter (Section 16.1) that operates between the nerve endings and muscles. It attaches itself to a specific receptor in the muscle end plate. This attachment transmits a signal to the muscle to contract; shortly thereafter, the muscle relaxes. A specific enzyme, acetylcholinesterase, then catalyzes the hydrolysis of the acetylcholine, removing it from the receptor site and preparing it for the next signal transmission—that is, the next contraction.

Succinylcholine is sufficiently similar to acetylcholine so that it, too, binds to the receptor of the muscle end plate. However, acetylcholinesterase can hydrolyze succinylcholine only very slowly. While it remains bound to the receptor, no new signal can reach the muscle to allow it to contract again. Thus the muscle stays relaxed for a long time.

This feature makes succinylcholine a good muscle relaxant during minor surgery, especially when a tube must be inserted into the bronchus (bronchoscopy). For example, after intravenous administration of 50 mg of succinylcholine, paralysis and respiratory arrest are observed within 30 seconds. While respiration is carried on artificially, the bronchoscopy can be performed within minutes.

15.2 How Are Enzymes Named and Classified?

Enzymes are commonly given names derived from the reaction that they catalyze and/or the compound or type of compound on which they act. For example, lactate dehydrogenase speeds up the removal of hydrogen from lactate (an oxidation reaction). Acid phosphatase catalyzes the hydrolysis of phosphate ester bonds under acidic conditions. As can be seen from these examples, the names of most enzymes end in "-ase." Some enzymes, however, have older names, which were assigned before their actions were clearly understood. Among these are pepsin, trypsin, and chymotrypsin—all enzymes of the digestive tract.

Enzymes can be classified into six major groups according to the type of reaction they catalyze (see also Table 15.1):

1. **Oxidoreductases** catalyze oxidations and reductions.
2. **Transferases** catalyze the transfer of a group of atoms, such as from one molecule to another.
3. **Hydrolases** catalyze hydrolysis reactions.
4. **Lyases** catalyze the addition of two groups to a double bond or the removal of two groups from adjacent atoms to create a double bond.
5. **Isomerases** catalyze isomerization reactions.
6. **Ligases,** or synthetases, catalyze the joining of two molecules.

TABLE 15.1 Classification of Enzymes

Class	Typical Example	Reaction Catalyzed	Section Number in This Book
1. Oxidoreductases	Lactate dehydrogenase		20.2
2. Transferases	Aspartate amino transferase or aspartate transaminase		20.8
3. Hydrolases	Acetylcholinesterase		16.3
4. Lyases	Aconitase		19.4
5. Isomerases	Phosphohexose isomerase		20.2
6. Ligases	Tyrosine-tRNA synthetase	$ATP + L\text{-tyrosine} + tRNA$ $\longrightarrow L\text{-tyrosyltRNA} + AMP + PP_i$	16.6

Oxidoreductases: $CH_3-CH-COO^- \longrightarrow CH_3-C-COO^-$ with OH and O groups. L-(+)-Lactate → Pyruvate

Transferases: Aspartate + α-Ketoglutarate → Oxaloacetate + Glutamate

Hydrolases: $CH_3-C-OCH_2CH_2\overset{+}{N}(CH_3)_3 + H_2O$ (Acetylcholine) $\longrightarrow CH_3COOH + HOCH_2CH_2\overset{+}{N}(CH_3)_3$ (Acetic acid, Choline)

Lyases: cis-Aconitate $+ H_2O \longrightarrow$ Isocitrate

Isomerases: Glucose 6-phosphate → Fructose 6-phosphate

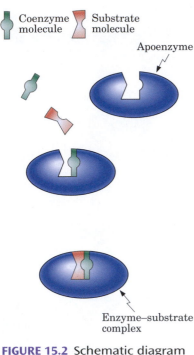

Coenzyme molecule   Substrate molecule

Apoenzyme

Enzyme–substrate complex

FIGURE 15.2 Schematic diagram of the active site of an enzyme and the participating components.

Cofactor The nonprotein part of an enzyme necessary for its catalytic function

Coenzyme A nonprotein organic molecule, frequently a B vitamin, that acts as a cofactor.

Active site A three-dimensional cavity of the enzyme with specific chemical properties that enable it to accommodate the substrate

Inhibitor A compound that binds to an enzyme and lowers its activity

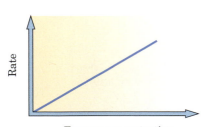

Rate

Enzyme concentration

FIGURE 15.3 The effect of enzyme concentration on the rate of an enzyme-catalyzed reaction. Substrate concentration, temperature, and pH are constant.

15.3 What Is the Terminology Used with Enzymes?

Some enzymes, such as pepsin and trypsin, consist of polypeptide chains only. Other enzymes contain nonprotein portions called **cofactors.** The protein (polypeptide) portion of the enzyme is called an **apoenzyme.**

The cofactors may be metallic ions, such as Zn^{2+} or Mg^{2+}, or organic compounds. Organic cofactors are called **coenzymes.** An important group of coenzymes are the B vitamins, which are essential to the activity of many enzymes (Section 19.3). Another important coenzyme is heme (Figure 14.16), which is part of several oxidoreductases as well as part of hemoglobin. In any case, an apoenzyme cannot catalyze a reaction without its cofactor, nor can the cofactor function without the apoenzyme. When a metal ion is a cofactor, it can bind directly to the protein or to the coenzyme, if the enzyme contains one.

The compound on which the enzyme works, and whose reaction it speeds up, is called the **substrate.** The substrate usually binds to the enzyme surface while it undergoes the reaction. The substrate binds to a specific portion of the enzyme during the reaction, called the **active site.** If the enzyme has coenzymes, they are located at the active site. Therefore, the substrate is simultaneously surrounded by parts of the apoenzyme, coenzyme, and metal ion cofactor (if any), as shown in Figure 15.2.

Activation is any process that initiates or increases the action of an enzyme. It can be the simple addition of a cofactor to an apoenzyme or the cleavage of a polypeptide chain of a proenzyme (Section 15.6B).

Inhibition is the opposite—any process that makes an active enzyme less active or inactive (Section 15.5). Inhibitors are compounds that accomplish this task, and there are many types of enzyme inhibition. **Competitive inhibitors** bind to the active site of the enzyme surface, thereby preventing the binding of substrate. **Noncompetitive inhibitors** bind to some other portion of the enzyme surface and sufficiently alter the tertiary structure of the enzyme so that its catalytic effectiveness is eliminated. That is, the enzyme cannot catalyze while the inhibitor is bound. Both competitive and noncompetitive inhibition are *reversible,* but some compounds alter the structure of the enzyme *permanently* and thus make it *irreversibly* inactive.

15.4 What Factors Influence Enzyme Activity?

Enzyme activity is a measure of how much reaction rates are increased. In this section, we examine the effects of concentration, temperature, and pH on enzyme activity.

A. Enzyme and Substrate Concentration

If we keep the concentration of substrate constant and increase the concentration of enzyme, the rate increases linearly (Figure 15.3). That is, if the enzyme concentration doubles, the rate doubles as well; if the enzyme concentration triples, the rate also triples. This is the case in practically all enzyme reactions, because the molar concentration of enzyme is almost always much lower than that of substrate (that is, many more molecules of substrate are typically present than molecules of enzyme).

Conversely, if we keep the concentration of enzyme constant and increase the concentration of substrate, we get an entirely different type of curve, called a saturation curve (Figure 15.4). In this case, the rate does not increase continuously. Instead, a point is reached after which the rate stays the same even if we increase the substrate concentration further. This happens because, at the saturation point, substrate molecules are bound to all available active sites of the enzymes. Because the reactions take place at the active sites, once they are all occupied, the reaction is proceeding at its maximum rate. Increasing the substrate concentration can no longer increase the rate because the excess substrate cannot find any active sites to which to bind.

B. Temperature

Temperature affects enzyme activity because it changes the conformation of the enzyme. In uncatalyzed reactions, the rate usually increases as the temperature increases. Changing the temperature has a different effect on enzyme-catalyzed reactions. When we start at a low temperature (Figure 15.5), an increase in temperature first causes an increase in rate. However, protein conformations are very sensitive to temperature changes. Once the optimal temperature is reached, any further increase in temperature alters the enzyme conformation. The substrate may then not fit properly onto the changed enzyme surface, so the rate of reaction actually *decreases*.

After a *small* temperature increase above the optimum, the decreased rate could be increased again by lowering the temperature because, over a narrow temperature range, changes in conformation are reversible. However, at some higher temperature above the optimum, we reach a point where the protein denatures (Section 14.12); the conformation is then altered irreversibly, and the polypeptide chain cannot refold to its native conformation. At this point, the enzyme is completely inactivated. The inactivation of enzymes at low temperatures is used in the preservation of food by refrigeration.

Most enzymes from bacteria and higher organisms have an optimal temperature around 37°C. However, the enzymes of organisms that live at the ocean floor at 2°C have an optimal temperature in that range. Other organisms live in ocean vents under extreme conditions, and their enzymes have optimal conditions at ranges from 90 to 105°C. The enzymes of these hyperthermophile organisms also have other extreme requirements, such as pressures up to 100 atm, and some of them have an optimal pH in the range of 1 to 4. Enzymes from these hyperthermophiles, especially polymerases that catalyze the polymerization of DNA (Section 17.6), have gained commercial importance.

C. pH

As the pH of its environment changes the conformation of a protein (Section 14.12), we would expect pH-related effects to resemble those observed when the temperature changes. Each enzyme operates best at a certain pH (Figure 15.6). Once again, within a narrow pH range, changes in enzyme activity are reversible. However, at extreme pH values (either acidic or basic), enzymes are denatured irreversibly, and enzyme activity cannot be restored by changing back to the optimal pH.

15.5 What Are the Mechanisms of Enzyme Action?

We have seen that the action of enzymes is highly specific for a substrate. What kind of mechanism can account for such specificity? About 100 years ago, Arrhenius suggested that catalysts speed up reactions by combining with the

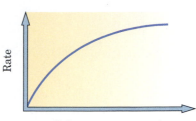

FIGURE 15.4 The effect of substrate concentration on the rate of an enzyme-catalyzed reaction. Enzyme concentration, temperature, and pH are constant.

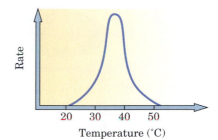

FIGURE 15.5 The effect of temperature on the rate of an enzyme-catalyzed reaction. Substrate and enzyme concentrations and pH are constant.

FIGURE 15.6 The effect of pH on the rate of an enzyme-catalyzed reaction. Substrate and enzyme concentrations and temperature are constant.

substrate to form some kind of intermediate compound. In an enzyme-catalyzed reaction, this intermediate is called the **enzyme–substrate complex.**

A. Lock-and-Key Model

To account for the high substrate specificity of most enzyme-catalyzed reactions, a number of models have been proposed. The simplest and most frequently referenced is the **lock-and-key model** (Figure 15.7). This model assumes that the enzyme is a rigid, three-dimensional body. The surface that contains the active site has a restricted opening into which only one kind of substrate can fit, just as only the proper key can fit exactly into a lock and turn it open.

According to the lock-and-key model, an enzyme molecule has its particular shape because that shape is necessary to maintain the active site in exactly the conformation required for that particular reaction. An enzyme molecule is very large (typically consisting of 100 to 200 amino acid residues), but the active site is usually composed of only two or a few amino acid residues, which may well be located at different places in the chain. The other amino acids—those that are not part of the active site—are located in the sequence in which we find them because that sequence causes the molecule as a whole to fold up in exactly the required way. This arrangement emphasizes that the shape and the functional groups on the surface of the active site are of utmost importance in recognizing a substrate.

The lock-and-key model was the first to explain the action of enzymes. For most enzymes, however, this model is too restrictive. Enzyme molecules are in a dynamic state, not a static one. There are constant motions within them, so that the active site has some flexibility. Also, while the lock-and-key model does a good job explaining why the enzyme binds to the substrate, if the fit is that perfect, there would be no reason for the reaction to occur, as the enzyme-substrate complex would be too stable.

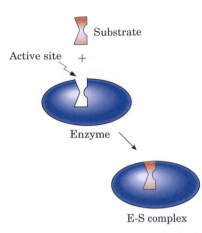

FIGURE 15.7 The lock-and-key model of the enzyme mechanism.

Lock-and-key model A model explaining the specificity of enzyme action by comparing the active site to a lock and the substrate to a key

B. Induced-Fit Model

From x-ray diffraction, we know that the size and shape of the active site cavity change when the substrate enters. To explain this phenomenon, an American biochemist, Daniel Koshland, introduced the **induced-fit model** (Figure 15.8), in which he compared the changes occurring in the shape of the cavity upon substrate binding to the changes in the shape of a glove when a hand is inserted. That is, the enzyme modifies the shape of the active site to accommodate the substrate. Recent experiments during actual catalysis have demonstrated that not only does the shape of the active site change with the binding of substrate, but even in the bound state both the backbone and the side chains of the enzyme are in constant motion.

Both the lock-and-key and the induced-fit models explain the phenomenon of competitive inhibition (Section 15.3). The inhibitor molecule fits into the active site cavity in the same way the substrate does (Figure 15.9), thereby preventing the substrate from entering. The result: Whatever reaction is supposed to take place on the substrate does not occur.

Many cases of noncompetitive inhibition can also be explained by the induced-fit model. In this case, the inhibitor does not bind to the active site but rather binds to a different part of the enzyme. Nevertheless, the binding causes a change in the three-dimensional shape of the enzyme molecule, which so alters the shape of the active site that once the substrate is bound, there can be no catalysis (Figure 15.10).

If we compare enzyme activity in the presence and the absence of an inhibitor, we can tell whether competitive or noncompetitive inhibition is taking place (Figure 15.11). The maximum reaction rate is the same without

Induced-fit model A model explaining the specificity of enzyme action by comparing the active site to a glove and the substrate to a hand

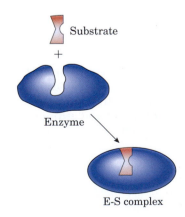

FIGURE 15.8 The induced-fit model of the enzyme mechanism.

Chemical Connections 15B

Enzymes and Memory

There are thousands of different enzymes in a cell, and we will study many of them in the following chapters. New information about the importance of enzymes is published every week in the scientific literature. One class of enzyme important in many metabolic processes is the class called **kinases** (a type of transferase; see Table 15.1). One such kinase, called protein kinase Mζ (PKMζ) (ζ is the symbol for zeta), has recently been implicated in the maintenance of long-term memory. Scientists created a drug called ZIP that blocks this enzyme. They gave rats saccharine-laced water and then induced nausea shortly afterwards. Control rats then had an aversion to saccharine-laced water for weeks afterwards. Humans have the same response: Normally a person who throws up shortly after eating a specific type of food will remember the experience and not want to consume the same food. Researchers then injected the cerebral cortex of test rats and found that they lost their aversion to saccharine within two hours. Since blocking the PKMζ eliminated the memory, here was a first indication that this specific enzyme is required

for long-term memory retention, a novel finding. The next step will be to determine if the drug ZIP eliminates all learning past a certain point or whether it could be used selectively. Researchers have been looking for ways to selectively block memories, such as the painful memories of trauma survivors.

Memory molecule. PKMζ sustains long-term memory in the cerebral coretex of rats

an inhibitor and in the presence of a competitive inhibitor. The only difference is that this maximum rate is achieved at a low substrate concentration with no inhibitor but at a high substrate concentration when an inhibitor is present. This is the true sign of competitive inhibition, because in this scenario the substrate and the inhibitor are competing for the same active site. If the substrate concentration is sufficiently increased, the inhibitor will be displaced from the active site by Le Chatelier's principle.

FIGURE 15.9 The mechanism of competitive inhibition. When a competitive inhibitor enters the active site, the substrate cannot enter.

FIGURE 15.10 Mechanism of noncompetitive inhibition. The inhibitor binds itself to a site other than the active site (allosterism), thereby changing the conformation of the active site. The substrate still binds but there is no catalysis.

FIGURE 15.11 Enzyme kinetics in the presence and the absence of inhibitors.

Chemical Connections 15C

Active Sites

The perception of the active site as either a rigid cavity (lock-and-key model) or a partly flexible template (induced-fit model) is an oversimplification. Not only is the geometry of the active site important, but so are the specific interactions that take place between the enzyme surface and the substrate. To illustrate, we take a closer look at the active site of pyruvate kinase. This enzyme catalyzes the transfer of the phosphate group from phosphoenol pyruvate (PEP) to ADP, an important step in glycolysis (Section 20.2).

Phosphoenol pyruvate   ADP   Pyruvate   ATP

The active site of the enzyme binds both substrates, PEP and ADP (see figure below left). The rabbit muscle pyruvate kinase has two cofactors, K^+ and either Mn^{2+} or Mg^{2+}. The divalent cation is coordinated to the carbonyl and carboxylate oxygen of the pyruvate substrate and to a glutamate 271 and aspartate 295 residue of the enzyme. (The numbers indicate the position of the amino acid in the sequence.) The nonpolar $=CH_2$ group lies in a hydrophobic pocket formed by an alanine 292, glycine 294, and threonine 327 residue. The K^+ on the other side of the active site is coordinated with the phosphate of the substrate and the serine 76 and asparagine 74 residue of the enzyme. Lysine 269 and arginine 72 are also part of the catalytic apparatus anchoring the ADP. This arrangement of the active site illustrates that specific folding into secondary and tertiary structures is required to bring important functional groups together. The residues of amino acids participating in the active sites are sometimes close in the sequence (asparagine 74 and serine 76), but mostly remain far apart (glutamate 271 and aspartate 295). The figure below right illustrates the secondary and tertiary structures providing such a stable active site.

The active site and substrates of pyruvate kinase.

Ribbon cartoon of pyruvate kinase. Pyruvate, Mg^{2+}, and K^+ are depicted as space-filling models.

If the inhibitor is noncompetitive, it cannot be displaced by addition of excess substrate because it is bound to a different site. In this case, the enzyme cannot be restored to its maximum activity, and the maximum rate of the reaction is lower than it would be in the absence of the inhibitor. With a noncompetitive inhibitor, it is always as though less enzyme were available. Competitive and noncompetitive inhibition are the two most common extremes of enzyme inhibition. Many other types of reversible inhibitors exist, but they are beyond the scope of this book.

Enzymes can also be inhibited irreversibly if a compound is bound covalently and permanently to or near the active site. Such inhibition occurs with penicillin, which inhibits the enzyme transpeptidase; this enzyme is necessary for cross-linking bacterial cell walls. Without cross-linking, the bacterial cytoplasm spills out, and the bacteria die (Chemical Connections 11B). Chemical Connections 15D describes two medical applications of inhibitors.

C. Catalytic Power of Enzymes

Both the lock-and-key model and the induced-fit model emphasize the shape of the active site. However, the chemistry at the active site is actually the most important factor. A survey of known active sites of enzymes shows that five amino acids participate in the active sites in more than 65% of all cases. They are, in order of their dominance, His > Cys > Asp > Arg > Glu. A quick glance at Table 14.1 reveals that most of these amino acids have either acidic or basic side chains. Thus acid–base chemistry often underlies the mode of catalysis. The example given in Chemical Connections 15C confirms this relationship. Out of the eleven amino acids in the catalytic site, two are Arg, one is Glu, one is Asp, and two are the related Asn.

We have said that enzymes cannot change the thermodynamic relationships between the substrates and the products of a reaction; rather, they speed up the reaction. But how do they really accomplish this feat? If we look at an energy diagram of a hypothetical reaction, there are reactants on one side and products on the other. The thermodynamic relationship is described by the height difference between the two, as shown in Figure 15.12(a). In any reaction that can be written as follows:

$$A + B \rightleftharpoons C + D$$

before A and B can become C and D, they must pass through a **transition state** where they are something in between. This situation is often thought of as being an "energy hill" that must be scaled. The energy required to climb this hill is the activation energy. Enzymes are powerful catalysts because they lower the energy hill, as shown in Figure 15.12(b). They reduce the activation energy.

How the enzyme reduces the activation energy is very specific to the enzyme and the reaction being catalyzed. As we noted, however, a few amino acids show up in most of the active sites. The specific amino acids in the active site and their exact orientation make it possible for the substrate(s) to bind to the active site and then react to form products. For example, papain is a protease, an enzyme that cleaves peptide bonds as we saw with trypsin. Two critical amino acids are present in the active site of papain (Figure 15.13). The histidine (shown in blue) helps attract the peptide and hold it in the correct orientation via hydrogen bonding (shown as red dashes). The sulfur on the cysteine side chain performs a type of reaction called a **nucleophilic attack** on the carbonyl carbon of the peptide bond, and the C—N bond is broken. Such nucleophilic attacks appear in the vast majority of enzyme mechanisms, and they are possible because of the precise arrangement of the amino acid side chains that can participate in this type of organic reaction.

(a)

(b)

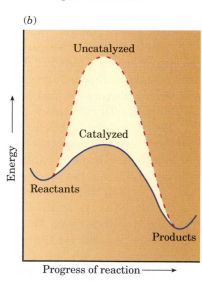

FIGURE 15.12 Activation energy profiles. (*a*) The activation energy profile for a typical reaction. (*b*) A comparison of the activation energy profiles for catalyzed and uncatalyzed reactions.

Nucleophilic attack A chemical reaction where an electron-rich atom, such as oxygen or sulfur, bonds to an electron-deficient atom, such as carbonyl carbon

Chemical Connections 15D

Medical Uses of Inhibitors

A key strategy in the treatment of acquired immunodeficiency syndrome (AIDS) has been to develop specific inhibitors that selectively block the actions of enzymes unique to the human immunodeficiency virus (HIV), which causes AIDS. Many laboratories are working on this approach to the development of therapeutic agents.

One of the most important targets is HIV protease, an enzyme essential to the production of new virus particles in infected cells. HIV protease is unique to this virus. It catalyzes the processing of viral proteins in an infected cell. Without these proteins, viable virus particles cannot be released to cause further infection. The structure of HIV protease, including its active site, was elucidated by x-ray crystallography. With this structure in mind, scientists then designed and synthesized competitive inhibitors to bind to the active site. Improvements were made in the drug design by obtaining structures of a series of inhibitors bound to the active site of HIV protease. These structures were also elucidated by x-ray crystallography. This process eventually led to several HIV protease inhibitors: saquinavir from Hoffmann-La Roche, ritonavir from Abbott, indinavir from Merck, viracept from Agouron Pharmaceuticals, and amprenavir from Vertex Pharmaceuticals. (These companies maintain highly informative home pages on the World Wide Web.)

Treatment of AIDS is most effective when a combination of drug therapies is used, and HIV protease inhibitors play an important role. Especially promising results (such as lowering of levels of the virus in the bloodstream) are obtained when HIV protease inhibitors are part of drug regimens for AIDS.

Sometimes the search for an inhibitor as part of drug design leads to unexpected results. Scientists have long sought better drugs to fight *angina* (chest pains due to poor blood flow to the heart) and *hypertension* (otherwise known as high blood pressure), a common ailment these days. Blood flow increases when the smooth muscles in the blood vessels relax. This relaxation is due to a decrease in intracellular Ca^{2+}, which is in turn triggered by an increase in cyclic GMP (cGMP, Chapter 17). Cyclic GMP is degraded by enzymes called phosphodiesterases. Scientists thought that if they could design an inhibitor

Active site of HIV-1 protease complexed with VX-478.

Structure of amprenavir (VX-478), an HIV protease inhibitor.

of these phosphodiesterases, the cGMP would last longer, the blood vessels would stay open longer, and blood pressure would decrease. Scientists developed a drug to mimic cGMP in the hopes of inhibiting phosphodiesterases. The structural name of the drug is sildenafil citrate, but a company called Pfizer marketed it under the name of Viagra.

Unfortunately, Viagra showed no significant benefits for reducing the pain of angina or decreasing blood pressure. However, some men in the clinical trials of the drug noted penile erections. Apparently the drug did work to inhibit the phosphodiesterases in the penile vascular tissue, leading to smooth muscle relaxation and increased blood flow. Despite the fact that the drug did not accomplish what it was intended for, this competitive inhibitor became a very big seller for the companies that produce it.

cGMP   Viagra

Note the structural similarity between cGMP (left) and Viagra.

FIGURE 15.13 Papain is a cysteine protease. A critical cysteine residue is involved in the nucleophilic attack on the peptide bond it hydrolyzes.

15.6 How Are Enzymes Regulated?

A. Feedback Control

Enzymes are often regulated by environmental conditions. **Feedback control** is an enzyme regulation process in which formation of a product inhibits an earlier reaction in the sequence. The reaction product of one enzyme may control the activity of another, especially in a complex system in which enzymes work cooperatively. For example, in such a system, each step is catalyzed by a different enzyme:

$$A \xrightarrow{E_1} B \xrightarrow{E_2} C \xrightarrow{E_3} D$$

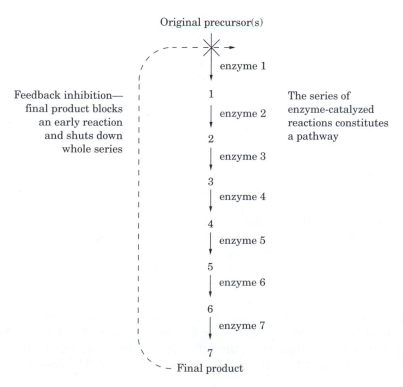

A schematic representation of a pathway showing feedback inhibition

Original precursor(s)

enzyme 1

Feedback inhibition—
final product blocks
an early reaction
and shuts down
whole series

1

enzyme 2

The series of
enzyme-catalyzed
reactions constitutes
a pathway

2

enzyme 3

3

enzyme 4

4

enzyme 5

5

enzyme 6

6

enzyme 7

7

Final product

The final product in the chain may inhibit the activity of the first enzyme (by competitive, noncompetitive, or some other type of inhibition). When the concentration of the final product is low, all of the reactions proceed rapidly. As the concentration increases, however, the action of enzyme 1 becomes inhibited and eventually stops. In this manner, the accumulation of the final product serves as a message that tells enzyme 1 to shut down because the cell has enough final product for its present needs. Shutting down enzyme 1 stops the entire process.

B. Proenzymes

Proenzyme (zymogen) A protein that becomes an active enzyme after undergoing a chemical change

Some enzymes are manufactured by the body in an inactive form. To make them active, a small part of their polypeptide chain must be removed. These inactive forms of enzymes are called **proenzymes** or **zymogens.** After the excess polypeptide chain is removed, the enzyme becomes active.

For example, trypsin is manufactured in the pancreas as the inactive molecule trypsinogen (a zymogen). When a fragment containing six amino acid residues is removed from the N-terminal end, the molecule becomes a fully active trypsin molecule. Removal of the fragment not only shortens the chain, but also changes the three-dimensional structure (the tertiary structure), thereby allowing the molecule to achieve its active form.

Why does the body go to so much trouble? Why not just make the fully active trypsin to begin with? The reason is very simple. Trypsin is a protease—it catalyzes the hydrolysis of peptide bonds (Figure 15.1)—and is, therefore, an important catalyst for the digestion of the proteins we eat. But it would not be good if it cleaved the proteins of which our own bodies are made! Therefore, the pancreas makes trypsin in an inactive form; only after trypsin enters the digestive tract does it become active.

C. Allosterism

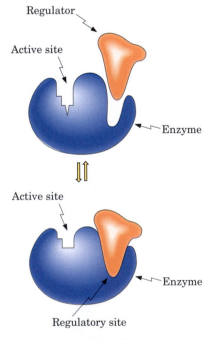

Regulator

Active site

Enzyme

⇅

Active site

Enzyme

Regulatory site

FIGURE 15.14 The allosteric effect. Binding of a regulator to a site other than the active site changes the shape of the active site.

Allosteric enzyme An enzyme in which the binding of a regulator on one site on the enzyme modifies the enzyme's ability to bind the substrate in the active site

Sometimes regulation takes place by means of an event that occurs at a site other than the active site but that eventually affects the active site. This type of interaction is called **allosterism,** and any enzyme regulated by this mechanism is called an **allosteric enzyme.** If a substance binds noncovalently and reversibly to a site *other than the active site,* it may affect the enzyme in either of two ways: It may inhibit enzyme action **(negative modulation)** or it may stimulate enzyme action **(positive modulation).**

The substance that binds to the allosteric enzyme is called a **regulator,** and the site to which it attaches is called a **regulatory site.** In most cases, allosteric enzymes contain more than one polypeptide chain (subunits); the regulatory site is on one polypeptide chain and the active site is on another.

Specific regulators can bind reversibly to the regulatory sites. For example, the enzyme depicted in Figure 15.14 is an allosteric enzyme. In this case, the enzyme has only one polypeptide chain, so it carries both the active site and the regulatory site at different points in this chain. The regulator binds reversibly to the regulatory site. As long as the regulator remains bound to the regulatory site, the total enzyme–regulator complex will be inactive. When the regulator is removed from the regulatory site, the enzyme becomes active. In this way, the regulator controls the allosteric enzyme action.

The concepts describing allosteric enzymes include a model of an enzyme that has two forms. One form is more likely to bind the substrate and produce the product than the other form. This more active form is referred to as the **R form,** where "R" stands for *relaxed*. The less active form

A dimeric protein that can exist in either of two states: R_0 or T_0. This protein can bind three ligands:

1. Substrate (S) ▼ : Binds only to R

2. Activator (A) ▲ : Binds only to R

3. Inhibitor (I) ➤ : Binds only to T

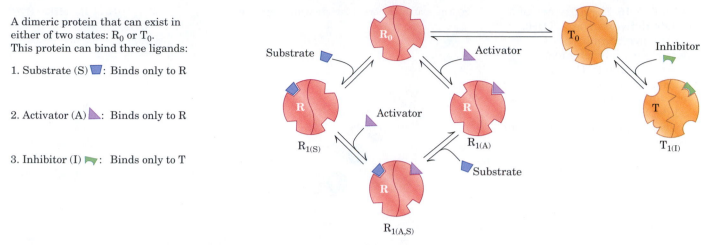

ACTIVE FIGURE 15.15 Effects of binding activators and inhibitors to allosteric enzymes. The enzyme has an equilibrium between the T form and the R form. An activator is anything that binds to the regulatory site and favors the R form. An inhibitor binds to the regulatory site and favors the T form. **Go to this book's companion website at www.cengage.com/chemistry/bettelheim to explore an interactive version of this figure.**

is referred to as the **T form,** where "T" stands for *taut* (Figure 15.15). There is an equilibrium between the T form and the R form. When the enzyme is in the R form, it will bind substrate well and catalyze the reaction. Allosteric regulators are seen to function by binding to the enzyme and favoring one form versus the other.

D. Protein Modification

The activity of an enzyme may also be controlled by **protein modification.** The modification is usually a change in the primary structure, typically by addition of a functional group covalently bound to the apoenzyme. The best-known example of protein modification is the activation or inhibition of enzymes by phosphorylation. A phosphate group is often bonded to a serine or tyrosine residue. In some enzymes, such as glycogen phosphorylase (Section 21.1), the phosphorylated form is the active form of the enzyme. Without it, the enzyme is less active or inactive.

The opposite example is the enzyme pyruvate kinase (PK, discussed in Chemical Connections 15C). Pyruvate kinase from the liver is inactive when it is phosphorylated. Enzymes that catalyze such phosphorylation go by the common name of *kinases*. When the activity of PK is not needed, it is phosphorylated (to PKP) by a protein kinase using ATP as a substrate as well as a source of energy (Section 19.3). When the system wants to turn on PK activity, the phosphate group, Pi, is removed by another enzyme, phosphatase, which renders PK active.

(*a*) The five tetramers of lactate dehydrogenase

(*b*)

Isozymes (Isoenzymes) Enzymes that perform the same function but have different combinations of subunits and thus different quaternary structures

E. Isoenzymes

Another type of regulation of enzyme activity occurs when the same enzyme appears in different forms in different tissues. Lactate dehydrogenase (LDH) catalyzes the oxidation of lactate to pyruvate, and vice versa (Figure 20.3, step 11). The enzyme has four subunits (tetramer). Two kinds of subunits, called H and M, exist. The enzyme that dominates in the heart is an H_4 enzyme, meaning that all four subunits are of the H type, although some M-type subunits are present as well. In the liver and skeletal muscles, the M type dominates. Other types of tetramer combinations exist in different tissues: H_3M, H_2M_2, and HM_3. These different forms of the same enzyme are called **isozymes** or **isoenzymes.**

The different subunits confer subtle, yet important differences to the function of the enzyme in relation to the tissue. The heart is a purely aerobic organ, except perhaps during a heart attack. LDH is used to convert lactate to pyruvate in the heart. The H_4 enzyme is allosterically inhibited by high levels of pyruvate (its product) and has a higher affinity for lactate (its substrate) than does the M_4 enzyme, which is optimized for the opposite reaction. The M_4 isozyme favors the production of lactate.

The distribution of LDH isozymes can be seen using the technique of electrophoresis, where samples are separated in a gel using an electric field. Besides their kinetic differences, the two subunits of LDH carry different charges. Therefore, each combination of subunits travels in the electric field at a different rate (Figure 15.16).

15.7 How Are Enzymes Used in Medicine?

Most enzymes are confined within the cells of the body. However, small amounts of them can also be found in body fluids such as blood, urine, and cerebrospinal fluid. The level of enzyme activity in these fluids can easily be monitored. This information can prove extremely useful: Abnormal activity (either high or low) of particular enzymes in various body fluids signals either the onset of certain diseases or their progression. Table 15.2 lists some enzymes used in medical diagnosis and their activities in normal body fluids.

TABLE 15.2 Enzyme Assays Useful in Medical Diagnosis

Enzyme	Normal Activity	Body Fluid	Disease Diagnosed
Alanine amino-transferase (ALT)	3–17 U/L*	Serum	Hepatitis
Acid phosphatase	2.5–12 U/L	Serum	Prostate cancer
Alkaline phospha-tase (ALP)	13–38 U/L	Serum	Liver or bone disease
Amylase	19–80 U/L	Serum	Pancreatic disease or mumps
Aspartate amino-transferase (AST)	7–19 U/L	Serum	Heart attack or hepatitis
	7–49 U/L	Cerebrospinal fluid	
Lactate dehydro-genase (LDH)	100–350 WU/mL	Serum	Heart attack
Creatine phospho-kinase (CPK)	7–60 U/L	Serum	
Phosphohexose isomerase (PHI)	15–75 U/L	Serum	

*U/L = International units per liter; WU/mL = Wrobleski units per milliliter.

For example, a number of enzymes are assayed (measured) during myocardial infarction to diagnose the severity of the heart attack. Dead heart muscle cells spill their enzyme contents into the serum. As a consequence, the level of creatine phosphokinase (CPK) in the serum rises rapidly, reaching a maximum within two days. This increase is followed by a rise in aspartate aminotransferase (AST; formerly called glutamate-oxaloacetate transaminase, or GOT). This second enzyme reaches a maximum two to three days after the heart attack. In addition to CPK and AST, lactate dehydrogenase (LDH) levels are monitored; they peak after five to six days. In infectious hepatitis, the alanine aminotransferase (ALT; formerly called glutamate-pyruvate transaminase, or GPT) level in the serum can rise to 10 times normal. There is also a concurrent increase in AST activity in the serum.

In some cases, administration of an enzyme is part of therapy. After duodenal or stomach ulcer operations, for instance, patients are advised to take tablets containing digestive enzymes that are in short supply in the stomach after surgery. Such enzyme preparations contain lipases, either alone or combined with proteolytic enzymes. They are sold under such names as Pancreatin, Acro-lase, and Ku-zyme.

15.8 What Are Transition-State Analogs and Designer Enzymes?

As we saw in Section 15.5C, an enzyme lowers the activation energy for a reaction, making the transition state more favorable. It does so by having an active site that actually fits best to the transition state rather than to the substrates or the products. This has been documented by the use of **transition-state analogs,** molecules with a shape that mimics the transition state of the substrate.

Proline racemase, for example, catalyzes a reaction that converts L-proline to D-proline. During this reaction, the α-carbon must change from a tetrahedral arrangement to a planar form, and then back to a tetrahedral form, but

Transition-state analog A molecule that mimics the transition state of a chemical reaction and that is used as an inhibitor of an enzyme

Chemical Connections 15E

Glycogen Phosphorylase: A Model of Enzyme Regulation

An excellent example of the subtle elegance of enzyme regulation can be seen in the enzyme glycogen phosphorylase, an enzyme that breaks down glycogen (Chapter 12) into glucose when the human body needs energy. Glycogen phosphorylase is a dimer controlled by modification and by allosterism.

Two forms of phosphorylase exist called phosphorylase *b* and phosphorylase *a* as shown in the figure. Phosphorylase *a* has a phosphate attached to each subunit, which was put there by the enzyme phosphorylase kinase. Phosphorylase *b* does not have the phosphate. The kinase is activated by hormonal signals that indicate a need for quick energy or a need for more blood glucose, depending on the tissue.

Phosphorylase is also controlled allosterically by a variety of regulators. The *b* form is converted to a more active form in the presence of AMP. Glucose-6-phosphate, glucose, and caffeine convert it to the less active form. The *a* form is also converted to the less active form by glucose and caffeine. In general, the equilibrium lies more toward the active form with phosphorylase *a* than with phosphorylase *b*.

This combination of regulation is very beneficial, because it leads to both quick changes and long-lasting ones. When you need quick action for the "fight or flight" response, the first few muscle contractions will cause ATP to be broken down. AMP will increase, which will, in much less than a second, convert some of the phosphorylase to the more active form (R state). At the same time, you will probably experience an adrenaline rush that will cause the activation of phosphorylase kinase and the subsequent phosphorylation of glycogen phosphorylase from the *b* form to the *a* form. This conversion will then cause even more phosphorylase to shift to the active R form (see right side of figure, arrow going down). The hormonal response is a little slower, taking seconds to minutes to take effect, but

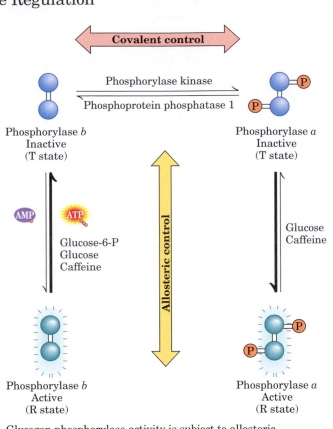

Glycogen phosphorylase activity is subject to allosteric control and covalent modification via phosphorylation. The phosphorylated form is more active. The enzyme that puts a phosphate group on phosphorylase is called phosphorylase kinase. **Go to this book's companion website at www.cengage.com/chemistry/bettelheim to explore an interactive version of this figure.**

it is more long-lasting because the equilibrium will stay shifted to the R form until other enzymes (phosphatases) remove the phosphates. Thus the combination of allosteric and covalent control gives us the best of both worlds.

with the orientation of two bonds reversed (Figure 15.17). An inhibitor of the reaction is pyrrole-2-carboxylate, a chemical that is structurally similar to what proline would look like at its transition state because it is always planar at the equivalent carbon. This inhibitor binds to proline racemase 160 times more strongly than proline does. Transition-state analogs have been used with many enzymes to help verify a suspected mechanism and structure of the transition state as well as to inhibit an enzyme selectively. They are currently being used for drug design when a specific enzyme is the cause of a disease.

In 1969, William Jencks proposed that an immunogen (a molecule that elicits an antibody response) would elicit antibodies (Chapter 23) with catalytic activity if the immunogen mimicked the transition state of the reaction. Richard Lerner and Peter Schultz, who created the first catalytic

Chemical Connections 15F

One Enzyme, Two Functions

The enzyme called prostaglandin enderoperoxide synthase (PGHS) catalyzes the conversion of arachidonic acid to PGH_2 (Section 13.12) in two steps:

Arachidonic acid

Cyclooxygenase

PGG₂

Peroxidase

PGH₂

In the process, PGHS inserts two molecules of oxygen into arachidonic acid.

The enzyme itself is a single protein molecule. It is associated with the cell membrane and has a heme coenzyme. It has two distinct functions. The enzyme's *cyclooxgenase activity* is to close a substituted cyclopentane ring. The *peroxidase activity* of the same enzyme yields the 15-hydroxy derivative prostaglandin, PGH_2.

Commercial painkillers act two ways in inhibiting the formation of prostaglandin, PGH_2. First, PGHS's cyclooxygenase activity is inhibited by aspirin and related NSAIDs (Chemical Connections 13H). Aspirin inhibits PGHS by acetylating serine in the active site; as a result, the active site is no longer able to accommodate arachidonic acid. Other NSAIDs also inhibit the cyclooxygenase activity by competitive inhibition, but they do not inhibit the peroxidase activity. Second, another class of inhibitor has antioxidant activity. For example, acetaminophen (Tylenol) acts as a painkiller by inhibiting the peroxidase activity of PGHS.

L-Proline

Planar transition state

D-Proline

Pyrrole-2-carboxylate
(inhibitor and transition-state analog)

FIGURE 15.17 The proline racemase reaction. Pyrrole-2-carboxylate mimics the planar transition state of the reaction.

antibodies, verified this hypothesis in 1986. Because an antibody is a protein designed to bind to specific molecules on the immunogen, the antibody will, in essence, serve as a fake active site. For example, the reaction of pyridoxal phosphate and an amino acid to form the corresponding α-ketoacid and pyridoxamine phosphate is a very important reaction in amino acid metabolism. The molecule $N^\alpha-(5'$-phosphopyridoxyl)-L-lysine serves as a transition-state analog for this reaction. When this antigen molecule was used to elicit antibodies, these antibodies, or **abzymes,** had catalytic

Abzyme An antibody that has catalytic ability because it was created by using a transition-state analog as an immunogen

Chemical Connections 15G

Catalytic Antibodies Against Cocaine

Many addictive drugs, including heroin, operate by binding to a particular receptor in the neurons, mimicking the action of a neurotransmitter. When a person is addicted to such a drug, a common way to attempt to treat the addiction is to use a compound that blocks the receptor and denies the drug access to it. Cocaine addiction has always been difficult to treat, due primarily to its unique modus operandi. As shown below, cocaine blocks the reuptake of the neurotransmitter dopamine. As a result, dopamine stays in the system longer, overstimulating neurons and leading to the reward signals in the brain that lead to addiction. Using a drug to block a receptor would be of no use with cocaine addiction and would probably just make removal of dopamine even more unlikely.

Cocaine (Section 8.2) can be degraded by a specific esterase, an enzyme that hydrolyzes an ester bond that is part of cocaine's structure. In the process of this hydrolysis, the cocaine must pass through a transition state that changes its shape. Catalytic antibodies to the transition state of the hydrolysis of cocaine have now been created. When administered to patients suffering from cocaine addiction, the antibodies successfully hydrolyze cocaine to two harmless degradation products—benzoic acid and ecgonine methyl ester. When degraded, the cocaine cannot block dopamine reuptake. No prolongation of the neuronal stimulus occurs, and the addictive effects of the drug vanish over time.

The mechanism of action of cocaine. (*a*) Dopamine acts as a neurotransmitter. It is released from the presynaptic neuron, travels across the synapse, and bonds to dopamine receptors on the postsynaptic neuron. It is later released and taken up into vesicles in the presynaptic neuron. (*b*) Cocaine increases the amount of time that dopamine is available to the dopamine receptors by blocking its uptake. (Adapted from *Immunotherapy for Cocaine Addiction,* by D. W. Landry, *Scientific American,* February 1997, pp 42–45.)

Degradation of cocaine by esterases or catalytic antibodies. Cocaine (*a*) passes through a transition state (*b*) on its way to being hydrolyzed to benzoic acid and ecgonine methyl ester (*c*). Transition-state analogs are used to generate catalytic antibodies for this reaction. (Adapted from *Immunotherapy for Cocaine Addiction,* by D. W. Landry, *Scientific American,* February 1997, pp 42–45.)

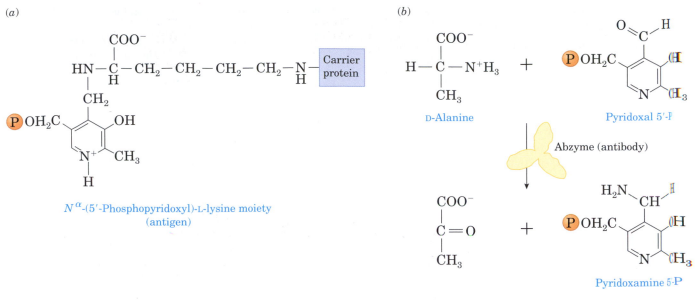

FIGURE 15.18 Abzymes. (*a*) N^α-(5'-phosphopyridoxyl)-L-lysine moiety is a transition-state analog for the reaction of an amino acid with pyridoxal 5'-phosphate. When this moiety is attached to a protein and injected into a host, it acts like an antigen and the host produces antibodies that have catalytic activity (abzymes). (*b*) The abzyme is then used to catalyze the reaction.

activity (Figure 15.18). Thus, in addition to helping us to verify the nature of the transition state or making an inhibitor, transition-state analogs now offer the possibility of creating designer enzymes to catalyze a wide variety of reactions.

Summary

OWL End-of-chapter problems identified in blue are assignable in **GOB OWL.**

Section 15.1 What Are Enzymes?

- **Enzymes** are macromolecules that catalyze chemical reactions in the body. Most enzymes are very specific—they catalyze only one particular reaction.
- The compound whose reaction is catalyzed by an enzyme is called the **substrate.**
- Most enzymes are proteins, although some are made of RNA.

Section 15.2 How Are Enzymes Named and Classified? Problem 15.10

- Enzymes are classified into six major groups according to the type of reaction they catalyze.
- Enzymes are typically named after the substrate and the type of reaction they catalyze, by adding the ending "-ase."

Section 15.3 What Is the Terminology Used with Enzymes?

- Some enzymes are made of polypeptide chains only. Others have, besides the polypeptide chain (the

apoenzyme), nonprotein **cofactors,** which are either organic compounds (**coenzymes**) or inorganic ions

- Only a small part of the enzyme surface, called the **active site,** participates in the actual catalysis of chemical reactions. Cofactors, if any, are part of the active site.
- Compounds that slow enzyme action are called **inhibitors.**
- A **competitive inhibitor** attaches itself to the active site. A **noncompetitive inhibitor** binds to other parts of the enzyme surface.

Section 15.4 What Factors Influence Enzyme Activity?

- The higher the enzyme and substrate concentrations, the higher the enzyme activity. At sufficiently high substrate concentrations, however, a saturation point is reached. After this point, increasing substrate concentration no longer increases the reaction rate.
- Each enzyme has an optimal temperature and pH at which it has its greatest activity.

Section 15.5 What Are the Mechanisms of Enzyme Action? Problem 15.31

- Two closely related mechanisms that seek to explain enzyme activity and specificity are the **lock-and-key model** and the **induced-fit model.**
- Enzymes lower the **activation energy** required for a biochemical reaction to occur.

Section 15.6 How Are Enzymes Regulated? Problem 15.39

- Enzyme activity is regulated by five mechanisms.
- In **feedback control,** the concentration of products influences the rate of the reaction.
- Some enzymes, called **proenzymes** or **zymogens,** must be activated by removing a small portion of the polypeptide chain.
- In **allosterism,** an interaction takes place at a position other than the active site but affects the active site, either positively or negatively.

- Enzymes can be activated or inhibited by **protein modification.**
- Enzyme activity is also regulated by **isozymes** (isoenzymes), which are different forms of the same enzyme.

Section 15.7 How Are Enzymes Used in Medicine? Problem 15.47

- Abnormal enzyme activity can be used to diagnose certain diseases.

Section 15.8 What Are Transition-State Analogs and Designer Enzymes? Problem 15.74

- The active site of an enzyme favors the production of the **transition state.**
- Molecules that mimic the transition state are called **transition-state analogs,** and they make potent enzyme inhibitors.

Problems

■ Indicates problems that are assignable in **GOB OWL**.

Blue numbered problems are applied.

Go to this book's companion website at **www.cengage. com/chemistry/bettelheim** for interactive versions of the How To tutorials and Active Figures, and to quiz yourself on this chapter.

Section 15.1 What Are Enzymes?

15.1 What is the difference between a *catalyst* and an *enzyme?*

15.2 What are ribozymes made of?

15.3 Would a lipase hydrolyze two triglycerides, one containing only oleic acid and the other containing only palmitic acid, with equal ease?

15.4 Compare the activation energy in uncatalyzed reactions and in enzyme-catalyzed reactions.

15.5 Why does the body need so many different enzymes?

15.6 Trypsin cleaves polypeptide chains at the carboxyl side of a lysine or arginine residue (Figure 15.1). Chymotrypsin cleaves polypeptide chains on the carboxyl side of an aromatic amino acid residue or any other nonpolar, bulky side chain. Which enzyme is more specific? Explain.

Section 15.2 How Are Enzymes Named and Classified?

15.7 Both lyases and hydrolases catalyze reactions involving water molecules. What is the difference in the types of reactions that these two enzymes catalyze?

15.8 Monoamine oxidases are important enzymes in brain chemistry. Judging from the name, which of

the following would be a suitable substrate for this class of enzymes:

(a) $HO-$

$$\overset{OH}{\underset{|}{CH}}-CH_2NH_2$$

(b) $CH_3-\overset{\overset{O}{\|}}{C}-N(CH_3)_2$

(c) $-NO_2$

15.9 On the basis of the classification given in Section 15.2, decide to which group each of the following enzymes belongs:

(a) Phosphoglyceromutase

$$^-OOC-\underset{\underset{OH}{|}}{CH}-CH_2-OPO_3{}^{2-}$$

3-Phosphoglycerate

$$\rightleftharpoons {}^-OOC-\underset{\underset{OPO_3{}^{2-}}{|}}{CH}-CH_2-OH$$

2-Phosphoglycerate

(b) Urease

$$H_2N-\overset{\overset{O}{\|}}{C}-NH_2 + H_2O \rightleftharpoons 2NH_3 + CO_2$$

Urea

(c) Succinate dehydrogenase

$$^-OOC-CH_2-CH_2-COO^- \quad + \quad FAD$$

Succinate   Coenzyme
(oxidized form)

Fumarate   Coenzyme
(reduced form)

(d) Aspartase

$$+ \quad NH_4^+$$

Fumarate

$$\rightleftharpoons ^-OOC-CH_2-CH-COO^-$$
$$NH_3^+$$

L-Asparate

15.10 ■ What kind of reaction does each of the following enzymes catalyze?

   (a) Deaminases   (b) Hydrolases

   (c) Dehydrogenases   (d) Isomerases

Section 15.3 What Is the Terminology Used with Enzymes?

15.11 What is the difference between a *coenzyme* and a *cofactor*?

15.12 In the citric acid cycle, an enzyme converts succinate to fumarate (see the reaction in Problem 15.9c). The enzyme consists of a protein portion and an organic molecule portion called FAD. What terms do we use to refer to (a) the protein portion and (b) the organic molecule portion?

15.13 What is the difference between reversible and irreversible noncompetitive inhibition?

Section 15.4 What Factors Influence Enzyme Activity?

15.14 In most enzyme-catalyzed reactions, the rate of reaction reaches a constant value with increasing substrate concentration. This relationship is described in a saturation curve diagram (Figure 15.4). If the enzyme concentration, on a molar basis, is twice the maximum substrate concentration, would you obtain a saturation curve?

15.15 At a very low concentration of a certain substrate, we find that, when the substrate concentration doubles, the rate of the enzyme-catalyzed reaction also doubles. Would you expect the same finding at a very high substrate concentration? Explain.

15.16 If we wish to double the rate of an enzyme-catalyzed reaction, can we do so by increasing the temperature by 10°C? Explain.

15.17 A bacterial enzyme has the following temperature-dependent activity.

   (a) Is this enzyme more or less active at normal body temperature than when a person has a fever?

   (b) What happens to the enzyme activity if the patient's temperature is lowered to 35°C?

15.18 The optimal temperature for the action of lactate dehydrogenase is 36°C. It is irreversibly inactivated at 85°C, but a yeast containing this enzyme can survive for months at −10°C. Explain how this can happen.

15.19 The activity of pepsin was measured at various pH values. When the temperature and the concentrations of pepsin and substrate were held constant, the following activities were obtained:

pH	Activity
1.0	0.5
1.5	2.6
2.0	4.8
3.0	2.0
4.0	0.4
5.0	0.0

   (a) Plot the pH dependence of pepsin activity.

   (b) What is the optimal pH?

   (c) Predict the activity of pepsin in the blood at pH 7.4.

15.20 How can the pH profile of an enzyme tell you something about the mechanism if you know the amino acids at the active site?

Section 15.5 What Are the Mechanisms of Enzyme Action?

15.21 ■ Urease can catalyze the hydrolysis of urea but not the hydrolysis of diethylurea. Explain why diethylurea is not hydrolyzed.

$$\underset{\text{Urea}}{H_2N-\overset{\overset{\displaystyle O}{\|}}{C}-NH_2} \qquad \underset{\text{Diethylurea}}{CH_3CH_2-NH-\overset{\overset{\displaystyle O}{\|}}{C}-NH-CH_2CH_3}$$

15.22 The following reaction may be represented by the cartoon figures:

Glucose + ATP ⇌ glucose 6-phosphate + ADP

In this enzyme-catalyzed reaction, Mg^{2+} is a cofactor, fluoroglucose is a competitive inhibitor, and Cd^{2+} is a noncompetitive inhibitor. Identify each component of the reaction by a cartoon figure and assemble them to show (a) the normal enzyme reaction, (b) a competitive inhibition, and (c) a noncompetitive inhibition.

15.23 Which amino acids appear most frequently in the active sites of enzymes?

15.24 What kind of chemical reaction occurs most frequently at the active site?

15.25 Which of the following is a correct statement describing the induced-fit model of enzyme action? Substrates fit into the active site
(a) because both are exactly the same size and shape.
(b) by changing their size and shape to match those of the active site.
(c) by changing the size and shape of the active site upon binding.

15.26 What is the maximum rate that can be achieved in competitive inhibition compared with noncompetitive inhibition?

15.27 Enzymes are long protein chains, usually containing more than 100 amino acid residues. Yet the active site contains only a few amino acids. Explain why the other amino acids of the chain are present and what would happen to the enzyme activity if the enzyme's structure were changed significantly.

15.28 Sucrose (common table sugar) is hydrolyzed to glucose and fructose. The reaction is catalyzed by the enzyme invertase. Using the following data, determine whether the inhibition by 2 M urea is competitive or noncompetitive.

Sucrose Concentration (M)	Velocity (arbitrary units)	Velocity + Inhibitor
0.0292	0.182	0.083
0.0584	0.265	0.119
0.0876	0.311	0.154
0.117	0.330	0.167
0.175	0.372	0.192

Section 15.6 How Are Enzymes Regulated?

15.29 The hydrolysis of glycogen to yield glucose is catalyzed by the enzyme phosphorylase. Caffeine, which is not a carbohydrate and not a substrate for the enzyme, inhibits phosphorylase. What kind of regulatory mechanism is at work?

15.30 Can the product of a reaction that is part of a sequence act as an inhibitor for another reaction in the sequence? Explain.

15.31 What is the difference between a *zymogen* and a *proenzyme*?

15.32 The enzyme trypsin is synthesized by the body in the form of a long polypeptide chain containing 235 amino acids (trypsinogen), from which a piece must be cut before the trypsin can be active. Why does the body not synthesize trypsin directly?

15.33 Give the structure of a tyrosyl residue of an enzyme modified by a protein kinase.

15.34 What is an *isozyme*?

15.35 The enzyme glycogen phosphorylase initiates the phosphorolysis of glycogen to glucose 1-phosphate. It comes in two forms: Phosphorylase *b* is less active, and phosphorylase *a* is more active. The difference between the *b* and *a* forms is the modification of the apoenzyme. Phosphorylase *a* has two phosphate groups added to the polypeptide chain. In analogy with the pyruvate kinase discussed in the text, give a scheme indicating the transition between the *b* and *a* forms. Which enzymes and which cofactors control this reaction?

15.36 How can you tell if an enzyme is allosteric by plotting velocity versus substrate?

15.37 Explain the nature of the two types of control of glycogen phosphorylase. What is the advantage to having both control types?

15.38 Which type of regulation discussed in Section 15.6 is the least reversible? Explain.

15.39 ■ The enzyme phosphofructokinase (PFK) (Chapter 20) has two types of subunits, M and L, for muscle and liver. These subunits combine to form a tetramer. How many isozymes of PFK exist? What are their designations?

15.40 If you separated PFK using electrophoresis, how would the isozymes migrate if the M subunit has a lower pI than the L subunit?

Section 15.7 How Are Enzymes Used in Medicine?

15.41 After a heart attack, the levels of certain enzymes rise in the serum. Which enzyme would you monitor within 24 hours following a suspected heart attack?

15.42 The enzyme formerly known as GPT (glutamate-pyruvate transaminase) has a new name: ALT (alanine aminotransferase). Looking at the equation in Section 20.9, which is catalyzed by this enzyme, what prompted this change of name?

15.43 If an examination of a patient indicated elevated levels of AST but normal levels of ALT, what would be your tentative diagnosis?

15.44 Which LDH isozyme is monitored in the case of a heart attack?

15.45 Chemists who have been exposed for years to organic vapors usually show higher-than-normal activity when given the alkaline phosphatase test. Which organ in the body do organic vapors affect?

15.46 Which enzyme preparation is given to patients after duodenal ulcer surgery?

15.47 ■ Chymotrypsin is secreted by the pancreas and passed into the intestine. The optimal pH for this enzyme is 7.8. If a patient's pancreas cannot manufacture chymotrypsin, would it be possible to supply it orally? What happens to chymotrypsin's activity during its passage through the gastrointestinal tract?

Section 15.8 What Are Transition-State Analogs and Designer Enzymes?

15.48 Explain why transition-state analogs are potent inhibitors.

15.49 How do transition-state analogs relate to the idea of the induced-fit model of enzymes?

15.50 Explain the relationship between transition-state analogs and abzymes.

Chemical Connections

15.51 (Chemical Connections 15A) Acetylcholine causes muscles to contract. Succinylcholine, a close relative, is a muscle relaxant. Explain the different effects of these related compounds.

15.52 (Chemical Connections 15A) An operating team usually administers succinylcholine before bronchoscopy. What is achieved by this procedure?

15.53 (Chemical Connections 15B) PKMζ is a type of enzyme called a kinase. Kinases are very important in metabolism. Look through the metabolism chapters (Chapters 19 and 20) and find two examples of kinases. What reactions do kinases catalyze?

15.54 (Chemical Connections 15B) Explain how researchers used the drug ZIP to test its effect on long-term memory. How did they know that food aversion was a long-term memory phenomenon?

15.55 (Chemical Connections 15B) Why would researchers want to be able to selectively block long-term memory?

15.56 (Chemical Connections 15C) What role does Mn^{2+} play in anchoring the substrate in the active site of protein kinase?

15.57 (Chemical Connections 15C) Which amino acids of the active site interact with the $=CH_2$ group of the phosphoenol pyruvate? Do these amino acids provide the same surface environment? What is the nature of the interaction?

15.58 (Chemical Connections 15D) What is the strategy in drug design to fight AIDS?

15.59 (Chemical Connections 15D) Why did scientists want to create a drug to inhibit cGMP diesterases?

15.60 (Chemical Connections 15E) Explain the difference between phosphorylase a and phosphorylase b. Which is more active and why?

15.61 (Chemical Connections 15E) What is the relationship between protein modification and allosteric control of glycogen phosphorylase?

15.62 (Chemical Connections 15F) Which activity of prostaglandin enderoperoxide synthase (PGHS) is inhibited by Tylenol, and which is inhibited by aspirin?

15.63 (Chemical Connections 15G) Explain how catalytic antibodies are produced to fight cocaine addiction.

15.64 (Chemical Connections 15G) Why can inhibitors not be used to block a cocaine receptor as is often done with other drug addictions?

15.65 (Chemical Connections 15G) How does cocaine work as a drug?

Additional Problems

15.66 Where can one find enzymes that are both stable and active at 90°C?

15.67 Food can be preserved by inactivation of enzymes that would cause spoilage—for example, by refrigeration. Give an example of food preservation in which the enzymes are inactivated (a) by heat and (b) by lowering the pH.

15.68 Why is enzyme activity during myocardial infarction measured in patients' serum rather than in their urine?

15.69 What is the common characteristic of the amino acids of which the carboxyl groups of the peptide bonds can be hydrolyzed by trypsin?

15.70 Many enzymes are active only in the presence of Zn^{2+}. What common term is used for ions such as Zn^{2+} when discussing enzyme activity?

15.71 An enzyme has the following pH dependence:

At what pH do you think this enzyme works best?

15.72 What enzyme is monitored in the diagnosis of infectious hepatitis?

15.73 The enzyme chymotrypsin catalyzes the following type of reaction:

On the basis of the classification given in Section 15.2, to which group of enzymes does chymotrypsin belong?

15.74 ■ Nerve gases operate by forming covalent bonds at the active site of cholinesterase. Is this an example of competitive inhibition? Can the nerve gas molecules be removed by simply adding more substrate (acetylcholine) to the enzyme?

15.75 What would be the appropriate name for an enzyme that catalyzes each of the following reactions:

(a) $CH_3CH_2OH \longrightarrow CH_3\overset{\displaystyle O}{\overset{\displaystyle \|}{C}}-H$

(b) $CH_3\overset{\displaystyle O}{\overset{\displaystyle \|}{C}}-O-CH_2CH_3 + H_2O$

$\longrightarrow CH_3\overset{\displaystyle O}{\overset{\displaystyle \|}{C}}-OH + CH_3CH_2OH$

15.76 In Section 21.5, a reaction between pyruvate and glutamate to form alanine and α-ketoglutarate is given. How would you classify the enzyme that catalyzes this reaction?

15.77 A liver enzyme is made of four subunits: 2A and 2B. The same enzyme, when isolated from the brain, has the following subunits: 3A and 1B. What would you call these two enzymes?

15.78 What is the function of a ribozyme?

15.79 Can an enzyme catalyze the forward reaction but not the backward reaction for its substrate-product pair(s)? Explain.

Looking Ahead

15.80 Caffeine is a stimulant that is taken by many people in the form of coffee, tea, chocolate, and cola beverages. It is also used by many athletes. Caffeine has many effects, including stimulating lipases. Given its effect on lipases and on glycogen phosphorylase, would you predict caffeine would be more effective as an aid to a runner in a 10 K race or in a 1 mile race?

15.81 Caffeine is also a diuretic, which means it increases the movement of water through the kidneys and into the urine. Why would this potentially offset its value to a distance athlete?

15.82 Until the discovery of thermophilic bacteria that live in conditions of extreme heat and pressure, it was impossible to have an automated system of DNA synthesis. Explain why this was so, given that it takes temperatures of around 90°C to separate strands of DNA.

15.83 What characteristics of RNA make it likely to have catalytic ability? Why is DNA less likely to have catalytic activity?

Chemical Communications: Neurotransmitters and Hormones

16

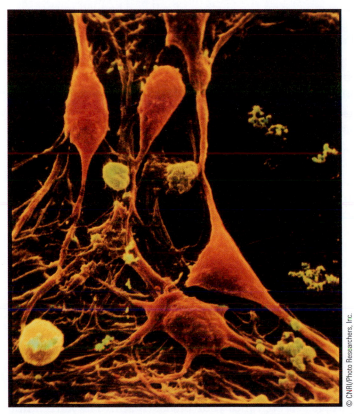

Nerve cells. Colored scanning electron micrograph (SEM) of neurons (nerve cells) from the cerebral cortex (the outer, heavily folded, gray matter of the brain). Neurons exist in varying sizes & shapes throughout the nervous system, but all have similar basic structure; a large central cell body (colored gold, containing nucleus) & processes of 2 types: a single axon (a nerve fiber) which is the effector part of the cell that terminates on other neurons (or organs), & 1 or more dendrites, smaller processes that act as sensory receptors. Similar types of neurons are arranged in layers in the cerebral cortex. Magnification: X 470 at 6×7cm size. Manification: X 235 at 35mm size.

© CNRI/Photo Researchers, Inc.

Key Questions

16.1 What Molecules Are Involved in Chemical Communications?

16.2 How Are Chemical Messengers Classified as Neurotransmitters and Hormones?

16.3 How Does Acetylcholine Act as a Messenger?

16.4 What Amino Acids Act as Neurotransmitters?

16.5 What Are Adrenergic Messengers?

16.6 What Is the Role of Peptides in Chemical Communication?

16.7 How Do Steroid Hormones Act as Messengers?

16.1 What Molecules Are Involved in Chemical Communications?

The importance of chemical communications in health becomes apparent from a quick glance at Table 16.1. This list contains only a small sample of drugs pertaining to this chapter and it is of utmost interest to health care

OWL Online homework for this chapter may be assigned in **GOB OWL**.

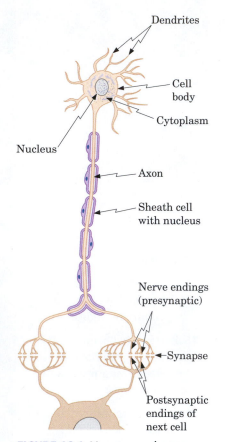

FIGURE 16.1 Neuron and synapse.

Neurotransmitter A chemical messenger between a neuron and another target cell: neuron, muscle cell, or cell of a gland

Hormone A chemical messenger released by an endocrine gland into the bloodstream and transported there to reach its target cell

providers. In fact, a vast number of drugs in the pharmacopoeia act, in one way or another, to influence chemical communications.

Each cell in the body is an isolated entity enclosed in its own membrane. Furthermore, within each cell of higher organisms, organelles, such as the nucleus or the mitochondrion, are enclosed by membranes separating them from the rest of the cell. If cells could not communicate with one another, the thousands of reactions in each cell would be uncoordinated. The same is true for organelles within a cell. Such communication allows the activity of a cell in one part of the body to be coordinated with the activity of cells in a different part of the body. There are three principal types of molecules for communications:

- **Receptors** are protein molecules that bind to ligands and effect some type of change. They may be on the surface of cells, imbedded in the membrane of subcellular organelles, or free in solution. Most of the receptors we will study are membrane bound.

- **Chemical messengers,** also called ligands, interact with the receptors. (Chemical messengers fit into the receptor sites in a manner reminiscent of the lock-and-key model mentioned in Section 15.5.)

- **Secondary messengers** in many cases carry the message from the receptor to the inside of the cell and amplify the message.

If your house is on fire and the fire threatens your life, external signals—light, smoke, and heat—register alarm at specific receptors in your eyes, nose, and skin. From there the signals are transmitted by specific compounds to nerve cells, or **neurons.** Nerve cells are present throughout the body and, together with the brain, constitute the nervous system. In the neurons, the signals travel as electric impulses along the axons (Figure 16.1). When they reach the end of the neuron, the signals are transmitted to adjacent neurons by specific compounds called **neurotransmitters.** Communication between the eyes and the brain, for example, is by neural transmission.

As soon as the danger signals are processed in the brain, other neurons carry messages to the muscles and to the endocrine glands. The message to the muscles is to run away or to take some other action in response to the fire (save the baby or run to the fire extinguisher, for example). To do so, the muscles must be activated. Again, neurotransmitters carry the necessary messages from the neurons to the muscle cells and the endocrine glands. The endocrine glands are stimulated, and a different chemical signal, called a **hormone**, is secreted into your bloodstream. "The adrenaline begins to flow." Adrenaline is a hormone that binds to specific receptors in muscle and liver cells. Once bound, it triggers the production of a second messenger, cyclic AMP (cAMP). The second messenger leads to modification of several enzymes involved in carbohydrate metabolism. The immediate result is that the cells produce quick energy so that the muscles can fire rapidly and often, allowing the organism to use its strength and speed in the moment of crisis. We will revisit second messengers in Section 16.6.

Without these chemical communicators, the whole organism—you—would not survive because there is a constant need for coordinated efforts to face a complex outside world. The chemical communications between different cells and different organs play a role in the proper functioning of our bodies. Its significance is illustrated by the fact that *a large percentage of the drugs we encounter in medical practice try to influence this communication.* The scope of these drugs covers all fields—from prescriptions against hypertension, to heart disease, to antidepressants, to painkillers, just to mention a few. There are several ways these drugs act in the body. A drug may affect either the messenger, the receptor, the secondary messenger, or any one of a host of enzymes that are activated or inhibited as part of a metabolic pathway see (Chapter 15).

TABLE 16.1 Drugs That Affect Nerve Transmission

Messenger	Drugs That Affect Receptor Sites		Drugs That Affect Available Concentration of the Neurotransmitter or Its Removal from Receptor Sites	
	Agonists (Activate Receptor Sites)	Antagonists (Block Receptor Sites)	Increase Concentration	Decrease Concentration
Acetylcholine (cholinergic)	Nicotine Succinylcholine	Curare Atropine	Malathion Nerve gases Succinylcholine Donepezil (Aricept)	*Clostridium botulinum* toxin
Calcium ion		Nifedipine (Adalat) Diltiazem (Cardiazem)	Digitoxin (Lanoxicaps)	
Epinephrine (α-adrenergic)	Terazosin (Hytrin)			
Norepinephrine (β-adrenergic)	Phenylephrine Epinephrine (Adrenalin)	Propranolol (Inderal)	Amphetamines	Reserpine Methyldopa (Aldomet) Metyrosine (Demser)
Dopamine (adrenergic)		Clozapine (Clorazil)	Entacapon (Comtan)	
Serotonin (adrenergic)		Ondansetron (Zofran)	Antidepressant Fluoxetine (Prozac)	
Histamine (adrenergic)	2-Methyl-histamine	Fexofenadine (Allegra) Diphenhydramine (Benadryl) Ranitidine (Zantac) Cimetidine (Tagamet)	Histidine	Hydrazino-histidine
Glutamic acid (amino acid)	*N*-Methyl-D-aspartate	Phenylcyclidine		
Enkephalin (peptidergic)	Opiate Morphine Heroin Meperidine (Demerol)		Naloxone (Narcan)	

1. An **antagonist** drug blocks the receptor and prevents its stimulation.

2. An **agonist** drug competes with the natural messenger for the receptor site. Once there, it stimulates the receptor.

3. Other drugs decrease the concentration of the messenger by controlling the release of messengers from their storage.

4. Other drugs increase the concentration of the messenger by inhibiting its removal from the receptors.

5. Still others act to inhibit or activate specific enzymes inside the cells.

Table 16.1 presents selected drugs and their modes of action that affect neurotransmission. We will take a closer look at chemical communication as it relates to enzyme control later.

16.2 How Are Chemical Messengers Classified as Neurotransmitters and Hormones?

As mentioned earlier, neurotransmitters are compounds that communicate between two nerve cells or between a nerve cell and another cell (such as a muscle cell). A nerve cell (Figure 16.1) consists of a main cell body from which projects a long, fiber-like part called an **axon.** Coming off the other side of the main body are hair-like structures called **dendrites.**

Typically, neurons do not touch each other. Between the axon end of one neuron and the cell body or dendrite end of the next neuron is a space filled with an aqueous fluid, called a **synapse.** If the chemical signal travels, say, from axon to dendrite, we call the nerve ends on the axon the **presynaptic** site. The neurotransmitters are stored at the presynaptic site in **vesicles,** which are small, membrane-enclosed packages. Receptors are located on the **postsynaptic** site of the cell body or the dendrite.

Synapse An aqueous small space between the tip of a neuron and its target cell

Hormones are diverse compounds secreted by specific tissues (the endocrine glands), released into the bloodstream, and then adsorbed onto specific receptor sites, usually relatively far from their source. (This is the physiological definition of a hormone.) Table 16.2 lists some of the principal hormones. Figure 16.2 shows the target organs of hormones secreted by the pituitary gland.

The distinction between hormones and neurotransmitters is physiological, not chemical. Whether a certain compound is considered to be a neurotransmitter or a hormone depends on whether it acts over a short distance across a synapse (2×10^{-6} cm), in which case it is a neurotransmitter, or over a long distance (20 cm) from the secretory gland through the bloodstream to

TABLE 16.2 The Principal Hormones and Their Actions

Gland	Hormone	Action	Structures Shown in
Parathyroid	Parathyroid hormone	Increases blood calcium Excretion of phosphate by kidney	
Thyroid	Thyroxine (T_4)	Growth, maturation, and metabolic rate	Section 14.5
	Triiodothyronine (T_3)	Metamorphosis	
Pancreatic islets			
Beta cells	Insulin	Hypoglycemic factor	Section 16.8
		Regulation of carbohydrates, fats, and proteins	Chemical Connections 8G
Alpha cells	Glucagon	Liver glycogenolysis	
Adrenal medulla	Epinephrine Norepinephrine	Liver and muscle glycogenolysis	Section 8.5
Adrenal cortex	Cortisol	Carbohydrate metabolism	Section 4.10
	Aldosterone	Mineral metabolism	Section 4.10
	Adrenal androgens	Androgenic activity (especially females)	
Kidney	Renin	Hydrolysis of blood precursor protein to yield angiotensin	
Anterior pituitary	Luteinizing hormone	Causes ovulation	
	Interstitial cell-stimulating hormone	Formation of testosterone and progesterone in interstitial cells	
	Prolactin	Growth of mammary gland	
	Mammotropin	Lactation Corpus luteum function	
Posterior pituitary	Vasopressin	Contraction of blood vessels Kidney reabsorption of water	Section 5.8
	Oxytocin	Stimulates uterine contraction and milk ejection	Section 5.8
Ovaries	Estradiol	Estrous cycle	Section 4.10
	Progesterone	Female sex characteristics	Section 4.10
Testes	Testosterone	Male sex characteristics	Section 4.10
	Androgens	Spermatogenesis	

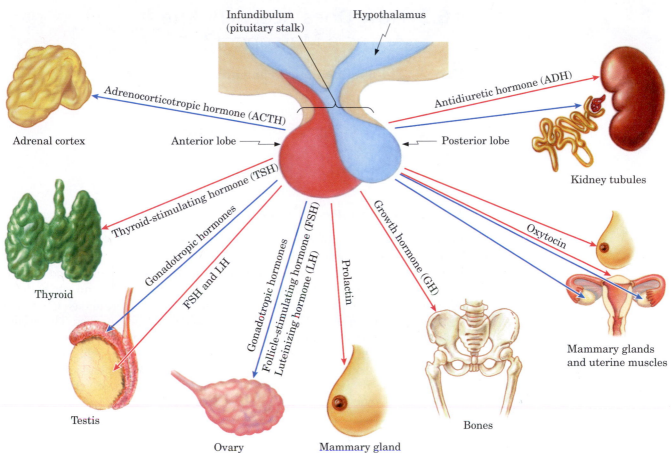

FIGURE 16.2 The pituitary gland is suspended from the hypothalamus by a stalk of neural tissue. The hormones secreted by the anterior and posterior lobes of the pituitary gland and the target tissues they act on are shown.

Modified from P. W. Davis and E. P. Solomon, *The World of Biology.* Philadelphia: Saunders College Publishing, 1986

the target cell, in which case it is a hormone. For example, epinephrine and norepinephrine are both neurotransmitters and hormones.

There are, broadly speaking, five classes of chemical messengers: *cholinergic, amino acid, adrenergic, peptidergic,* and *steroid* messengers. This classification is based on the chemical nature of the important messenger in each group. Neurotransmitters can belong to all five classes, and hormones can belong to the last three classes.

Messengers can also be classified according to how they work. Some of them—epinephrine, for example—*activate enzymes.* Others affect the *synthesis of enzymes and proteins* by turning on the genes that produce them (Section 18.2). Steroid hormones (Section 13.10) work in this manner. Finally, some affect the *permeability of membranes;* acetylcholine, and insulin belong to this class.

Yet another way of classifying messengers is according to their potential to *act directly* or through a *secondary messenger.* The steroid hormones act directly. They can penetrate the cell membrane and pass through the membrane of the nucleus. For example, estradiol stimulates uterine growth.

Other chemical messengers act through secondary messengers. For example, epinephrine, glucagon, luteinizing hormone, norepinephrine, and vasopressin use cAMP as a secondary messenger (details in Section 16.5C).

In the following sections, we will sample the mode of communication within each of the five chemical categories of messengers.

16.3 How Does Acetylcholine Act as a Messenger?

The main **cholinergic neurotransmitter** is acetylcholine:

Acetylcholine

A. Cholinergic Receptors

There are two kinds of receptors for this messenger. We will look at one that exists on the motor end plates of skeletal muscles or in the sympathetic ganglia. The nerve cells that bring messages contain stored acetylcholine in the vesicles in their axons. The receptor on the muscle cells or neurons is also known as nicotinic receptor because nicotine (see Chemical Connections 8B) inhibits the neurotransmission of these nerves. The receptor itself is a *transmembrane protein* (Figure 13.2) made of five different subunits. The central core of the receptor is an ion channel through which, when open, Na^+ and K^+ ions can pass (Figure 16.3). When the ion channels are closed, the K^+ ion concentration is higher inside the cell than outside; the reverse is true for the Na^+ ion concentration.

B. Storage of Messengers

Events begin when a message is transmitted from one neuron to the next by neurotransmitters. The message is initiated by calcium ions (see Chemical Connections 16A). When the concentration in a neuron reaches a certain level (more than 0.1 μM), the vesicles containing acetylcholine fuse with the presynaptic membrane of the nerve cells. Then they empty the neurotransmitters into the synapse. The messenger molecules travel across the synapse and are adsorbed onto specific receptor sites.

FIGURE 16.3 Acetylcholine in action. The receptor protein has five subunits. When two molecules of acetylcholine bind to the two α subunits, a channel opens to allow the passage of Na^+ and K^+ ions by facilitated transport (Chemical Connections 13C).

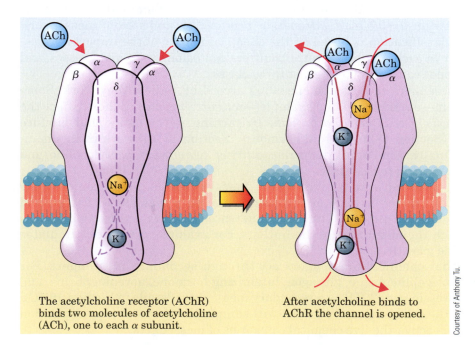

The acetylcholine receptor (AChR) binds two molecules of acetylcholine (ACh), one to each α subunit.

After acetylcholine binds to AChR the channel is opened.

Courtesy of Anthony Tu.

Chemical Connections 16A

Calcium as a Signaling Agent (Secondary Messenger)

The message delivered to the receptors on the cell membranes by neurotransmitters or hormones must be delivered intracellularly to various locations within the cell. The most universal yet most versatile signaling agent is the cation Ca^{2+}.

Calcium ions in the cells come from either extracellular sources or intracellular stores, such as the endoplasmic reticulum. If the ions come from the outside, they enter the cells through specific calcium channels. Calcium ions control our heart beats, our movements through the action of skeletal muscles, and, through the release of neurotransmitters in our neurons, learning and memory. They are also involved in signaling the beginning of life at fertilization and its end at death. Calcium ion signaling controls these functions via two mechanisms: (1) increased concentration (forming sparks and puffs) and (2) duration of the signals.

In the resting state of the neuron, the Ca^{2+} concentration is about 0.1 μM. When neurons are stimulated, this level may increase to 0.5 μM. However, to elicit a fusion between the synaptic vesicles and the plasma membrane of the neuron, much higher concentrations are needed (10–25 μM).

An increase in calcium ion concentration may take the form of sparks or puffs. The source of calcium ions may be external (calcium influx caused by the electric signal of nerve transmission) or internal (calcium released from the stores of the endoplasmic reticulum). Upon receiving the signal of calcium puffs, the vesicles storing acetylcholine travel to the membrane of the presynaptic cleft. There they fuse with the membrane and empty their contents into the synapse.

Calcium ions can also control signaling by controlling the duration of the signal. The signal in arterial smooth muscle lasts for 0.1 to 0.5 s. The wave of Ca^{2+} in the liver lasts 10 to 60 s. The calcium wave in the human egg lasts 1 to 35 min after fertilization. Thus, by combining the concentration, localization, and duration of the signal, calcium ions can deliver messages to perform a variety of functions.

The effects of Ca^{2+} are modulated through specific calcium-binding proteins. In all nonmuscle cells and in smooth muscles, calmodulin serves as the calcium-binding protein. Calmodulin-bound calcium activates an enzyme, protein kinase II, which then phosphorylates an appropriate protein substrate. In this way, the signal is translated into metabolic activity.

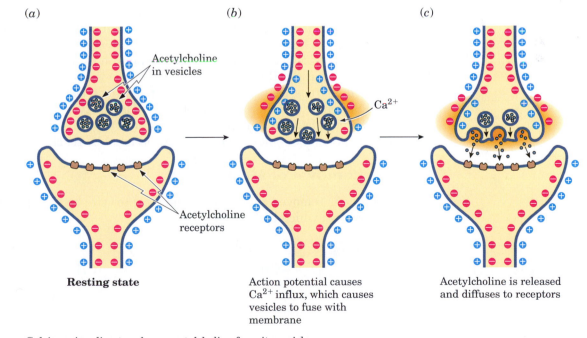

(a)

Acetylcholine in vesicles

Acetylcholine receptors

Resting state

(b)

Ca^{2+}

Action potential causes Ca^{2+} influx, which causes vesicles to fuse with membrane

(c)

Acetylcholine is released and diffuses to receptors

Calcium signaling to release acetylcholine from its vesicles.

Chemical Connections 16B

Botulism and Acetylcholine Release

When meat or fish is improperly cooked or preserved, a deadly type of food poisoning, called botulism, may result. The culprit is the bacterium *Clostridium botulinum,* whose toxin prevents the release of acetylcholine from the presynaptic vesicles. Therefore, no neurotransmitter reaches the receptors on the surface of muscle cells, and the muscles no longer contract. Left untreated, the affected person may die.

Surprisingly, the botulin toxin has a valuable medical use. It is used in the treatment of involuntary muscle spasms—for example, in facial tics. These tics are caused by the uncontrolled release of acetylcholine. Controlled administration of the toxin, when applied locally to the facial muscles, stops the uncontrolled contractions and relieves the facial distortions.

"Facial distortion" has multiple meanings in the cosmetics industry. "Frown lines" and other wrinkles can also be removed by temporarily paralyzing facial muscles. The U.S. Food and Drug Administration has approved Botox (botulin toxin) for cosmetic use, meaning that its manufacturer can now advertise its rejuvenating appeal. Its

Clostridium botulinum is a food-poisoning bacterium.

use is spreading fast. Botox has been used on an off-label basis for this application for a number of years, especially in Hollywood. Indeed, several movie directors have complained that some actors have used so much Botox that they can no longer show a variety of facial expressions.

C. The Action of Messengers

The presence of acetylcholine molecules at the postsynaptic receptor site triggers a conformational change (Section 14.10) in the receptor protein. This opens the *ion channel* and allows ions to cross membranes freely. Na^+ ions have a higher concentration outside the neuron than do K^+ ions; thus more Na^+ enters the cell than K^+ leaves. Because it involves ions, which carry electric charges, this process is translated into an electric signal. After a few milliseconds, the channel closes again. The acetylcholine still occupies the receptor. For the channel to reopen and transmit a new signal, the acetylcholine must be removed and the neuron must be reactivated.

D. The Removal of Messengers

Acetylcholine is removed rapidly from the receptor site by the enzyme *acetylcholinesterase,* which hydrolyzes it.

This rapid removal enables the nerves to transmit more than 100 signals per second. By this means, the message moves from neuron to neuron until it is finally transmitted, again by acetylcholine molecules, to the muscles or endocrine glands that are the ultimate target of the message.

The action of the acetylcholinesterase enzyme is essential to the entire process. When this enzyme is inhibited, the removal of acetylcholine is incomplete, and nerve transmission ceases.

Chemical Connections 16C

Alzheimer's Disease and Chemical Communication

Alzheimer's disease is the name given to the symptoms of severe memory loss and other senile behavior that afflict about 1.5 million people in the United States.

People with Alzheimer's disease are forgetful, especially about recent events. As the disease advances, they become confused and, in severe cases, lose their ability to speak; at that point, they need total care. As yet, there is no cure for this disease. Postmortem identification of this disease focuses on two pathological hallmarks in the brain: (1) buildup of protein deposits known as β-amyloid plaques outside the nerve cells and (2) neurofibrillar tangles composed of tau proteins. Controversy exists as to whether one of the two is the primary cause of the neurodegeneration observed in Alzheimer's disease. Each has its advocates. Those who favor the primacy of the plaque are nicknamed β-aptists, and the other camp sports the name of tauists.

Tau protein binds to microtubules, one of the major cytoskeleton proteins. Genetic mutation of tau protein or environmental factors such as hyperphosphorylation may alter the ability of tau to bind to microtubules. These altered tau proteins form tangles in the cytoplasm of neurons. Neurofibrillar tangles have been found in the brains of patients with Alzheimer's disease in the absence of plaque as well, which suggests that tau abnormality may be sufficient to cause neurodegeneration.

In most brains affected by Alzheimer's disease, the dominant feature is plaques. These plaques consist of protein fibers, some 7–10 nm thick, that are mixed with small peptides called β-amyloid peptides. These peptides originate from a water-soluble precursor, amyloid precursor protein (APP); this transmembrane protein has an unknown function. Certain enzymes, called presenilins, cut peptides containing 38, 40, and 42 amino acids from the transmembrane region of APP. Mutations of APP proteins in patients with Alzheimer's disease cause preferential accumulation of the 42-amino-acid peptide, forming β-pleated sheets, which precipitates creating plaques.

In Alzheimer's disease, nerve cells in the cerebral cortex die, the brain becomes smaller, and part of the cortex atrophies. The depression of the folds on the brain surface becomes deeper.

While many researchers focus on the β-amyloid plaques and the tau proteins, it is not clear whether these are the real culprits that cause neuron death. Another chemical messenger, Ca^{2+}, may also be involved. Current research is leading to the conclusion that calcium flux into neurons is disrupted in Alzheimer's patients. The β-amyloid proteins are believed to form channels in the neuron outer membrane leading to higher-than-normal levels

(a)

(b)

Courtesy of Anatomical Collection, National Museum of Health and Medicine, Armed Forces Institute of Pathology, Washington, D.C.

(a) Normal brain. (b) Brain of a person with Alzheimer's disease.

of intracellular calcium. The presenilins may also play a role in the way calcium ion is released from intracellular stores, primarily from the endoplasmic reticulum (ER). Mutant presenilins from Alzheimer's patients are thought to provide a leak from the ER into the cytosol, as well as possibly affect the protein called SERCA that is supposed to clear calcium ion from the cytosol. While β-amyloid and tau proteins are the most notable and obvious characteristics of brain tissue from the disease, many believe that an overload of calcium ion actually causes the cell death. Patients with Alzheimer's disease also have significantly diminished acetylcholine transferase activity in their brains. This enzyme synthesizes acetylcholine by transferring the acetyl group from acetyl-CoA to choline:

Chemical Connections 16C

Alzheimer's Disease and Chemical Communication (continued)

The diminished concentration of acetylcholine can be partially compensated for by inhibiting the enzyme acetylcholinesterase, which decomposes acetylcholine. Certain drugs, which act as acetylcholinesterase inhibitors, have been shown to improve memory and other cognitive functions in some people with the disease. Drugs such as donepezil (Aricept), rivastigmine (Exelon), and galantamine (Razadyne) belong to this category; they all moderate the symptoms of Alzheimer's disease. The alkaloid huperzine A, an active ingredient of Chinese herb tea that has been used for centuries to improve memory, is also a potent inhibitor of acetylcholinesterase.

E. Control of Neurotransmission

Acetylcholinesterase is inhibited reversibly by succinylcholine (Chemical Connections 15A) and decamethonium bromide.

Decamethonium bromide

Succinylcholine and decamethonium bromide resemble the choline end of acetylcholine and, therefore, act as competitive inhibitors of acetylcholinesterase. In small doses, these reversible inhibitors relax the muscles temporarily, making them useful as muscle relaxants in surgery. In large doses, they are deadly.

The inhibition of acetylcholinesterase is but one way in which cholinergic neurotransmission is controlled. Another way is to modulate the action of the receptor. Because acetylcholine enables the ion channels to open and propagate signals, this mode of action is called *ligand-gated ion channels*. The attachment of the ligand to the receptor is critical in signaling. Nicotine given in low doses is a stimulant; it is an agonist because it prolongs the receptor's biochemical response. When given in large doses, however, nicotine becomes an antagonist and blocks the action on the receptor. As such, it may cause convulsions and respiratory paralysis. Succinylcholine, besides being a reversible inhibitor of acetylcholinesterase, also has this concentration-dependent agonist/antagonist effect on the receptor. A strong antagonist, which blocks the receptor completely, can interrupt the communication between neuron and muscle cell. The venom of a number of snakes, such as cobratoxin, exerts its deadly influence in this manner. The plant extract curare, which was used in poisoned arrows by the Amazon Indians, works in the same way. In small doses, curare is used as a muscle relaxant.

Finally, the supply of the acetylcholine messenger can influence the proper nerve transmission. If the acetylcholine messenger is not released from its storage, as in botulism (Chemical Connections 16B), or if its synthesis is impaired, as in Alzheimer's disease (Chemical Connections 16C), the concentration of acetylcholine is reduced, and nerve transmission is impaired.

16.4 What Amino Acids Act as Neurotransmitters?

A. Messengers

Amino acids are distributed throughout the neurons individually or as parts of peptides and proteins. They can also act as neurotransmitters. Some of them, such as glutamic acid, aspartic acid, and cysteine, act as **excitatory neurotransmitters** similar to acetylcholine and norepinephrine. Others, such as glycine, β-alanine, taurine (Section 13.11), and mainly γ-aminobutyric acid (GABA), are **inhibitory neurotransmitters;** they reduce neurotransmission. Note that some of these neurotransmitter amino acids are not found in proteins.

$$^+H_3NCH_2CH_2SO_3^- \qquad ^+H_3NCH_2CH_2COO^- \qquad ^+H_3NCH_2CH_2CH_2COO^-$$

Taurine $\qquad\qquad$ β-Alanine $\qquad\qquad$ γ-Aminobutyric acid
(GABA)
(IUPAC name:
4-aminobutanoic acid)

B. Receptors

Each of these amino acids has its own receptor; in fact, glutamic acid has at least five subclasses of receptors. The best known is the N-methyl-D-aspartate (NMDA) receptor. This ligand-gated ion channel is similar to the nicotinic cholinergic receptor discussed in Section 16.3:

N-Methyl-D-aspartate

When glutamic acid binds to this receptor, the ion channel opens, Na^+ and Ca^{2+} flow into the neuron, and K^+ flows out of the neuron. The same thing happens when NMDA, being an agonist, stimulates the receptor. The gate of this channel is closed by a Mg^{2+} ion.

Phencyclidine (PCP), an antagonist of this receptor, induces hallucination. PCP, known by the street name "angel dust," is a controlled substance; it causes bizarre psychotic behavior and long-term psychological problems.

C. Removal of Messengers

In contrast to acetylcholine, there is no enzyme that would degrade glutamic acid and thereby remove it from its receptor once the signaling has occurred. Glutamic acid is removed by **transporter** molecules, which bring it back through the presynaptic membrane into the neuron. This process is called **reuptake.**

Transporter A protein molecule that carries small molecules, such as glucose or glutamic acid, across a membrane

16.5 What Are Adrenergic Messengers?

A. Monoamine Messengers

The third class of neurotransmitters/hormones, the adrenergic messengers, includes such monoamines as epinephrine, serotonin, dopamine, and histamine. (Structures of these compounds can be found later in this section and in Chemical Connections 16D.) These monoamines transmit signals by a mechanism whose beginning is similar to the action of acetylcholine. That is, they are adsorbed on a receptor.

B. Signal Transduction

Signal transduction A cascade of events through which the signal of a neurotransmitter or hormone delivered to its receptor is carried inside the target cell and amplified into many signals that can cause protein modifications, enzyme activation, and the opening of membrane channels

Once a hormone or neurotransmitter binds to a receptor, some mechanism must propagate the signal to the cell. The process by which the initial signal is spread and amplified throughout the cell is called signal transduction. The process involves intermediate compounds that pass the signal on to the ultimate targets. Eventually enzymes are modified to alter their activity or membrane channels are opened or closed. The term "signal transduction" and much of the pioneering research on this topic come from the work of Martin Rodbell (1925–1998) of the National Institutes of Health, winner of the 1994 Nobel Prize in physiology and medicine.

The action of monoamine neurotransmitters is a prime example. Once the monoamine neurotransmitter/hormone (for example, norepinephrine) is adsorbed onto the receptor site, the signal will be amplified inside the cell. In the example shown in Figure 16.4, the receptor has an associated protein called G-protein. This protein is the key to the cascade that produces many signals inside the cell (amplification). The active G-protein has an associated nucleotide, guanosine triphosphate (GTP). It is an analog of adenosine triphosphate (ATP), in which the aromatic base adenine is substituted by guanine (Section 17.2). The G-protein becomes inactive when its associated nucleotide is hydrolyzed to guanosine diphosphate (GDP). Signal transduction starts with the active G-protein, which activates the enzyme adenylate cyclase.

G-protein also participates in another signal transduction cascade, which involves inositol-based compounds (Section 13.6) as signaling molecules. Phosphatidylinositol diphosphate (PIP_2) mediates the action of hormones and neurotransmitters. These messengers can stimulate the phosphorylation of enzymes, in a manner similar to the cAMP cascade. They also play an important role in the release of calcium ions from their storage areas in the endoplasmic reticulum (ER) or sarcoplasmic reticulum (SR).

C. Secondary Messengers

Adenylate cyclase produces a secondary messenger inside the cell, cyclic AMP (cAMP). The manufacture of cAMP activates processes that result in the transmission of an electrical signal. The cAMP is manufactured by adenylate cyclase from ATP:

Adenosine triphosphate
(ATP)

Cyclic adenosine monophosphate
(cAMP)

Pyrophosphate
(PPi)

ACTIVE FIGURE 16.4   The sequence of events in the postsynaptic membrane when norepinephrine is absorbed onto the receptor site. (*a*) The active G-protein hydrolyzes GTP. The energy of hydrolysis of GTP to GDP activates adenylate cyclase. A molecule of cAMP forms when adenylate cyclase cleaves ATP into cAMP and pyrophosphate. (*b*) Cyclic AMP activates a protein kinase by dissociating the regulatory (R) unit from the catalytic unit (C). A second molecule of ATP, shown in (*b*), has phosphorylated the catalytic unit and been converted to ADP. (*c*) The catalytic unit phosphorylates the ion-translocating protein that blocked the channel for ion flow. The phosphorylated ion-translocating protein changes its shape and position and opens the ion gates. **Go to this book's companion website at www. cengage.com/chemistry/bettelheim to explore an interactive version of this figure.**

The activation of adenylate cyclase accomplishes two important goals:

1. It converts an event occurring at the outer surface of the target cell (adsorption onto receptor site) to a change inside the target cell (formation of cAMP). Thus the primary messenger (neurotransmitter or hormone) does not have to cross the membrane.

2. It amplifies the signal. One molecule adsorbed on the receptor triggers the adenylate cyclase to make many cAMP molecules. In this way, the signal is amplified many thousands of times.

D. Removal of Signal

How does this signal amplification stop? When the neurotransmitter or hormone dissociates from the receptor, the adenylate cyclase halts the manufacture of cAMP. The cAMP already produced is destroyed by the enzyme phosphodiesterase, which catalyzes the hydrolysis of the phosphoric ester bond, yielding AMP.

The amplification through the secondary messenger (cAMP) is a relatively slow process. It may take from 0.1 s to a few minutes. Therefore, in cases where the transmission of signals must be fast (milliseconds or seconds), a neurotransmitter such as acetylcholine acts on membrane permeability directly, without the mediation of a secondary messenger.

E. Control of Neurotransmission

The G-protein–adenylate cyclase cascade in transduction signaling is not limited to monoamine messengers. A wide variety of peptide hormones and neurotransmitters (Section 16.6) use this signaling pathway. Included among them are glucagon, vasopressin, luteinizing hormone, enkephalins, and P-protein. Neither is the opening of ion channels, depicted in Figure 16.4, the only target of this signaling. A number of enzymes can be phosphorylated by protein kinases, and the phosphorylation controls whether these enzymes will be active or inactive (Section 15.6).

The fine control of the G-protein–adenylate cyclase cascade is essential for health. Consider the toxin of the bacterium *Vibrio cholerae,* which permanently activates G-protein. The result is the symptoms of cholera—namely, severe dehydration as a result of diarrhea. This problem arises because the activated G-proteins overproduce cAMP. This excess, in turn, opens the ion channels, which leads to a large outflow of ions and accompanying water from the epithelial cells to the intestines. Therefore, the first measure taken in treating cholera victims is to replace the lost water and salt.

F. Removal of Neurotransmitters

The inactivation of the adrenergic neurotransmitters differs somewhat from the inactivation of the cholinergic transmitters. While acetylcholine is decomposed by acetylcholinesterase, most of the adrenergic neurotransmitters are inactivated in a different way. *The body inactivates monoamines by oxidizing them to aldehydes.* Enzymes that catalyze these reactions, called monoamine oxidases (MAOs), are very common in the body. For example, one MAO converts both epinephrine and norepinephrine to the corresponding aldehyde:

Epinephrine (salt form)

Norepinephrine (salt form)

Many drugs that are used as antidepressants or antihypertensive agents are MAO inhibitors—for example, Marplan and Nardil. They prevent MAOs from converting monoamines to aldehydes, thereby increasing the concentration of the active adrenergic neurotransmitters.

There is also an alternative way to remove adrenergic neurotransmitters. Shortly after adsorption onto the postsynaptic membrane, the neurotransmitter comes off the receptor site and is reabsorbed through the presynaptic membrane and stored again in the vesicles.

Chemical Connections 16D

Parkinson's Disease: Depletion of Dopamine

Parkinson's disease is characterized by spastic motion of the eyelids as well as rhythmic tremors of the hands and other parts of the body, often when the patient is at rest. The posture of the patient changes to a forward, bent-over position; walking becomes slow, with shuffling footsteps. The cause of this degenerative nerve disease is unknown, but genetic factors and environmental effects, such as exposure to pesticides or high concentrations of metals such as Mn^{2+} ion, have been implicated.

The neurons affected contain, under normal conditions, mostly dopamine as a neurotransmitter. People with Parkinson's disease have depleted amounts of dopamine in their brains, but the dopamine receptors are not affected. Thus the first line of remedy is to *increase the concentration of dopamine.* Dopamine cannot be administered directly, because it cannot penetrate the blood–brain barrier and therefore does not reach the tissue where its action is needed. L-dopa, by contrast, is transported through the arterial wall and converted to dopamine in the brain:

(S)-3,4-Dihydroxyphenylalanine
(L-Dopa)

Dopamine

When L-dopa is administered, many patients with Parkinson's disease are able to synthesize dopamine and resume normal nerve transmission. In these individuals, L-dopa reverses the symptoms of their disease, although the respite is only temporary. In other patients, the L-dopa regimen provides little benefit.

Another way to increase dopamine concentration is to *prevent its metabolic elimination.* The drug entacapon (Comtan) inhibits an enzyme that is instrumental in clearing dopamine from the brain. The enzyme (catechol-O-methyl transferase, COMT) converts dopamine to 3-methoxy-4-hydroxy-L-phenylalanine, which is then eliminated. Entacapon is usually administered together with L-dopa. Another drug, (R)-selegiline (L-Deprenyl), is a monoamine oxidase (MAO) inhibitor. L-Deprenyl, which is also given in combination with L-dopa, can reduce the symptoms of Parkinson's disease and even increase the life span of patients. It increases the level of dopamine by *preventing its oxidation by MAOs.*

Other drugs may treat only the symptoms of Parkinson's disease: the spastic motions and the tremors. These drugs, such as benztropine (Cogentin), are similar to atropine and act on cholinergic receptors, thereby preventing muscle spasms.

The real cure for Parkinson's disease may lie in transplanting human embryonic dopamine neurons. In preliminary work, such grafts have been functionally integrated in patients' brains producing dopamine. In the most successful cases, patients have been able to resume normal, independent life following the transplant.

Certain drugs designed to affect one neurotransmitter may also affect another. An example is the drug methylphenidate (Ritalin). In high doses, this drug enhances the dopamine concentration in the brain and acts as a stimulant. In small doses, it is prescribed to calm hyperactive children or to minimize ADD (attention deficit disorder). It seems that in smaller doses Ritalin raises the concentration of serotonin. This neurotransmitter decreases hyperactivity without affecting the dopamine levels of the brain.

Serotonin

The close connection between two monoamine neurotransmitters, dopamine and serotonin, is also evident in their roles in controlling the nausea and vomiting that often follow general anesthesia and chemotherapy. Blockers of dopamine receptors in the brain, such as promethazine (Phenergan), can alleviate the symptoms after anesthesia. A blocker of serotonin receptors in the brain as well as on the terminals of the vagus nerve in the stomach, such as ondansetrone (Zofran), is the drug of choice for preventing chemotherapy-induced vomiting.

Synthesis and degradation of dopamine are not the only way that the brain keeps its concentration at a steady state. The concentration is also controlled by specific proteins, called *transporters,* that ferry the used dopamine from the receptor back across the synapse into the original neuron for reuptake. Cocaine addiction works through such a transporter. Cocaine binds to the dopamine transporter, like a reversible inhibitor, thereby preventing the reuptake of dopamine. As a consequence, dopamine is not transported back to the original neuron and stays in the synapse, increasing the continuous firing of signals, which is the psychostimulatory effect associated with a cocaine "high."

Chemical Connections 16E

Nitric Oxide as a Secondary Messenger

The toxic effects of the simple gaseous molecule NO have long been recognized. Therefore, it came as a surprise to find that this compound plays a major role in chemical communications. This simple molecule is synthesized in the cells when arginine is converted to citrulline. (These two compounds appear in the urea cycle; see Section 20.8.) Nitric oxide is a relatively nonpolar molecule, and shortly after it has been produced in the nerve cell, it quickly diffuses across the lipid bilayer membrane. During its short half-life (4–6 s), it can reach a neighboring cell. Because NO passes through membranes, it does not need extracellular receptors to deliver its message. NO is very unstable, so there is no need for a special mechanism to carry out its destruction.

NO acts as an intercellular messenger between the endothelial cells surrounding the blood vessels and the smooth muscles encompassing these cells. It relaxes the muscle cells, thereby dilating the blood vessels. The outcome is less-restricted blood flow and a drop in blood pressure. This reason also explains why nitroglycerin (Chemical Connections 4D) is effective against angina: It produces NO in the body.

Another role of NO in dilating blood vessels lies in remedying impotence. The impotence-relieving drug Viagra enhances the activity of NO by inhibiting an enzyme (phosphodiesterase) that otherwise would reduce NO's effect on smooth muscles. When the NO concentration is sufficiently high, the blood vessels dilate, allowing enough blood to flow to provide an erection. In most cases, this happens within an hour after taking the pill.

Sometimes the dilation of blood vessels is not so beneficial. Headaches are caused by dilated arteries in the head. NO-producing compounds in food—nitrites in smoked and cured meats and sodium glutamate in seasoning—can cause such headaches. Nitroglycerin itself often induces headaches.

Nitric oxide is toxic. This toxicity is used by our immune system (Section 23.2B) to fight infections caused by viruses.

Monosodium Glutamate (MSG) is a flavor enhancer. However, as a NO-producing molecule, it can also cause headaches. Because many Chinese restaurants used to use MSG, people who got headaches after eating Chinese food were often said to have the "Chinese Restaurant Syndrome."

The toxic effect of NO is also evident in strokes. In a stroke, a blocked artery restricts the blood flow to certain parts of the brain; the oxygen-starved neurons then die. Next, neurons in the surrounding area, 10 times larger than the place of the initial attack, release glutamic acid, which stimulates other cells. They, in turn, release NO, which kills all cells in the area. Thus the damage to the brain is spread tenfold. A concentrated effort is under way to find inhibitors of the NO-producing enzyme, nitric oxide synthase, that can be used as antistroke drugs. For the discovery of NO and its role in blood pressure control, three pharmacologists—Robert Furchgott, Louis Ignarro, and Ferid Murad—received the 1998 Nobel Prize in physiology.

G. Histamines

The neurotransmitter histamine is present in mammalian brains. It is synthesized from the amino acid, histidine, by decarboxylation:

The action of histamine as a neurotransmitter is very similar to that of other monoamines. There are two kinds of receptors for histamine. One receptor, H_1, can be blocked by antihistamines such as dimenhydrinate (Dramamine) and diphenhydramine (Benadryl). The other receptor, H_2, can be blocked by ranitidine (Zantac) and cimetidine (Tagamet).

H_1 receptors are found in the respiratory tract. They affect the vascular, muscular, and secretory changes associated with hay fever and asthma. Therefore, antihistamines that block H_1 receptors relieve these symptoms. The H_2 receptors are found mainly in the stomach and affect the secretion of HCl. Cimetidine and ranitidine, both H_2 blockers, reduce acid secretion and, therefore, are effective drugs for ulcer patients. The main culprit in the formation of most ulcers, however, is the bacteria *Helicobacter pylori*. Sir James W. Black of the United Kingdom received the 1988 Nobel Prize in medicine for the invention of cimetidine and such other drugs as propranolol that kill the ulcer-causing bacteria (Table 16.1).

Antihistamines block the H_1 receptor for histamine.

Example 16.1 Identifying Enzymes in the Adrenergic Pathway

Three enzymes in the adrenergic neurotransmission pathway affect the transduction of the signals. Identify them and describe how they affect the neurotransmission.

Solution

Adenylate cyclase amplifies the signal by producing cAMP secondary messengers. Phosphatase terminates the signal by hydrolyzing cAMP. Monoamine oxidase (MAO) reduces the frequency of signals by oxidizing the monoamine neurotransmitters to the corresponding aldehydes.

Problem 16.1

What is the functional difference between G-protein and GTP?

16.6 What Is the Role of Peptides in Chemical Communication?

A. Messengers

Many of the most important hormones affecting metabolism belong to the peptidergic messengers group. Among them are insulin (Section 14.8 and Chemical Connections 16F) and glucagon, hormones of the pancreatic islets, and vasopressin and oxytocin (Section 14.8) products of the posterior pituitary gland.

In the last few years, scientists have isolated a number of brain peptides that have affinity for certain receptors and, therefore, act as if they were neurotransmitters. Some 25 or 30 such peptides are now known.

The first brain peptides isolated were the **enkephalins.** These pentapeptides are present in certain nerve cell terminals. They bind to specific pain receptors and seem to control pain perception. Because they bind to the receptor site that also binds the pain-killing alkaloid morphine, it is assumed that the N-terminal end of the pentapeptide fits the receptor (Figure 16.5).

Even though morphine remains the most effective agent in reducing pain, its clinical use is limited because of its side effects. These include respiratory depression, constipation, and, most significantly, addiction. The clinical use of enkephalins has yielded only modest relief. The challenge is to develop analgesic drugs that do not involve the opiate receptors in the brain.

Another brain peptide, **neuropeptide Y,** affects the hypothalamus, a region that integrates the body's hormonal and nervous systems. Neuropeptide

Morphine

Methionine enkephalin

FIGURE 16.5 Similarities between the structure of morphine and that of the brain's own pain regulators, the enkephalins.

Y is a potent orexic (appetite-stimulating) agent. When its receptors are blocked (for example, by leptin, the "thin" protein), appetite is suppressed. Leptin is an anorexic agent.

Yet another peptidergic neurotransmitter is **substance P** (*P* for "pain"). This 11-amino-acid peptide is involved in transmission of pain signals. In injury or inflammation, sensory nerve fibers transmit signals from the peripheral nervous system (where the injury occurred) to the spinal cord, which processes the pain. The peripheral neurons synthesize and release substance P, which bonds to receptors on the surface of the spinal cord. Substance P, in its turn, removes the magnesium block at the *N*-methyl-D-aspartate (NMDA) receptor. Glutamic acid, an excitatory amino acid, can then bind to this receptor. In doing so, it amplifies the pain signal going to the brain.

B. Secondary Messengers and Control of Metabolism

All peptidergic messengers, hormones, and neurotransmitters act through secondary messengers. Glucagon, luteinizing hormone, antidiuretic hormone, angiotensin, enkaphalin, and substance P use the G-protein–adenylate cyclase cascade that we saw in the previous section.

Glucagon is a peptide hormone that is critical for maintaining blood glucose levels. When the pancreas senses that blood glucose is dropping, it releases glucagon. When glucagon is released, it binds to receptors on liver cells and acts through a series of reactions to raise blood glucose. The method of action is far from simple, however. When glucagon binds to its receptor and activates the G-protein cascade, the second messenger, cAMP, activates protein kinase, an enzyme that phosphorylates many target enzymes. As shown in Figure 16.6, protein kinase phosphorylates two key enzymes in carbohydrate metabolism, fructose bisphosphatase 2 (FBP-2) and phosphfructokinase 2 (PFK-2). Phosphorylating these two enzymes has opposite effects. The kinase is inactivated and the phosphatase is activated. This lowers the intracellular concentration of fructose 2,6 bisphosphate, a key metabolic regulator. The reduced level of the regulator increases the level of the pathway called **gluconeogenesis** (Chapter 21) and reduces the level of the pathway called **glycolysis** (Chapter 20). Gluconeogenesis produces glucose and glycolysis uses it. Thus by turning on gluconeogenesis and turning off glycolysis, the liver produces more glucose for the blood.

Insulin is another peptide hormone produced by the pancreas, but its overall effect is roughly the opposite of glucagon's. Insulin binds to insulin receptors on liver and muscle cells, as shown in Figure 16.7. The receptor is an example of a protein called a tyrosine kinase. A specific tyrosine residue becomes phosphorylated on the receptor, initiating its kinase *activity*. A target protein called IRS (insulin receptor substrate) is then phosphorylated by the

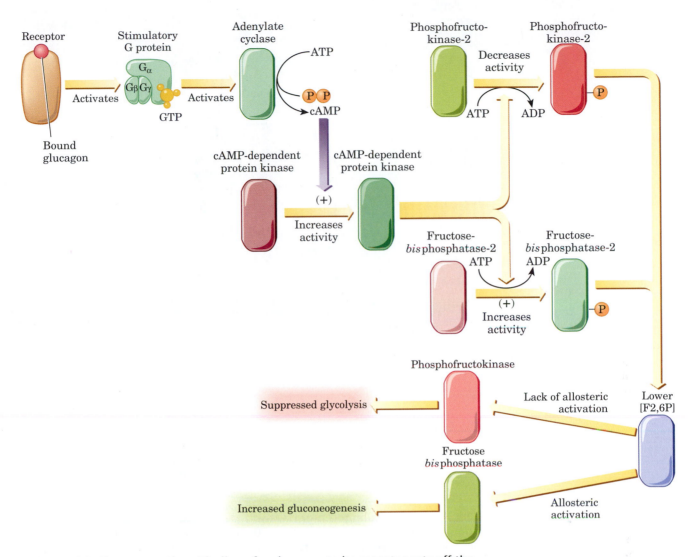

FIGURE 16.6 Glucagon action. Binding of a glucagon to its receptor sets off the chain of events that leads to the activation of a cAMP-dependent protein kinase. The enzymes phosphorylated in this case are phosphofructokinase-2, which is inactivated, and fructose-*bis* phosphate-2, which is activated. The combined result of phosphorylating these two enzymes is to lower the concentration of fructose-2, 6-*bis* phosphate (F2,6P). A lower concentration of F2,6P leads to lack of allosteric activation of phosphofructokinase 1 and lowered glycolysis while also leading to allosteric activation of fructose bisphosphatase 1 and increased gluconeogenesis.

active tyrosine kinase. The phosphorylated IRS acts as the second messenger. It causes the phosphorylation of many target enzymes in the cell. The effect is to reduce the level of glucose in the blood by increasing the rate of pathways that use glucose and slowing the rate of pathways that make glucose.

16.7 How Do Steroid Hormones Act as Messengers?

In Section 13.10, we saw that a large number of hormones possess steroid ring structures. These hormones, which include the sex hormones, are hydrophobic; therefore, they can cross plasma membranes of the cell by diffusion. There is no need for special receptors embedded in the membrane for these hormones. It has been shown, however, that **steroid hormones**

From Lehninger, *Principles of Biochemistry*, Second Edition, by David L. Nelson and Michael M. Cox. © 1982, 1992 by Worth Publishers. Used with permission of W. H. Freeman and Company.

FIGURE 16.7 The insulin receptor has two types of subunits, α and β. The α-subunit is on the extracellular side of the membrane, and it binds to insulin. The β-subunit spans the membrane. When insulin binds to the α-subunit, the β-subunits autophosphorylate on tyrosine residues. These then phosphorylate target proteins called insulin receptor substrates (IRS). The IRSs act as the second messengers in the cells.

interact inside the cell with protein receptors. Most of these receptors are localized in the nucleus of the cell, but small amounts also exist in the cytoplasm. When they interact with steroids, they facilitate their migration through the aqueous cytoplasm; the proteins themselves are hydrophilic.

Progesterone

Once inside the nucleus, the steroid–receptor complex can either bind directly to the DNA or combine with a **transcription factor,** a protein that binds to DNA and alters the expression of a gene (Section 18.2), influencing the synthesis of a certain key protein. Thyroid hormones, which also have large hydrophobic domains, have protein receptors as well, which facilitates their transport through aqueous media.

Chemical Connections 16F

Diabetes

The disease diabetes mellitus affects over 20 million people in the United States. In a normal person, the pancreas, a large gland behind the stomach, secretes insulin and several other hormones. Diabetes usually results from low insulin secretion. Insulin is necessary for glucose molecules to penetrate such cells as brain, muscle, and fat cells, where they can be used. It accomplishes this task by being adsorbed onto the receptors in the target cells. This adsorption triggers the manufacture of cyclic GMP (not cAMP); this secondary messenger, in turn, increases the transport of glucose molecules into the target cells.

In the resulting cascade of events, the first step is the self-(auto)phosphorylation of the receptor molecule itself, on the cytoplasmic side. The phosphorylated insulin receptor activates enzymes and regulatory proteins by phosphorylating them. As a consequence, glucose transporter molecules (GLUT4) that are stored inside the cells migrate to the plasma membrane. Once there, they facilitate the movement of glucose across the membrane. This transport relieves the accumulation of glucose in blood serum and makes it available for metabolic activity inside the cells. The glucose can then be used as an energy source, stored as glycogen, or even diverted to enter fat and other molecular biosynthetic pathways.

In diabetic patients, the glucose level rises to 600 mg/100 mL of blood or higher (normal is 80 to 100 mg/100 mL). Two kinds of diabetes exist. In insulin-dependent diabetes, patients do not manufacture enough of this hormone in the pancreas. So-called Type I disease develops early, before the age of 20, and must be treated with daily injections of insulin. Even with daily injections of insulin, the blood sugar level fluctuates, which may cause other disorders, such as cataracts, retinal dystrophy leading to blindness, kidney disease, heart attack, and nervous disorders.

One way to counteract these fluctuations is to monitor the blood sugar and, as the glucose level rises, to administer insulin. Such monitoring requires pricking the finger six times per day, an invasive regimen that few diabetics follow faithfully. Recently, noninvasive monitoring techniques have been developed. One of the most promising employs contact lenses. The blood sugar's fall and rise is mimicked by the glucose content of tears. A fluorescent sensor in the contact lens monitors these glucose fluctuations; its data can be read via a hand-held fluorimeter. Thus the patterns of glucose fluctuation can be obtained noninvasively and, if the level rises to the danger zone, it can be counteracted by insulin intake.

The delivery of insulin also has undergone a revolution. The tried-and-true methods of injections and insulin pumps are still widely used, but new delivery is available orally or by nasal sniffing.

In Type II (non–insulin-dependent) diabetes, patients have enough insulin in the blood but cannot utilize it properly because the target cells have an insufficient number of receptors. Such patients typically develop the disease after age 40 and are likely to be obese. Overweight people usually have a lower-than-normal number of insulin receptors in their adipose (fat) cells.

Oral drugs can help patients with Type II diabetes in several ways. For example, sulfonyl urea compounds, such as tolbutamide (Orinase), increase insulin secretion. In addition, insulin concentration in the blood can be increased by enhancing its release from the β-cells of pancreatic islets. The drug repaglinide (Prandin) blocks the K^+-ATP channels of the β-cells, facilitating Ca^{2+} influx, which induces the release of insulin from the cells.

The oral drugs seem to control the symptoms of diabetes, but fluctuations in insulin levels may turn high blood sugar (hyperglycemia), into low blood sugar (hypoglycemia), which is just as dangerous. Other drugs for Type II diabetes that do not elicit hypoglycemia attempt to control the glucose level at its source. Miglitol (Glyset), an anti-α-glucosidase drug, inhibits the enzyme that converts glycogen or dietary starch into glucose. The drug metformin (Glucophage) decreases glucose production in the liver, carbohydrate absorption in the intestines, and glucose uptake by fat cells.

Tolbutamide
(Orinase)

Metformin
(Glucophage is the
hydrochloride of metformin)

Chemical Connections 16G

Hormones and Biological Pollutants

Hormones are some of the most powerful chemicals when we consider their effects on development and metabolism and their low concentrations: They act at the level of parts per billion. In the last 20 years, people have begun to worry about many types of biological pollutants that affect development. For example, many people prefer to eat organically raised chickens to avoid pesticides ingested by factory-farmed the animals. They also prefer to eat hormone-free meat and poultry, fearing the effects of the hormones on their children. It is known that the age of onset of puberty has been dropping in the last 30 years, and many believe this drop is due to the effects of biological pollutants that mimic human hormones.

One such chemical is called bisphenol A (BPA).

BPA is an estrogen-like compound used to make polycarbonate bottles, such as baby bottles, as well as the linings of food cans and Tupperware. Small amounts of BPA can leach out into food, and many people have detectable levels of the compound in their blood, although the levels found are below the maximum safe dose set by the Environmental Protection Agency. In 1997, a reproductive biologist found that very low levels of BPA fed to pregnant mice caused enlargement of the prostate of their male offspring. Other studies found increases in chromosomal abnormalities in the eggs of mice after BPA leached out of plastic mouse cages. Epidemiological studies have linked human health problems to BPA, such as breast cancer and early onset of puberty. The National Toxicology Program formed a twelve-member panel to study the effects of BPA, but so far the results have been inconclusive. While they concluded that the dangers of BPA were negligible, they did have some concern about possible risk to fetuses and children. Reproductive biologist Dr. John Vandenbergh of North Carolina State University, a member of the NTP panel, was quoted as saying, "I think there is a human risk. What we are trying to do here is define what the risk is."

Depending on who you listen to, BPA is either perfectly harmless or a serious health threat. Even research scientists do not agree on the answer. An Internet search for bisphenol A will lead to many websites dedicated to discussing the possible dangers. One thing is clear, however. Hormones are critical to human physiology and even tiny perturbations in their levels may affect our health.

Polycarbonate used for plastic bowls and baby bottles contains bisphenol A (BPA), a substance of concern to the Environmental Protection Agency for its link to human health problems.

The steroid hormonal response through protein synthesis is not fast. In fact, it takes hours to occur. Steroids can also act at the cell membrane, influencing ligand-gated ion channels. Such a response would take only seconds. An example of such a fast response occurs in fertilization. The sperm head contains proteolytic enzymes, which act on the egg to facilitate its penetration. These enzymes are stored in acrosomes, organelles found on the sperm head. During fertilization, progesterone originating from the follicle cells surrounding the egg acts on the acrosome outer membrane, which disintegrates within seconds and releases the proteolytic enzymes.

The same steroid hormones depicted in Figure 13.6 act as neurotransmitters, too. These neurosteroids are synthesized in the brain cells in both neurons and glia, and they affect receptors—mainly the NMDA and GABA receptors (Section 16.4). Progesterone and progesterone metabolites in brain cells can induce sleep, have analgesic and anticonvulsive effects, and can even serve as natural anesthetics.

Summary

OWL End-of-chapter problems identified in blue are assignable in **GOB OWL.**

Section 16.1 What Molecules Are Involved in Chemical Communications?

- Cell-to-cell communications are carried out by three kinds of molecules.
- **Receptors** are protein molecules embedded in the membranes of cells.
- **Chemical messengers,** or ligands, interact with receptors.
- **Secondary messengers** carry and amplify the signals from the receptor to inside the cell.

Section 16.2 How Are Chemical Messengers Classified as Neurotransmitters and Hormones?

- **Neurotransmitters** send chemical messengers across a short distance—the **synapse** between two neurons or between a neuron and a muscle or endocrine gland cell. This communication occurs in milliseconds.
- **Hormones** transmit their signals more slowly and over a longer distance, from the source of their secretion (endocrine gland), through the bloodstream, into target cells.
- **Antagonists** block receptors; **agonists** stimulate receptors.
- Five kinds of chemical messengers exist: **cholinergic, amino acid, adrenergic, peptidergic,** and **steroid.** Neurotransmitters may belong to all five classes, hormones to the last three classes. Acetylcholine is cholinergic, glutamic acid is an amino acid, epinephrine (adrenaline) and norepinephrine are adrenergic, enkephalins are peptidergic, and progesterone is a steroid.

Section 16.3 How Does Acetylcholine Act as a Messenger? Problem 16.10

- Nerve transmission starts with the neurotransmitters, such as acetylcholine packaged in **vesicles** in the **presynaptic end** of neurons.
- When neurotransmitters are released, they cross the membrane and the synapse and are adsorbed onto receptor sites on the **postsynaptic** membranes. This adsorption triggers an electrical response.
- Some neurotransmitters act directly, whereas others act through a secondary messenger, **cyclic AMP.**

- After the electrical signal is triggered, the neurotransmitter molecules must be removed from the postsynaptic end. In the case of acetylcholine, this removal is done by an enzyme called acetylcholinesterase.

Section 16.4 What Amino Acids Act as Neurotransmitters?

- Amino acids, many of which differ from the amino acids found in proteins, bind to their receptors, which are ligand-gated ion channels.
- Removal of amino acid messengers takes place by **reuptake** through the presynaptic membrane, rather than by hydrolysis.

Section 16.5 What Are Adrenergic Messengers?

- The mode of action of monoamines such as epinephrine, serotonin, dopamine, and histamine is similar to that of acetylcholine, in the sense they start with binding to a receptor.
- Cyclic AMP is an important secondary messenger.
- The mode of removal of monoamines differs from the hydrolysis of acetylcholine. In the case of monoamines, enzymes (MAOs) oxidize them to aldehydes.

Section 16.6 What Is the Role of Peptides in Chemical Communication? Problem 16.40

- Peptides and proteins bind to receptors on the target cell membrane and use secondary messengers to exert their influence.
- **Signal transduction** is the process that occurs after a ligand binds to its receptor. In this process, the signal is carried inside the cell and is amplified.

Section 16.7 How Do Steroid Hormones Act as Messengers? Problems 16.42, 16.64, 16.71

- Steroids penetrate the cell membrane, and their receptors are found in the cytoplasm. Together with their receptors, they penetrate the cell nucleus.
- Steroid hormones can act in three ways: (1) They activate enzymes, (2) they affect the gene transcription of an enzyme or protein, and (3) they change membrane permeability.
- The same steroids can also act as neurotransmitters, when synthesized in neurons.

Problems

■ Indicates problems that are assignable in **GOB OWL**.

Blue numbered problems are applied.

Go to this book's companion website at **www.cengage. com/chemistry/bettelheim** for interactive versions of the How To tutorials and Active Figures, and to quiz yourself on this chapter.

Section 16.1 What Molecules Are Involved in Chemical Communications?

16.2 What kind of signal travels along the axon of a neuron?

16.3 What is the difference between a *chemical messenger* and a *secondary messenger?*

Section 16.2 How Are Chemical Messengers Classified as Neurotransmitters and Hormones?

16.4 Define the following:
(a) Synapse
(b) Receptor
(c) Presynaptic
(d) Postsynaptic
(e) Vesicle

16.5 What is the role of Ca^{2+} in releasing neurotransmitters into the synapse?

16.6 Which signal takes longer: (a) a neurotransmitter or (b) a hormone? Explain.

16.7 Which gland controls lactation?

16.8 To which of the three groups of chemical messengers do these hormones belong?
(a) Norepinephrine
(b) Thyroxine
(c) Oxytocin
(d) Progesterone

Section 16.3 How Does Acetylcholine Act as a Messenger?

16.9 How does acetylcholine transmit an electric signal from neuron to neuron?

16.10 ■ Which end of the acetylcholine molecule fits into the receptor site?

16.11 Cobra venom and botulin are both deadly toxins, but they affect cholinergic neurotransmission differently. How does each cause paralysis?

16.12 Different ion concentrations across a membrane generate a potential (voltage). We call such a membrane *polarized.* What happens when acetylcholine is adsorbed on its receptor?

Section 16.4 What Amino Acids Act as Neurotransmitters?

16.13 List two features by which taurine differs from the amino acids found in proteins.

16.14 How is glutamic acid removed from its receptor?

16.15 What is unique in the structure of GABA that distinguishes it from the amino acids that are present in proteins?

16.16 What is the structural difference between NMDA, an agonist of a glutamic acid receptor, and L-aspartic acid?

Section 16.5 What Are Adrenergic Messengers?

16.17 (a) Identify two monoamine neurotransmitters in Table 16.1.
(b) Explain how they act.
(c) Which medication controls the diseases caused by the lack of monoamine neurotransmitters?

16.18 What bond is hydrolyzed and what bond is formed in the synthesis of cAMP?

16.19 How is the catalytic unit of protein kinase activated in adrenergic neurotransmission?

16.20 The formation of cyclic AMP is described in Section 16.5. Show by analogy how cyclic GMP is formed from GTP.

16.21 By analogy to the action of MAO on epinephrine, write the structural formula of the product of the corresponding oxidation of dopamine.

16.22 The action of protein kinase is the next-to-last step in the G-protein–adenylate cyclase cascade signal transduction. What kind of effects can elicit the phosphorylation by this enzyme?

16.23 Explain how adrenergic neurotransmission is affected by (a) amphetamines and (b) reserpine. (See Table 16.1.)

16.24 Which step in the events depicted in Figure 16.3 provides an electrical signal?

16.25 What kind of product is the MAO-catalyzed oxidation of epinephrine?

16.26 How is histamine removed from the receptor site?

16.27 Cyclic AMP affects the permeability of membranes for ion flow.
(a) What blocks the ion channel?
(b) How is this blockage removed?
(c) What is the direct role of cAMP in this process?

16.28 Dramamine and cimetidine are both antihistamines. Would you expect Dramamine to cure ulcers and cimetidine to relieve the symptoms of asthma? Explain.

Section 16.6 What Is the Role of Peptides in Chemical Communication?

16.29 What is the chemical nature of enkephalins?

16.30 What is the mode of action of Demerol as a painkiller? (See Table 16.1.)

16.31 What enzyme catalyzes the formation of inositol-1,4,5-triphosphate from inositol-1,4-diphosphate? Give the structures of the reactant and the product.

16.32 How is the secondary messenger inositol-1,4,5-triphosphate inactivated?

16.33 What second messenger is formed in response to glucagon binding to its receptor?

16.34 Which organ produces glucagon and why?

16.35 What is the direct target of the second messenger produced when glucagon binds to its receptor?

16.36 In the course of the glucagon effect, what does protein kinase A do?

16.37 Why does glucagon lead to the activation of gluconeogenesis and the inhibition of glycolysis?

16.38 How is fructose 2,6 bisphosphate involved in glucose metabolism?

16.39 Describe the signaling pathway involving insulin.

16.40 ■ Does insulin use a G-protein signaling pathway? What is the nature of the insulin receptor?

Section 16.7 How Do Steroid Hormones Act as Messengers?

16.41 Where are the receptors for steroid hormones located—on the cell surface or elsewhere?

16.42 ■ Do steroid hormones affect protein synthesis? If so, does this effect have any implications for the time the hormonal response can take?

16.43 Can steroid hormones act as neurotransmitters?

Chemical Connections

16.44 (Chemical Connections 16A) What is the difference between calcium sparks (puffs) and calcium waves?

16.45 (Chemical Connections 16A) What is the role of calmodulin in signaling by Ca^{2+} ions?

16.46 (Chemical Connections 16A) To enable a fusion between the synaptic vesicle and the plasma membrane, the calcium concentration is increased. How many-fold of an increase in Ca^{2+} concentration is needed?

16.47 (Chemical Connections 16B) What is the mode of action of botulinum toxin?

16.48 (Chemical Connections 16B) How can a deadly botulinum toxin contribute to the facial beauty of Hollywood actors and actresses?

16.49 (Chemical Connections 16C) What are the neurofibrillar tangles in the brains of patients with Alzheimer's disease made of? How do they affect the cell structure?

16.50 (Chemical Connections 16C) What are the plaques in the brains of patients with Alzheimer's disease made of?

16.51 (Chemical Connections 16C) Alzheimer's disease causes loss of memory. What kind of drugs may provide some relief for—if not cure—this disease? How do they act?

16.52 (Chemical Connections 16C) How are the β-amyloid proteins and the presenilins involved with calcium flux in brain cells?

16.53 (Chemical Connections 16D) Why would a dopamine pill be ineffective in treating Parkinson's disease?

16.54 (Chemical Connections 16D) What is the mechanism by which cocaine stimulates the continuous firing of signals between neurons?

16.55 (Chemical Connections 16D) Parkinson's disease is due to a paucity of dopamine neurons, yet its symptoms are relieved by drugs that block cholinergic receptors. Explain.

16.56 (Chemical Connections 16D) In certain cases, embryonic dopamine neurons transplanted into the brains of patients with advanced Parkinson's disease resulted in complete remission. How was this result possible?

16.57 (Chemical Connections 16E) How can NO cause headaches?

16.58 (Chemical Connections 16E) How is NO synthesized in the cells?

16.59 (Chemical Connections 16E) How is the toxicity of NO detrimental in strokes?

16.60 (Chemical Connections 16F) Tolbutamide is called a sulfonyl urea compound. Identify the sulfonyl urea moiety in the structure of the drug.

16.61 (Chemical Connections 16F) What is the difference between insulin-dependent and non–insulin-dependent diabetes?

16.62 (Chemical Connections 16F) How does insulin facilitate the absorption of glucose from blood serum into adipocytes (fat cells)?

16.63 (Chemical Connections 16F) Diabetic patients must frequently monitor the fluctuation of glucose levels in their blood. What is the advantage of the latest technique of monitoring the glucose content in tears over the older technique of obtaining frequent blood samples?

16.64 ■ (Chemical Connections 16G) Why type of compound is bisphenol A and where does it come from?

16.65 (Chemical Connections 16G) What are the possible biological effects of ingesting bisphenol A?

16.66 What experimental evidence led scientists to be concerned about bisphenol A?

Additional Problems

16.67 Considering its chemical nature, how does aldosterone (Section 13.10) affect mineral metabolism (Table 16.2)?

16.68 What is the function of the ion-translocating protein in adrenergic neurotransmission?

16.69 Decamethonium acts as a muscle relaxant. If an overdose of decamethonium occurs, can paralysis be prevented by administering large doses of acetylcholine? Explain.

16.70 Endorphin, a potent painkiller, is a peptide containing 22 amino acids; among them are the same five N-terminal amino acids found in the enkephalins. Does this explain endorphin's pain-killing action?

16.71 ■ How do alanine and beta alanine differ in structure?

16.72 Where is a G-protein located in adrenergic neurotransmission?

16.73 (Chemical Connections 16E) List a number of effects that are caused when NO, acting as a secondary messenger, relaxes smooth muscles.

16.74 (a) In terms of their action, what do the hormone vasopressin and the neurotransmitter dopamine have in common?

(b) What is the difference in their modes of action?

16.75 What is the difference in the models of action between acetylcholinesterase and acetylcholine transferase?

16.76 How does cholera toxin exert its effect?

16.77 Give the formulas for the following reaction:

$$GTP + H_2O \rightleftharpoons GDP + P_i$$

16.78 Insulin is a hormone that, when it binds to a receptor, enables glucose molecules to enter the cell and be metabolized. If you have a drug that is an agonist, how would the glucose level in the serum change upon administering the drug?

16.79 (Chemical Connections 16D) Ritalin is used to alleviate hyperactivity in attention deficit disorder of children. How does this drug work?

16.80 The pituitary gland releases luteinizing hormone (LH), which enhances the production of progesterone in the uterus. Classify these two messengers and discuss how each delivers its message.

Tying It Together

16.81 Why are receptors proteins rather than any other kind of molecule?

16.82 Why is it useful for organisms to have several different classes of neurotransmitters and hormones?

16.83 What relationship do adrenergic messengers have to amino acid messengers, and what does this relationship say about the biochemical origin of adrenergic messengers?

16.84 What functional groups are found in the structures of chemical messengers? What do these structural features imply about the active sites of the enzymes that process these messengers?

Looking Ahead

16.85 Why is insulin not administered orally in treatment of insulin-dependent diabetes?

16.86 One of the challenges in treating cholera is that of preventing dehydration. What can make this a double challenge? (*Hint:* See Chapter 22; cholera is frequently a water-borne disease.)

16.87 Do any chemical messengers have a *direct* effect on the synthesis of nucleic acids?

16.88 Would you expect the role of chemical messengers to have any bearing on the body's requirements for energy?

Challenge Problems

16.89 Do all chemical messengers require the same time to elicit a response? If there are differences, how do the underlying response mechanisms differ?

16.90 A number of agricultural pesticides are acetylcholinesterase inhibitors. Why is their use carefully controlled?

16.91 What benefit is it to an organism to have two different enzymes for the synthesis and breakdown of acetylcholine—acetylcholine transferase and acetylcholinesterase, respectively?

16.92 Which would be a better form of therapy for cocaine addiction—an inhibitor for the dopamine transporter or a substance that degrades cocaine?

Nucleotides, Nucleic Acids, and Heredity

AP/Wide World Photos.

While these dogs might appear to be a normal mother and puppy, the latter is really the first cloned dog, Snuppy. The larger dog is a male Afghan whose DNA was used to create the clone.

17.1 What Are the Molecules of Heredity?

Each cell of our bodies contains thousands of different protein molecules. Recall from Chapter 14 that all of these molecules are made up of the same 20 amino acids, just arranged in different sequences. That is, the hormone insulin has a different amino acid sequence from the globin of the red blood cell. Even the same protein—for example, insulin—has a different sequence in different species (Section 14.8). Within the same species, individuals may have some differences in their proteins, although the differences are much less dramatic than those seen between species. This variation is most obvious in cases where people have such conditions as hemophilia, albinism, or color-blindness because they lack certain proteins that normal people have

OWL Online homework for this chapter may be assigned in **GOB OWL**.

Human chromosomes magnified about 8000 times.

Gene The unit of heredity; a DNA segment that codes for one protein or one type of RNA

or because the sequence of their amino acids is somewhat different (see Chemical Connections 14D).

After scientists became aware of the differences in amino acid sequences, their next quest was to determine how cells know which proteins to synthesize out of the extremely large number of possible amino acid sequences. The answer is that an individual gets the information from its parents through *heredity*. Heredity is the transfer of characteristics, anatomical as well as biochemical, from generation to generation. We all know that a pig gives birth to a pig and a mouse gives birth to a mouse.

It was easy to determine that the information is obtained from the parent or parents, but what form does this information take? During the last 60 years, revolutionary developments have enabled us to answer this question—the transmission of heredity occurs on the molecular level.

From about the end of the nineteenth century, biologists suspected that the transmission of hereditary information from one generation to another took place in the nucleus of the cell. More precisely, they believed that structures within the nucleus, called **chromosomes,** have something to do with heredity. Different species have different numbers of chromosomes in the nucleus. The information that determines external characteristics (red hair, blue eyes) and internal characteristics (blood group, hereditary diseases) was thought to reside in **genes** located inside the chromosomes.

Chemical analysis of nuclei showed that they are largely made up of special basic proteins called *histones* and a type of compound called *nucleic acids*. By 1940, it became clear through the work of Oswald Avery (1877–1955) that, of all the material in the nucleus, only a nucleic acid called deoxyribonucleic acid (DNA) carries the hereditary information. That is, the genes are located in the DNA. Other work in the 1940s by George Beadle (1903–1989) and Edward Tatum (1909–1975) demonstrated that each gene controls the manufacture of one protein, and that external and internal characteristics are expressed through this gene. Thus the expression of the gene (DNA) in terms of an enzyme (protein) led to the study of protein synthesis and its control. *The information that tells the cell which proteins to manufacture is carried in the molecules of DNA.* We now know that not all genes lead to the production of protein, but all genes do lead to the production of another type of nucleic acid, called ribonucleic acid (RNA).

17.2 What Are Nucleic Acids Made Of?

Two kinds of nucleic acids are found in cells: **ribonucleic acid (RNA)** and **deoxyribonucleic acid (DNA).** Each has its own role in the transmission of hereditary information. As we just saw, DNA is present in the chromosomes of the nuclei of eukaryotic cells. RNA is not found in the chromosomes, but rather is located elsewhere in the nucleus and even outside the nucleus, in the cytoplasm. As we will see in Section 17.4, there are six types of RNA, all with specific structures and functions.

Both DNA and RNA are polymers. Just as proteins consist of chains of amino acids, and polysaccharides consist of chains of monosaccharides, so nucleic acids are also chains. The building blocks (monomers) of nucleic acid chains are *nucleotides*. Nucleotides themselves, however, are composed of three simpler units: a base, a monosaccharide, and a phosphate. We will look at each of these components in turn.

A. Bases

Bases Purines and pyrimidines, which are components of nucleotides, DNA, and RNA

The **bases** found in DNA and RNA are chiefly those shown in Figure 17.1. All of them are basic because they are heterocyclic aromatic amines (Section 8.1). Two of these bases—adenine (A) and guanine (G)—are purines;

FIGURE 17.1 The five principal bases of DNA and RNA. Note how the rings are numbered. The hydrogens shown in blue are lost when the bases bond to monosaccharides.

the other three—cytosine (C), thymine (T), and uracil (U)—are pyrimidines. The two purines (A and G) and one of the pyrimidines (C) are found in both DNA and RNA, but uracil (U) is found only in RNA, and thymine (T) is found only in DNA. Note that thymine differs from uracil only in the methyl group in the 5 position. Thus both DNA and RNA contain four bases: two pyrimidines and two purines. For DNA, the bases are A, G, C, and T; for RNA, the bases are A, G, C, and U.

B. Sugars

The sugar component of RNA is D-ribose (Section 12.1C). In DNA, it is 2-deoxy-D-ribose (hence the name deoxyribonucleic acid).

The full name of β-D-ribose is β-D-ribofuranose and that of β-2-deoxy-D-ribose is β-2-deoxy-D-ribofuranose (see Section 12.2A).

The combination of sugar and base is known as a **nucleoside.** The purine bases are linked to C-1 of the monosaccharide through N-9 (the nitrogen at position 9 of the five-membered ring) by a β-N-glycosidic bond:

Nucleoside A compound composed of ribose or deoxyribose and a base

The nucleoside made of guanine and ribose is called **guanosine.** Table 17.1 gives the names of the other nucleosides.

The pyrimidine bases are linked to C-1 of the monosaccharide through their N-1 by a β-N-glycosidic bond.

Uridine

C. Phosphate

Nucleotide A nucleoside bonded to one, two, or three phosphate groups

The third component of nucleic acids is phosphoric acid. When this group forms a phosphate ester (Section 11.5) bond with a nucleoside, the result is a compound known as a **nucleotide.** For example, adenosine combines with phosphate to form the nucleotide adenosine 5′-monophosphate (AMP):

The ′ sign in adenosine 5′-monophosphate is used to distinguish which molecules the phosphate is bound to. Numbers without primes refer to positions on the purine or pyrimidine base. Numbers on the sugar are denoted with primes.

Table 17.1 gives the names of the other nucleotides. Some of these nucleotides play important roles in metabolism. They are part of the structure of key coenzymes, cofactors, and activators (Sections 19.3 and 21.2). Most notably, adenosine 5′-triphosphate (ATP) serves as a common currency into which the energy gained from food is converted and stored. In ATP, two more phosphate groups are joined to AMP in phosphate anhydride bonds (Section 11.5). In adenosine 5′-diphosphate (ADP), only one phosphate group is bonded to the AMP. All other nucleotides have important multiphosphorylated forms. For example, guanosine exists as GMP, GDP, and GTP.

TABLE 17.1 The Eight Nucleosides and Eight Nucleotides in DNA and RNA

Base	Nucleoside	Nucleotide
		DNA
Adenine (A)	Deoxyadenosine	Deoxyadenosine 5′-monophosphate (dAMP)*
Guanine (G)	Deoxyguanosine	Deoxyguanosine 5′-monophosphate (dGMP)*
Thymine (T)	Deoxythymidine	Deoxythymidine 5′-monophosphate (dTMP)*
Cytosine (C)	Deoxycytidine	Deoxycytidine 5′-monophosphate (dCMP)*
		RNA
Adenine (A)	Adenosine	Adenosine 5′-monophosphate (AMP)
Guanine (G)	Guanosine	Guanosine 5′-monophosphate (GMP)
Uracil (U)	Uridine	Uridine 5′-monophosphate (UMP)
Cytosine (C)	Cytidine	Cytidine 5′-monophosphate (CMP)

*The *d* indicates that the sugar is deoxyribose.

In Section 17.3, we will see how DNA and RNA are chains of nucleotides. In summary:

A nucleoside = Base + Sugar

A nucleotide = Base + Sugar + Phosphate

A nucleic acid = A chain of nucleotides

Example 17.1 Nucleotide Structure

GTP is an important store of energy. Draw the structure of guanosine triphosphate.

Strategy

When drawing nucleotides, there are three things to consider. First, determine if the sugar is ribose or deoxyribose. Then attach the correct base to the C1 position of the sugar. Last, put in the correct number of phosphates.

Solution

The base guanine is linked to a ribose unit by a *β-N*-glycosidic linkage. The triphosphate is linked to C-5′ of the ribose by an ester bond.

Problem 17.1

Draw the structure of UMP.

Chemical Connections 17A

Anticancer Drugs

A major difference between cancer cells and most normal cells is that cancer cells divide much more rapidly. Rapidly dividing cells require a constant new supply of DNA. One component of DNA is the nucleoside deoxythymidine, which is synthesized in the cell by the methylation of the uracil base.

Fluorouracil

If fluorouracil is administered to a cancer patient as part of chemotherapy, the body converts it to fluorouridine, a compound that irreversibly inhibits the enzyme that manufactures thymidine from uridine, greatly decreasing DNA synthesis. Because this inhibition affects the rapidly dividing cancer cells more than the healthy cells, the growth of the tumor and the spread of the cancer are arrested. Unfortunately, chemotherapy with fluorouracil or other anticancer drugs also weakens the body, because it interferes with DNA synthesis in normal cells.

Chemotherapy is used intermittently to give the body time to recover from the side effects of the drug. During the period after chemotherapy, special precautions must be taken so that bacterial infections do not debilitate the already-weakened body.

17.3 What Is the Structure of DNA and RNA?

In Chapter 14, we saw that proteins have primary, secondary, and higher-order structures. Nucleic acids, which are chains of monomers, also have primary, secondary, and higher-order structures.

A. Primary Structure

Nucleic acids are polymers of nucleotides, as shown schematically in Figure 17.2. Their primary structure is the sequence of nucleotides. Note that it can be divided into two parts: (1) the backbone of the molecule and (2) the bases that are the side-chain groups. The backbone in DNA consists of alternating deoxyribose and phosphate groups. Each phosphate group is linked to the 3′ carbon of one deoxyribose unit and simultaneously to the 5′ carbon of the next deoxyribose unit (Figure 17.3). Similarly, each monosaccharide unit forms a phosphate ester at the 3′ position and another at the 5′ position. The primary structure of RNA is the same except that each sugar is ribose (so an —OH group appears in the 2′ position) rather than deoxyribose and U is present instead of T.

Thus the backbone of the DNA and RNA chains has two ends: a 3′ —OH end and a 5′ —OH end. These two ends have roles similar to those of the C-terminal and N-terminal ends in proteins. The backbone provides structural stability for the DNA and RNA molecules.

As noted earlier, the bases that are linked, one to each sugar unit, are the side chains. They carry all of the information necessary for protein synthesis. Through analysis of the base composition DNA molecules from many different species, Erwin Chargaff (1905–2002) showed that the quantity of adenine (in moles) is always approximately equal to the quantity of thymine, and the quantity of guanine is always approximately equal to the quantity of cytosine, although the adenine/guanine ratio varies widely from

Nucleic acid A polymer composed of nucleotides

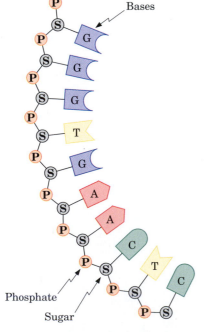

FIGURE 17.2 Schematic diagram of a nucleic acid molecule. The four bases of each nucleic acid are arranged in various specific sequences.

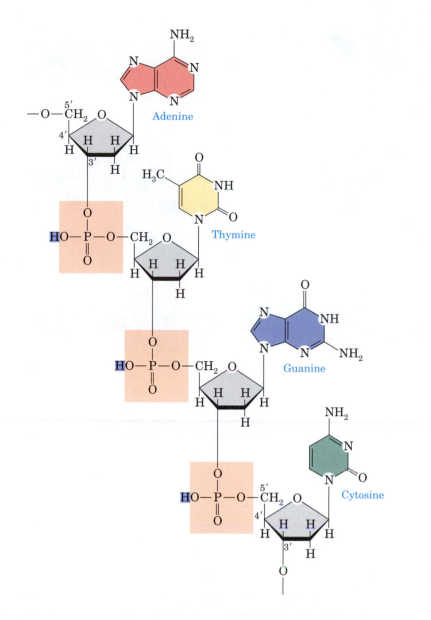

FIGURE 17.3 Primary structure of the DNA backbone. The hydrogens shown in blue cause the acidity of nucleic acids. In the body, at neutral pH, the phosphate groups carry a charge of −1 and the hydrogens are replaced by Na$^+$ and K$^+$.

species to species (see Table 17.2). This important information helped to establish the secondary structure of DNA, as we will soon see.

Just as the order of the amino acid residues of protein side chains determines the primary structure of the protein (for example, —Ala—Gly—Glu—Met—), the order of the bases (for example, —ATTGAC—) provides the primary structure of DNA. As with proteins, we need a convention to tell us which end to start with when we write the sequence of bases. For nucleic acids, the convention is to begin the sequence with the nucleotide that has the free 5′ terminus. Thus the sequence AGT means that adenine is the base at the 5′ terminus and thymine is the base at the 3′ terminus.

TABLE 17.2 Base Composition and Base Ratio in Two Species

	Base Composition (mol %)				Base Ratio	
Organism	A	G	C	T	A/T	G/C
Human	30.9	19.9	19.8	29.4	1.05	1.01
Wheat germ	27.3	22.7	22.8	27.1	1.01	1.00

FIGURE 17.4 Three-dimensional structure of the DNA double helix.

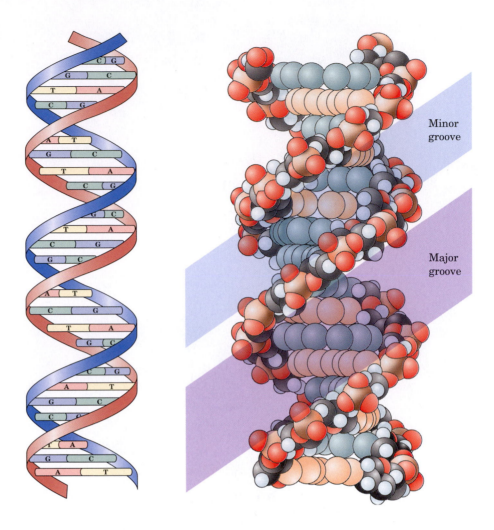

Minor groove

Major groove

B. Secondary Structure of DNA

In 1953, James Watson (1928–) and Francis Crick (1916–2004) established the three-dimensional structure of DNA. Their work is a cornerstone in the history of biochemistry. The model of DNA developed by Watson and Crick was based on two important pieces of information obtained by other workers: (1) the Chargaff rule that (A and T) and (G and C) are present in equimolar quantities and (2) x-ray diffraction photographs obtained by Rosalind Franklin (1920–1958) and Maurice Wilkins (1916–2004). By the clever use of these facts, Watson and Crick concluded that DNA is composed of two strands entwined around each other in a **double helix,** as shown in Figure 17.4.

In the DNA double helix, the two polynucleotide chains run in opposite directions (which is called antiparallel). Thus, at each end of the double helix, there is one 5′ — OH and one 3′ — OH terminus. The sugar–phosphate backbone is on the outside, exposed to the aqueous environment, and the bases point inward. The bases are hydrophobic, so they try to avoid contact with water. Through their hydrophobic interactions, they stabilize the double helix. The bases are paired according to Chargaff's rule: For each adenine on one chain, a thymine is aligned opposite it on the other chain; each guanine on one chain has a cytosine aligned with it on the other chain. *The bases so paired form hydrogen bonds with each other, two for A—T and three for G—C, thereby stabilizing the double helix* (Figure 17.5). A—T and G—C are **complementary base pairs.**

The important thing, as Watson and Crick realized, is that only adenine could fit with thymine and only guanine could fit with cytosine. Let us consider the other possibilities. Can two purines (AA, GG, or AG) fit opposite each

Double helix The arrangement in which two strands of DNA are coiled around each other in a screw-like fashion

Watson, Crick, and Wilkins were awarded the 1962 Nobel Prize in medicine for their discovery. Franklin died in 1958. The Nobel Committee does not award the Nobel Prize posthumously.

Thymine   Adenine

AT pair

Cytosine   Guanine

GC pair

FIGURE 17.5 A and T pair up by forming two hydrogen bonds; G and C pair up by forming three hydrogen bonds.

other? Figure 17.6 shows that they would overlap. How about two pyrimidines (TT, CC, or CT)? As shown in Figure 17.6, they would be too far apart. *There must be a pyrimidine opposite a purine.* But could A fit opposite C, or G opposite T? Figure 17.7 shows that the hydrogen bonding would be much weaker.

The entire action of DNA—and of the heredity mechanism—depends on the fact that, *wherever there is an adenine on one strand of the helix, there must be a thymine on the other strand because that is the only base that fits and forms strong hydrogen bonds with adenine, and similarly for G and C.* The entire heredity mechanism rests on these aligned hydrogen bonds (Figure 17.5), as we will see in Section 17.6.

Two purines overlap

Gap between two pyrimidines

FIGURE 17.6 The bases of DNA cannot stack properly in the double helix if a purine is opposite a purine or if a pyrimidine is opposite a pyrimidine.

Rosalind Franklin (1920–1958).

Watson and Crick with their model of the DNA molecule.

FIGURE 17.7 Only one hydrogen bond is possible for TG or CA. These combinations are not found in DNA. Compare this figure with Figure 17.5.

Thymine   Guanine

Cytosine   Adenine

The form of the DNA double helix shown in Figure 17.4 is called B-DNA. It is the most common and most stable form. Other forms become possible where the helix is wound more tightly or more loosely, or is wound in the opposite direction. With B-DNA, a distinguishing feature is the presence of a **major groove** and a **minor groove,** which arise because the two strands are not equally spaced around the helix. Interactions of proteins and drugs with the major and minor grooves of DNA serve as an active area of research.

C. Higher-Order Structures of DNA

If a human DNA molecule were fully stretched out, its length would be perhaps 1 m. However, the DNA molecules in the nuclei are not stretched out, but rather coiled around basic protein molecules called **histones.** The acidic DNA and the basic histones attract each other by electrostatic (ionic) forces, combining to form units called **nucleosomes.** In a nucleosome, eight histone molecules form a core, around which a 147-base-pair DNA double helix is wound. Nucleosomes are further condensed into **chromatin** when a 30-nm-wide fiber forms in which nucleosomes are wound in a **solenoid** fashion, with six nucleosomes forming a repeating unit (Figure 17.8). Chromatin fibers are organized still further into loops, and loops are arranged into bands to provide the superstructure of chromosomes.

The beauty of establishing the three-dimensional structure of the DNA molecule was that the knowledge of this structure immediately led to the explanation for the transmission of heredity—how the genes transmit traits from one generation to another. Before we look at the mechanism of DNA replication (in Section 17.6), let us summarize the three differences in structure between DNA and RNA:

Chromatin The DNA complexed with histone and nonhistone proteins that exists in eukaryotic cells between cell divisions

Solenoid A coil wound in the form of a helix

DNA double helix

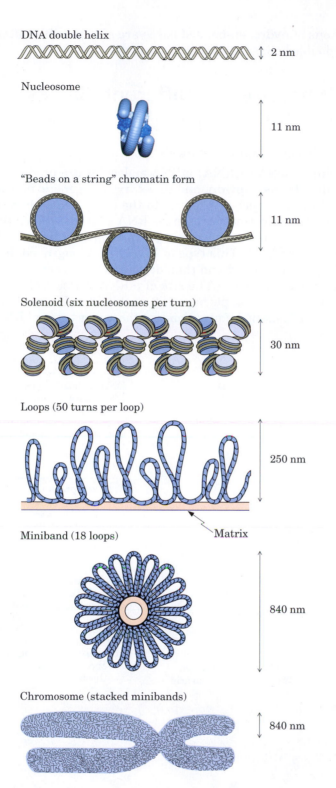

2 nm

Nucleosome

11 nm

"Beads on a string" chromatin form

11 nm

Solenoid (six nucleosomes per turn)

30 nm

Loops (50 turns per loop)

250 nm

Matrix

Miniband (18 loops)

840 nm

Chromosome (stacked minibands)

840 nm

FIGURE 17.8 Superstructure of chromosomes. In nucleosomes, the bandlike DNA double helix winds around cores consisting of eight histones. Solenoids of nucleosomes form 30 nm filament. Loops and minibands are other substructures.

1. DNA has four bases: A, G, C, and T. RNA has three of these bases —A,G, and C—but its fourth base is U, not T.

2. In DNA, the sugar is 2-deoxy-D-ribose. In RNA, it is D-ribose.

3. DNA is almost always double-stranded, with the helical structure shown in Figure 17.4.

There are several kinds of RNA (as we will see in Section 17.4); None of them has a repetitive double-stranded structure like DNA, although base-pairing can occur within a chain (see, for example, Figure 17.10). When it does, adenine pairs with uracil because thymine is not present. Other combinations of hydrogen-bonded bases are also possible outside the confines of a double helix, and Chargaff's rule does not apply.

17.4 What Are the Different Classes of RNA?

We previously noted that there are six types of RNA.

Messenger RNA (mRNA) The RNA that carries genetic information from DNA to the ribosome and acts as a template for protein synthesis

1. **Messenger RNA (mRNA)** mRNA molecules are produced in the process called **transcription,** and they carry the genetic information from the DNA in the nucleus directly to the cytoplasm, where most of the protein is synthesized. Messenger RNA consists of a chain of nucleotides whose sequence is exactly complementary to that of one of the strands of the DNA. This type of RNA is not long-lived, however. It is synthesized as needed and then degraded, so its concentration at any given time is rather low. The size of mRNA varies widely, with the average unit containing perhaps 750 nucleotides. Figure 17.9 shows the basic flow of genetic information and the major types of RNA.

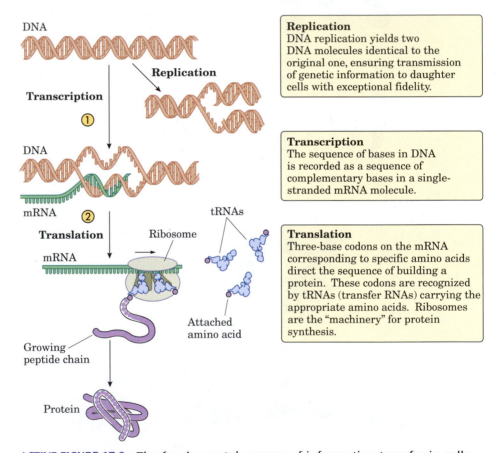

Replication
DNA replication yields two DNA molecules identical to the original one, ensuring transmission of genetic information to daughter cells with exceptional fidelity.

Transcription
The sequence of bases in DNA is recorded as a sequence of complementary bases in a single-stranded mRNA molecule.

Translation
Three-base codons on the mRNA corresponding to specific amino acids direct the sequence of building a protein. These codons are recognized by tRNAs (transfer RNAs) carrying the appropriate amino acids. Ribosomes are the "machinery" for protein synthesis.

ACTIVE FIGURE 17.9 The fundamental process of information transfer in cells. (1) Information encoded in the nucleotide sequence of DNA is transcribed through synthesis of an RNA molecule whose sequence is dictated by the DNA sequence. (2) As the sequence of this RNA is read (as groups of three consecutive nucleotides) by the protein synthesis machinery, it is translated into the sequence of amino acids in a protein. This information transfer system is encapsulated in what is known as the central dogma of molecular biology: DNA ⟶ RNA ⟶ protein. **Go to this book's companion website at www.cengage.com/chemistry/bettelheim to explore an interactive version of this figure.**

FIGURE 17.10 Structure of tRNA. (*a*) Two-dimensional simplified cloverleaf structure. (*b*) Three-dimensional structure.

2. Transfer RNA (tRNA) Containing from 73 to 93 nucleotides per chain, tRNAs are relatively small molecules. There is at least one different tRNA molecule for each of the 20 amino acids from which the body makes its proteins. The three-dimensional tRNA molecules are L-shaped, but they are conventionally represented as a cloverleaf in two dimensions. Figure 17.10 shows a typical structure. Transfer RNA molecules contain not only cytosine, guanine, adenine, and uracil, but also several other modified nucleotides, such as 1-methylguanosine.

Transfer RNA (tRNA) The RNA that transports amino acids to the site of protein synthesis in ribosomes

1-Methylguanosine

3. Ribosomal RNA (rRNA) Ribosomes, which are small spherical bodies located in the cells but outside the nuclei, contain rRNA. They consist of about 35% protein and 65% ribosomal RNA (rRNA). These large molecules have molecular weights up to 1 million. As discussed in Section 17.5, protein synthesis takes place on the ribosomes.

Dissociation of ribosomes into their components has proved to be a useful way of studying their structure and properties. A particularly important endeavor has been to determine both the number and the kind of RNA and protein molecules that make up ribosomes. This approach has helped elucidate the role of ribosomes in protein synthesis. In both prokaryotes and eukaryotes, a ribosome consists of two subunits, one larger than the other. In turn, the smaller subunit consists of one large RNA molecule and about 20 different proteins; the larger subunit

Ribosomal RNA (rRNA) The RNA complexed with proteins in ribosomes

Ribosome Small spherical bodies in the cell made of protein and RNA; the site of protein synthesis

Splicing The removal of an internal RNA segment and the joining of the remaining ends of the RNA molecule

Whole cells

Lysis and fractionation

70S shape

⊢——~200 Å——⊣
Prokaryote ribosome
(thousands per cell)

Dissociation
10^{-4} M Mg^{2+}
Elevating Mg^{2+}
to 10^{-2} M is
sufficient to
reverse this step

30S subunit   50S subunit

Both about 2/3 RNA and 1/3 protein

Detergent   Dissociation of subunits into component parts

16S rRNA and 21 different proteins

23S rRNA and 5S rRNA and 34 different proteins

FIGURE 17.11 The structure of a typical prokaryotic ribosome. The individual components can be mixed, producing functional subunits. Reassociation of subunits gives rise to an intact ribosome. The designation S refers to Svedberg, a unit of relative size determined when molecules are separated by centrifugation.

consists of two RNA molecules in prokaryotes (three in eukaryotes) and about 35 different proteins in prokaryotes (about 50 in eukaryotes) (Figure 17.11). The subunits are easily dissociated from one another in the laboratory by lowering the Mg^{2+} concentration of the medium. Raising the Mg^{2+} concentration to its original level reverses the process, and active ribosomes can be reconstituted by this method.

4. **Small Nuclear RNA (snRNA)** A recently discovered RNA molecule is snRNA, which is found, as the name implies, in the nucleus of eukaryotic cells. This type of RNA is small, about 100 to 200 nucleotides long, but it is neither a tRNA molecule nor a small subunit of rRNA. In the cell, it is complexed with proteins to form **small nuclear ribonucleoprotein particles, snRNPs,** pronounced "snurps." Their function is to help with the processing of the initial mRNA transcribed from DNA into a mature form that is ready for export out of the nucleus. This process is often referred to as **splicing,** and it is an active area of research. While studying splicing, scientists realized that part of the splicing reaction involved catalysis by the RNA portion of a snRNP and not the protein portion. This recognition led to the discovery of **ribozymes,** RNA-based enzymes, for which Thomas Cech received the Nobel Prize. Splicing will be discussed further in Chapter 18.

5. **Micro RNA (miRNA)** A very recent discovery is another type of small RNA, miRNA. These RNAs are only 20–22 nucleotides long but are important in the timing of an organism's development. They play important roles in cancer, stress respsonses, and viral infections. They inhibit translation of mRNA into protein and promote the degradation of mRNA. It was recently discovered, however, that these versatile RNAs can also stimulate protein production in cells when the cell cycle has been arrested.

6. **Small Interfering RNA (siRNA)** The process called RNA interference was heralded as the breakthrough of the year in 2002 by *Science* magazine. Short stretches of RNA (20–30 nucleotides long), called small interfering RNA, have been found to have an enormous control over gene expression. This process serves as a protective mechanism in many species, with the siRNAs being used to eliminate expression of an undesirable gene, such as one that causes uncontrolled cell growth or one that came from a virus. siRNAs lead to the degradation of specific mRNA molecules. Scientists who wish to study gene expression are also using these small RNAs. In what has become an explosion of new biotechnology, many companies have been created to produce and market designer siRNAs to knock out hundreds of known genes. This technology also has medical applications, as siRNA has been used to protect mouse liver from hepatitis and to help clear infected liver cells of the disease.

Table 17.3 summarizes the basic types of RNA.

TABLE 17.3 The Roles of Different Kinds of RNA

RNA Type	Size	Function
Transfer RNA	Small	Transports amino acids to site of protein synthesis
Ribosomal RNA	Several kinds— variable in size	Combines with proteins to form ribosomes, the site of protein synthesis
Messenger RNA	Variable	Directs amino acid sequence of proteins
Small nuclear RNA	Small	Processes initial mRNA to its mature form in eukaryotes
Micro RNA	Small	Affects gene expression; important in growth and development
Small interfering RNA	Small	Affects gene expression; used by scientists to knock out a gene being studied

17.5 What Are Genes?

A gene is a stretch of DNA, containing a few hundred nucleotides, that carries one particular message—for example, "make a globin molecule" or "make a tRNA molecule." One DNA molecule may have between 1 million and 100 million bases. Therefore, many genes are present in one DNA molecule. In bacteria, this message is continuous; in higher organisms, it is not. That is, stretches of DNA that spell out (encode) the amino acid sequence to be assembled are interrupted by long stretches that seemingly do not code for anything. The coding sequences are called **exons,** short for "expressed sequences," and the noncoding sequences are called **introns,** short for "intervening sequences."

For example, the globin gene has three exons broken up by two introns. Because DNA contains both exons and introns, the mRNA transcribed from it also contains both exons and introns. The introns are spliced out by ribozymes, and the exons are spliced together before the mRNA is used to synthesize a protein. In other words, the introns function as spacers and, in rare instances, as enzymes, catalyzing the splicing of exons into mature mRNA. Figure 17.12 shows the difference between prokaryotic and eukaryotic production of proteins.

In prokaryotes, the genes on a stretch of DNA are next to each other. These are turned into a sequence of mRNA, which are then translated by ribosomes to make proteins, all of which happens simultaneously. In eukaryotes, the genes are separated by introns and the processes take place in different compartments. The DNA is turned into RNA in the nucleus, but then the initial mRNA, containing introns, is transported to the cytosol where the exons are spliced out. The final mRNA is then translated to protein. The process of making RNA and protein is the subject of Chapter 18.

In humans, only 3% of the DNA codes for proteins or RNA with clear functions. Introns are not the only noncoding DNA sequences, however. **Satellites** are DNA molecules in which short nucleotide sequences are repeated hundreds or thousands of times. Large satellite stretches appear at the ends and centers of chromosomes and provide stability for the chromosomes.

Smaller repetitive sequences, called **mini-satellites** or **microsatellites,** are associated with cancer when they mutate.

Exon Nucleotide sequence in DNA or mRNA that codes for a protein

Intron A nucleotide sequence in DNA or mRNA that does not code for a protein

17.6 How Is DNA Replicated?

The DNA in the chromosomes carries out two functions: (1) It reproduces itself and (2) it supplies the information necessary to make all the RNA and proteins in the body, including enzymes. The second function is covered in Chapter 18. Here we are concerned with the first, **replication.**

Each gene is a section of a DNA molecule that contains a specific sequence of the four bases A, G, T, and C, typically comprising about 1000 to 2000 nucleotides. The base sequence of the gene carries the information necessary to produce one protein molecule. If the sequence is changed (for example, if one A is replaced by a G, or if an extra T is inserted), a different protein is produced, which might have an impaired function, as in sickle cell anemia (Chemical Connections 14D).

Consider the monumental task that must be accomplished by the organism. When an individual is conceived, the egg and sperm cells unite to form the zygote. This cell contains only a small amount of DNA, but it nevertheless provides all the genetic information that the individual will ever

Replication The process by which copies of DNA are made during cell division

Prokaryotes:

Eukaryotes:

Exons are protein-coding regions that must be joined by removing introns, the noncoding intervening sequences. The process of intron removal and exon joining is called splicing.

ACTIVE FIGURE 17.12 The properties of mRNA molecules in prokaryotic versus eukaryotic cells during transcription and translation. **Go to this book's companion website at www.cengage.com/chemistry/bettelheim to explore an interactive version of this figure.**

have. In a human cell, some 3 billion base pairs must be duplicated at each cell cycle, and a fully grown human being may contain more than 1 trillion cells. Each cell contains the same amount of DNA as the original single cell. Furthermore, cells are constantly dying and being replaced. Thus there must be a mechanism by which DNA molecules can be copied over and over again without error. In Section 17.7, we will see that such errors sometimes do occur and can have serious consequences. Here, however, we want to examine this remarkable mechanism that takes place every day in billions of organisms, from microbes to whales, and has been taking place for billions of years—with only a tiny percentage of errors.

Replication begins at a point in the DNA called an **origin of replication.** In human cells, the average chromosome has several hundred origins of replication where the copying occurs simultaneously. The DNA double helix has two strands running in opposite directions. The point on the DNA where replication is proceeding is called the **replication fork.** (See Figure 17.13.)

Chemical Connections 17B

Telomeres, Telomerase, and Immortality

Every person has a genetic makeup consisting of about 3 billion pairs of nucleotides, distributed over 46 chromosomes. Telomeres are specialized structures at the ends of chromosomes. In vertebrates, telomeres are TTAGGG sequences that are repeated hundreds to thousands of times. In normal **somatic cells** that divide in a cyclic fashion throughout the life of the organism (via mitosis), chromosomes lose about 50 to 200 nucleotides from their telomeres at each cell division.

DNA polymerase, the enzyme that links the fragments, does not work at the end of linear DNA. This fact results in the shortening of the telomeres at each replication. The telomere shortening acts as a clock by which the cells count the number of times they have divided. After a certain number of divisions, the cells stop dividing, having reached the limit of the aging process.

In contrast to somatic cells, all immortal cells (germ cells in proliferative stem cells, normal fetal cells, and cancer cells) possess an enzyme, telomerase, that can extend the shortened telomeres by synthesizing new chromosomal ends. Telomerase is a ribonucleoprotein; that is, it is made of RNA and protein. The activity of this enzyme seems to confer immortality to the cells.

(*b*) **A mechanism by which telomerase may work. (In this case, RNA of the telomerase acts as a template for reverse transcription)**

Extension of DNA on the RNA of telomerase

Telomerase

RNA template in telomerase

3' end of the leading strand is elongated.

RNA primer   Lagging strand extended by polymerase

Removal of the primer shortens the DNA, but it is now longer by one repeat unit.

The telomerase extension cycle is repeated until there is an adequate number of DNA repeats for the end of the chromosome to survive.

(*a*) **Replication at the end of a linear template**

Template strand

End of linear chromosome

RNA primer   Synthesis (lagging strand)

This portion of the end of the chromosome will be lost when the primer is removed.

If the unwinding of the double helix begins in the middle, the synthesis of new DNA molecules on the old templates continues in both directions until the entire molecule is duplicated. Alternatively, the unwinding can start at one end and proceed in one direction until the entire double helix is unwound.

Replication is bidirectional and takes place at the same speed in both directions. An interesting detail of DNA replication is that the two daughter strands are synthesized in different ways. One of the syntheses is continuous along the 3' to 5' strand (see Figure 17.13). It is called the **leading strand.** Along the other strand that runs in the 5' ⟶ 3' direction, the synthesis is discontinuous. It is called the **lagging strand.**

The replication process is called **semiconservative** because each daughter molecule has one parental strand (conserved) and one newly synthesized one.

Replication always proceeds from the 5' to the 3' direction from the perspective of the chain that is being synthesized. The actual reaction occurring is a nucleophilic attack by the 3' hydroxyl of the deoxyribose of one nucleotide

ACTIVE FIGURE 17.13 General features of the replication of DNA. The two strands of the DNA double helix are shown separating at the replication fork. **Go to this book's companion website at www.cengage.com/chemistry/ bettelheim to explore an interactive version of this figure.**

FIGURE 17.14 The addition of a nucleotide to a growing DNA chain. The 3'-hydoxyl group at the end of the growing DNA chain is a nucleophile. It attacks at the phosphorus adjacent to the sugar in the nucleotide, which will be added to the growing chain.

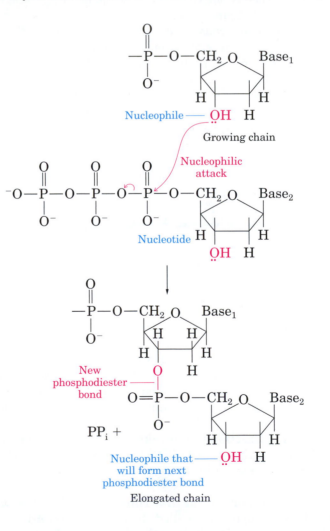

against the first phosphate on the 5' carbon of the incoming nucleoside triphosphate, as shown in Figure 17.14.

One of the more interesting aspects of DNA replication is that the basic reaction of synthesis always requires an existing chain with a nucleotide

TABLE 17.4 Components of Replisomes and Their Functions

Component	Function
Helicase	Unwinds the DNA double helix
Primase	Synthesizes short oligonucleotides (primers)
Clamp protein	Allows the leading strand to be threaded through
DNA polymerase	Joins the assembled nucleotides
Ligase	Joins Okazaki fragments in the lagging strand

that has a free 3′-hydroxyl to do the nucleophilic attack. DNA replication cannot begin without this preexisting chain to latch onto. We call this chain a **primer.** In all known forms of replication, the primer is made out of RNA, not DNA.

Replication is a very complex process involving a number of enzymes and binding proteins. A growing body of evidence indicates that these enzymes assemble their products in "factories" through which the DNA moves. Such factories may be bound to membranes in bacteria. In higher organisms, the replication factories are not permanent structures. Instead, they may be disassembled and their parts reassembled in ever-larger factories. These assemblies of enzyme "factories" go by the name of **replisomes,** and they contain key enzymes such as polymerases, helicases, and primases (Table 17.4). The primases can shuttle in and out of the replisomes. Other proteins, such as clamp loaders and clamp proteins, through which the newly synthesized primer is threaded, are also parts of the replisomes.

The replication of DNA occurs in a number of distinct steps. A few of the salient features are enumerated here:

1. **Opening up the superstructure** During replication, the very condensed superstructure of chromosomes must be opened so that it becomes accessible to enzymes and other proteins. A complicated signal transduction mechanism accomplishes this feat. One notable step of the signal transduction is the acetylation and deacetylation of key lysine residues of histones. When histone acetylase, an enzyme, puts acetyl groups on key lysine residues, some positive charges are eliminated and the strength of the DNA–histone interaction is weakened:

$$\text{Histone}-(\text{CH}_2)_4-\text{NH}_3^+ + \text{CH}_3-\text{COO}^- \underset{\text{deacetylation}}{\overset{\text{acetylation}}{\rightleftharpoons}} \text{Histone}-(\text{CH}_2)_4-\overset{\displaystyle H}{\underset{\displaystyle |}{N}}-\overset{\displaystyle O}{\overset{\displaystyle \|}{C}}-\text{CH}_3 + \text{H}_2\text{O}$$

   This process allows the opening up of key regions on the DNA molecule. When another enzyme, histone deacetylase, removes these acetyl groups, the positive charges are reestablished. That, in turn, facilitates regaining the highly condensed structure of **chromatin.**

2. **Relaxation of Higher-Order Structures of DNA** Topoisomerases (also called gyrases) are enzymes that facilitate the relaxation of supercoiling in DNA. They do so during replication by temporarily introducing either single- or double-strand breaks in DNA. The transient break forms a phosphodiester linkage between a tyrosyl residue of the enzyme and either the 5′ or 3′ end of a phosphate on the DNA. Once the supercoiling is relaxed, the broken strands are joined together, and the topoisomerase diffuses from the location of the replicating fork. Topoisomerases are also involved in the untangling of the replicated chromosomes, before cell division can occur.

3. **Unwinding the DNA Double Helix** The replication of DNA molecules starts with the unwinding of the double helix, which can occur at

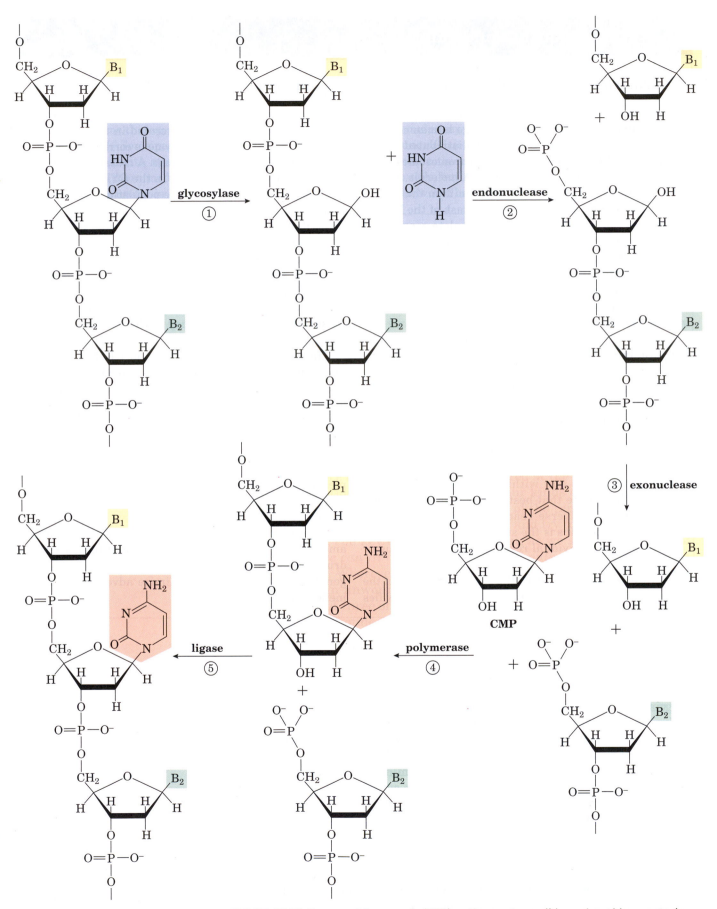

FIGURE 17.15 Base excision repair (BER) pathway. A uracil is replaced by a cytosine.

The sugar–phosphate backbone is still intact. At the **AP site** (*ap*urinic or *ap*yrimidinic site) created in this way, the backbone is cleaved by a second enzyme, endonuclease (2). A third enzyme, exonuclease (3), liberates the sugar–phosphate unit of the damaged site.

2. In the synthesis step, the enzyme DNA polymerase (4) inserts the correct nucleotide, cytidine, and the enzyme DNA ligase seals (5) the backbone to complete the repair.

A second repair mechanism removes not a single mismatched base but rather whole nucleotides—as many as 25 to 32 residue oligonucleotides. Known as NER, for *n*ucleotide *e*xcision *r*epair, it similarly involves a number of repair enzymes.

Any defect in the repair mechanisms may lead to harmful or even lethal mutations. For example, individuals with inherited XP (xeroderma pigmentosa), a condition in which one or another enzyme in the NER repair pathway is missing or defective, have a 1000 times greater risk of developing skin cancer than do normal individuals.

17.8 How Do We Amplify DNA?

To study DNA for basic or applied scientific purposes, we must have enough of it to work with. There are several ways of amplifying DNA. One approach is to allow a rapidly growing organism, like bacteria, to replicate DNA for us. This process, which is usually referred to as **cloning,** will be discussed further in Chapter 18. Millions of copies of selected DNA fragments can also be made within a few hours with high precision by a technique called **polymerase chain reaction (PCR),** which was discovered by Kary B. Mullis (1944–), who shared the 1993 Nobel Prize in chemistry for this achievement.

PCR techniques can be used if the sequence of a gene to be copied is known, or at least a sequence bordering the desired DNA is known. In such a case, one can synthesize two primers that are complementary to the ends of the gene or to the bordering DNA. The primers are polynucleotides consisting of 12 to 16 nucleotides. When added to a target DNA segment, they hybridize with the end of each strand of the gene.

$$5'CATAGGACAGC—OH \qquad \text{Primer}$$
$$3'TACGTATCCTGTCGTAGG— \qquad \text{Gene}$$

In cycle 1 (Figure 17.16), the polymerase extends the primers in each direction as individual nucleotides are assembled and connected on the template DNA. In this way, two new copies are created. The two-step process is repeated (cycle 2) when the primers are **hybridized** with the new strands, and the primers are extended again. At that point, four new copies have been created. The process continues, and in 25 cycles, 2^{25} or some 33 million copies can be made. In practice, only a few million are produced, which is sufficient for the isolation of a gene.

This fast process is practical because of the discovery of heat-resistant polymerases isolated from bacteria that live in hot thermal vents on the sea floor (Section 15.4B). A temperature of 95°C is needed because the double helix must be unwound to hybridize the primer to the target DNA. Once single strands of DNA have been exposed, the mixture is cooled to 70°C. The primers are hybridized and subsequent extensions take place. The 95°C and 70°C cycles are repeated over and over. No new enzyme is required because the polymerase is stable at both temperatures.

Hybridization The process in which two strands of nucleic acids or segments of nucleic acid strands form a double stranded structure through hydrogen bonding of complementary base pairs

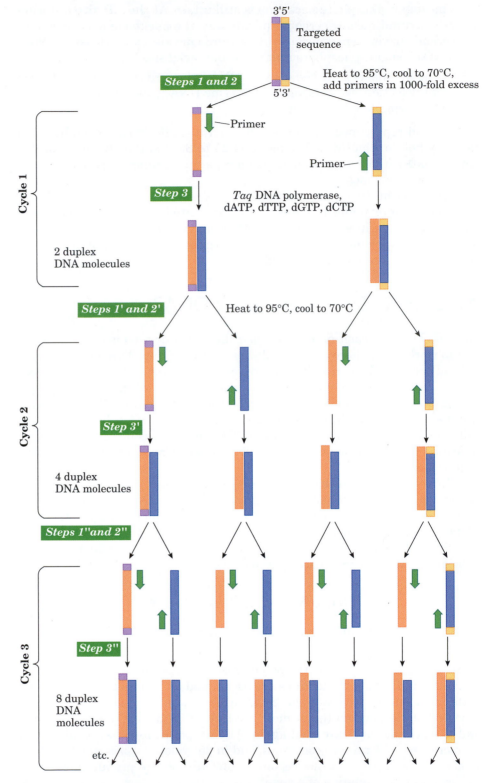

ACTIVE FIGURE 17.16 Polymerase chain reaction (PCR). Oligonucleotides complementary to a given DNA sequence prime the synthesis of only that sequence. Heat-stable *Taq* DNA polymerase survives many cycles of heating. Theoretically, the amount of the specific primed sequence is doubled in each cycle. **Go to this book's companion website at www.cengage.com/chemistry/bettelheim to explore an interactive version of the figure.**

PCR techniques are routinely used when a gene or a segment of DNA must be amplified from a few molecules. It is used in studying genomes (Chemical Connections 17E), in obtaining evidence from a crime scene (Chemical Connections 17C), and even in obtaining the genes of long-extinct species found fossilized in amber.

Summary

OWL End-of-chapter problems identified in blue are assignable in **GOB OWL.**

Section 17.1 What Are the Molecules of Heredity?

- Heredity is based on genes located in chromosomes.
- Genes are sections of DNA that encode specific RNA molecules.

Section 17.2 What Are Nucleic Acids Made Of?

- **Nucleic acids** are composed of sugars, phosphates, and organic bases.
- Two kinds of nucleic acids exist: **ribonucleic acid (RNA)** and **deoxyribonucleic acid (DNA).**
- In DNA, the sugar is the monosaccharide 2-deoxy-D-ribose; in RNA, it is D-ribose.
- In DNA, the heterocyclic amine bases are adenine (A), guanine (G), cytosine (C), and thymine (T).
- In RNA, they are A, G, C, and uracil (U).
- Nucleic acids are giant molecules with backbones made of alternating units of sugar and phosphate. The bases are side chains joined by β-N-glycosidic bonds to the sugar units.

Section 17.3 What Is the Structure of DNA and RNA?

- DNA is made of two strands that form a double helix. The sugar–phosphate backbone runs on the outside of the double helix, and the hydrophobic bases point inward.
- **Complementary pairing** of the bases occurs in the double helix, such that each A on one strand is hydrogen-bonded to a T on the other strand, and each G is hydrogen-bonded to a C. No other pairs fit.
- DNA is coiled around basic protein molecules called **histones.** Together they form **nucleosomes,** which are further condensed into chromatin.
- The DNA molecule carries, in the sequence of its bases, all the information necessary to maintain life. When cell division occurs and this information is passed from parent cell to daughter cells, the sequence of the parent DNA is copied.

Section 17.4 What Are the Different Classes of RNA? Problems 17.34, 17.40

- There are six kinds of RNA: **messenger RNA (mRNA), transfer RNA (tRNA), ribosomal RNA (rRNA), small nuclear RNA (snRNA), micro RNA (miRNA),** and **small interfering RNA (siRNA).**
- mRNA, tRNA, and rRNA are involved in all protein synthesis.
- Small nuclear RNA is involved in splicing reactions and has been found in some cases to have catalytic activity.
- RNA with catalytic activity is called a **ribozyme.**

Section 17.5 What Are Genes?

- A **gene** is a segment of a DNA molecule that carries the sequence of bases that directs the synthesis of one particular protein or RNA molecule.
- DNA in higher organisms contains sequences, called **introns,** that do not code for proteins.
- The sequences that do code for proteins are called **exons.**

Section 17.6 How Is DNA Replicated? Problems 17.54, 17.59

- DNA replication occurs in several distinct steps.
- The superstructures of chromosomes are initially loosened by acetylation of histones. **Topoisomerases** relax the higher structures. **Helicases** at the replication fork separate the two strands of DNA.
- RNA primers and primases are needed to start the synthesis of daughter strands. The **leading strand** is synthesized continuously by **DNA polymerase.** The **lagging strand** is synthesized discontinuously as **Okazaki fragments.**
- **DNA ligase** seals the nicks and the Okazaki fragments.

Section 17.7 How Is DNA Repaired?

- An important **DNA repair** mechanism is BER, or single base excision repair.

Section 17.8 How Do We Amplify DNA?

- The **polymerase chain reaction (PCR)** technique can make millions of copies with high precision in a few hours.

Problems

■ Indicates problems that are assignable in **GOB OWL**.

Blue numbered problems are applied.

Go to this book's companion website at **www.cengage. com/chemistry/bettelheim** for interactive versions of the How To tutorials and Active Figures, and to quiz yourself on this chapter.

Section 17.1 What Are the Molecules of Heredity?

17.2 What structures of the cell, visible in a microscope, contain hereditary information?

17.3 Name one hereditary disease.

17.4 What is the basic unit of heredity?

Section 17.2 What Are Nucleic Acids Made Of?

17.5 (a) Where in a cell is the DNA located?
(b) Where in a cell is the RNA located?

17.6 What are the components of (a) a nucleotide and (b) a nucleoside?

17.7 What are the differences between DNA and RNA?

17.8 Draw the structures of ADP and GDP. Are these structures parts of nucleic acids?

17.9 What is the difference in structure between thymine and uracil?

17.10 Which DNA and RNA bases contain a carbonyl group?

17.11 Draw the structures of (a) cytidine and (b) deoxycytidine.

17.12 Which DNA and RNA bases are primary amines?

17.13 What is the difference in structure between D-ribose and 2-deoxy-D-ribose?

17.14 What is the difference between a nucleoside and a nucleotide?

17.15 RNA and DNA refer to nucleic *acids*. Which part of the molecule is acidic?

17.16 What type of bond exists between the ribose and the phosphate in AMP?

17.17 What type of bond exists between the two phosphates in ADP?

17.18 What type of bond connects the base to the ribose in GTP?

Section 17.3 What Is the Structure of DNA and RNA?

17.19 In RNA, which carbons of the ribose are linked to the phosphate group and which are linked to the base?

17.20 What constitutes the backbone of DNA?

17.21 Draw the structures of (a) UDP and (b) dAMP.

17.22 In DNA, which carbon atoms of 2-deoxy-D-ribose are bonded to the phosphate groups?

17.23 The sequence of a short DNA segment is ATGGCAATAC.
(a) What name do we give to the two ends (terminals) of a DNA molecule?
(b) In this segment, which end is which?
(c) What would be the sequence of the complementary strand?

17.24 Chargaff showed that, in samples of DNA taken from many different species, the molar quantity of A was always approximately equal to the molar quantity of T; the same is true for C and G. How did this information help to establish the structure of DNA?

17.25 How many hydrogen bonds can form between uracil and adenine?

17.26 How many histones are present in a nucleosome?

17.27 What is the nature of the interaction between histones and DNA in nucleosomes?

17.28 What are chromatin fibers made of?

17.29 What constitutes the superstructure of chromosomes?

17.30 What is the primary structure of DNA?

17.31 What is the secondary structure of DNA?

17.32 What is the major groove of a DNA helix?

17.33 What are the higher-order structures of DNA that eventually make up a chromosome?

Section 17.4 What Are the Different Classes of RNA?

17.34 ■ Which type of RNA has enzyme activity? Where does it function mostly?

17.35 Which has the longest chains: tRNA, mRNA, or rRNA?

17.36 Which type of RNA contains modified nucleotides?

17.37 Which type of RNA has a sequence exactly complementary to that of DNA?

17.38 Where is rRNA located in the cell?

17.39 What kind of functions do ribozymes, in general, perform?

17.40 ■ Which of the RNA types are always involved in protein synthesis?

17.41 What is the purpose of small nuclear RNA?

17.42 What is the purpose of siRNA?

17.43 What is the difference between miRNA and siRNA?

Section 17.5 What Are Genes?

17.44 Define:
(a) Intron   (b) Exon

17.45 Does mRNA also have introns and exons? Explain.

17.46 (a) What percentage of human DNA codes for proteins?

(b) What is the function of the rest of the DNA?

17.47 Do satellites code for a particular protein?

17.48 Do all genes code for a protein? If not, what do they code for?

Section 17.6 How Is DNA Replicated?

17.49 A DNA molecule normally replicates itself millions of times, with almost no errors. What single fact about the structure is most responsible for this fidelity of replication?

17.50 Which functional groups on the bases form hydrogen bonds in the DNA double helix?

17.51 Draw the structures of adenine and thymine, and show with a diagram the two hydrogen bonds that stabilize A—T pairing in DNA.

17.52 Draw the structures of cytosine and guanine, and show with a diagram the three hydrogen bonds that stabilize C—G pairing in nucleic acids.

17.53 How many different bases are present in a DNA double helix?

17.54 ■ What is a replication fork? How many replication forks may exist simultaneously on an average human chromosome?

17.55 Why is replication called semiconservative?

17.56 How does the removal of some positive charges from histones enable the opening of the chromosomal superstructure?

17.57 Write the chemical reaction for the deacetylation of acetyl-histone.

17.58 What is the quaternary structure of helicases in eukaryotes?

17.59 ■ What are helicases? What is their function?

17.60 Can dATP serve as a source for a primer?

17.61 What are the side products of the action of primase in forming primers?

17.62 What do we call the enzymes that join nucleotides into a DNA strand?

17.63 In which direction is the DNA molecule synthesized continuously?

17.64 What kind of bond formation do polymerases catalyze?

17.65 Which enzyme catalyzes the joining of Okazaki fragments?

17.66 What is the nature of the chemical reaction that joins nucleotides together?

17.67 From the perspective of the chain being synthesized, in which direction does DNA synthesis proceed?

Section 17.7 How Is DNA Repaired?

17.68 As a result of damage, a few of the guanine residues in a gene are methylated. What kind of mechanism could the cell use to repair the damage?

17.69 What is the function of endonuclease in the BER repair mechanism?

17.70 When cytosine is deaminated, uracil is formed. Uracil is a naturally occurring base. Why would the cell use base excision repair to remove it?

17.71 Which bonds are cleaved by glycosylase?

17.72 What are AP sites? Which enzyme creates them?

17.73 Why are patients with xeroderma pigmentosa 1000 times more likely to develop skin cancer than normal individuals are?

Section 17.8 How Do We Amplify DNA?

17.74 What is the advantage of using DNA polymerase from thermophilic bacteria that live in hot thermal vents in PCR?

17.75 What 12-nucleotide primer would you use in the PCR technique when you want to amplify a gene whose end is as follows: 3′TACCGTCATCCGGTG5′?

Chemical Connections

17.76 (Chemical Connections 17A) Draw the structure of the fluorouridine nucleoside that inhibits DNA synthesis.

17.77 (Chemical Connections 17A) Give an example of how anticancer drugs work in chemotherapy.

17.78 (Chemical Connections 17B) What sequence of nucleotides is repeated many times in telomeres?

17.79 (Chemical Connections 17B) Why are as many as 200 nucleotides lost at each replication?

17.80 (Chemical Connections 17B) How does telomerase make a cancer cell immortal?

17.81 (Chemical Connections 17B) Why is DNA loss with replication not a problem for bacteria? (Hint: Bacteria have a circular genome.)

17.82 (Chemical Connections 17C) After having been cut by restriction enzymes, how are DNA fragments separated from each other?

17.83 (Chemical Connections 17C) How is DNA fingerprinting used in paternity suits?

17.84 (Chemical Connections 17C) Why is it easier to exclude someone via DNA fingerprinting than it is to prove that he or she is the person whose sample is being tested?

17.85 (Chemical Connections 17C) What is the principle behind paternity testing via DNA fingerprinting?

17.86 (Chemical Connection 17D) What would be some advantages to having your own genome sequenced?

17.87 (Chemical Connection 17D) How could the information from sequencing your genome be used against you theoretically if it fell into the wrong hands?

17.88 (Chemical Connection 17D) Why might a person make bad personal or lifestyle choices if she had knowledge of her genome?

17.89 (Chemical Connections 17E) What is the function of cytochrome P-450?

■ Problems assignable in GOB OWL

17.90 (Chemical Connections 17E) How does knowledge of the human genome enable patients to be screened for individual drug tolerance?

Additional Problems

17.91 What is the active site of a ribozyme?

17.92 Why is it important that a DNA molecule be able to replicate itself millions of times without error?

17.93 ■ Draw the structures of (a) uracil and (b) uridine.

17.94 How would you classify the functional groups that bond together the three different components of a nucleotide?

17.95 Which nucleic acid molecule is the largest?

17.96 What kind of bonds are broken during replication? Does the primary structure of DNA change during replication?

17.97 In sheep DNA, the mol % of adenine (A) was found to be 29.3. Based on Chargaff's rule, what would be the approximate mol % of G, C, and T?

Looking Ahead

17.98 DNA is the blueprint for the cell, but not all genes in DNA lead to protein. Gene expression is the study of how genes are used to make their particular product. What are some examples of gene products that do not lead to proteins?

17.99 In a process similar to DNA replication, RNA is produced via the process called transcription. The enzyme used is RNA polymerase. When RNA is synthesized, what is the direction of the synthesis reaction?

17.100 The Human Genome Project showed that human DNA is not considerably bigger than much simpler organisms, with about 30,000 total genes. However, humans make over 100, 000 different proteins. How is this possible? (*Hint:* think about splicing.)

17.101 One of the biggest differences between DNA replication and transcription is that RNA polymerase does not require a primer. How does this fact relate to the theory that primordial life was based on RNA and not DNA?

17.102 How could life have evolved if DNA leads to RNA which leads to protein, but it takes many proteins to replicate DNA and to transcribe DNA into RNA?

17.103 When DNA is heated sufficiently, the strands separate. The energy that it takes to separate the DNA is related to the amount of guanine and cytosine bases. Why is this so?

17.104 If you wanted to amplify DNA using a technique similar to PCR but you had no source of a heat-stable DNA polymerase, what would you have to do to get the amplification?

17.105 Why do you think that DNA synthesis has evolved to have extensive proofreading and repair mechanisms while RNA synthesis has far fewer?

Gene Expression and Protein Synthesis

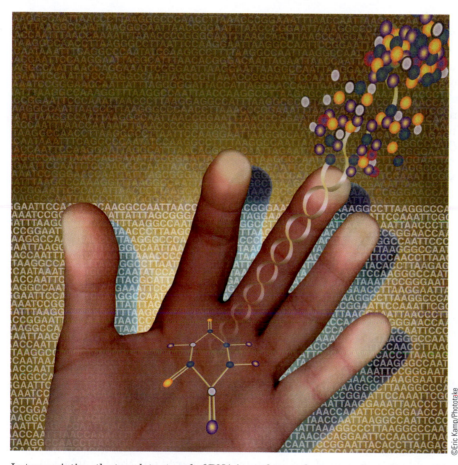

In transcription, the template strand of DNA is used to produce a complementary strand of RNA. Transcription is the most controlled and best understood part of gene regulation.

©Eric Kamp/Phototake

Key Questions

18.1 How Does DNA Lead to RNA and Protein?

18.2 How Is DNA Transcribed into RNA?

18.3 What Is the Role of RNA in Translation?

18.4 What Is the Genetic Code?

18.5 How Is Protein Synthesized?

18.6 How Are Genes Regulated?

18.7 What Are Mutations?

18.8 How and Why Do We Manipulate DNA?

18.9 What Is Gene Therapy?

18.1 How Does DNA Lead to RNA and Protein?

We have seen that the DNA molecule is a storehouse of information. We can compare it to a loose-leaf cookbook, each page of which contains one recipe. The pages are the genes. To prepare a meal, we use a number of recipes.

OWL Online homework for this chapter may be assigned in **GOB OWL**.

FIGURE 18.1 The central dogma of molecular biology. The yellow arrows represent the general cases, and the blue arrows represent special cases in RNA viruses.

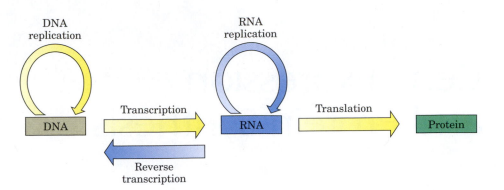

Similarly, to provide a certain inheritable trait, a number of genes (Chapter 17)—segments of DNA—are needed.

Of course, the recipe itself is not the meal. The information in the recipe must be expressed in the proper combination of food ingredients. Similarly, the information stored in DNA must be expressed in the proper combination of amino acids representing a particular protein. The way this expression works is now so well established that it is called the **central dogma of molecular biology.** The dogma states that *the information contained in DNA molecules is transferred to RNA molecules, and then from the RNA molecules the information is expressed in the structure of proteins.* **Gene expression** is the turning on or activation of a gene. Transmission of information occurs in two steps: transcription and translation.

Figure 18.1 shows the central dogma of gene expression. In some viruses (shown in blue), gene expression does not work this way. In some viruses with RNA genomes, replication proceeds from RNA to RNA. In retroviruses, RNA is reverse transcribed to DNA.

Gene expression The activation of a gene to produce a specific protein; it involves both transcription and translation

Transcription

Transcription The process in which information encoded in a DNA molecule is copied into an mRNA molecule

Because the information (that is, the DNA) is in the nucleus of a eukaryotic cell and the amino acids are assembled outside the nucleus, the information must first be carried out of the nucleus. This step is analogous to copying a recipe from a cookbook. All the necessary information is copied, albeit in a slightly different format, as if we were converting the printed page into handwriting. On the molecular level, this task is accomplished by transcribing the information from the DNA molecule into a molecule of messenger RNA, so named because it carries the message from the nucleus to the site of protein synthesis. Other RNAs are similarly transcribed. rRNA is needed to form ribosomes, and tRNA is required to carry out the translation into protein language (Chapter 17). The transcribed information on the different RNA molecules is then carried out of the nucleus.

Translation

Translation The process in which information encoded in an mRNA molecule is used to assemble a specific protein

The mRNA serves as a template on which the amino acids are assembled in the proper sequence. To complete the assembly, the information that is written in the language of nucleotides must be translated into the language of amino acids. The translation is done by another type of RNA, transfer RNA (Section 17.4). An exact, word-to-word translation occurs. Each amino acid in the protein language has a corresponding word in the RNA language. Each word in the RNA language is a sequence of three bases. This correspondence between three bases and one amino acid is called the genetic code (we will discuss the code in Section 18.4).

In higher organisms (eukaryotes), transcription and translation occur sequentially. The transcription takes place in the nucleus. After RNA leaves the nucleus and enters the cytoplasm, the translation takes place there. In lower organisms (prokaryotes), there is no nucleus and thus transcription and translation occur simultaneously in the cytoplasm. This extended form of the "central dogma" was challenged in 2001, when it was found that even in eukaryotes about 15% of the proteins are produced in the nucleus itself. Clearly, some simultaneous transcription and translation do occur even in higher organisms.

We know more about bacterial transcription and translation because they are simpler than the processes operating in higher organisms and have been studied for a longer time. Nevertheless, we will concentrate on studying gene expression and protein synthesis in the eukaryotic systems because they are more relevant to human health care.

18.2 How Is DNA Transcribed into RNA?

Transcription starts when the DNA double helix begins to unwind at a point near the gene that is to be transcribed (Figure 18.2). As we saw in Section 17.3C, nucleosomes form chromatin and higher condensed structures in the chromosomes. To make the DNA available for transcription, these superstructures change constantly. Specific **binding proteins** bind to the nucleosomes, making the DNA become less dense and more accessible. Only then can the enzyme **helicase,** a ring-shaped complex of six protein subunits, unwind the double helix.

Only one strand of the DNA molecule is transcribed. The strand that serves as a template for the formation of RNA has several names, including the **template strand,** the **(−) strand,** and the **antisense strand.** The other strand, while not used as a template, actually has a sequence that matches the RNA that will be produced. This strand is called the **coding strand,** the **(+) strand,** and the **sense strand.** Of these names, coding strand and template strand are the most commonly used.

Ribonucleotides assemble along the unwound DNA strand in the complementary sequence. Opposite each C on the DNA is a G on the growing mRNA; the other complementary bases follow the patterns G ⟶ C, A ⟶ U, and T ⟶ A. The ribonucleotides, when aligned in this way, are linked to form the appropriate RNA.

In eukaryotes, three kinds of **polymerases** catalyze transcription. RNA polymerase I (pol I) catalyzes the formation of most of the rRNA; Pol II catalyzes mRNA formation; and Pol III catalyzes tRNA formation as well as one ribosomal subunit and other small regulatory RNA types, like snRNA. Each enzyme is a complex of 10 or more subunits. Some subunits are unique to each kind of polymerase, whereas other subunits appear in all three polymerases. Figure 18.3 shows the architecture of yeast RNA polymerase II.

Template strand The strand of DNA that serves as the template during RNA synthesis

Coding strand The strand of DNA with a sequence that matches the RNA produced during transcription

ACTIVE FIGURE 18.2 Transcription of a gene. The information in one DNA strand is transcribed to a strand of RNA. The termination site is the locus of termination of transcription. **Go to this book's companion website at www.cengage. com/chemistry/bettelheim to explore an interactive version of this figure.**

FIGURE 18.3 Architecture of yeast RNA polymerase II. Transcription of DNA (helical structure) into RNA (red) is shown. The template strand of DNA is shown in blue and the coding strand in green. Transcription takes place in the clamp region of the active site, shown at center right. The jaws that hold DNA downstream of the active site are shown at lower left.

Courtesy of Professor Roger Kornberg, Stanford University.

The eukaryotic gene has two major parts: the **structural gene** itself, which is transcribed into RNA, and a **regulatory** portion that controls the transcription. The structural gene is made of exons and introns (Figure 18.4). The regulatory portion is not transcribed, but rather has control elements.

One such control element is a **promoter.** On the DNA strand, there is always a sequence of bases that the polymerase recognizes as an **initiation signal,** saying, in essence, "Start here." The promoter is unique to each gene. Besides unique nucleotide sequences, promoters contain **consensus sequences,** such as the TATA box, which gets its name from the sequence beginning, TATAAT. A TATA box lies approximately 26 base pairs upstream—that is, before the beginning of the transcription process (see Figure 18.4). By convention, all sequences of DNA used to describe transcription are given from the perspective of the coding strand. TATA boxes are common to all eukaryotes. All three RNA polymerases interact with their promoter regions via **transcription factors** that are binding proteins.

Another type of control element is an enhancer, a DNA sequence that can be far removed from the promoter region. Enhancers also bind to transcription factors, enhancing transcription above the basal level that would be seen without such binding. Enhancers will be discussed in Section 18.6.

After initiation, the RNA polymerase zips up the complementary bases by forming a phosphate ester bond (Section 11.5) between each ribose and the next phosphate group. This process is called **elongation.**

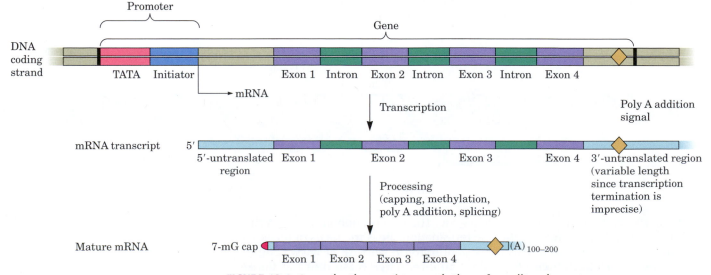

FIGURE 18.4 Organization and transcription of a split eukaryote gene.

At the end of the gene is a **termination sequence** that tells the enzyme, "Stop the transcription." The enzyme Pol II has two different forms. At the C-terminal domain, Pol II has serine and threonine repeats that can be phosphorylated. When Pol II starts the initiation, the enzyme is in its unphosphorylated form. Upon phosphorylation, it performs the elongation process. After termination of the transcription, Pol II is dephosphorylated by a phosphatase. In this manner, Pol II is constantly recycled between its initiation and elongation roles.

The enzyme synthesizes the mRNA molecule from the 5′ to the 3′ end (the zipper can move in only one direction). Because complementary nucleotide chains (RNA and DNA) run in opposite directions, however, the enzyme moves along the DNA template strand in the 3′ ⟶ 5′ direction (Figure 18.2). As the RNA is synthesized, it moves away from the DNA template, which then rewinds to the original double-helix form. Transfer RNA and ribosomal RNA are also synthesized on DNA templates in this manner.

The RNA products of transcription are not necessarily the functional RNAs. Previously, we have seen that in higher organisms mRNA contains exons and introns (Section 17.5). To ensure that mRNA is functional, the transcribed product is capped at both ends. The 5′ end acquires a methylated guanine (7-mG cap) and the 3′ end has a poly-A tail that may contain 100 to 200 adenine residues. Once the two ends are capped, the introns are spliced out in a **post-transcription process** (Figure 18.4). Similarly, a transcribed tRNA must be trimmed and capped, and some of its nucleotides methylated, before it becomes functional tRNA. Functional rRNA also undergoes post-transcriptional methylation.

Example 18.1 DNA Polymerases

Polymerase II both initiates the transcription and performs the elongation. What are the two forms of the enzyme in these processes? What chemical bond formation occurs in the conversion between these two forms?

Solution

The phosphorylated form of Pol II performs the elongation and the unphosphorylated form initiates the transcription. The chemical bond formed in the phosphorylation is the phosphoric ester between the —OH of serine and threonine residues of the enzyme and phosphoric acid.

Problem 18.1

DNA is highly condensed in the chromosomes. What is the sequence of events that enables the transcription of a gene to begin?

18.3 What Is the Role of RNA in Translation?

Translation is the process by which the genetic information preserved in the DNA and transcribed into the mRNA is converted to the language of proteins—that is, the amino acid sequence. Three types of RNA (mRNA, rRNA, and tRNA) participate in the process.

The synthesis of proteins takes place on the ribosomes (Section 17.4). These spheres dissociate into two parts—a larger and a smaller body. Each of these bodies contains rRNA and some polypeptide chains that act as

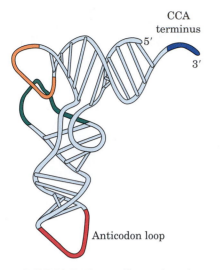

CCA
terminus

5'

3'

Anticodon loop

FIGURE 18.5 Three-dimensional structure of tRNA.

Codon The sequence of three nucleotides in messenger RNA that codes for a specific amino acid

Anticodon A sequence of three nucleotides on tRNA complementary to the codon in mRNA

Genetic code The sequence of triplets of nucleotides (codons) that determines the sequence of amino acids in a protein

enzymes, speeding up the synthesis. In higher organisms, including humans, the larger ribosomal body is called the 60S ribosome, and the smaller one is called the 40S ribosome. The designation "S" refers to *Svedberg,* a measure of density used in centrifugation. In prokaryotes, the corresponding ribosomal subunits are called the 50S and the 30S, respectively. Messenger RNA is attached to the smaller ribosomal body and later joined by the larger body. Together they form a unit on which the mRNA is stretched out. Triplets of bases on the mRNA are called **codons.** After the mRNA is attached to the ribosome in this way, the 20 amino acids are brought to the site, each carried by its own particular tRNA molecule.

The most important segments of the tRNA molecule are (1) the site to which enzymes attach the amino acids and (2) the recognition site. Figure 18.5 shows that the 3′ terminus of the tRNA molecule is single-stranded; this end carries the amino acid.

As we have said, each tRNA is specific for one amino acid only. How does the body make sure that alanine, for example, attaches only to the one tRNA molecule that is specific for alanine? The answer is that each cell carries specific enzymes for this purpose. These **aminoacyl-tRNA synthetases** recognize specific tRNA molecules and amino acids. The enzyme then attaches the amino acid to the terminal group of the tRNA, forming an ester bond.

The second important segment of the tRNA molecule carries the **codon recognition site,** which is a sequence of three bases called an **anticodon** located at the opposite end of the molecule in the three-dimensional structure of tRNA (see Figure 18.5). This triplet of bases is complementary to the sequence of the codon and allows the tRNA to align with the mRNA. Thus, the mRNA and tRNA are antiparallel at the point of contact.

18.4 What Is the Genetic Code?

By 1961, it was apparent that the order of bases in a DNA molecule corresponds to the order of amino acids in a particular protein, but the code was unknown. Obviously, it could not be a one-to-one correspondence. There are only four bases, so if A coded for glycine, G for alanine, C for valine, and T for serine, there would be 16 amino acids that could not be coded.

In 1961, Marshall Nirenberg (1927–) and his co-workers attempted to break the code in a very ingenious way. They made a synthetic molecule of mRNA consisting of uracil bases only. They put this molecule into a cell-free system that synthesized proteins and then supplied the system with all 20 amino acids. The only polypeptide produced was a chain consisting solely of the amino acid phenylalanine. This experiment showed that the code for phenylalanine must be UUU or some other multiple of U.

A series of similar experiments by Nirenberg and other workers followed, and by 1967 the entire genetic code had been broken. *Each amino acid is coded for by a sequence of three bases,* called a *codon.* Table 18.1 shows the complete code.

The first important aspect of the **genetic code** is that it is almost universal. In virtually every organism, from a bacterium to an elephant to a human, the same sequence of three bases codes for the same amino acid. The universality of the genetic code implies that all living matter on Earth arose from the same primordial organisms. This finding is perhaps the strongest evidence supporting Darwin's theory of evolution.

Some exceptions to the genetic code in Table 18.1 occur in mitochondrial DNA. Because of that fact and other evidence, it is thought that the mitochondrion may have been an ancient free-living entity. During evolution,

TABLE 18.1 The Genetic Code

First Position (5'-end)	Second Position								Third Position (3'-end)
	U		**C**		**A**		**G**		
U	UUU	Phe	UCU	Ser	UAU	Tyr	UGU	Cys	U
	UUC	Phe	UCC	Ser	UAC	Tyr	UGC	Cys	C
	UUA	Leu	UCA	Ser	UAA	Stop	UGA	Stop	A
	UUG	Leu	UCG	Ser	UAG	Stop	UGG	Trp	G
C	CUU	Leu	CCU	Pro	CAU	His	CGU	Arg	U
	CUC	Leu	CCC	Pro	CAC	His	CGC	Arg	C
	CUA	Leu	CCA	Pro	CAA	Gln	CGA	Arg	A
	CUG	Leu	CCG	Pro	CAG	Gln	CGG	Arg	G
A	AUU	Ile	ACU	Thr	AAU	Asn	AGU	Ser	U
	AUC	Ile	ACC	Thr	AAC	Asn	AGC	Ser	C
	AUA	Ile	ACA	Thr	AAA	Lys	AGA	Arg	A
	AUG*	Met	ACG	Thr	AAG	Lys	AGG	Arg	G
G	GUU	Val	GCU	Ala	GAU	Asp	GGU	Gly	U
	GUC	Val	GCC	Ala	GAC	Asp	GGC	Gly	C
	GUA	Val	GCA	Ala	GAA	Glu	GGA	Gly	A
	GUG	Val	GCG	Ala	GAG	Glu	GGG	Gly	G

*AUG also serves as the principal initiation codon.

it developed a symbiotic relationship with eukaryotic cells. For example, some of the respiratory enzymes located on the cristae of the mitochondrion (see Section 19.2) are encoded in the mitochondrial DNA, and other members of the same respiratory chain are encoded in the nucleus of the eukaryotic cell.

There are 20 amino acids in proteins, but 64 possible combinations of four bases into triplets. All 64 codons (triplets) have been deciphered. Three of them—UAA, UAG, and UGA—are "stop signs." They terminate protein synthesis. The remaining 61 codons code for amino acids. Because there are only 20 amino acids, there must be more than one codon for each amino acid. Indeed, some amino acids have as many as six codons. Leucine, for example, is coded by UUA, UUG, CUU, CUC, CUA, and CUG.

Just as there are three stop signs in the code, there is also an initiation sign. The initiation sign is AUG, which is also the codon for the amino acid methionine. This means that, in all protein synthesis, the first amino acid initially put into the protein is always methionine. Methionine can also be put into the middle of the chain.

Although all protein synthesis starts with methionine, most proteins in the body do not have a methionine residue at the N-terminus of the chain. In most cases, the initial methionine is removed by an enzyme before the polypeptide chain is completed. The code on the mRNA is always read in the 5' $\longrightarrow$ 3' direction, and the first amino acid to be linked to the initial methionine is the N-terminal end of the translated polypeptide chain.

The genetic code is said to be continuous and unpunctuated. If the mRNA is AUGGGCCAA, then the AUG is one codon and specifies the first amino acid. The GGC is the second codon and specifies the second amino acid. The CCA is the third codon and specifies the third amino acid. There are no overlapping codons and no nucleotides interspersed.

Example 18.2 The Genetic Code

Which amino acid is represented by the codon CGU? What is its anticodon?

Solution

Looking at Table 18.1, we find that CGU corresponds to arginine; the anticodon is GCA (read 3′ to 5′ to show how the codon and anticodon match up).

Problem 18.2

What are the codons for histidine? What are the anticodons?

18.5 How Is Protein Synthesized?

So far we have met the molecules that participate in protein synthesis (Section 18.3) and the dictionary of the translation, the genetic code. Now let us look at the actual mechanism by which the polypeptide chain is assembled.

There are four major stages in protein synthesis: activation, initiation, elongation, and termination. At each stage, a number of molecular entities participate in the process (Table 18.2). We will look specifically at prokaryotic translation because it has been studied longer, and we have more complete information on it. The details of eukaryotic translation are very similar, however.

A. Activation

Each amino acid is first activated by reacting with a molecule of ATP:

TABLE 18.2 Molecular Components of Reactions at Four Stages of Protein Synthesis

Stage	Molecular Components
Activation	Amino acids, ATP, tRNAs, aminoacyl-tRNA synthetases
Initiation	fMet–tRNA^{fMet}, 30S ribosome, initiation factors, mRNA with Shine–Dalgarno sequence, 50S ribosome, GTP
Elongation	30S and 50S ribosomes, aminoacyl-tRNAs, elongation factors, mRNA, GTP
Termination	Release factors, GTP

The activated amino acid is then bound to its own particular tRNA molecule with the aid of an enzyme (a synthetase) that is specific for that particular amino acid and that particular tRNA molecule:

The different synthetases recognize their substrates by stretches of nucleotide sequences on the tRNA. The specific recognition by an enzyme, aminoacyl-tRNA synthetase, of its proper tRNA and amino acid is often referred to as the **second genetic code.** This step is very important because once the amino acid is on the tRNA, there is no other opportunity to check for the correct pairing. In other words, the anticodon of the tRNA will match up with its correct codon on the mRNA regardless of whether it is carrying the correct amino acid, so the aminoacyl-tRNA synthetases have to get it right.

B. Initiation

The initiation stage consists of three steps:

1. **Forming the pre-initiation complex** To initiate the protein synthesis a unique tRNA is used, designated as **tRNA^{fMet}**. This tRNA carries a formylated methionine (fMet) residue, but it is used solely for the initiation step. It is attached to the 30S ribosomal body and forms the pre-initiation complex, along with GTP, [(Figure 18.6)(1)]. Just as in transcription, each step in translation is aided by a number of factors; these proteins are called **initiation factors.**

2. **Migration to mRNA** Next, the pre-initiation complex binds to the mRNA (2). The ribosome is aligned on the mRNA by recognizing a special RNA sequence called the **Shine-Dalgarno** sequence, which is complementary to a sequence on the 30S ribosomal subunit. The anticodon of the fMet–tRNA^{fMet}, UAC, lines up against the start codon, AUG.

3. **Forming the full ribosomal complex** The 50S ribosomal body joins the 30S ribosomal complex (3). The complete ribosome carries three sites. The one shown in the middle in Figure 18.6 is called the **P site,** because the growing peptide chain will bind there. The one next to it on the right is called the **A (acceptor) site,** because it accepts the incoming tRNA bringing the next amino acid. As the full initiation complex is completed, the initiation factors dissociate and the GTP is hydrolyzed to GDP.

C. Elongation

1. **Binding to the A site** At this point, the A site is vacant, and each of the aminoacycl-tRNA molecules can try to fit itself in. However, only one of the tRNAs carries the right anticodon that corresponds to the next codon on the mRNA. This is an alanine tRNA in Figure 18.6. The binding

ACTIVE FIGURE 18.6 The formation of an initiation complex. The 30S ribosomal subunit binds to mRNA and the fMet-tRNA^{fMet} in the presence of GTP and the three initiation factors, forming the 30S initiation complex (Step 1). The 50S ribosomal subunit is added, forming the full initiation complex (Step 2). **Go to this book's companion website at www.cengage.com/chemistry/bettelheim to explore an interactive version of this figure.**

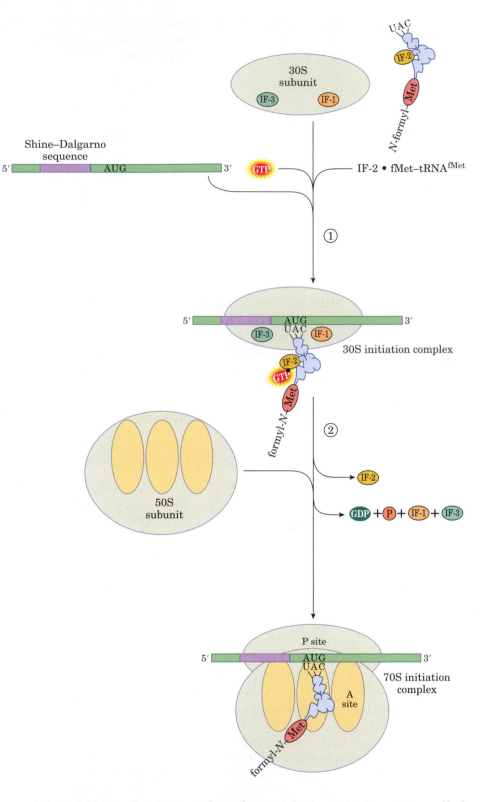

of this tRNA to the A site takes place with the aid of proteins called **elongation factors** and GTP [Figure 18.7 (2)].

2. **Forming the first peptide bond** At the A site, the new amino acid, alanine (Ala), is linked to the fMet in a peptide bond by the enzyme **peptidyl transferase.** The empty tRNA remains on the P site [Figure 18.7 (3)].

3. **Translocation** In the next phase of elongation, the whole ribosome moves one codon along the mRNA. Simultaneously with this move, the

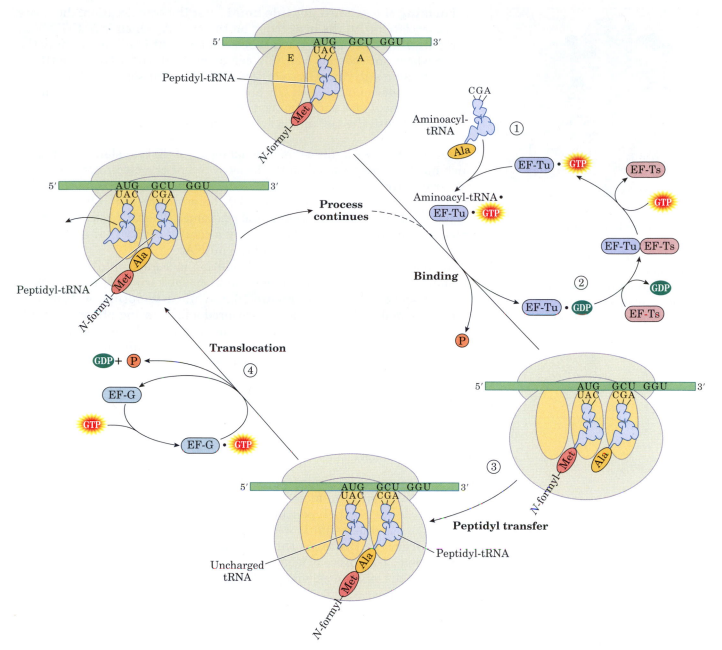

ACTIVE FIGURE 18.7 The steps in chain elongation. (1) An aminoacyl-tRNA is bound to the A site on the ribosome. Elongation factors and GTP are required. The P site on the ribosome is already occupied. (2) Elongation factors are recycled to prepare another incoming tRNA, and GTP is hydrolyzed. (3) The peptide bond is formed, leaving an uncharged tRNA at the P site. (4) In the translocation step, the uncharged tRNA is pushed into the E site and more GTP is hydrolyzed. The A site is now over the next codon on the mRNA. **Go to this book's companion website at www.cengage.com/chemistry/bettelheim to explore an interactive version of this figure.**

dipeptide is **translocated** from the A site to the P site (4). The empty tRNA is moved to the E site. When this cycle occurs one more time, the empty tRNA will be ejected and go back to the pool of tRNA that is available for activation with an amino acid.

FIGURE 18.8 Ribosome in action. The lower yellow half represents the 30S ribosome; the upper blue half represents the 50S ribosome. The yellow and green twisted cones are tRNAs, and the chain of beads stand for mRNA. The elongation factors are in dark blue.

Courtesy of Dr. J. Frank, Wadsworth Center, Albany, New York

4. **Forming the second peptide bond** After the translocation, the A site is associated with the next codon on the mRNA, which is 5′ GGU 3′ in Figure 18.7. Once again, each tRNA can try to fit itself in, but only the one whose anticodon is 5′ACC 3′ can align itself with GGU. This tRNA, which carries glycine (Gly), now comes in. The transferase establishes a new peptide bond between Gly and Ala, moving the dipeptide from the P site to the A site and forming a tripeptide. These elongation steps are repeated until the last amino acid is attached.

Figure 18.8 shows a three-dimensional model of the translational process, which has been constructed on the basis of recent cryoelectron microscopy and x-ray diffraction studies. This model shows how the elongation factor proteins (in dark blue) fit into a cleft between the 50S (blue) and the 30S (pale yellow) bodies of prokaryotic ribosomes. The tRNAs on the P site (green) and on the A site (yellow) occupy a central cavity in the ribosomal complex. The orange beads represent the mRNA.

The mechanism of peptide bond formation is a nucleophilic attack by the amino group of the A site amino acid upon the carbonyl group of the P site amino acid, as shown in Figure 18.9. While attempting to study this mechanism in detail, researchers discovered a fascinating phenomenon. It turns out that in the vicinity of the nucleophilic attack, there is no protein in the ribosome that could catalyze such a reaction. The only chemical groups nearby that could catalyze a reaction are on a purine of the ribosomal RNA. Thus the ribosome is a ribozyme. Previously, catalytic RNA had been found only in some RNA splicing reactions, but here is a scenario in which RNA catalyzes one of the principal reactions of life.

D. Termination

After the final translocation, the next codon reads "stop" (UAA, UGA, or UAG). At this point, no more amino acids can be added. Releasing factors then cleave the polypeptide chain from the last tRNA via a GTP-requiring mechanism that is not yet fully understood. The tRNA itself is released from the P site. At the end, the whole mRNA is released from the ribosome. This process is shown in Figure 18.10. While the mRNA is attached to the ribosomes, many polypeptide chains are synthesized on it simultaneously.

FIGURE 18.9 Peptide bond formation in protein synthesis. Nucleophilic attack by the amino group of the A-site aminoacyl-tRNA on the carbonyl carbon of the P-site peptidyl-tRNA is facilitated when a purine moiety of the rRNA abstracts a proton.

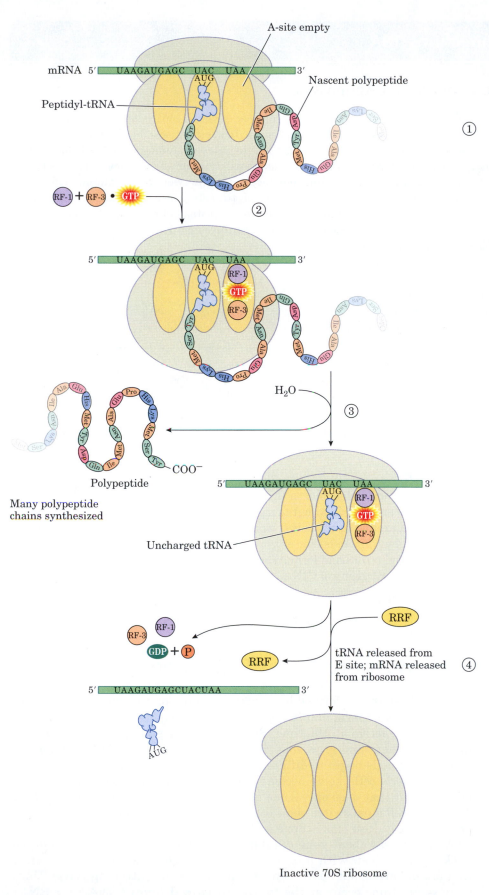

FIGURE 18.10 The events in peptide chain termination. As the ribosome moves along the mRNA, it encounters a stop codon, such as UAA (Step 1). Release factors and GTP bind to the A-site (Step 2). The peptide is hydrolyzed from the tRNA (Step 3). Finally, the entire complex dissociates, and the ribosome, mRNA, and other factors can be recycled (Step 4).

Chemical Connections 18A

Breaking the Dogma: The Twenty-First Amino Acid

Many amino acids, such as citrulline and ornithine found in the urea cycle (Chapter 20), are not building blocks of proteins. Other nonstandard amino acids such as hydroxyproline (Chapter 14) are formed after translation by post-translational modification. When discussing amino acids and translation in the past, the magic number was always 20. That is, only 20 standard amino acids were put onto tRNA molecules for protein synthesis. In the late 1980s, another amino acid was found in proteins from eukaryotes and prokaryotes alike, including humans. It is selenocysteine, a cysteine residue that has the sulfur replaced by selenium.

Selenocysteine is formed by placing a serine onto a special tRNA molecule called tRNA^{sec}. Once bound, the oxygen in the serine side chain is replaced by selenium.

This tRNA molecule has an anticodon that matches the UGA "stop" codon. In special cases, the UGA is not read as a "stop"; rather, the selenocysteine-tRNA^{sec} is loaded into the A site and translation continues. Some are, therefore, calling selenocysteine the twenty-first amino acid. The methods by which the cell knows when to put selenocysteine into the protein instead of reading UGA as a "stop" codon remain under investigation.

Selenocysteine

Example 18.3 Translation

A tRNA has an anticodon, 5′ AAG 3′. Which amino acid will this tRNA carry? What are the steps necessary for the amino acid to bind to the tRNA?

Solution

The amino acid is leucine, as the codon is 5′ CUU 3′. Remember that sequences are read from left to right as 5′ ⟶ 3′, so you have to flip the anticodon around to see how it would bind to the codon. Leucine has to be activated by ATP. A specific enzyme, leucine-tRNA synthetase, catalyzes the carboxyl ester bond formation between the carboxyl group of leucine and the —OH group of tRNA.

Problem 18.3

What are the reactants in the reaction forming valine-tRNA?

18.6 How Are Genes Regulated?

Every embryo that is formed by sexual reproduction inherits its genes from both the parent sperm and the egg cells. But the genes in its chromosomal DNA are not active all the time. Rather, they are switched on and off during development and growth of the organism. Soon after formation of the embryo, the cells begin to differentiate. Some cells become neurons, some become muscle cells, some become liver cells, and so on. Each cell is a specialized unit that uses only some of the many genes it carries in its DNA. Thus each cell must switch some of its genes on and off—either permanently or temporarily. How this is done is the subject of **gene regulation.**

We know less about gene regulation in eukaryotes than in the simpler prokaryotes. Even with our limited knowledge, however, we can state that organisms do not have a single, unique way of controlling genes. Many gene regulations occur at the **transcriptional level** (DNA ⟶ RNA). Others operate at the **translational level** (mRNA ⟶ protein). A few of these processes are listed here as examples.

Gene regulation The control process by which the expression of a gene is turned on or off. Because RNA synthesis proceeds in one direction (5′ ⟶ 3′), the gene (DNA) to be transcribed runs from 3′ to 5′. Thus the control sites are in front of, or upstream of, the 3′ end of the structural gene.

Chemical Connections 18B

Viruses

Nucleic acids are essential for life as we know it. No living thing can exist without them because they carry the information necessary to make protein molecules. The smallest forms of life, the viruses, consist only of a molecule of nucleic acid surrounded by a "coat" of protein molecules. In some viruses, the nucleic acid is DNA; in others, it is RNA. No virus has both. Whether viruses can be considered a true form of life became a topic of debate recently. In the summer of 2002, a group of scientists at the State University of New York, Stony Brook, reported that they had synthesized the polymyelitis virus in the laboratory from fragments of DNA. This new "synthetic" virus caused the same polio symptoms and death as the wild virus.

The shapes and sizes of viruses vary greatly, as shown in the figure. Because their structures are so simple, viruses cannot reproduce themselves in the absence of other organisms. They carry DNA or RNA but do not have the nucleotides, enzymes, amino acids, and other molecules necessary to replicate their nucleic acid (Section 17.6) or to synthesize proteins (Section 18.5). Instead, viruses invade the cells of other organisms and cause those cells (the hosts) to do these tasks for them. Typically, the protein coat of a virus remains outside the host cell, attached to the cell membrane, while the DNA or RNA is pushed inside. Once the viral nucleic acid is inside the cell, the cell stops replicating its own DNA and making its own proteins. Instead, it replicates the viral nucleic acid and synthesizes the viral protein, according to the instructions on the viral nucleic acid. One host cell can make many copies of the virus.

In many cases, the cell bursts when a large number of new viruses have been synthesized, sending the new viruses out into the intercellular material, where they can infect other cells. This kind of process causes the host organism to get sick, and perhaps to die. Among the many human diseases caused by viruses are measles, hepatitis, mumps, influenza, the common cold, rabies, and smallpox. There is no cure for most viral diseases. Antibiotics, which can kill bacteria, have no effect on viruses. So far, the best defense against these diseases has been immunization (Chemical Connections 23B), which under the proper circumstances can work spectacularly well. Smallpox, once one of the most dreaded diseases, has been eradicated from this planet by many years of vaccination, and comprehensive programs of vaccination against such diseases as polio and measles have greatly reduced the incidence of these diseases.

Lately, a number of antiviral agents have been developed. They completely stop the reproduction of viral nucleic acids (DNA or RNA) inside infected cells without preventing the DNA of normal cells from replicating. One such drug is called vidarabine, or Ara-A, and is sold under the trade name Vira-A.

Herpes virus

Poliomyelitis virus

Influenza virus

The shape of three viruses. (Illustration by Irving Geis, from *Scientific American*, January 1963. Rights owned by Howard Hughes Medical Institute. Not to be reproduced without permission.)

Vidarabine

Antiviral agents often act like anticancer drugs in that they have structures similar to one of the nucleotides necessary for the synthesis of nucleic acids. Vidarabine is the same as adenosine, except that the sugar is arabinose instead of ribose. Vidarabine is used to fight a life-threatening viral illness, herpes encephalitis. It is also effective in neonatal herpes infection and chickenpox. However, like many other anticancer and antiviral drugs, vidarabine is toxic, causing nausea and diarrhea. In some cases, it has caused chromosomal damage.

A. Control at the Transcriptional Level

In eukaryotes, transcription is regulated by three entities: promoters, enhancers, and response elements.

1. Promoters of a gene are located adjacent to the transcription site. They are defined by an initiator and conserved sequences, such as a TATA box (see also Section 18.2 and Figure 18.4) or one or more copies of other sequences, such as the GGGCGG sequence called a GC box. In eukaryotes, the enzyme RNA-polymerase has little affinity for binding to DNA. Instead, different transcription factors, or binding proteins, bind to the different modules of the promoter.

   There are two basic types of transcription factors. The first is called a **general transcription factor (GTF).** These proteins form a complex with RNA polymerase and the DNA and help to position the RNA polymerase correctly and stimulate the initiation of transcription. For transcription of genes that will lead to mRNA (that is, Pol II transcription), there are six GTFs, all named as TFII and then a letter, for Pol II transcription factor. All of these transcription factors are necessary to establish the initiation of transcription. As can be seen in Figure 18.11, the events in the initiation of Pol II transcription are very complicated. Six transcription factors must bind to the DNA and RNA polymerase to initiate transcription. They first form the **pre-initiation complex.** The critical event in starting the transcription is the conversion to the **open complex,** which involves the phosphorylation of the C-terminal end of the RNA polymerase. Only when the open complex has formed can transcription take place. During elongation, three transcription factors (B, E, and H) are released. Transcription factor F remains bound to Pol II with D bound to the TATA box. Only factor F continues with the polymerase.

   With the aid of these transcription factors, the promoter functions to control the transcription on a steady, normal level. Transcription factors may allow the synthesis of mRNA (and from there the target protein) to vary by a factor of 1 million. This wide variation in eukaryotic cells is exemplified by the α-A-crystallin gene, which can be expressed in the lens of the eye, at a rate a millionfold higher than that in the liver cell of the same organism.

2. Another group of transcription factors speeds up the transcription process by binding to DNA sequences that may be located several thousand nucleotides away from the transcription site. These sequences are known as **enhancers.** To stimulate transcription, an enhancer is brought to the vicinity of the promoter by the formation of a loop. Figure 18.12 shows how the transcription factor binds to the enhancer element and forms a bridge to the basal transcription unit. This complex then allows the RNA polymerase II to speed up the transcription when higher-than-normal production of proteins is needed.

   Other DNA sequences bind transcription factors but have the opposite effect—they slow down transcription. These are called **silencers.**

3. The third type of transcription control involves a type of enhancer called **response elements.** These enhancers are activated by their transcription factors in response to an outside stimulus. The stimulus may be heat shock, heavy metal toxicity, or simply a hormonal signal, such as the binding of a steroid hormone to its receptor. The response element of steroids is in front of, and 260 base pairs upstream from, the starting point of transcription. Only the receptor with the bound steroid hormone can interact with its response element, thereby initiating transcription.

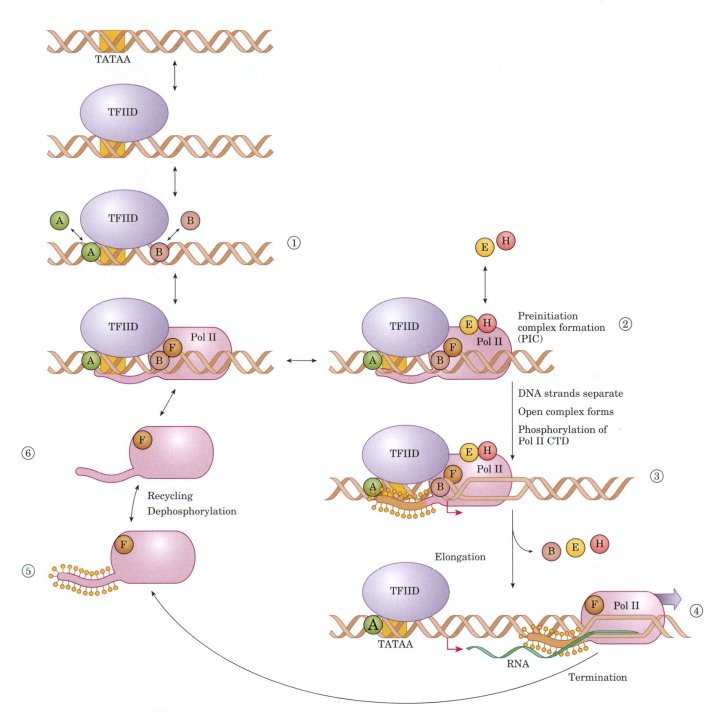

ACTIVE FIGURE 18.11 A schematic representation of the order of events of Pol II transcription. General Transcription Factor TFIID binds to the TATA box on the DNA and recruits TFIIA and TFIIB (Step 1). RNA Polymerase II carrying TFIIF binds to the DNA, followed by TFIIE and TFIIH to form the preinitiation complex (PIC) (Step 2). The C-terminal domain of Pol II is phosphorylated and the DNA strands are separated to form the open complex (Step 3). TFIIB, TFIIE, and TFIIH are released as the polymerase synthesizes RNA in the process of elongation (Step 4). Transcription is terminated when the mRNA is complete and the Pol II is released (Step 5). The Pol II is dephosphorylated and is ready to be recycled for another round of transcription (Step 6). **Go to this book's companion website at www.cengage.com/chemistry/bettelheim to explore an interactive version of this figure.**

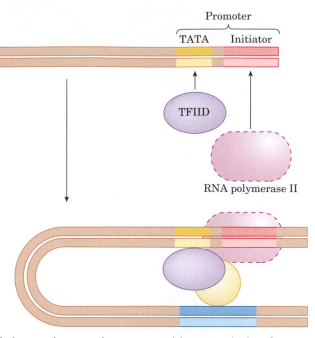

FIGURE 18.12 DNA looping brings enhancers in contact with transcription factors and RNA polymerase.

FIGURE 18.13 Cys$_2$His$_2$ zinc finger motifs. (*a*) The coordination between zinc and cysteine and histidine residues. (*b*) The secondary structure. (Adapted from Evans R. M. and Hollenberg, S. M., *Cell, 52:1, 1988,* Figure 1).

The difference between an enhancer and a response element is largely a matter of our own understanding of the system. We call something a response element when we understand the bigger picture of how controlling the gene is related to a pattern of metabolism. Many response elements may be controlling a particular process, and a given gene may be under the control of more than one response element.

4. Transcription does not occur at the same rate throughout the cell's entire life cycle. Instead, it is accelerated or slowed down as the need arises. The signal to speed up transcription may originate from outside of the cell. One such signal, in the GTP–adenylate cyclase–cAMP pathway (Section 16.5B), produces **phosphorylated protein kinase.** This enzyme enters the nucleus, where it phosphorylates transcription factors, which aid in the transcription cascade.

How do these transcription factors find the specific gene control sequences into which they fit, and how do they bind to them? The interaction between the protein and DNA involves nonspecific electrostatic interactions (positive ions attracting negative ions and repelling other positive ions) as well as more specific hydrogen bonding. The transcription factors find their targeted sites by twisting their protein chains so that a certain amino acid sequence is present at the surface. One such conformational twist is provided by **metal-binding fingers** (Figure 18.13). These finger shapes are created by ions, which form covalent bonds with the amino acid side chains of the protein.

The zinc fingers interact with specific DNA (or sometimes RNA) sequences. The recognition comes by hydrogen bonding between a nucleotide (for example, guanine) and the side chain of a specific amino acid (for example, arginine). Zinc fingers allow the proteins to bind in the major groove of DNA, as shown in Figure 18.14.

Besides metal-binding fingers, at least two other prominent transcription factors exist: **helix-turn-helix** and **leucine zipper.**

B. Control on the Post-transcriptional Level

In the spring of 2000, scientists were eagerly awaiting the results of the Human Genome Project and an accurate count of the number of genes in the human genome. The odds-on favorite would have been 100,000 to 150,000 genes. After all, it was known that humans produce 90,000 different proteins. The dogma stated that "one gene leads to one mRNA leads to one protein." The only exceptions to this rule were thought to occur in the production of antibodies and other immunoglobulin-based proteins. These proteins were known to undergo a type of post-transcriptional modification called alternative splicing, whereby the primary mRNA transcript can be spliced in different ways to give multiple mature mRNAs, and therefore multiple proteins.

It came as a big shock, therefore, when the data revealed that humans have about 30,000 genes, a number close to that of the roundworm or corn. If 30,000 genes can lead to 90,000 proteins, there must be far more alternative splicing to account for it. Scientists now believe that splicing RNA in different ways is a very important process that leads to the differences in species that are otherwise similar. Chimpanzees and humans, for example, share 99% of their DNA. They also produce very similar protein complements. However, significant differences have been found in some tissues, most notably the brain, where certain human genes are more active and others generate different proteins by alternative splicing.

Figure 18.15 summarizes the various ways that alternative splicing can produce many different proteins. Exons may be included in all products, or they may be present on only some of them. Different splicing sites can appear either on the 5′ or 3′ side. In some cases, introns may even be retained in the final product.

Alternative splicing provides another powerful technique for controlling gene regulation. Within the same cell or the same organism, different genes can be spliced in different ways at different times, controlling what the eventual products of the gene are.

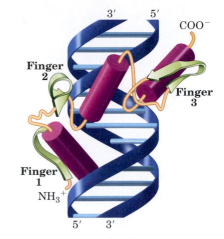

FIGURE 18.14 Zinc finger proteins follow the major groove of DNA. (Adapted with permission from Pavletich N. and Pabo C. O., *Science* 252: 809, 1991 Figure 2. Copyright © 1991 AAAS.)

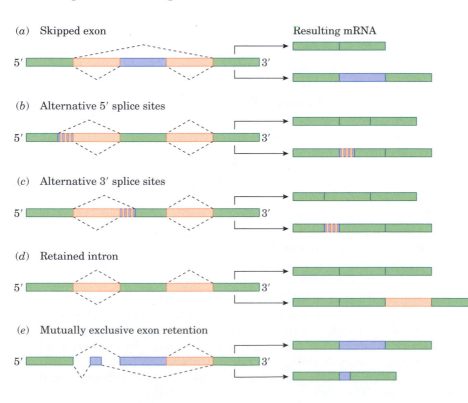

(a) Skipped exon

(b) Alternative 5′ splice sites

(c) Alternative 3′ splice sites

(d) Retained intron

(e) Mutually exclusive exon retention

Resulting mRNA

■ Exon always spliced in ■ Exon alternatively spliced ■ Intron

FIGURE 18.15 Alternate splicing. A gene's primary transcript can be edited in several different ways where splicing activity is indicated by dashed lines. An exon may be left out (*a*). Splicing machinery may recognize alternative 5′ splice sites for an intron (*b*) or alternative 3′ splice sites (*c*). An intron may be retained in the final mRNA transcript (*d*). Exons may be retained on a mutually exclusive basis (*e*). (*Scientific American*).

C. Control on the Translational Level

During translation, a number of mechanisms ensure quality control.

1. **The specificity of a tRNA for its unique amino acid** First, the attachment of the proper amino acid to the proper tRNA must be achieved. The enzyme that catalyzes this reaction, aminoacyl-tRNA synthetase (AARS), is specific for each amino acid. For those amino acids that have more than one type of tRNA, the same synthetase catalyzes the reaction for all of the tRNA types for that amino acid. The AARS enzymes recognize their tRNAs by specific nucleotide sequences. Furthermore, the active site of the enzyme has two **sieving portions.** For example, in isoleucyl-tRNA synthetases, the first sieve excludes any amino acids that are larger than isoleucine. If a similar amino acid such as valine, which is smaller than isoleucine, arrives at the active site, the second sieve eliminates it. The second sieving site is thus a "proofreading" site.

2. **Recognition of the stop codon** Another quality-control measure occurs at the termination. The stop codons must be recognized by release factors, leading to the release of the polypeptide chain and allowing the recycling of the ribosomes. Otherwise, a longer polypeptide chain may be toxic. The release factor combines with GTP and binds to the ribosomal A site when that site is occupied by the termination codon. Both the GTP and the peptidyl-tRNA ester bond are hydrolyzed. This hydrolysis releases the polypeptide chain and the deacylated tRNA. Finally, the ribosome dissociates from the mRNA. As we saw in Chemical Connections 18A, sometimes the stop codon is used to continue translating by inserting a very uncommon amino acid, such as selenocysteine.

3. **Post-translational Controls**

   (a) *Removal of methionine.* In most proteins, the methionine residue at the N-terminus, which was added in the initiation step, is removed. A special enzyme, methionine aminopeptidase, cleaves the peptide bond. In the case of prokaryotes, when the N-terminus is supposed to be methionine, another enzyme cleaves off the formyl group.

   (b) *Chaperoning.* The tertiary structure of a protein is largely determined by the amino acid sequence (primary structure). Proteins begin to fold even as they are synthesized on ribosomes. Nevertheless, misfolding may occur due to mutation in a gene, lack of fidelity in transcription, or translational errors. All of these errors may lead to aggregation of misfolded proteins that can be detrimental to the cell—for example, as seen in amyloid diseases, such as Alzheimer's disease or Jakob-Creutzfeldt disease. Certain proteins in living cells, called **chaperones,** help the newly synthesized polypeptide chains to fold properly. They recognize hydrophobic regions exposed on unfolded proteins and bind to them. Chaperones then shepherd the proteins to the biologically desirable folding as well as to their local destinations within the cell.

   (c) *Degradation of misfolded proteins.* A third post-translational control exists in the form of **proteasomes.** These cylindrical assemblies include a number of protein subunits with proteolytic activity in the core of the cylinder. If the rescue by chaperones fails, proteases may degrade the misfolded protein first by targeting it with ubiquitination (see Chemical Connections 20E) and finally by proteolysis.

18.7 What Are Mutations?

In Section 17.6, we saw that the base-pairing mechanism provides an almost perfect way to copy a DNA molecule during replication. The key word here is "almost." No machine, not even the copying mechanism of DNA replication, is totally without error. It has been estimated that, on average, one error occurs for every 10^{10} bases (that is, one in 10 billion). An error in the copying of a sequence of bases is called a **mutation.** Mutations can occur during replication. Base errors can also occur during transcription (a non-inheritable error).

These errors may have widely varying consequences. For example, the codon for valine in mRNA can be GUA, GUG, GUC, or GUU. In DNA, these codons correspond to GTA, GTG, GTC, and GTT, respectively. Assume that the original codon in the DNA is GTA. If a mistake is made during replication and GTA is spelled as GTG in the copy, there will be no harmful mutation. Instead, when a protein is synthesized, the GTG will appear on the mRNA as GUG, which also codes for valine. Therefore, although a mutation has occurred, the same protein is manufactured.

All sequences on DNA are given as coding strand sequences. Thus the codon, which is on the mRNA, has the same sequence as the coding strand DNA, except that T is replaced by U.

Chemical Connections 18C

Mutations and Biochemical Evolution

We can trace the genetic relationship of different species through the variability of their amino acid sequences in different proteins. For example, the blood of all mammals contains hemoglobin, but the amino acid sequences of the hemoglobins are not identical. In the table below, we see that the first 10 amino acids in the β-globin of humans and gorillas are exactly the same. In fact, there is only one amino acid difference, at position 104, between us and apes. The β-globin of the pig differs from ours at 10 positions, of which 2 are in the N-terminal decapeptide. That of the horse differs from ours in 26 positions, of which 4 are in this decapeptide. β-Globin seems to have gone through many mutations during the evolutionary process, because only 26 of the 146 sites are invariant—that is, exactly the same in all species studied so far.

The relationship between different species can also be established by finding similarities in their mRNA primary structures. Because the mutations actually occurred on the original DNA molecule and then were perpetuated in the progeny by the mutant DNA, it is instructive to learn how a point mutation may occur in different species. Looking at position 4 of the β-globin molecule, we see a change from serine to threonine. The code for serine is AGU or AGC, whereas that for threonine is ACU or ACC (Table 18.1). Thus a change from G to C in the second position of the codon created the divergence between the β-globins of humans and horses. The genes of closely related species, such as humans and apes, have very similar primary structures, presumably because these two species diverged on the evolutionary tree only recently. In contrast, species far removed from each other diverged long ago and have undergone more mutations, which show up as differences in the primary structures of their DNA, mRNA, and consequently proteins.

The *number* of amino acid substitutions is significant in the evolutionary process caused by mutation, but the *kind* of substitution is even more important. If the substitution involves an amino acid with physicochemical properties similar to those of the amino acid in the ancestor protein, the mutation is most probably viable. For example, in human and gorilla β-globin, position 4 is occupied by threonine, but it is occupied by serine in the pig and horse. Both amino acids provide an —OH carrying side chain.

| | Amino Acid Sequence of the N-Terminal Decapeptides of β-Globin in Different Species | | | | | | | | | |
| | Position | | | | | | | | | |
Species	1	2	3	4	5	6	7	8	9	10
Human	Val	His	Leu	Thr	Pro	Glu	Glu	Lys	Ser	Ala
Gorilla	Val	His	Leu	Thr	Pro	Glu	Glu	Lys	Ser	Ala
Pig	Val	His	Leu	Ser	Ala	Glu	Glu	Lys	Ser	Ala
Horse	Val	Glu	Leu	Ser	Gly	Glu	Glu	Lys	Ala	Ala

Chemical Connections 18D

Silent Mutations

A silent mutation is a mutation that changes DNA but not the amino acid incorporated. For example, if the DNA coding strand has a TTC, the mRNA will be UUC and it will code for phenylalanine. If a mutation in the DNA changes the sequence to TTT, the DNA has undergone a silent mutation because in the resultant mRNA, UUU and UUC both code for the same amino acid. At least that is what scientists believed for decades. Recent evidence, however, has shown that this is not always true. Researchers at the National Cancer Institute were studying a gene called *MDR1*, which is named for its association with multiple drug resistance in tumor cells. They had sequences of this gene and knew that there were some common silent mutations. Interestingly, they discovered that there was a response to silent mutations of this gene that influenced patients' response to certain drugs. A silent mutation leading to an observable change was striking, as a silent mutation should have no effect on the final product.

Apparently, not all codons are translated equally. Different codons may require alternate versions of the tRNA for a particular amino acid. Even though the amino acid incorporated is the same, the pace with which the ribosome is able to incorporate the amino acid differs depending on which codon it is. As shown below in the figure, translation kinetics can affect the form of the final protein. If the wild type codon is used, translation proceeds normally and produces the normal conformation of the protein. However, if a silent mutation changes the pace of the movement of the ribosome, folding differences, will result in the creation of an abnormal protein conformation.

Reprinted by permission of *Science* magazine, from SNPs, Silent but not Invisible by Anton A. Komar, *Science* 315, 466-467 [2007].

Translation kinetics and protein folding. Unaffected translation kinetics results in a correctly folded protein. Abnormal kinetics, caused by the ribosome moving faster or slower through certain mRNA regions, can produce a different final protein conformation. Abnormal kinetics may arise from a silent single nucleotide polymorphism (SNP) in a gene that creates a codon synonymous to the wild type codon. However, this synonymous codon substitution may lead to different kinetics of mRNA translation, thus yielding a proetin with a different final structure and function.

Now assume that the original sequence in the gene's DNA is GAA, which will also be GAA in the mRNA and codes for glutamic acid. If a mutation occurs during replication and GAA becomes TAA, a very serious mutation will have occurred. The TAA on the DNA coding strand would be UAA on the mRNA, which does not code for any amino acid but rather is a stop signal. Thus, instead of continuing to build a protein chain with glutamic acid, the synthesis will stop altogether. An important protein will not be manufactured, or at least will be manufactured incorrectly, and the organism may be sick or even die. As we saw in Chemical Connections 14D, sickle cell anemia is caused by a single base mutation that causes a glutamic acid to be replaced with valine.

Ionizing radiation (X rays, ultraviolet light, gamma rays) can cause mutations. Furthermore, a large number of organic compounds can induce mutations by reacting with DNA. Such compounds are called **mutagens.** Many changes

Chemical Connections 18E

p53: A Central Tumor Suppressor Protein

There are some 36 known **tumor suppressor genes,** the products of which are proteins controlling cell growth. None of them is more important than the protein with a molar mass of 53,000, simply named **p53.** This protein responds to a variety of cellular stresses, including DNA damage, lack of oxygen (hypoxia), and aberrantly activated oncogenes. In about 40% of all cancer cases, the tumor contains p53 that underwent mutation. Mutated p53 protein can be found in 55% of lung cancers, about half of all colon and rectal cancers, and some 40% of lymphomas, stomach cancers, and pancreatic cancers. In addition, in one third of all soft tissue sarcomas, p53 is inactive, even though it did not undergo mutation.

These statistics indicate that the normal function of the p53 protein is to suppress tumor growth. When it is mutated or otherwise not present in sufficient or active form, it is unable to perform this protective function and cancer spreads. The p53 protein binds to specific sequences of double-stranded DNA. When X rays or γ-rays damage DNA, an increase in p53 protein concentration is observed. The increased binding of p53 controls the cell cycle by holding it between cell division and DNA replication. The time gained in this arrested cell cycle allows the DNA to repair its damage. If that fails, the p53 protein triggers apoptosis, the programmed death of the injured cell.

Recently, it was reported that p53 performs a finely tuned function in cells. It suppresses tumor growth, but if p53 is overexpressed (that is, its concentration is high), it contributes to the premature aging of the organism.

In those conditions, p53 arrests the cell cycle not only of damaged cells but also of stem cells. These cells normally differentiate into various types (muscle cells, nerve cells, and so forth) and replace those cells that die with aging. Excess p53 slows down this differentiation. In mice, an abundance of p53 protein made the animal cancer-free but at a cost: The mice lost weight and muscle, their bones became brittle, and their wounds took longer to heal. All in all, their average life span was 20% shorter than that of normal mice.

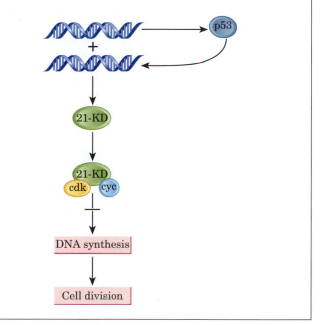

caused by radiation and mutagens do not become mutations because the cell has repair mechanisms, such as nucleotide excision repair (NER), which can prevent mutations by cutting out damaged areas and resynthesizing them (see Section 17.7 for a description of repair mechanisms). Despite this defense mechanism, certain errors in copying that result in mutations do slip by. Many compounds (both synthetic and natural) are mutagens, and some can cause cancer when introduced into the body. These substances are called **carcinogens** (Chemical Connections 4B). One of the main tasks of the U.S. Food and Drug Administration and the Environmental Protection Agency is to identify carcinogens and eliminate them from our food, drugs, and environment. Even though most carcinogens are mutagens, the reverse is not true.

Not all mutations are harmful. Some are beneficial because they enhance the survival rate of the species. For example, mutation is used to develop new strains of plants that can withstand pests.

If a mutation is harmful, it results in an inborn genetic disease. This condition may be carried as a recessive gene from generation to generation with no individual demonstrating the symptoms of the disease. When both parents carry recessive genes, however, an offspring has a 25% chance of inheriting the disease. If the defective gene is dominant, on the other hand, every carrier will develop symptoms.

Human Diversity and Transcription Factors

Research on the human genome showed that there was not enough diversity in the base genome of structural genes to explain the vast differences between species or even between individuals of a species. However, when we add in genes for transcription factors and the DNA sequences of enhancers and silencers, we have a much wider range of possible differences. One example can be seen in the human population. There is a particular protein usually found on the surface of red blood cells called Duffy. The DNA that encodes Duffy is regulated by a specific enhancer. It has been found that nearly 100% of West

Africans lack the Duffy proteins on their red blood cells. The lack of the Duffy protein has been traced to a single nucleotide mutation in the enhancer region of the Duffy gene. It turns out that the Duffy protein is a docking site for malaria-carrying parasites, and red blood cells lacking Duffy are resistant to malaria. This is an example of human evolution in progress. There would be significant evolutionary pressure in favor of the mutation that disrupts the synthesis of the Duffy protein in areas like West Africa, where malaria is present.

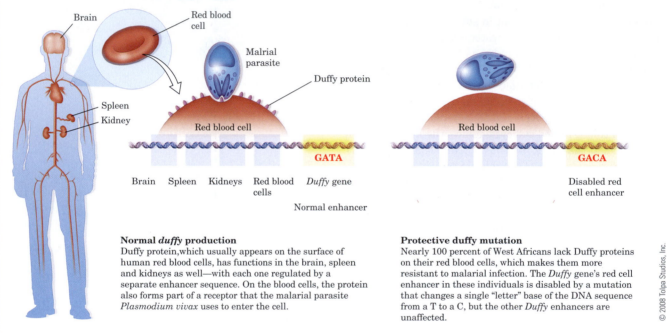

The human genome, like those of flies and fish, also displays evidence of evolution through changes to enhancer DNA. One example is the adaptive loss of a protein called Duffy on red blood cells in a West African population living in areas where malaria is endemic.

Normal *duffy* production
Duffy protein, which usually appears on the surface of human red blood cells, has functions in the brain, spleen and kidneys as well—with each one regulated by a separate enhancer sequence. On the blood cells, the protein also forms part of a receptor that the malarial parasite *Plasmodium vivax* uses to enter the cell.

Protective duffy mutation
Nearly 100 percent of West Africans lack Duffy proteins on their red blood cells, which makes them more resistant to malarial infection. The *Duffy* gene's red cell enhancer in these individuals is disabled by a mutation that changes a single "letter" base of the DNA sequence from a T to a C, but the other *Duffy* enhancers are unaffected.

© 2008 Tolpa Studios, Inc.

18.8 How and Why Do We Manipulate DNA?

There are no cures for the inborn genetic diseases discussed in Section 18.7. The best we can do is detect the carriers and, through genetic counseling of prospective parents, try not to perpetuate the defective genes. However, **recombinant DNA techniques** give us some hope for the future. At this time, these DNA techniques are being used mostly in bacteria, plants, and test animals (such as mice), but they are slowly being applied to humans as well, as will be described in Section 18.9.

One example of recombinant DNA techniques begins with certain circular DNA molecules found in the cells of the bacterium *Escherichia coli*. These molecules, called **plasmids** (Figure 18.16), consist of double-stranded

FIGURE 18.16 Plasmids from a bacterium used in the recombinant DNA technique.

© Thomas Broker

DNA arranged in a ring. Certain highly specific enzymes called **restriction endonucleases** cleave DNA molecules at specific locations (a different location for each enzyme). For example, one of these enzymes may split a double-stranded DNA as follows:

We use "B" to indicate the remaining DNA in a bacterial plasmid.

We use "H" to indicate a human gene.

The enzyme is so programmed that whenever it finds this specific sequence of bases in a DNA molecule, it cleaves it as shown. Because a plasmid is circular, cleaving it in this way produces a double-stranded chain with two ends (Figure 18.17). These are called "sticky ends" because on one strand each has several free bases that are ready to pair up with a complementary section if they can find one.

The next step is to give the strands such a section. This is done by adding a gene from some other species. The gene is a strip of double-stranded DNA that contains the necessary base sequence. For example, we can put in the human gene that manufactures insulin, which we can get in two ways:

1. It can be made in a laboratory by chemical synthesis; that is, chemists can combine the nucleotides in the proper sequence to make the gene.

2. We can cut a human chromosome with the same restriction enzyme. Because it is the same enzyme, it cuts the human gene so as to leave the same sticky ends:

The human gene must be cut at two places so that a piece of DNA that carries two sticky ends is freed. To splice the human gene into the plasmid, the two are mixed in the presence of DNA ligase, and the sticky ends come together:

$$\text{H—G} \quad + \quad \text{AATTC—B} \xrightarrow{\text{DNA ligase}} \text{H—GAATTC—B}$$
$$\text{H—CTTAA} \qquad \text{G—B} \qquad \text{H—CTTAAG—B}$$

This reaction takes place at both ends of the human gene, turning the plasmid into a circle once again (Figure 18.17).

The modified plasmid is then put back into a bacterial cell, where it replicates naturally every time the cell divides. Bacteria multiply quickly, so soon we have a large number of bacteria, all containing the modified plasmid. All these cells now manufacture human insulin by transcription and translation. In this way, we can use bacteria as a factory to manufacture specific proteins. This new industry has tremendous potential for lowering the price of drugs that are currently manufactured by isolation from human or animal tissues (for example, human interferon, a molecule that fights infection). Not only bacteria but also viruses can be used to create recombinant DNA (Figure 18.18).

Example 18.4 Restriction Endonucleases

Two different restriction endonucleases act on the following sequence of a double-stranded DNA:

$$\text{AATGAATTCGAGGC}$$
$$\text{TTACTTAAGCTCCG}$$

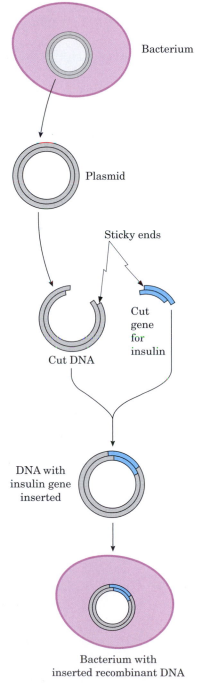

FIGURE 18.17 The recombinant DNA technique can be used to turn a bacterium into an insulin "factory."

One endonuclease, EcoRI, recognizes the sequence GAATTC and cuts the sequence between G and A. The other endonuclease, TaqI, recognizes the sequence TCGA and cuts the sequence between T and C. What are the sticky ends that each of these endonucleases will create?

Solution

EcoRI ⌇⌇⌇⌇ AATG   AATTCGAGGC⌇⌇⌇
        ⌇⌇⌇ TTACTTAA   GCTCCG ⌇⌇

TaqI   ⌇⌇⌇⌇ AATGAATT   CGAGGC⌇⌇⌇
      ⌇⌇⌇ TTACTTAAGC   TCCG ⌇⌇

Problem 18.4

Show the sticky end of the following double-stranded DNA sequence that is cut by TaqI:

⌇⌇⌇ CCTCGATTG ⌇⌇⌇
 ⌇⌇⌇ GGAGCTAAC ⌇⌇⌇

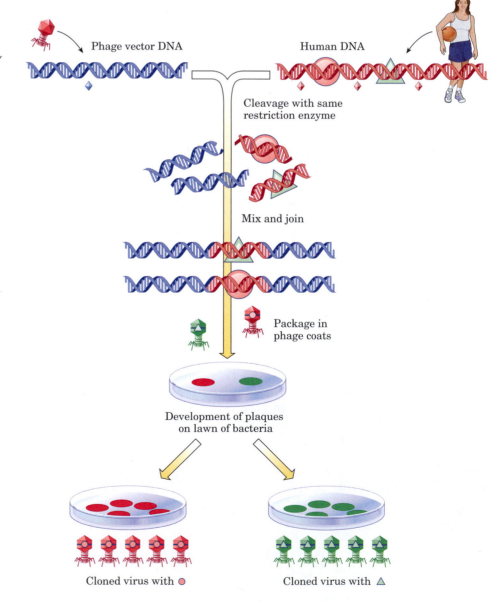

FIGURE 18.18 The cloning of human DNA fragments with a viral vector. (Adapted with permission from *Dealing with Genes: The Language of Heredity* by Paul Berg and Maxine Singer, 1992 © by University Science Books).

Phage is another word for a virus that infects bacteria.

Phage vector DNA   Human DNA

Cleavage with same restriction enzyme

Mix and join

Package in phage coats

Development of plaques on lawn of bacteria

Cloned virus with ●   Cloned virus with △

18.9 What Is Gene Therapy?

While viruses have traditionally been seen as problems for humans, there is one field where they are now being used for good. Viruses can be used to alter somatic cells, where a genetic disease is treated by the introduction of a gene for a missing protein. This process is called **gene therapy.**

The most successful form of gene therapy to date involves the gene for **adenosine deaminase (ADA),** an enzyme involved in purine catabolism (Section 17.8). If this enzyme is missing, dATP builds up in tissues, inhibiting the action of the enzyme ribonucleotide reductase. The result is a deficiency of the other three deoxyribonucleoside triphosphates (dNTPs). The dATP (in excess) and the other three dNTPs (deficient) are precursors for DNA synthesis. This imbalance particularly affects DNA synthesis in lymphocytes, on which much of the immune response depends (Chapter 23). Individuals who are homozygous for adenosine deaminase deficiency develop **severe combined immune deficiency (SCID),** the "bubble boy" syndrome. They are prone to infection because of their highly compromised immune systems. The ultimate goal of the planned gene therapy is to take bone marrow cells from affected individuals, introduce the gene for adenosine deaminase into the cells using a virus as a vector, and then reintroduce the bone marrow cells in the body, where they will produce the desired enzyme. The first clinical trials for a cure to ADA-SCID were simple enzyme replacement therapies begun in 1982. The patients in these trials were given injections of ADA. Later clinical trials sought to correct the gene in mature T cells. In 1990, transformed T cells were given to recipients via transfusions. In trials at the National Institutes of Health (NIH), two girls, aged 4 and 9 years at the start of treatment, showed improvement to the extent that they could attend regular public schools and had no more than the average number of infections. Administration of bone marrow stem cells in addition to T cells was the next step; clinical trials of this procedure were undertaken with two infants, aged 4 months and 8 months in the year 2000. After 10 months, the children were healthy and had restored immune systems.

There are two types of delivery methods in human gene therapy. The first, called **ex vivo,** is the type used to combat SCID. *Ex vivo* means that somatic cells are removed from the patient, altered with the gene therapy, and then returned to the patient. The most common vector for this approach is **Maloney murine leukemia virus (MMLV).** Figure 18.19 shows how the virus is used for gene therapy. Some of the MMLV is altered to remove certain genes, thereby rendering the virus unable to replicate. These genes are replaced with an **expression cassette,** which contains the gene being administered, such as the ADA gene, along with a suitable promoter. This mutated virus is used to infect a packaging cell line. Normal MMLV is also used to infect the packaging cell line. The normal MMLV will not replicate in the packaging cell line, but it will restore the mutated virus's ability to replicate, albeit only in this cell line. These controls are necessary to keep mutant viruses from escaping to other tissues. The mutated virus particles are collected from the packaging cell line and used to infect the target cells—bone marrow cells, in the case of SCID. MMLV is a retrovirus, so it infects the target cell and produces DNA from its RNA genome; this DNA can then become incorporated into the host genome, along with the promoter and ADA gene. In this way, the target cells that were collected are transformed, and they will produce ADA. These cells are then put back into the patient.

In the second delivery method, called **in vivo,** the virus is used to directly infect the patient's tissues. The most common vector for this delivery is the DNA virus, **adenovirus.** A particular vector can be chosen based on specific receptors on the target tissue. Adenovirus has receptors in lung and liver cells, and it has been used in clinical trials for gene therapy of cystic fibrosis

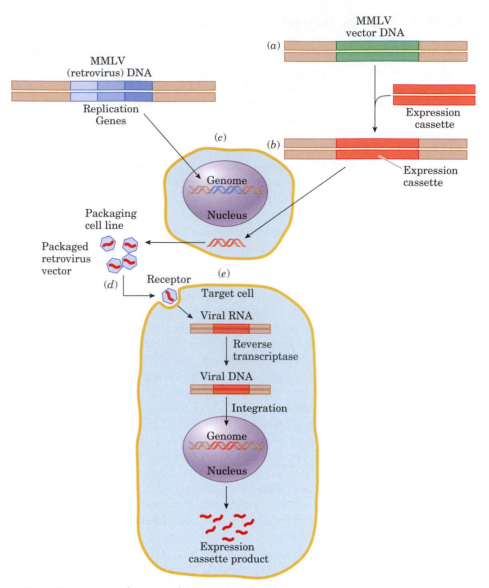

FIGURE 18.19 Gene therapy via retroviruses. The Maloney murine leukemia virus (MMLV) is used for ex vivo gene therapy. Replication genes are removed from the virus and replaced with an expression cassette containing the gene being replaced with gene therapy. The altered virus is grown in a packaging cell line that will allow replication. Viruses are collected and used to infect cultured target cells from the host needing the gene therapy. The altered virus produces RNA, which in turn produces DNA via reverse transcription. The DNA becomes integrated in the host cell's genome, and the host cells produce the desired protein. The cultured cells are returned to the host.

and ornithine transcarbamoylase deficiency. Clinical trials using gene therapy to combat cystic fibrosis and certain tumors in humans are now under way. In mice, gene therapy has been successful in fighting diabetes.

The field of gene therapy is exciting and full of promise, but many obstacles to its success in humans remain. There are also many risks, such as a dangerous immunological response to the vector carrying the gene, or the danger of a gene becoming incorporated into the host chromosome at a location that activates a cancer-causing gene. Both undesirable outcomes have happened in a limited number of human cases.

Gene therapy has been approved in humans only for the manipulation of somatic cells. It is illegal to tamper with human gametes in an attempt to create a heritable change in the human genome.

Summary

Section 18.1 How Does DNA Lead to RNA and Protein?

- A **gene** is a segment of a DNA molecule that carries the sequence of bases that directs the synthesis of a specific RNA molecule. When the RNA is mRNA, it specifies the synthesis of a specific protein.
- The information stored in the DNA is transcribed onto RNA and then expressed in the synthesis of a protein molecule. This process involves two steps: **transcription** and **translation.**

Section 18.2 How Is DNA Transcribed into RNA?

- In transcription, the information is copied from DNA onto mRNA by complementary base pairing. There are also start and stop signals.
- The enzyme that synthesizes RNA is called RNA polymerase. In eukaryotes, three kinds of polymerase are used for the different types of RNA.

Section 18.3 What Is the Role of RNA in Translation?

- mRNA is strung out along the ribosomes.
- Transfer RNA carries the individual amino acids, with each tRNA going to a specific site on the mRNA.
- A sequence of three bases (a triplet) on mRNA constitutes a **codon.** It spells out the particular amino acid that the tRNA brings to this site.
- Each tRNA has a recognition site, the **anticodon,** that pairs up with the codon.
- When two tRNA molecules are aligned at adjacent sites, the amino acids that they carry are linked by an enzyme, forming a peptide bond.
- The translation process continues until the entire protein is synthesized.

Section 18.4 What Is the Genetic Code?

- The **genetic code** provides the correspondence between a codon and an amino acid.
- In most cases, there is more than one codon for each amino acid, but the opposite is not true: A given codon will specify only one amino acid.

Section 18.5 How Is Protein Synthesized?
Problem 18.25

- Protein synthesis takes place in four stages: activation, initiation, elongation, and termination.
- Several of the steps of translation require an input of energy in the form of GTP.
- Ribosomes have three sites: the A site, the P site, and the E site.
- No protein is found in the area where the peptide synthesis is catalyzed. Thus the ribosome is a ribozyme.

Section 18.6 How Are Genes Regulated?
Problem 18.33

- Most of human DNA (96 to 98%) does not code for proteins.
- A number of mechanisms for gene regulation exist on both the transcriptional level and the translational level.
- **Promoters** have initiator and conserved sequences.
- **Transcription factors** bind to the promoter, thereby regulating the rate of transcription.
- **Enhancers** are nucleotide sequences far removed from the transcription sites. They bind to transcription factors and increase the level of transcription.
- Some translational controls, such as **release factors,** act during translation; others, such as **chaperones,** act after translation is completed.

Section 18.7 What Are Mutations?

- A change in the sequence of bases is called a **mutation.**
- Mutations can be caused by an internal mistake or induced by chemicals or radiation. In fact, a change in just one base can cause a mutation.
- A mutation may be harmful or beneficial, or it may cause no change in the amino acid sequence. If a mutation is very harmful, the organism may die.
- Chemicals that cause mutations are called **mutagens.** Chemicals that cause cancer are called **carcinogens.** Many carcinogens are mutagens, but the reverse is not true.

Section 18.8 How and Why Do We Manipulate DNA? **Problem 18.48**

- With the discovery of restriction enzymes that can cut DNA molecules at specific points, scientists have found ways to splice DNA segments together.
- A human gene (for example, the one that codes for insulin) can be spliced into a bacterial plasmid. The bacteria, when multiplied, can then transmit this new information to the daughter cells so that the ensuing generations of bacteria can manufacture human insulin. This powerful method is called the **recombinant DNA technique.**
- Genetic engineering is the process by which genes are inserted into cells.

Section 18.9 What Is Gene Therapy?

- Gene therapy is a technique whereby a missing gene is replaced using a viral vector.
- In **ex vivo** gene therapy, cells are removed from a patient, given the missing gene, and then the cells are given back to the patient.
- In **in vivo** therapy, the patient is given the virus directly.

Problems

■ Indicates problems that are assignable in **GOB OWL**.

Blue numbered problems are applied.

Go to this book's companion website at **www.cengage. com/chemistry/bettelheim** for interactive versions of the How To tutorials and Active Figures, and to quiz yourself on this chapter.

Section 18.1 How Does DNA Lead to RNA and Protein?

18.5 Does the term *gene expression* refer to

(a) transcription, (b) translation, or (c) transcription plus translation?

18.6 In what part of the eukaryote cell does transcription occur?

18.7 Where does most of the translation occur in a eukaryote cell?

Section 18.2 How Is DNA Transcribed into RNA?

18.8 What is the function of RNA polymerase?

18.9 What is the role of helicase in transcription?

18.10 Where is an initiation signal located?

18.11 Which end of the DNA contains the termination signal?

18.12 What would happen to the transcription process if a drug added to a eukaryote cell inhibits the phosphatase?

18.13 Where is the methyl group located in the guanine cap?

18.14 How are the adenine nucleotides linked together in the polyA tails?

Section 18.3 What Is the Role of RNA in Translation?

18.15 Where are the codons located?

18.16 What are the two most important sites on tRNA molecules?

18.17 What are the ribosomal subunits for eukaryotic translation?

Section 18.4 What Is the Genetic Code?

18.18 (a) If a codon is GCU, what is the anticodon? (b) For which amino acid does this codon code?

18.19 If a segment of DNA is 981 units long, how many amino acids appear in the protein encoded by this DNA segment? (Assume that the entire segment is used to code for the protein and that there is no methionine at the N-terminal end of the protein.)

18.20 In what sense does the universality of the genetic code support the theory of evolution?

18.21 Which amino acids have the most possible codons? Which have the fewest?

18.22 Using the first column of Table 18.1, explain how changing the second base of the codon is more detrimental to a protein than changing the first or the third base.

Section 18.5 How Is Protein Synthesized?

18.23 To which end of the tRNA is the amino acid bonded? Where does the energy come from to form the tRNA–amino acid bond?

18.24 There are three sites on the ribosome, each participating in the translation. Identify them and describe what is happening at each site.

18.25 ■ What is the main role of (a) the 40S ribosome and (b) the 60S ribosome?

18.26 What are the prokaryotic equivalents of the eukaryotic ribosomal subunits?

18.27 What is the function of elongation proteins?

18.28 What are the stages of protein synthesis?

18.29 Explain the nature of the tRNA used to initiate translation.

18.30 Explain what happens to the fMet initially put at the N-terminus.

18.31 Explain why scientists now refer to the ribosome as a ribozyme.

18.32 Why is amino acid activation called the second genetic code?

Section 18.6 How Are Genes Regulated?

18.33 ■ Which molecules are involved in gene regulation at the transcriptional level?

18.34 Where are enhancers located? How do they work?

18.35 Where are the sieving portions of AARS enzymes located? What is their function?

18.36 What are the two types of transcription factors, and how do they work?

18.37 What is the difference between an enhancer and a response element?

18.38 How does alternative splicing lead to protein diversity?

18.39 What is the function of proteosomes in quality control?

18.40 What kind of interactions exist between metal-binding fingers and DNA?

Section 18.7 What Are Mutations?

18.41 Using Table 18.1, give an example of a mutation that (a) does not change anything in a protein molecule and (b) might cause fatal changes in a protein.

18.42 How do cells repair mutations caused by X rays?

18.43 Can a harmful mutation-causing genetic disease exist from generation to generation without exhibiting the symptoms of the disease? Explain.

18.44 Are all mutagens also carcinogens?

Section 18.8 How and Why Do We Manipulate DNA?

18.45 How do restriction endonucleases operate?

18.46 What are sticky ends?

18.47 A new genetically engineered corn has been approved by the Food and Drug Administration. This new corn shows increased resistance to a destructive insect called a corn borer. What is the difference, in principle, between this genetically engineered corn and one that developed insect resistance by mutation (natural selection)?

18.48 ■ EcoRI restriction endonuclease recognizes the sequence GAATTC and cuts it between G and A. What will be the sticky ends of the following double-helical sequence when EcoRI acts on it?

CAAAGAATTCG
GTTTCTTAAGC

Chemical Connections

18.49 Why can it be argued that the discovery of restriction enzymes was the key to the beginning of modern molecular biology?

18.50 (Chemical Connections 18A) Why is selenocysteine called the twenty-first amino acid? Why were amino acids such as hydroxyproline and hydroxylysine not counted as additional amino acids?

18.51 (Chemical Connections 18B) What is a viral "coat"?

18.52 (Chemical Connections 18B) Where do the ingredients—amino acids, enzymes, and so forth—necessary to synthesize the viral coat come from?

18.53 (Chemical Connections 18C) What is an invariant site?

18.54 (Chemical Connections 18D) What is a silent mutation?

18.55 (Chemical Connections 18D) If a mRNA codon has the sequence UCU, can there be a mutation in the third base that is not a silent mutation? Why or why not?

18.56 (Chemical Connections 18D) If a mRNA codon has the sequence UAU, which mutations of the third base would be the worst? Why?

18.57 (Chemical Connections 18D) Why can a silent mutation sometimes lead to a different protein product?

18.58 (Chemical Connections 18D) How was the gene MDR1 involved in the discovery that silent mutations could lead to observable changes?

18.59 (Chemical Connections 18E) What is p53? Why is its mutated form associated with cancer?

18.60 (Chemical Connections 18E) How does p53 promote DNA repair?

18.61 (Chemical Connections 18F) What is the Duffy protein and how is it important in the epidemiology of malaria?

18.62 (Chemical Connections 18F) What is the nature of the mutation by which West Africans do not produce the Duffy protein?

18.63 (Chemical Connections 18F) Consider a gene X that makes a protein Y. Give several examples of mutations that could affect the production of protein Y.

18.64 (Chemical Connections 18F) How is the Duffy protein thought to be related to human evolution?

Additional Problems

18.65 ■ In both the transcription and the translation steps of protein synthesis, a number of different molecules come together to act as a factor unit. What are these units of (a) transcription and (b) translation?

18.66 In the tRNA structure, there are stretches where complementary base pairing is necessary and other areas where it is absent. Describe two functionally critical areas (a) where base pairing is mandatory and (b) where it is absent.

18.67 Is there any way to prevent a hereditary disease? Explain.

18.68 How does the cell ensure that a specific amino acid (say, valine) attaches itself only to the one tRNA molecule that is specific for valine?

18.69 (a) What is a plasmid?
(b) How does it differ from a gene?

18.70 Why do we call the genetic code *degenerate*?

18.71 Glycine, alanine, and valine are classified as nonpolar amino acids. Compare their codons. What similarities do you find? What differences do you find?

18.72 ■ Looking at the multiplicity (degeneracy) of the genetic code, you may get the impression that the third base of the codon is irrelevant. Point out where this is not the case. Out of the 16 possible combinations of the first and second bases, in how many cases is the third base irrelevant?

18.73 Which polypeptide is coded for by the mRNA sequence 5′-GCU-GAA-GUC-GAG-GUG-UGG-3′?

18.74 A new endonuclease is found. It cleaves double-stranded DNA at every location where C and G are paired on opposite strands. Could this enzyme be used in producing human insulin by the recombinant DNA technique? Explain.

Bioenergetics: How the Body Converts Food to Energy

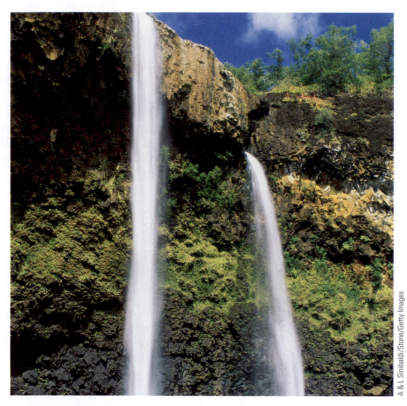

A & L Sinibaldi/Stone/Getty Images

Wailua Falls, Hawaii, is a natural demonstration of two pathways ending in a common pool.

19.1 What Is Metabolism?

Living cells are in a dynamic state, which means that compounds are constantly being synthesized and then broken down into smaller fragments. Thousands of different reactions take place at the same time. **Metabolism** is the sum total of all the chemical reactions involved in maintaining the dynamic state of the cell.

OWL Online homework for this chapter may be assigned in **GOB OWL**.

In general, we can classify metabolic reactions into two broad groups: (1) those in which molecules are broken down to provide the energy needed by cells and (2) those that synthesize the compounds needed by cells—both simple and complex. **Catabolism** is the process of breaking down molecules to supply energy. The process of synthesizing (building up) molecules is **anabolism.** The same compounds may be synthesized in one part of a cell and broken down in a different part of the cell.

In spite of the large number of chemical reactions, a mere handful dominate cell metabolism. In this chapter and Chapter 20, we focus our attention on the catabolic pathways that yield energy. A **biochemical pathway** is a series of consecutive biochemical reactions. We will see the actual reactions that enable the chemical energy stored in our food to be converted to the energy we use every minute of our lives—to think, to breathe, and to use our muscles to walk, write, eat, and everything else. In Chapter 21, we will look at some synthetic (anabolic) pathways.

The food we eat consists of many types of compounds, largely the ones we discussed in earlier chapters: carbohydrates, lipids, and proteins. All of them can serve as fuel, and we derive our energy from them. To convert those compounds to energy, the body uses a different pathway for each type of compound. *All of these diverse pathways converge to one* **common catabolic pathway,** which is illustrated in Figure 19.1. In the figure, the diverse pathways are shown as different food streams. The small molecules produced from the original large molecules in food drop into an imaginary collecting funnel that represents the common catabolic pathway. At the end of the funnel appears the energy carrier molecule, adenosine triphosphate (ATP).

The purpose of catabolic pathways is to convert the chemical energy in foods to molecules of ATP. In the process, foods also yield metabolic intermediates, which the body can use for synthesis. In this chapter, we deal with the common catabolic pathway. In Chapter 20, we discuss the ways in which the different types of food (carbohydrates, lipids, and proteins) feed molecules into the common catabolic pathway.

Common catabolic pathway A series of chemical reactions through which foodstuffs are oxidized to yield energy in the form of ATP; the common catabolic pathway consists of (1) the citric acid cycle (Section 19.4) and (2) oxidative phosphorylation (Sections 19.5 and 19.6)

19.2 What Are Mitochondria, and What Role Do They Play in Metabolism?

A typical animal cell has many components, as shown in Figure 19.2. Each serves a different function. For example, the replication of DNA (Section 17.6) takes place in the **nucleus; lysosomes** remove damaged cellular components and some unwanted foreign materials; and **Golgi bodies** package and process proteins for secretion and delivery to other cellular compartments. The specialized structures within cells are called **organelles.**

The **mitochondria,** which possess two membranes (Figure 19.3), are the organelles in which the common catabolic pathway takes place in higher organisms. The enzymes that catalyze the common pathway are all located in these organelles. Because these enzymes are synthesized in the cytosol, they must be imported through the two membranes. They cross the outer membrane through translocator outer membrane (TOM) channels and are accepted in the intermembrane space by chaperone-like translocator inner membrane (TIM) complexes, which also insert them into the inner membrane.

Because the enzymes are located inside the inner membrane of mitochondria, the starting materials of the reactions in the common pathway must

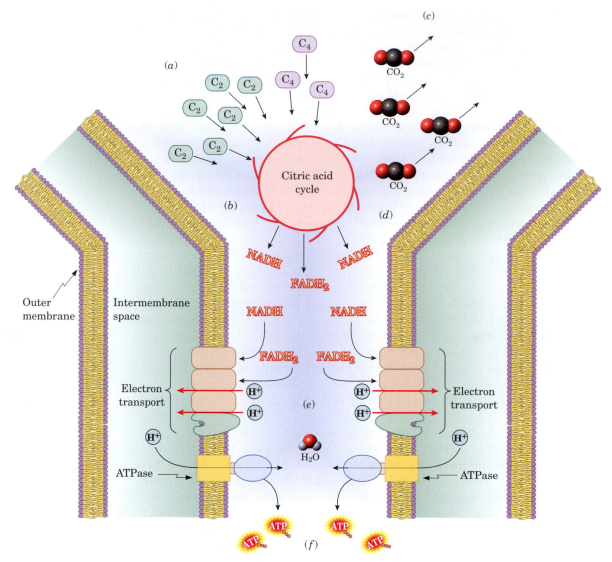

FIGURE 19.1 In this simplified schematic diagram of the common catabolic pathway, an imaginary funnel represents what happens in the cell. (*a*) The diverse catabolic pathways drop their products into the funnel of the common catabolic pathway, mostly in the form of C_2 fragments (Section 19.4). (The source of the C_4 fragments will be shown in Section 20.9.) (*b*) The spinning wheel of the citric acid cycle breaks these molecules down further. (*c*) The carbon atoms are released in the form of CO_2, and (*d*) the hydrogen atoms and electrons are picked up by special compounds such as NAD^+ and FAD. (*e*) Then the reduced NADH and $FADH_2$ cascade down into the stem of the funnel, where the electrons are transported inside the walls of the stem and the H^+ ions are expelled to the outside. (*f*) In their drive to get back, the H^+ ions form the energy carrier ATP. Once back inside, they combine with the oxygen that picked up the electrons and produce water.

pass through the two membranes to enter the mitochondria. Products must leave the same way.

The inner membrane of a mitochondrion is quite resistant to the penetration of any ions and of most uncharged molecules. (*Mitochondria* is the plural form; *mitochondrion* is the singular form.) However, ions and molecules can still get through the membrane—they are transported across it by the numerous protein molecules embedded in it (Figure 13.2). The outer membrane, by contrast, is quite permeable to small molecules and ions and does not have transporting membrane proteins.

An animal cell

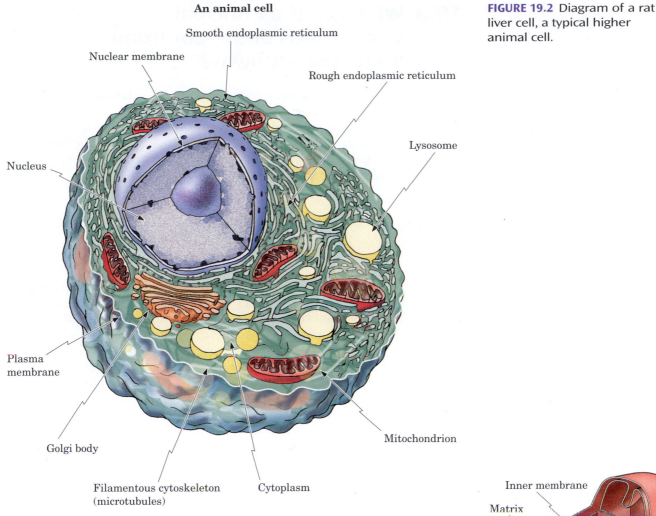

Nuclear membrane

Smooth endoplasmic reticulum

Rough endoplasmic reticulum

Nucleus

Lysosome

Plasma membrane

Golgi body

Filamentous cytoskeleton (microtubules)

Cytoplasm

Mitochondrion

FIGURE 19.2 Diagram of a rat liver cell, a typical higher animal cell.

The **matrix** is the inner nonmembranous portion of a mitochondrion (Figure 19.3). The inner membrane is highly corrugated and folded. On the basis of electron microscopic studies, the Romanian-born American cell biologist George Palade (1912–2008) proposed his baffle model of the mitochondrion in 1952. The baffles, which are called **cristae,** project into the matrix like the bellows of an accordion. The enzymes of the oxidative phosphorylation cycle are localized on the cristae. The space between the inner and outer membranes is the **intermembrane space.** This classical baffle model of mitochondria underwent some changes in the late 1990s after three-dimensional pictures were obtained by the new technique called electron-microscope tomography. The 3-D images indicate that the cristae have narrow tubular connections to the inner membrane. These tubular connections may control the diffusion of metabolites from the inside to the intermembrane space. Furthermore, the space between the outer and the inner membrane varies during metabolism, possibly controlling the rate of reactions.

The enzymes of the citric acid cycle are located in the matrix. We will soon see in detail how the specific sequence of these enzymes causes the chain of events in the common catabolic pathway. Furthermore, we will discuss the ways in which nutrients and reaction products move into and out of the mitochondria.

Inner membrane

Matrix

Outer membrane

(a)

Crista

(b)

© Dennis Kunkel/Phototake

FIGURE 19.3 (a) Schematic of a mitochondrion cut to reveal the internal organization. (b) Colorized transmission electron micrograph of a mitochondrion in a heart muscle cell.

19.3 What Are the Principal Compounds of the Common Metabolic Pathway?

The common catabolic pathway has two parts: the **citric acid cycle** (also called the tricarboxylic acid cycle or the Krebs cycle) and the **electron transport chain** and **phosphorylation,** together called the **oxidative phosphorylation pathway.** To understand what actually happens in these reactions, we must first introduce the principal compounds participating in the common catabolic pathway.

A. Agents for Storage of Energy and Transfer of Phosphate Groups

The most important of these agents are three rather complex compounds: **adenosine monophosphate (AMP), adenosine diphosphate (ADP),** and **adenosine triphosphate (ATP)** (Figures 19.4 and 19.5). All three of these molecules contain the heterocyclic amine adenine (Section 17.2) and the sugar D-ribose (Section 12.2) joined together by a β-N-glycosidic bond, forming adenosine (Section 17.2).

AMP, ADP, and ATP all contain adenosine connected to phosphate groups. The only difference between the three molecules is the number of phosphate groups. As you can see from Figure 19.5, each phosphate is attached to the next by an anhydride bond (Section 19.5A). ATP contains three phosphates—one phosphoric ester and two phosphoric anhydride bonds. In all three molecules, the first phosphate is attached to the ribose by a phosphoric ester bond (Section 11.5B).

A phosphoric anhydride bond contains more chemical energy (7.3 kcal/mol) than a phosphoric ester linkage (3.4 kcal/ mol). Thus, when ATP and ADP are hydrolyzed to yield phosphate ion (Figure 19.5), they release more energy per phosphate group than does AMP. When one phosphate group is hydrolyzed from each, the following energy yields are obtained: AMP = 3.4 kcal/mol; ADP = 7.3 kcal/mol; ATP = 7.3 kcal/mol. (The PO_4^{3-} ion is generally called inorganic phosphate.) Conversely, when inorganic phosphate bonds to AMP or ADP, greater amounts of energy are added to the chemical bond than when it bonds to adenosine. ADP and ATP contain *high-energy* phosphoric anhydride bonds.

ATP releases the most energy and AMP releases the least energy when each gives up one phosphate group. This property makes ATP a very useful compound for energy storage and release. The energy gained in the oxidation of food is stored in the form of ATP, albeit only for a short while. ATP molecules in the cells normally do not last longer than about 1 min.

FIGURE 19.4 Adenosine 5'- monophosphate (AMP).

FIGURE 19.5 Hydrolysis of ATP produces ADP plus dihydrogen phosphate ion plus energy.

They are hydrolyzed to ADP and inorganic phosphate to yield energy that drives other processes, such as muscle contraction, nerve signal conduction, and biosynthesis. As a consequence, ATP is constantly being formed and decomposed. Its turnover rate is very high. Estimates suggest that the human body manufactures and degrades as much as 40 kg (approximately 88 lb) of ATP every day. Even with these considerations, the body is able to extract only 40 to 60% of the total caloric content of food.

B. Agents for Transfer of Electrons in Biological Oxidation–Reduction Reactions

Two other actors in this drama are the coenzymes (Section 15.3) NAD^+ (nicotinamide adenine dinucleotide) and FAD (flavin adenine dinucleotide), both of which contain an ADP core (Figure 19.6). (The + in NAD^+ refers to the positive charge on the nitrogen.) In NAD^+, the operative part of the coenzyme is the nicotinamide part. In FAD, the operative part is the flavin portion. In both molecules, ADP is the handle by which the apoenzyme holds on to the coenzyme; the other end of the molecule carries out the actual chemical reaction. For example, when NAD^+ is reduced, the nicotin-amide part of the molecule gets reduced:

The reduced form of NAD^+ is called NADH. The same reduction happens on two nitrogens of the flavin portion of FAD:

The reduced form of FAD is called $FADH_2$. We view NAD^+ and FAD coenzymes as the **hydrogen ion** and **electron-transporting molecules.**

FIGURE 19.6 The structures of NAD$^+$ and FAD.

FAD
*F*lavin *a*denine *d*inucleotide

C. Agent for Transfer of Acetyl Groups

The final principal compound in the common catabolic pathway is **coenzyme A** (CoA; Figure 19.7), which is the **acetyl (CH_3CO—)-transporting molecule**. Coenzyme A also contains ADP, but here the next structural unit is pantothenic acid, another B vitamin. Just as ATP can be viewed as an ADP molecule to which a —PO_3^{2-}— group is attached by a high-energy bond, so **acetyl coenzyme A** can be considered a CoA molecule linked to an acetyl group by a high-energy thioester bond, for which the energy of hydrolysis is 7.51 kcal/mol. The active part of coenzyme A is the mercaptoethylamine. The acetyl group of acetyl coenzyme A is attached to the SH group:

Acetyl group The group CH_3CO—

$$CH_3-\overset{\overset{\textstyle O}{\|}}{C}-S-CoA$$

Acetyl coenzyme A

| Mercaptoethylamine | Pantothenic acid | Phosphorylated ADP |

FIGURE 19.7 The structure of coenzyme A.

19.4 What Role Does the Citric Acid Cycle Play in Metabolism?

The common catabolism of carbohydrates and lipids begins when they have been broken down into pieces of two carbon atoms each. The two-carbon fragments are the acetyl portions of acetyl coenzyme A. The acetyl is now fragmented further in the citric acid cycle.

Figure 19.8 gives the details of the citric acid cycle. A good way to gain an insight into this cycle is to use Figure 19.8 in conjunction with the simplified schematic diagram shown in Figure 19.9, which shows only the carbon balance.

We will now follow the two carbons of the acetyl group through each step in the citric acid cycle. The circled numbers correspond to those in Figure 19.8.

Step 1 Acetyl coenzyme A enters the cycle by combining with a C_4 compound called oxaloacetate:

The first thing that happens is the addition of the —CH_3 group of the acetyl CoA to the C=O of the oxaloacetate, catalyzed by the enzyme citrate synthase. This event is followed by hydrolysis of the thioester to produce the C_6 compound citrate ion and CoA. Therefore, Step ① is a building-up rather than a breaking-down process. In Step ⑧ we will see where the oxaloacetate comes from.

Step 2 The citrate ion is dehydrated to *cis*-aconitate, after which the *cis*-aconitate is hydrated, but this time to isocitrate instead of citrate:

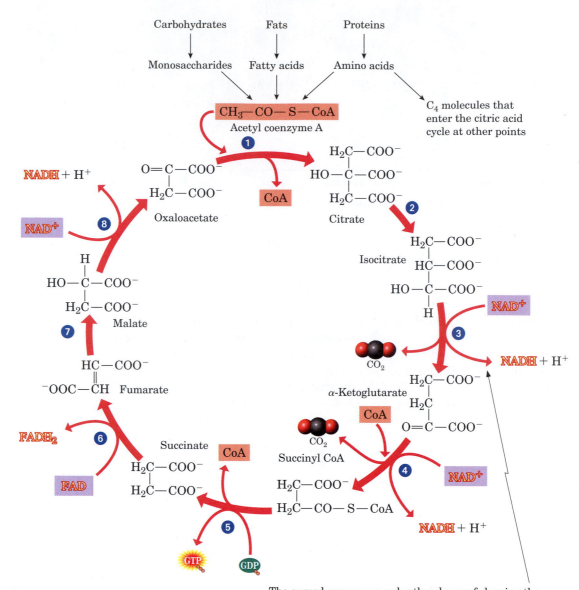

The curved arrows are a shorthand way of showing the reactants and products. For example, in step ❸ NAD^+ reacts with isocitrate to produce α-ketoglutarate, CO_2, NADH, and H^+. The last two then leave the site of the reaction.

FIGURE 19.8 The citric acid (Krebs) cycle. The numbered steps are explained in detail in the text. [Hans Krebs (1900–1981), Nobel laureate in 1953, established the relationships among the different components of the cycle.]

In citrate, the alcohol is a tertiary alcohol. We learned in Section 5.2 that tertiary alcohols cannot be oxidized. The alcohol in the isocitrate is a secondary alcohol, which upon oxidation yields a ketone.

Decarboxylation The process that leads to the loss of CO_2 from a —COOH group.

Step 3 The isocitrate is now oxidized and **decarboxylated** at the same time:

Oxidation:

$$\underset{\text{Isocitrate}}{\begin{array}{l} H_2C-COO^- \\ | \\ HC-COO^- \\ | \\ HO-C-COO^- \\ | \\ H \end{array}} + \text{NAD}^+ \xrightarrow[\text{dehydrogenase}]{\text{Isocitrate}} \underset{\text{Oxalosuccinate}}{\begin{array}{l} H_2C-COO^- \\ | \\ HC-COO^- \\ | \\ O=C-COO^- \end{array}} + \text{NADH} + H^+$$

Decarboxylation:

In oxidizing the secondary alcohol to a ketone, the oxidizing agent NAD^+ removes two hydrogens. One of the hydrogens is added to NAD^+ to produce NADH. (Recall that NAD^+ and NADH are the oxidized and reduced forms, respectively, of nicotinamide adenine dinucleotide [Figure 19.6]). The other hydrogen replaces the COO^- that goes into making CO_2. Note that the CO_2 given off comes from the original oxaloacetate and not from the two carbons of the acetyl CoA. Both of these carbons are still present in the α-ketoglutarate. Also note that we are now down to a C_5 compound, α-ketoglutarate.

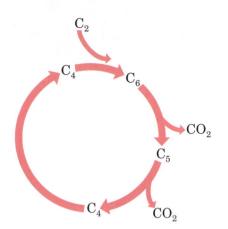

FIGURE 19.9 A simplified view of the citric acid cycle, showing only the carbon balance.

Steps 4 and 5 Next, a complex system removes another CO_2 once again from the original oxaloacetate portion rather than from the acetyl CoA portion:

(Recall again that NAD^+ and NADH are the oxidized and reduced forms, respectively, of nicotinamide adenine dinucleotide [Figure 19.6]). The P_i in this equation is the usual notation for inorganic phosphate.

We are now down to a C_4 compound, succinate. This oxidative decarboxylation is more complex than the first. It occurs in many steps and requires a number of cofactors. For our purpose, it is sufficient to know that, during this second oxidative decarboxylation, a high-energy compound called **guanosine triphosphate (GTP)** is also formed.

GTP is similar to ATP, except that guanine replaces adenine. Otherwise, the linkages of the base to ribose and the phosphates are exactly the same as in ATP. The function of GTP is also similar to that of ATP—namely, it stores energy in the form of high-energy phosphoric anhydride bonds (chemical energy). The energy from the hydrolysis of GTP drives many important biochemical reactions—for example, the signal transduction in neurotransmission (Section 16.5).

As a final note on the decarboxylation steps, the CO_2 molecules given off in Steps ③ and ④ are the ones we exhale.

Step 6 In this step, succinate is oxidized by FAD, which removes two hydrogens to give fumarate (the double bond in this molecule is *trans*):

This reaction cannot be carried out in the laboratory, but with the aid of an enzyme catalyst, the body does it easily. (Recall that FAD and $FADH_2$ are the oxidized and reduced forms, respectively, of flavin adenine dinucleotide [Figure 19.6]).

Step 7 The fumarate is now hydrated to give the malate ion:

Step 8 In the final step of the cycle, malate is oxidized to give oxaloacetate:

(Recall that NAD^+ and NADH are the oxidized and reduced forms, respectively, of nicotinamide adenine dinucleotide [Figure 19.6]). Thus the final product of the Krebs cycle is oxaloacetate, which is the compound with which we started in Step ①.

In this process, the original two acetyl carbons of acetyl CoA were added to the C_4 oxaloacetate to produce a C_6 unit, which then lost two carbons in the form of CO_2 to produce, at the end of the process, the C_4 unit oxaloacetate. The net effect is that one two-carbon acetyl group enters the cycle and two carbon dioxides are given off.

How does the citric acid cycle produce energy? We have already learned that one step in the process produces a high-energy molecule of GTP. However, most of the energy is produced in the other steps that convert NAD^+ to NADH and FAD to $FADH_2$. These reduced coenzymes carry the H^+ and electrons that eventually will provide the energy for the synthesis of ATP (discussed in detail in Sections 19.5 and 19.6).

This stepwise degradation and oxidation of acetate in the citric acid cycle results in the most efficient extraction of energy. Rather than being generated in one burst, the energy is released in small packets that are carried away step by step in the form of NADH and $FADH_2$.

The cyclic nature of this acetate degradation has other advantages besides maximizing energy yield:

1. The citric acid cycle components provide raw materials for amino acid synthesis as the need arises (Chapter 21). For example, α-ketoglutaric acid is used to synthesize glutamic acid.

2. The many-component cycle provides an excellent method for regulating the speed of catabolic reactions.

The regulation can occur at many different parts of the cycle, so that feedback information can be used at many points to speed up or slow down the process as necessary.

The following equation represents the overall reactions in the citric acid cycle:

$$GDP + P_i + CH_3-CO-S-CoA + 2H_2O + 3NAD^+ + FAD$$
$$\longrightarrow CoA + GTP + 2CO_2 + 3NADH + FADH_2 + 3H^+ \quad \text{(Eq.19.1)}$$

The citric acid cycle is controlled by feedback mechanisms. When the essential product of this cycle, $NADH + H^+$, and the end product of the common catabolic pathway, ATP, accumulate, they inhibit some of the enzymes in the cycle. Citrate synthase (Step ①), isocitrate dehydrogenase (Step ③), and α-ketoglutarate dehydrogenase (part of the complex enzyme system in

Step ④) are inhibited by ATP and/or by NADH + H^+. This inhibition slows down or shuts off the cycle. Conversely, when the feed material, acetyl CoA, is in abundance, the cycle accelerates. The enzyme isocitrate dehydrogenase (Step ③) is stimulated by ADP and NAD^+, which are the essential reactants from which the end products of the cycle are derived.

19.5 How Do Electron and H^+ Transport Take Place?

The reduced coenzymes NADH and $FADH_2$ are end products of the citric acid cycle. They carry hydrogen ions and electrons and, therefore, have the potential to yield energy when these combine with oxygen to form water:

$$4H^+ + 4e^- + O_2 \longrightarrow 2H_2O + \text{energy}$$

This simple exothermic reaction is carried out in many steps. The oxygen in this reaction is the oxygen we breathe.

A number of enzymes are involved in this reaction, all of which are embedded in the inner membrane of the mitochondria. These enzymes are situated in a particular *sequence* in the membrane so that the product from one enzyme can be passed on to the next enzyme, in a kind of assembly line. The enzymes are arranged in order of increasing affinity for electrons, so electrons flow through the enzyme system (Figure 19.10).

The sequence of the electron-carrying enzyme systems starts with complex I. The largest complex, it contains some 40 subunits, among them a flavoprotein and several FeS clusters. **Coenzyme Q** (CoQ; also called ubiquinone) is associated with complex I, which oxidizes the NADH produced in the citric acid cycle and reduces the CoQ:

$$NADH + H^+ + CoQ \rightarrow NAD^+ + CoQH_2$$

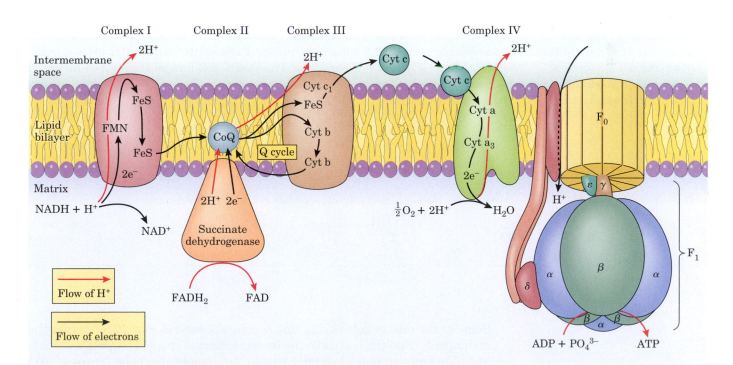

FIGURE 19.10 Schematic diagram of the electron and H^+ transport chain and subsequent phosphorylation. The combined processes are also known as oxidative phosphorylation. **Go to this book's companion website at www.cengage.com/chemistry/bettelheim to explore an interactive version of this figure.**

Chemical Connections 19A

Uncoupling and Obesity

The health concerns that surround the growing number of obese people in developed countries have led to research into the causes and alleviation of obesity. A number of weight-reducing drugs exist. Some of them operate as uncouplers of electron transport and oxidative phosphorylation.

The discovery of a role for uncouplers in weight reduction occurred more or less by accident. During World War I, many ammunition workers were exposed to 2,4-dinitrophenol (DNP), a compound used to prepare the explosive picric acid, which is structurally related to the well-known explosive trinitrotoluene (TNT). After it was observed that these workers lost weight, DNP was used as a weight-reducing drug during the 1920s. Unfortunately, DNP eliminated not only the fat but sometimes also the dieter, and its use as a diet pill was discontinued after 1929.

Today we know why DNP works as a weight-reducing drug: It is an effective protonophore—a compound that transports ions through cell membranes passively, without the expenditure of energy. As noted earlier, H^+ ions accumulate in the intermembrane space of mitochondria and, under normal conditions, drive the synthesis of ATP while they are going back to the inside. This process is Mitchell's chemiosmotic principle in action. When DNP is ingested, it transfers the H^+ back to the mitochondrion easily, and no ATP is manufactured. The energy of the electron separation is dissipated as heat and is not built in as chemical energy in ATP. The loss of this energy-storing compound makes the utilization of food much less efficient, resulting in weight loss.

A similar mechanism provides heat in hibernating bears. The bears have brown fat; its color is derived from

The role of brown fat in hibernation may be related to obesity in humans.

the numerous mitochondria in the tissue. The brown fat also contains an uncoupling protein called thermogenin, a protonophore that allows the ions to stream back into the mitochondrial matrix without manufacturing ATP. The heat generated in this manner keeps the animal alive during cold winter days. In similar fashion, an uncoupling protein is known to be involved in obesity, but it is not known what relationship, if any, exists between this protein and hibernation. The question of human obesity and its prevention is important enough, however, to make uncoupling in brown fat a point of departure for obesity research.

DNP

TNT

Picric acid

Some of the energy released in this reaction is used to move $2H^+$ across the membrane, from the matrix to the intermembrane space. The CoQ is soluble in lipids and can move laterally within the membrane. (The figure of $2H^+$ transported across the membrane is the minimum number that allows the overall oxidation process to take place. Some researchers would say that the number of protons transported by some of these respiratory complexes should be higher.)

Complex II also catalyzes the transfer of electrons to CoQ. The source of the electrons is the oxidation of succinate in the citric acid cycle, producing $FADH_2$. The final reaction is

$$FADH_2 + CoQ \longrightarrow FAD + CoQH_2$$

The energy of this reaction is not sufficient to pump two protons across the membrane, however, nor is there an appropriate channel for such a transfer.

Complex III delivers the electrons from $CoQH_2$ to **cytochrome c.** This integral membrane complex contains 11 subunits, including cytochrome b, cytochrome c_1, and FeS clusters. (The letters used to designate the cytochromes were given in order of their discovery.) Each cytochrome has an iron-ion–containing heme center (Section 14.11) embedded in its own protein. Complex III has two channels through which two H^+ ions are pumped from the $CoQH_2$ into the intermembrane space. The process is very complicated. To simplify matters, we can imagine that it occurs in two distinct steps, as does the electron transfer. Because each cytochrome c can pick up only one electron, two cytochrome c units are needed:

$$CoQH_2 + 2 \text{ cytochrome } c(\text{oxydized})$$
$$\longrightarrow CoQ + 2H^+ + 2 \text{ cytochrome } c \text{ (reduced)}$$

Cytochrome c is also a mobile carrier of electrons—it can move laterally in the intermembrane space.

Complex IV, known as cytochrome oxidase, contains 13 subunits—most importantly, cytochrome a_3, a heme that has an associated copper center. Complex IV is an integral membrane protein complex. The electron movement flows from cytochrome c to cytochrome a to cytochrome a_3. There, the electrons are transferred to the oxygen molecule, and the O—O bond is cleaved. The oxidized form of the enzyme takes up two H^+ ions from the matrix for each oxygen atom. The water molecule formed in this way is released into the matrix:

$$\tfrac{1}{2}O_2 + 2H^+ + 2e^- \longrightarrow H_2O$$

During this process, two more H^+ ions are pumped out of the matrix and into the intermembrane space. Although the mechanism of pumping out protons from the matrix is not known, the energy driving this process is derived from the energy of water formation. This final pumping into the intermembrane space makes a total of six H^+ ions per NADH + H^+ and four H^+ ions per $FADH_2$ molecule.

19.6 What Is the Role of the Chemiosmotic Pump in ATP Production?

How do the electron and H^+ transports produce the chemical energy of ATP? In 1961, Peter Mitchell (1920–1992), an English chemist, proposed the **chemiosmotic theory** to answer this question: The energy in the electron transfer chain creates a proton gradient. A **proton gradient** is a continuous variation in the H^+ concentration along a given region. In this case, there is a higher concentration of H^+ in the intermembrane space than inside the mitochondrion. The driving force, which is the result of the spontaneous flow of ions from a region of high concentration to a region of low concentration, propels the protons back to the mitochondrion through a complex

Chemiosmotic theory Mitchell's proposal that electron transport is accompanied by an accumulation of protons in the intermembrane space of the mitochondrion, which in turn creates osmotic pressure; the protons driven back to the mitochondrion under this pressure generate ATP.

known as **proton-translocating ATPase.** This compound is located on the inner membrane of the mitochondrion (Figure 19.10) and is the active enzyme that catalyzes the conversion of ADP and inorganic phosphate to ATP (the reverse of the reaction shown in Figure 19.5):

$$ADP + P_i \xrightleftharpoons{ATPase} ATP + H_2O$$

Subsequent studies have confirmed this theory, and Mitchell received the Nobel Prize in chemistry in 1978.

The proton-translocating ATPase is a complex "rotor engine" made of 16 different proteins. The F_0 sector, which is embedded in the membrane, contains the **proton channel** (Figure 19.10). The 12 subunits that form this channel rotate every time a proton passes from the cytoplasmic side (intermembrane) to the matrix side of the mitochondrion. This rotation is transmitted to a "rotor" in the F_1 sector. F_1 contains five kinds of polypeptides. The rotor (γ and ε subunits) is surrounded by the catalytic unit (made of α and β subunits) that synthesizes the ATP. The catalytic unit converts the mechanical energy of the rotor into chemical energy of the ATP molecule. The last unit, the "stator," containing the δ subunit, stabilizes the whole complex. The proton-translocating ATPase can catalyze the reaction in both directions. When protons that have accumulated on the outer surface of the mitochondrion stream inward, the enzyme manufactures ATP and stores the electrical energy (due to the flow of charges) in the form of chemical energy. In the reverse reaction, the enzyme hydrolyzes ATP and, as a consequence, pumps out H^+ from the mitochondrion. Each pair of protons that is translocated gives rise to the formation of one ATP molecule. Only when the two parts of the proton-translocating ATPase F_1 and F_0 are linked is energy production possible. When the interaction between F_1 and F_0 is disrupted, the energy transduction is lost.

The protons that enter a mitochondrion combine with the electrons transported through the electron transport chain and with oxygen to form water. The net result of the two processes (electron/H^+ transport and ATP formation) is that each oxygen molecule we inhale combines with four H^+ ions and four electrons to give two water molecules. The four H^+ ions and four electrons come from the NADH and $FADH_2$ molecules produced in the citric acid cycle. The oxygen, therefore, has two functions:

• It oxidizes NADH to NAD^+ and $FADH_2$ to FAD so that these molecules can go back and participate in the citric acid cycle.

• It provides energy for the conversion of ADP to ATP.

The latter function is accomplished indirectly, not through the reduction of O_2 to H_2O. The entrance of the H^+ ions into the mitochondrion drives the ATP formation, but the H^+ ions enter the mitochondrion because the O_2 depleted the H^+ ion concentration when water was formed. This rather complex process involves the transport of electrons along a whole series of enzyme molecules (which catalyze all these reactions); however, the cell cannot utilize the O_2 molecules without it and eventually will die.

The electron and H^+ transport chain and subsequent phosphorylation process are collectively known as oxidative phosphorylation. The following equations represent the overall reactions in oxidative phosphorylation:

$$NADH + 3\ ADP + \tfrac{1}{2}O_2 + 3P_i + H^+ \longrightarrow NAD^+ + 3\ ATP + H_2O \quad \text{(Eq.19.2)}$$

$$FADH_2 + 2\ ADP + \tfrac{1}{2}O_2 + 2P_i \longrightarrow FAD + 2\ ATP + H_2O \quad \text{(Eq.19.3)}$$

19.7 What Is the Energy Yield Resulting from Electron and H$^+$ Transport?

The energy released during electron transport is finally captured in the chemical energy of the ATP molecule. Therefore, it is instructive to look at the energy yield in the universal biochemical currency: the number of ATP molecules.

Each pair of protons entering a mitochondrion results in the production of one ATP molecule. For each NADH molecule, three pairs of protons are pumped into the intermembrane space in the electron transport process. Therefore, for each NADH molecule, we get three ATP molecules, as can be seen in Equation 19.2. For each FADH$_2$ molecule, only four protons are pumped out of the mitochondrion. Therefore, only two ATP molecules are produced for each, as seen in Equation 19.3. Note that ATP production is reported to the nearest whole number. The process is complex, and these numbers represent the least complicated way of looking at it.

Now we can produce the energy balance for the entire common catabolic pathway (citric acid cycle and oxidative phosphorylation combined). For each C$_2$ fragment entering the citric acid cycle, we obtain three NADH and one FADH$_2$ (Equation 19.1) plus one GTP, which is equivalent in energy to one ATP. Thus the total number of ATP molecules produced per C$_2$ fragment is

$$3 \text{ NADH} \times 3 \text{ ATP/NADH} = 9 \text{ ATP}$$
$$1 \text{ FADH}_2 \times 2 \text{ ATP/FADH}_2 = 2 \text{ ATP}$$
$$1 \text{ GTP} = 1 \text{ ATP}$$
$$\overline{ = 12 \text{ ATP}}$$

Each C$_2$ fragment that enters the cycle produces 12 ATP molecules and uses up two O$_2$ molecules. The total effect of the energy-production chain of reactions discussed in this chapter (the common catabolic pathway) is to oxidize one C$_2$ fragment with two molecules of O$_2$ to produce two molecules of CO$_2$ and 12 molecules of ATP

$$C_2 + 2O_2 + 12ADP + 12P_i \longrightarrow 12ATP + 2CO_2$$

The important thing is not the waste product, CO$_2$, but rather the 12 ATP molecules. These molecules will now release their energy when they are converted to ADP.

19.8 How Is Chemical Energy Converted to Other Forms of Energy?

As mentioned in Section 19.3, the storage of chemical energy in the form of ATP lasts only a short time. Usually, within a minute, the ATP is hydrolyzed (an exothermic reaction) and releases its chemical energy. How does the body use this chemical energy? To answer this question, let us look at the different forms in which energy is needed in the body.

A. Conversion to Other Forms of Chemical Energy

The activity of many enzymes is controlled and regulated by phosphorylation. For example, the enzyme phosphorylase, which catalyzes the breakdown of glycogen (Chemical Connections 20B), occurs in an inactive form,

Thick filament
(myosin)

Thin filament
(actin)

(*a*) Relaxed muscle

(*b*) Contracted muscle

FIGURE 19.11 Schematic diagram of muscle contraction.

phosphorylase *b*. When ATP transfers a phosphate group to a serine residue, the enzyme becomes active. Thus the chemical energy of ATP is used
in the form of chemical energy to activate phosphorylase *b* so that glycogen
can be utilized. We will see several other examples of this energy conversion
in Chapters 20 and 21.

B. Electrical Energy

The body maintains a high concentration of K^+ ions inside the cells despite
the fact that the K^+ concentration is low outside the cells. The reverse is
true for Na^+. So that K^+ does not diffuse out of the cells and Na^+ does
not enter them, special transport proteins in the cell membranes constantly
pump K^+ into and Na^+ out of the cells. This pumping requires energy, which
is supplied by the hydrolysis of ATP to ADP. Because of this pumping, the
charges inside and outside the cell are unequal, which generates an electric
potential. Thus the chemical energy of ATP is transformed into electrical
energy, which operates in neurotransmission (Section 16.2).

C. Mechanical Energy

ATP is the immediate source of energy in muscle contraction. In essence, muscle contraction takes place when thick and thin filaments slide past each other
(Figure 19.11). The thick filament is myosin, an ATPase enzyme (that is, one
that hydrolyzes ATP). The thin filament, actin, binds strongly to myosin in
the contracted state. However, when ATP binds to myosin, the actin–myosin
complex dissociates, and the muscle relaxes. When myosin hydrolyzes ATP, it
interacts with actin once more, and a new contraction occurs. In this way, the
hydrolysis of ATP drives the alternating association and dissociation of actin
and myosin and, consequently, the contraction and relaxation of the muscle.

D. Heat Energy

One molecule of ATP upon hydrolysis to ADP yields 7.3 kcal/mol. Some of
this energy is released as heat and used to maintain body temperature. If we
estimate that the specific heat of the body is about the same as that of water,
a person weighing 60 kg would need to hydrolyze approximately 99 moles
(approximately 50 kg) of ATP to raise the temperature of the body from room
temperature, 25°C, to 37°C. Not all body heat is derived from ATP hydrolysis;
some other exothermic reactions in the body also make heat contributions.

Summary of Key Questions

OWL End-of-chapter problems identified in blue are assignable in **GOB OWL.**

Section 19.1 What Is Metabolism?

- The sum total of all the chemical reactions involved in maintaining the dynamic state of cells is called **metabolism.**
- The breaking down of molecules is **catabolism;** the building up of molecules is **anabolism.**

Section 19.2 What Are Mitochondria, and What Role Do They Play in Metabolism?
Problems 19.3, 19.6

- Many metabolic activities in cells take place in specialized structures called **organelles.**
- **Mitochondria** are the organelles in which the reactions of the **common catabolic pathway** take place.

Section 19.3 What Are the Principal Compounds of the Common Metabolic Pathway?

- The common metabolic pathway oxidizes a two-carbon C_2 fragment (acetyl) from different foods. The products of C_2 oxidation are water and carbon dioxide.
- The energy from oxidation is built into the high-chemical-energy-storing molecule **ATP.** As the C_2 fragments are oxidized, protons (H^+) and electrons are released and passed along to carriers.
- The principal carriers in the common catabolic pathway are as follows: ATP is a phosphate carrier; **CoA** is the C_2 fragment carrier; and **NAD$^+$** and **FAD** carry the hydrogen ions (protons) and electrons. The unit common to all of these carriers is **ADP.** The nonactive end of the carriers acts as a handle that fits into the active sites of the enzymes.

Section 19.4 What Role Does the Citric Acid Cycle Play in Metabolism? **Problem 19.22**

- In the **citric acid cycle,** the C_2 fragment first combines with a C_4 fragment (oxaloacetate) to yield a C_6 fragment (citrate). An oxidative decarboxylation yields a C_5 fragment. One CO_2 is released, and one $NADH + H^+$ is passed to the **electron transport chain** for further oxidation.
- Another oxidative decarboxylation provides a C_4 fragment. Once again, a CO_2 is released, and another $NADH + H^+$ is passed to the electron transport chain.
- The enzymes of the citric acid cycle are located in the mitochondrial matrix. Control of this cycle takes place by a feedback mechanism.

Section 19.5 How Do Electron and H^+ Transport Take Place? **Problem 19.40**

- The electrons of NADH enter the electron transport chain at the complex I stage. The coenzyme Q (CoQ) of this complex picks up the electrons and the H^+ and becomes $CoQH_2$. The energy of this reduction reaction is used to expel two H^+ ions from the matrix into the intermembrane space.

- Complex II also has CoQ. Electrons and H^+ are passed to this complex, and complex II catalyzes the transfer of electrons from $FADH_2$. However, no H^+ ions are pumped into the intermembrane space at this point.
- Electrons are passed along by $CoQH_2$ to complex III of the electron transport chain. At complex III, the two H^+ ions from the $CoQH_2$ are expelled into the intermembrane space. Cytochrome c of complex III transfers electrons to complex IV through redox reactions.
- As the electrons are transported from cytochrome c to complex IV, two more H^+ ions are expelled from the matrix of the mitochondrion to the intermembrane space.
- For each NADH, six H^+ ions are expelled. For each $FADH_2$, four H^+ ions are expelled.
- The electrons passed to complex IV return to the matrix, where they combine with oxygen and H^+ to form water.

Section 19.6 What Is the Role of the Chemiosmotic Pump in ATP Production?

- Both the citric acid cycle and **oxidative phosphorylation** take place in the mitochondria. The enzymes of the citric acid cycle are found in the mitochondrial matrix, whereas the enzymes of the electron transport chain and oxidative phosphorylation are located on the inner mitochondrial membrane. Some of them project into the intermembrane space.
- When the H^+ ions expelled by electron transport stream back into the mitochondrion, they drive a complex enzyme called **proton-translocating ATPase,** which makes one ATP molecule for each two H^+ ions that enter the mitochondrion.
- The proton-translocating ATPase is a complex "rotor engine." The proton channel part (F_0) is embedded in the membrane, and the catalytic unit (F_1) converts mechanical energy to chemical energy of the ATP molecule.

Section 19.7 What Is the Energy Yield Resulting from Electron and H^+ Transport?

- For each $NADH + H^+$ coming from the citric acid cycle, three ATP molecules are formed. For each $FADH_2$, two ATP molecules are formed. The overall result: For each C_2 fragment that enters the citric acid cycle, 12 ATP molecules are produced.

Section 19.8 How Is Chemical Energy Converted to Other Forms of Energy? **Problem 19.70**

- Chemical energy is stored in ATP only for a short time—ATP is quickly hydrolyzed, usually within a minute.
- This chemical energy is used to do chemical, mechanical, and electrical work in the body and to maintain body temperature.

Problems

■ Indicates problems that are assignable in **GOB OWL**.

Blue numbered problems are applied.

Go to this book's companion website at **www.cengage. com/chemistry/bettelheim** for interactive versions of the How To tutorials and Active Figures, and to quiz yourself on this chapter.

Section 19.1 What Is Metabolism?

19.1 To what end product is the energy of foods converted in the catabolic pathways?

19.2 (a) How many sequences are in the common catabolic pathway?

(b) Name these sequences.

Section 19.2 What Are Mitochondria, and What Role Do They Play in Metabolism?

19.3 ■ (a) How many membranes do mitochondria have?

(b) Which membrane is permeable to ions and small molecules?

19.4 How do the enzymes of the common pathway find their way into the mitochondria?

19.5 What are cristae, and how are they related to the inner membrane of mitochondria?

19.6 ■ (a) Where are the enzymes of the citric acid cycle located?

(b) Where are the enzymes of oxidative phosphorylation located?

Section 19.3 What Are the Principal Compounds of the Common Metabolic Pathway?

19.7 How many high-energy phosphate bonds are in the ATP molecule?

19.8 What are the products of the following reaction? Complete the equation.

$$AMP + H_2O \xrightarrow{H^+}$$

19.9 Which yields more energy, (a) the hydrolysis of ATP to ADP or (b) the hydrolysis of ADP to AMP?

19.10 How much ATP is needed for normal daily activity in humans?

19.11 What kind of chemical bond exists between the ribitol and the phosphate group in FAD?

19.12 When NAD^+ is reduced, two electrons enter the molecule, together with one H^+ ion. Where in the product will the two electrons be located?

19.13 Which atoms in the flavin portion of FAD are reduced to yield $FADH_2$?

19.14 NAD^+ has two ribose units in its structure; FAD has a ribose and a ribitol. What is the relationship between these molecules?

19.15 In the common catabolic pathway, a number of important molecules act as carriers (transfer agents).

(a) Which is the carrier of phosphate groups?

(b) Which are the coenzymes transferring hydrogen ions and electrons?

(c) What kind of groups does coenzyme A carry?

19.16 The ribitol in FAD is bound to phosphate. What is the nature of this bond? On the basis of the energies of the different bonds in ATP, estimate how much energy (in kcal/mol) would be obtained from the hydrolysis of this bond.

19.17 What kind of chemical bond exists between the pantothenic acid and mercaptoethylamine in the structure of CoA?

19.18 Name the vitamin B molecules that are part of the structure of (a) NAD^+, (b) FAD, and (c) coenzyme A.

19.19 In both NAD^+ and FAD, the vitamin B portion of the molecule is the active part. Is this also true for CoA?

19.20 What type of compound is formed when coenzyme A reacts with acetate?

19.21 The fats and carbohydrates metabolized by our bodies are eventually converted to a single compound. What is it?

Section 19.4 What Role Does the Citric Acid Cycle Play in Metabolism?

19.22 ■ The first step in the citric acid cycle is abbreviated as

$$C_2 + C_4 = C_6$$

(a) What do these symbols stand for?

(b) What are the common names of the three compounds involved in this reaction?

19.23 What is the only C_5 compound in the citric acid cycle?

19.24 Identify by number those steps of the citric acid cycle that are not redox reactions.

19.25 Which substrate in the citric acid cycle is oxidized by FAD? What is the oxidation product?

19.26 In Steps ③ and ⑤ of the citric acid cycle, the compounds are shortened by one carbon each time. What is the form of this one-carbon compound? What happens to it in the body?

19.27 According to Table 15.1, to what class of enzymes does fumarase belong?

19.28 List all the enzymes or enzyme systems of the citric acid cycle that could be classified as oxidoreductases.

19.29 Is ATP directly produced during any step of the citric acid cycle? Explain.

19.30 There are four dicarboxylic acid compounds, each containing four carbons, in the citric acid cycle. Which is (a) the least oxidized and (b) the most oxidized?

19.31 Why is a many-step cyclic process more efficient in utilizing energy from food than a single-step combustion?

19.32 Did the two CO_2 molecules given off in one turn of the citric acid cycle originate from the entering acetyl group?

19.33 Which intermediates of the citric acid cycle contain $C=C$ double bonds?

19.34 The citric acid cycle can be regulated by the body; that is, it can be slowed down or speeded up. What mechanism controls this process?

19.35 Oxidation is defined as loss of electrons. When oxidative decarboxylation occurs, as in Step ④, of the citric acid cycle, where do the electrons of the α-ketoglutarate go?

Section 19.5 How Do Electron and H^+ Transport Take Place?

19.36 What is the main function of oxidative phosphorylation (the electron transport chain)?

19.37 What are the mobile electron carriers of the oxidative phosphorylation?

19.38 In each complex of the electron transport system, the redox reaction occurs mostly around Fe ions.
(a) Identify the compounds that contain such Fe centers.
(b) Identify the compounds that contain ion centers other than iron.

19.39 What kind of motion is set up in the proton-translocating ATPase by the passage of H^+ from the intermembrane space into the matrix?

19.40 ■ The following reaction is a reversible reaction:

$$NADH \rightleftharpoons NAD^+ + H^+ + 2e^-$$

(a) Where does the forward reaction occur in the common catabolic pathway?
(b) Where does the reverse reaction occur?

19.41 In oxidative phosphorylation, water is formed from H^+, e^-, and O_2. Where does this take place?

19.42 At what points in oxidative phosphorylation are the H^+ ions and the electrons separated from each other?

19.43 How many ATP molecules are generated (a) for each H^+ translocated through the ATPase complex and (b) for each C_2 fragment that goes through the complete common catabolic pathway?

19.44 When H^+ is pumped out into the intermembrane space, is the pH there increased, decreased, or unchanged compared with that in the matrix?

Section 19.6 What Is the Role of the Chemiosmotic Pump in ATP Production?

19.45 What is the channel through which ions reenter the matrix of mitochondria?

19.46 The proton gradient accumulated in the intermembrane area of a mitochondrion drives the ATP manufacturing enzyme, ATPase. Why do you think Mitchell called this concept the "chemiosmotic theory"?

19.47 Which part of the proton-translocating ATPase machinery is the catalytic unit? What chemical reaction does it catalyze?

19.48 When the interaction between the two parts of proton-translocating ATPase, F_0 and F_1, are disrupted, no energy production is possible. Which subunits maintain connections between F_0 and F_1, and what names are designated for these subunits?

Section 19.7 What Is the Energy Yield Resulting from Electron and H^+ Transport?

19.49 If each mole of ATP yields 7.3 kcal of energy upon hydrolysis, how many kilocalories of energy would you get from 1 g of CH_3COO^- entering the citric acid cycle?

19.50 A hexose (C_6) enters the common metabolic pathway in the form of two C_2 fragments.
(a) How many molecules of ATP are produced from one hexose molecule?
(b) How many O_2 molecules are used up in the process?

Section 19.8 How Is Chemical Energy Converted to Other Forms of Energy?

19.51 (a) How do muscles contract?
(b) Where does the energy in muscle contraction come from?

19.52 Give an example of the conversion of the chemical energy of ATP to electrical energy.

19.53 How is the enzyme phosphorylase activated?

Chemical Connections

19.54 (Chemical Connections 19A) What is a protonophore?

19.55 (Chemical Connections 19A) Oligomycin is an antibiotic that allows electron transport to continue but stops phosphorylation in both bacteria and humans. Would you use it as an antibacterial drug for people? Explain.

Additional Problems

19.56 (a) What is the difference in structure between ATP and GTP?
(b) Compared with ATP, would you expect GTP to carry more, less, or about the same amount of energy?

19.57 How many grams of CH_3COOH (from acetyl CoA) molecules must be metabolized in the common metabolic pathway to yield 87.6 kcal of energy?

19.58 What is the basic difference in the functional groups between citrate and isocitrate?

19.59 The passage of ions from the cytoplasmic side into the matrix generates mechanical energy. Where is this energy of motion exhibited first?

19.60 What kind of reaction occurs in the citric acid cycle when a C_6 compound is converted to a C_5 compound?

19.61 What structural characteristics do citric acid and malic acid have in common?

19.62 Two ketoacids are important in the citric acid cycle. Identify them, and tell how they are manufactured.

19.63 Which filament of muscles is an enzyme, catalyzing the reaction that converts ATP to ADP?

19.64 One of the end products of food metabolism is water. How many molecules of H_2O are formed from the entry of each molecule of (a) NADH + H^+ and (b) $FADH_2$? (*Hint:* Use Figure 19.10.)

19.65 How many stereocenters are in isocitrate?

19.66 Acetyl CoA is labeled with radioactive carbon as shown: CH_3*CO—S—CoA. This compound enters the citric acid cycle. If the cycle is allowed to progress to only the α-ketoglutarate level, will the CO_2 expelled by the cell be radioactive?

19.67 Where is the H^+ ion channel located in the proton-translocating ATPase complex?

19.68 Is the passage of H^+ ion through the channel converted directly into chemical energy?

19.69 Does all the energy used in ATP synthesis come from the mechanical energy of rotation?

19.70 ■ (a) In the citric acid cycle, how many steps can be classified as decarboxylation reactions?
(b) In each case, what is the concurrent oxidizing agent? (*Hint:* See Table 15.1.)

19.71 What is the role of succinate dehydrogenase in the citric acid cycle?

19.72 How many stereocenters are in malate?

19.73 What is the source of carbon dioxide that we exhale?

19.74 Does oxygen combine directly with carbon-containing molecules to produce carbon dioxide?

19.75 Some soft drinks contain citric acid as flavoring. Is it a good nutrient?

19.76 Is mitochondrial ATPase an integral membrane protein? Explain your answer.

19.77 Do all complexes of the electron transport chain generate enough energy to produce ATP?

19.78 Why is mitochondrial ATPase considered a motor protein?

Tying It Together

19.79 Why does citrate isomerize to isocitrate before any oxidation steps take place in the citric acid cycle?

19.80 Why is the material in this chapter called the common catabolic pathway, rather than giving that designation to any other metabolic reactions?

19.81 What are two ways in which iron is part of the structure of the proteins of the electron transport pathway?

19.82 Why is it necessary for the proteins of the electron transport chain to be integral membrane proteins?

19.83 Why is it necessary to have mobile electron carriers as part of the electron transport chain?

19.84 Why does the loss of CO_2 make the citric acid cycle irreversible?

Looking Ahead

19.85 Why is the citric acid cycle central to biosynthetic pathways as well as to catabolism?

19.86 Is there a significant difference in the energy yield of the central catabolic pathway if FAD is used as an electron carrier rather than NAD^+?

19.87 Are biosynthetic pathways likely to involve oxidation, like the common catabolic pathway, or reduction? Why?

19.88 Are biosynthetic pathways likely to release energy, like the common catabolic pathway, or require energy? Why?

Challenge Problems

19.89 In a typical human, body weight fluctuates very little during the course of a day. How can this statement be consistent with the estimate that the human body manufactures as much as 40 kg of ATP every day?

19.90 When the electron transport pathway was first studied, researchers used inhibitors to block the flow of electrons in their work. Why is it likely that such inhibitors could help establish the order of carriers?

19.91 Oxygen does not appear in any of the reactions of the citric acid cycle, but it is considered part of aerobic metabolism. Why?

19.92 Is it likely that some of the important molecules for transfer of phosphate groups, electrons, and acetyl groups will appear in other metabolic pathways discussed in future chapters?

Specific Catabolic Pathways: Carbohydrate, Lipid, and Protein Metabolism

20

© Dennis Degnan/Corbis

The ballet dancer derives energy from catabolism of nutrients.

20.1 What Is the General Outline of Catabolic Pathways?

The food we eat serves two main purposes: (1) It fulfills our energy needs, and (2) it provides the raw materials to build the compounds our bodies need. Before either of these processes can take place, food—carbohydrates, fats, and proteins—must be broken down into small molecules that can be absorbed through the intestinal walls. We will deal with most of the details of digestion

OWL Online homework for this chapter may be assigned in **GOB OWL**.

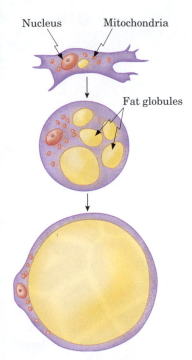

Nucleus   Mitochondria

Fat globules

FIGURE 20.1 Storage of fat in a fat cell. As more and more fat droplets accumulate in the cytoplasm, they coalesce to form a very large globule of fat. Such a fat globule may occupy most of the cell, pushing the cytoplasm and the organelles to the periphery. (Modified from C. A. Villee, E. P. Solomon, and P. W. Davis, *Biology,* Philadelphia, Saunders College Publishing, 1985.)

Amino acid pool   The free amino acids found both inside and outside cells throughout the body

Glycolysis   The biochemical pathway that breaks down glucose to pyruvate, which yields chemical energy in the form of ATP and reduced coenzymes

in Chapter 22. In this chapter, along with the preceding and following chapters, we will keep to our main train of thought: the chemical aspects of metabolism.

A. Carbohydrates

Complex carbohydrates (di- and polysaccharides) in the diet are broken down by enzymes and stomach acid to produce monosaccharides, the most important of which is glucose (Section 22.3). Glucose also comes from the enzymatic breakdown of glycogen that is stored in the liver and muscles until needed. Once monosaccharides are produced, they can be used either to build new oligo- and polysaccharides or to provide energy. The specific pathway by which energy is extracted from monosaccharides is called glycolysis (Sections 20.2 and 20.3).

B. Lipids

Ingested fats are hydrolyzed by lipases to glycerol and fatty acids or to monoglycerides, which are absorbed through the intestine (Section 20.4). In a similar fashion, complex lipids are hydrolyzed to smaller units before their absorption. As with carbohydrates, these smaller molecules (fatty acids, glycerol, and so on) can be used to build the complex molecules needed in membranes; they can be oxidized to provide energy; or they can be stored in **fat storage depots** (Figure 20.1). The stored fats can later be hydrolyzed to glycerol and fatty acids whenever they are needed as fuel.

The specific pathway by which energy is extracted from glycerol involves the same glycolysis pathway as that used for carbohydrates (Section 20.4). The specific pathway used by the cells to obtain energy from fatty acids is called β-oxidation (Section 20.5).

C. Proteins

As you might expect from your knowledge of their structures, proteins are hydrolyzed by HCl in the stomach and by digestive enzymes in the stomach (pepsin) and intestines (trypsin, chymotrypsin, and carboxypeptidases) to produce their constituent amino acids. The amino acids absorbed through the intestinal wall enter the **amino acid pool.** They serve as building blocks for proteins as needed and, to a smaller extent (especially during starvation), as a fuel for energy. In the latter case, the nitrogen of the amino acids is catabolized through oxidative deamination and the urea cycle and is expelled from the body as urea in the urine (Section 20.8). The carbon skeletons of the amino acids enter the common catabolic pathway (Chapter 19) as either α-ketoacids (pyruvic, oxaloacetic, and α-ketoglutaric acids) or acetyl coenzyme A (Section 20.9).

In all cases, *the specific pathways of carbohydrate, triglyceride (fat), and protein catabolism converge to the common catabolic pathway* (Figure 20.2). In this way, the body needs fewer enzymes to get energy from diverse food materials. Efficiency is achieved because a minimal number of chemical steps are required and because the energy-producing factories of the body are localized in the mitochondria.

20.2 What Are the Reactions of Glycolysis?

Glycolysis is the specific pathway by which the body gets energy from monosaccharides. The detailed steps in glycolysis are shown in Figure 20.3, and the most important features are shown schematically in Figure 20.4.

A. Glycolysis of Glucose

In the first steps of glucose metabolism, energy is consumed rather than released. At the expense of two molecules of ATP (which are converted to ADP), glucose is phosphorylated. First, glucose 6-phosphate is formed in Step ①, then, after isomerization to fructose 6-phosphate in Step ②, a second

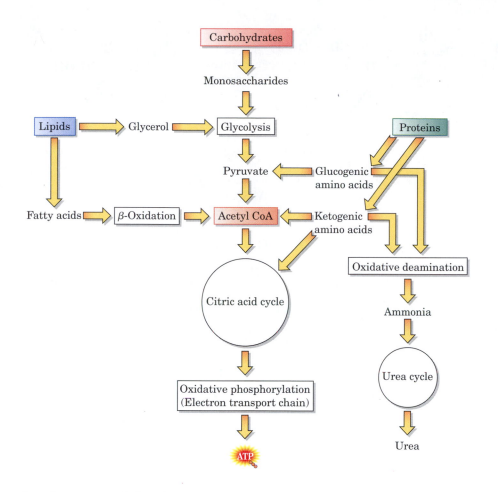

FIGURE 20.2 The convergence of the specific pathways of carbohydrate, fat, and protein catabolism into the common catabolic pathway, which is made up of the citric acid cycle and oxidative phosphorylation.

phosphate group is bonded to yield fructose 1,6-bisphosphate in Step ③. We can consider these steps to be the activation process.

In the second stage, the C_6 compound, fructose 1,6-bisphosphate, is broken into two C_3 fragments in Step ④. The two C_3 fragments, glyceraldehyde 3-phosphate and dihydroxyacetone phosphate, are in equilibrium (they can be converted to each other). Only glyceraldehyde 3-phosphate is oxidized in glycolysis, but as this species is removed from the equilibrium mixture, the equilibrium shifts and dihydroxyacetone phosphate is converted to glyceraldehyde 3-phosphate.

In the third stage, glyceraldehyde 3-phosphate is oxidized to 1,3-bisphosphoglycerate in Step ⑤. The hydrogen of the aldehyde group is removed by the NAD^+ coenzyme. In Step ⑥, the phosphate from the carboxyl group is transferred to ADP, yielding ATP and 3-phosphoglycerate. The latter compound, after isomerization in Step ⑦ and dehydration in Step ⑧, is converted to phosphoenolpyruvate, which loses its remaining phosphate in Step ⑨ and yields pyruvate and another ATP molecule. [In Step ⑨, after hydrolysis of the phosphate, the resulting enol of pyruvic acid tautomerizes to the more stable keto form (Section 9.5).] Step ⑨ is also the "payoff" step, as the two ATP molecules produced here (one for each C_3 fragment) represent the net yield of ATPs in glycolysis. Step ⑨ is catalyzed by an enzyme, pyruvate kinase, whose active site was depicted in Chemical Connections 15C. This enzyme plays a key role in the regulation of glycolysis. For example, pyruvate kinase is inhibited by ATP and activated by AMP. Thus, when plenty of ATP is available, glycolysis is shut down; when ATP is scarce and AMP levels are high, the glycolytic pathway is speeded up.

All of these glycolysis reactions occur in the cytoplasm outside the mitochondria. Because they occur in the absence of O_2, they are also called reactions of the **anaerobic pathway.** As indicated in Figure 20.4, the end product of glycolysis, pyruvate, does not accumulate in the body. In certain

FIGURE 20.3 Glycolysis, the pathway of glucose metabolism. (Steps ⑩, ⑫, and ⑬ are shown in Figure 20.4.) Some of the steps are reversible, but equilibrium arrows are not shown (they appear in Figure 20.4).

bacteria and yeast, pyruvate undergoes decarboxylation in Step ⑩ to produce ethanol. In some bacteria, and in mammals in the absence of oxygen, pyruvate is reduced to lactate in Step ⑪. The reactions that produce ethanol in organisms capable of alcoholic fermentation operate in reverse when

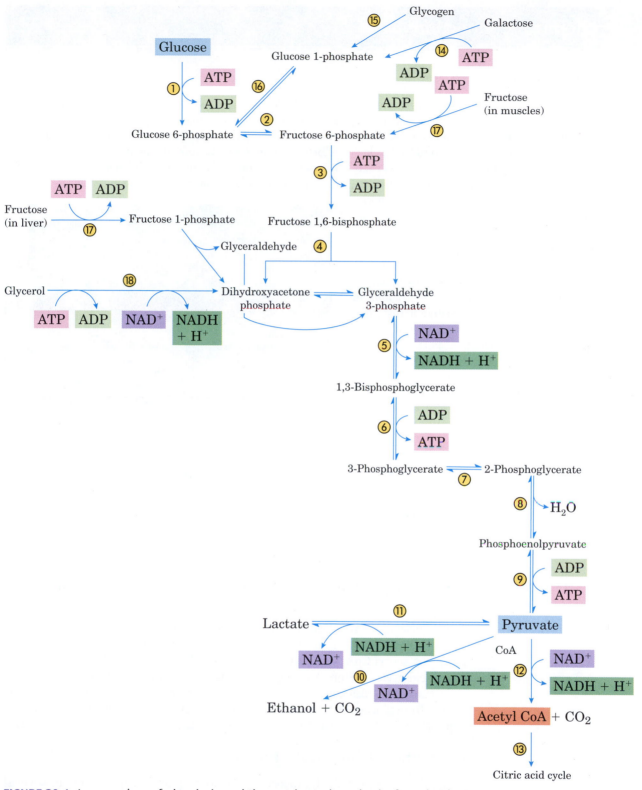

FIGURE 20.4 An overview of glycolysis and the entries to it and exits from it. The equilibrium arrows represent reversible steps. A change in conditions can, and does, affect the relative amounts of starting material metabolized by glycolysis, or the fate of the pyruvate produced.

humans metabolize ethanol. Acetaldehyde (Section 9.2), which is a product of one of these reactions, is a toxic substance that's responsible for much of the damage in fetal alcohol syndrome. Transfer of nutrients and oxygen to the fetus is depressed, with tragic consequences.

Chemical Connections 20A

Lactate Accumulation

Many athletes suffer muscle cramps when they engage in strenuous exercise (see Chapter 7). This problem results from a shift from normal glucose catabolism (glycolysis: citric acid cycle: oxidative phosphorylation) to that of lactate production (see Step ⑪ in Figure 20.4). During exercise, oxygen is used up rapidly, which slows down the rate of the common catabolic pathway. The demand for energy makes anaerobic glycolysis proceed at a high rate, but because the aerobic (oxygen-demanding) pathways are slowed down, not all the pyruvate produced in glycolysis can enter the citric acid cycle. The excess pyruvate ends up as lactate, which causes painful muscle contractions.

The same shift in catabolism occurs in heart muscle when coronary thrombosis leads to cardiac arrest. The blockage of the artery leading to the heart muscles cuts off the oxygen supply. The common catabolic pathway and its ATP production are consequently shut off. Glycolysis proceeds at an accelerated rate, causing lactate to accumulate. The heart muscle contracts, producing a cramp. Just as in skeletal muscle, massage of heart muscles can relieve the cramp and start the heart beating. Even if heartbeat is restored within 3 minutes (the amount of time the brain can survive without being damaged), acidosis may develop as a result of the cardiac arrest. Therefore, at the same time that efforts are under way to start the heart beating by chemical, physical, or electrical means, an intravenous infusion of 8.4% bicarbonate solution is given to combat acidosis.

Knowledge of biochemistry is a big help in treating cardiac arrest.

B. Entrance to the Citric Acid Cycle

Pyruvate is not the end product of glucose metabolism. Most importantly, pyruvate goes through an oxidative decarboxylation in the presence of coenzyme A in Step ⑫ to produce acetyl CoA:

$$NAD^+ + CH_3 - \overset{\overset{\displaystyle O}{\|}}{C} - COO^- + CoA - SH \longrightarrow CH_3 - \overset{\overset{\displaystyle O}{\|}}{C} - S - CoA + CO_2 + NADH + H^+$$

Pyruvate   Acetyl coenzyme A

This reaction is catalyzed by a complex enzyme system, pyruvate dehydrogenase, that sits on the inner membrane of the mitochondrion. The reaction produces acetyl CoA, CO_2, and NADH + H$^+$. The acetyl CoA then enters the citric acid cycle in Step ⑬ and goes through the common catabolic pathway.

In summary, after converting complex carbohydrates to glucose, the body gets energy from glucose by converting it to acetyl CoA (by way of pyruvate) and then using the acetyl CoA as a starting material for the common catabolic pathway.

C. Pentose Phosphate Pathway

As we saw in Figure 20.4, glucose 6-phosphate plays a central role in several different entries into the glycolytic pathway. However, glucose 6-phosphate can also be used by the body for other purposes, not just for the production of energy in the form of ATP. Most importantly, glucose 6-phosphate can be shunted to the **pentose phosphate pathway** in Step ⑲ (Figure 20.5). This pathway has the capacity to produce NADPH and ribose in Step ⑳ as well as energy.

NADPH is needed in many biosynthetic processes, including synthesis of unsaturated fatty acids (Section 21.3), cholesterol, and amino acids as well as photosynthesis (Chemical Connections 21A) and the reduction of ribose to

Pentose phosphate pathway
The biochemical pathway that produces ribose and NADPH from glucose 6-phosphate or, alternatively, releases energy

FIGURE 20.5 Simplified schematic representation of the pentose phosphate pathway, also called a shunt. Steps 19 and 21 in this figure represent multiple steps in the actual pathway.

Nicotinamide adenine dinucleotide phosphate ($NADP^+$)

deoxyribose for DNA. Ribose is needed for the synthesis of RNA (Section 17.3). Therefore, when the body needs these synthetic ingredients more than energy, glucose 6-phosphate is used in the pentose phosphate pathway. When energy is needed, glucose 6-phosphate remains in the glycolytic pathway—even ribose 5-phosphate can be channeled back to glycolysis through glyceraldehyde 3-phosphate. Through this reversible reaction, the cells can also obtain ribose directly from the glycolytic intermediates. In addition, NADPH is badly needed in red blood cells as a defense against oxidative damages. Glutathione is the main agent used to keep the hemoglobin in its reduced form. It is regenerated by NADPH, so an insufficient supply of NADPH leads to the destruction of red blood cells, causing severe anemia.

20.3 What Is the Energy Yield from Glucose Catabolism?

Using Figure 20.4, let us sum up the energy derived from glucose catabolism in terms of ATP production. First, however, we must take into account the fact that glycolysis takes place in the cytoplasm, whereas oxidative phosphorylation occurs in the mitochondria. Therefore, the NADH + H$^+$ produced in glycolysis in the cytoplasm must be converted to NADH in the mitochondria before it can be used in oxidative phosphorylation.

NADH is too large to cross the mitochondrial membrane. Two routes are available to get the electrons into the mitochondria; they have different efficiencies. In glycerol 3-phosphate transport, which operates in muscle and nerve cells, only two ATP molecules are produced for each NADH + H$^+$. In the other transport route, which operates in the heart and the liver, three ATP molecules are produced for each NADH + H$^+$ produced in the cytoplasm, just as is the case in the mitochondria (Section 19.7). Because most energy production takes place in skeletal muscle cells, when we construct the energy balance sheet, we use two ATP molecules for each NADH + H$^+$ produced in the cytoplasm. (Muscles attached to bones are called skeletal muscles; cardiac muscle is in a different category.)

TABLE 20.1 ATP Yield from Complete Glucose Metabolism

Step Numbers in Figure 20.4	Chemical Steps	Number of ATP Molecules Produced
① ② ③	Activation (glucose ⟶ 1,6-fructose bisphosphate)	−2
⑤	Phosphorylation 2 (glyceraldehyde 3-phosphate ⟶ 1,3-bisphosphoglycerate), producing 2 (NADH + H⁺) in cytosol	4
⑥ ⑨	Dephosphorylation 2 (1,3-bisphosphoglycerate ⟶ pyruvate)	4
⑫	Oxidative decarboxylation 2 (pyruvate ⟶ acetyl CoA), producing 2 (NADH + H⁺) in mitochondrion	6
⑬	Oxidation of two C_2 fragments in citric acid cycle and oxidative phosphorylation common pathways, producing 12 ATP for each C_2 fragment	24
	Total	36

Armed with this knowledge, we are ready to calculate the energy yield of glucose in terms of ATP molecules produced in skeletal muscles. Table 20.1 shows this calculation. In the first stage of glycolysis (Steps ①, ②, and ③), two ATP molecules are used up, but this loss is more than compensated for by the production of 14 ATP molecules in Steps ⑤, ⑥, ⑨, and ⑫, and in the conversion of pyruvate to acetyl CoA. The net yield of these steps is 12 ATP molecules. As we saw in Section 19.7, the oxidation of one acetyl CoA molecule produces 12 ATP molecules, and one glucose molecule provides two acetyl CoA molecules. Therefore, the total net yield from metabolism of one glucose molecule in skeletal muscle is 36 molecules of ATP, or 6 ATP molecules per carbon atom.

$$C_6H_{12}O_6 + 6O_2 \longrightarrow 6CO_2 + 6H_2O$$

If the same glucose is metabolized in the heart or liver, the electrons of the two NADH molecules produced in glycolysis are transported into the mitochondrion by the malate-aspartate shuttle. Through this shuttle, the two NADH molecules yield a total of 6 ATP molecules, so that, in this case, 38 ATP molecules are produced for each glucose molecule. It is instructive to note that most of the energy (in the form of ATP) from glucose is produced in the common metabolic pathway. Recent investigations suggest that approximately 30 to 32 ATP molecules are actually produced per glucose molecule (2.5 ATP/NADH and 1.5 ATP/FADH₂). Further research into the complexity of the oxidative phosphorylation pathway is needed to verify these numbers. We alluded to this point briefly in Chapter 19 when we said that the ATP yields were reported to the nearest whole number. Here we see how the complexity of oxidative phosphorylation affects energy metabolism.

Glucose is not the only monosaccharide that can be used as an energy source. Other hexoses, such as galactose (Step ⑭) and fructose (Step ⑰), enter the glycolysis pathway at the stages indicated in Figure 20.4. They also yield 36 molecules of ATP per hexose molecule. Furthermore, glycogen stored in liver and muscle cells and elsewhere can be converted by enzymatic breakdown and phosphorylation to glucose 1-phosphate (Step ⑮). This compound, in turn, isomerizes to glucose 6-phosphate, providing an entry into the glycolytic pathway (Step ⑯). The pathway in which glycogen breaks down to glucose is called **glycogenolysis.**

Now that we have seen the catabolic reactions of carbohydrates, we turn our attention to another major source of energy, the catabolism of lipids.

Glycogenolysis The biochemical pathway for the breakdown of glycogen to glucose

Recall that for triglycerides, which are the main storage form of the chemical energy of lipids, we have to consider two parts, glycerol and fatty acids.

20.4 How Does Glycerol Catabolism Take Place?

The glycerol hydrolyzed from fats or complex lipids (Chapter 13) can also be a rich energy source. The first step in glycerol utilization is an activation step. The body uses one ATP molecule to form glycerol 1-phosphate, which is the same as glycerol 3-phosphate:

The glycerol phosphate is oxidized by NAD^+ to dihydroxyacetone phosphate, yielding $NADH + H^+$ in the process. Dihydroxyacetone phosphate then enters the glycolysis pathway (Step ⑱ in Figure 20.4) and is isomerized to glyceraldehyde 3-phosphate. A net yield of 20 ATP molecules is produced from each glycerol molecule, or 6.7 ATP molecules per carbon atom.

20.5 What Are the Reactions of β-Oxidation of Fatty Acids?

As early as 1904, Franz Knoop, working in Germany, proposed that the body utilizes fatty acids as an energy source by breaking them down into fragments. Prior to fragmentation, the β-carbon (the second carbon atom from the COOH group) is oxidized:

$$-\text{C}-\text{C}-\text{C}-\overset{\beta}{\text{C}}-\overset{\alpha}{\text{C}}-\text{COOH}$$

The name **β-oxidation** has its origin in Knoop's prediction. It took about 50 years to establish the mechanism by which fatty acids are utilized as an energy source.

Figure 20.6 depicts the overall process of fatty acid metabolism. As is the case with the other foods we have seen, the first step involves activation. In the general case of lipid catabolism, this activation occurs in the cytosol, where the fat was previously hydrolyzed to glycerol and fatty acids. It converts ATP to AMP and inorganic phosphate (Step ①), which is equivalent to the cleavage of two high-energy phosphate bonds. The chemical energy derived from the hydrolysis of ATP is built into the compound acyl CoA, which forms when the fatty acid combines with coenzyme A. The fatty acid oxidation occurs inside the mitochondrion, so the acyl group of acyl CoA must pass through the mitochondrial membrane. Carnitine is the acyl group transporter. The enzyme system that catalyzes the process is carnitine acyltransferase.

Once the fatty acid in the form of acyl CoA is inside the mitochondrion, the β-oxidation starts. In the first oxidation (dehydrogenation; Step ②), two hydrogens are removed, creating a *trans* double bond between the alpha and beta carbons of the acyl chain. The hydrogens and electrons are picked up by FAD.

In Step ③, the double bond is hydrated. An enzyme specifically places the hydroxyl group on C-3, the beta carbon. The second oxidation (dehydrogenation;

β-oxidation The biochemical pathway that degrades fatty acids to acetyl CoA by removing two carbons at a time and yielding energy

ACTIVE FIGURE 20.6
The β-oxidation spiral of fatty acids. Each loop in the spiral contains two dehydrogenations, one hydration, and one fragmentation. At the end of each loop, one acetyl CoA molecule is released. **Go to this book's companion website at www. cengage.com/chemistry/ bettelheim to explore an interactive version of this figure.**

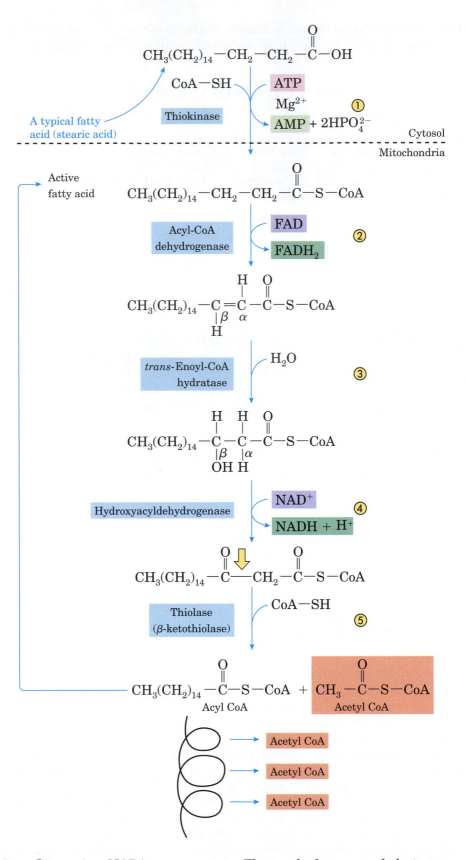

Step ④) requires NAD^+ as a coenzyme. The two hydrogens and electrons removed are transferred to the NAD^+ to form $NADH + H^+$. In the process, a secondary alcohol is oxidized to a ketone at the beta carbon. In Step ⑤, the enzyme thiolase cleaves the terminal C_2 fragment (an acetyl CoA) from the chain, and the rest of the molecule is bonded to a new molecule of coenzyme A.

Chemical Connections 20B

Effects of Signal Transduction on Metabolism

The ligand-binding/G-protein/adenylate cyclase cascade, which activates proteins by phosphorylation, has a wide range of effects other than opening or closing ion-gated channels (Figure 16.4). An example of such a target is the glycogen phosphorylase enzyme that participates in the breakdown of glycogen stored in muscles. This enzyme cleaves units of glucose 1-phosphate from glycogen, which enters into the glycolytic pathway and yields quick energy (Section 20.2). The active form of the enzyme, phosphorylase a, is phosphorylated. When it is dephosphorylated (phosphorylase b), it is inactive. When epinephrine signals danger to a muscle cell, phosphorylase is activated through the cascade, and quick energy is produced. In this way, the signal is converted to a metabolic event, allowing the muscles to contract rapidly and enabling the person in danger to fight or run away.

Not all phosphorylation of enzymes results in activation. Consider glycogen synthase. In this case, the phosphorylated form of the enzyme is inactive, and the dephosphorylated form is active. This enzyme participates in glycogenesis, the conversion of glucose to, and storage in the form of, glycogen. The action of glycogen synthase is the opposite of phosphorylase. Nature provides a beautiful balance, inasmuch as the danger signal of the epinephrine hormone has a dual target: It activates phosphorylase to get quick energy, but simultaneously it inactivates glycogen synthase so that the available glucose will be used solely for energy and will not be stored.

The deer can run away from danger because a hormonal signal leads to metabolism of carbohydrates, providing energy.

The cycle then starts again with the remaining acyl CoA, which is now two carbon atoms shorter. At each turn of the cycle, one acetyl CoA is produced. Most fatty acids contain an even number of carbon atoms. The cyclic spiral continues until it reaches the last four carbon atoms. When this fragment enters the cycle, two acetyl CoA molecules are produced in the fragmentation step.

The β-oxidation of unsaturated fatty acids proceeds in the same way. An extra step is involved, in which the *cis* double bond is isomerized to a *trans* bond, but otherwise the spiral is the same.

20.6 What Is the Energy Yield from Stearic Acid Catabolism?

To compare the energy yield from fatty acids with that of other foods, let us select a typical and quite abundant fatty acid—stearic acid, the C_{18} saturated fatty acid.

We start with the initial step, in which energy is used up rather than produced. The reaction breaks two high-energy phosphoric anhydride bonds:

$$ATP \longrightarrow AMP + 2P_i + energy$$

This reaction is equivalent to hydrolyzing two molecules of ATP to ADP. In each cycle of the spiral, we obtain one $FADH_2$, one $NADH + H^+$, and one acetyl CoA. Stearic acid (C_{18}) goes through seven cycles in the spiral before it reaches the final stage. In the last (eighth) cycle, one $FADH_2$, one $NADH + H^+$, and two acetyl CoA molecules are produced. Now we can add up the energy. Table 20.2 shows that, for a C_{18} fatty acid, we obtain a total of 146 ATP molecules.

TABLE 20.2 ATP Yield from Complete Stearic Acid Metabolism

Step Number in Figure 20.6	Chemical Steps	Happens	Number of ATP Molecules Produced
①	Activation (stearic acid ⟶ stearyl CoA	Once	−2
②	Dehydrogenation (acyl CoA ⟶ *trans-enoyl* CoA), producing $FADH_2$	8 times	16
④	Dehydrogenation (hydroxyacyl CoA ⟶ ketoacyl CoA), producing $NADH + H^+$	8 times	24
	C_2 fragment (acetyl CoA ⟶ common catabolic pathway), producing 12 ATP for each C_2 fragment	9 times	108
		Total	146

It is instructive to compare the energy yield from fats with that from carbohydrates, as both are important constituents of the diet. In Section 20.2 we saw that glucose produces 36 ATP molecules—that is, 6 ATP molecules for each carbon atom. For stearic acid, there are 146 ATP molecules and 18 carbons, or 146/18 = 8.1 ATP molecules per carbon atom. The ATP produced from the glycerol portion of fats adds to the total. Fatty acids have a higher caloric value than carbohydrates.

20.7 What Are Ketone Bodies?

In spite of the high caloric value of fats, the body preferentially uses glucose as an energy supply. When an animal is well fed (plenty of sugar intake), fatty acid oxidation is inhibited, and fatty acids are stored in the form of neutral fat in fat depots. When physical exercise demands energy, when the glucose supply dwindles (as in fasting or starvation), or when glucose cannot be utilized (as in the case of diabetes), the β-oxidation pathway of fatty acid metabolism is mobilized. In some pathological conditions, glucose may not be available at all, giving added importance to this point.

Unfortunately, low glucose supply also slows down the citric acid cycle. This lag happens because some oxaloacetate is essential for the continuous operation of the citric acid cycle (Figure 19.8). Oxaloacetate is produced from malate, but it is also produced by the carboxylation of phosphoenolpyruvate (PEP):

If there is no glucose, there will be no glycolysis, no PEP formation, and, therefore, greatly reduced oxaloacetate production.

Thus, even though the fatty acids are oxidized, not all of the resulting fragments (acetyl CoA) can enter the citric acid cycle because not enough oxaloacetate is present. As a result, acetyl CoA builds up in the body, with the following consequences.

Test kit for the presence of ketone bodies in the urine.

Charles D. Winters/Cengage Learning

The liver is able to condense two acetyl CoA molecules to produce aceto-acetyl CoA:

When the acetoacetyl CoA is hydrolyzed, it yields acetoacetate, which can be reduced to form β-hydroxybutyrate:

These two compounds, together with smaller amounts of acetone, are collectively called **ketone bodies.** Under normal conditions, the liver sends these compounds into the bloodstream to be carried to tissues and utilized there as a source of energy via the common catabolic pathway. The brain, for example, normally uses glucose as an energy source. During periods of starvation, however, ketone bodies may serve as the major energy source for the brain. Normally, the concentration of ketone bodies in the blood is low. During starvation and in untreated diabetes mellitus, however, ketone bodies accumulate in the blood and can reach high concentrations. When this buildup occurs, the excess is secreted in the urine. A check of urine for ketone bodies is used in the diagnosis of diabetes.

Ketone bodies A collective name for acetone, acetoacetate and β-hydroxybutyrate; compounds produced from acetyl CoA in the liver that are used as a fuel for energy production by muscle cells and neurons

Example 20.1 Counting ATPs

Ketone bodies are a source of energy, especially during dieting and starvation. If acetoacetate is metabolized through β-oxidation and the common pathway, how many ATP molecules would be produced?

Strategy

First determine from Section 20.7 that two acetyl CoA molecules are produced. Then consult Table 20.2, final step.

Solution

In Step ①, activation of acetoacetate to acetoacetyl-CoA requires 2 ATPs. Step ⑤ yields two acetyl CoA molecules, which enter the common catabolic pathway, yielding 12 ATPs for each acetyl CoA, for a total of 24 ATPs. The net yield, therefore, is 22 ATP molecules.

Problem 20.1

Which fatty acid yields more ATP molecules per carbon atom: (a) stearic acid or (b) lauric acid?

Chemical Connections 20C

Ketoacidosis in Diabetes

In untreated diabetes, the glucose concentration in the blood is high because a lack of insulin prevents utilization of glucose by the cells. Regular injections of insulin can remedy this situation. However, in some stressful conditions, **ketoacidosis** can still develop.

A typical case was a diabetic patient admitted to the hospital in a semi-comatose state. He showed signs of dehydration, his skin was inelastic and wrinkled, his urine showed high concentrations of glucose and ketone bodies, and his blood contained excess glucose and had a pH of 7.0, a drop of 0.4 pH unit from normal, which is an indication of severe acidosis. The patient's urine also contained the bacterium *Escherichia coli*. This indication of urinary tract infection explained why the normal doses of insulin were insufficient to prevent ketoacidosis.

The stress of infection can upset the normal control of diabetes by changing the balance between administered insulin and other hormones produced in the body. This imbalance happened during the patient's infection, and his body started to produce ketone bodies in large quantities. Both glucose and ketone bodies appear in the blood before they show up in the urine.

The acidic nature of ketone bodies (acetoacetic acid and β-hydroxybutyric acid) lowers the blood pH. A large drop in pH is prevented by the bicarbonate/carbonic acid buffer, but even a drop of 0.3 to 0.5 pH unit is sufficient to decrease the Na^+ concentration. Such a decrease of Na^+ ions in the interstitial fluids draws out K^+ ions from the cells. This, in turn, impairs brain function and leads to coma. During the secretion of ketone bodies and glucose in the urine, a lot of water is lost, the body becomes dehydrated, and the blood volume shrinks. As a consequence, the blood pressure drops, and the pulse rate increases to compensate for it. Smaller quantities of nutrients reach the brain cells, which can also cause coma.

The patient mentioned here was infused with physiological saline solution to remedy his dehydration. Extra doses of insulin restored his glucose level to normal, and antibiotics cured the urinary infection.

Devices that allow diabetics to monitor blood glucose levels are widely marketed.

20.8 How Is the Nitrogen of Amino Acids Processed in Catabolism?

The proteins of our foods are hydrolyzed to amino acids in digestion. These amino acids are primarily used to synthesize new proteins. Unlike carbohydrates and fats, however, they cannot be stored, so excess amino acids are catabolized for energy production. Section 20.9 explains what happens to the carbon skeletons of the amino acids. Here we discuss the catabolic fate of the nitrogen. Figure 20.7 gives an overview of the entire process of protein catabolism.

In the tissues, amino groups ($-NH_2$) freely move from one amino acid to another. The enzymes that catalyze these reactions are the transaminases. In essence, nitrogen catabolism in the liver occurs in three stages: transamination, oxidative deamination, and the urea cycle.

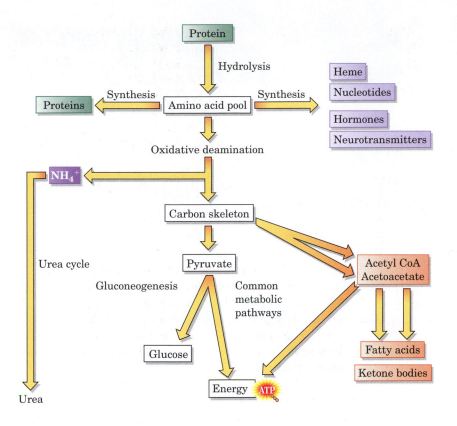

FIGURE 20.7 The overview of pathways in protein catabolism.

A. Transamination

In the first stage, **transamination,** amino acids transfer their amino groups to α-ketoglutarate:

Transamination The exchange of the amino group of an amino acid and a keto group of an α-ketoacid

The carbon skeleton of the amino acid remains behind as an α-ketoacid, the catabolism of which is discussed in the next section.

B. Oxidative Deamination

The second stage of nitrogen catabolism is the **oxidative deamination** of glutamate, which occurs in the mitochondrion:

Oxidative deamination The reaction in which the amino group of an amino acid is removed and an α-ketoacid is formed

FIGURE 20.8 The urea cycle.

The oxidative deamination yields NH_4^+ and regenerates α-ketoglutarate, which can again participate in the first stage (transamination). The NADH + H^+ produced in the second stage enters the oxidative phosphorylation pathway and eventually produces three ATP molecules. The body must get rid of NH_4^+ because both it and NH_3 are toxic.

C. Urea Cycle

Urea cycle A cyclic pathway that produces urea from ammonia and carbon dioxide

In the third stage, the NH_4^+ is converted to urea through the **urea cycle** (Figure 20.8). In Step ①, NH_4^+ is condensed with CO_2 in the mitochondrion to form an unstable compound, carbamoyl phosphate. This condensation occurs at the expense of two ATP molecules. In Step ②, carbamoyl phosphate is condensed with ornithine, a basic amino acid similar in structure to lysine, but which does not occur in proteins, to produce citrulline.

The resulting citrulline diffuses out of the mitochondrion into the cytoplasm.

A second condensation reaction in the cytoplasm takes place between citrulline and aspartate, forming argininosuccinate (Step ③):

The energy for this reaction comes from the hydrolysis of ATP to AMP and pyrophosphate (PP_i).

In Step ④, the argininosuccinate is split into arginine and fumarate:

In Step ⑤, the final step, arginine is hydrolyzed to urea and ornithine:

The final product of the three stages is urea, which is excreted in the urine of mammals. The ornithine reenters the mitochondrion, completing the cycle. It is then ready to pick up another carbamoyl phosphate. An important aspect of carbamoyl phosphate's role as an intermediate is that it can be used for synthesis of nucleotide bases (Chapter 17). Furthermore, the urea cycle is linked to the citric acid cycle in that both involve fumarate. In fact, Hans Krebs, who elucidated the citric acid cycle, was also instrumental in establishing the urea cycle.

Not all organisms dispose of metabolic nitrogen in the form of urea. Bacteria and fish, for example, release ammonia directly into the surrounding water. Ammonia is toxic in high concentrations, but the surrounding water dilutes the ammonia enough for these organisms to excrete nitrogen in this form. Birds and reptiles secrete nitrogen in the form of uric acid, the concentrated white solid so familiar in bird droppings.

D. Other Pathways of Nitrogen Catabolism

The urea cycle is not the only way that the body can dispose of the toxic NH_4^+ ions. The oxidative deamination process, which produced the NH_4^+ in the first place, is reversible. Therefore, the buildup of glutamate from α-ketoglutarate and NH_4^+ is always possible. A third possibility for disposing of NH_4^+ is the ATP-dependent amidation of glutamate to yield glutamine:

Glutamate   Glutamine

20.9 How Are the Carbon Skeletons of Amino Acids Processed in Catabolism?

After transamination of amino acids (Section 20.8A) to glutamate, the alpha amino group is removed from glutamate by oxidative deamination (Section 20.8B). The remaining carbon skeletons are used as an energy source (Figure 20.8). We will not study the pathways involved except to point out the eventual fate of the skeletons. Not all of the carbon skeletons of amino acids are used as fuel. Some may be degraded up to a certain point, and the resulting intermediate may then be used as a building block to construct another needed molecule.

For example, if the carbon skeleton of an amino acid is catabolized to pyruvate, the body has two possible choices: (1) use the pyruvate as an energy supply via the common catabolic pathway or (2) use it as a building block to synthesize glucose (Section 21.1). Those amino acids that yield a carbon skeleton that is degraded to pyruvate or another intermediate capable of conversion to glucose (such as oxaloacetate) are called **glucogenic.**

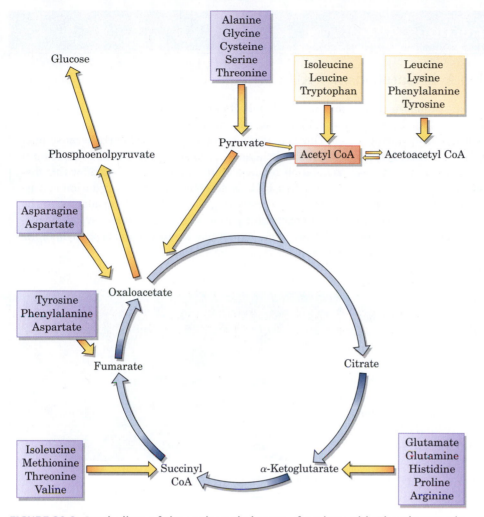

FIGURE 20.9 Catabolism of the carbon skeletons of amino acids. The glucogenic amino acids are in the purple boxes; the ketogenic ones in the gold boxes.

One example is alanine (Figure 20.9). When alanine reacts with α-ketoglutaric acid, the transamination produces pyruvate directly:

In contrast, many amino acids are degraded to acetyl CoA and acetoacetic acid. These compounds cannot form glucose but are capable of yielding ketone bodies; they are called **ketogenic.** Leucine is an example of a ketogenic amino acid. Some amino acids are both glucogenic and ketogenic—for example, phenylalanine.

Both glucogenic and ketogenic amino acids, when used as an energy supply, enter the citric acid cycle at some point (Figure 20.9) and are eventually oxidized to CO_2 and H_2O. The oxaloacetate (a C_4 compound) produced in this manner enters the citric acid cycle, adding to the oxaloacetate produced from PEP and in the cycle itself.

Chemical Connections 20D

Hereditary Defects in Amino Acid Catabolism: PKU

Many hereditary diseases involve missing or malfunctioning enzymes that catalyze the breakdown of amino acids. The oldest known of such diseases is cystinuria, which was described as early as 1810. In this disease, cystine shows up as flat hexagonal crystals in the urine. Stones form because of cystine's low solubility in water. This problem leads to blockage in the kidneys or the ureters and requires surgery to resolve it. One way to reduce the amount of cystine secreted is to remove as much methionine as possible from the diet. Beyond that, an increased fluid intake increases the volume of the urine, reducing the solubility problem. In addition, penicillamine can prevent cystinuria.

An even more important genetic defect is the absence of the enzyme phenylalanine hydroxylase, which causes a disease called phenylketonuria (PKU). In normal catabolism, this enzyme helps degrade phenylalanine by converting it to tyrosine. If the enzyme is defective, phenylalanine is converted to phenylpyruvate (see the discussion of the conversion of alanine to pyruvate in Section 20.9). Phenylpyruvate (an α-ketoacid) accumulates in the body and inhibits the conversion of pyruvate to acetyl CoA, thereby depriving the cells of energy via the common catabolic pathway. This effect is most important in the brain, which gets its energy from the utilization of glucose. PKU results in mental retardation.

This genetic defect can be detected early because phenylpyruvic acid appears in the urine and blood. A federal regulation requires that all infants be tested for this disease. When PKU is detected, mental retardation can be prevented by restricting the intake of phenylalanine in the diet. In particular, patients with PKU should avoid the artificial sweetener aspartame because it yields phenylalanine when hydrolyzed in the stomach.

A newborn being tested for PKU.

Courtesy of U.S. Air Force photo/Staff Sgt Eric T. Sheler

20.10 What Are the Reactions of Catabolism of Heme?

Carbohydrates, lipids, and proteins are the principal sources of energy in catabolism. Other cellular components contribute far less energy when they are catabolized. Their breakdown products can, however, affect the body. We will use the catabolism of heme as an example of an easily visible result of degradation.

Red blood cells are continuously being manufactured in the bone marrow. Their life span is relatively short—about four months. Aged red blood cells are destroyed in the phagocytic cells. (Phagocytes are specialized blood cells that destroy foreign bodies.) When a red blood cell is destroyed, its hemoglobin is metabolized: The globin (Section 14.11) is hydrolyzed to amino acids, and the heme is first oxidized to biliverdin and finally reduced to bilirubin (Figure 20.10). The color change observed in bruises signal the redox reactions occurring in heme catabolism: Black and blue are due to the congealed blood, green to the biliverdin, and yellow to the bilirubin. The iron is preserved in ferritin, an iron-carrying protein, and reused. The bilirubin enters the liver via the blood and is then transferred to the gallbladder, where it is stored in the bile and finally excreted via the small intestine. The color of feces is provided by urobilin, an oxidation product of bilirubin.

FIGURE 20.10 Heme degradation from heme to biliverdin to bilirubin.

Postscript

It is useful to summarize the main points of catabolic pathways by showing how they are related. Figure 20.11 shows how all catabolic pathways lead to the citric acid cycle, producing ATP by the reoxidation of NADH and FADH$_2$. We saw the common metabolic pathway in Chapter 19, and here we see how it is related to all of catabolism.

FIGURE 20.11 A summary of catabolism showing the role of the common metabolic pathway. Note that the end products of the catabolism of carbohydrates, lipids, and amino acids all appear. (TA is transamination; → → → is a pathway with many steps.)

Summary

OWL End-of-chapter problems identified in blue are assignable in **GOB OWL**.

Section 20.1 What Is the General Outline of Catabolic Pathways?

- The foods we eat consist of carbohydrates, lipids, and proteins.
- There are specific breakdown pathways for each kind of nutrient.

Section 20.2 What Are the Reactions of Glycolysis?

- The specific pathway of carbohydrate catabolism is **glycolysis.**
- Hexose monosaccharides are activated by ATP and eventually converted to two C_3 fragments, dihydroxyacetone phosphate and glyceraldehyde phosphate.
- The glyceraldehyde phosphate is further oxidized and eventually ends up as pyruvate. All of these reactions occur in the cytosol.
- Pyruvate is converted to acetyl CoA, which is further catabolized in the common pathway.
- When the body needs intermediates for synthesis rather than energy, the glycolytic pathway can be shunted to the **pentose phosphate pathway.** NADPH, which is necessary for reduction, is obtained in this way.
- The pentose phosphate pathway also yields ribose, which is necessary for synthesis of RNA.

Section 20.3 What Is the Energy Yield from Glucose Catabolism? Problem 20.17

- When completely metabolized, a hexose molecule yields the energy of 36 ATP molecules.

Section 20.4 How Does Glycerol Catabolism Take Place?

- Fats are broken down to glycerol and fatty acids.
- Glycerol is catabolized in the glycolysis pathway and yields 20 ATP molecules.

Section 20.5 What Are the Reactions of β-Oxidation of Fatty Acids? Problem 20.24

- Fatty acids are broken down into fragments in the **β-oxidation** spiral.

- At each turn of the spiral, one acetyl CoA is released along with one $FADH_2$ and one $NADH + H^+$. These products go through the common catabolic pathway.

Section 20.6 What Is the Energy Yield from Stearic Acid Catabolism?

- Stearic acid, a C_{18} compound, yields 146 molecules of ATP.

Section 20.7 What Are Ketone Bodies?

- In starvation and under certain pathological conditions, not all of the acetyl CoA produced in the β-oxidation of fatty acids enters the common catabolic pathway.
- Some acetyl CoA forms acetoacetate, β-hydroxybutyrate, and acetone, commonly called **ketone bodies.**
- Excess ketone bodies in the blood are secreted in urine.

Section 20.8 How Is the Nitrogen of Amino Acids Processed in Catabolism? Problem 20.37

- Proteins are broken down to amino acids. The nitrogen of the amino acids is first transferred to glutamate.
- Glutamate is **oxidatively deaminated** to yield ammonia.
- Mammals get rid of the toxic ammonia by converting it to urea in the **urea cycle;** urea is secreted in urine.

Section 20.9 How Are the Carbon Skeletons of Amino Acids Processed in Catabolism?

- The carbon skeletons of amino acids are catabolized via the citric acid cycle.
- Some amino acids, called **glucogenic amino acids,** enter as pyruvate or other intermediates of the citric acid cycle.
- Other amino acids are incorporated into acetyl CoA or ketone bodies and are called **ketogenic amino acids.**

Section 20.10 What Are the Reactions of Catabolism of Heme?

- Heme is catabolized to bilirubin, which is excreted in the feces.

Problems

■ Indicates problems that are assignable in **GOB OWL**.

Blue numbered problems are applied.

Go to this book's companion website at **www.cengage. com/chemistry/bettelheim** for interactive versions of the How To tutorials and Active Figures, and to quiz yourself on this chapter.

Section 20.1 What Is the General Outline of Catabolic Pathways?

20.2 What are the products of lipase-catalyzed hydrolysis of fats?

20.3 What is the main use of amino acids in the body?

Section 20.2 What Are the Reactions of Glycolysis?

20.4 Although catabolism of a glucose molecule eventually produces a lot of energy, the first step uses up energy. Explain why this step is necessary.

20.5 In one step of the glycolysis pathway, a chain is broken into two fragments, only one of which can be further degraded in the glycolysis pathway. What happens to the other fragment?

20.6 Kinases are enzymes that catalyze the addition (or removal) of a phosphate group to (or from) a substance. ATP is also involved. How many kinases are in glycolysis? Name them.

20.7 (a) Which steps in glycolysis of glucose need ATP?

(b) Which steps in glycolysis yield ATP directly?

20.8 At which intermediate of the glycolytic pathway does oxidation—and hence energy production—begin? In what form is the energy produced?

20.9 At what point in glycolysis can ATP act as an inhibitor? What kind of enzyme regulation occurs in this inhibition?

20.10 The end product of glycolysis, pyruvate, cannot enter as such into the citric acid cycle. Which process converts this C_3 compound to a C_2 compound?

20.11 What essential compound is produced in the pentose phosphate pathway that is needed for synthesis as well as for defense against oxidative damages?

20.12 Which of the following steps yields energy and which consumes energy?

(a) Pyruvate $\longrightarrow$ lactate

(b) Pyruvate $\longrightarrow$ acetyl CoA + CO_2

20.13 How many moles of lactate are produced from 3 moles of glucose?

20.14 How many moles of net NADH + H^+ are produced from 1 mole of glucose going to

(a) acetyl CoA? (b) Lactate?

Section 20.3 What Is the Energy Yield from Glucose Catabolism?

20.15 Of the 36 molecules of ATP produced by the complete metabolism of glucose, how many are produced directly in glycolysis alone—that is, before the common pathway?

20.16 How many net ATP molecules are produced in the skeletal muscles for each glucose molecule

(a) In glycolysis alone (up to pyruvate)?

(b) In converting pyruvate to acetyl CoA?

(c) In the total oxidation of glucose to CO_2 and H_2O?

20.17 ■ (a) If fructose is metabolized in the liver, how many moles of net ATP are produced from each mole during glycolysis?

(b) How many moles are produced if the same thing occurs in a muscle cell?

20.18 In Figure 20.3, Step ⑤ yields one NADH. Yet in Table 20.1, the same step indicates a yield of 2 NADH + H^+. Is there a discrepancy between these two statements? Explain.

Section 20.4 How Does Glycerol Catabolism Take Place?

20.19 Based on the names of the enzymes participating in glycolysis, what would be the name of the enzyme catalyzing the activation of glycerol?

20.20 Which yields more energy upon hydrolysis, ATP or glycerol 1-phosphate? Why?

Section 20.5 What Are the Reactions of β-Oxidation of Fatty Acids?

20.21 Two enzymes participating in β-oxidation have the word "thio" in their names.

(a) Name the two enzymes.

(b) To which chemical group does this name refer?

(c) What is the common feature in the action of these two enzymes?

20.22 (a) Which part of the cells contains the enzymes needed for β-oxidation of fatty acids?

(b) How does the activated fatty acid get there?

20.23 Assume that lauric acid (C_{12}) is metabolized through β-oxidation. What are the products of the reaction after three turns of the spiral?

20.24 ■ Is the β-oxidation of fatty acid (without the subsequent metabolism of C_2 fragments via the common metabolic pathway) more efficient with a short-chain fatty acid than with a long-chain fatty acid? Is more ATP produced per carbon atom in a short-chain fatty acid than in a long-chain fatty acid during β-oxidation?

Section 20.6 What Is the Energy Yield from Stearic Acid Catabolism?

20.25 Calculate the number of ATP molecules obtained in the β-oxidation of myristic acid, $CH_3(CH_2)_{12}COOH$.

20.26 Assume that the *cis-trans* isomerization in the β-oxidation of unsaturated fatty acids does not require energy. Which fatty acid yields the greater amount of energy, saturated (stearic acid) or monounsaturated (oleic acid)? Explain.

20.27 Assuming that both fats and carbohydrates are available, which does the body preferentially use as an energy source?

20.28 If equal weights of fats and carbohydrates are eaten, which will give more calories? Explain.

Section 20.7 What Are Ketone Bodies?

20.29 Acetoacetate is the common source of acetone and β-hydroxybutyrate. Name the type of reactions that yield these ketone bodies from acetoacetate.

20.30 Do ketone bodies have nutritional value?

20.31 What happens to the oxaloacetate produced from carboxylation of phosphoenolpyruvate?

Section 20.8 How Is the Nitrogen of Amino Acids Processed in Catabolism?

20.32 What kind of reaction is the following, and what is its function in the body?

$$
\begin{array}{c}
\text{H}_3\text{C} \quad \text{H} \quad\quad \overset{\displaystyle \text{COO}^-}{\underset{\displaystyle \text{C}=\text{O}}{|}} \\
\text{CH}-\overset{\displaystyle |}{\underset{\displaystyle |}{\text{C}}}-\text{COO}^- + \text{CH}_2 \\
\text{H}_3\text{C} \quad \text{NH}_3{}^+ \quad\quad \overset{\displaystyle \text{CH}_2}{\underset{\displaystyle \text{COO}^-}{|}}
\end{array}
$$

$$
\longrightarrow
\begin{array}{c}
\text{H}_3\text{C} \\
\quad \text{CH}-\text{C}-\text{COO}^- + \\
\text{H}_3\text{C} \quad\quad \text{O}
\end{array}
\quad
\begin{array}{c}
\text{COO}^- \\
\text{H}-\text{C}-\text{NH}_3{}^+ \\
\text{CH}_2 \\
\text{CH}_2 \\
\text{COO}^-
\end{array}
$$

20.33 Write an equation for the oxidative deamination of alanine.

20.34 Ammonia, NH_3, and ammonium ion, $NH_4{}^+$, are both soluble in water and could easily be excreted in the urine. Why does the body convert them to urea rather than excreting them directly?

20.35 What are the sources of the nitrogen in urea?

20.36 What compound is common to both the urea and citric acid cycles?

20.37 ■ (a) What is the toxic product of the oxidative deamination of glutamate?

(b) How does the body get rid of it?

20.38 If the urea cycle is inhibited, in what other ways can the body get rid of $NH_4{}^+$ ions?

Section 20.9 How Are the Carbon Skeletons of Amino Acids Processed in Catabolism?

20.39 The metabolism of the carbon skeleton of tyrosine yields pyruvate. Why is tyrosine a glucogenic amino acid?

Section 20.10 What Are the Reactions of Catabolism of Heme?

20.40 Why is a high bilirubin content in the blood an indication of liver disease?

20.41 When hemoglobin is fully metabolized, what happens to the iron in it?

20.42 Describe which groups on the biliverdin (Figure 20.10) are the oxidation products and which are the reduction products in the degradation of heme.

Chemical Connections

20.43 (Chemical Connections 20A) What causes cramps of the muscles when a person is fatigued?

20.44 (Chemical Connections 20B) How does the signal of epinephrine result in depletion of glycogen in the muscle?

20.45 (Chemical Connections 20C) What system counteracts the acidic effect of ketone bodies in the blood?

20.46 (Chemical Connections 20C) The patient whose condition is described in Chemical Connections 20C was transferred to a hospital in an ambulance. Could a nurse in the ambulance tentatively diagnose his diabetic condition without running blood and urine tests? Explain.

20.47 (Chemical Connections 20D) Draw structural formulas for each reaction component, and complete the following equation:

$$\text{Phenylalanine} \longrightarrow \text{Phenylpyruvate} + ?$$

Additional Problems

20.48 If you receive a laboratory report showing the presence of a high concentration of ketone bodies in the urine of a patient, which disease would you suspect?

20.49 Which compounds give the sequence in coloration of bruises from black and blue to green and yellow?

20.50 ■ (a) At which step of the glycolysis pathway does NAD^+ participate (see Figures 20.3 and 20.4)?

(b) At which step does $NADH + H^+$ participate?

(c) As a result of the overall pathway, is there a net increase of NAD^+, of $NADH + H^+$, or of neither?

20.51 What is the net energy yield in moles of ATP produced when yeast converts one mole of glucose to ethanol?

20.52 Can the intake of alanine, glycine, and serine relieve hypoglycemia caused by starvation? Explain.

20.53 How can glucose be utilized to produce ribose for RNA synthesis?

20.54 Write the products of the transamination reaction between alanine and oxaloacetate:

20.55 Phosphoenolpyruvate (PEP) has a high-energy phosphate bond that has more energy than the anhydride bonds in ATP. Which step in glycolysis suggests that this is so?

20.56 Suppose that a fatty acid labeled with radioactive carbon-14 is fed to an experimental animal. Where would you look for the radioactivity?

20.57 Which functional groups are present in carbamoyl phosphate?

20.58 Is the urea cycle an energy-producing or an energy-consuming pathway?

20.59 Which intermediate of the glycolytic pathway can replenish oxaloacetate in the citric acid cycle?

20.60 ■ How many turns of the spiral are there in the β-oxidation of (a) lauric acid and (b) palmitic acid?

Tying It Together

20.61 The equations of glycolysis indicate that there is a net gain of two ATP molecules for each molecule of glucose processed. Why is it that we see the figure of 36 ATP molecules in Table 20.1?

20.62 What reactions can pyruvate undergo once it is formed? Are these reactions aerobic, anaerobic, or both?

20.63 Is lactate a dead-end product of metabolism, or does it play some role in generating (or regenerating) some useful compound?

20.64 Why do ketone bodies occur in the blood of people who are on severely restricted diets?

20.65 Can amino acids be catabolized to yield energy?

20.66 Suggest a reason why the carbon skeletons and nitrogen-containing portions of amino acids are catabolized separately.

Looking Ahead

20.67 Put the following words into two related groups: energy-yielding, oxidative, anabolism, reductive, energy-requiring, catabolism.

20.68 Would you expect the biosynthesis of a protein from the constituent amino acids to require energy or to release energy? Explain.

20.69 In what ways can the production of glucose from CO_2 and H_2O in photosynthesis be considered the exact reversal of complete aerobic catabolism of glucose? In what ways is it different?

20.70 Why is the citric acid cycle *the* central pathway in metabolism?

Challenge Problems

20.71 With their oxygen-containing functional groups, sugars are more oxidized than the hydrocarbon side chains of fatty acids. Does this fact have any bearing on the energy yield of carbohydrates compared to that of fats?

20.72 Many soft drinks contain citric acid to add flavor. Is it likely to be a good nutrient?

20.73 The intermediates of glycolysis carry phosphate groups, which are charged. The intermediates of the citric acid cycle are not phosphorylated. Suggest a reason for this difference. (*Hint:* In what parts of the cell do these pathways occur?)

20.74 One occasionally hears diet advice that proteins and carbohydrates should not be eaten at the same meal. Does this advice make sense to you in light of Figure 20.11?

20.75 The production of ATP is not shown explicitly in Figure 20.11. What point in this figure indicates that ATP production does indeed take place?

20.76 Many metabolic pathways, including those of catabolism, are long and complex. Suggest a reason for this observation.

Biosynthetic Pathways

Algae on mudflats.

21.1 What Is the General Outline of Biosynthetic Pathways?

In the human body, and in most other living tissues, the pathways by which a compound is synthesized (anabolism) are usually different from the pathways by which it is degraded (catabolism). (Anabolic pathways are also called biosynthetic pathways, and we will use these terms interchangeably.) There are several reasons why it is biologically advantageous for anabolic and catabolic pathways to be different. We will give two of them here:

1. **Flexibility** If the normal **biosynthetic pathway** is blocked, the body can often use the reverse of the degradation pathway (recall that most

steps in degradation are reversible), thereby providing another way to make the necessary compounds.

2. **Overcoming the effect of Le Chatelier's principle** This point can be illustrated by the cleavage of a glucose unit from a glycogen molecule, an equilibrium process:

$$\underset{\text{Glycogen}}{(\text{Glucose})_n} + P_i \underset{\text{phosphorylase}}{\rightleftharpoons} \underset{\substack{\text{Glycogen} \\ \text{(one unit smaller)}}}{(\text{Glucose})_{n-1}} + \text{Glucose 1-phosphate} \quad (21.1)$$

Phosphorylase catalyzes not only glycogen degradation (the forward reaction), but also glycogen synthesis (the reverse reaction). However, the body contains a large excess of inorganic phosphate, P_i. This excess would drive the reaction, on the basis of Le Chatelier's principle, to the right, which represents glycogen degradation. To provide a method for the synthesis of glycogen even in the presence of excess inorganic phosphate, a different pathway is needed in which P_i is not a reactant. Thus the body uses the following synthetic pathway:

$$\underset{\text{Glycogen}}{(\text{Glucose})_{n-1}} + \text{UDP-glucose} \longrightarrow \underset{\substack{\text{Glycogen} \\ \text{(one unit larger)}}}{(\text{Glucose})_n} + \text{UDP} \quad (21.2)$$

Not only do the synthetic pathways differ from the catabolic pathways, but the energy requirements are also different, as are the pathways' locations. Most catabolic reactions occur in the mitochondria, whereas anabolic reactions generally take place in the cytoplasm. We will not describe the energy balances of the biosynthetic processes in detail as we did for catabolism. However, keep in mind that, while energy (in the form of ATP) is *obtained* in the degradative processes, biosynthetic processes *consume* energy.

21.2 How Does the Biosynthesis of Carbohydrates Take Place?

We discuss the biosynthesis of carbohydrates by looking at three examples:

- Conversion of atmospheric CO_2 to glucose in plants
- Synthesis of glucose in animals and humans
- Conversion of glucose to other carbohydrate molecules in animals and humans

A. Conversion of Atmospheric Carbon Dioxide to Glucose in Plants

The most important biosynthesis of carbohydrates takes place in plants, green algae, and cyanobacteria, with the last two representing an important part of the marine food web. In the process of **photosynthesis,** the energy of the sun is built into chemical bonds of carbohydrates. The overall reaction is

Photosynthesis The process in which plants synthesize carbohydrates from CO_2 and H_2O with the help of sunlight and chlorophyll

$$6H_2O + 6CO_2 \xrightarrow[\text{chlorophyll}]{\substack{\text{energy in} \\ \text{the form of} \\ \text{sunlight}}} \underset{\text{Glucose}}{C_6H_{12}O_6} + 6O_2 \quad (21.3)$$

Although the primary product of photosynthesis is glucose, it is largely converted to other carbohydrates, mainly cellulose and starch. The very complicated process of glucose biosynthesis takes place in large protein–cofactor complexes (Chemical Connections 21A). We will not discuss it further here except to note that the carbohydrates of plants—starch, cellulose, and other mono- and polysaccharides—serve as the basic carbohydrate supply of all animals, including humans.

B. Synthesis of Glucose in Animals

Gluconeogenesis The process by which glucose is synthesized in the body

In Chapter 20, we saw that when the body needs energy, carbohydrates are broken down via glycolysis. When energy is not needed, glucose can be synthesized from the intermediates of the glycolytic and citric acid pathways. This process is called **gluconeogenesis.** As shown in Figure 21.1, a large number of intermediates—pyruvate, lactate, oxaloacetate, malate, and several amino acids (the glucogenic amino acids we met in Section 20.9)—can serve as starting compounds. Gluconeogenesis proceeds in the reverse order from glycolysis, and many of the enzymes of glycolysis also catalyze gluconeogenesis. At four points, however, unique enzymes (marked in Figure 21.1) catalyze only gluconeogenesis and not the breakdown reactions. These four enzymes make *gluconeogenesis a pathway that is distinct from glycolysis.* Note that ATP is used up in gluconeogenesis and produced in glycolysis, another difference between the two pathways.

During periods of strenuous exercise, the body needs to replenish its carbohydrate supply. The Cori cycle makes use of lactate produced in glycolysis (Section 20.2) as the starting point for gluconeogenesis. Lactate produced in

FIGURE 21.1 Gluconeogenesis. All reactions take place in cytosol, except for those shown in the mitochondria.

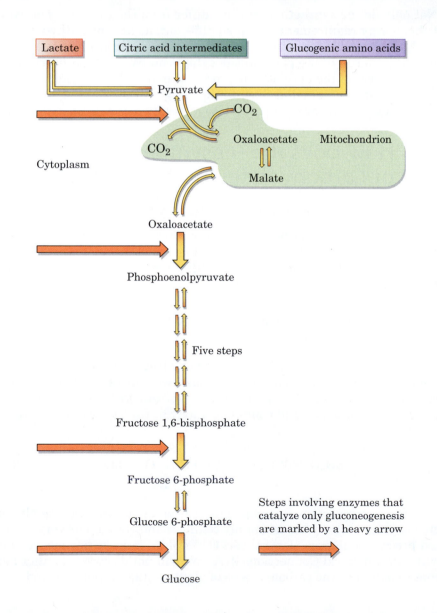

Chemical Connections 21A

Photosynthesis

Photosynthesis requires sunlight, water, CO_2, and pigments found in plants, mainly chlorophyll. The overall reaction shown in Equation 21.3 actually occurs in two distinct steps. First, light interacts with the pigments that are located in highly membranous organelles of plants, called **chloroplasts.** Chloroplasts resemble mitochondria (Section 19.2) in many respects: They contain a whole chain of oxidation–reduction enzymes similar to the cytochrome and iron–sulfur complexes of mitochondrial membranes, and they contain a proton-translocating ATPase. In a manner similar to mitochondria, the proton gradient accumulated in the intermembrane region drives the synthesis of ATP in chloroplasts (see the discussion of the chemiosmotic pump in Section 19.6).

The chlorophyll is the central part of a complex machinery called photosystem I and II. The detailed structure of photosystem I was elucidated in 2001. The photosystem consists of three monomeric entities, which are designated as I, II and III. Each monomer contains 12 different proteins, 96 chlorophyll molecules, and 30 cofactors that include iron clusters, lipids, and Ca^{2+} ions. Its most important feature shows that a central Mg^{2+} is bound to the sulfur of a methionine residue of a surrounding protein. This Mg-S linkage makes this whole assembly a powerful oxidizing agent, meaning that it can readily accept electrons.

Courtesy of Dr. Petra Fromme

Side view of monomer III in photosystem I.

Inside of a chloroplast.

working muscle is then transported in the bloodstream to the liver, where gluconeogenesis converts it to glucose (Figure 21.2). The newly formed glucose is then transported back to the muscle by the blood, where it can fuel further exercise. Note that the two different pathways, glycolysis and gluconeogenesis, take place in different organs. This division of labor ensures that both pathways are not simultaneously active in the same tissues, which is highly inefficient.

C. Conversion of Glucose to Other Carbohydrates in Animals

The third important biosynthetic pathway for carbohydrates is the conversion of glucose to other hexoses and hexose derivatives and the

Chemical Connections 21A

Photosynthesis *(continued)*

The chlorophyll itself, buried in a complex protein that traverses the chloroplast membranes, is a molecule similar to the heme we have already encountered in hemoglobin (Figure 14.18). In contrast to heme, however, chlorophyll contains Mg^{2+} instead of Fe^{2+}.

Chlorophyll a

The reactions in photosynthesis, collectively called the light reactions, are the ones in which chlorophyll captures the energy of sunlight and, with its aid, strips the electrons and protons from water to form oxygen, ATP, and $NADPH + H^+$ (see Section 20.2C):

$$H_2O + ADP + P_i + NADP^+ + sunlight$$
$$\longrightarrow \tfrac{1}{2}O_2 + ATP + NADPH + H^+$$

Another group of reactions, called the dark reactions because they do not need light, in essence converts CO_2 to carbohydrates:

$$CO_2 + ATP + NADPH + H^+$$
$$\longrightarrow (CH_2O)_n + ADP + P_i + NADP^+$$
Carbohydrates

Energy, now in the form of ATP, is used to help $NADPH + H^+$ reduce carbon dioxide to carbohydrates. Thus the protons and electrons stripped in the light reactions are added to the carbon dioxide in the dark reactions. This reduction takes place in a multistep cyclic process called the **Calvin cycle,** named after its discoverer, Melvin Calvin (1911–1997), who was awarded the 1961 Nobel Prize in chemistry for his work. In this cycle, CO_2 is first attached to a C_5 fragment and breaks down to two C_3 fragments (triose phosphates). Through a complex series of steps, these fragments are converted to a C_6 compound and, eventually, to glucose.

$$CO_2 + C_5 = 2C_3 = C_6$$

The critical step in the dark reactions (Calvin cycle) is the attachment of CO_2 to ribulose 1,5-bisphosphate, a compound derived from ribulose (Table 12.2). The enzyme that catalyzes this reaction, ribulose-1,5-bisphosphate carboxylase-oxygenase, nicknamed RuBisCO, is one of the slowest in nature. As in traffic, the slowest-moving vehicle determines the overall flow, so RuBisCO is the main factor in the low efficiency of the Calvin cycle. Because of the low efficiency of this enzyme, most plants convert less than 1% of the absorbed radiant energy into carbohydrates. To overcome its inefficiency, plants must synthesize large quantities of this enzyme. More than half of the soluble proteins in plant leaves are RuBisCO enzymes, whose synthesis requires a large energy expenditure.

synthesis of di-, oligo-, and polysaccharides. The common step in all of these processes is the activation of glucose by uridine triphosphate (UTP) to form UDP-glucose:

Uridine diphosphate (UDP)
Uridine diphosphate glucose (UDP-glucose)

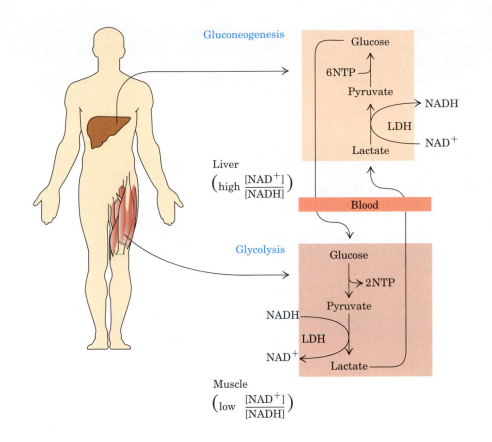

UDP is similar to ADP except that the base is uracil instead of adenine. UTP, an analog of ATP, contains two high-energy phosphate anhydride bonds. For example, when the body has excess glucose and wants to store it as glycogen (a process called **glycogenesis**), the glucose is first converted to glucose 1-phosphate, but then a special enzyme catalyzes the reaction:

Glycogenesis The conversion of glucose to glycogen

$$\text{glucose 1-phosphate} + \boxed{\text{UTP}} \longrightarrow \text{UDP-glucose} + \begin{array}{c} \text{O} \quad\quad \text{O} \\ \parallel \quad\quad \parallel \\ {}^-\text{O}-\text{P}-\text{O}-\text{P}-\text{O}^- \\ | \quad\quad | \\ \text{O}^- \quad\quad \text{O}^- \end{array}$$

$$\underset{\text{Glycogen}}{\text{UDP-glucose} + (\text{glucose})_n} \longrightarrow \boxed{\text{UDP}} + \underset{\substack{\text{Glycogen} \\ \text{(one unit larger)}}}{(\text{glucose})_{n+1}}$$

The biosynthesis of many other di- and polysaccharides and their derivatives also uses the common activation step: forming the appropriate UDP compound.

21.3 How Does the Biosynthesis of Fatty Acids Take Place?

The body can synthesize all the fatty acids it needs except for linoleic and linolenic acids (essential fatty acids; see Section 13.2). The source of carbon in this synthesis is acetyl CoA. Because acetyl CoA is also a degradation product of the β-oxidation spiral of fatty acids (Section 20.5), we might expect the synthesis to be the reverse of the degradation. This is not the case.

FIGURE 21.3 The biosynthesis of fatty acids. The ACP (central blue sphere) has a long side chain (⌐) that carries the growing fatty acid (∿). The ACP rotates counterclockwise, and its side chain sweeps over a multienzyme system (empty spheres). As each cycle is completed, a C_2 fragment is added to the growing fatty acid chain.

For one thing, the majority of fatty acid synthesis occurs in the cytoplasm, whereas degradation takes place in the mitochondria. Fatty acid synthesis is catalyzed by a multienzyme system.

However, one aspect of fatty acid synthesis is the same as in fatty acid degradation: Both processes involve acetyl CoA, so both proceed in units of two carbons. Fatty acids are built up two carbons at a time, just as they are broken down two carbons at a time (Section 20.5).

Most of the time, fatty acids are synthesized when excess food is available. That is, when we eat more food than we need for energy, our bodies turn the excess acetyl CoA (produced by catabolism of carbohydrates; see Section 20.2) into fatty acids and then to fats. The fats are stored in the fat depots, which are specialized fat-carrying cells (see Figure 20.1)

The key to fatty acid synthesis is an **acyl carrier protein (ACP).** It can be looked upon as a merry-go-round—a rotating protein molecule to which the growing chain of fatty acids is bonded. As the growing chain rotates with the ACP, it sweeps over the multienzyme complex; at each enzyme, one reaction of the chain is catalyzed (Figure 21.3).

At the beginning of this cycle, the ACP picks up an acetyl group from acetyl CoA and delivers it to the first enzyme, fatty acid synthase, here called synthase for short:

Acetyl CoA   Acetyl ACP

$$CH_3\overset{\overset{O}{\|}}{C}-S-ACP + \text{synthase}-SH \longrightarrow CH_3\overset{\overset{O}{\|}}{C}-S-\text{synthase} + HS-ACP$$

The —SH group is the site of acyl binding as a thioester.

The C_2 fragment on the synthase is condensed with a C_3 fragment attached to the ACP in a process in which CO_2 is given off:

Malonyl-ACP

Acetoacetyl-ACP

The result is a C_4 fragment, which is reduced twice and dehydrated before it becomes a fully saturated C_4 group. This event marks the end of one cycle of the merry-go-round. These three steps are the reverse of what we saw in the β-oxidation of fatty acids in Section 20.5.

In the next cycle, the fragment is transferred to the synthase, and another malonyl-ACP (C_3 fragment) is added. When CO_2 splits out, a C_6 fragment is obtained. The merry-go-round continues to turn. At each turn, another C_2 fragment is added to the growing chain. Chains up to C_{16} (palmitic acid) can be obtained in this process. If the body needs longer fatty acids—for example, stearic (C_{18})—another fragment is added to palmitic acid by a different enzyme system.

Chemical Connections 21B

The Biological Basis of Obesity

Obesity has long been associated with several known disease states, such as diabetes and even cancer, making it a hot topic in modern society. Geneticists and enzymologists have found some research results that may be of use in understanding and treating obesity.

Recently, researchers were able to identify the first gene that showed a clear relationship with a person's tendency toward obesity. The gene has been labeled *FTO*. Interestingly, although this gene is positively correlated with obesity, nobody yet knows what it does. A British team of scientists studied samples from over 4000 individuals and identified the FTO gene, which was shown to be related to body mass index (BMI). A specific variant of FTO with a single nucleotide change was found. Individuals who had two copies of the variant were 1.67 times more likely to be obese than individuals lacking any copies of the variant. The researchers said that even though they currently do not know the function of the gene, its high correlation with obesity has people racing to understand it.

The second area of research on obesity concerns an enzyme that controls the level of a key intermediate in fatty acid biosynthesis, malonyl-CoA . This intermediate has two very important functions in metabolism. First, it is converted to fatty acids and not to any other compound in biosynthesis. Second, it strongly inhibits the enzyme that helps transfer fatty acids to the mitochondrion and therefore inhibits fatty-acid oxidation. The level of malonyl-CoA in the cytosol can determine whether the cell will be oxidizing fats or storing fats. The enzyme that produces malonyl-CoA is acetyl-CoA carboxylase, or ACC. There are two forms of this enzyme, each encoded by separate genes. ACC1 is found in the liver and adipose tissue, while ACC2 is found in cardiac and skeletal muscle. High glucose concentrations and high insulin concentrations lead to stimulation of ACC2. Exercise has the opposite effect. During exercise, an AMP-dependent protein kinase phosphorylates ACC2 and inactivates it.

Some recent studies looked at the nature of weight gain and weight loss with respect to ACC2. The investigators created a strain of mice lacking the gene for ACC2. These mice ate more than their wild-type counterparts but had significantly lower stores of lipids (30%–40% less in skeletal muscle and 10% less in cardiac muscle). Even the adipose tissue, which still had ACC1, showed a reduction in stored triacylglycerols of up to 50%. The mice showed no other abnormalities. They grew and reproduced normally and had normal life spans. The investigators concluded that reduced pools of malonyl-CoA due to the lack of ACC2 lead to two results: increased β-oxidation via removal of the block on transfer of fatty acids to the mitochondrion, and a decrease in fatty acid synthesis. They speculate that ACC2 would be a good target for drugs used to combat obesity.

A number of factors contribute to obesity.

© Jeff Mitchell/Getty Images

Unsaturated fatty acids are obtained from saturated fatty acids by an oxidation step in which hydrogen is removed and combined with O_2 to form water:

$$R-CH_2-CH_2-(CH_2)_n COOH + O_2 + \boxed{NADPH} + H^+$$

$$\xrightarrow{\text{enzyme}} \begin{array}{c} H \qquad\quad H \\ \diagdown \quad\, / \\ C = C \\ / \qquad \diagdown \\ R \qquad\quad (CH_2)_n COOH \end{array} + 2H_2O + \boxed{NADP^+}$$

An example of such elongation and unsaturation is docosahexenoic acid, a 22-carbon fatty acid with 6 *cis* double bonds (22:6). Docosahexenoic acid is part of the glycerophospholipids prevalent in the membranes where the visual pigment rhodopsin resides. Its presence is necessary to provide fluidity to the membrane so that it can process the light signals reaching our retina.

We have already seen that lipids are a highly efficient form of energy storage. They accumulate all too easily when we take in excess energy in the form of nutritional "calories." Obesity-related health questions are becoming all too common in developed countries, leading to research on ways to address the situation (Chemical Connections 21B).

21.4 How Does the Biosynthesis of Membrane Lipids Take Place?

The various membrane lipids (Sections 13.6 to 13.8) are assembled from their constituents. We just saw how fatty acids are synthesized in the body. These fatty acids are activated by CoA, forming acyl CoA. Glycerol 1-phosphate, which is obtained from the reduction of dihydroxyacetone phosphate (a C_3 fragment of glycolysis; see Figure 20.4), is the second building block of glycerophospholipids. This compound combines with two acyl CoA molecules, which can be the same or different:

We encountered glycerol 1-phosphate as a vehicle for transporting electrons in and out of mitochondria (Section 20.3). To complete the molecule, an activated serine or choline or ethanolamine is added to the $-OPO_3^{2-}$ group, forming phosphate esters (see the structures in Section 13.6; Figure 21.4 shows a model of phosphatidylcholine). Choline is activated by cytidine triphosphate (CTP), yielding CDP-choline. This process is similar to the activation of glucose by UTP (Section 21.2C) except that the base is cytosine rather than uracil (Section 17.2). Sphingolipids (Section 13.7) are similarly built up from smaller molecules. An activated phosphocholine is added to the sphingosine part of ceramide (Section 13.7) to make sphingomyelin.

The glycolipids are constructed in a similar fashion. Ceramide is assembled as described above, and the carbohydrate is added one unit at a time in the form of activated monosaccharides (UDP-glucose and so on).

Cholesterol, the molecule that controls the fluidity of membranes and is a precursor of all steroid hormones and bile salts, is also synthesized by the human body. It is assembled in the liver from fragments that come from the acetyl group of acetyl CoA. All of the carbon atoms of cholesterol come from the carbon atoms of acetyl CoA molecules (Figure 21.5). Cholesterol in the brain is synthesized in nerve cells themselves; its presence is necessary to form synapses. Cholesterol from our diets and cholesterol that is synthesized in the liver circulate in the plasma as LDL (see Section 13.9) and are not available for synapse formation because LDL cannot cross the blood–brain barrier.

FIGURE 21.4 A model of phosphatidylcholine, commonly called lecithin.

FIGURE 21.5 Biosynthesis of cholesterol. The circled carbon atoms come from the —CH$_3$ group, and the other carbon atoms come from the —CO— group of the acetyl group of acetyl CoA.

Cholesterol synthesis starts with the sequential condensation of three acetyl CoA molecules to form a compound 3-hydroxy-3-methylglutaryl CoA (HMGCoA):

A key enzyme, HMG-CoA reductase, controls the rate of cholesterol synthesis. It reduces the thioester of HMGCoA to a primary alcohol, yielding CoA in the process. The resulting compound, mevalonate, undergoes phosphorylation and decarboxylation to yield a C$_5$ compound, isopentenyl pyrophosphate:

From this basic C$_5$ unit, the other multiple-C$_5$ compounds are formed that eventually lead to cholesterol synthesis. These intermediates are geranyl, C$_{10}$, and farnesyl, C$_{15}$, pyrophosphates:

Finally, cholesterol is synthesized from the condensation of two farnesyl pyrophosphate molecules.

The statin drugs, such as lovastatin, competitively inhibit the key enzyme HMG-CoA reductase and, thereby, the biosynthesis of cholesterol. They are frequently prescribed to control the cholesterol level in the blood so as to prevent atherosclerosis (Section 13.9E).

Note that the intermediates in cholesterol synthesis, the geranyl and farnesyl pyrophosphates, are made of isoprene units; we discussed these C$_5$ compounds in Section 3.5. The C$_{10}$ and C$_{15}$ compounds are also used to enable protein molecules to be dispersed in the lipid bilayers of membranes. When these multiple-isoprene units are attached to a protein in a process called **prenylation,** the protein becomes more hydrophobic and is able to move laterally within the bilayer with greater ease. (The name *prenylation* originates from isoprene, the five-carbon unit from which the cholesterol intermediates C$_{10}$, C$_{15}$, and C$_{30}$ are made.) Prenylation marks proteins to be associated with membranes and to perform other cellular functions, such as signal transduction of G-protein (Section 16.5B).

21.5 How Does the Biosynthesis of Amino Acids Take Place?

The human body needs 20 different amino acids to make its protein chains—all 20 are found in a normal diet. Some of the amino acids can be synthesized from other compounds; these are the nonessential amino acids. Others cannot be synthesized by the human body and must be supplied in the diet; these are the **essential amino acids** (see Section 22.6). Most nonessential amino acids are synthesized from some intermediate of either glycolysis (Section 20.2) or the citric acid cycle (Section 19.4). Glutamate plays a central role in the synthesis of five nonessential amino acids. Glutamate itself is synthesized from α-ketoglutarate, one of the intermediates in the citric acid cycle:

$$\boxed{\text{NADH}} + \text{H}^+ + \text{NH}_4^+ + \quad \begin{array}{c} \text{COO}^- \\ | \\ \text{C}=\text{O} \\ | \\ \text{CH}_2 \\ | \\ \text{CH}_2 \\ | \\ \text{COO}^- \end{array} \quad \rightleftharpoons \quad \begin{array}{c} \text{COO}^- \\ | \\ \text{CH}-\text{NH}_3^+ \\ | \\ \text{CH}_2 \\ | \\ \text{CH}_2 \\ | \\ \text{COO}^- \end{array} \quad + \boxed{\text{NAD}^+} + \text{H}_2\text{O}$$

α-Ketoglutarate   Glutamate

The forward reaction is the synthesis, and the reverse reaction is the oxidative deamination (degradation) reaction we encountered in the catabolism

Chemical Connections 21C

Essential Amino Acids

The biosynthesis of proteins requires the presence of all of the protein's constituent amino acids. If any of the 20 amino acids is missing or in short supply, protein biosynthesis is inhibited.

Some organisms, including bacteria, can synthesize all the amino acids that they need. Other species, including humans, must obtain some amino acids from dietary sources. The essential amino acids in human nutrition are listed in Table 21.1. The body can synthesize some of these amino acids, but not in sufficient quantities for its needs, especially in the case of growing children (particularly children's requirement for arginine and histidine).

Amino acids are not stored (except in proteins), so dietary sources of essential amino acids are needed at regular intervals. Protein deficiency—especially a prolonged deficiency

in sources that contain essential amino acids—leads to the disease **kwashiorkor.** The problem in this disease, which is particularly severe in growing children, is not simply starvation, but the breakdown of the body's own proteins.

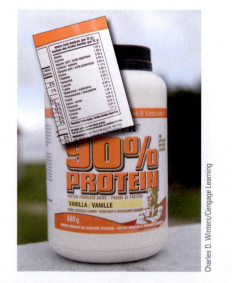

The label on this supplement lists the amino acid content and points out which ones are essential amino acids.

TABLE 21.1 Amino Acid Requirements in Humans

Essential		Nonessential	
Arginine	Methionine	Alanine	Glutamine
Histidine	Phenylalanine	Asparagine	Glycine
Isoleucine	Threonine	Aspartate	Proline
Leucine	Tryptophan	Cysteine	Serine
Lysine	Valine	Glutamate	Tyrosine

of amino acids (Section 20.8B). In this case, the synthetic and degradative pathways are exactly the reverse of each other.

Glutamate can serve as an intermediate in the synthesis of alanine, serine, aspartate, asparagine, and glutamine. For example, the transamination reaction we saw in Section 20.8A leads to alanine formation:

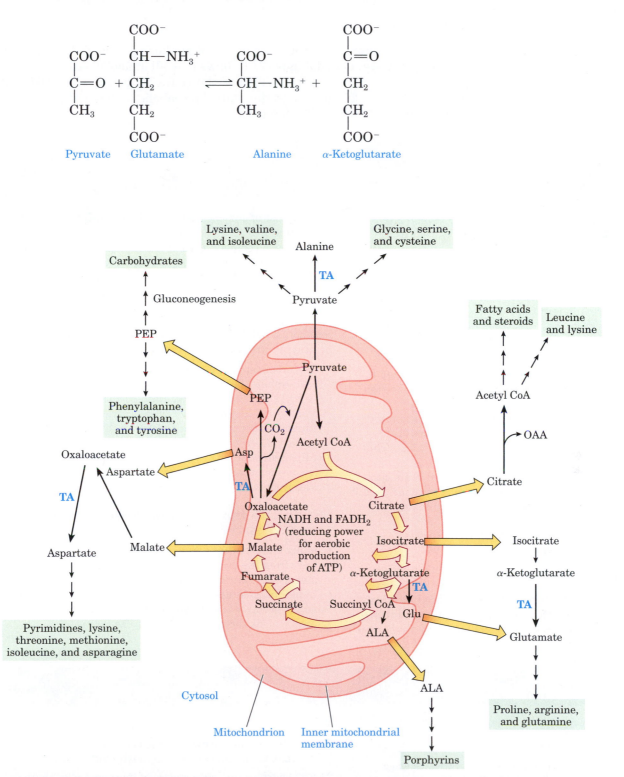

Pyruvate   Glutamate   Alanine   α-Ketoglutarate

FIGURE 21.6 A summary of anabolism showing the role of the central metabolic pathway. Note that carbohydrates, lipids, and amino acids all appear as products. (OAA is oxaloacetate; ALA is a derivative of succinyl CoA; TA is transamination; ⟶ ⟶ ⟶ is a pathway with many steps.)

Besides being the building blocks of proteins, amino acids serve as intermediates for a large number of biological molecules. We have already seen that serine is needed in the synthesis of membrane lipids (Section 21.4). Certain amino acids are also intermediates in the synthesis of heme and of the purines and pyrimidines that are the raw materials for DNA and RNA (Chapter 17).

Postscript

It is useful to summarize the main points of anabolic pathways by considering how they are related. Figure 21.6 shows how all anabolic pathways start at the citric acid cycle, using ATP and the reducing power of NADH and FADH$_2$. Chapter 20 introduced the central metabolic pathway, and here we see how it is related to all of anabolism.

Summary

OWL End-of-chapter problems identified in blue are assignable in **GOB OWL.**

Section 21.1 What Is the General Outline of Biosynthetic Pathways?

- For most biochemical compounds, the biosynthetic pathways are different from the degradation pathways.

Section 21.2 How Does the Biosynthesis of Carbohydrates Take Place? **Problems 21.7, 21.9**

- In **photosynthesis,** carbohydrates are synthesized in plants from CO$_2$ and H$_2$O, using sunlight as an energy source.
- Glucose can be synthesized by animals from the intermediates of glycolysis, from the intermediates of the citric acid cycle, and from glucogenic amino acids. This process is called **gluconeogenesis.**
- When glucose or other monosaccharides are built into di-, oligo-, and polysaccharides, each monosaccharide unit in its activated form is added to a growing chain.

Section 21.3 How Does the Biosynthesis of Fatty Acids Take Place? **Problem 21.15**

- Fatty acid biosynthesis is accomplished by a multienzyme system.
- The key to fatty acid biosynthesis is the **acyl carrier protein (ACP),** which acts as a merry-go-round transport system: It carries the growing fatty acid chain over

a number of enzymes, each of which catalyzes a specific reaction.
- With each complete turn of the merry-go-round, a C$_2$ fragment is added to the growing fatty acid chain.
- The source of the C$_2$ fragment is malonyl-ACP, a C$_3$ compound bonded to ACP. It becomes C$_2$ with the loss of CO$_2$.

Section 21.4 How Does the Biosynthesis of Membrane Lipids Take Place? **Problem 21.26**

- Glycerophospholipids are synthesized from glycerol 1-phosphate, fatty acids that are activated by conversion to acyl CoA, and activated alcohols such as choline.
- Cholesterol is synthesized from acetyl CoA. Three C$_2$ fragments are condensed to form C$_6$, hydroxymethylglutaryl CoA.
- After reduction and decarboxylation, isoprene C$_5$ units are formed that condense to the C$_{10}$ and C$_{15}$ intermediates from which cholesterol is built.

Section 21.5 How Does the Biosynthesis of Amino Acids Take Place?

- Many nonessential amino acids are synthesized in the body from the intermediates of glycolysis or the intermediates of the citric acid cycle.
- In half of these cases, glutamate is the donor of the amino group in transamination.
- Amino acids serve as building blocks for proteins.

Problems

■ Indicates problems that are assignable in **GOB OWL.**

Blue numbered problems are applied.

Go to this book's companion website at **www.cengage. com/chemistry/bettelheim** for interactive versions of the How To tutorials and Active Figures, and to quiz yourself on this chapter.

21.1 Why are the pathways that the body uses for anabolism and catabolism mostly different?

21.2 How does the large excess of inorganic phosphate in a cell affect the amount of glycogen present? Explain.

21.3 Glycogen can be synthesized in the body by the same enzymes that degrade it. Why is this process utilized in glycogen synthesis only to a small extent, while most glycogen biosynthesis occurs via a different synthetic pathway?

21.4 Do most anabolic and catabolic reactions take place in the same location?

Section 21.2 How Does the Biosynthesis of Carbohydrates Take Place?

21.5 What is the difference in the overall chemical equations for photosynthesis and for respiration?

21.6 In photosynthesis, what are the sources of (a) carbon, (b) hydrogen, and (c) energy?

21.7 ■ Name a compound that can serve as a raw material for gluconeogenesis and is (a) from the glycolytic pathway, (b) from the citric acid cycle, and (c) an amino acid.

21.8 How is glucose activated for glycogen synthesis?

21.9 ■ Glucose is the only carbohydrate compound that the brain can use for energy. Which pathway is mobilized to supply the need of the brain during starvation: (a) glycolysis, (b) gluconeogenesis, or (c) glycogenesis? Explain.

21.10 Are the enzymes that combine two C_3 compounds into a C_6 compound in gluconeogenesis the same as or different from those that cleave the C_6 compound into two C_3 compounds in glycolysis?

21.11 Devise a scheme in which maltose is formed starting with UDP-glucose.

21.12 Glycogen is written as $(glucose)_n$.
(a) What does n stand for?
(b) What is the approximate value of n?

21.13 What are the constituents of UTP?

Section 21.3 How Does the Biosynthesis of Fatty Acids Take Place?

21.14 What is the source of carbon in fatty acid synthesis?

21.15 ■ (a) Where in the body does fatty acid synthesis occur?
(b) Does fatty acid degradation occur in the same location?

21.16 Is ACP an enzyme?

21.17 In fatty acid biosynthesis, which compound is added repeatedly to the synthase?

21.18 (a) What is the name of the first enzyme in fatty acid synthesis?
(b) What does it do?

21.19 From which compound is CO_2 released in fatty acid synthesis?

21.20 What are the common functional groups in CoA, ACP, and synthase?

21.21 In the synthesis of unsaturated fatty acids, $NADPH + H^+$ is converted to $NADP^+$, yet this is an oxidation step and not a reduction step. Explain.

21.22 Which of these fatty acids can be synthesized by the multienzyme fatty acid synthesis complex alone?
(a) Oleic (b) Stearic
(c) Myristic (d) Arachidonic
(e) Lauric

21.23 Some enzymes can use NADH as well as NADPH as a coenzyme. Other enzymes use one or the other exclusively. Which features would prevent NADPH from fitting into the active site of an enzyme that otherwise can accommodate NADH?

21.24 Are fatty acids for energy, in the form of fat, synthesized in the same way as fatty acids for the lipid bilayer of membrane?

21.25 Linoleic and linolenic acids cannot be synthesized in the human body. Does this mean that the human body cannot make an unsaturated fatty acid from a saturated one?

Section 21.4 How Does the Biosynthesis of Membrane Lipids Take Place?

21.26 ■ When the body synthesizes the following membrane lipid, from which building blocks is it assembled?

21.27 Name the activated constituents necessary to form the glycolipid glucoceramide.

21.28 Why is HMG-CoA reductase a key enzyme in cholesterol synthesis?

21.29 Describe by carbon skeleton designation how a C_2 compound ends up as a C_5 compound.

Section 21.5 How Does the Biosynthesis of Amino Acids Take Place?

21.30 Which reaction is the reverse of the synthesis of glutamate from α-ketoglutarate, ammonia, and $NADH + H^+$?

21.31 Which amino acid will be synthesized by the following process?

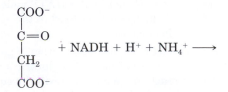

21.32 Draw the structure of the compound needed to synthesize asparagine from glutamate by transamination.

21.33 Name the products of the following transamination reaction:

$$(CH_3)_2CH-\overset{\overset{\displaystyle O}{\|}}{C}-COO^-$$

$$+ \ ^-OOC-CH_2-CH_2-\underset{\underset{\displaystyle NH_3^+}{|}}{CH}-COO^- \longrightarrow$$

Chemical Connections

21.34 (Chemical Connections 21A) Photosystem I and II are complex factories of proteins, chlorophyll, and many cofactors. Where are these photosystems located in plants, and in which reaction of photosynthesis do they participate?

21.35 (Chemical Connections 21A) Which coenzyme reduces CO_2 in the Calvin cycle?

21.36 (Chemical Connections 21B) What is the metabolic importance of malonyl-CoA?

21.37 (Chemical Connections 21B) What enzyme provides a possible target for drugs to treat obesity?

21.38 (Chemical Connections 21C) What is the result of eating only proteins that do not contain all 20 amino acids?

Additional Problems

21.39 In the structure of $NADP^+$, what are the bonds that connect nicotinamide and adenine to the ribose units?

21.40 Which C_3 fragment carried by ACP is used in fatty acid synthesis?

21.41 When glutamate transaminates phenylpyruvate, which amino acid is produced?

$$C_6H_5-CH_2-\overset{\overset{\displaystyle O}{\|}}{C}-COO^-$$

$$+ \ ^-OOC-CH_2-CH_2-\underset{\underset{\displaystyle NH_3^+}{|}}{CH}-COO^- \longrightarrow$$

21.42 Name three compounds based on isoprenoid units that play a role in cholesterol biosynthesis.

21.43 Each activation step in the synthesis of complex lipids occurs at the expense of one ATP molecule. How many ATP molecules are used in the synthesis of one molecule of lecithin?

21.44 Consider the fact that the deamination of glutamic acid and its synthesis from α-ketoglutaric acid are equilibrium reactions. Which way will the equilibrium shift when the human body is exposed to cold temperature?

21.45 Which compound reacts with glutamate in a transamination process to yield serine?

21.46 ■ What are the names of the C_{10} and C_{15} intermediates in cholesterol biosynthesis?

21.47 Which is carbon 1 in HMG-CoA (3-hydroxy-3-methylglutaryl CoA)?

21.48 In most biosynthetic processes, the reactant is reduced to obtain the desired product. Verify whether this statement holds for the overall reaction of photosynthesis.

21.49 What is the major difference in structure between chlorophyll and heme?

21.50 Can the complex enzyme system participating in every fatty acid synthesis manufacture fatty acids of any length?

Tying It Together

21.51 How does fatty acid biosynthesis differ from catabolism of fatty acids?

21.52 How does the energy source differ in carbohydrate biosynthesis in plants and in animals?

21.53 The enzyme that catalyzes carbon dioxide fixation in photosynthesis is one of the least efficient known. What does this fact imply about energy requirements in photosynthesis?

Looking Ahead

21.54 A vegan diet is one that excludes all animal products. Is it possible to get all essential nutrients from such a diet? Will it be more difficult or easier to achieve this goal with a diet that allows animal products?

21.55 Many key proteins in the immune system are glycoproteins (proteins that incorporate sugars in their structure). Would you expect the biosynthesis of such proteins to be affected by a lack of essential amino acids, a low-carbohydrate diet, or both? Explain.

Challenge Problems

21.56 The foods that we eat supply carbohydrates, fats, and proteins. Based on what you have learned in this chapter, which would you predict that we could do without? Explain.

21.57 In general, catabolic and biosynthetic processes do not take place in the same part of the cell. Why is this separation advantageous?

21.58 Would you expect feedback inhibition to play a role in long biosynthetic pathways? Give the reason for your answer.

21.59 If laboratory rats are fed all the amino acids except one of the essential ones and then fed the missing amino acid four hours later, what will be the effect on protein synthesis and why?

21.60 Do humans possess all the anabolic pathways shown in Figure 21.6? If any are missing in humans, what are they likely to be?

Nutrition

Foods high in fiber include whole grains, legumes, fruits, and vegetables.

22.1 How Do We Measure Nutrition?

In Chapters 19 and 20, we saw what happens to the food that we eat in its final stages—after the proteins, lipids, and carbohydrates have been broken down into their components. In this chapter, we discuss the earlier stages—nutrition and diet—and then the digestive processes that break down these large molecules into the small ones that undergo metabolism. Food provides energy and new molecules to replace those that the body uses. This synthesis of new molecules is especially important for the period during which a child becomes an adult.

The components of food and drink that provide for growth, replacement, and energy are called **nutrients.** Not all components of food are nutrients. Some components of food and drinks, such as those that provide flavor, color, or aroma, enhance our pleasure in the food but are not themselves nutrients.

Nutritionists classify nutrients into six groups:

1. Carbohydrates
2. Lipids
3. Proteins
4. Vitamins
5. Minerals
6. Water

Digestion The process in which the body breaks down large molecules into smaller ones that can be absorbed and metabolized

Discriminatory curtailment diet A diet that avoids certain food ingredients that are considered harmful to the health of an individual—for example, low-sodium diets for people with high blood pressure

For food to be used in our bodies, it must be absorbed through the intestinal walls into the bloodstream or lymph system. Some nutrients, such as vitamins, minerals, glucose, and amino acids, can be absorbed directly. Others, such as starch, fats, and proteins, must first be broken down into smaller components before they can be absorbed. This breakdown process is called **digestion.**

A healthy body needs the proper intake of all nutrients. However, nutrient requirements vary from one person to another. For example, more energy is needed to maintain the body temperature of an adult than that of a child. For this reason, nutritional requirements are usually given per kilogram of body weight. Furthermore, the energy requirements of a physically active individual are greater than those of a person in a sedentary occupation. Therefore, when average values are given, as in **Dietary Reference Intakes (DRI)** and in the former guidelines called **Recommended Daily Allowances (RDA),** one should be aware of the wide range that these average values represent.

The public interest in nutrition and diet changes with time and geography. Seventy or eighty years ago, the main nutritional interest of most Americans was getting enough food to eat and avoiding diseases caused by vitamin deficiencies, such as scurvy or beriberi. This issue is still the main concern of the large majority of the world's population. In affluent societies such as the industrialized nations, however, today's nutritional message is no longer "eat more," but rather "eat less and discriminate more in your selection of food." Dieting to reduce body weight is a constant effort in a sizable percentage of the American population. Many people discriminate in their selection of food to avoid cholesterol (Section 13.9E) and saturated fatty acids to reduce the risk of heart attacks.

Along with such **discriminatory curtailment diets** came many faddish diets. **Diet faddism** is an exaggerated belief in the effects of nutrition upon health and disease. This phenomenon is not new; it has been prevalent for many years. Many times it is driven by visionary beliefs, but backed by little science. In the nineteenth century, Dr. Kellogg (of cornflakes fame) recommended a largely vegetarian diet based on his belief that meat produces sexual excess. Eventually his religious fervor withered and his brother made a commercial success of his inventions of grain-based food. Another fad is raw food, which bans any application of heat higher than 118°F to food. Heat, these faddists maintain, depletes the nutritional value of proteins and vitamins and concentrates pesticides in food. Obviously raw food diet is vegetarian, as it excludes meat and meat products.

A recommended food is rarely as good, and a condemned food is rarely as bad, as faddists claim. Each food contains a large variety of nutrients. For example, a typical breakfast cereal lists the following items as its ingredients: milled corn, sugar, salt, malt flavoring, and vitamins A, B, C, and D, plus flavorings and preservatives. U.S. consumer laws require that most packaged food be labeled in a uniform manner to show the nutritional values of the food. Figure 22.1 shows a typical label of the type found on almost every can, bottle, or box of food that we buy.

Such labels must list the percentages of **Daily Values** for four key vitamins and minerals: vitamins A and C, calcium, and iron. If other vitamins or minerals have been added, or if the product makes a nutritional claim about other nutrients, their values must be shown as well. The percent daily values on

Nutrition Facts

Serving size 1 Bar (28g)
Servings per Container 6

Amount Per Serving

Calories 120	Calories from Fat 35

	% Daily Value*
Total Fat 4g	6%
Saturated Fat 2g	10%
Cholesterol 0mg	0%
Sodium 45mg	2%
Potassium 100mg	3%
Total Carbohydrate 19g	6%
Dietary fiber 2g	8%
Sugars 13g	
Protein 2g	

Vitamin A 15%	•	Vitamin C 15%
Calcium 15%	•	Iron 15%
Vitamin D 15%	•	Vitamin E 15%
Thiamin 15%	•	Riboflavin 15%
Niacin 15%	•	Vitamin B6 15%
Folate 15%	•	Vitamin B12 15%
Biotin 10%	•	Pantothenic Acid 10%
Phosphorus 15%	•	Iodine 2%
Magnesium 4%	•	Zinc 4%

*Percent Daily Values are based on a 2,000 calorie diet. Your daily values may be higher or lower depending on your calorie needs.

	Calories:	2,000	2,500
Total Fat	Less than	65g	80g
Sat Fat	Less than	20g	25g
Cholesterol	Less than	300mg	300mg
Sodium	Less than	2,400mg	2,400mg
Potassium		3,500mg	3,500mg
Total Carbohydrate		300g	375g
Dietary Fiber		25g	30g

FIGURE 22.1 A food label for a peanut butter crunch bar. The portion at the bottom (following the asterisk) gives the same categories of information on all labels that carry it.

the labels are based on a daily intake of 2000 Cal. For anyone who eats more than that amount, the actual percentage figures would be lower (and higher for those who eat less). Note that each label specifies the serving size; the percentages are based on that portion, not on the contents of the entire package. The section at the bottom of the label is exactly the same on all labels, no matter what the food; it shows the daily amounts of nutrients recommended by the government, based on consumption of either 2000 or 2500 Cal. Some food packages are allowed to carry shorter labels, either because they have only a few nutrients or because the package has limited label space. The uniform labels make it much easier for consumers to know exactly what they are eating.

In 1992, the U.S. Department of Agriculture (USDA) issued a set of guidelines regarding what constitutes a healthy diet, depicted in the form of a pyramid (Figure 22.2). These guidelines considered the basis of a healthy

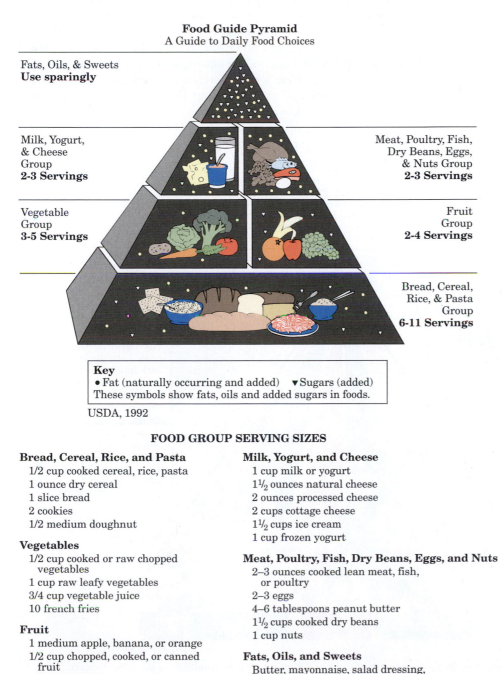

Food Guide Pyramid
A Guide to Daily Food Choices

Fats, Oils, & Sweets
Use sparingly

Milk, Yogurt, & Cheese Group
2-3 Servings

Meat, Poultry, Fish, Dry Beans, Eggs, & Nuts Group
2-3 Servings

Vegetable Group
3-5 Servings

Fruit Group
2-4 Servings

Bread, Cereal, Rice, & Pasta Group
6-11 Servings

Key
• Fat (naturally occurring and added)   ▼ Sugars (added)
These symbols show fats, oils and added sugars in foods.

USDA, 1992

FIGURE 22.2 The Food Guide Pyramid developed by the U.S. Department of Agriculture as a general guide to a healthful diet.

FOOD GROUP SERVING SIZES

Bread, Cereal, Rice, and Pasta
1/2 cup cooked cereal, rice, pasta
1 ounce dry cereal
1 slice bread
2 cookies
1/2 medium doughnut

Vegetables
1/2 cup cooked or raw chopped vegetables
1 cup raw leafy vegetables
3/4 cup vegetable juice
10 french fries

Fruit
1 medium apple, banana, or orange
1/2 cup chopped, cooked, or canned fruit
3/4 cup fruit juice
1/4 cup dried fruit

Milk, Yogurt, and Cheese
1 cup milk or yogurt
$1\frac{1}{2}$ ounces natural cheese
2 ounces processed cheese
2 cups cottage cheese
$1\frac{1}{2}$ cups ice cream
1 cup frozen yogurt

Meat, Poultry, Fish, Dry Beans, Eggs, and Nuts
2–3 ounces cooked lean meat, fish, or poultry
2–3 eggs
4–6 tablespoons peanut butter
$1\frac{1}{2}$ cups cooked dry beans
1 cup nuts

Fats, Oils, and Sweets
Butter, mayonnaise, salad dressing, cream cheese, sour cream, jam, jelly

Chemical Connections 22A

The New Food Guide Pyramid

Over time, scientists began questioning some aspects of the original food pyramid shown in Figure 22.2. For example, certain types of fat are known to be essential to health and actually reduce the risk of heart disease. Also, little evidence exists to back up the claim that a high intake of carbohydrates is beneficial, although for certain sports it is essential. The original pyramid glorified carbohydrates while casting all fats as the "bad guys." In fact, plenty of evidence does link the consumption of saturated fat with high cholesterol and risk of heart disease—but mono- and polyunsaturated fats have the opposite effect. While many scientists recognized the distinction between the various types of fats, they felt that the average person would not understand them, so the original pyramid was designed to just send a simple message: "Fat is bad." The implied corollary to "fat is bad" was that "carbohydrates are good." However, after years of study, no evidence proved that a diet in which 30% or fewer of the calories come from fat is healthier than one with a higher level of fat consumption.

In an attempt to reconcile the latest nutritional data while presenting them in a form understandable to the average person, the USDA has created a website at www.mypyramid.gov. This interactive website allows visitors to take a brief tutorial about the new pyramid as well as to calculate their ideal amount of the various food types. In essence, the new pyramid starts with the old pyramid and tips it on its side. This view does not suggest that one food group is better than another; rather, it shows that proper nutrition is a blend of all the food groups. The nutrients are classified into six groups: grains, vegetables, fruits, oils, milk, and meat and beans. For each group, the pyramid describes the amounts and varieties that should be consumed. A big difference between this pyramid and the original version is that this one includes a section dedicated to exercise. There is also an expanded section on foods whose intakes should be limited, such as certain fats, sugars, and salts.

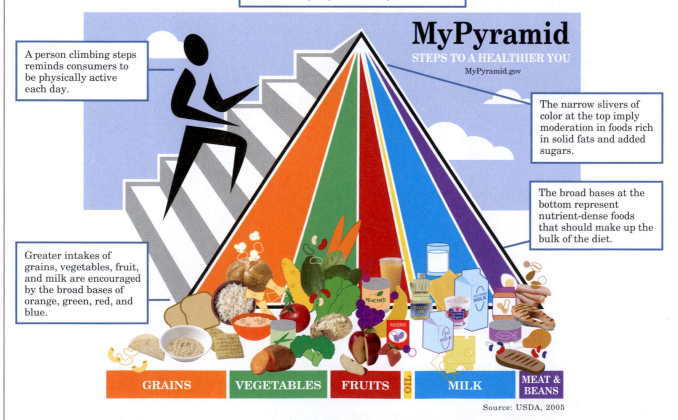

The colors of the pyramid illustrate variety: each color represents one of the five food groups, plus one for oils. Different band widths suggest the proportional contribution of each food group to a healthy diet.

A person climbing steps reminds consumers to be physically active each day.

The narrow slivers of color at the top imply moderation in foods rich in solid fats and added sugars.

The broad bases at the bottom represent nutrient-dense foods that should make up the bulk of the diet.

Greater intakes of grains, vegetables, fruit, and milk are encouraged by the broad bases of orange, green, red, and blue.

MyPyramid
STEPS TO A HEALTHIER YOU
MyPyramid.gov

GRAINS   VEGETABLES   FRUITS   OIL   MILK   MEAT & BEANS

Source: USDA, 2005

The new Food Guide Pyramid. This version of the recommended amounts of different food types comes from the latest research compiled by the USDA. See its website at www.mypyramid.gov for a tutorial. (USDA, 2005) diet to be the foods richest in starch (bread, rice, and so on), plus lots of fruits and vegetables (which are rich in vitamins and minerals). Protein-rich foods (meat, fish, dairy products) were to be consumed more sparingly, and fats, oils, and sweets were not considered necessary at all. The pyramid shape demonstrated the relative importance of each type of food group, with the most important forming the base and the least important or unnecessary appearing at the top. This pictorial description has been used in many textbooks and taught in schools to children of all ages since its initial publication. However, the USDA has recently revised the information and appearance of the food pyramid. Chemical Connections 22A discusses the most recent version of the pyramid.

An important non-nutrient in some foods is **fiber,** which generally consists of the indigestible portion of vegetables and grains. Lettuce, cabbage, celery, whole wheat, brown rice, peas, and beans are all high in fiber. Chemically, fiber is made of cellulose, which, as we saw in Section 12.5C, cannot be digested by humans. Although we cannot digest it, fiber is necessary for proper operation of the digestive system; without it, constipation may result. In more serious cases, a diet lacking sufficient fiber may lead to colon cancer. The DRI recommendation is to ingest 35 g/day for men age 50 years and younger, and 25 g/day for women of the same age.

Fiber The cellulose-based non-nutrient component in our food

Basal caloric requirement The caloric requirement for a resting body

22.2 Why Do We Count Calories?

The largest part of our food supply goes to provide energy for our bodies. As we saw in Chapters 19 and 20, this energy comes from the oxidation of carbohydrates, fats, and proteins. The energy derived from food is usually measured in calories. One nutritional calorie (Cal) equals 1000 cal or 1 kcal. Thus, when we say that the average daily nutritional requirement for a young adult male is 3000 Cal, we mean the same amount of energy needed to raise the temperature of 3000 kg of water by 1°C or of 30 kg (64 lb) of water by 100°C. A young adult female needs 2100 Cal/day. These are peak requirements—children and older people, on average, require less energy. Keep in mind that these energy requirements apply to active people. For bodies completely at rest, the corresponding energy requirement for young adult males is 1800 Cal/day, and that for females is 1300 Cal/day. The requirement for a resting body is called the **basal caloric requirement.**

An imbalance between the caloric requirement of the body and the caloric intake creates health problems. Chronic caloric starvation exists in many parts of the world where people simply do not have enough food to eat because of prolonged drought, the devastation of war, natural disasters, or overpopulation. Famine particularly affects infants and children. Chronic starvation, called **marasmus,** increases infant mortality by as much as 50%. It results in arrested growth, muscle wasting, anemia, and general weakness. Even if starvation is later alleviated, it leaves permanent damage, insufficient body growth, and lowered resistance to disease.

At the other end of the caloric spectrum is excessive caloric intake. It results in *obesity,* or the accumulation of body fat. Obesity is becoming an epidemic in the U.S. population, with important consequences: It increases the risk of hypertension, cardiovascular disease, and diabetes. Obesity is defined by the National Institutes of Health as applying to a person who has a body mass index (BMI) of 30 or greater. The BMI is a measure of body fat based on height and weight that applies to both adult men and women. For example, a person 70 inches tall is *normal* (BMI less than 25) if he or she weighs 174 lb or less. A person of the same height is *overweight* if he or she weighs more than 174 lb but less than 209 lb; an individual is *obese*

Charles D. Winters/Cengage Learning

Different activity levels are associated with different caloric needs.

if he or she weighs more than 209 lb. More than 200 million Americans are overweight or obese.

Reducing diets aim at decreasing caloric intake without sacrificing any essential nutrients. A combination of exercise and lower caloric intake can eliminate obesity, but usually these diets must achieve their goal over an extended period. Crash diets give the illusion of quick weight loss, but most of this decrease is due to loss of water, which can be regained very quickly. To reduce obesity, we must lose body fat, not water. Achieving this goal takes a lot of effort, because fats contain so much energy. A pound of body fat is equivalent to 3500 Cal. Thus, to lose 10 lb, it is necessary either to consume 35,000 fewer Cal, which can be achieved if one reduces caloric intake by 350 Cal every day for 100 days (or by 700 Cal daily for 50 days) or uses up, through exercise, the same number of food calories.

22.3 How Does the Body Process Dietary Carbohydrates?

Carbohydrates are the major source of energy in the diet. They also furnish important compounds for the synthesis of cell components (Chapter 21). The main dietary carbohydrates are the polysaccharide starch, the disaccharides lactose and sucrose, and the monosaccharides glucose and fructose. Before the body can absorb carbohydrates, it must break down di-, oligo-, and polysaccharides into monosaccharides, because only monosaccharides can pass into the bloodstream.

The monosaccharide units are connected to each other by glycosidic bonds. Glycosidic bonds are cleaved by hydrolysis. In the body, this hydrolysis is catalyzed by acids and by enzymes. When a metabolic need arises, storage polysaccharides—amylose, amylopectin, and glycogen—are hydrolyzed to yield glucose and maltose.

This hydrolysis is aided by a number of enzymes:

- **α-Amylase** attacks all three storage polysaccharides at random, hydrolyzing the α-1,4-glycosidic bonds,

- **β-Amylase** also hydrolyzes the α-1,4-glycosidic bonds but in an orderly fashion, cutting disaccharidic maltose units one by one from the nonreducing end of a chain.

- The **debranching enzyme** hydrolyzes the α-1,6-glycosidic bonds (Figure 22.3).

In acid-catalyzed hydrolysis, storage polysaccharides are cut at random points. At body temperature, acid catalysis is slower than the enzyme-catalyzed hydrolysis.

The digestion (hydrolysis) of starch and glycogen in our food supply starts in the mouth, where α-amylase is one of the main components of saliva. Hydrochloric acid in the stomach and other hydrolytic enzymes in the intestinal tract hydrolyze starch and glycogen to produce mono- and disaccharides (D-glucose and maltose).

D-glucose enters the bloodstream and is carried to the cells to be utilized (Section 20.2). For this reason, D-glucose is often called blood sugar. In healthy people, little or none of this sugar ends up in the urine except for short periods of time (binge eating). In diabetes mellitus, however, glucose is not completely metabolized and does appear in the urine. As a consequence, it is necessary to test the urine of diabetic patients for the presence of glucose (Chemical Connections 12C).

The latest DRI guideline, issued by the National Academy of Sciences in 2002, recommends a minimum carbohydrate intake of 130 g/day. Most

FIGURE 22.3 The action of different enzymes on glycogen and starch.

Chemical Connections 22B

Why Is It So Hard to Lose Weight?

One of the great tragedies of being human is that it is far too easy to gain weight and far too difficult to lose it. If we had to analyze the specific chemical reactions involved in this reality, we would look very carefully at the citric acid cycle, especially the decarboxylation reactions. Of course, all foods consumed in excess can be stored as fat. This is true for carbohydrates, proteins, and, of course, fats. In addition, these molecules can be interconverted, with the exception that fats cannot give a net yield of carbohydrates. Why can fats not yield carbohydrates? The only way that a fat molecule could make glucose would be to enter the citric acid cycle as acetyl-CoA and then be drawn off as oxaloacetate for gluconeogenesis (Section 21.2). Unfortunately, the two carbons that enter (in the citric acid cycle) are effectively lost by the decarboxylations (Section 19.4). This leads to an imbalance in the catabolic versus anabolic pathways.

All roads lead to fats, but fats cannot lead back to carbohydrates. Humans are very sensitive to glucose levels in the blood because so much of our metabolism is geared toward protecting our brain cells, which prefer glucose as a fuel. If we eat more carbohydrates than we need, the excess carbohydrates will turn to fats. As we know, it is very easy to put on fat, especially as we age.

What about the reverse? Why don't we just stop eating? Won't that reverse the process? Yes and no. When we start eating less, fat stores become mobilized for energy production. Fat is an excellent source of energy because it forms acetyl-CoA and gives a steady influx for the citric acid cycle. Thus we can lose some weight by reducing caloric intake. Unfortunately, our blood sugar will also drop as soon as our glycogen stores run out. We have very little stored glycogen that could be tapped to maintain our blood glucose levels.

When the blood glucose drops, we become depressed, sluggish, and irritable. We start having negative thoughts like, "this dieting thing is really stupid. I should eat a pint of Oreo cookie ice cream." If we continue the diet, and given that we cannot turn fats into carbohydrates, where will the blood glucose come from? Only one source is left—proteins. Proteins will be degraded to amino acids, and eventually

be converted to pyruvate for gluconeogenesis. Thus we will begin to lose muscle as well as fat.

There is a bright side to this process, however. Using our knowledge of biochemistry, we can see that there is a better way to lose weight than dieting—exercise! If you exercise correctly, you can train your body to use fats to supply acetyl-CoA for the citric acid cycle. If you consume a normal diet, you will maintain your blood glucose and not degrade proteins for that purpose; your ingested carbohydrates will be sufficient to maintain both blood glucose and carbohydrate stores. With the proper ratio of exercise to food intake, and the proper balance of the right types of nutrients, we can increase the breakdown of fats without sacrificing carbohydrate stores or proteins. In essence, it is easier and healthier to train off the weight than to diet off the weight. This fact has been known for a long time. Now we are in a position to see why it is so, biochemically.

Working out is far superior to dieting if you want to lose weight.

people exceed this value. Artificial sweeteners (Chemical Connections 22C) can be used to reduce mono- and disaccharide intake.

Not everyone considers this recommendation to be the final word. Certain diets, such as the Atkins diet, recommend reducing carbohydrate intake to force the body to burn stored fat for energy supply. For example, during the introductory two-week period in the Atkins diet, only 20 g of carbohydrates per day are recommended in the form of salad and vegetables; no fruits or starchy vegetables are allowed. For the longer term, an additional 5 g of carbohydrates per day is added in the form of fruits. This restriction induces ketosis, the production of ketone bodies (Section 20.7) that may create muscle weakness and kidney problems.

Many fad diets exist today. The Atkins diet was preceded by the Zone Diet and the Sugar Buster's Diet, both of which attempted to limit carbohydrate intake. Another diet suggests that you match the foods you eat to your ABO blood type. To date, little scientific evidence supports any of these approaches, although some aspects of many diets have merit.

22.4 How Does the Body Process Dietary Fats?

Fats are the most concentrated source of energy. About 98% of the lipids in our diet are fats and oils (triglycerides); the remaining 2% consist of complex lipids and cholesterol.

The lipids in the food we eat must be hydrolyzed into smaller components before they can be absorbed into the blood or lymph system through the intestinal walls. The enzymes that promote this hydrolysis are located in the small intestine and are called *lipases.* However, because lipids are insoluble in the aqueous environment of the gastrointestinal tract, they must be dispersed into fine colloidal particles before the enzymes can act on them.

Bile salts (Section 13.11) perform this important function. Bile salts are manufactured in the liver from cholesterol and stored in the gallbladder. From there, they are secreted through the bile ducts into the intestine. Lipases act on the emulsion produced by bile salts and dietary fats, breaking the fats into glycerol and fatty acids and the complex lipids into fatty acids, alcohols (glycerol, choline, ethanolamine, sphingosine), and carbohydrates. These hydrolysis products are then absorbed through the intestinal walls.

Only two fatty acids are essential in higher animals, including humans: linolenic and linoleic acids (Section 13.3). Nutritionists occasionally list arachidonic acid as an **essential fatty acid.** In reality, our bodies can synthesize arachidonic acid from linoleic acid.

22.5 How Does the Body Process Dietary Protein?

Although the proteins in our diet can be used for energy (Section 20.9), their main use is to furnish amino acids from which the body synthesizes its own proteins (Section 18.5).

The digestion of dietary proteins begins with cooking, which denatures proteins. (Denatured proteins are hydrolyzed more easily by hydrochloric acid in the stomach and by digestive enzymes than are native proteins.) *Stomach acid* contains about 0.5% HCl. This HCl both denatures the proteins and hydrolyzes the peptide bonds randomly. *Pepsin,* the proteolytic enzyme of stomach juice, hydrolyzes peptide bonds on the amino side of the aromatic amino acids: tryptophan, phenylalanine, and tyrosine (see Figure 22.4).

Most protein digestion occurs in the small intestine. There, the enzyme *chymotrypsin* hydrolyzes internal peptide bonds at the same amino acids as does pepsin, except it does so on the other side, leaving these amino

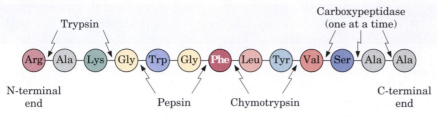

FIGURE 22.4 Different enzymes hydrolyze peptide chains in different but specific ways. Note that both chymotrypsin and pepsin would hydrolyze all of the same amino acids, but they are shown here hydrolyzing separate ones for comparison of the side they hydrolyze.

acids as the carboxyl termini of their fragments. Another enzyme, *trypsin,* hydrolyzes them only on the carboxyl side of arginine and lysine. Other enzymes, such as *carboxypeptidase,* hydrolyze amino acids one by one from the C-terminal end of the protein. The amino acids and small peptides are then absorbed through the intestinal walls.

The human body is incapable of synthesizing ten of the amino acids in sufficient quanities needed to make proteins. These ten **essential amino acids** must be obtained from our food; they are shown in Table 21.1. The body hydrolyzes food proteins into their amino acid constituents and then puts the amino acids together again to make body proteins. For proper nutrition, the human diet should contain about 20% protein.

A dietary protein that contains all of the essential amino acids is called a **complete protein.** Casein, the protein of milk, is a complete protein, as are most other animal proteins—those found in meat, fish, and eggs. People who eat adequate quantities of meat, fish, eggs, and dairy products get all the amino acids they need to keep healthy. About 50 g/day of complete proteins constitutes an adequate quantity.

An important animal protein that is not complete is gelatin, which is made by denaturing collagen (Section 14.12). Gelatin lacks tryptophan and is low in several other amino acids, including isoleucine and methionine. Many people on quick-reducing diets consume "liquid protein." This substance is simply denatured and partially hydrolyzed collagen (gelatin). Therefore, if it is the only protein source in the diet, some essential amino acids will be lacking.

Most plant proteins are incomplete. For example, corn protein lacks lysine and tryptophan; rice protein lacks lysine and threonine; wheat protein lacks lysine; and legumes are low in methionine and cysteine. Even soy protein, one of the best plant proteins, is very low in methionine. Adequate amino acid nutrition is possible with a vegetarian diet, but only if a wide range of vegetables is eaten. **Protein complementation** is one such diet. In protein complementation, two or more foods complement the others' deficiencies. For example, grains and legumes complement each other, with grains being low in lysine but high in methionine. Over time, such protein complementation in vegetarian diets became the staple in many parts of the world—corn tortillas and beans in Central and South America, rice and lentils in India, and rice and tofu in China and Japan.

In many developing countries, protein deficiency diseases are widespread because the people get their protein mostly from plants. Among these diseases is **kwashiorkor,** whose symptoms include a swollen stomach, skin discoloration, and retarded growth.

Proteins are inherently different from carbohydrates and fats when it comes to their relationship to the diet. Unlike the other two fuel sources, proteins have no storage form. If you eat a lot of carbohydrate, you will

Essential amino acid An amino acid that the body cannot synthesize in the required amounts and so must be obtained in the diet

Chemical Connections 22C

Dieting and Artificial Sweeteners

Obesity is a serious problem in the United States. Many people would like to lose weight but are unable to control their appetites enough to do so, which explains why artificial sweeteners are so popular. These substances have a sweet taste but do not add calories. Many people restrict their sugar intake. Some are forced to do so by diseases such as diabetes; others, by the desire to lose weight. Because most of us like to eat sweet foods, artificial sweeteners are added to many foods and drinks for those who must (or want to) restrict their sugar intake.

Four noncaloric artificial sweeteners are approved by the U.S. Food and Drug Administration (FDA). The oldest of these substances, saccharin, is 450 times sweeter than sucrose. Saccharin has been in use for about 100 years. Unfortunately, some tests have shown that this sweetener, when fed in massive quantities to rats, caused some of these rats to develop cancer. Other tests have given negative results. Recent research has shown that this cancer in rats is not relevant to human consumption. Saccharin continues to be sold on the market.

A newer artificial sweetener, aspartame, does not have the slight aftertaste that saccharin does. Aspartame is the methyl ester of a dipeptide, Asp-Phe (aspartylphenylalanine). Its sweetness was discovered in 1969. After extensive biological testing, it was approved by the FDA in 1981 for use in cold cereals, drink mixes, gelatins, and as tablets or powder to be used as a sugar substitute. Aspartame is 100 to 150 times sweeter than sucrose. It is made from natural amino acids, so both the aspartic acid and the phenylalanine have the L configuration. Other possibilities have also been synthesized: the L-D, D-L and D-D configurations. However, they are all bitter rather than sweet.

Aspartame is sold under such brand names as Equal and NutraSweet. A newer version of aspartame, neotame, was approved by the FDA in 2002. Twenty-five times sweeter than aspartame, it is essentially the same compound but with one —H substituted by a —CH$_2$—CH$_2$—C(CH$_3$)$_3$ group at the amino terminus.

Acesulfame-K

A third artificial sweetener, acesulfame-K, is 200 times sweeter than sucrose and is used (under the name Sunette) mostly in dry mixes. The fourth option, Sucralose, is used in sodas, baked goods, and tabletop packets. This trichloro derivative of a disaccharide is 600 times sweeter than sucrose and has no aftertaste. Neither Sucralose nor acesulfame-K is metabolized in the body; that is, they pass through unchanged.

Sucralose

Of course, calories in the diet do not come solely from sugar. Dietary fat is an even more important source (Section 22.4). It has long been hoped that some kind of artificial fat, which would taste the same but have no (or only a few) calories, would help in losing weight. Procter & Gamble Company has developed such a product, called Olestra. Like natural fats, this molecule is a carboxylic ester; instead of glycerol, however, the alcohol component is sucrose (Section 12.4A). All eight of sucrose's OH groups are converted to ester groups; the carboxylic acids are long-chain fatty acids similar to those found in normal triglycerides. The one illustrated here is made with the C$_{10}$ decanoic (capric) acid.

Saccharin   Aspartame

Olestra

Chemical Connections 22C

Dieting and Artificial Sweeteners *(continued)*

Although Olestra has a chemical structure similar to a fat, the human body cannot digest it because the enzymes that digest ordinary fats are not designed to work with the particular size and shape of this molecule. Therefore, it passes through the digestive system unchanged, and we derive no calories from it. Olestra can be used instead of an ordinary fat in cooking such items as cookies and potato chips. People who eat these foods will then be consuming fewer calories.

On the downside, Olestra can cause diarrhea, cramps, and nausea in some people—effects that increase with the amount consumed. Individuals who are susceptible to such side effects will have to decide if the potential to lose weight is worth the discomfort involved. Furthermore, Olestra dissolves and sweeps away some vitamins and nutrients in other foods being digested at the same time. To counter this effect, food manufacturers must add some of these nutrients (vitamins A, D, E, and K) to the product. FDA requires that all packages of Olestra-containing foods carry a warning label explaining these side effects.

---

store glucose in the form of glycogen. If you eat a lot of anything, you will store fat. However, if you eat a lot of protein (more than required for your needs), there is no place to store extra protein. Protein in excess will be metabolized to other substances, such as fat. For this reason, you must consume adequate protein every day. This requirement is especially critical in athletes and growing children. If an athlete works out intensely one day but eats incomplete protein, he or she cannot repair the damaged muscles. The fact that the athlete may have eaten an excess of a complete protein the day before will not help.

22.6 What Is the Importance of Vitamins, Minerals, and Water?

Vitamins and **minerals** are essential for good nutrition. Animals maintained on diets that contain sufficient carbohydrates, fats, and proteins and provided with an ample water supply cannot survive on these alone; they also need the essential organic components called vitamins and the inorganic ions called minerals. Many vitamins, especially those in the B group, act as coenzymes and inorganic ions as cofactors in enzyme-catalyzed reactions (Table 22.1). Table 22.2 lists the structures, dietary sources, and functions of the vitamins and minerals. Deficiencies in vitamins and minerals lead to many nutritionally controllable diseases (one example is shown in Figure 22.5); these conditions are also listed in Table 22.2.

The recent trend in vitamin appreciation is connected to their general role rather than to any specific action they have against a particular disease. For example, today the role of vitamin C in prevention of scurvy is barely mentioned but it is hailed as an important antioxidant. Similarly, other antioxidant vitamins or vitamin precursors dominate the medical literature. As an example, it has been shown that consumption of carotenoids (other than β-carotene) and vitamins E and C contributes significantly to respiratory health. The most important of the three is vitamin E. Furthermore, the loss of vitamin C during hemodialysis contributes significantly to oxidative damage in patients, leading to accelerated atherosclerosis.

Courtesy of Drs. P. G. Bullogh and V. J. Vigorita and the Gower Medical Publishing Company

FIGURE 22.5 Symptoms of rickets, a vitamin D deficiency in children. The nonmineralization of the bones of the radius and the ulna results in prominence of the wrist.

TABLE 22.1 Vitamins and Trace Elements as Coenzymes and Cofactors

Vitamin/Trace Element	Form of Coenzyme	Representative Enzyme	Reference
B$_1$, thiamine	Thiamine pyrophosphate, TPP	Pyruvate dehydrogenase	Step 12, Section 20.2
B$_2$, riboflavin	Flavin adenine dinucleotide, FAD	Succinate dehydrogenase	Step 6, Section 19.4
Niacin	Nicotinamide adenine dinucleotide, NAD$^+$	D-Glyceraldehyde-3-phosphate dehydrogenase	Step 5, Section 20.2
Pantothenic acid	Coenzyme A, CoA	Fatty acid synthase	Step 1, Section 21.3
B$_6$, pyridoxal	Pyridoxal phosphate, PLP	Aspartate amino transferase	Class 2, Section 15.2
B$_{12}$, cobalamine		Ribose reductase	Step 1, Section 17.2
Biotin	N-carboxybiotin	Acetyl-CoA carboxylase	Malonyl-CoA, Section 21.3
Folic acid		Purine biosynthesis	Section 17.2
Mg		Pyruvate kinase	Chemical Connections 15C
Fe		Cytochrome oxidase	Section 19.5
Cu		Cytochrome oxidase	Section 19.5
Zn		DNA polymerase	Section 17.6
Mn		Arginase	Step 5, Section 20.8
K		Pyruvate kinase	Chemical Connections 15C
Ni		Urease	Section 15.1
Mo		Nitrate reductase	
Se		Glutathione peroxidase	Chemical Connections 14A

Chemical Connections 22D

Iron: An Example of a Mineral Requirement

Iron, whether in the form Fe(II) or Fe(III), is usually found in the body in association with proteins. Little or no iron can be found "free" in the blood. Because iron-containing proteins are ubiquitous, there is a dietary requirement for this mineral. Severe deficits can lead to iron-deficiency anemia.

Iron usually occurs as the Fe(III) form in food. This is also the form released from iron pots when food is cooked in them. However, iron must be in the Fe(II) state to be absorbed. Reduction from Fe(III) to Fe(II) can be accomplished by ascorbate (vitamin C) or by succinate. Factors that affect absorption include the solubility of a given compound of iron, the presence of antacids in the digestive tract, and the source of the iron. To give some examples, iron may form insoluble complexes with phosphate or oxalate, and the presence of antacids in the digestive tract may decrease iron absorption. Iron from meats is more easily absorbed than iron from plant sources.

Requirements for iron vary according to age and gender. Infants and adult men need 10 mg per day; infants are born with a three- to six-month supply. Children and women (ages 16 through 50) need 15 to 18 mg per day. Women lose 20 to 23 mg of iron during each menstrual period. Pregnant and lactating women need more than 18 mg per day. After a blood loss, anyone, regardless of age or gender, needs more than these amounts. Distance runners, particularly marathoners, are also at risk of becoming anemic through loss of blood cells in the feet caused by the pounding of the thousands of foot falls that occur during a long run. People with iron deficiencies may experience a craving for nonfood items like clay, chalk, and ice.

Iron, in the form of ferrous sulfate, is a common over-the-counter mineral supplement.

TABLE 22.2 Vitamins and Minerals: Sources, Functions, Deficiency Diseases, and Daily Requirements

Name	Structure	Best Food Source	Function	Deficiency Symptoms and Diseases	Recommended Dietary Allowance[a]
Fat-Soluble Vitamins					
A		Liver, butter, egg yolk, carrots, spinach, sweet potatoes	Vision; to heal eye and skin injuries	Night blindness; blindness; keratinization of epithelium and cornea	800 μg (1500 μg)[b]
D		Salmon, sardines, cod liver oil, cheese, eggs, milk	Promotes calcium and phosphate absorption and mobilization	Rickets (in children): pliable bones; osteomalacia (in adults); fragile bones	5–10 μg; exposure to sunlight
E		Vegetable oils, nuts, potato chips, spinach	Antioxidant	In cases of malabsorption such as in cystic fibrosis: anemia. In premature infants: anemia	8–10 mg
K		Spinach, potatoes, cauliflower, beef liver	Blood clotting	Uncontrolled bleeding (mostly in newborn infants)	65–80 μg
Water-Soluble Vitamins					
B$_1$ (thiamine)		Beans, soybeans, cereals, ham, liver	Coenzyme in oxidative decarboxylation and in pentose phosphate shunt	Beriberi. In alcoholics: heart failure; pulmonary congestion	1.1 mg
B$_2$ (riboflavin)		Kidney, liver, yeast, almonds, mushrooms, beans	Coenzyme of oxidative processes	Invasion of cornea by capillaries; cheilosis; dermatitis	1.4 mg

Vitamin	Amount	Function	Deficiency disease	Food sources
Nicotinic acid (niacin)	15–18 mg	Coenzyme of oxidative processes	Pellagra	Chickpeas, lentils, prunes, peaches, avocados, figs, fish, meat, mushrooms, peanuts, bread, rice, beans, berries
B$_6$ (pyridoxal)	1.6–2.2 mg	Coenzyme in transamination; heme synthesis	Convulsions; chronic anemia; peripheral neuropathy	Meat, fish, nuts, oats, wheat germ, potato chips
Folic acid	400 μg	Coenzyme in methylation and in DNA synthesis	Anemia	Liver, kidney, eggs, spinach, beets, orange juice, avocados, cantaloupe
B$_{12}$	1–3 μg	Part of methyl-removing enzyme in folate metabolism	Patchy demyelination; degradation of nerves, spinal cord, and brain	Oysters, salmon, liver, kidney

TABLE 22.2 Vitamins and Minerals: Sources, Functions, Deficiency Diseases, and Daily Requirements *(continued)*

Name	Structure	Best Food Source	Function	Deficiency Symptoms and Diseases	Recommended Dietary Allowance[a]
Water-Soluble Vitamins					
Pantothenic acid	$HOCH_2C(CH_3)_2-CHOH-C(=O)-NHCH_2CH_2COOH$	Peanuts, buckwheat, soybeans, broccoli, lima beans, liver kidney, brain, heart	Part of CoA; fat and carbohydrate metabolism	Gastrointestinal disturbances; depression	4–7 mg
Biotin	(biotin ring structure) $(CH_2)_4COOH$	Yeast, liver, kidney, nuts, egg yolk	Synthesis of fatty acids	Dermatitis; nausea; depression	30–100 μg
C (ascorbic acid)	(ascorbic acid ring structure)	Citrus fruit, berries, broccoli, cabbage, peppers, tomatoes	Hydroxylation of collagen, wound healing; bond formation; antioxidant	Scurvy; capillary fragility	60 mg
Minerals					
Potassium		Apricots, bananas, dates, figs, nuts, raisins, beans, chickpeas, cress, lentils	Provides membrane potential	Muscle weakness	3500 mg
Sodium		Meat, cheese, cold cuts, smoked fish, table salt	Osmotic pressure	None	2000–2400 mg
Calcium		Milk, cheese, sardines, caviar	Bone formation; hormonal function; blood coagulation; muscle contraction	Muscle cramps; osteoporosis; fragile bones	800–1200 mg
Chloride		Meat, cheese, cold cuts, smoked fish, table salt	Osmotic pressure	None	1700–5100 mg

Mineral	Sources	Function	Deficiency/Excess	Amount
Phosphorus	Lentils, nuts, oats, grain flours, cocoa, egg yolk, cheese, meat (brain, sweetbreads)	Balancing calcium in diet	Excess causes structural weakness in bones	800–1200 mg
Magnesium	Cheeses, cocoa, chocolate, nuts, soybeans, beans	Cofactor in enzymes	Hypocalcemia	280–350 mg
Iron	Raisins, beans, chickpeas, parsley, smoked fish, liver, kidney, spleen, heart, clams, oysters	Oxidative phosphorylation; hemoglobin	Anemia	15 mg
Zinc	Yeast, soybeans, nuts, corn, cheese, meat, poultry	Cofactor in enzymes, insulin	Retarded growth; enlarged liver	12–15 mg
Copper	Oysters, sardines, lamb, liver	Oxidative enzymes cofactor	Loss of hair pigmentation, anemia	1.5–3 mg
Manganese	Nuts, fruits, vegetables, whole-grain cereals	Bone formation	Low serum cholesterol levels; retarded growth of hair and nails	2.0–5.0 mg
Chromium	Meat, beer, whole wheat and rye flours	Glucose metabolism	Glucose not available to cells	0.05–0.2 mg
Molybdenum	Liver, kidney, spinach, beans, peas	Protein synthesis	Retarded growth	0.075–0.250 mg
Cobalt	Meat, dairy products	Component of vitamin B_{12}	Pernicious anemia	0.05 mg (20–30 mg)[b]
Selenium	Meat, seafood	Fat metabolism	Muscular disorders	0.05–0.07 mg (2.4–3.0 mg)[b]
Iodine	Seafood, vegetables, meat	Thyroid glands	Goiter	150–170 μg (1000 μg)[b]
Fluorine	Fluoridated water; fluoridated toothpaste	Enamel formation	Tooth decay	1.5–4.0 mg (8–20 mg)[b]

[a]The U.S. RDAs are set by the Food and Nutrition Board of the National Research Council. The numbers given here are based on the latest recommendations (*National Research Council Recommended Dietary Allowances*, 10th ed., 1989, National Academy Press, Washington, D.C.). The RDA varies with age, sex, and level of activity; the numbers given are average values for both sexes between the ages of 18 and 54.

[b]Toxic if doses above the level shown in parentheses are taken.

Chemical Connections 22E

Food for Performance Enhancement

Athletes do whatever they can to enhance their performance. While the press focuses on those illegal methods employed by some athletes, such as using steroids or erythropoietin (EPO), many athletes continue to seek legal ways of enhancing their performance through diet and diet supplements. Any substance that aids performance is called an **ergogenic aid.**

After vigorous exercise lasting 30 minutes or more, performance typically declines because the glycogen stores in the muscle are depleted. After 90 minutes to 2 hours, liver glycogen stores also become depleted. There are two ways to combat this outcome. First, one can start the event with a full load of glycogen in the muscle and liver. This is why many athletes load up on carbohydrates in the form of pasta or other high-carbohydrate meals in the days before the event. Second, one can maintain the blood glucose levels during the event, so that liver glycogen does not have to be used for this purpose and so that ingested sugars can help support the athlete's energy needs, sparing some of the muscle and liver glycogen. This is why athletes often consume energy bars and sport drinks during an event.

The most often used ergogenic aid, although many athletes might not realize it, is caffeine. Caffeine works in two different ways. First, it acts as a general stimulant of the central nervous system, giving the athlete the sensation of having a lot of energy. Second, it stimulates the breakdown of fatty acids for fuel via its role as an activator of the lipases that break triacyclglycerols down (Section 22.4). Caffeine is a "double-edged sword," however, because it can also cause dehydration and actually inhibits the breakdown of glycogen.

In recent years, a new performance-enhancing food has appeared on the market and quickly become a best-seller: creatine. It is sold over the counter. Creatine is a naturally occurring amino acid in the muscles, which store energy in the form of high-energy phosphocreatine. During a short and strenuous bout of exercise, such as a 100-meter sprint, the muscles first use up the ATP obtainable from the reaction of phosphocreatine with ADP; only then do they rely on the glycogen stores.

Creatine is sold over the counter.

Both creatine and carbohydrates are natural food and body components and, therefore, cannot be considered equivalent to banned performance enhancers such as anabolic steroids or "andro" (Chemical Connections 13F). They are beneficial in improving performance in sports where short bursts of energy are needed, such as weight lifting, jumping, and sprinting. Lately, creatine has been used experimentally to preserve muscle neurons in degenerative diseases such as Parkinson's disease, Huntington's disease, and muscular dystrophy. Creatine also has few known hazards, even with long-term use. It is a highly nitrogenated compound, however, and overuse of creatine leads to the same problems as eating a diet characterized by excessive protein. The molecule must be hydrated, so water can be tied up with creatine that should be hydrating the body. The kidneys must also deal with excretion of extra nitrogen.

While athletes spend a lot of time and money on commercial ergogenic aids, the most important ergogenic aid remains water, the elixir of life that has been almost forgotten as it is replaced by its more expensive sports-drink cousins. A 1% level of dehydration during an event can adversely affect athletic performance. For a 150-pound athlete, this means losing 1.5 pounds of water as sweat. An athlete could easily lose much more than that in running a 10-kilometer race on all but the coldest days. To make things more difficult, performance is affected before thirst becomes noticeable. Only by drinking often during a long event can dehydration be prevented, and the athlete should begin drinking before noticing that she or he is thirsty. Water must be consumed either alone or as a part of a sports drink before you notice that you are thirsty.

Phosphocreatine + ADP ⟶ Creatine + ATP

Chemical Connections 22F

Organic Food—Hope or Hype?

The organic food industry is growing rapidly around the world. Organic markets like Whole Foods are flourishing along with associated organic restaurants. For a food to be labeled organic in the United States, it must be certified by the USDA as being grown in an environment without synthetic pesticides and fertilizers. Organic meat and poultry must be raised without hormones on feed grown organically. While it may seem intuitive to prefer such organically grown products, the price is often discouraging, between 10% and 100% more expensive than its nonorganic counterparts. The true benefits of organic food depend on many variables. For example, research shows fewer benefits to eating organic bananas as most pesticides in nonorganic bananas are thrown away with the peel. Other fruits, such as peaches, strawberries, apples, and pears, are the opposite. A 2002 study by the USDA showed that 98% of these nonorganic fruits tested had measurable levels of pesticides.

Researchers agree that pregnant women and children have the highest risk from pesticides. Pesticides have been shown to cross the placenta during pregnancy, and one study found that there was a positive link between pesticides present in some New York apartments and impaired fetal growth. A University of Washington study found that preschoolers fed nonorganic diets had six times the levels of certain pesticides as children fed organic food. High doses of certain pesticides have been shown to cause neurological and reproductive disorders in children exposed to them. With young bodies less able to rid themselves of toxins, eating organic food would be most important for pregnant and breast-feeding mothers. For those concerned about nonorganic meat and poultry, the issue is the use of antibiotics and growth hormones in raising the animals. Antibiotics are a worry as they can help produce resistant strains of bacteria. Some studies have suggested a link between certain types of cancer and the use of growth hormones.

Besides the potential dangers of the pesticides comes the question of the quality of the nutrients in the food. This is a more difficult parameter to measure and the results have been less conclusive. Recent studies on wheat have shown no significant differences in the quality of the wheat's nutrients when grown on synthetic pesticides compared to organic processes. Subjectively, however, many people prefer the taste of organically grown fruits, vegetables, and grains, despite the bitter taste the price tag might leave. If the rapid growth of the organic food industry is any indication, the general public believes the price tag is worth it.

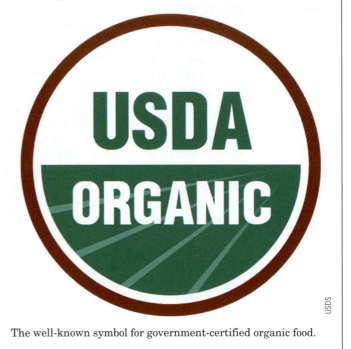

The well-known symbol for government-certified organic food.

Some vitamins have unusual effects besides their normal participation as coenzymes in many metabolic pathways or action as antioxidants in the body. Among these are the well-known effect of riboflavin, vitamin B_2, as a photosensitizer. Niacinamide, the amide form of vitamin B (niacin), is used in megadoses (2 g/day) to treat an autoimmune disease called bullous pemphigoid, which causes blisters on the skin. Conversely, the same megadoses in healthy patients may cause harm.

Although the concept of RDA has been used since 1940 and periodically updated as new knowledge dictated, a new concept is being developed in the field of nutrition. The DRI, currently in the process of completion, is designed to replace the RDA and is tailor-made to different ages and genders. It gives a set of two to four values for a particular nutrient in RDI:

- The estimated average requirement
- The recommended dietary allowance

- The adequate intake level
- The tolerable upper intake level

For example, the RDA for vitamin D is 5–10 μg, the adequate intake listed in the DRI for vitamin D for a person between age 9 and 50 years is 5 μg, and the tolerable upper intake level for the same age group is 50 μg.

A third set of standards appears on food labels in the United States, the Daily Values discussed in Section 22.2. Each gives a single value for each nutrient and reflects the need of an average healthy person eating 2000 to 2500 calories per day. The Daily Value for vitamin D, as it appears on vitamin bottles, is 400 International Units, which is 10 μg, the same as the RDA.

Water makes up 60% of our body weight. Most of the compounds in our body are dissolved in water, which also serves as a transporting medium to carry nutrients and waste materials. We must maintain a proper balance between water intake and water excretion via urine, feces, sweat, and exhalation of breath. A normal diet requires about 1200 to 1500 mL of water per day, in addition to the water consumed as part of our foods. Public drinking water systems in the United States are regulated by the Environmental Protection Agency (EPA), which sets minimum standards for protection of public health. Public water supplies are treated with disinfectants (typically chlorine) to kill microorganisms. Chlorinated water may have both an aftertaste and an odor. Private wells are not regulated by EPA standards.

Bottled water may come from springs, streams, or public water sources. Because bottled water is classified as a food, it comes under the FDA supervision, which requires it to meet the same standards for purity and sanitation as tap water. Most bottled water is disinfected with ozone, which leaves no flavor or odor in water.

Summary

⚙WL End-of-chapter problems identified in blue are assignable in **GOB OWL.**

Section 22.1 How Do We Measure Nutrition?

- **Nutrients** are components of foods that provide for growth, replacement, and energy.
- Nutrients are classified into six groups: carbohydrates, lipids, proteins, vitamins, minerals, and water.
- Each food contains a variety of nutrients. The largest part of our food intake is used to provide energy for our bodies.

Section 22.2 Why Do We Count Calories?

- A typical young adult needs 3000 Cal (male) or 2100 Cal (female) as an average daily caloric intake.
- **Basal caloric requirements,** the energy needed when the body is completely at rest, are less than the normal requirements.
- An imbalance between energy needs and caloric intake may create health problems. For example, chronic starvation increases infant mortality, whereas obesity leads to hypertension, cardiovascular disease, and diabetes.

Section 22.3 How Does the Body Process Dietary Carbohydrates?

- Carbohydrates are the major source of energy in the human diet.
- Monosaccharides are directly absorbed in the intestines, while oligo- and polysaccharides, such as starch, are digested with the aid of stomach acid, α- and β-amylases, and debranching enzymes.

Section 22.4 How Does the Body Process Dietary Fats? **Problem 22.20**

- Fats are the most concentrated source of energy.
- Fats are emulsified by bile salts and digested by lipases before being absorbed as fatty acids and glycerol in the intestines.
- Essential fatty and amino acids are needed as building blocks because the human body cannot synthesize them.

Section 22.5 How Does the Body Process Dietary Protein?

- Proteins are hydrolyzed by stomach acid and further digested by enzymes such as pepsin and trypsin before being absorbed as amino acids.

- There is no storage form for proteins, so good protein sources must be consumed in the diet every day.

Section 22.6 What Is the Importance of Vitamins, Minerals, and Water? **Problems 22.28, 22.32, 22.60, 22.61**

- Vitamins and minerals are essential constituents of the diet that are needed in small quantities.

- The fat-soluble vitamins are A, D, E, and K.
- Vitamins C and the B group are water-soluble vitamins.
- Most of the B vitamins are essential coenzymes.
- The most important dietary minerals are Na^+, Cl^-, K^+, $PO_4{}^{3-}$, Ca^{2+}, Fe^{2+}, and Mg^{2+}, but trace minerals are also necessary.
- Water makes up 60% of body weight.

Problems

■ Indicates problems that are assignable in **GOB OWL**.

Blue numbered problems are applied.

Go to this book's companion website at **www.cengage. com/chemistry/bettelheim** for interactive versions of the How To tutorials and Active Figures, and to quiz yourself on this chapter.

Section 22.1 How Do We Measure Nutrition?

22.1 Are nutrient requirements uniform for everyone?

22.2 Is banana flavoring, isopentyl acetate, a nutrient?

22.3 If sodium benzoate, a food preservative, is excreted as such, and if calcium propionate, another food preservative, is metabolized to CO_2 and H_2O, would you consider either of these preservatives to be a nutrient? If so, why?

22.4 Is corn grown solely with organic fertilizers more nutritious than that grown with chemical fertilizers?

22.5 Which part of the Nutrition Facts label found on food packages is the same for all labels carrying it?

22.6 Of which kinds of food does the U.S. government recommend that we have the most servings each day?

22.7 What is the importance of fiber in the diet?

22.8 Can a chemical that, in essence, goes through the body unchanged be an essential nutrient? Explain.

Section 22.2 Why Do We Count Calories?

22.9 A young adult female needs a caloric intake of 2100 Cal/day. Her basal caloric requirement is only 1300 Cal/day. Why is the extra 800 Cal needed?

22.10 What ill effects may obesity bring?

22.11 Assume that you want to lose 20 lb of body fat in 60 days. Your present dietary intake is 3000 Cal/day. What should your caloric intake be, in Cal/day, to achieve this goal, assuming no change in exercise habits?

22.12 What is marasmus?

22.13 Diuretics help to secrete water from the body. Would diuretic pills be a good way to reduce body weight?

Section 22.3 How Does the Body Process Dietary Carbohydrates?

22.14 Humans cannot digest wood; termites do so with the aid of bacteria in their digestive tract. Is there a basic difference in the digestive enzymes present in humans and termites?

22.15 What is the product of the reaction when α-amylase acts on amylose?

22.16 Does HCl in the stomach hydrolyze both the 1,4- and 1,6-glycosidic bonds?

22.17 Beer contains maltose. Can beer consumption be detected by analyzing the maltose content of a blood sample?

Section 22.4 How Does the Body Process Dietary Fats?

22.18 Which nutrient provides energy in its most concentrated form?

22.19 What is the precursor of arachidonic acid in the body?

22.20 ■ How many (a) essential fatty acids and (b) essential amino acids do humans need in their diets?

22.21 Do lipases degrade (a) cholesterol or (b) fatty acids?

22.22 What is the function of bile salts in digestion of fats?

Section 22.5 How Does the Body Process Dietary Protein?

22.23 Is it possible to get a sufficient supply of nutritionally adequate proteins by eating only vegetables?

22.24 Suggest a way to cure kwashiorkor.

22.25 What is the difference between protein digestion by trypsin and by HCl?

22.26 Which one will be digested faster: (a) a raw egg or (b) a hard-boiled egg? Explain.

Section 22.6 What Is the Importance of Vitamins, Minerals, and Water?

22.27 In a prison camp during a war, the prisoners are fed plenty of rice and water but nothing else. What would be the result of such a diet in the long run?

22.28 ■ (a) How many milliliters of water per day does a normal diet require?

(b) How many calories does this amount of water contribute?

22.29 Why did British sailing ships carry a supply of limes?

22.30 What are the symptoms of vitamin A deficiency?

22.31 What is the function of vitamin K?

22.32 ■ (a) Which vitamin contains cobalt?

(b) What is the function of this vitamin?

22.33 Vitamin C is recommended in megadoses by some people for prevention of all kinds of diseases, ranging from colds to cancer. What disease has been scientifically proven to be prevented when sufficient daily doses of vitamin C are in the diet?

22.34 Why is the Recommended Dietary Allowance (RDA) being phased out in favor of the Daily Reference Intake (DRI)?

22.35 What are the nonspecific effects of vitamin E, C, and carotenoids?

22.36 What are the best dietary sources of calcium, phosphorus, and cobalt?

22.37 Which vitamins contain a sulfur atom?

22.38 What are the symptoms of vitamin B_{12} deficiency?

Chemical Connections

22.39 (Chemical Connections 22A) What is the difference between the original Food Guide Pyramid published in 1992 and the revised version presented here?

22.40 (Chemical Connections 22A) With respect to carbohydrates and fats, why are there multiple locations in the new Food Guide Pyramid for these nutrients?

22.41 (Chemical Connections 22B) Explain what is meant by this statement: "All nutrients in excess can turn into fat, but fat cannot be turned into carbohydrate."

22.42 (Chemical Connections 22B) What does blood glucose have to do with dieting?

22.43 (Chemical Connections 22B) What is the most effective weight loss method?

22.44 How can the difference between weight loss through dieting and through exercise be explained by biochemistry?

22.45 (Chemical Connections 22B) Plants have a pathway that humans lack, called the glyoxylate pathway. It allows acetyl-CoA to bypass the two decarboxylation steps of the Citric Acid Cycle. How would dieting be different if humans had this pathway?

22.46 (Chemical Connections 22C) Describe the difference between the structure of aspartame and the methyl ester of phenylalanylaspartic acid.

22.47 (Chemical Connections 22C)

(a) Which artificial sweeteners are not metabolized in the body?

(b) What could be products of digested aspartame?

22.48 (Chemical Connections 22C) What is common in the structures of Olestra and Sucralose?

22.49 (Chemical Connections 22D) Why is there a requirement for iron in the diet?

22.50 (Chemical Connections 22D) What is the form of iron found in the body?

22.51 (Chemical Connections 22D) What factors influence the absorption of iron from the digestive system?

22.52 (Chemical Connections 22D) What factors influence a person's requirement for iron?

22.53 (Chemical Connections 22E) Looking in Table 14.1, which lists the common amino acids found in proteins, which amino acid most resembles creatine?

22.54 (Chemical Connections 22E) Which single compound will have the greatest effect on athletic performance?

22.55 (Chemical Connections 22E) Identify two ways carbohydrates are used for athletic performance.

22.56 (Chemical Connections 22E) Why is creatine an effective ergogenic aid? For which types of competitions is it effective?

22.57 (Chemical Connections 22E) How is caffeine used as an ergogenic aid? What are possible downsides to caffeine use?

22.58 (Chemical Connections 22F) What is meant by "organic" as it relates to food?

22.59 (Chemical Connections 22F) What are the major considerations surrounding organic vs. non organic foods?

Additional Problems

22.60 ■ Which two chemicals are used most frequently to disinfect public water supplies?

22.61 ■ Which vitamin is part of coenzyme A (CoA)? List a step (or the enzyme) that has CoA as a coenzyme in (a) glycolysis and (b) fatty acid synthesis.

22.62 Which vitamin is prescribed in megadoses to combat autoimmune blisters?

22.63 Why is it necessary to have protein in our diets?

22.64 Which chemical processes take place during digestion?

22.65 According to the U.S. government's Food Guide Pyramid, are there any foods that we can completely omit from our diets and still be healthy?

22.66 Does the debranching enzyme help in digesting amylose?

22.67 As an employee of a company that markets walnuts, you are asked to provide information for an ad that would stress the nutritional value of walnuts. What information would you provide?

22.68 In diabetes, insulin is administered intravenously. Explain why this hormone protein cannot be taken orally.

22.69 Egg yolk contains a lot of lecithin (a phosphoglyceride). After ingesting a hard-boiled egg, would you find an increase in the lecithin level of your blood? Explain.

22.70 What would you call a diet that scrupulously avoids phenylalanine-containing compounds? Could aspartame be used in such a diet?

22.71 What kind of supplemental enzyme would you recommend for a patient after a peptic ulcer operation?

22.72 In a trial, a woman was accused of poisoning her husband by adding arsenic to his meals. Her attorney stated that this supplementation was done to promote her husband's health, as arsenic is an essential nutrient. Would you accept this argument?

Immunochemistry

Two natural killer (NK) cells, shown in yellow-orange, attacking a leukemia cell, shown in red.

© Meckes/Ottawa/Photo Researchers, Inc.

23.1 How Does the Body Defend Itself from Invasion?

When you were in elementary school, you may have had chickenpox along with many other children. The viral disease passed from one person to the next and ran its course, but after the children recovered, they never had chickenpox again. Those who were infected became *immune* to this disease. Humans and other vertebrates possess a highly developed, complex immune system that defends the body against foreign invaders. Figure 23.1, which presents an overview of this complex system, could serve as a road map as you read the different sections in this chapter. Referring back to

FIGURE 23.1 Overview of the immune system: its components and their interactions.

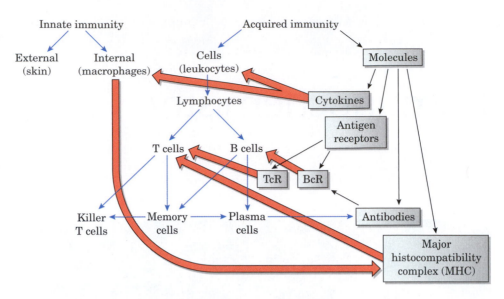

Figure 23.1, you will find the precise location of the topic under discussion and its relationship to the immune system as a whole. As you can see, the immune system is very involved with multiple layers of protection against invading organisms. In this section, we will briefly introduce the major parts of the immune system. These topics will then be expanded in the sections that follow.

A. Innate Immunity

When one considers the tremendous number of bacteria, viruses, parasites, and toxins that our bodies encounter, it is a wonder that we are not continually sick. Most students learn about antibodies in high school, and these days everyone learns about T cells due to their relationship to AIDS. When discussing immunity, however, many more weapons of defense exist other than T cells and antibodies. In reality, you only discover that you are sick once a pathogen has managed to beat the front-line defense, which is called **innate immunity.**

Innate immunity includes several components. One part, called **external innate immunity,** includes physical barriers, such as skin, mucus, and tears. All of these barriers act to hinder penetration by pathogens and do not require specialized cells to fight a pathogen. If a pathogen—whether a bacterium, virus, or parasite—is able to breach this outer layer of defense, the cellular warriors of the innate system come into play. The cells of the innate immune system that we will discuss are **dendritic cells, macrophages, and natural killer (NK) cells.** Among the first and most important cells to join the fight are the dendritic cells, so called due to their long, tentacle-like projections (Figure 23.2).

Natural killer cells function as the police officers of **internal innate immunity**. When they encounter cancerous cells, cells infected by a virus, or any other suspicious cells, they attach themselves to those abnormal cells (see the chapter-opening photo). Other nonspecific responses include the proliferation of macrophages, which engulf and digest bacteria and reduce inflammation. In the inflammatory response to injury or infection, the capillaries dilate to allow greater flow of blood to the site of the injury, enabling agents of the internal innate immunity system to congregate there.

B. Adaptive Immunity

Vertebrates have a second line of defense, called **adaptive** or **acquired immunity.** We refer to this type of immunity when we talk colloquially about the **immune system.** The key features of the immune system are *specificity*

Innate immunity The natural nonspecific resistance of the body against foreign invaders, which has no memory

Eye of Science/Photo Researchers, Inc.

FIGURE 23.2 Dendritic cells get their name from their tentacle-like arms. The one shown here is from a human.

and *memory.* The key cellular components of acquired immunity are **T cells** and **B cells.** The immune system uses antibodies and cell receptors designed specifically for each type of invader. In a second encounter with the same invader, the response is more rapid, more vigorous, and more prolonged than it was in the first case, because the immune system remembers the nature of the invader from the first encounter.

The invaders may be bacteria, viruses, molds, or pollen grains. A body with no defense against such invaders could not survive. In a rare genetic disease, a person is born without a functioning immune system. Attempts have been made to bring up such children in an enclosure totally sealed from the environment. While in this environment, they can survive; when the environment is removed, however, such people always die within a short time. The severity of this disease, called severe combined immunodeficiency, explains why it was the first disease treated with gene therapy (Section 18.9). The disease AIDS (Section 23.8) slowly destroys the immune system, particularly a type of T cell, leaving its victims to die from some invading organism that a person without AIDS would easily be able to fight off.

As we shall see, the beauty of the body's immune system lies in its flexibility. The system is capable of making millions of potential defenders, so it can almost always find just the right one to counter the invader, even when it has never seen that particular organism before.

C. Components of the Immune System

Foreign substances that invade the body are called **antigens.** The immune system is made of both cells and molecules. Two types of **white blood cells,** called lymphocytes, fight against the invaders: (1) T cells kill the invader by contact and (2) B cells manufacture **antibodies,** which are soluble immunoglobulin molecules that immobilize antigens.

The basic molecules of the immune system belong to the **immunoglobulin superfamily.** All molecules of this class have a certain portion of the molecule that can interact with antigens, and all are glycoproteins. The polypeptide chains in this superfamily have two domains: a constant region and a variable region. The constant region has the same amino acid sequence in each of the same class of molecules. In contrast, the variable region is antigen-specific, which means that the amino acid sequence in this region is unique for each antigen. The variable regions are designed to recognize only one specific antigen.

There are three representatives of the immunoglobulin superfamily in the immune system:

1. Antibodies are soluble immunoglobulins secreted by the plasma cells.

2. Receptors on the surface of T cells (**TcR**) recognize and bind antigens presented to them.

3. Molecules that present antigens also belong to this superfamily. They reside inside the cells. These protein molecules are known as major histocompatibility complexes (MHC).

When a cell is infected by an antigen, MHC molecules interact with it and bring a characteristic portion of the antigen to the surface of the cell. Such a surface presentation then marks the diseased cell for destruction. This can happen in a cell that was infected by a virus, and it can happen in macrophages that engulf and digest bacteria and viruses.

D. The Speed of the Immune Response

The process of finding and making just the right immunoglobulin to fight a particular invader is relatively slow, compared to the actions of the chemical messengers discussed in Chapter 16. While neurotransmitters act within a

Antigen A substance foreign to the body that triggers an immune response

Antibody A glycoprotein molecule that interacts with an antigen

Immunoglobulin superfamily Glycoproteins that are composed of constant and variable protein segments having significant homologies to suggest that they evolved from a common ancestry

Macrophage ingesting bacteria (the rod-shaped structures). The bacteria will be pulled inside the cell within a membrane-bound vesicle and quickly killed.

© David M. Phillips/Photo Researchers, Inc.

millisecond and hormones within seconds, minutes, or hours, immunoglobulins respond to an antigen over a longer span of time—weeks and months.

Although the immune system can be considered another form of chemical communication (Chapter 16), it is much more complex than neurotransmission because it involves molecular signals and interplay between various cells. In these constant interactions, the major elements are as follows: (1) the cells of the immune system, (2) the antigens and their perception by the immune system, (3) the antibodies (immunoglobulin molecules) that are designed to immobilize antigens, (4) the receptor molecules on the surface of the cells that recognize antigens, and (5) the cytokine molecules that control these interactions. Because the immune system is the cornerstone of the body's defenses, its importance for students in health-related sciences is undeniable.

23.2 What Organs and Cells Make Up the Immune System?

The blood plasma circulates in the body and comes in contact with the other body fluids through the semipermeable membranes of the blood vessels. Therefore, blood can exchange chemical compounds with other body fluids and, through them, with the cells and organs of the body (Figure 23.3).

A. Lymphoid Organs

The lymphatic capillary vessels drain the fluids that bathe the cells of the body. The fluid within these vessels is called **lymph.** Lymphatic vessels circulate throughout the body and enter certain organs, called **lymphoid organs,** such as the thymus, spleen, tonsils, and lymph nodes (Figure 23.4). The cells primarily responsible for the functioning of the immune system are the specialized white blood cells called **lymphocytes.** As their name implies, these cells are mostly found in the lymphoid organs. Lymphocytes may be either specific for a given antigen or nonspecific.

T cells are lymphocytes that originate in the bone marrow but mature in the thymus gland. B cells are lymphocytes that originate and mature in the bone marrow. Both B and T cells are found mostly in the lymph, where they circulate looking for invaders. Small numbers of lymphocytes are also found in the blood. To get there, they must squeeze through tiny openings between the endothelial cells. This process is aided by signaling molecules called **cytokines** (Section 23.6). The sequence of the response of the body to a foreign invader is depicted schematically in Table 23.1.

FIGURE 23.3 Exchange of compounds among three body fluids: blood, interstitial fluid, and lymph. (After Holum, J. R., *Fundamentals of General, Organic and Biological Chemistry,* New York: John Wiley & Sons, 1978, p. 569)

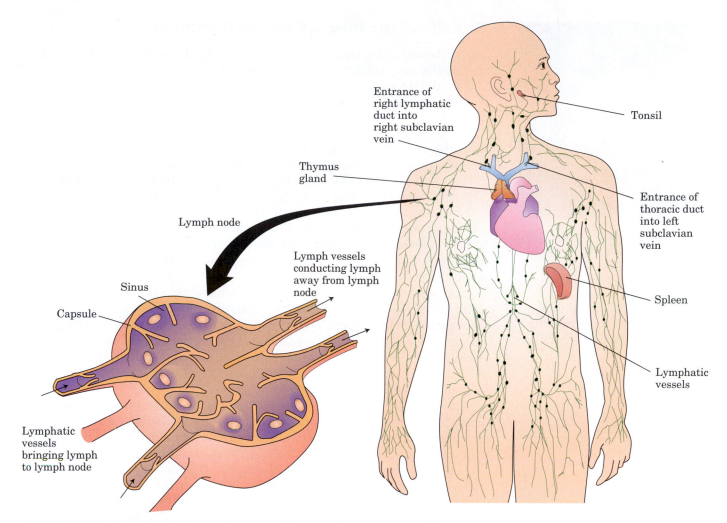

FIGURE 23.4 The lymphatic system is a web of lymphatic vessels containing a clear fluid, called lymph, and various lymphatic tissues and organs located throughout the body. Lymph nodes are masses of lymphatic tissue covered with a fibrous capsule. Lymph nodes filter the lymph. In addition, they are packed with macrophages and lymphocytes.

TABLE 23.1 Interactions Among the Different Cells of the Immune System

Courtesy of Steve Haley, William Bowers, and Richard Hunt, University of South Carolina Medical School

FIGURE 23.5 Dendritic cells and the other cells of the immune system. This figure shows a rat dendritic cell interacting with a T cell. Through these interactions, the dendritic cells teach the acquired immunity system what to attack.

B. Cells of the Internal Innate Immunity

As mentioned in Section 23.1A, the major cells of innate immunity are dendritic cells, macrophages, and natural killer cells.

Dendritic cells are found in the skin, mucus membranes, the lungs, and the spleen. They are the first cells of the innate system that will have a crack at any virus or bacterium that wanders into their path. Using suction-cup like receptors, they grab onto invaders and then engulf them by endocytosis. These cells then chop up the devoured pathogens and bring parts of their proteins to the surface. Here the protein fragments are displayed on a protein called a **major histocompatibility complex (MHC).** The dendritic cells travel through the lymph to the spleen, where they present these antigens to other cells of the immune system, the **helper T cells (T_H cells)** as shown in Figure 23.5. Dendritic cells are members of a class of cells referred to as **antigen presenting cells (APCs),** and they are the starting point in most of the responses that are traditionally associated with the immune system.

Macrophages are the first cells in the blood that encounter an antigen. They belong to the internal innate immune system; inasmuch as they are nonspecific, macrophages attack virtually anything that is not recognized as part of the body, including pathogens, cancer cells, and damaged tissues. Macrophages engulf an invading bacterium or virus and kill it. The "magic bullet" in this case is the NO molecule, which is both toxic and can act as a secondary messenger (Chemical Connections 16E).

The NO molecule is short-lived and must be constantly manufactured anew. When an infection begins, the immune system manufactures the protein interferon. It, in turn, activates a gene that produces an enzyme, nitric oxide synthase. With the aid of this enzyme, the macrophages, endowed with NO, kill the invading organisms. Macrophages then digest the engulfed antigen and display a small portion of it on their surface.

Natural killer (NK) cells target abnormal cells. Once in physical contact with such cells, NK cells release proteins, aptly called perforins, that perforate the target cell membranes and create pores. The membrane of the target cell becomes leaky, allowing hypotonic liquid from the surroundings to enter the cell, which swells and eventually bursts.

C. Cells of Adaptive Immunity: T and B Cells

T cells interact with the antigens presented by APCs and produce other T cells that are highly specific to the antigen. When these T cells differentiate, some of them become **killer T cells,** also called **cytotoxic T cells (T_c cells),** which kill the invading foreign cells by cell-to-cell contact. Killer T cells, like NK cells, act through perforins, which attach themselves to the target cell, in effect punching holes in its membranes. Through these holes water rushes into the target cell; it swells and eventually bursts.

Other T cells become **memory cells.** They remain in the bloodstream, so that if the same antigen enters the body again, even years after the primary infection, the body will not need to build up its defenses anew but is ready to kill the invader instantly.

A third type of T cell is the **helper T cell (T_H cell).** These cells do not kill other cells directly, but rather are involved in recognizing antigens on APCs and recruiting other cells to help fight the infection.

The production of antibodies is the task of **plasma cells.** These cells are derived from B cells after the B cells have been exposed to an antigen.

The lymphatic vessels, in which most of the antigen attacking takes place, flow through a number of lymph nodes (Figure 23.4). These nodes

Both natural killer cells and killer T cells act the same way, through perforin. The T_c cells attack specific targets; the NK cells attack all suspicious targets.

are essentially filters. Most plasma cells reside in lymph nodes, so most antibodies are produced there. Each lymph node is also packed with millions of other lymphocytes. More than 99% of all invading bacteria and foreign particles are filtered out in the lymph nodes. As a consequence, the outflowing lymph is almost free of invaders and is packed with antibodies produced by the plasma cells. All lymphocytes derive from stem cells in the bone marrow. Stem cells are undifferentiated cells that can become many different cell types. As shown in Figure 23.6, they can differentiate into T cells in the thymus or B cells in the bone marrow.

23.3 How Do Antigens Stimulate the Immune System?

A. Antigens

Antigens are foreign substances that elicit an immune response; for this reason, they are also called immunogens. Three features characterize an antigen. The first condition is foreignness—molecules of your own body should not elicit an immune response. The second condition is that the antigen must be of molecular weight greater than 6000. The third condition is that the molecule must have sufficient complexity. A polypeptide made of lysine only, for example, is not immunogenic.

Antigens can be proteins, polysaccharides, or nucleic acids, as all of these substances are large molecules. Antigens may be soluble in the cytoplasm, or they may be found at the surface of cells, either embedded in the membrane or just absorbed on it. An example of polysaccharidic antigenicity is ABO blood groups (Chemical Connections 12D).

In protein antigens, only part of the primary structure is needed to cause an immune response. Between 5 and 7 amino acids are needed to interact with an antibody, and between 10 and 15 amino acids are necessary to bind to a receptor on a T cell. The smallest unit of an antigen capable of binding with an antibody is called the **epitope.** The amino acids in an epitope do not have to be in sequence in the primary structure, as folding and secondary structures may bring amino acids that are not in sequence into each other's proximity. For example, the amino acids in positions 20 and 28 may form part of an epitope. *Antibodies can recognize all types of antigens, but T-cell receptors recognize only peptide antigens.*

As noted earlier, antigens may be in the interior of an infected cell or on the surface of a virus or bacterium that penetrated the cell. To elicit an immune reaction, the antigen or its epitope must be brought to the surface of the infected cell. Similarly, after a macrophage swallows up and partially digests an antigen, the macrophage must bring the epitope back to its surface to elicit an immune response from T cells (Table 23.1).

B. Major Histocompatibility Complexes

The task of bringing an antigen's epitope to the infected cell's surface is performed by the major histocompatibility complex (MHC). The name derives from the fact that its role in the immune response was first discovered in organ transplants. MHC molecules are transmembrane proteins belonging to the immunoglobulin superfamily. Two classes of MHC molecules exist (Figure 23.7), both of which have peptide-binding variable domains. Class I MHC is made of a single polypeptide chain, whereas class II MHC is a dimer. Class I MHC molecules seek out antigen molecules that have been *synthesized inside a virus-infected cell.* Class II MHC molecules pick up exogenous

FIGURE 23.6 The development of lymphocytes. All lymphocytes are ultimately derived from the stem cells of the bone marrow. In the thymus, T cells develop into helper T cells and killer T cells. B cells develop in the bone marrow.

There are rare exceptions: In an autoimmune disease, the body mistakes its own protein as foreign

Epitope The smallest number of amino acids on an antigen that elicits an immune response

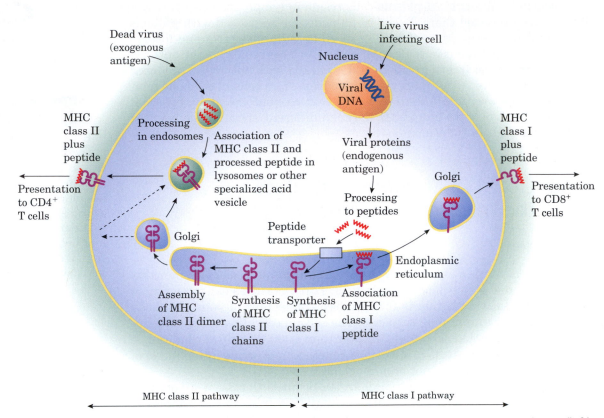

FIGURE 23.7 Differential processing of antigens in the MHC class II pathway (left) or MHC class I pathway (right). Cluster determinants (CD) are parts of the T-cell receptor complex (see Section 23.5B).

"dead" antigens. In each case, the epitope attached to the MHC is brought to the cell surface to be presented to T cells.

For example, if a macrophage engulfed and digested a virus, the result would be dead antigens. The digestion occurs in several steps. First, the antigen is processed in lysosomes, special organelles of cells that contain proteolytic enzymes. An enzyme called GILT (gamma-interferon inducible lysosomal thiol reductase) breaks the disulfide bridges of the antigen by reduction. The reduced peptide antigen unfolds and is exposed to proteolytic enzymes that hydrolyze it to smaller peptides. These peptides serve as epitopes that are recognized by class II MHC. The difference between MHC I and MHC II becomes significant when we look at the functions of T cells. Antigens bound to MHC I will interact with killer T cells, while those bound to MHC II will interact with helper T cells.

23.4 What Are Immunoglobulins?

A. Classes of Immunoglobulins

Immunoglobulins are glycoproteins—that is, carbohydrate-carrying protein molecules. Not only do the different classes of immunoglobulins vary in molecular weight and carbohydrate content, but their concentration in the blood differs dramatically as well (Table 23.2). The IgG and IgM antibodies are the most important antibodies in the blood. They interact with antigens and trigger the swallowing up (phagocytosis) of these cells by phagocytes. Inside the phagocytes, the antigens are destroyed inside the lysosomes. Antigens bound to antibodies are also destroyed in the blood system by

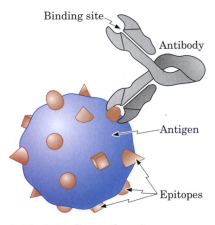

Antibody binding to the epitope of an antigen.

Chemical Connections 23A

The Mayapple and Chemotherapy Agents

Among the life-threatening diseases in the United States, heart disease remains the number one killer. Collectively cancers represent our second leading cause of death and it is estimated that, within the next several decades, they will head the list. Reducing risk factors is a major focus in the war against heart disease. Because risk factors for cancers are not as clearly recognized, however, treatment by surgery, radiation therapy, and chemotherapy remains the major focus in this war. In chemotherapy, a chemical or combination of chemicals is introduced into the body to destroy rapidly dividing cells. While we are only now beginning to understand how these agents work within the body, many such agents have been known for over a century. Among them is the antitumor effect of an extract of the common mayapple, *Podophyllum peltatum,* described in 1861 by Robert Bentley of Kings College, London. The active principle of mayapple extract was identified twenty

years later and given the name picropodophyllin. It was not until 1946 that the mechanism of action of the drug was discovered. We now know that it suppresses tumor growth by inhibiting the formation of the mitotic spindle, which holds cells undergoing nuclear division. The structure of picropodophyllin was determined in 1954 and in the 1970s, scientists at the Sandoz Company succeeded in synthesizing several analogs of picropodophyllin that are even more effective than their mayapple parent drug. Among them is etoposide, which is effective in treating small-cell lung cancer, testicular cancer, lymphomas, leukemias, and several types of brain tumors. Etoposide also inhibits topoisomerase II, an important DNA-regulating enzyme.

Picropodophyllin

Etoposide

a complicated procedure called the complement pathway. The IgA molecules are found mostly in secretions: tears, milk, and mucus. Therefore, these immunoglobulins attack the invading material before it gets into the bloodstream. The IgE molecules play a part in such allergic reactions as asthma and hay fever, and are involved in the body's defense against parasites.

B. Structure of Immunoglobulins

Each immunoglobulin molecule is made of four polypeptide chains: two identical light chains and two identical heavy chains. The four polypeptide chains are arranged symmetrically, forming a Y shape (Figure 23.8).

TABLE 23.2 Immunoglobulin Classes

Class	Molecular Weight (MW)	Carbohydrate Content (%)	Concentration in Serum (mg/100 mL)
IgA	200,000–700,000	7–12	90–420
IgD	160,000	<1	1–40
IgE	190,000	10–12	0.01–0.1
IgG	150,000	2–3	600–1800
IgM	950,000	10–12	50–190

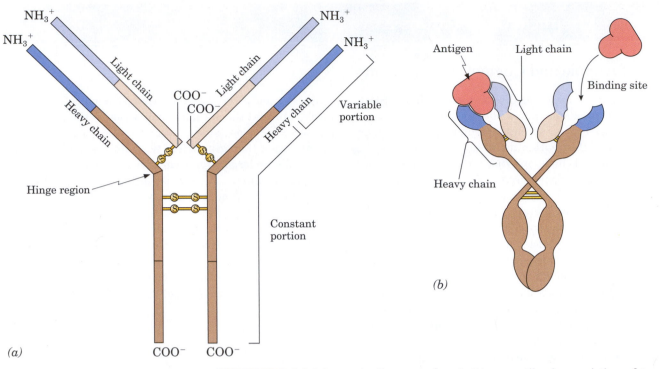

FIGURE 23.8 (*a*) Schematic diagram of an IgG-type antibody consisting of two heavy chains and two light chains connected by disulfide bonds. The amino terminal end of each chain has the variable portion. (*b*) Model showing how an antibody bonds to an antigen.

FIGURE 23.9 (*a*) An antigen–antibody complex. The antigen (shown in green) is lysozyme. The heavy chain of the antibody is shown in blue; the light chain in yellow. The most important amino acid residue (glutamine in the 121 position) on the antigen is the one that fits into the antibody groove (shown in red). (*b*) The antigen–antibody complex has been pulled apart. Note how they fit each other.

Courtesy of Dr. R. J. Poljak, Pasteur Institute, Paris, France

Four disulfide bridges link the four chains into a unit. Both light and heavy chains have constant and variable regions. The constant regions have the same amino acid sequences in different antibodies, and the variable regions have different amino acid sequences in different antibodies.

The variable regions of the antibody recognize the foreign substance (the antigen) and bind to it (Figure 23.9). Because each antibody contains two variable regions, it can bind two antigens, thereby forming a large aggregate, as shown in Figure 23.10.

The binding of the antigen to the variable region of the antibody occurs not by covalent bonds but rather by much weaker intermolecular forces such as London dispersion forces, dipole–dipole interactions, and hydrogen bonds. This binding is similar to the way in which substrates bind to enzymes or hormones and neurotransmitters bind to a receptor site. That is, the antigen must fit into the antibody surface. Humans have more than 10,000 different antibodies circulating at measurable levels, which enables our bodies to fight a large number of foreign invaders. However, the potential number of antibodies that can be created by the available genes is in the millions.

C. B Cells and Antibodies

Each B cell synthesizes only one unique immunoglobulin antibody, and that antibody contains a unique antigen-binding site to one epitope. Before encountering an antigen, these antibodies are inserted in the plasma membrane of the B cells, where they serve as receptors. When an antigen interacts with its receptor, it stimulates the B cell to divide and differentiate into plasma cells. These daughter cells secrete soluble antibodies that have the same antigen-binding sites as the original antibody/receptor. The soluble secreted antibodies appear in the serum (the noncellular part of blood) and can react with the antigen. Thus, an immunoglobulin produced in B cells acts both as a receptor to be stimulated by the antigen and as a secreted messenger ready to neutralize and eventually destroy the antigen (Figure 23.11).

D. How Does the Body Acquire the Diversity Needed to React to Different Antigens?

From the moment of conception, an organism contains all of the DNA it will ever have, including that DNA that will lead to antibodies and T-cell receptors. Thus the organism is born with a repertoire of genes necessary to fight infections. During B-cell development, the variable regions of the heavy chains are assembled by a process called V(J)D recombination. A number of exons are present in each of three different areas—V, J, and D—of the immunoglobin gene. Combining one exon from each area yields a *new V(J)D gene*. This process creates a great diversity because of the large number of ways that this combination can be performed (Figure 23.12). For one type of antibody light chain, called kappa, there are roughly 40 V genes and 5 J genes. That alone gives rise to 40 × 5 or 200 combinations of V and J. For another type of light chain, called lambda, about 120 combinations are possible. With the heavy chains, there is even greater diversity: about 50 V genes, 27 D genes, and 6 J genes. By the time one gets done calculating the possible combinations of all of the combinations of V, J, D, and the C regions of both the heavy and light chains, there are more than 2 million possible combinations.

However, that is merely the first step. A second level of diversity is created by mutation of the V(J)D genes in somatic cells. As cells proliferate in response to recognizing an antigen, such mutations can cause a 1000-fold

FIGURE 23.10 An antigen–antibody reaction forms a precipitate. An antigen, such as a bacterium or virus, typically has several binding sites for antibodies. Each variable region of an antibody (each prong of the Y) can bind to a different antigen. The aggregate thus formed precipitates and is attacked by phagocytes and the complement system.

B cell

Maturing B cells make antibodies that bind to the antigens present in the bloodstream; B cells proliferate

Plasma cell releases antibodies into bloodstream

FIGURE 23.11 B cells have antibodies on their surfaces, which allow them to bind to antigens. The B cells with antibodies for the antigens present grow and develop. When B cells develop into plasma cells, they release circulating antibodies into the bloodstream. (Adapted from "How the Immune System Develops," by Irving L. Weissman and Max D. Cooper; illustrated by Jared Schneidman, *Scientific American*, September 1993.) **Go to this book's companion website at www.cengage.com/chemistry/bettelheim to explore an interactive version of this figure.**

FIGURE 23.12 Diversification of immunoglobulins by V(J)D recombination. Exons of three genes—the V(ariable) (A, B, C, D); J(oin) (1, 2, 3, 4); and D(iverse) (a, b, c, d) genes—combine to form new V(J)D genes that are transcribed to the corresponding mRNAs. The expression of these new genes results in a wide variety of immunoglobulins having different variable regions on their heavy chains.

increase in the binding affinity of an antigen to an antibody. This process is called **affinity maturation.**

There are three ways to create mutations. Two affect the variable regions, and one alters the constant region

1. Somatic hypermutation (SHM) creates a point mutation (only one nucleotide). The resultant protein of the mutation is able to bind more strongly with the antigen.

2. In a gene conversion (GC) mutation, stretches of nucleotide sequences are copied from a pseudogene V and entered into the V(J)D. This also enables the proteins synthesized from the GC mutation to make stronger contact with the antigen than without the mutation.

3. Mutations on the constant region of the chain can be achieved by class switch recombination (CSR). In this case, exons of the constant regions are swapped between highly repetitive regions.

The diversity of antibodies created by the V(J)D recombination is greatly amplified and finely tuned by the mutations on these genes. Because the response to the antigen occurs on the gene level, it is easily preserved and transmitted from one generation of cells to the next.

While the possible combination of genes leading to antibody diversity seems limitless, it is important to remember that the basis of antibody diversity is the genetic blueprint that the organism was given. Antibodies do not appear because they are needed; rather, they are selected and proliferated because they already existed in small quantities before they were stimulated by recognition of an antigen.

E. Monoclonal Antibodies

When an antigen is injected into an organism (for example, human lysozyme into a rabbit), the initial response is quite slow. It may take from one to two weeks before the anti-lysozyme antibody shows up in the rabbit's

Chemical Connections 23B

Monoclonal Antibodies Wage War on Breast Cancer

Breast cancer is currently the second leading cause of cancer-related deaths in North America, but this status is likely to change in the near future. The survival rate for women diagnosed with breast cancer has been rising for the last ten years. Among the combination of contributing factors are increasing breast cancer awareness, leading to earlier detection, and the development of many new drugs and techniques to battle the disease.

Cancers result from a wide variety of complicated errors in metabolism. To combat cancer, scientists first identify specific differences between cancer cells and normal cells and then look for ways of stopping the change from normal cell to cancer cell or of attacking the cancer cell once it has formed. Many drugs used to combat breast cancer, as well as other cancers, work by directing monoclonal antibodies against specific cell surface proteins that have been identified as being active in cancer. One protein found in many breast cancers is Human Epidermal Growth Factor 2 (HER2), a member of a larger class of epidermal growth factors that are implicated in many cancer types. These proteins are receptors that bind to specific ligands, causing rapid cell growth. Studies reveal that many breast cancer types show increased levels of HER2.

In breast cancer, HER2 causes aggressive tumor growth, so any drug that can stop its action can be a potent anticancer agent. One such potent weapon is a monoclonal antibody called trastuzumab, approved for use in 1998; it significantly increases the life expectancy of patients in both early-stage and metastatic breast cancer. The success of trastuzumab led to the creation of newer drugs, such as pertuzumab, which attacks the HER2 protein at a different site and also keeps it from interacting with other receptors that have been linked to cancer.

Several strategies use monoclonal antibodies to combat breast cancer, as shown in the figure. The antibody (shown in orange) may bind directly to the chemical growth factor before it binds to its receptor, as shown at the top of the figure. The antibody may also block the binding site on the receptor so the growth factor cannot bind. Many cellular effects are initiated by the dimerization of two cell receptors (Section 15.6D), and monoclonal antibodies can also stop that process. Some of the cell receptors that can lead to cancer are based on a tyrosine kinase (Section 15.6D), and monoclonal antibodies have been created that act as inhibitors of this activity (Chapter 15). Finally, new technologies are being developed that link a monoclonal antibody to a specific toxin. When the antibody binds to a critical cell receptor on a cancerous cell, the toxin is carried inside the cell and kills it.

Much progress is being made in development of individualized therapy in which profiling of a patient's specific cells lets doctors know which cell proteins are the culprits. Once they identify a specific protein target, they can use the proper combination of drugs. This ability is already making a significant impact in the survival rates of breast cancer patients, and we can expect even more progress in the years to come.

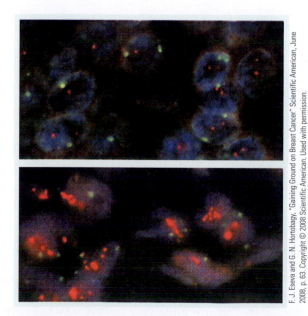

Red fluorescent tags spot the HER2 receptor. In cancer cells the proton is duplicated many times (bottom) compared to a control cell (top).

F. J. Eseva and G. N. Hortobagy, "Gaining Ground on Breast Cancer." Scientific American, June 2008, p. 63. Copyright © 2008 Scientific American. Used with permission.

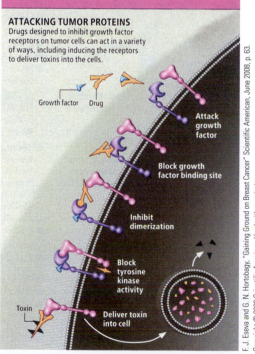

ATTACKING TUMOR PROTEINS
Drugs designed to inhibit growth factor receptors on tumor cells can act in a variety of ways, including inducing the receptors to deliver toxins into the cells.

Growth factor Drug

Attack growth factor

Block growth factor binding site

Inhibit dimerization

Block tyrosine kinase activity

Toxin

Deliver toxin into cell

F. J. Eseva and G. N. Hortobagy, "Gaining Ground on Breast Cancer." Scientific American, June 2008, p. 63. Copyright © 2008 Scientific American. Used with permission.

serum. Those antibodies, however, are not uniform. The antigen may have many epitopes, and the antisera contains a mixture of immunoglobulins with varying specificity for all the epitopes. Even antibodies to a single epitope usually have a variety of specificities.

Each B cell (and each progeny plasma cell) produces only one kind of antibody. In principle, each such cell should represent a potential source of a supply of homogeneous antibody by cloning. This is not possible in practice because lymphocytes do not grow continuously in culture. In the late 1970s, Georges Köhler and César Milstein developed a method to circumvent this problem, a feat for which they received the Nobel Prize in physiology in 1984. Their technique requires fusing lymphocytes that make the desired antibody with mouse myeloma cells. The resulting **hybridoma** (hybrid myeloma), like all cancer cells, can be cloned in culture (Figure 23.13) and produces the desired antibody. Because the clones are the progeny of a single cell, they produce homogeneous **monoclonal antibodies.** With this technique, it becomes possible to produce antibodies to almost any antigen in quantity. Monoclonal antibodies can, for instance, be used to assay for biological substances that can act as antigens. A striking example of their usefulness is in testing blood for the presence of HIV; this procedure has become routine to protect the public blood supply. Monoclonal antibodies are also commonly used in cancer treatment, as described in Chemical Connections 23B.

23.5 What Are T Cells and T-Cell Receptors?

A. T-Cell Receptors

Like B cells, T cells carry on their surface unique receptors that interact with antigens. We noted earlier that T cells respond only to protein antigens. An individual has millions of different T cells, each of which carries on its surface a unique T-cell receptor (TcR) that is specific for one antigen only. The TcR is a glycoprotein made of two subunits cross-linked by disulfide bridges. Like immunoglobulins, TcRs have constant (C) and variable (V)

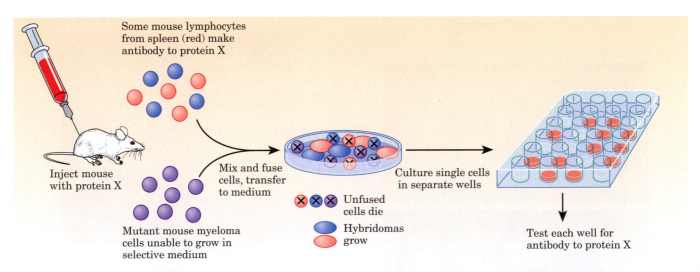

FIGURE 23.13 A procedure for producing monoclonal antibodies against a protein antigen X. A mouse is immunized against the antigen X, and some of its spleen lymphocytes produce antibody. The lymphocytes are fused with mutant myeloma cells that cannot grow in a given medium because they lack an enzyme found in the lymphocytes. Unfused cells die because lymphocytes cannot grow in culture and the mutant myeloma cells cannot survive in this medium. The individual cells are grown in culture in separate wells and tested for antibody to protein X.

regions. The antigen binding occurs on the variable region. The similarity in amino acid sequence between immunoglobulins (Ig) and TcR, as well as the organization of the polypeptide chains, makes TcR molecules members of the immunoglobulin superfamily.

There are, however, some fundamental differences between immunoglobulins and TcRs. For instance, immunoglobulins have four polypeptide chains, whereas TcRs contain only two subunits. Immunoglobulins can interact directly with antigens, but TcRs can interact with them only when the epitope of an antigen is presented by an MHC molecule. Lastly, immunoglobulins can undergo somatic mutation. This kind of mutation can occur in all body cells except the ones involved in sexual reproduction. Thus Ig molecules can increase their diversity by somatic mutation; TcRs cannot.

B. T-Cell Receptor Complex

A TcR is anchored in the membrane by hydrophobic transmembrane segments (Figure 23.14). TcR alone, however, is not sufficient for antigen binding. Also needed are other protein molecules that act as coreceptors and/or signal transducers. These molecules go under the name of CD3, CD4, and CD8, where "CD" stands for **cluster determinant.** TcR and CD together form the **T-cell receptor complex.**

The CD3 molecule adheres to the TcR in the complex not through covalent bonds but rather by intermolecular forces. It is a signal transducer because, upon antigen binding, CD3 becomes phosphorylated. This event sets off a signaling cascade inside the cell, which is carried out by different kinases. We saw a similar signaling cascade in neurotransmission (Section 16.5).

The CD4 and CD8 molecules act as **adhesion molecules** as well as signal transducers. A T cell has either a CD4 or a CD8 molecule to help bind the

Adhesion molecules Various protein molecules that help to bind an antigen to the T-cell receptor and dock the T cell to another cell via an MHC

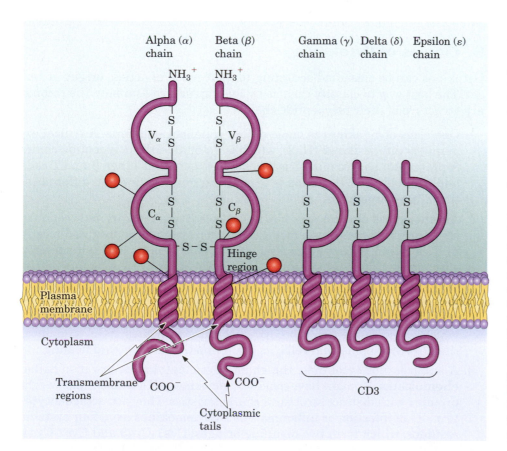

FIGURE 23.14 Schematic structure of a TcR complex. The TcR consists of two chains: α and β. Each has two extracellular domains: an amino-terminal V domain and a carboxyl-terminal C domain. Domains are stabilized by intrachain disulfide bonds between cysteine residues. The α and β chains are linked by an interchain disulfide bond near the cell membrane (hinge region). Each chain is anchored on the membrane by a hydrophobic transmembrane segment and ends in the cytoplasm with a carboxyl-terminal segment rich in cationic residues. Both chains are glycosylated (red spheres). The cluster determinant (CD) coreceptor consists of three chains: γ, δ, and ε. Each is anchored in the plasma membrane by a hydrophobic transmembrane segment. Each is also cross-linked by a disulfide bridge, and the carboxylic terminal is located in the cytoplasm.

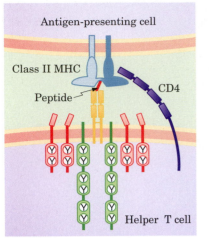

FIGURE 23.15 Interaction between helper T cells and antigen-presenting cells. Foreign peptides are displayed on the surface by MHC II proteins. These bind to the T-cell receptor of a helper T cell. A docking protein called CD4 helps link the two cells together.

Cytokine A glycoprotein that traffics between cells and alters the function of a target cell

antigen to the receptor and to dock the T cell to an APC or B cell (Figure 23.15). A unique characteristic of the CD4 molecule is that it binds strongly to a special glycoprotein that has a molecular weight of 120,000 (gp120). This glycoprotein exists on the surface of the human immunodeficiency virus (HIV). Through this binding to CD4, HIV can enter and infect helper T cells and cause AIDS. Helper T-cells die as a result of HIV infection, which depletes the T cell population so drastically that the immune system can no longer function. As a consequence, the body succumbs to opportunistic pathogen infections (Section 23.8).

23.6 How Is the Immune Response Controlled?

A. Nature of Cytokines

Cytokines are glycoprotein molecules that are produced by one cell but alter the function of another cell. They have no antigen specificity. Cytokines transmit intercellular communications between different types of cells at diverse sites in the body. They are short-lived and are not stored in cells.

Cytokines facilitate a coordinated and appropriate inflammatory response by controlling many aspects of the immune reaction. They are released in bursts, in response to all manner of insult or injury (real or perceived). They travel and bind to specific cytokine receptors on the surface of macrophages and B and T cells and induce cell proliferation.

One set of cytokines are called **interleukins** (**IL**s) because they communicate between and coordinate the actions of leukocytes (all kinds of white blood cells). Macrophages secrete IL-1 upon a bacterial infection. The presence of IL-1 then induces shivering and fever. The elevated body temperature both reduces the bacterial growth and speeds up the mobilization of the immune system. One leukocyte may make many different cytokines, and one cell may be the target of many cytokines.

B. Classes of Cytokines

Cytokines can be classified according to their mode of action, origin, or target. The best way to classify them is by their structure—namely, the secondary structure of their polypeptide chains.

1. One class of cytokines is made of four α-helical segments. A typical example is interleukin-2 (IL-2), which is a 15,000-MW polypeptide chain. A prominent source of IL-2 is T cells. IL-2 activates other B and T cells and macrophages. Its action is to enhance proliferation and differentiation of the target cells.

2. Another class of cytokines has only β-pleated sheets in its secondary structure. Tumor necrosis factor (TNF), for example, is produced mostly by T cells and macrophages. Its name comes from its ability to destroy susceptible tumor cells through lysis, after it binds to receptors on the tumor cell.

3. A third class of cytokines has both α-helix and β-sheet secondary structures. A representative of this class is epidermal growth factor (EGF), which is a cysteine-rich protein. As its name indicates, EGF stimulates the growth of epidermal cells; its main function is in wound healing.

4. A subgroup of cytokines, the chemotactic cytokines, are also called **chemokines.** Humans have some 40 chemokines, all low-molecular-weight proteins with distinct structural characteristics. They attract leukocytes to a site of infection or inflammation. All chemokines have four cysteine residues, which form two disulfide bridges: Cys1-Cys3 and Cys2-Cys4.

Chemokine A low-molecular-weight polypeptide that interacts with special receptors on target cells and alters their functions

Chemical Connections 23C

Immunization

Smallpox was a scourge over many centuries, with each outbreak leaving many people dead and others maimed by deep pits on the face and body. A form of immunization was practiced in ancient China and the Middle East by intentionally exposing people to scabs and fluids of lesions of victims of smallpox. This practice was known as variolation in the Western world, where the disease was called variola. Variolation was introduced to England and the American Colonies in 1721.

Edward Jenner, an English physician, noted that milkmaids who had contracted cowpox from infected cows seemed to be immune to smallpox. Cowpox was a mild disease, whereas smallpox could be lethal. In 1796, Jenner performed a potentially deadly experiment: He dipped a needle into the pus of a cowpox-infected milkmaid and then scratched a boy's hand with the needle. Two months later, Jenner injected the boy with a lethal dose of smallpox-carrying agent. The boy survived and did not develop any symptoms of the disease. The word spread, and Jenner was soon established in the immunization business. When the news reached France, skeptics there coined a derogatory term, *vaccination*, which literally means "encowment." The derision did not last long, and the practice was soon adopted worldwide.

A century later, in 1879, Louis Pasteur found that tissue infected with rabies has much weakened virus in it. When injected into patients, it elicits an immune response that protects against rabies. Pasteur named these attenuated, protective antigens *vaccines* in honor of Jenner's work. Today, immunization and vaccination are synonymous.

Vaccines are available for several diseases, including polio, measles, and smallpox, to name a few. A vaccine may be made of either dead viruses or bacteria or weakened ones. For example, the Salk polio vaccine is a polio virus that has been made harmless by treatment with formaldehyde; it is given by intramuscular injection. In contrast, the Sabin polio vaccine is a mutated form of the wild virus; the mutation makes the virus sensitive to temperature. The mutated live virus is taken orally. The body temperature and the gastric juices render it harmless before it penetrates the bloodstream.

Many cancers have specific carbohydrate cell-surface tumor markers. If such antigens could be introduced by injection without endangering the individual, they could provide an ideal vaccine. Obviously, one cannot use live or even attenuated cancer cells for vaccination. However, the expectation is that a synthetic analog of a natural tumor marker will elicit the same immune reaction that the original cancer-surface markers do. Thus injecting such innocuous synthetic compounds may prompt the body to produce immunoglobulins that can provide a cure for the cancer—or at least prevent the occurrence of it. A compound called 12:13 dEpoB, a derivative of the macrolide epothilon B, is currently being explored for its potential to become an anticancer vaccine.

Vaccines change lymphocytes into plasma cells that produce large quantities of antibodies to fight any invading antigens. This is only the immediate, short-term response, however. Some lymphocytes become memory cells rather than plasma cells. These memory cells do not secrete antibodies, but instead store them to serve as a detecting device for future invasion of the same foreign cells. In this way, long-term immunity is conferred. If a second invasion occurs, these memory cells divide directly into both antibody-secreting plasma cells and more memory cells. This time the response is faster because it does not have to go through the process of activation and differentiation into plasma cells, which usually takes two weeks.

Smallpox, which was once one of humanity's worst-scourges, has been totally wiped out, and smallpox vaccination is no longer required. Because smallpox is a potential weapon of bioterrorism, the U.S. government has recently begun gearing up its smallpox vaccine production.

This school-age boy is receiving an intramuscular vaccine.

Courtesy of the CDC.

Chemokines have a variety of names, such as interleukin-8 (IL-8) and monocyte chemotactic proteins (MCP-1 to MCP-4). Chemokines interact with specific receptors, which consist of seven helical segments coupled to GTP-binding proteins.

C. Mode of Action of Cytokines

When a tissue is injured, leukocytes are rushed to the inflamed area. Chemokines help leukocytes migrate out of the blood vessels to the site of injury. There leukocytes, in all their forms—neutrophils, monocytes, lymphocytes—accumulate and attack the invaders by engulfing them (phagocytosis) and later killing them. Other phagocytic cells, the macrophages, which reside in the tissues and thus do not have to migrate, do the same. These activated phagocytic cells destroy their prey by releasing endotoxins that kill bacteria and/or by producing such highly reactive oxygen intermediates as superoxide, singlet oxygen, hydrogen peroxide, and hydroxyl radicals.

Chemokines are also major players in chronic inflammation, autoimmune diseases, asthma, and other forms of allergic inflammation, and even in transplant rejection.

> Singlet oxygen is a form of O_2 where the outer electrons are raised to a higher energy level state. There are two common forms of singlet oxygen and both are reactive oxygen species.

23.7 How Does the Body Distinguish "Self" from "Nonself"?

One major problem facing body defenses is how to recognize the foreign body as "not self" and thereby avoid attacking the "self"—that is, the healthy cells of the organism.

A. Selection of T and B Cells

The members of the adaptive immunity system, B cells and T cells, are all specific and have memory, so they target only truly foreign invaders. The T cells mature in the thymus gland. During the maturation process, those T cells that fail to recognize and interact with MHC and thus cannot respond to foreign antigens are eliminated through a selection process. They essentially die by neglect. T cells that express receptors (TcR) that are prone to interact with normal self antigens are also eliminated through the selection process (Figure 23.16). Thus the activated T cells leaving the thymus gland carry TcRs that can respond to foreign antigens. Even if some T cells prone to react with self antigen escape the selection detection, they can be deactivated through the signal transduction system that, among other functions, performs tyrosine kinase activation and phosphatase deactivation, similar to those processes seen in adrenergic neurotransmitter signaling (Section 16.5).

Similarly, the maturation of B cells in the bone marrow depends on the engagement of their receptors, BcR, with antigen. Those B cells that are prone to interact with self antigen are also eliminated before they leave the bone marrow. As with T cells, many signaling pathways control the proliferation of B cells. Among them, the activation by tyrosine kinase and the deactivation by phosphatase provide a secondary control.

B. Discrimination of the Cells of the Innate Immunity System

The first line of defense is the innate immunity system, in which cells such as natural killer cells or macrophages have no specific targets and no memory of which epitope signals danger. Nevertheless, these cells must somehow

FIGURE 23.16 A two-stage process leads to the growth and differentiation of T cells. (*a*) In the absence of antigen, proliferation of T cells does not take place. Those T-cell lines die by neglect. (*b*) In the presence of antigen alone, the T-cell receptor binds to antigen presented on the surface of a macrophage cell by the MHC protein. There is still no proliferation of T cells because the second signal is missing. In this way the body can avoid an inappropriate response to its own antigens. This process occurs early in the development of T cells, effectively eliminating those cells that would otherwise be activated by the self antigens. (*c*) When an infection occurs, a B7 protein is produced in response to the infection. The B7 protein on the surface of the infected cell binds to a CD28 protein on the surface of the immature T cell, giving the second signal that allows it to grow and proliferate. (Adapted from "How the Immune System Recognizes Invaders," by Charles A. Janeway, Jr.; illustration by Ian Warpole, *Scientific American*, September 1993.) **Go to this book's companion website at www.cengage.com/chemistry/bettelheim to explore an interactive version of this figure.**

discriminate between normal and abnormal cells to identify targets. The mechanism by which this identification is accomplished has only recently been explored and is not yet fully understood. The main point is that the cells of innate immunity have two kinds of receptors on their surface: an **activating receptor** and an **inhibitory receptor.** When a healthy cell of the body encounters a macrophage or a natural killer cell, the inhibitory receptor on the latter's surface recognizes the epitope of the normal cell, binds to it, and prevents the activation of the killer cell or macrophage. Conversely, when a macrophage encounters a bacterium with a foreign antigen on its surface, the antigen binds to the activating receptor of the macrophage. This ligand binding prompts the macrophage to engulf the bacterium by phagocytosis. Such foreign antigens may be the lipopolysaccharides of gram-negative bacteria or the peptidoglycans of gram-positive bacteria.

When a cell is infected, damaged, or transformed into a malignant cell, the epitopes that signaled healthy cells diminish greatly, and unusual epitopes

Chemical Connections 23D

Antibiotics: A Double-Edged Sword

Living in the modern world, you have undoubtedly taken advantage of antibiotics. Indeed, many of the diseases of the past have been all but eradicated by these drugs, which can stop a bacterial life cycle in its tracks. Common infections that may have proved fatal at the beginning of the 1900s are often treated successfully today with penicillin or another common antibiotic, such as erythromycin or cephalosporin.

However, antibiotics can also cause several problems. Many people are allergic to penicillin and its derivatives, and these antibiotic allergies can be very potent. A person may take an antibiotic once and have no symptoms. A subsequent use of the same antibiotic may cause a severe skin rash or hives. A third exposure could be fatal. Indiscriminate use of antibiotics can also be harmful. Many diseases are caused by viruses, which do not respond to treatment with antibiotics. However, patients do not want to hear that there is nothing they can do except wait out the disease, so they often are given antibiotics. Antibiotics are also prescribed before the exact nature of the infection is known. This indiscriminate use of antibiotics is the major cause of the increasing incidence of drug-resistant microorganisms.

One disease that has flourished due to misuse of antibiotics is gonorrhea. One strain of *Neisseria gonorrhoeae* produces β-lactamase, an enzyme that degrades penicillin. These strains are referred to as PPNG, penicillinase-producing *N. gonorrhoeae*. Before 1976, almost no cases of PPNG were reported in the United States. Today, thousands of cases occur nationwide. The source of the problem has been traced to military bases in the Philippines, where soldiers acquired the disease from prostitutes. Among prostitutes, there is a common practice of using low doses of antibiotics in an attempt to prevent the spread of sexually transmitted diseases. In reality, constant use of antibiotics has just the opposite effect—it causes development of drug-resistant strains.

An antibiotic commonly given to young children for earaches is amoxicillin, a penicillin derivative. Severe and repeated earaches can cause hearing loss, so parents are often quick to put their children on antibiotics. There are two downsides, however, to overuse of this antibiotic. First, it has minimal efficiency because the bacteria causing the earaches are localized inside the ear, where the antibiotic has little access. Second, overuse of the antibiotic affects the inside of the growing teeth, causing them to remain soft and leading to future dental problems.

When a person with a bacterial infection uses antibiotics early in the infection, he or she never has the opportunity to elicit a true immune response. For this reason, the person will be susceptible to the same disease again and again. This issue is seen nowadays with diseases such as strep throat, which many people have almost yearly. Some doctors are trying to avoid prescribing antibiotics until their patients have had a chance to fight the disease on their own. Some patients are purposely avoiding using antibiotics for the same reason. While such a strategy is attractive and intuitive, if we take a closer look at strep throat, we can see yet another side to this story.

Rheumatic fever is a complication of untreated strep throat. It is characterized by fever and widespread inflammation of the joints and heart. These effects are produced by the body's immune response to the M protein of the group A streptococci. The M protein resembles a major protein in heart tissue. As a result, the antibodies attack the heart valves as well as the bacterial M protein. About 3% of individuals who do not get treatment with antibiotics when they have strep throat develop rheumatic fever. About 40% of patients with rheumatic fever develop heart valve damage, which may not become evident for 10 years or more. The best way to avoid this complication is early treatment of strep throat with antibiotics.

In summary, antibiotics are a very important weapon in our arsenal against disease, but they should not be used indiscriminately. If they are used, the entire course of treatment should be taken to completion. The last thing you would want to do is kill off most—but not all—of the bacteria you were infected with. This would leave behind a few "superbacteria," which would then be drug-resistant.

Cipro is a common antibiotic, often used to fight strep throat.

are presented on the surface of these cancer cells. The effect is that fewer inhibitory receptors of macrophages or natural killer cells can bind to the surface of the target cell and more activating receptors find inviting ligands. As a consequence, the balance shifts in favor of activation and the macrophages or killer cells will do their job.

C. Autoimmune Diseases

In spite of the safeguards in the body intended to prevent acting against "self," in the form of healthy cells, many diseases exist in which some part of a pathway in the immune system goes awry. The skin disease psoriasis is thought to be a T-cell–mediated disease in which cytokines and chemokines play an essential role. Other autoimmune diseases, such as myasthenia gravis, rheumatoid arthritis, multiple sclerosis (Chemical Connections 13D), and insulin-dependent diabetes (Chemical Connections 16F), also involve cytokines and chemokines. Allergies are another example of malfunctioning of the immune system. Pollens and animal furs are allergens that can provoke asthma attacks. Some people are so sensitive to certain food allergens that even a knife used in smearing peanut butter may prove fatal in a person known to be allergic to peanuts.

The major drug treatment for autoimmune diseases involves the glucocorticoids, the most important of which is cortisol (Section 13.10A). It is a standard therapy for rheumatoid arthritis, asthma, inflammatory bone diseases, psoriasis, and eczema. The beneficial effects of glucocorticoids are overshadowed, however, by their undesirable side effects, which include osteoporosis, skin atrophy, and diabetes. Glucocorticoids regulate the synthesis of cytokines either directly, by interacting with their genes, or indirectly, through transcription factors.

Macrolide drugs are used to suppress the immune system during tissue transplantation or in the case of certain autoimmune diseases. Drugs such as cyclosporin A or rapamycin bind to receptors in the cytosol and, through secondary messengers, inhibit the entrance of nuclear factors into the nucleus. Normally those nuclear factors signal a need for transcription, so their absence prevents the transcription of cytokines—for example, interleukin-2.

> Macrolide drugs are a class of drug, mostly antibiotics, that all have a large macrocyclic lactone ring system. Common examples are erythromycin and clarithromycin

23.8 How Does the Human Immunodeficiency Virus Cause AIDS?

Human immunodeficiency virus (HIV) is the most infamous of the retroviruses, as it is the causative agent of acquired immunodeficiency syndrome (AIDS). This disease affects more than 40 million people worldwide and has continually thwarted attempts to eradicate it. The best medicine today can slow it down, but nothing has been able to stop AIDS.

The HIV genome is a single-stranded RNA that has a number of proteins packed around it, including the virus-specific reverse transcriptase and protease. A protein coat surrounds the RNA–protein assemblage, giving the overall shape of a truncated cone. Finally, a membrane envelope encloses the protein coat. The envelope consists of a phospholipid bilayer formed from the plasma membrane of cells infected earlier in the life cycle of the virus, as well as some specific glycoproteins, such as gp41 and gp120, as shown in Figure 23.17.

HIV offers a classic example of the mode of operation of retroviruses. The HIV infection begins when the virus particle binds to receptors on the

(a)

(b)

gp120
gp41
Lipid membrane
p17 matrix
p24 capsid
Reverse transcriptase
RNA in p7 nucleocapsid
Protease
Integrase

Eye of Science/Photo Researchers, Inc.

FIGURE 23.17 The architecture of HIV. (*a*) The RNA genome is surrounded by P7 nucleocapsid proteins and by several viral enzymes—namely, reverse transcriptase, integrase, and protease. The truncated cone consists of P24 capsid protein subunits. The P17 matrix (another layer of protein) lies inside the envelope, which consists of a lipid bilayer and glycoproteins such as gp41 and gp120. (*b*) An electron micrograph shows both mature virus particles, in which the core (the truncated cone) is visible, and immature virus particles, in which it is not.

surface of a cell (Figure 23.18). The viral core is inserted into the cell and partially disintegrates. The reverse transcriptase catalyzes the production of DNA from the viral RNA. The viral DNA is integrated into the DNA of the host cell. The DNA, including the integrated viral DNA, is transcribed to RNA. Smaller RNAs are produced first, specifying the amino acid sequences of viral regulatory proteins. Larger RNAs, which specify the amino acid sequences of viral enzymes and coat proteins, are made next. The viral protease assumes particular importance in the budding of new virus particles. Both the viral RNA and viral proteins are included in the budding virus, as is some of the membrane of the infected cell.

A. HIV's Ability to Confound the Immune System

Why is HIV so deadly and so hard to stop? Many viruses, such as adenovirus, cause nothing more than the common cold; others, such as the virus that causes severe accute respiratory syndrome (SARS), are deadly. At the same time, we have seen the complete eradication of the deadly SARS virus, whereas adenovirus is still with us. HIV has several characteristics that lead to its persistence and eventual lethality. Ultimately, it is deadly because of its targets, the helper T cells. The immune system is under constant attack by the virus, and millions of helper T cells and killer T cells are called up to fight billions of virus particles. Through degradation of the T-cell membrane via budding and the activation of enzymes that lead to cell death, the T-cell count diminishes to the point that the infected person is no longer able to mount a suitable immune response. As a result, the individual eventually succumbs to pneumonia or another opportunistic disease.

There are many reasons that the disease is so persistent. For example, it is slow acting. SARS was eradicated quickly because the virus was quick to act, making it easy to find infected people before they had a chance to spread the disease. In contrast, HIV-infected individuals can go years before they become aware that they have the disease. However, this is only a small part of what makes HIV so difficult to kill.

FIGURE 23.18 HIV infection begins when the virus particle binds to CD4 receptors on the surface of the cell (Step 1). The viral core is inserted into the cell and partially disintegrates (Step 2). The reverse transcriptase catalyzes the production of DNA from the viral RNA. The viral DNA is integrated into the DNA of the host cell (Step 3). The DNA, including the integrated viral DNA, is transcribed to RNA (Step 4). Smaller RNAs are produced first, specifying the amino acid sequence of viral regulatory proteins (Step 5). Larger RNAs, which specify the amino acid sequences of viral enzymes and coat proteins, are made next (Step 6). The viral protease assumes particular importance in the budding of new virus particles (Step 7). Both the viral RNA and viral proteins are included in the budding virus, as is some of the membrane of the infected cell (Step 8).

HIV is difficult to kill because it is difficult to find. For an immune system to fight a virus, it needs to locate specific macromolecules that can be bound to antibodies or T-cell receptors. The reverse transcriptase of HIV is very inaccurate in replication. The result is rapid mutation of HIV, a situation that presents a considerable challenge to those who want to devise treatments for AIDS. The virus mutates so rapidly that multiple strains of HIV may be present in a single individual.

Another trick the virus plays is a conformational change of the gp120 protein when it binds to the CD4 receptor on a T cell. The normal shape of the gp120 monomer may elicit an antibody response, but these antibodies are largely ineffective. The gp120 forms a complex with gp41 and changes shape when it binds to CD4. It also binds to a secondary site on the T cell that normally binds to a cytokine. This change exposes a part of gp120 that was previously hidden and, therefore, cannot elicit antibodies.

HIV is also adept at evading the innate immunity system. Natural killer cells attempt to attack the virus, but HIV binds a particular cell protein, called cyclophilin, to its capsid, which blocks the antiviral agent restriction factor-1. Another of HIV's proteins blocks the viral inhibitor called CEM-15, which normally disrupts the viral life cycle.

Lastly, HIV hides from the immune system by cloaking its outer membrane in sugars that are very similar to the natural sugars found on most of its host's cells, rendering the immune system blind to it.

B. The Search for a Vaccine

The attempt to find a vaccine for HIV is akin to the search for the Holy Grail, and it has met with about as much success. One strategy for using a vaccine to stimulate the body's immunity to HIV is shown in Figure 23.19. DNA for a unique HIV gene, such as the *gag* gene, is injected into muscle. The *gag* gene leads to the Gag protein, which is taken up by antigen-presenting cells and displayed on their cell surfaces. This then elicits the cellular immune response, stimulating killer and helper T cells. It also stimulates the humoral immune response, spurring production of antibodies. Figure 23.19 also shows a second phase of the treatment, a booster shot of an altered adenovirus that carries the *gag* gene.

Unfortunately, most attempts at making antibodies have proved unsuccessful to date. The most thorough attempt was made by the VaxGen Company, which carried the research through the third stage of clinical trials, testing the vaccine on more than 1000 high-risk people and comparing them to 1000 individuals who did not receive the vaccine. In the study, 5.7% of the people who received the vaccine eventually became infected, compared to 5.8% of the placebo group. Many people analyzed the data and, despite attempts to show a better response in certain ethnic groups, the trials had to be declared a failure. The vaccine, called AIDSVAX, was based on gp120.

C. Antiviral Therapy

While the search for an effective vaccine continued with little to minor success, pharmaceutical companies flourished by designing drugs that would inhibit retroviruses. By 1996, there were 16 drugs used to inhibit either the HIV reverse transcriptase or the protease. Several others are in clinical trials, including drugs that target gp41 and gp120 in an attempt to prevent entry of the virus. A combination of drugs to inhibit retroviruses has been dubbed **highly active antiretroviral therapy (HAART).** Initial attempts at HAART were very successful, driving the viral load almost to the point of being undetectable, with the concomitant rebounding of the CD4 cell population. However, as always seems to be the case with HIV, it later turned out that while the virus was knocked down, it was not knocked out. HIV remained in hiding in the body and would bounce back as soon as the therapy was stopped. Thus the best-case scenario for an AIDS patient was a lifetime of expensive drug therapies. In addition, long-term exposure to HAART was found to cause constant nausea and anemia, as well as diabetes symptoms, brittle bones, and heart disease.

D. A Second Chance for Antibodies

In the wake of the realization that patients could not stay on the HAART program indefinitely, several researchers attempted to combine HAART with vaccination. Even though most of the vaccines were not found to be effective alone, they proved more effective in combination with HAART. In addition, once on the vaccine, patients were able to take a rest from the other drugs, giving their bodies and minds time to recover from the side effects of the antiviral therapy.

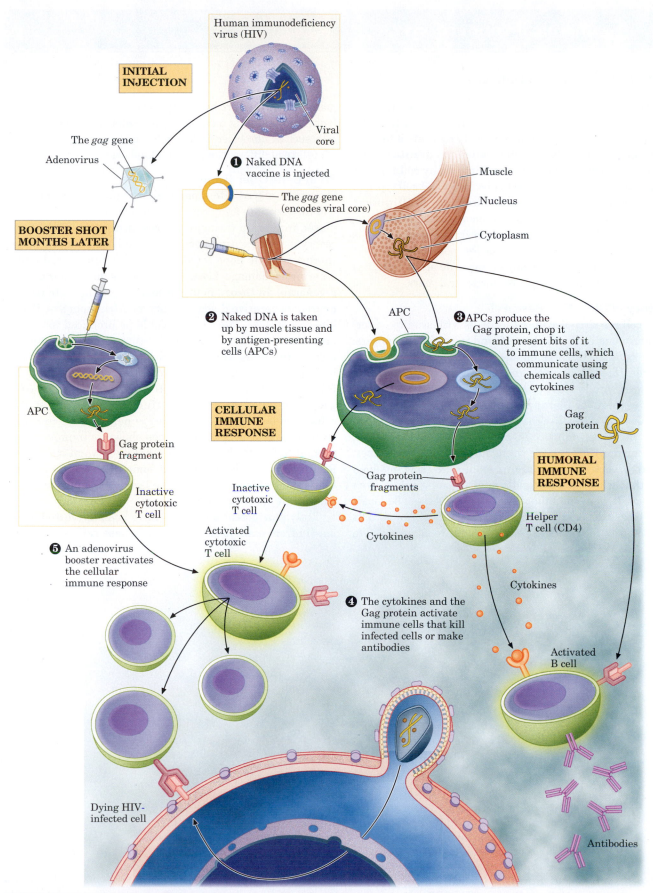

FIGURE 23.19 One strategy for an AIDS vaccine. (Reprinted by permission from Ezzell C., "Hope in a Vial," *Sci. Am.* 39–45 June 2002.)

Why Are Stem Cells Special?

Stem cells are the precursors of all other cell types, including T and B lymphocytes. These undifferentiated cells have the ability to form any cell type as well as to replicate into more stem cells. Stem cells are often called **progenitor cells** due to their ability to differentiate into many cell types. A **pluripotent** stem cell is able to give rise to all cell types in an embryo or an adult. Some cells are called **multipotent** because they can differentiate into more than one cell type, but not into all cell types. The further from a zygote a cell is in the course of its development, the less potent the cell type. The use of stem cells, especially **embryonic stem cells,** has been an exciting field of research for several years.

History of Stem Cell Research

Stem cell research began in the 1970s with studies on teratocarcinoma cells, which are found in testicular cancers. These cells are bizarre blends of differentiated and undifferentiated cells. Such **embryonal carcinoma cells (EC)** were found to be pluripotent, which led to the idea of using them for therapy. However, this line of research was suspended because the cells came from tumors, which made their use dangerous, and they were **aneuploid,** which means they had the wrong number of chromosomes.

Early work with embryonic stem cells (ES) involved cells that were grown in culture after being taken from embryos. It was found that these stem cells could be maintained for long periods. In contrast, most differentiated cells will not grow for extended periods in culture. Stem cells are maintained in culture by addition of certain factors, such as leukemia-inhibiting factor (LIF) or feeder cells (nonmitotic cells such as fibroblasts). Once released from these controls, ES cells will differentiate into all kinds of cells.

Stem Cells Offer Hope

Stem cells placed into a particular tissue, such as blood, will differentiate and grow into blood cells. Others, when placed into brain tissue, will grow brain cells. This discovery is very exciting because it had been believed that there was little hope for patients with spinal cord and other nerve damage, because these cells do not normally regenerate. In theory, neurons could be produced to treat neurodegenerative diseases, such as Alzheimer's or Parkinson's disease. Muscle cells could be produced to treat muscular dystrophies and heart disease. In one study, mouse stem cells were injected into a mouse heart that had undergone a myocardial infarction. The cells spread from an unaffected region into the infarcted zone and began to grow new heart tissue. Human pluripotent stem cells have been used to regenerate nerve tissue in rats with nerve injuries and have been shown to improve motor and cognitive ability in rats that underwent a stroke. Results such as these have led some scientists to claim that stem cell technology will be the most important advancement in science since cloning.

Truly pluripotent stem cells have been harvested primarily from embryonic tissue, and these cells show the

Pluripotent embryonic stem cells can be grown in cell culture. They can be maintained in an undifferentiated state by growing them on certain feeder cells, such as fibroblasts or by using leukemia inhibitory factor (LIF). When removed from the feeder cells or when the LIF is removed, they begin to differentiate in a wide variety of tissue types, which could then be harvested and grown for tissue therapy. (Taken from Donovan P. J., and Gearhart J., *Nature,* 414:92–97, 2001)

Chemical Connections 23E

Why Are Stem Cells Special? *(continued)*

greatest ability to differentiate into various tissues and to reproduce in cell culture. Stem cells have also been taken from adult tissues, as some stem cells are always present in an organism, even at the adult stage. These cells are usually multipotent, as they can form several different cell types, but are not as versatile as ES cells. Many scientists believe that the ES cells represent a better source for tissue therapy than adult stem cells for this reason.

The acquisition and use of stem cells can be related to a technique called cell reprogramming, which is a necessary component of whole-mammal cloning, such as the process that created the world's most famous sheep, Dolly. Most somatic cells in an organism contain the same genes, but the cells develop as different tissues with wildly different patterns of gene expression. A mechanism that alters expression of genes without changing the actual DNA sequence is called an **epigenetic** mechanism. An epigenetic state of the DNA in a cell is a heritable trait that allows a "molecular memory" to exist in the cells. In essence, a liver cell remembers where it came from and will continue to divide and remain a liver cell. These epigenetic states involve methylation of cytosine–guanine dinucleotides and interactions with proteins of chromatin (Section 17.3C). Mammalian genes possess an additional level of epigenetic information called **imprinting,** which allows the DNA to retain a molecular memory of its germ-line origin. The paternal DNA is imprinted differently than the maternal DNA. In normal development, only DNA that came from both parents would be able to combine and lead to a viable offspring.

The epigenetic states of somatic cells are generally locked in a way that the differentiated tissues remain stable. The key to whole-organism cloning was the ability to erase the epigenetic state and return to the state of a fertilized egg, which has the potential to produce all cell types. If the nucleus of a somatic cell is injected into a recipient oocyte, the epigenetic state of the DNA can be reprogrammed, or at least "partially" reprogrammed. The molecular memory is erased, and the cell begins to behave like a true zygote. This development can be used to derive pluripotent stem cells or to transfer a blastocyst into a mother-carrier for growth and development. In November 2001, the first human cloned blastocyst was created in this way, with the aim of growing enough cells to harvest pluripotent stem cells for research.

Currently, debate continues to rage worldwide over the use of embryonic stem cells. The issue at hand is one of ethics and the definition of life. Embryonic stem cells come from many sources, including aborted fetuses, umbilical cords, and embryos from in vitro fertilization clinics. The report of the cloned human embryonic cells merely added to the controversy. The U.S. government has banned funding for stem cell research, but allows research to continue on all existing embryonic cell lines. Some big questions that people will have to answer are the following: Do a few cells created by therapeutic cloning of your own somatic cells constitute life? If these cells do constitute life, do they have the same rights as a human being conceived naturally? If it were possible, should someone be allowed to grow his or her own therapeutic clone into an adult?

E. The Future of Antibody Research

Early attempts at creating a vaccine appear to have failed because the vaccine elicited too many useless antibodies. What patients need is a **neutralizing antibody,** one capable of completely eliminating its target. Researchers discovered a patient who had HIV for six years, but never developed AIDS. They then studied his blood and found a rare antibody, which they labeled **b12.** In laboratory trials, **b12** was found to stop most strains of HIV. What made b12 different from the other antibodies? Structural analysis revealed that this antibody has a different shape from a normal immunoglobulin. It has sections of long tendrils that fit into a fold in gp120. This fold in gp120 cannot mutate very much; otherwise, the protein will not be able to dock properly with the CD4 receptor.

Another antibody was found in a different patient who seemed resistant to HIV. This antibody was actually a dimer and had a shape more like an "I" than the traditional "Y". This antibody, called **2G12,** recognizes some of the sugars on HIV's outer membrane that are unique to HIV.

By identifying a few such antibodies, researchers have been able to search for a vaccine in the opposite direction from the normal way. In **retrovaccination,** researchers have the antibody and need to find a vaccine to elicit it, instead of injecting vaccines and noting which antibodies they produce.

Summary

⭕WL End-of-chapter problems identified in blue are assignable in **GOB OWL.**

Section 23.1 How Does the Body Defend Itself from Invasion?

- The human immune system protects us against foreign invaders. It consists of two parts: (1) the natural resistance of the body, called innate immunity, and (2) adaptive or acquired immunity.
- **Innate immunity** is nonspecific. **Macrophages** and **natural killer (NK) cells** are cells of innate immunity that function as police officers.
- **Adaptive** or **acquired immunity** is highly specific, being directed against one particular invader.
- Acquired immunity (known as the immune system) also has memory, unlike innate immunity.

Section 23.2 What Organs and Cells Make Up the Immune System? Problem 23.7

- The principal cellular components of the immune system are white blood cells, or **leukocytes**. The specialized leukocytes in the lymph system are called **lymphocytes**. They circulate mostly in the **lymphoid organs**.
- The **lymph** system is a collection of vessels extending throughout the body and connected to the interstitial fluid on the one hand and to the blood vessels on the other hand.
- Lymphocytes that mature in the bone marrow and produce soluble immunoglobulins are **B cells**. Lymphocytes that mature in the thymus gland are **T cells**.

Section 23.3 How Do Antigens Stimulate the Immune System?

- **Antigens** are large, complex molecules of foreign origin. An antigen may be a bacterium, a virus, or a toxin.
- An antigen may interact with antibodies, T-cell receptors (TcR), or major histocompatibility complex (MHC) molecules. All three types of molecules belong to the **immunoglobulin superfamily.**
- An **epitope** is the smallest part of an antigen that binds to antibodies, TcRs, and MHCs.

Section 23.4 What Are Immunoglobulins?
Problems 23.19, 23.22

- Antibodies are **immunoglobulins**. These water-soluble glycoproteins are made of two heavy chains and two light chains. All four chains are linked together by disulfide bridges.

- Immunoglobulins contain variable regions in which the amino acid composition of each antibody is different. These regions interact with antigens to form insoluble large aggregates.
- A large variety of antibodies are synthesized by a number of processes in the body.
- During B-cell development, the **variable regions** of the heavy chains are assembled by a process called V(J)D recombination.
- Mutations on these new genes create even greater diversity. Somatic hypermutation (SHM) that creates a point mutation (only one nucleotide) is one way. Patches of mutation introduced into the V(J)D gene constitute gene conversion (GC) mutation.
- Immunoglobulins respond to an antigen over a long span of time, lasting for weeks and even months.
- All antigens—whether proteins, polysaccharides, or nucleic acids—interact with soluble immunoglobulins produced by B cells.

Section 23.5 What Are T Cells and T-Cell Receptors?

- Protein antigens interact with T cells. The binding of the epitope to the TcR is facilitated by MHC, which carries the epitope to the T-cell surface, where it is presented to the receptor.
- Upon epitope binding to the receptor, the T cell is stimulated. It proliferates and can differentiate into (1) killer T cells, (2) memory cells, or (3) helper T cells.
- The TcR has a number of helper molecules, such as CD4 or CD8, that enable it to bind the epitope tightly and to bind to other cells via MHC proteins.
- CD (**cluster determinant**) molecules also belong to the immunoglobulin superfamily.
- Antibodies can recognize all types of antigens, but TcRs recognize only peptide antigens.

Section 23.6 How Is the Immune Response Controlled? Problem 23.37

- The control and coordination of the immune response are handled by **cytokines**, which are small protein molecules.
- Chemotactic cytokines, the **chemokines**, such as interleukin-8, facilitate the migration of leukocytes from blood vessels into the site of injury or inflammation. Other cytokines activate B and T cells and macrophages, enabling them to engulf the foreign body, digest it, or destroy it by releasing special toxins.

- Some cytokines, such as tumor necrosis factor (TNF), can lyse tumor cells.

Section 23.7 How Does the Body Distinguish "Self" from "Nonself"?

- A number of mechanisms in the body ensure that the body recognizes "self."
- In adaptive immunity, T and B cells that are prone to interact with "self" antigens are eliminated.
- In innate immunity, two kinds of receptors exist on the surfaces of T and B cells: an **activating receptor** and an **inhibitory receptor**. The inhibitory receptor recognizes the epitope of a normal cell, binds to it, and prevents the activation of the killer T cell or macrophage.
- Many autoimmune diseases are T-cell–mediated diseases, in which cytokines and chemokines play an essential role.

- The standard drug treatment for autoimmune diseases involves glucocorticoids, which prevent the transcription and hence the synthesis of cytokines.

Section 23.8 How Does the Human Immunodeficiency Virus Cause AIDS?
Problem 23.54

- HIV is a retrovirus that enters helper T cells.
- The virus weakens the immune system by destroying these helper T cells through damage to their cell membranes and activation of enzymes that cause apoptosis.
- HIV has been studied for 25 years in an attempt to find a cure, but none has been discovered as yet. The virus hides from the host's immune system and mutates so frequently that no effective antibody response can be mounted.
- A combination of therapies combining enzyme inhibitors and antibodies has achieved the most success.

Problems

■ Indicates problems that are assignable in **GOB OWL**.

Blue numbered problems are applied.

Go to this book's companion website at **www.cengage.com/chemistry/bettelheim** for interactive versions of the How To tutorials and Active Figures, and to quiz yourself on this chapter.

Section 23.1 How Does the Body Defend Itself from Invasion?

23.1   Give two examples of external innate immunity in humans.

23.2   Which form of immunity is characteristic of vertebrates only?

23.3   How does the skin fight bacterial invasion?

23.4   T-cell receptors and MHC molecules both interact with antigens. What is the difference in the mode of interaction between the two?

23.5   What differentiates innate immunity from adaptive (acquired) immunity?

Section 23.2 What Organs and Cells Make Up the Immune System?

23.6   Where in the body do you find the largest concentration of antibodies as well as T cells?

23.7   ■ Where do T and B cells mature and differentiate?

23.8   What are memory cells? What is their function?

23.9   What are the favorite targets of macrophages? How do they kill the target cells?

Section 23.3 How Do Antigens Stimulate the Immune System?

23.10   Would a foreign substance, such as aspirin (MW 180), be considered an antigen by the body?

23.11   What kind of antigen does a T cell recognize?

23.12   What is the smallest unit of an antigen that is capable of binding to an antibody?

23.13   How does the body process antigens to be recognized by class II MHC?

23.14   What role do MHC molecules play in the immune response of the ABO blood groups?

23.15   To which class of compounds do MHCs belong? Where would you find them?

23.16   What is the difference in function between class I and class II MHC molecules?

Section 23.4 What are Immunoglobulins?

23.17   When a foreign substance is injected in a rabbit, how long does it take to find antibodies against the foreign substance in the rabbit serum?

23.18   Distinguish among the roles of the IgA, IgE, and IgG immunoglobulins.

23.19   ■ (a) Which immunoglobulin has the highest carbohydrate content and the lowest concentration in the serum?

   (b) What is its main function?

23.20   Chemical Connections 12D states that the antigen in the red blood cells of a person with B-type blood is a galactose unit. Show schematically how the antibody of a person with A-type blood would aggregate the red blood cells of a B-type person if such a transfusion were made by mistake.

23.21   In the immunoglobulin structure, the "hinge region" joins the stem of the Y to the arms. The hinge region can be cleaved by a specific enzyme to yield one F_c fragment (the stem of the Y) and two F_{ab} fragments

---

■   Problems assignable in **GOB OWL**

(the two arms). Which of these two kinds of fragments can interact with an antigen? Explain.

23.22 ■ How are the light and heavy chains of an antibody held together?

23.23 What do we mean by the term *immunoglobulin superfamily?*

23.24 If you could isolate two monoclonal antibodies from a certain population of lymphocytes, in what sense would they be similar to each other and in what sense would they differ?

23.25 What kind of interaction takes place between an antigen and an antibody?

23.26 How is a new protein created on the variable portion of a heavy chain by V(J)D recombination?

23.27 What accounts for antibody diversity?

Section 23.5 What Are T Cells and T-Cell Receptors?

23.28 T-cell receptor molecules are made of two polypeptide chains. Which part of the chain acts as a binding site, and what binds to it?

23.29 What is the difference between a T-cell receptor (TcR) and a TcR complex?

23.30 What kind of tertiary structure characterizes the TcR?

23.31 What are the components of the TcR complex?

23.32 By what chemical process does CD3 transduct signals inside the cell?

23.33 Which adhesion molecule in the TcR complex helps HIV infect a leukocyte?

23.34 Three kinds of molecules in the T cell belong to the immunoglobulin superfamily. List them and indicate briefly their functions.

23.35 What functions do CD4 and CD8 serve in the immune response?

Section 23.6 How Is the Immune Response Controlled?

23.36 What kind of molecules are cytokines?

23.37 ■ With what do cytokines interact? Do they bind to antigens?

23.38 As in most biochemistry literature, one encounters a veritable alphabet soup when reading about cytokines. Identify these cytokines by their full names: (a) TNF, (b) IL, and (c) EGF.

23.39 What are chemokines? How do they deliver their message?

23.40 What is the characteristic chemical signature in the structure of chemokines?

23.41 What are the chemical characteristics of cytokines that allow their classification?

23.42 Which amino acid appears in all chemokines?

Section 23.7 How Does the Body Distinguish "Self" from "Nonself"?

23.43 How does the body prevent T cells from being active against a "self" antigen?

23.44 What makes a tumor cell different from a normal cell?

23.45 Name a signaling pathway that controls the maturation of B cells and prevents those with an affinity for self antigen from becoming active.

23.46 How does the inhibitory receptor on a macrophage prevent an attack on normal cells?

23.47 Which components of the immune system are principally involved in autoimmune diseases?

23.48 How do glucocorticoids make individuals with autoimmune disease feel more comfortable?

Section 23.8 How Does the Human Immunodeficiency Virus Cause AIDS?

23.49 Which cells are attacked by HIV?

23.50 How does HIV gain entry into the cells it attacks?

23.51 How does HIV confound the human immune system?

23.52 What types of therapy are used to fight AIDS?

23.53 Why have vaccines been relatively unsuccessful in stopping AIDS?

23.54 ■ What are the structural features of the two types of neutralizing antibodies that have been the most successful at combating AIDS? What makes these antibodies more effective?

Chemical Connections

23.55 (Chemical Connections 23A) How are the mayapple and chemotherapy related?

23.56 (Chemical Connection 23B) What is contributing to the higher survival rate of women with breast cancer?

23.57 (Chemical Connection 23B) Why are monoclonal antibodies a good choice for a weapon against breast cancer?

23.58 (Chemical Connection 23B) Explain a situation in which a monoclonal antibody would be superior to a polyclonal antibody as a cancer drug.

23.59 (Chemical Connection 23B) What type of evidence suggests that the HER2 protein is important in many breast cancers?

23.60 (Chemical Connection 23B) How are monoclonal antibodies used to fight cancer?

23.61 (Chemical Connection 23B) What do tyrosine kinases have to do with cancer?

23.62 (Chemical Connections 23C) What made Edward Jenner "the father of immunization"? In your opinion, could one do such an experiment today legally?

23.63 (Chemical Connections 23C) What was the observation that led Edward Jenner to attempt his experiment?

23.64 (Chemical Connections 23C) What is the derivation of the word *vaccination*?

23.65 (Chemical Connections 23D) Why are allergies to antibiotics dangerous?

23.66 (Chemical Connections 23D) What do we mean by the term "indiscriminate use of antibiotics"?

23.67 (Chemical Connections 23D) Why has the sexually transmitted disease (STD) gonorrhea benefited from indiscriminate use of antibiotics?

23.68 (Chemical Connections 23D) What are the downsides to use of amoxicillin to combat earaches in children?

23.69 (Chemical Connections 23D) Why can strep throat be a serious condition apart from the problems directly associated with the sore throat?

23.70 (Chemical Connections 23E) What are the different types of stem cells?

23.71 (Chemical Connections 23E) Why are stem cells special? Why do scientists think they offer so much hope?

23.72 (Chemical Connections 23E) What are epigenetic states? Why do scientists want to be able to manipulate the epigenetic state of stem cells?

Additional Problems

23.73 Which immunoglobulins form the first line of defense against invading bacteria?

23.74 Which cells of the innate immunity system are the first to interact with an invading pathogen?

23.75 Which compound or complex of compounds of the immune system is mostly responsible for the proliferation of leukocytes?

23.76 Name a process beside V(J)D recombination that can enhance immunoglobulin diversity in the variable region.

23.77 Name a tumor cell marker, a synthetic analog of which may be the first anticancer vaccine.

23.78 Is the light chain of an immunoglobulin the same as the V region?

23.79 Where are TNF receptors located?

23.80 The variable regions of immunoglobulins bind the antigens. How many polypeptide chains carry variable regions in one immunoglobulin molecule?

Exponential Notation

The **exponential notation** system is based on powers of 10 (see table). For example, if we multiply $10 \times 10 \times 10 = 1000$, we express this as 10^3. The 3 in this expression is called the **exponent** or the **power**, and it indicates how many times we multiplied 10 by itself and how many zeros follow the 1.

There are also negative powers of 10. For example, 10^{-3} means 1 divided by 10^3:

$$10^{-3} = \frac{1}{10^3} = \frac{1}{1000} = 0.001$$

Numbers are frequently expressed like this: 6.4×10^3. In a number of this type, 6.4 is the **coefficient** and 3 is the exponent, or power of 10. This number means exactly what it says:

$$6.4 \times 10^3 = 6.4 \times 1000 = 6400$$

Similarly, we can have coefficients with negative exponents:

$$2.7 \times 10^{-5} = 2.7 \times \frac{1}{10^5} = 2.7 \times 0.00001 = 0.000027$$

For numbers greater than 10 in exponential notation, we proceed as follows: *Move the decimal point to the left,* to just after the first digit. The (positive) exponent is equal to the number of places we moved the decimal point.

Exponential notation is also called scientific notation.

For example, 10^6 means a one followed by six zeros, or 1,000,000, and 10^2 means 100.

APP. 1.1 Examples of Exponential Notation

10,000	$= 10^4$
1000	$= 10^3$
100	$= 10^2$
10	$= 10^1$
1	$= 10^0$
0.1	$= 10^{-1}$
0.01	$= 10^{-2}$
0.001	$= 10^{-3}$

Example

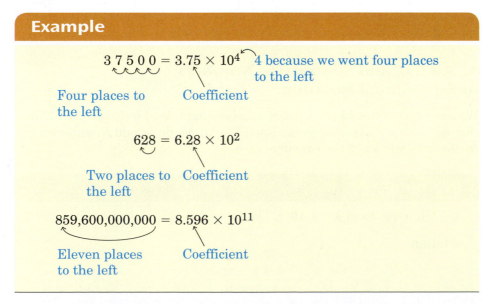

$$3\,7\,5\,0\,0 = 3.75 \times 10^4 \quad \text{4 because we went four places to the left}$$

Four places to the left   Coefficient

$$628 = 6.28 \times 10^2$$

Two places to the left   Coefficient

$$859{,}600{,}000{,}000 = 8.596 \times 10^{11}$$

Eleven places to the left   Coefficient

We don't really have to place the decimal point after the first digit, but by doing so we get a coefficient between 1 and 10, and that is the custom.

Using exponential notation, we can say that there are 2.95×10^{22} copper atoms in a copper penny. For large numbers, the exponent is always *positive*. Note that we do not usually write out the zeros at the end of the number.

For small numbers (less than 1), we move the decimal point *to the right,* to just after the first nonzero digit, and use a *negative exponent*.

Example

$$0.00346 = 3.46 \times 10^{-3}$$

Three places to
the right

$$0.000004213 = 4.213 \times 10^{-6}$$

Six places to
the right

In exponential notation, a copper atom weighs 2.3×10^{-25} pounds.

To convert exponential notation into fully written-out numbers, we do the same thing backward.

Example

Write out in full: (a) 8.16×10^7 (b) 3.44×10^{-4}

Solution

(a) $8.16 \times 10^7 = 81,600,000$ (b) $3.44 \times 10^{-4} = 0.000344$

Seven places to the right   Four places to the left
(add enough zeros)

When scientists add, subtract, multiply, and divide, they are always careful to express their answers with the proper number of digits, called significant figures. This method is described in Appendix II.

A. Adding and Subtracting Numbers in Exponential Notation

We are allowed to add or subtract numbers expressed in exponential notation *only if they have the same exponent*. All we do is add or subtract the coefficients and leave the exponent as it is.

Example

Add 3.6×10^{-3} and 9.1×10^{-3}.

Solution

$$
\begin{aligned}
3.6 &\times 10^{-3} \\
+\ 9.1 &\times 10^{-3} \\
\hline
12.7 &\times 10^{-3}
\end{aligned}
$$

The answer could also be written in other, equally valid ways:

$$12.7 \times 10^{-3} = 0.0127 = 1.27 \times 10^{-2}$$

When it is necessary to add or subtract two numbers that have different exponents, we first must change them so that the exponents are the same.

A calculator with exponential notation changes the exponent automatically.

Example

Add 1.95×10^{-2} and 2.8×10^{-3}.

Solution

To add these two numbers, we make both exponents -2. Thus, $2.8 \times 10^{-3} = 0.28 \times 10^{-2}$. Now we can add:

$$
\begin{array}{r}
1.95 \times 10^{-2} \\
+\ 0.28 \times 10^{-2} \\
\hline
2.23 \times 10^{-2}
\end{array}
$$

B. Multiplying and Dividing Numbers in Exponential Notation

To multiply numbers in exponential notation, we first multiply the coefficients in the usual way and then algebraically *add* the exponents.

Example

Multiply 7.40×10^5 by 3.12×10^9.

Solution

$$7.40 \times 3.12 = 23.1$$

Add exponents:

$$10^5 \times 10^9 = 10^{5+9} = 10^{14}$$

Answer:

$$23.1 \times 10^{14} = 2.31 \times 10^{15}$$

Example

Multiply 4.6×10^{-7} by 9.2×10^4

Solution

$$4.6 \times 9.2 = 42$$

Add exponents:

$$10^{-7} \times 10^4 = 10^{-7+4} = 10^{-3}$$

Answer:

$$42 \times 10^{-3} = 4.2 \times 10^{-2}$$

To divide numbers expressed in exponential notation, the process is reversed. We first divide the coefficients and then algebraically *subtract* the exponents.

Example

Divide: $\dfrac{6.4 \times 10^8}{2.57 \times 10^{10}}$

Solution

$$6.4 \div 2.57 = 2.5$$

Subtract exponents:

$$10^8 \div 10^{10} = 10^{8-10} = 10^{-2}$$

Answer:

$$2.5 \times 10^{-2}$$

Example

Divide: $\dfrac{1.62 \times 10^{-4}}{7.94 \times 10^7}$

Solution

$$1.62 \div 7.94 = 0.204$$

Subtract exponents:

$$10^{-4} \div 10^7 = 10^{-4-7} = 10^{-11}$$

Answer:

$$0.204 \times 10^{-11} = 2.04 \times 10^{-12}$$

Scientific calculators do these calculations automatically. All that is necessary is to enter the first number, press $+$, $-$, $\times$, or $\div$, enter the second number, and press $=$. (The method for entering numbers of this form varies; consult the instructions that come with the calculator.) Many scientific calculators also have a key that will automatically convert a number such as 0.00047 to its scientific notation form (4.7×10^{-4}), and vice versa.

Significant Figures

If you measure the volume of a liquid in a graduated cylinder, you might find that it is 36 mL, to the nearest milliliter, but you cannot tell if it is 36.2, or 35.6, or 36.0 mL because this measuring instrument does not give the last digit with any certainty. A buret gives more digits, and if you use one you should be able to say, for instance, that the volume is 36.3 mL and not 36.4 mL. But even with a buret, you could not say whether the volume is 36.32 or 36.33 mL. For that, you would need an instrument that gives still more digits. This example should show you that *no measured number can ever be known exactly.* No matter how good the measuring instrument, there is always a limit to the number of digits it can measure with certainty.

We define the number of **significant figures** as the number of digits of a measured number that have uncertainty only in the last digit.

What do we mean by this definition? Assume that you are weighing a small object on a laboratory balance that can weigh to the nearest 0.1 g, and you find that the object weighs 16 g. Because the balance weighs to the nearest 0.1 g, you can be sure that the object does not weigh 16.1 g or 15.9 g. In this case, you would write the weight as 16.0 g. To a scientist, there is a difference between 16 g and 16.0 g. Writing 16 g says that you don't know the digit after the 6. Writing 16.0 g says that you do know it: It is 0. However, you don't know the digit after that. Several rules govern the use of significant figures in reporting measured numbers.

A. Determining the Number of Significant Figures

In Section 1.3, we saw how to determine the number of significant figures in a reported number. We summarize those guidelines here:

1. Nonzero digits are always significant.
2. Zeros at the beginning of a number are never significant.
3. Zeros between nonzero digits are always significant.
4. Zeros at the end of a number that contains a decimal point are always significant.
5. Zeros at the end of a number that contains no decimal point may or may not be significant.

We use periods as decimal points throughout this text to indicate the significant figures in numbers with trailing zeros. For example, 1000. mL has four significant figures; 20. m has two significant figures.

B. Multiplying and Dividing

The rule in multiplication and division is that the final answer should have the *same* number of significant figures as there are in the number with the *fewest* significant figures.

Example

Do the following multiplications and divisions:

(a) 3.6×4.27

(b) 0.004×217.38

(c) $\dfrac{42.1}{3.695}$

(d) $\dfrac{0.30652 \times 138}{2.1}$

Solution

(a) 15 (3.6 has two significant figures)

(b) 0.9 (0.004 has one significant figure)

(c) 11.4 (42.1 has three significant figures)

(d) 2.0×10^1 (2.1 has two significant figures)

C. Adding and Subtracting

In addition and subtraction, the rule is completely different. The number of significant figures in each number doesn't matter. The answer is given to the *same number of decimal places* as the term with the fewest decimal places.

Example

Add or subtract:

(a) 320.084
    80.47
   200.23
    20.0
   _____
   620.8

(b) 61.4532
    13.7
    22
     0.003
   _____
    97

(c) 14.26
    −1.05041
   _____
    13.21

Solution

In each case, we add or subtract in the normal way but then round off so that the only digits that appear in the answer are those in the columns in which every digit is significant.

D. Rounding Off

When we have too many significant figures in our answer, it is necessary to round off. In this book we have used the rule that if *the first digit dropped* is 5, 6, 7, 8, or 9, we raise *the last digit kept* to the next number; otherwise, we do not.

Example

In each case, drop the last two digits:

(a) 33.679   (b) 2.4715   (c) 1.1145   (d) 0.001309   (e) 3.52

Solution

(a)   33.679 = 33.7
(b)   2.4715 = 2.47
(c)   1.1145 = 1.11
(d)   0.001309 = 0.0013
(e)   3.52 = 4

E. Counted or Defined Numbers

All of the preceding rules apply to *measured* numbers and **not** to any numbers that are *counted* or *defined*. Counted and defined numbers are known exactly. For example, a triangle is defined as having 3 sides, not 3.1 or 2.9. Here, we treat the number 3 as if it has an infinite number of zeros following the decimal point.

Example

Multiply 53.692 (a measured number) $\times$ 6 (a counted number).

Solution

$$322.15$$

Because 6 is a counted number, we know it exactly, and 53.692 is the number with the fewest significant figures. All we really are doing is adding 53.692 six times.

Answers

Chapter 1 Organic Chemistry

1.1 Following are Lewis structures showing all valence electrons and with all bond angles labeled.

(a)

(b)

1.2 Of the four alcohols with the molecular formula $C_4H_{10}O$, two are 1°, one is 2°, and one is 3°. For the Lewis structures of the 3° alcohol and one of the 1° alcohols, some C—CH₃ bonds are drawn longer to avoid crowding in the Lewis structures.

CH₃CH₂CH₂CH₂OH
Primary (1°)

CH₃CH₂CHCH₃
Secondary (2°)

CH₃CHCH₂OH
Primary (1°)

CH₃COH
Tertiary (3°)

1.3 The three secondary (2°) amines with the molecular formula $C_4H_{11}N$ are

CH₃CH₂CH₂NHCH₃   CH₃CHNHCH₃   CH₃CH₂NHCH₂CH₃
                    |
                   CH₃

1.4 The three ketones with the molecular formula $C_5H_{10}O$ are

CH₃CH₂CH₂CCH₃   CH₃CH₂CCH₂CH₃   CH₃CCHCH₃
                                          |
                                         CH₃

1.5 Following are the two carboxylic acids with the molecular formula $C_4H_8O_2$. The second structure drawn for each shows the fully condensed $-CO_2H$ group.

CH₃CH₂CH₂COH or CH₃CH₂CH₂CO₂H   and

CH₃CHCOH or CH₃CHCO₂H
    | |
   CH₃   CH₃

1.6 The four esters with the molecular formula $C_4H_8O_2$ are

HCOCH₂CH₂CH₃

(1)

HCOCHCH₃
       |
      CH₃

(2)

CH₃COCH₂CH₃

(3)

CH₃CH₂COCH₃

(4)

1.7 (a) T (b) T (c) F (d) F

1.9 Assuming each is pure, there are no differences in chemical or physical properties.

1.11 Wöhler heated ammonium chloride and silver cyanate, both inorganic compounds, and obtained urea, an organic compound.

1.13 The four most common elements that make up organic compounds and the number of bonds each typically forms are
H: forms one bond
C: forms four bonds
O: forms two bonds
N: forms three bonds.

1.15 Following are Lewis dot structures for each element. Under each is the number of electrons in its valence shell.

(a) ·Ċ· (b) ·Ö· (c) ·Ṅ· (d) :F̈·
   (4)   (6)   (5)   (7)

1.17 (a) H—Ö—Ö—H

      Hydrogen peroxide

(b) H—N̈—N̈—H
     | |
     H H

      Hydrazine

(c) H—C—Ö—H with H above and below C

      Methanol

(d) H—C—S̈—H with H above and below C

      Methanethiol

(e) H—C—N̈—H with H above and below C and H below N

      Methylamine

(f) H—C—C̈l: with H above and below C

      Chloromethane

1.19 Following is a Lewis structure for each ion.

(a) H—Ö—C(=O)—Ö:⁻

(b) ⁻:Ö—C(=O)—Ö:⁻

(c) CH_3-C(=O)—Ö:⁻

(d) :C̈l:⁻

1.21 To use the VSEPR model to predict bond angles and the geometry about atoms of carbon, nitrogen, and oxygen (1) write the Lewis structure for the target molecule showing all valence electrons; (2) determine the number of regions of electron density around an atom of C, N, or O. (3) if you find

four regions of electron density, predict bond angles of 109.5°; If you find three regions, predict bond angles of 120° if you find two regions, predict bond angles of 180°.

1.23 You would find two regions of electron density around oxygen and, therefore, predict 180° for the C—O—H bond angle. If only two regions of electron density are shown, predict 180° for the C—O—H bond angle

1.25 (a) 120° about C and 109.5° about O.
(b) 109.5° about N.
(c) 120° about N.
(d) This is a molecular model of (c) and shows the 120° bond angle about N.

1.27 A functional group is a part of an organic molecule that undergoes a predictable set of chemical reactions.

1.29

(d) —N̈—H with H below;
(e) —C(=O)—Ö—

1.31 When applied to alcohols, tertiary (3°) means that the carbon bearing the −OH group is bonded to three other carbon atoms.

1.33 When applied to amines, tertiary (3°) means that the amine nitrogen is bonded to three carbon groups.

1.35 (a) The four primary (1°) alcohols with the molecular formula $C_5H_{12}O$ are

$CH_3CH_2CH_2CH_2CH_2OH$ $CH_3CH_2CHCH_2OH$
                                       CH_3

        CH_3
CH_3CCH_2OH $CH_3CHCH_2CH_2OH$
        CH_3 CH_3

(b) The three secondary (2°) alcohols with the molecular formula $C_5H_{12}O$ are

OH   OH   OH
$CH_3CHCH_2CH_2CH_3$ $CH_3CH_2CHCH_2CH_3$ $CH_3CHCHCH_3$
                                                    CH_3

(c) The one tertiary (3°) alcohol with the molecular formula $C_5H_{12}O$ is

1.37 The eight carboxylic acids with the molecular formula $C_6H_{12}O_2$ are:

a six-carbon chain	a five-carbon chain with a one-carbon branch	a four-carbon chain with two carbons as branches	a six-carbon chain	a five-carbon chain with a one-carbon branch	a four-carbon chain with two carbons as branches
CH₃CH₂CH₂CH₂CH₂CO₂H	CH₃CHCH₂CH₂CO₂H (with CH₃ branch)	CH₃CHCHCO₂H (with CH₃ branches)	CH₃CH₂CH₂CH₂CH₂CHO	CH₃CHCH₂CH₂CHO (with CH₃ branch)	CH₃CHCHCHO (with CH₃ branches)

1.39 Taxol was discovered by a survey of indigenous plants sponsored by the National Cancer Institute with the goal of discovering new chemicals for fighting cancer.

1.41 Arrows point to atoms and show bond angles about each atom.

(a) CH₃—CH₂—CH₂—ÖH   109.5°

(b) CH₃—CH₂—C(=O)—H   109.5°, 120°

(c) CH₃—C=CH₂   109.5°, 120°

(d) CH₃—C≡C—CH₃   109.5°, 180°

(e) CH₃—C(=O)—Ö—CH₃   109.5°, 120°, 109.5°

(f) CH₃—N(CH₃)—CH₃   109.5°

1.43 Predict 109.5° for C-P-C bond angle.

CH₃—P̈—CH₃ (with CH₃ below)   109.5°

1.45 The eight aldehydes with the molecular formula C₆H₁₂O are below. The aldehyde functional group is written CHO.

1.47 (a) nonpolar covalent (b) nonpolar covalent
(c) nonpolar covalent (d) polar covalent (e) polar covalent
(f) polar covalent (g) polar covalent (h) polar covalent

1.49 Under each formula is given the difference in electronegativity between the atoms of the most polar bond.

(a) H—C—O—H   O—H (3.5–2.1 = 1.4)

(b) H—C—N—H   N—H (3.0–2.1 = 0.9)

(c) H—S—C—C—N—N   N—H (3.0–2.1 = 0.9)

(d) H—C—C—C—H   C=O (3.5–2.5 = 1.0)

(e) H₂C=O   C=O (3.5–2.5 = 1.0)

(f) H—C—C—O—H   O—H (3.5–2.1 = 1.4)

1.51 Following is one structural formula for each part. More than one answer is possible for parts a, b and c.

(a) CH₃CH₂CH₂C(=O)—OH

(b) CH₃CH₂—C(=O)—O—CH₃

(c) CH₃—CH(OH)—C(=O)—CH₃

(d) H—C(=O)—C(CH₃)—OH

(e) CH₂=CH—CH₂OH

Chapter 2 Alkanes

2.1 This compound is octane, and its molecular formula is C_8H_{18}.

2.2 (a) constitutional isomers (b) the same compound.

2.3 Following are structural formulas and line-angle formulas for the three constitutional isomers with the molecular formula C_5H_{12}.

2.4 (a) 5-isopropyl-2-methyloctane. Its molecular formula is $C_{12}H_{26}$.
(b) 4-isopropyl-4-propyloctane. Its molecular formula is $C_{14}H_{30}$.

2.5 (a) isobutylcyclopentane, C_9H_{18}
(b) *sec*-butylcycloheptane, $C_{11}H_{22}$
(c) 1-ethyl-1-methylcyclopropane, C_6H_{12}

2.6 The structure with the three methyl groups equatorial is

2.7 Cycloalkanes (a) and (c) show *cis-trans* isomerism.

(a) or

cis-1,3-Dimethylcyclopentane

trans-1,3-Dimethylcyclopentane

(c) or

cis-1,3-Dimethylcyclohexane

trans-1,3-Dimethylcyclohexane

2.8 In order of increasing boiling point, they are
(a) 2,2-dimethylpropane (9.5°C), 2-methylbutane (27.8°C), pentane (36.1°C)

(b) 2,2,4-trimethylhexane, 3,3-dimethylheptane, nonane

2.9 The two chloroalkanes with their IUPAC and common names are

1-Chloropropane   2-Chloropropane
(Propyl chloride)   (Isopropyl chloride)

2.11 (a) A hydrocarbon is a compound that contains only carbon and hydrogen.
(b) An alkane is a saturated hydrocarbon.
(c) A saturated hydrocarbon contains only C—C and C—H single bonds.

2.13 In a line-angle formula, each line terminus and vertex represents a carbon atom. Bonds are represented by combinations of one, two, or three parallel lines.

2.15 (a) $C_{10}H_{22}$ (b) C_8H_{18} (c) $C_{11}H_{24}$

2.17 (a) T (b) T (c) F (d) F

2.19 None represent the same compound. There are three sets of constitutional isomers. Compounds (a), (d), and (e) have the molecular formula C_4H_8O and are one set, compounds (c) and (f) have the molecular formula $C_5H_{10}O$ and are a second set, and compounds (g) and (h) have the molecular formula $C_6H_{10}O$ and are a third set.

2.21 (a) T (b) T (c) T

2.23 2-methylpropane and 2-methylbutane.

2.25 (a) T (b) F (c) T

2.27

2.29 A condensed structural formula shows only the order of bonding of the atoms in the compound. There is no attempt in it to show bond angles or molecular shape.

2.31 (a) F (b) F (c) T (d) T (e) F

2.33 No

2.35 Structural formulas for the six cycloalkanes with the molecular formula C_5H_{10} are

Cyclopentane   Methylcyclobutane   1,1-Dimethylcyclopropane

trans-1,2-Dimethyl-   cis-1,2-Dimethyl-   Ethylcyclopropane
cyclopropane   cyclopropane

2.37 (a) T (b) F (c) F (d) F (e) T (f) T

2.39 Heptane, C_7H_{16}, has a boiling point of 98°C and a molecular weight of 100. Its molecular weight is approximately 5.5 times that of water. Although they are considerably smaller, water molecules associate in the liquid phase by relatively strong hydrogen bonds whereas the much larger heptane molecules associate only by relatively weak London dispersion forces.

2.41 Alkanes are insoluble in water.

2.43 Boiling points of unbranched alkanes are related to their surface area; the larger the surface area, the greater the strength of dispersion forces, and the higher the boiling point. The relative increase in molecular size per CH_2 group is greatest between CH_4 and CH_3CH_3 and becomes progressively smaller as the molecular weight increases. Therefore, increase in boiling point per added CH_2 group is greatest between CH_4 and CH_3CH_3 and becomes progressively smaller for higher alkanes.

2.45 (a) F (b) T (c) T

2.47 The heat of combustion of methane is 212 kcal/mol or 212/16 = 13.3 kcal/gram. The heat of combustion of propane is 530 kcal/mol or 530/44 = 12.0 kcal/gram. Therefore, the heat energy per gram is greater for methane.

2.49

1-Chloropentane   2-Chloropentane   3-Chloropentane

2.51 (a) One ring contains only carbon atoms.
(b) One ring contains two nitrogen atoms.
(c) One ring contains two oxygen atoms.

2.53 Octane will produce more engine knocking than heptane.

2.55 The Freons are a class of chlorofluorocarbons. They were considered ideal for use as heat transfer agents in refrigeration systems because they are nontoxic, not corrosive, and nonflammable and odorless. (c) The two Freons used for this purpose were Freon 11(CCl_3F) and Freon-12 (CCl_2F_2).

2.57 They are hydrofluorocarbons and hydrochlorofluorocarbons. These compounds are much more chemically reactive in the atmosphere than the original Freons and are destroyed before they reach the stratosphere.

2.59 (a) The longest chain is pentane. Its IUPAC name is 2-methylpentane.
(b) The pentane chain is numbered incorrectly. Its IUPAC name is 2-methylpentane.
(c) The longest chain is pentane. Its IUPAC name is 3-ethyl-3-methylpentane.
(d) The longest chain is hexane. Its IUPAC name is 3,4-dimethylhexane.
(e) The longest chain is heptane. Its IUPAC name is 4-methylheptane.
(f) The longest chain is octane. Its IUPAC name is 3-ethyl-3-methyloctane.
(g) The ring is numbered incorrectly. Its IUPAC name is 1-ethyl-3-methylcyclohexane.
(h) The ring is numbered incorrectly. Its IUPAC name is 1-ethyl-3-methylcyclohexane.

2.61 Tetradecane is a liquid at room temperature.

2.63 On the left is the planar hexagon representation. On the right is the more stable chair conformation of this isomer.

2-Isopropyl-5-methylcyclohexanol

In this chair conformation all groups on the ring are in equatorial positions

2.65 Following is the alternative representation of 2-deoxy-D-ribose.

Chapter 3 Alkenes and Alkynes

3.1 (a) 3,3-dimethyl-1-pentene (b) 2,3-dimethyl-2-butene
(c) 3,3-dimethyl-1-butyne

3.2 (a) *trans*-3,4-dimethyl-2-pentene
(b) *cis*-4-ethyl-3-heptene

3.3 (a) 1-isopropyl-4-methylcyclohexene
(b) cyclooctene (c) 4-*tert*-butylcyclohexene

3.4 Line-angle formulas for the other two heptadienes are

cis,trans-2,4-Heptadiene   *cis,cis*-2,4-Heptadiene

3.5 Four stereoisomers are possible (two pairs of *cis-trans* isomers).

3.6

3.7 Propose a two-step mechanism similar to that for the addition of HCl to propene.
Step 1: Reaction of H^+ with the carbon–carbon double bond gives a 3° carbocation intermediate.

A 3° carbocation intermediate

Step 2: Reaction of the 3° carbocation intermediate with bromide ion completes the valence shell of carbon and gives the product.

3.8 The product from each acid-catalyzed hydration is the same alcohol.

3.9 Propose a three-step mechanism similar to that for the acid-catalyzed hydration of propene.

Step 1: Reaction of the carbon-carbon double bond with H⁺ gives a 3° carbocation intermediate.

A 3° carbocation intermediate

Step 2: Reaction of the 3° carbocation intermediate with water completes the valence shell of carbon and gives an oxonium ion.

An oxonium ion

Step 3: Loss of H⁺ from the oxonium ion completes the reaction and generates a new H⁺ catalyst.

3.10

(a) CH₃—C(CH₃)(H₃C)—CH(Br)—CH₂Br   (b)

3.11 (a) F (b) F (c) F (d) T

3.13 A saturated hydrocarbon contains only carbon–carbon and carbon–hydrogen single bonds. An unsaturated hydrocarbon contains one or more carbon–carbon double or triple bonds or an aromatic ring (Chapter 13).

3.15

(a) 109.5° 120°   (b) 120° CH₂OH

(c) HC≡C—CH=CH₂ 180° 120°   (d) 120°

3.17

(a)   (b) CH₃CHC≡CCH₂CH₃ with CH₃

(c) CH₂=CCH₂CH₃ with CH₃   (d) HC≡CCCH₂CH₃ with CH₃ and CH₂CH₃

(e) CH₃C=CCH₂CH₃ with H₃C and CH₃

3.19 (a) 2,5-dimethyl-1-hexene
(b) 1,3-dimethylcyclopentene
(c) 2-methyl-1-butene
(d) 2-propyl-1-pentene

3.21 (a) The longest chain is four carbon atoms. The correct name is 2-methyl-1-butene.
(b) The ring is numbered incorrectly. The correct name is 4-isopropylcyclohexene.
(c) The longest chain is six carbon atoms. The correct name is 3-methyl-2-hexene.
(d) The longest chain containing the C=C is five carbon atoms. The correct name 2-ethyl-3-methyl-1-pentene
(e) The ring is numbered incorrectly. The correct name is 3, 3-dimethylcyclohexene.
(f) The longest chain is seven carbon atoms. The correct name is 3-methyl-3-heptene.

3.23 Only (b) 2-hexene, (c) 3-hexene, and (e) 3-methyl-2-hexene show *cis-trans* isomerism.

3.25 Arachidonic acid is the all *cis* isomer.

Arachidonic acid

3.27 Only parts (b) and (d) show *cis-trans* isomerism.

(b)   (d)

3.29 Following is a structural formula for β-ocimene.

β-Ocimene

3.31 (a) T (b) T (c) F (d) F

3.33 The four isoprene units are shown in bold.

Vitamin A (retinol)

3.35 In an alkene addition reaction, one bond of the carbon–carbon double bond breaks and single bonds to two new atoms or groups of atoms form in its place.

$$CH_3CH{=}CH_2 + H_2O \xrightarrow{H_2SO_4} CH_3\overset{\overset{\displaystyle OH}{|}}{C}HCH_3$$

3.37

(a)

(b)

(c) $CH_3(CH_2)_5CH{=}CH + HI \longrightarrow CH_3(CH_2)_5\overset{\overset{\displaystyle I}{|}}{C}HCH_3$

(d)

(e) $CH_3CH{=}CHCH_2CH_3 + H_2O \xrightarrow{H_2SO_4}$

$CH_3\overset{\overset{\displaystyle OH}{|}}{C}HCH_2CH_2CH_3 + CH_3CH_2\overset{\overset{\displaystyle OH}{|}}{C}HCH_3CH_3$

(f) $CH_2{=}CHCH_2CH_2CH_3 + H_2O \xrightarrow{H_2SO_4}$

$CH_3\overset{\overset{\displaystyle OH}{|}}{C}HCH_2CH_2CH_3$

3.39

(a) $CH_3\overset{\overset{\displaystyle CH_3}{|}}{\underset{\underset{\displaystyle Cl}{|}}{C}}CH_2CH_2CH_3$ (b) $CH_3\overset{\overset{\displaystyle CH_3}{|}}{\underset{\underset{\displaystyle OH}{|}}{C}}CH_2CH_2CH$

3.41 (a) (b)

(c)

3.43

(a) $CH_2{=}\overset{\overset{\displaystyle CH_3}{|}}{C}CH_2CH_3$ or $CH_3\overset{\overset{\displaystyle CH_3}{|}}{C}{=}CHCH_3$

(b) $CH_3\overset{\overset{\displaystyle CH_3}{|}}{C}HCH{=}CH_2$ (c) $CH_2{=}CHCH_2CH_2CH_3$

3.45 Reaction involves acid-catalyzed hydration of each double bond. There are two *cis-trans* isomers. The structural formula for terpin hydrate is shown below in the most stable chair conformation with the —$(CH_3)_2$CHOH group equatorial.

Limonene   Terpin

3.47 Compound A is 2-methyl-1,3-butadiene.
3.49 The reagents are shown over the arrows.

(a)   (c)

(b)   (d)

3.51 Ethylene is a natural ripening agent for fruits.
3.53 Its molecular formula is $C_{16}H_{30}O_2$.
3.55 Rods are used for peripheral and night vision. Cones function in bright light and are used for color vision.
3.57 The most common consumer items made of high-density polyethylene (HDPE) are milk and water jugs, grocery bags, and squeeze bottles. The most common consumer items made of low-density polyethylene (LDPE) are packaging for baked goods, vegetables, and other produce, as well as trash bags. Currently, only HDPE materials are recyclable.
3.59 There are five compounds with the molecular formula C_4H_8. All are constitutional isomers. The only *cis-trans* isomers are *cis*-2-butene and *trans*-2-butene.

Cyclobutane   Methyl-   1-Butene   *cis*-2-   *trans*-2-
                cyclopropane   Butene   Butene

3.61 (a) The carbon skeleton of lycopene can be divided into eight isoprene units, here shown in bold bonds.

(b) Eleven of the 13 double bonds have the possibility for *cis-trans* isomerism. The double bonds at either end of the molecule cannot show *cis-trans* isomerism.
3.63

(a)   (b)   —CH_3 (c)   (d)

3.65 Each alkene hydration reaction follows Markovnikov's rule. H adds preferentially to carbon-3 and OH adds to carbon-4 to give 3-hexanol. Each carbon of the double bond again has the same pattern of substitution, so 3-hexanol is the only product.

3.67 Reagents are shown over the arrows.

3.69 Oleic acid has one double bond about which *cis-trans* isomerism is possible, so $2^1 = 2$ isomers are possible (one pair of *cis-trans* isomers).
Linoleic acid has two double bonds about which *cis-trans* isomerism is possible, so $2^2 = 4$ isomers are possible (two pairs of *cis-trans* isomers).
Linolenic acid has 3 double bonds about which *cis-trans* isomerism is possible, so $2^3 = 8$ isomers are possible (four pairs of *cis-trans* isomers).

Chapter 4 Benzene and Its Derivatives

4.1 (a) 2,4,6-tri-*tert*-butylphenol
(b) 2,4-dichloroaniline
(c) 3-nitrobenzoic acid
4.3 A saturated compound contains only single covalent bonds. An unsaturated compound contains one or more double or triple bonds or aromatic rings. The most common double bonds are C=C, C=O, and C=N. The most common triple bond is C≡C.
4.5 The members of each class of hydrocarbons contain fewer hydrogens than an alkane or cycloalkane with the same number of carbon atoms. Alternatively, each class of hydrocarbons contains one or more carbon-carbon double or triple bonds.
4.7 No.
4.9 (a) CH_4 (b) $CH_2\!=\!CH_2$

Methane   Ethene
             (Ethylene)

(c) $HC\!\equiv\!CH$ (d)

Ethyne
(Acetylene)

Benzene

4.11 Benzene consists of six carbons, each surrounded by three regions of electron density, which gives 120° for all bond angles. The presence of only trigonal planar ring carbons means that all ring substituents are coplanar (lie in the same plane) and *cis-trans* isomerism is not possible. Cyclohexane,

on the other hand, consists of carbons each surrounded by four regions of electron density, which gives 109.5° for all bond angles. It is the tetrahedral nature of the ring atoms in cyclohexane that allows for up and down substituents and hence *cis-trans* isomerism (because no C—C rotation is possible in such a cyclic system.

4.13 (a) T (b) F (c) F (d) T (e) F (f) F
4.15

4.17 Only cyclohexene will react with a solution of bromine in dichloromethane. A solution of Br_2/CH_2Cl_2 is reddish-purple, whereas 1,2-dibromocyclohexane is colorless. To tell which bottle contains which compound, place a small quantity of each compound in a test tube and to each add a few drops of Br_2/CH_2Cl_2 solution. If the red color disappears, the compound is cyclohexene, which has been converted to 1,2-dibromocyclohexane. If the reddish-purple color remains, the compound is benzene because in the absence of a catalyst, aromatic compounds do not react with Br_2 in dichloromethane.

4.21 (a) Nitration using HNO_3/H_2SO_4 followed by sulfonation using H_2SO_4. The order of the steps may be reversed.
(b) Bromination using $Br_2/FeCl_3$ followed by chlorination using $Cl_2/FeCl_3$. The order of the steps may also be reversed.
4.23 (a) T (b) T (c) F (d) F (e) F (f) T (g) T
(h) T (i) T
4.25 Autoxidation is the reaction of a C—H group with oxygen, O_2, to form a hydroperoxide C—O—O—H.
4.27 Vitamin E participates in one of the chain propagation steps of autoxidation and forms a stable radical, which breaks the cycle of chain propagation steps.
4.29 By definition, a carcinogen is a cancer-causing substance. The most important carcinogens present in cigarette smoke belong to the class of compounds called polynuclear aromatic hydrocarbons (PAHs).
4.31 Cyclonite (RDX) has the largest percentage of its molecular weight contributed by nitro groups.

Explosive	Mol Wt	NO₂ groups	% NO₂ groups
TNT	227.1	138	60.77
Nitroglycerine	227.1	138	60.77
Cyclonite	222.1	138	62.13
PETN	316.1	184	58.21

4.33 The functional groups most responsible for the water solubility of these dyes are the two ionic $—SO_3^-Na^+$ groups.
4.35 Capsaicin is isolated from various species of peppers (*Capsicum* and *Solanaecae*).
4.37 Following are the three possible resonance-contributing structures for naphthalene.

4.39 BHT participates in one of the chain propagation steps of autooxidation. It forms a stable radical and thus terminates the radical chain reaction.
4.41 Styrene reacts with bromine by addition to the carbon–carbon double bond.

Chapter 5 Alcohols, Ethers, and Thiols

5.1 (a) 2-Heptanol
(b) 2,2-Dimethyl-1-propanol
(c) *cis*-3-Isopropylcyclohexanol
5.2 (a) Primary (b) Secondary
(c) Primary (d) Tertiary
5.3 The structure of the major alkene product from each reaction is enclosed in a box.
In each case, the major product contains the more substituted double bond.

5.4

2-Methyl-   1-Methyl-   1-Methyl-
cyclohexanol   cyclohexene (C) cyclohexanol (D)

5.5 Each secondary alcohol is oxidized to a ketone.

5.6 (a) Ethyl isobutyl ether   (b) Cyclopentyl methyl ether
5.7 (a) 3-Methyl-1-butanethiol   (b) 3-Methyl-2-butanethiol
5.9 The difference is in the number of carbon atoms bonded to the carbon bearing the OH group. For primary alcohols, there is one; for a secondary alcohol, there are two and for a tertiary alcohol, there are three.
5.11

5.13 (a) Prednisone contains three ketones, one primary alcohol, one tertiary alcohol, one disubstituted carbon–carbon double bond, and one trisubstituted carbon–carbon double bond.
(b) Estradiol contains one secondary alcohol and one disubstituted phenol.
5.15 Low-molecular-weight alcohols form hydrogen bonds with water molecules through both the oxygen and hydrogen atoms of their —OH groups. Low-molecular-weight ethers form hydrogen bonds with water molecules only through the oxygen atom of their —O— group. The greater extent of hydrogen bonding between alcohol and water molecules makes the low-molecular-weight alcohols more soluble in water than the low-molecular-weight ethers.
5.17 Both types of hydrogen bonding are shown in the following illustration.

5.19 In order of increasing boiling point they are

$$CH_3CH_2CH_3 \quad CH_3CH_2OH$$
$$-42°C \qquad\qquad 78°C$$

$$CH_3CH_2CH_2CH_2OH \quad HOCH_2CH_2OH$$
$$117°C \qquad\qquad\qquad 198°C$$

5.21 The thickness (viscosity) of these three liquids is related to the degree of hydrogen bonding between their molecules in the liquid state. Hydrogen bonding is strongest between molecules of glycerol, weaker between molecules of ethylene glycol, and weakest between molecules of ethanol.

5.23 In order of decreasing solubility in water, they are:
(a) ethanol > diethyl ether > butane
(b) 1,2-hexanediol > 1-hexanol > hexane

5.25 (a) T (b) T (c) F (d) F (e) F (f) F (g) T (h) F (i) T (j) F

5.27 Phenols are weak acids, with pK_a values approximately equal to 10. Alcohols are considerably weaker acids and have about the same acidity as water (pK_a values around 16).

5.29

(a) $CH_3CH_2CH_2CH_2OH \xrightarrow[\text{heat}]{H_2SO_4} CH_3CH_2CH=CH_2 + H_2O$

(b) $CH_3CH_2CH_2CH_2OH \xrightarrow[H_2SO_4]{K_2Cr_2O_7} CH_3CH_2CH_2CO_2H$

5.31

(b) $HOCH_2CH_2CH_2CH_2OH \xrightarrow[H_2SO_4]{K_2Cr_2O_7} HOCCH_2CH_2COH$

5.33 (a) H_2SO_4, heat (b) H_2O/H_2SO_4
(c) $K_2Cr_2O_7/H_2SO_4$ (d) HBr (e) Br_2
(f) H_2/Pd (g) $K_2Cr_2O_7/H_2SO_4$
(h) $K_2Cr_2O_7/H_2SO_4$ (i) $K_2Cr_2O_7/H_2SO_4$

5.35 2-Propanol (isopropyl alcohol) and glycerin (glycerol) are derived from propene. 2-Propanol is the alcohol in rubbing alcohol. Major uses of glycerol are in skin care products and cosmetics. It is also the starting material for the synthesis of nitroglycerin.

5.37 (a) Dicyclopentyl ether (b) Dipentyl ether
(c) Diisopropyl ether

5.39 (a) 2-Butanethiol (b) 1-Butanethiol
(c) Cyclohexanethiol

5.41 Because 1-butanol molecules associate by hydrogen bonding in the liquid state, it has the higher boiling point

(117°C). There is little polarity to an S—H bond. The only interactions among 1-butanethiol molecules in the liquid state are the considerably weaker London dispersion forces. For this reason, 1-butanethiol has the lower boiling point (98°).

5.43 (a) T (b) T (c) T (d) T (e) T (f) T

5.45 Nobel discovered that diatomaceous earth absorbs nitroglycerin so that it will not explode without a fuse.

5.47 Dichromate ion is reddish-orange, chromium (III) ion is green. When breath containing ethanol passes through a solution containing dichromate ion, ethanol is oxidized and dichromate ion is reduced to green chromium (III) ion.

5.49 Normal bond angles about tetrahedral carbon and divalent oxygen are 109.5°. In ethylene oxide the C—C—O and C—O—C bond angles are compressed to approximately 60°, which results in angle strain within the molecule.

5.51 The molecular formula of each is $C_3H_2ClF_5O$. They have the same molecular formula but a different connectivity of their atoms.

Enflurane   Isoflurane

5.53 $CH_3CH_2OH + 3O_2 \longrightarrow 2CO_2 + 3H_2O$

5.55 The eight isomeric alcohols with the molecular formula $C_5H_{12}O$ are

5.57 Ethylene glycol has two —OH groups by which each molecule participates in hydrogen bonding, whereas 1-propanol has only one. The stronger intermolecular forces of attraction between molecules of ethylene glycol give it the higher boiling point.

5.59 Arranged in order of increasing boiling point, they are

$$CH_3CH_2CH_2CH_2CH_2CH_3 \quad CH_3CH_2CH_2CH_2CH_2OH$$

Hexane   1-Pentanol
(Insoluble)   (2.3 g/100 mL)

$$HOCH_2CH_2CH_2CH_2OH$$
1,4-Butanediol
(Infinitely soluble)

5.61 Each is prepared from 2-methyl-1-propanol (circled) as shown in this flow chart.

5.63 The three functional groups are a thiol, a primary amine, and a carboxyl group.
Oxidation of the thiol gives a disulfide.

Cystine

Chapter 6 Chirality: The Handedness of Molecules

6.1 The enantiomers of each part are drawn with two groups in the plane of the paper, a third group toward you in front of the plane, and the fourth group away from you behind the plane.

6.2 The group of higher priority in each set is circled.

(a) [—CH$_2$OH] and —CH$_2$CH$_2\overset{\text{O}}{\overset{\|}{\text{C}}}$OH

(b) [—CH$_2$NH$_2$] and —CH$_2\overset{\text{O}}{\overset{\|}{\text{C}}}$OH

6.3 The configuration is R and the compound is (R)-glyceraldehyde.

6.4 (a) Structures 1 and 3 are one pair of enantiomers. Structures 2 and 4 are a second pair of enantiomers.
(b) Compounds 1 and 2, 1 and 4, 2 and 3, and 3 and 4 are diastereomers.

6.5 Four stereoisomers are possible for 3-methylcyclohexanol. The *cis* isomer is one pair of enantiomers; the *trans* isomer is a second pair of enantiomers.

6.6 Each stereocenter is marked by an asterisk and the number of stereoisomers possible is shown under the structural formula.

(a)

$2^1 = 2$

(b) $CH_2\!\!=\!\!CHCHCH_2CH_3$

$2^1 = 2$

(c)

$2^2 = 4$

6.7 (a) T (b) T (c) T (d) F (e) T (f) T (g) T
(h) T (i) T

6.9 An achiral object has no handedness; it is an object whose mirror image is superposable on the original. An example is methane, CH_4.

6.11 Both constitutional isomers and stereoisomers have the same molecular formula. Whereas stereoisomers have the same connectivity, constitutional isomers have a different connectivity of their atoms.

6.13 2-Pentanol has a stereocenter (carbon 2). 3-Pentanol has no stereocenter.

6.15 The carbon of a carbonyl group has only three groups bonded to it. To be a stereocenter, a carbon must have four different groups bonded to it.

6.17 Compounds (b), (c), and (d) contain stereocenters, here marked by asterisks, and are chiral

6.19 Following are the mirror images of each.

(d)

6.21 (a) T (b) F (c) T (d) F (e) T (f) T
6.23 Only parts (b) and(c) contain stereocenters.

(b) (c)

6.25 Stereocenters are marked with an asterisk. Under each is the number of stereoisomers possible.

(a)

$(2^2 = 4)$

(Two pairs of enantiomers)

(b)

2

(cis + trans)

(c)

$(2^1 = 2)$

(one pair of enantiomers)

(d)

$(2^2 = 4)$

(two pair of enantiomers)

6.27 The specific rotation of its enantiomer is +41°.
6.29 (a) T (b) T (c) F (d) T
6.31 Of the eight alcohols with the molecular formula $C_5H_{12}O_2$, only three are chiral.

2-Pentanol   2-Methyl-1-butanol   3-Methyl-2-butanol

6.33 Of the eight carboxylic acids with the molecular formula $C_6H_{12}O_2$, only three are chiral.

6.35 Amoxicillin has four stereocenters.

Amoxicillian

6.37 Share your findings with others. You will find it interesting to compare them.
6.39 This molecule has eight stereocenters, here marked with asterisks. There are $2^8 = 256$ possible stereoisomers.

Triamcinolone acetonide

Chapter 7 Acids and Bases

7.1 acid reaction: $HPO_4^{2-} + H_2O \rightleftharpoons PO_4^{3-} + H_3O^+$;
base reaction: $HPO_4^{2-} + H_2O \rightleftharpoons H_2PO_4^- + OH^-$
7.2 (a) toward the left;

$$H_3O^+ + I^- \rightleftharpoons H_2O + HI$$
Weaker acid   Weaker base   Stronger base   Stronger acid

(b) toward the right;

$$CH_3COO^- + H_2S \rightleftharpoons CH_3COOH + HS^-$$
Weaker base   Weaker acid   Stronger acid   Stronger base

7.3 pK_a is 9.31
7.4 (a) ascorbic acid (b) aspirin
7.5 1.0×10^{-2}
7.6 (a) 2.46 (b) 7.9×10^{-5}, acidic
7.7 pOH = 4, pH = 10
7.8 0.0960 M
7.9 (a) 9.25 (b) 4.74
7.10 9.44
7.11 7.7
7.13 (a) $HNO_3(aq)+H_2O(\ell) \longrightarrow NO_3^-(aq)+H_3O^+(aq)$
(b) $HBr(aq)+H_2O(\ell) \longrightarrow Br^-(aq)+H_3O^+(aq)$
(c) $H_2SO_3(aq)+H_2O(\ell) \longrightarrow HSO_3^-(aq)+H_3O^+(aq)$
(d) $H_2SO_4(aq)+H_2O(\ell) \longrightarrow HSO_4^-(aq)+H_3O^+(aq)$
(e) $HCO_3^-(aq)+H_2O(\ell) \longrightarrow CO_3^{2-}(aq)+H_3O^+(aq)$
(f) $NH_4^+(aq) + H_2O(\ell) \longrightarrow NH_3(aq)+H_3O^+(aq)$
7.15 (a) weak (b) strong (c) weak (d) strong
(e) weak (f) weak
7.17 (a) false (b) false (c) true (d) false
(f) true (g) false (h) false
7.19 (a) A Brønsted-Lowry acid is a proton donor.
(b) A Brønsted-Lowry base is a proton acceptor.
7.21 (a) HPO_4^{2-} (b) HS^- (c) CO_3^{2-}
(d) $CH_3CH_2O^-$ (e) OH^-
7.23 (a) H_3O^+ (b) $H_2PO_4^-$ (c) $CH_3NH_3^+$
(d) HPO_4^{2-} (e) NH_4^+
7.25 The equilibrium favors the side with the weaker acid−weaker base combination. Equilibria (b) and (c) lie to the left; equilibrium (a) lies to the right.

$$C_6H_5OH + C_2H_5O^- \rightleftharpoons C_6H_5O^- + C_2H_5OH$$

Stronger acid   Stronger base   Weaker base   Weaker acid

$$HCO_3^- + H_2O \rightleftharpoons H_2CO_3 + OH^-$$

Weaker base   Weaker acid   Stronger acid   Stronger base

$$CH_3COOH + H_2PO_4^- \rightleftharpoons CH_3COO^- + H_3PO_4$$

Weaker acid   Weaker base   Stronger base   Stronger acid

7.27 (a) the pK_a of a weak acid (b) the K_a of a strong acid
7.29 (a) $0.10\,M$ HCl (b) $0.10\,M$ H_3PO_4 (c) $0.010\,M$ H_2CO_3
(d) $0.10\,M$ NaH_2PO_4 (e) $0.10\,M$ aspirin
7.31 Only (b) is a redox reaction. The others are acid–base reactions.
(a) $Na_2CO_3 + 2HCl \longrightarrow 2NaCl + CO_2 + H_2O$
(b) $Mg + 2HCl \longrightarrow MgCl_2 + H_2$
(c) $NaOH + HCl \longrightarrow NaCl + H_2O$
(d) $Fe_2O_3 + 6HCl \longrightarrow 2FeCl_3 + 3H_2O$
(e) $NH_3 + HCl \longrightarrow NH_4Cl$
(f) $CH_3NH_2 + HCl \longrightarrow CH_3NH_3Cl$
(g) $NaHCO_3 + HCl \longrightarrow NaCl + H_2O + CO_2$
7.33 (a) $10^{-3}\,M$ (b) $10^{-10}\,M$ (c) $10^{-7}\,M$ (d) $10^{-15}\,M$
7.35 (a) pH = 8 (basic) (b) pH = 10 (basic) (c) pH = 2 (acidic) (d) pH = 0 (acidic) (e) pH = 7 (neutral)
7.37 (a) pH = 8.5 (basic) (b) pH = 1.2 (acidic)
(c) pH = 11.1 (basic) (d) pH = 6.3 (acidic)
7.39 (a) pOH = 1.0, $[OH^-]$ = 0.10 M
(b) pOH = 2.4, $[OH^-]$ = $4.0 \times 10^{-3}\,M$
(c) pOH = 2.0, $[OH^-]$ = $1.0 \times 10^{-2}\,M$
(d) pOH = 5.6, $[OH^-]$ = $2.5 \times 10^{-6}\,M$
7.41 0.348 M
7.43 (a) 12 g of NaOH diluted to 400 mL of solution:

$$400\text{ mL sol}\left(\frac{1\text{ L sol}}{1000\text{ mL sol}}\right)\left(\frac{0.75\text{ mol NaOH}}{1\text{ L sol}}\right)$$
$$\times \left(\frac{40.0\text{ g NaOH}}{1\text{ mol NaOH}}\right) = 12\text{ g NaOH}$$

(b) 12 g of $Ba(OH)_2$ diluted to 1.0 L of solution:

$$\left(\frac{0.071\text{ mol Ba(OH)}_2}{1\text{ L sol}}\right)\left(\frac{171.4\text{ Ba(OH)}_2}{1\text{ mol Ba(OH)}_2}\right) = 12\text{ g Ba(OH)}_2$$

(c) 2.81 g KOH diluted to 500 mL
(d) 49.22 g sodium acetate diluted to 2 liters
7.45 5.66 mL
7.47 3.30×10^{-3} mol
7.49 The point at which the observed change occurs during a titration. It is usually so close to the equivalence point that the difference between the two becomes insignificant.
7.51 (a)
$$H_3O^+ + CH_3COO^- \rightleftharpoons CH_3COOH + H_2O \text{ (removal of } H_3O^+)$$
(b)
$$HO^- + CH_3COOH \rightleftharpoons CH_3COO^- + H_2O \text{ (removal of } OH^-)$$

7.53 Yes, the conjugate acid becomes the weak acid and the weak base becomes the conjugate base.

7.55 The pH of a buffer can be changed by altering the weak acid/conjugate base ratio, according to the Henderson-Hasselbalch equation. The buffer capacity can be changed without a change in pH by increasing or decreasing the amount of weak acid/conjugate base mixture while keeping the ratio of the two constant.
7.57 This would occur in a couple of cases. One is very common: You are using a buffer, such as Tris with a pK_a of 8.3, but you do not want the solution to have a pH of 8.3. If you wanted a pH of 8.0, for example, you would need unequal amounts of the conjugate acid and base, with there being more conjugate acid. Another case might be a situation where you are performing a reaction that you know will generate H^+ but you want the pH to be stable. In that situation, you might start with a buffer that was initially set to have more of the conjugate base so that it could absorb more of the H^+ that you know will be produced.
7.59 No. 100 mL of 0. 1 M phosphate at pH 7.2 has a total of 0.01 moles of weak acid and conjugate base with equimolar amounts of each. 20 mL of 1 M NaOH has 0.02 moles of base, so there is more total base than there is buffer to neutralize it. This buffer would be ineffective.
7.61 (a) According to the Henderson-Hasselbalch equation, no change in pH will be observed as long as the weak acid/conjugate base ratio remains the same.
(b) The buffer capacity increases with increasing amounts of weak acid/conjugate base concentrations; therefore, 1.0-mol amounts of each diluted to 1 L would have a greater buffer capacity than 0.1 mol of each diluted to 1 L.
7.63 From the Henderson-Hasselbach equation,
$$pH = pK_a + \log(A^-/HA)$$
$$A^-/HA = 10, \log(A^-/HA) = 1 \text{ since } 10^1 = 10$$
$$pH = pK_a + 1$$
7.65 When 0.10 mol of sodium acetate is added to 0.10 M HCl, the sodium acetate completely neutralizes the HCl to acetic acid and sodium chloride. The pH of the solution is determined by the incomplete ionization of acetic acid.
$$K_a = \frac{[CH_3COO^-][H_3O^+]}{[CH_3COOH]} \quad [H_3O^+]=[CH_3COO^-] = x$$
$$\sqrt{x^2} = \sqrt{K_a[CH_3COOH]} = \sqrt{(1.8 \times 10^{-5})(0.10)}$$
$$x = [H_3O^+] = 1.34 \times 10^{-3}\,M$$
$$pH = -\log[H_3O^+] = 2.9$$
7.67 $TRIS\text{-}H^+ + NaOH \longrightarrow TRIS + H_2O + Na^+$
7.69 The only parameter you need to know about a buffer is its pK_a. Choosing a buffer involves identifying the acid form that has a pK_a within one unit of the desired pH.
7.71 Choosing a buffer involves identifying the acid form that has a pK_a within one unit of the desired pH (a pH of 8.15). The TRIS buffer with a pK_a = 8.3 best fits this criteria.
7.73 $Mg(OH)_2$ is a weak base used in flame-retardant plastics.
7.75 (a) Respiratory acidosis is caused by hypoventilation, which occurs due to a variety of breathing difficulties, such as a windpipe obstruction, asthma, or pneumonia. (b) Metabolic acidosis is caused by starvation or heavy exercise.
7.77 Sodium bicarbonate is the weak base form of one of the blood buffers. It tends to raise the pH of blood, which is the purpose of the sprinter's trick, so that the person can absorb more H^+ during the event. By putting $NaHCO_3$ into the system, the following reaction will occur:
$HCO_3^- + H^+ \rightleftharpoons H_2CO_3$. The loss of H^+ means that the blood pH will rise.

7.79 (a) Benzoic acid is soluble in aqueous NaOH.
$C_6H_5COOH + NaOH \rightleftharpoons C_6H_5COO^- + H_2O$
$pK_a = 4.19$ $\qquad\qquad pK_a = 15.56$
(b) Benzoic acid is soluble in aqueous $NaHCO_3$.
$C_6H_5COOH + NaHCO_3 \rightleftharpoons CH_3C_6H_4O^- + H_2CO_3$
$pK_a = 4.19$ $\qquad\qquad pK_a = 6.37$
(c) Benzoic acid is soluble in aqueous Na_2CO_3.
$C_6H_5COOH + CO_3^{2-} \rightleftharpoons CH_3C_6H_4O^- + HCO_3^-$
$pK_a = 4.19$ $\qquad\qquad pK_a = 10.25$

7.81 The strength of an acid is not important to the amount of NaOH that would be required to hit a phenolphthalein end-point. Therefore, the more concentrated acid, the acetic acid, would require more NaOH.

7.83 $3.70 \times 10^{-3} M$

7.85 $0.9 M$

7.87 Yes, a pH of 0 is possible. A 1.0 M solution of HCl has $[H_3O^+] = 1.0 M$. $pH = -\log[H_3O^+] = -\log[1.0 M] = 0$

7.89 The qualitative relationship between acids and their conjugate bases states that the stronger the acid, the weaker its conjugate base. This can be quantified in the equation $K_b \times K_a = K_w$ or $K_b + 1.0 \times 10^{-14}/K_a$, where K_b is the base dissociation equilibrium constant for the conjugate base, K_a is the acid dissociation equilibruim constant for the acid, and K_w is the ionization equilibrium constant for water.

7.91 Yes. The strength of the acid is irrelevant. Both acetic acid and HCl have one H^+ to give up, so equal moles of either will require equal moles of NaOH to titrate to an end point.

7.93 You would need a ratio of 0.182 parts of the conjugate base to 1 part of the conjugate acid.

7.95 An equilibrium will favor the side of the weaker acid/weaker base. The larger the pK_a value, the weaker the acid.

7.97 (a) $HCOO^- + H_3O^+ \rightleftharpoons HCOOH + H_2O$
(b) $HCOOH + OH^- \rightleftharpoons HCOO^- + H_2O$

7.99 (a) 0.050 mol (b) 0.0050 mol (c) 0.50 mol

7.101 According to the Henderson-Hasselbalch equation,
$$pH = 7.21 + \log\frac{[HPO_4^{2-}]}{[H_2PO_4^-]}$$
As the concentration of $H_2PO_4^-$ increases, the $\log\frac{[HPO_4^{2-}]}{[H_2PO_4^-]}$ becomes negative, lowering the pH and becoming more acidic.

7.103 No. A buffer will have a pH equal to its pK_a only if equimolar amounts of the conjugate acid and base forms are present. If this is the basic form of Tris, then just putting any amount of it into water will give a pH much higher than the pK_a value.

7.105 (a) $pH = 7.1$, $[H_3O^+] = 7.9 \times 10^{-8} M$, basic
(b) $pH = 2.0$, $[H_3O^+] = 1.0 \times 10^{-2} M$, acidic
(c) $pH = 7.4$, $[H_3O^+] = 4.0 \times 10^{-8} M$, basic
(d) $pH = 7.0$, $[H_3O^+] = 1.0 \times 10^{-7} M$, neutral
(e) $pH = 6.6$, $[H_3O^+] = 2.5 \times 10^{-7} M$, acidic
(f) $pH = 7.4$, $[H_3O^+] = 4.0 \times 10^{-8} M$, basic
(g) $pH = 6.5$, $[H_3O^+] = 3.2 \times 10^{-7} M$, acidic
(h) $pH = 6.9$, $[H_3O^+] = 1.3 \times 10^{-7} M$, acidic

7.107 $4.9:1$, or $5:1$ to one significant figure

Chapter 8 Amines

8.1 Pyrrolidine has nine hydrogens; its molecular formula is C_4H_9N. Purine has four hydrogens; its molecular formula is $C_5H_4N_4$.

8.2

8.3

(a) HO—NH₂ (b) [diphenylamine structure]

(c) [diisopropylamine structure]

8.4 The stronger base is circled.

8.5 The product of each reaction is an ammonium salt.

(a) $(CH_3CH_2)_3\overset{+}{N}HCl^-$
Triethylammonium chloride

(b) [piperidinium acetate structure]
Piperidinium acetate

8.7 Each compound has a six-membered ring with three double bonds.

8.9 Following is a structural formula for each amine.

(a) [structure with NH₂] (b) $CH_3(CH_2)_6CH_2NH_2$

(c) [structure with NH₂] (d) $H_2N(CH_2)_5NH_2$

(e) [structure with NH₂ and Br] (f) $(CH_3CH_2CH_2CH_2)_3N$

8.11 For this molecular formula, there are four primary amines, three secondary amines, and one tertiary amine. Only 2-butanamine is chiral.

1° amines (4)

[structure] —NH₂
1-Butanamine
(Butylamine)

[structure] —NH₂
2-Butanamine
(*sec*-Butylamine)

2-Methyl-1-
propanamine
(Isobutylamine)

2-Methyl-2-
propanamine
(*tert*-Butylamine)

2° amines (3)

Methylpropyl-
amine

Isopropylmethyl-
amine

Diethylamine

3° amine (1)

Ethyldimethylamine

8.13 (a) F (b) T (c) T

8.15 The association of 1-butanol molecules by hydrogen bonding is stronger than the association of 1-butanamine molecules by hydrogen bonding because of the greater polarity of an O—H bond compared with the polarity of an N—H bond.

8.17 Low-molecular-weight amines are polar molecules and are soluble in water because they form relatively strong hydrogen bonds with water molecules. Hydrocarbons are nonpolar molecules and do not interact with water molecules.

8.19 Amines are more basic than alcohols because nitrogen is less electronegative than oxygen and, therefore, more willing to donate its unshared pair of electrons to H⁺ in an acid–base reaction to form a salt.

8.21 (a) Ethylammonium chloride
(b) Diethylammonium chloride
(c) Anilinium hydrogen sulfate

8.23 The form of amphetamine present at both pH 1.0 and pH 7.4 is its conjugate acid.

Conjugate acid
of amphetamine

8.25

(a) CH_3COH + ⟶ CH_3CO^-

(b) ... NH_2 + HCl ⟶ ... $NH_3{}^+Cl^-$

(c)

8.27 (a) Tamoxifen contains three aromatic (benzene) rings, one carbon–carbon double bond, one ether, and one tertiary amine.
(b) The amine is tertiary.
(c) Two stereoisomers are possible, one pair of *cis-trans* isomers.
(d) Insoluble in water and in blood.

8.29 Possible negative effects are long periods of sleeplessness, loss of weight, and paranoia.

8.31 Both coniine and nicotine have one stereocenter; two stereoisomers (one pair of enantiomers) are possible for each. The *S* enantiomer of each is shown below.

(*S*)-Coniine

(*S*)-Nicotine

8.33 The four stereocenters of cocaine hydrochloride are marked with asterisks. Following is the structural formula of the salt formed by reaction of cocaine with HCl.

Cocaine•HCl

8.35 Neither Librium nor Valium is chiral. Each is achiral (without handedness).

8.37 No. No unreacted HCl is present.

8.39 The amino group in each compound is secondary. Each compound has one benzene ring, one phenolic —OH group, a secondary alcohol on a carbon bonded to the benzene ring, and the same configuration at its single stereocenter. The amine of epinephrine is substituted with a methyl group, whereas hat on albuterol has a tertiary butyl group.

8.41 In order of decreasing ability to form intermolecular hydrogen bonds, they are $CH_3OH > (CH_3)_2NH > CH_3SH$. An O—H bond is more polar than an N—H bond, which is in turn, more polar than an S—H bond.

8.43 Butane, the least polar molecule, has the lowest boiling point; 1-propanol, the most polar molecule, has the highest boiling point.

$CH_3CH_2CH_2CH_3$ $CH_3CH_2CH_2NH_2$ $CH_3CH_2CH_2OH$
−0.5°C   7.2°C   77.8°C

8.45 (a) Following is a structural formula for 1-phenyl-2-amino-1-propanol.

1-Phenyl-2-amino-1-propanol

(b) This molecule has two stereocenters. $2^2=4$ stereoisomers are possible.

8.47 (a) The secondary aliphatic amine is the more basic nitrogen.

(b) The three stereocenters are marked by asterisks.

the 2° aliphatic amine is the more basic nitrogen

8.49 The structural formula of alanine is drawn on the left showing a free amino group ($-NH_2$) and a free carboxyl group ($-CO_2H$). An acid–base reaction between these two groups gives the internal salt shown in the right.

an internal salt

Chapter 9 Aldehydes and Ketones

9.1 (a) 3,3-dimethylbutanal (b) cyclopentanone
(c) 1-phenyl-1-propanone

9.2 Following are line-angle formulas for the eight aldehydes with the molecular formula $C_6H_{12}O$. In the three that are chiral, the stereocenter is marked by an asterisk.

Hexanal   4-Methylpentanal

3-Methylpentanal   2-Methylpentanal

2,3-Dimethylbutanal   3,3-Dimethylbutanal

2,2-Dimethylbutanal   2-Ethylbutanal

9.3 (a) 2,3-dihydroxypropanal (b) 2-aminobenzaldehyde
(c) 5-amino-2-pentanone

9.4 Each aldehyde is oxidized to a carboxylic acid.

(a)

Hexanedioic acid
(Adipic acid)

(b)

3-Phenylpropanoic acid

9.5 Each primary alcohol comes from the reduction of an aldehyde. Each secondary alcohol comes from the reduction of a ketone.

(a)

(b) CH_3O—⟨⟩—CH_2CH

(c)

9.6 Shown first is the hemiacetal and then the acetal.

Benzaldehyde   A hemiacetal

An acetal

9.7 (a) A hemiacetal derived from 3-pentanone (a ketone) and ethanol.
(b) Neither a hemiacetal nor an acetal. This compound is the dimethyl ether of ethylene glycol.
(c) An acetal derived from 5-hydroxypentanal and methanol.

9.8 Following is the keto form of each enol.

(a)   (b)

(c)

9.9 (a) T (b) T (c) T (d) F

9.11 In an aromatic aldehyde, the —CHO group is bonded to an aromatic ring. In an aliphatic aldehyde, it is bonded to a tetrahedral carbon atom.

9.13 Compounds (b), (c), (d), and (f) contain a carbonyl group.

9.15 Of the four aldehydes with the molecular formula C$_5$H$_{10}$O, only one is chiral. The stereocenter in it is marked by an asterisk.

Pentanal

3-Methylbutanal

2-Methylbutanal

2,2-Dimethylpropanal

9.17

(a) H—C—H (with O double bond)

(b) CH$_3$CH$_2$CH (with O double bond)

(c) [branched structure]—CHO

(d) CH$_3$(CH$_2$)$_8$CHO

(e) HO—[ring]—CHO

(f) HOCH$_2$CHCHO with OH

9.19 (a) 4-heptanone (b) 2-methylcyclopentanone
(c) *cis*-2-methyl-2-butenone (d) 2-hydroxypropanal
(e) 1-phenyl 2-propanone (f) hexanedial
9.21 (a) T (b) T (c) T (d) T
9.23 The carbonyl group of acetone forms hydrogen bonds with water. These hydrogen bonds are sufficient to make acetone soluble in water in all proportions. 4-Heptanone contains a carbonyl which, through its hydrogen bonding with water molecules, promotes water solubility. It also contains two three-carbon hydrocarbon groups bonded to the carbonyl carbon, which inhibit water solubility. In 4-heptanone, the combined hydrophobic effect of the two hydrocarbon groups is greater than the hydrophilic effect of the single carbonyl group, making 4-heptanone insoluble in water.
9.25 Pentane is a nonpolar hydrocarbon and the only attractive forces between its molecules in the liquid state are the very weak London dispersion forces. Pentane, therefore, has the lowest boiling point. Pentanal and 1-butanol are both polar molecules. Because 1-butanol has a polar OH group, its molecules can associate by hydrogen bonding. The intermolecular attraction between molecules of 1-butanol are greater than those between molecules of pentanal. 1-Butanol, therefore, has a higher boiling point than pentanal.
9.27 Acetone molecules have no O—H or N—H group through which to form intramolecular hydrogen bonds.
9.29 Only an aldehyde is oxidized by Tollens' reagent. Under the basic conditions of the reaction, the oxidation product is a sodium salt of a carboxylic acid. Upon neutralization with aqueous HCl, the oxidation product is isolated as a carboxylic acid. Each oxidation product is shown as it would be before treatment with HCl.

(a) CH$_3$CH$_2$CH$_2$CO$_2^-$Na$^+$ (b) [ring]—CO$_2^-$Na$^+$

(c) No reaction (d) No reaction

9.31 Liquid aldehydes are very susceptible to oxidation by atmospheric oxygen. To prevent this oxidation, they are generally stored under an atmosphere of nitrogen.
9.33 These experimental conditions reduce an aldehyde to a primary alcohol and a ketone to a secondary alcohol. Products (a) and (c) are chiral.

(a) CH$_3$CHCH$_2$CH$_3$ with OH, * (b) CH$_3$(CH$_2$)$_4$CH$_2$OH

(c) [cyclopentane ring with OH and CH$_3$, * marks] (d) [benzene ring with CH$_2$OH and OH]

9.35

(a) HOCH$_2$—C—CH$_2$OH (with O double bond) (b) Soluble

(c) HOCH$_2$CHCH$_2$OH with OH

9.37

(a) [ring]—CHCH$_3$ with OH (b) [ring]—CHCH$_3$ with OH

(c) No reaction (d) No reaction
9.39 Only compounds (a), (b), (d), and (f) will undergo keto-enol tautomerism because each has an H on an (α-carbon.
9.41 Following are the keto forms of each enol.

(a) [cyclopentanone] (b) CH$_3$CCH$_2$CH$_2$CH$_2$CH$_3$ (with O double bond)

(c) [ring]—CH$_2$CCH$_3$ (with O double bond)

9.43 Compounds (a), (c), (d), and (e) are acetals. Compound (b) is a hemiacetal. Compound (f) is neither an acetal nor a hemiacetal.
9.45 Following are structural formulas for the products of each hydrolysis.

(a) CH$_3$CH$_2$CCH$_2$CH$_3$ + HOCH$_2$CH$_2$OH (with O double bond)

(b) [ring]—CH with O double bond, OCH$_3$ + 2CH$_3$OH

(c) [cyclohexane]=O + 2CH$_3$OH

(d) [ring with OH and CHO] + CH$_3$OH

9.47 *Hydration* refers to the addition of one or more molecules of water to a substance. An example of hydration is the acid-catalyzed hydration of propene to give 2-propanol. *Hydrolysis* refers to the reaction of a substance with water with breaking (lysis) of one or more bonds in the substance. An example of hydrolysis is the acid-catalyzed reaction of an acetal with a molecule of water to give an aldehyde or ketone and two molecules of alcohol.

9.49 (a) To reduce the ketone to an alcohol, use either $NaBH_4$ followed by H_2O or H_2/M.
(b) To bring about dehydration of the alcohol to an alkene, use H_2SO_4 and heat.
(c) To add HBr to the carbon–carbon double bond, use concentrated HBr.
(d) To reduce the carbon–carbon double bond, use H_2/Pd.
(e) To add bromine to the carbon–carbon double bond, use a solution of Br_2 in CH_2Cl_2.

9.51 Compounds (a), (b), and (d) can be formed by the reduction of an aldehyde or a ketone.

9.53

C6H5CH=CHCH3

9.55 (a) Each compound is insoluble in water. Treat each with dilute aqueous HCl. Aniline, an aromatic amine, reacts with HCl to form a water-soluble salt. Cyclohexanone does not react with this reagent and is insoluble in aqueous HCl.
(b) Treat each with a solution of Br_2/CH_2Cl_2. Cyclohexene reacts to discharge the red color of Br_2 and to form 1,2-dibromocyclohexane, a colorless compound. Cyclohexanol does not react with this reagent.
(c) Treat each with a solution of Br_2/CH_2Cl_2. Cinnamaldehyde, which contains a carbon–carbon double bond, reacts to discharge the red color of Br_2 and to form 2,3-dibromo-3-phenylpropanal, a colorless compound. Benzaldehyde does not react with this reagent.

9.57 Each aldehyde or ketone in each will be reduced to an alcohol.

9.59

9.61 1-Propanol has the higher boiling point because of the greater attraction between its molecules due to hydrogen bonding through its hydroxyl group.

9.63 (a) The hydroxyaldehyde is first redrawn to show the OH group nearer the CHO group. Closing the ring in the hemiacetal formation gives the cyclic hemiacetal.
(b) 5-Hydroxyhexanal has one stereocenter, and two stereoisomers (one pair of enantiomers) are possible.
(c) The cyclic hemiacetal has two stereocenters, and four stereoisomers (two pairs of enantiomers) are possible.

CH3*CHCH2CH2CH2CH → Redraw to show OH near CHO →

OH

5-Hydroxyhexanal

The cyclic hemiacetal

9.65 Given first is the alkene that undergoes acid-catalyzed hydration to give the desired alcohol, then the aldehyde or ketone that undergoes reduction to give the desired alcohol.

(c) $CH_2\!\!=\!\!CHCH_3$ $CH_3\overset{\overset{\displaystyle O}{\|}}{C}CH_3$

(d)

9.67 Carbon 4 provides the —OH group and carbon 1 provides the —CHO group.
Following is a structural formula for the free aldehyde.

9.69

(a)

(b)

(c)

(d)

(e)

Chapter 10 Carboxylic Acids

10.1 (a) 2,3-dihydroxypropanoic acid
(b) 3-aminopropanoic acid
(c) 3,5-dihydroxy-3-methylpentanoic acid
10.2 Each acid is converted to its ammonium salt. Given are both the IUPAC and common names of each acid and its ammonium salt.

(a) $COOH + NH_3 \longrightarrow$ $COO^-NH_4^+$

Butanoic acid (Butyric acid)
Ammonium butanoate (Ammonium butyrate)

(b)

OH   OH

$COOH + NH_3 \longrightarrow$ $COOH^-NH_4^+$

(s)-2-Hydroxypropanoic acid
[(s)-Lactic acid]
Ammonium (s)-2-hydroxypropanoate
[Ammonium (s)-lactate]

10.3

(a)

(b)

10.5 (a) 3,4-dimethylpentanoic acid
(b) 2-aminobutanoic acid
(c) hexanoic acid

10.7

(a)

O_2N —— $COOH$

(b) $H_2NCH_2CH_2CH_2COOH$

(c)

$COOH$

(d)

$COOH$
$COOH$

10.9

(a)

(b) $CH_3\overset{\overset{\displaystyle O}{\|}}{C}O^-Li^+$ (c) $CH_3\overset{\overset{\displaystyle O}{\|}}{C}O^-NH_4^+$

(d)

(e)

(f) $(CH_3CH_2CH_2COO^-)_2Ca^{2+}$

10.11 Oxalic acid (IUPAC name: ethanedioic acid) is a dicarboxylic acid. In monopotassium oxalate, one of its carboxyl groups is present as its carboxylic anion, giving a charge of −1.

Monopotassium oxalate

10.13 The dimer drawn here shows two hydrogen bonds.

Hydrogen bonds

10.15 The carboxyl group contributes to water solubility; the hydrocarbon chain prevents water solubility.

10.17 In order of increasing boiling point, they are heptanal, 1-heptanol, and heptanoic acid.

$$CH_3CH_2CH_2CH_2CH_2CH_2CH$$

O
||

Heptanal
(bp 153°C)

$$CH_3CH_2CH_2CH_2CH_2CH_2CH_2OH$$

1-Heptanol
(bp 176°C)

$$CH_3CH_2CH_2CH_2CH_2CH_2COH$$

O
||

Heptanoic acid
(bp 223°C)

10.19 In order of increasing solubility in water, they are decanoic acid, pentanoic acid, and acetic acid.

10.21 (a) T (b) T (c) F (d) F (e) T (f) F
(g) T (h) F (i) T

10.23 In order of increasing acidity, they are benzyl alcohol, phenol, and benzoic acid.

10.25 Following are completed equations for these acid–base reactions.

(a)

(b)

(c)

(d)

10.27 Dividing both sides of the K_a equation by $[H_3O^+]$ gives the desired relationship.

10.29 The pK_a of lactic acid is 4.07. At this pH, lactic acid is present as 50% $CH_3CH(OH)COOH$ and 50% $CH_3CH(OH)COO^-$. At pH 7.45, which is more basic than pH 4.07, lactic acid is present primarily as the anion, $CH_3CH(OH)COO^-$.

10.31 In part (a), the —COOH group is a stronger acid than the —NH_3^+ group.

(a) $CH_3CHCOOH + NaOH \longrightarrow$
     |
     NH_3^+

$CH_3CHCOO^-Na^+ + H_2O$
     |
     NH_3^+

(b) $CH_3CHCOO^-Na^+ + NaOH \longrightarrow$
     |
     NH_3^+

$CH_3CHCOO^-Na^+ + H_2O$
     |
     NH_2

10.33 In part (a), the —NH_2 group is a stronger base than the —COO^- group.

(a) $CH_3CHCOO^-Na^+ + HCl \longrightarrow$
     |
     NH_2

$CH_3CHCOO^-Na^+ + NaCl$
     |
     NH_3^+

(b) $CH_3CHCOO^-Na^+ + HCl \longrightarrow CH_3CHCOOH + NaCl$
     | |
     NH_3^+ NH_3^+

10.35 Following is a structural formula for the ester formed in each reaction.

(a)

(b) CH_3CO-

(c)

10.37 Following is a structural formula for methyl 2-hydoxybenzoate.

Methyl 2-hydroxybenzoate
(Methyl salicylate)

10.39 Following are the expected organic products.

(a)

(b)

(c)

(d)

(e) No reaction

(f)

(g) No reaction

10.41 Step 1: Treatment of benzoic acid with HNO_3/H_2SO_4 brings about nitration of the aromatic ring.
Step 2: Treatment of 4-nitrobenzoic acid with H_2/M brings about catalytic reduction of the $-NO_2$ group to an $-NH_2$ group.

10.43 Each starting material is difunctional and can form esters at both ends. The following equation shows reaction of two molecules of adipic acid and two molecules of ethylene glycol to form a triester. The product has a free carboxyl group at one end and a free hydroxyl group at the other and ester formation can continue at each end of the chain.

Chapter 11 Carboxylic Anhydrides, Esters, and Amides

11.1

(a) CH_3CNH- ⬡   (b) ⬡$-CNH_2$

11.2 Under basic conditions, as in part (a), each carboxyl group is present as a carboxylate anion. Under acidic conditions, as in part (b), each carboxyl group is present in its un-ionized form.

(a)   + 2NaOH $\xrightarrow{H_2O}$

  + 2CH₃OH

(b) + H_2O $\xrightarrow{HCl}$

  + CH_3CH_2OH

11.3 In aqueous NaOH, each carboxyl group is present as a carboxylate anion, and each amine is present in its unprotonated form.

(a) $CH_3CN(CH_3)_2$ + NaOH $\xrightarrow[\text{heat}]{H_2O}$

$$CH_3CO^-Na^+ + (CH_3)_2NH$$

(b)   + NaOH $\xrightarrow[\text{heat}]{H_2O}$ H_2N ⌇ O^-Na^+

11.5 (a) benzoic anhydride
(b) methyl decanoate
(c) N-methylhexanamide
(d) 4-aminobenzamide or p-aminobenzamide
(e) cyclopentyl ethanoate or cyclopentyl acetate
(f) ethyl 3-hydroxybutanoate

11.7 Each reaction brings about hydrolysis of the amide. Each product is shown as it would exist under the specified reaction conditions.

(a) $-CNH_2$ + NaOH $\xrightarrow{H_2O}$

  $-CO^-Na^+ + NH_3$

(b)

11.9 The product is an amide.

11.11 (a) Aspartame is chiral. It has two stereocenters. Four stereoisomers are possible.

Aspartame

(b) Aspartame contains one carboxylate anion, one 1° ammonium ion, on amide group and one ester group.
(c) The net charge is zero.
(d) It is an internal salt. Therefore, expect it to be soluble in water.
(e) Following are the products of hydrolysis of the ester and amide bonds.

Hydrolysis in NaOH

+

+ CH₃OH

Hydrolysis in HCl

+ CH₃OH

11.13 Following are sections of two parallel nylon-66 chains, with hydrogen bonds between N—H and C═O groups indicted by dashed lines.

11.15 In the anhydrides of carboxylic acids, the functional group is two carbonyl (C═O) groups bonded to an oxygen atom. In an anhydride of phosphoric acid, the functional group is two phosphoryl (P═O) groups bonded to an oxygen atom.
11.17 The phosphate ester group is shown here in its doubly ionized form and has a net charge of −2.

Dihydroxyacetone phosophate

11.19 The box on the left encloses the COO of the ester group. On the right is the structural formula of chrysanthemic acid.

Pyrethrin I

Chrysanthemic acid

11.21 (a) The *cis/trans* ratio refers to *the cis-trans* relationship between the ester group and the two-carbon chain on the cyclopropane ring. Specifically the repellant in the commercial preparation consists of a minimum of 35% of the *cis* isomer and a maximum of 65% of the *trans* isomer.
(b) Permethrin has two stereocenters, and four stereoisomers (two pairs of enantiomers) are possible. The designation "(+/−)" refers to the fact that the *cis* isomer is present as a racemic mixture, as is the *trans* isomer.
11.23 The compound is salicin. Hydrolysis of the glucose unit and oxidation of the primary alcohol to a carboxyl group gives salicylic acid.
11.25 Both aspirin and ibuprofen contain a carboxylic acid and a benzene ring. Naproxen also contains a carboxylic acid and a benzene ring; in naproxen, the benzene ring is a part of a naphthalene ring.
11.27 A *sunblock* prevents all ultraviolet radiation from reaching protected skin by reflecting it away from the skin.

A *sunscreen,* absorbs UV radiation and then reiradiates the energy as heat.

11.29 The portion derived from urea contains the atoms —NH—CO—NH—.

11.31 Following is the structural formula of benzocaine.

Benzocaine
Ethyl 4-aminobenzote

11.33 The reaction involves hydrolysis of a phosphate ester and a mixed anhydride of a carboxylic acid and a phosphate acid. Two equivalents of water are required.

11.35 (a) Both lidocaine and carbocaine have an amide group a tertiary aliphatic amine, and an aromatic ring.
(b) Both are derived from 2,6-dimethylaniline and a 2-alkylamino-substituted carboxylic acid.

11.37 Hydrolysis gives hydrogen phosphate ion and the enol of pyruvate. The enol then undergoes rapid keto-enol tautomerism to give pyruvate ion. Recall from Chapter 9 that the keto form generally predominates in cases of keto-enol tautomerism.

Chapter 12 Carbohydrates

12.1 Following are Fischer projections for the four 2-ketopentoses. They consist of two pairs of enantiomers.

12.2 D-Mannose differs in configuration from D-glucose only at carbon 2. One way to arrive at the structures of the α and β forms of D-mannopyranose is to draw the corresponding α and β forms of D-glucopyranose, and then invert the configuration in each at carbon 2.

12.3 D-Mannose differs in configuration from D-glucose only at carbon 2.

12.4 Following is a Haworth projection and a chair conformation for this glycoside.

12.5 The β-glycosidic bond is between carbon 1 of the left unit and carbon 3 of the right unit.

12.7 The carbonyl group in an aldose is an aldehyde. In a ketose, it is a ketone. An aldopentose is an aldose that contains five carbon atoms. An aldoketose is a ketose that contains five carbon atoms.

12.9 The three most abundant hexoses in the biological world are D-glucose, D-galactose, and D-fructose. The first two are aldohexoses. The third is a 2-ketohexose.

12.11 To say that they are enantiomers means that they are nonsuperposable mirror images.

12.13 The D or L configuration in an aldopentose is determined by its configuration at carbon 4.

12.15 Compounds (a) and (c) are D-monosaccharides. Compound (b) is an L-monosaccharide.

12.17 A 2-ketoheptose has four stereocenters and 16 possible stereoisomers. Eight of these are D-2-ketoheptoses and eight are L-2-ketoheptoses. Following is one of the eight possible D-2-ketoheptoses.

12.19 In an amino sugar, one or more —OH groups are replaced by —NH_2 groups. The three most abundant amino sugars in the biological world are D-glucosamine, D-galactosamine, and N-acetyl-D-glucosamine.

12.21 (a) A pyranose is a six-membered cyclic hemiacetal form of a monosaccharide.
(b) A furanose is a five-membered cyclic hemiacetal form of a monosaccharide.

12.23 Yes, they are anomers. No, they are not enantiomers; that is, they are not mirror images. They differ in configuration only at carbon 1 and, therefore, are diastereomers.

12.25 A Haworth projection shows the six-membered ring as a planar hexagon. In reality, the ring is puckered and its

most stable conformation is a chair conformation with all bond angles approximately 109.5°.

12.27 Compound (a) differs from D-glucose only in the configuration at carbon 4. Compound (b) differs only at carbon 3.

12.29 The specific rotation of α-L-glucose is −112.2°.

12.31 A glycoside is a cyclic acetal of a monosaccharide. A glycosidic bond is the bond from the anomeric carbon to the —OR group of the glycoside.

12.33 No, glycosides cannot undergo mutarotation because the anomeric carbon is not free to interconvert between α and β configurations via the open-chain aldehyde or ketone.

12.35 Following are Fischer projections of D-glucose and D-sorbitol. The configurations at the four stereocenters of D-glucose are not affected by this reduction.

12.37 Ribitol is the reduction product of D-ribose. β-D-ribose 1-phosphate is the phosphoric ester of the OH group on the anomeric carbon of β-D-ribofuranose.

β-D-Ribofuranose 1-phosphate
(β-D-Ribose 1-phosphate)

Ribitol

12.39 To say that it is a β-1,4-glycosidic bond means that the configuration at the anomeric carbon (carbon 1 in this problem) of the monosaccharide unit forming the glycosidic bond is beta and that it is bonded to carbon 4 of the second monosaccharide unit. To say that it is an α-1,6-glycosidic bond means that the configuration at the anomeric carbon (carbon 1 in this problem) of the monosaccharide unit forming the glycosidic bond is alpha and that it is bonded to carbon 6 of the second monosaccharide unit.

12.41 (a) Both monosaccharide units are D-glucose.
(b) They are joined by a β-1,4-glycosidic bond.
(c) It is a reducing sugar and
(d) it undergoes mutarotation.

Units of D-glucose

A reducing sugar because this carbon is a hemiacetal and in equilibrium with the open-chain aldehyde

β-1,4-Glycosidic bond

12.43 An oligosaccharide contains approximately six to ten monosaccharide units. A polysaccharide contains more—generally many more—than ten monosaccharide units.

12.45 The difference lies in the degree of chain branching. Amylose is composed of unbranched chains, whereas amylopectin is a branched network with the branches started by α-1, 6-glycosidic bonds.

12.47 Cellulose fibers are insoluble in water because the strength of hydrogen bonding of a cellulose molecule in the fiber with surface water molecules is not sufficient to overcome the intermolecular forces that hold it in the fiber.

12.49 (a) In these structural formulas, the CH₃CO (the acetyl group) is abbreviated Ac.

Following are Haworth and chair structures for this repeating disaccharide.

β-1,4-Glycosidic bond

12.51 Its lubricating power decreases.

12.53 With maturation, children develop an enzyme capable of metabolizing galactose. Thus they are able to tolerate galactose as they mature. Until these children develop the ability to metabolize galactose, substituting sucrose for lactose replaces the galactose in lactose with fructose in sucrose.

12.55 L-Ascorbic acid is oxidized (there is loss of two hydrogen atoms) when it is converted to L-dehydroascorbic acid. L-Ascorbic acid is a biological reducing agent.

12.57 Types A, B, and O have in common D-galactose and L-fucose. Only type A has N-acetyl-D-glucosamine.

12.59 Mixing types A and B blood will result in coagulation.

12.61 Consult Table 12.1 for the structural formula of D-altrose and draw it. Then replace the —OH groups on carbons 2 and 6 with hydrogens.

$$
\begin{array}{c}
\text{CHO} \\
\text{HO} \!\!-\!\!\!-\!\! \text{H} \\
\text{H} \!\!-\!\!\!-\!\! \text{OH} \\
\text{H} \!\!-\!\!\!-\!\! \text{OH} \\
\text{H} \!\!-\!\!\!-\!\! \text{OH} \\
\text{CH}_2\text{OH}
\end{array}
$$

D-Altrose

$$
\begin{array}{c}
\text{CHO} \\
\text{H} \!-\!\!^2\!\!-\! \text{H} \\
\text{H} \!\!-\!\!\!-\!\! \text{OH} \\
\text{H} \!\!-\!\!\!-\!\! \text{OH} \\
\text{H} \!\!-\!\!\!-\!\! \text{OH} \\
^6\text{CH}_3
\end{array}
$$

2,6-Dideoxy-D-Altrose
(D-Digitoxose)

12.63 The monosaccharide unit of salicin is D-glucose.

12.65 Chitosan can be obtained from the shells of crustaceans such as shrimp and lobsters.

12.67 The five-membered ring of fructose is nearly planar, so a Haworth projection is a good representation of its structure.

12.69 In starch, α-glycosidic bonds join one glucose moiety to another. Cellulose has β-glycosidic bonds. This difference means that humans and other animals can digest starch, but not cellulose.

12.71 Amino sugars play an important structural role in polysaccharides, such as chitin, the hard shell of crabs, shrimp, and lobsters. Amino sugars also play a role in the structures of the blood group antigens.

12.73 The intermediate in this transformation is an enediol formed by keto-enol tautomerism of dihydroxyacetone phosphate. Keto-enol tautomerism of this intermediate gives D-glyceraldehyde-3-phosphate.

12.75 (a) Coenzyme A is chiral. It has five stereocenters.
(b) The functional groups, starting from the left, are a thiol (—SH), two amides, a secondary alcohol, a phosphate ester, a phosphate anhydride, a phosphate ester, another phosphate ester, a unit of 2-deoxyribose, and a (β-glycosidic bond to adenine, a heterocyclic amine.
(c) Yes, it is soluble in water because of the presence of a number of polar C=O groups, one —OH group and three phosphate groups, all of which will interact with water molecules by hydrogen bonding.
(d) Following are the products of hydrolysis of all amide, ester, and glycosidic bonds.

2,3-Dihydroxy-3-methyl-butanoic acid
(Pantothenic acid)

Adenine β-2-Deoxy-D-ribose   Dihydrogen phosphate

Chapter 13 Lipids

13.1 (a) It is an ester of glycerol and contains a phosphate group; therefore it is a glycerophospholipid. Besides glycerol and phosphate, it has a myristic acid and a linoleic acid component. The other alcohol is serine. Therefore, it belongs to the subgroup of cephalins.
(b) The components present are glycerol, myristic acid, linoleic acid, phosphate, and serine.

13.3 *Hydrophobic means* "water hating." If the body did not have such molecules, there could be no structure because the water would dissolve everything.

13.5 The melting point would increase. The *trans* double bonds would fit more in the packing of the long hydrophobic tails, creating more order and therefore more interaction between chains. This would require more energy to disrupt, and hence a higher melting point.

13.7 The diglycerides with the highest melting points will be the ones with two stearic acids (a saturated fatty acid). The lowest melting points will be the ones with two oleic acids (a monounsaturated fatty acid).

13.9 (b), because its molecular weight is higher.

13.11 lowest (c); then (b); highest (a)

13.13 The more long-chain groups, the lower the solubility; lowest (a); then (b); highest (c).

13.15 glycerol, sodium palmitate, sodium stearate, sodium linolenate

13.17 Complex lipids can be classified into two groups: phospholipids and glycolipids. Phospholipids contain an alcohol, two fatty acids, and a phosphate group. There are two types: glycerophospholipids and sphingolipids. In glycerophospholipids, the alcohol is glycerol. In sphingolipids, the alcohol is sphingosine. Glycolipids are complex lipids that contain carbohydrates.

13.19 The presence of *cis* double bonds in fatty acids produces greater fluidity because they cannot pack together as closely as saturated fatty acids.

13.21 Integral membrane proteins are embedded in the membrane. Peripheral membrane proteins are found on membrane surfaces.

13.23 A phosphatidyl inositol containing oleic acid and arachidonic acid:

13.25 Complex lipids that contain ceramides include sphingomyelin, sphingolipids, and the cerebroside glycolipids.

13.27 The hydrophilic functional groups of (a) glucocerebroside: carbohydrate; hydroxyl and amide groups of the cerebroside. (b) Sphingomyelin: phosphate group; choline; hydroxyl and amide of ceramide.

13.29 Cholesterol crystals may be found in (1) gallstones, which are sometimes pure cholesterol, and (2) joints of people suffering from bursitis.

13.31 The carbon of the steroid D ring to which the acetyl group is bonded in progesterone undergoes the most substitution.

13.33 LDL from the bloodstream enters the cells by binding to LDL receptor proteins on the surface. After binding, the LDL is transported inside the cells, where cholesterol is released by enzymatic degradation of the LDL.

13.35 Removing lipids from the triglyceride cores of VLDL particles increases the density of the particles and converts them from VLDL to LDL particles.

13.37 When serum cholesterol concentration is high, the synthesis of cholesterol in the liver is inhibited and the synthesis of LDL receptors in the cell is increased. Serum cholesterol levels control the formation of cholesterol in the liver by regulating enzymes that synthesize cholesterol.

13.39 Estradiol (E) is synthesized from progesterone (P) through the intermediate testosterone (T). First the D-ring acetyl group of P is converted to a hydroxyl group and T is produced. The methyl group in T, at the junction of the rings A and B, is removed and ring A becomes aromatic. The keto group in P and T is converted to a hydroxyl group in E.

13.41 Steroid structures are shown in Section 13.10. The major structural differences are at carbon 11. Progesterone has no substituents except hydrogen, cortisol has a hydroxyl group, cortisone has a keto group, and RU-486 has a large *p*-aminophenyl group. The functional group at carbon 11 apparently has little importance in receptor binding.

13.43 They have a steroid ring structure, they have a methyl group at carbon 13, they have a triply bonded group at carbon 17, and all have some unsaturation in the A ring, the B ring, or both.

13.45 Bile salts help solubilize fats. They are oxidation products of cholesterol themselves, and they bind to cholesterol, forming complexes that are eliminated in the feces.

13.47 (a) Glycocholate:

(b) Cortisone:

(c) PGE₂:

(d) Leukotriene B4:

13.49 Aspirin slows the synthesis of thromboxanes by inhibiting the COX enzyme. Because thromboxanes enhance the blood clotting process, the result is that strokes caused by blood clots in the brain will occur less often.

13.51 Waxes consist primarily of esters of long-chain saturated acids and alcohols. Because of the saturated components, wax molecules pack more tightly than those of triglycerides, which frequently have unsaturated components.

13.53 The transporter is a helical transmembrane protein. The hydrophobic groups on the helices are turned outward and interact with the membrane. The hydrophilic groups of the helices are on the inside and interact with the hydrated chloride ions.

13.55 (a) Sphingomyelin acts as an insulator.
(b) The insulator is degraded, impairing nerve conduction.

13.57 α-D-galactose, β-D-glucose, β-D-glucose

13.59 They prevent ovulation.

13.61 It inhibits prostaglandin formation by preventing ring closure.

13.63 NSAIDs inhibit cyclooxygenases (COX enzymes) that are needed for ring closure. Leukotrienes have no ring in their structure; therefore they are not affected by COX inhibitors.

13.65 (See Figure 13.2.) Polar molecules cannot penetrate thebilayer. They are insoluble in lipids. Nonpolar molecules can interact with the interior of the bilayer ("like dissolves like").

13.67 Both groups are derived from a common precursor, PGH₂, in a process catalyzed by the COX enzymes.

13.69 Coated pits are concentrations of LDL receptors on the surface of cells. They bind LDL and by endocytosis transfer it inside the cell.

13.71 In facilitated transport, a membrane protein assists in the movement of a molecule through the membrane with no requirement for energy. In active transport, a membrane protein assists in the process, but energy is required. ATP hydrolysis usually supplies the needed energy.

13.73 Aldosterone has an aldehyde group at the junction of the C and D rings. The other steroids have methyl groups.

13.75 The formula weight of the triglyceride is about 800 g/mol. This is 0.125 mol (100 g ÷ 800 g/mol = 0.125 mol). One mole of hydrogen is required for each mole of double bonds in the triglyceride. There are three double bonds, so the moles of hydrogen required for each 100 g = 0.125 mol × 3 = 0.375 mol of hydrogen gas. Converting to grams of hydrogen, 0.375 × 2 g/mol = 0.750 g hydrogen gas.

13.77   This lipid is a ceramide, a kind of sphingolipid.

13.79   Some proteins that are associated with membranes associate exclusively with one side of the membrane rather than the other.

13.81   Statements (c) and (d) are consistent with what is known about membranes. Covalent bonding between lipids and proteins [statement (e)] is not widespread. Proteins "float" in the lipid bilayers rather than being sandwiched between them [statement (a)]. Bulkier molecules tend to be found in the outer lipid layer [statement (b)].

13.83   Statement (c) is correct. Transverse diffusion is only rarely observed [statement (b)]. Proteins are bound to the inside and outside of the membrane [statement (a)].

13.85   Both lipids and carbohydrates contain carbon, hydrogen, and oxygen. Carbohydrates have aldehyde and ketone groups, as do some steroids. Carbohydrates have a number of hydroxyl groups, which lipids do not have to a great extent. Lipids have major components that are hydrocarbon in nature. These structural features imply that carbohydrates tend to be significantly more polar than lipids.

13.87   primarily lipid: olive oil and butter; primarily carbohydrate: cotton and cotton candy

13.89   The amounts are the key point here. Large amounts of sugar can provide energy. Fat burning due to the presence of taurine plays a relatively minor role because of the small amount.

13.91   The other ends of the molecules involved in the ester linkages in lipids, such as fatty acids, tend not to form long chains of bonds with other molecules.

13.93   The bulkier molecules tend to be found on the exterior of the cell because the curvature of the cell membrane provides more room for them.

13.95   The charges tend to cluster on membrane surfaces. Positive and negative charges attract each other. Two positive or two negative charges repel each other, so unlike charges do not have this repulsion.

Chapter 14 Proteins

14.1

Valine
(Val)

Phenylalanine
(Phe)

Valylphenylalanine
(Val-Phe)

14.2   a salt bridge

14.3   (a) storage   (b) movement

14.5   protection

14.7   Tyrosine has an additional hydroxyl group on the phenyl side chain.

14.9   arginine

14.11

Pyrrolidines
(heterocyclic
aliphatic amines)

14.13   They supply most of the amino acids we need in our bodies.

14.15   These structures are similar except that one of the hydrogens in the side chain of alanine has been replaced with a phenyl group in phenylalanine.

14.17   Amino acids are zwitterions; therefore they all have positive and negative charges. These molecules are very strongly attracted to each other, so they are solids at low temperatures.

14.19   All amino acids have a carboxyl group with a pK_a around 2 and an amino group with a pK_a between 8 and 10. One group is significantly more acidic and one is more basic. To have an un-ionized amino acid, the hydrogen would have to be on the carboxyl group and have vacated the amino group. Given that the carboxyl group is the stronger acid, this would never happen.

14.21

$$H_3\overset{+}{N}-\underset{\underset{COOH}{\overset{|}{CH_2}}}{\overset{\overset{H}{|}}{C}}-COO^-$$

14.23

14.25   the side-chain imidazole

14.27   The side chain of histidine is an imidazole with a nitrogen that reversibly binds to a hydrogen. When dissociated, it is neutral; when associated, it is positive. Therefore, chemically it is a base, even though it does have a pK_a in the acidic range.

14.29   histidine, arginine, and lysine

14.31   Serine may be obtained by the hydroxylation of alanine. Tyrosine is obtained by the hydroxylation of phenyl-alanine.

14.33   Thyroxine is a hormone that controls overall metabolic rate. Both humans and animals sometimes suffer from low levels of thyroxine, causing lack of energy and tiredness.

14.35

Alanylglutamine
(Ala-Gln)

Glutaminylalanine
(Gln-Ala)

14.37

NH₃⁺—CH—C—N—CH—C—N—CH—COO⁻

14.39   Only the peptide backbone contains polar units.

14.41

(a)

NH₃⁺—CH—C—N—CH—C—N—CH—COOH

(b) pH 2 is shown above. At pH 7.0 would look like:

NH₃⁺—CH—C—N—CH—C—N—CH—COO⁻

(c) At pH 10:

NH₂—CH—C—N—CH—C—N—CH—COO⁻

14.43   It would acquire a net positive charge and become more water-soluble.

14.45   (a) 256   (b) 160,000

14.47   valine or isoleucine

14.49   (a) secondary   (b) quaternary   (c) quaternary
(d) primary

14.51   Above pH 6.0, the COOH groups are converted to COO⁻ groups. The negative charges repel each other, disrupting the compact a-helix and converting it to a random coil.

14.53   (1) C-terminal end (2) N-terminal end
(3) pleated sheet (4) random coil
(5) hydrophobic interaction (6) disulfide bridge (7) α-helix
(8) salt bridge (9) hydrogen bonds

14.55   (a) Fetal hemoglobin has fewer salt bridges between the chains.
(b) Fetal hemoglobin has a higher affinity for oxygen.
(c) Fetal hemoglobin has an oxygen saturation curve that is in between myoglobin and maternal hemoglobin, so the graph would look like the figure below:

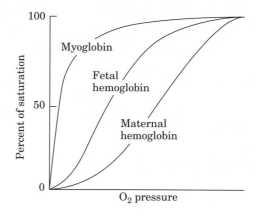

14.57   The heme and the polypeptide chain form the quaternary structure of cytochrome c. This is a conjugated protein.

14.59   the intramolecular hydrogen bonds between the peptide backbone carbonyl group and the N—H group.

14.61   cysteine

14.63   Ions of heavy metals like silver denature bacterial proteins by reacting with cysteine —SH groups. The proteins, denatured by formation of silver salts, form insoluble precipitates.

14.65   (Chemical Connections 14A) Nutrasweet contains phenylalanine. People suffering from the genetic disease phenylketonuria must avoid phenylalanine as they cannot metabolize it, and buildup in the body will have severe effects.

14.67   The symptoms of hunger, sweating, and poor coordination accompany diabetes when hypoglycemia is coming on.

14.69   The abnormal form has a higher percentage of β-pleated sheet compared to the normal form.

14.71   The oxygen-binding behavior of myoglobin is hyperbolic, while that of hemoglobin is sigmoidal.

14.73 The two most common are prion diseases and Alzheimer's disease.

14.75 Even if it is feasible, it is not completely correct to call the imaginary process that converts α-keratin to β-keratin "denaturation." Any process that changes a protein from α to β requires at least two steps: (1) conversion from the α form to a random coil, and (2) conversion from the random coil to the β form. The term "denaturation" describes only the first half of the process (Step 1). The second step would be called "renaturation." The overall process is called denaturation followed by renaturation. If we assume that the imaginary process actually occurs without passing through a random coil, then the term "denaturation" does not apply.

14.77 a quaternary structure because subunits are cross-linked

14.79 (a) hydrophobic (b) salt bridge (c) hydrogen bond (d) hydrophobic

14.81 glycine

14.83 one positive charge on the amino group

14.85 These amino acids have side chains that can catalyze organic reactions. They are polar or sometimes charged, and the ability to make hydrogen bonds or salt bridges can help catalyze the reaction.

14.87 Proteins can be denatured when the temperature is only slightly higher than a particular optimum. For this reason, the health of a warm-blooded animal is dependent on the body temperature. If the temperature is too high, proteins could denature and lose function.

14.89 Even if you know all of the genes in an organism, not all the genes code for proteins, nor are all genes expressed all the time.

14.91 A diet supplement full of collagen may help a person lose weight, but it would be of little use for repairing muscle tissue as collagen is not a good protein source. One third of its amino acids are glycine and another third are proline. Muscle repair requires high-quality protein to be effective.

Chapter 15 Enzymes

15.1 A catalyst is any substance that speeds up the rate of a reaction and is not itself changed by the reaction. An enzyme is a biological catalyst, which is either a protein or an RNA molecule.

15.3 Yes. Lipases are not very specific

15.5 because enzymes are very specific and thousands of reactions must be catalyzed in an organism

15.7 Lyases add water across a double bond or removes water from a molecule, thereby generating a double bond. Hydrolases use water to break an ester or amide bond, thereby generating two molecules.

15.9 (a) isomerase (b) hydrolase (c) oxidoreductase (d) lyase

15.11 *Cofactor* is more generic; it means a nonprotein part of an enzyme. A *coenzyme* is an organic cofactor.

15.13 In reversible inhibition, the inhibitor can bind and then be released. With noncompetitive inhibition, once the inhibitor is bound, no catalysis can occur. With irreversible inhibition, once the inhibitor is bound, the enzyme would be effectively dead, as the inhibitor could not be removed and no catalysis could occur.

15.15 No, at high substrate concentration the enzyme surface is saturated, and doubling of the substrate concentration will produce only a slight increase in the rate of the reaction or no increase at all.

15.17 (a) less active at normal body temperature (b) The activity decreases.

15.19 (a)

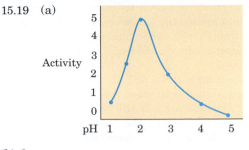

(b) 2

(c) zero activity

15.21 The active site of an enzyme is very specific for the size and shape of the substrate molecules. Urea is a small molecule and the urease active site is specific for it. Diethylurea has the two ethyl groups attached. It is unlikely that diethylurea would fit into an active site specific for urea.

15.23 The amino acid residues most often found at enzyme active sites are His, Cys, Asp, Arg, and Glu.

15.25 The correct answer is (c). Initially the enzyme does not have exactly the right shape for strongly binding a substrate, but the shape of the active site changes to better accommodate the substrate molecule.

15.27 Amino acid residues in addition to those at an enzyme active site are present to help form a three-dimensional pocket where the substrate binds. These amino acids act to make the size, shape, and environment (polar or nonpolar) of the active site just right for the substrate.

15.29 Caffeine is an allosteric regulator.

15.31 There is no difference. They are the same.

15.33

15.35

15.37 Glycogen phosphorylase is controlled by allosteric regulation and by phosphorylation. The allosteric controls are very fast, so that when the level of ATP drops, for example, there is an immediate response to the enzyme allowing more energy to be produced. The covalent modification by phosphorylation is triggered by hormone responses. They are a bit slower, but more long lasting and ultimately more effective.

15.39   Just as with lactate dehydrogenase, there are five isozymes of PFK: M_4, M_3L, M_2L_2, ML_3, and L_4.

15.41   Two enzymes that increase in serum concentration following a heart attack are creatine phosphokinase and aspartate aminotransferase. Creatine phosphokinase peaks earlier than aspartate aminotransferase and would be the best choice in the first 24 hours.

15.43   Serum levels of the enzymes AST and ALT are monitored for diagnosis of hepatitis and heart attack. Serum levels of AST are increased after a heart attack, but ALT levels are normal. In hepatitis, both enzymes are elevated. The diagnosis, until further testing, would indicate the patient may have had a heart attack.

15.45   Chemicals present in organic vapors are detoxified in the liver. The enzyme alkaline phosphatase is monitored to diagnose liver problems.

15.47   It is not possible to administer chymotrypsin orally. The stomach would treat it just as it does all dietary proteins: degrade it by hydrolysis to free amino acids. Even if whole, intact molecules of the enzyme were present in the stomach, the low pH in the region would not allow activity for the enzyme, which prefers an optimal pH of 7.8.

15.49   A transition-state analog is built to mimic the transition state of the reaction. It is not the same shape as the substrate or the product, but rather something in between. The potency with which such analogs can act as inhibitors lends credence to the theory of induced fit.

15.51   Succinylcholine has a chemical structure similar to that of acetylcholine, so both can bind to the acetylcholine receptor of the muscle end plate. The binding of either choline causes a muscle contraction. However, the enzyme acetylcholinesterase hydrolyzes succinylcholine only very slowly compared to acetylcholine. Muscle contraction does not occur as long as succinylcholine is still present and acts as a relaxant.

15.53   The most common reactions of kinases that we study in this book are the ones that involve using ATP to phosphorylate another molecule, be it an enzyme or a metabolite of a pathway. One example would be glycogen phosphorylase kinase. This enzyme catalyzes the following reaction as described in Chemical Connections 15E:

$$\text{Phosphorylase + ATP} \longrightarrow \text{Phosphorylase-P + ADP}$$

Another example is hexokinase from the pathway called glycolysis (chapter 28). Hexokinase catalyzes the following reaction:

$$\text{Glucose + ATP} \longrightarrow \text{glucose 6-P + ADP}$$

15.55   Many people have suffered psychological traumas that haunt them for many years or even their entire lives. If long-term memories could be selectively blocked, it would offer relief to patients suffering from something in their past.

15.57   In the enzyme pyruvate kinase, the $=CH_2$ of the substrate phosphoenolpyruvate sits in a hydrophobic pocket formed by the amino acids Ala, Gly, and Thr. The methyl group on the side chain of Thr, rather than the hydroxyl group, is in the pocket. Hydrophobic interactions are at work here to hold the substrate into the active site.

15.59 Researchers were trying to inhibit phosphodiesterases because cGMP acts to relax constricted blood vessels. This approach was hoped to help treat angina and high blood pressure.

15.61   Phosphorylase exists in a phosphorylated form and an unphosphorylated form, with the former being more active.

Phosphorylase is also controlled allosterically by several compounds, including AMP and glucose. While the two act semi-independently, they are related to some degree. The phosphorylated form has a higher tendency to assume the R state, which is more active, and the unphosphorylated form has a higher tendency to be in the less active T state.

15.63   In the processing of cocaine by specific esterase enzymes, the cocaine molecule passes through an intermediate state. A molecule was designed that mimics this transition state. This transition-state analog can be given to a host animal, which then produces antibodies to the analog. When these antibodies are given to a person, they act like an enzyme and degrade cocaine.

15.65   Cocaine blocks the reuptake of the neurotransmitter dopamine, leading to overstimulation of the nervous system.

15.67   (a) Vegetables such as green beans, corn, and tomatoes are heated to kill microorganisms before they are preserved by canning. Milk is preserved by the heating process, pasteurization. (b) Pickles and sauerkraut are preserved by storage in vinegar (acetic acid).

15.69   The amino acid residues (Lys and Arg) that are cleaved by trypsin have basic side chains; thus they are positively charged at physiological pH.

15.71   This enzyme works best at a pH of about 7.

15.73   a hydrolase

15.75   (a) The enzyme is called ethanol dehydrogenase or, more generally, alcohol dehydrogenase. It could also be called ethanol oxidoreductase.

(b) ethyl acetate esterase or ethyl acetate hydrolase

15.77   isozymes or isoenzymes

15.79   No, the direction a reaction goes is determined by the thermodynamics of the reaction, including the concentration of substrates and products. A reaction may only go in the forward direction in a metabolic pathway due to an overwhelming concentration of the substrates along with immediate removal of the products. However, the enzyme that catalyzes the reaction would catalyze the reaction in either direction if it is thermodynamically possible.

15.81   The athlete may benefit from the stimulatory effect of caffeine, but in a long race, the athlete would also become dehydrated from the diuretic effect on the kidneys. One of the most important factors to endurance performance is hydration, so any substance that causes dehydration would be detrimental to performance in a long-distance event.

15.83   The structure of RNA makes it more likely to be able to adopt a wider range of tertiary structures, so it can fold up to form globular molecules similar to protein-based enzymes. It also has an extra oxygen, which gives it an additional reactive group to use in catalysis or an electronegative group, useful in hydrogen bonding.

Chapter 16 Chemical Communications: Neurotransmitters and Hormones

16.1   G-protein is an enzyme; it catalyzes the hydrolysis of GTP to GDP. GTP, therefore, is a substrate.

16.3   A chemical messenger operates between cells; secondary messengers signal inside a cell in the cytoplasm.

16.5   The concentration of Ca^{2+} in neurons controls the process. When it reaches 10^{-4} M, the vesicles release the neurotransmitters into the synapse.

16.7   anterior pituitary gland

16.9 Upon binding of acetylcholine, the conformation of the proteins in the receptor changes and the central core of the ion channel opens.

16.11 The cobra toxin causes paralysis by acting as a nerve system antagonist. It blocks the receptor and interrupts the communication between neuron and muscle cell. The botulin toxin prevents the release of acetylcholine from presynaptic vesicles.

16.13 Taurine is a β-amino acid; its acidic group is $-SO_2OH$ instead of $-COOH$.

16.15 The amino group in GABA is in the gamma position; proteins contain only alpha amino acids.

16.17 (a) norepinephrine and histamine
(b) They activate a secondary messenger, cAMP, inside the cell.
(c) amphetamines and histidine

16.19 It is phosphorylated by an ATP molecule.

16.21 Product from the MAO-catalyzed oxidation of dopamine:

16.23 (a) Amphetamines increase and (b) reserpine decreases the concentration of the adrenergic neurotransmitter.

16.25 the corresponding aldehyde

16.27 (a) the ion-translocating protein itself
(b) It gets phosphorylated and changes its shape.
(c) It activates the protein kinase that does the phosphorylation of the ion-translocating protein.

16.29 They are pentapeptides.

16.31 The enzyme is a kinase. The reaction is the phosphorylation of inositol-1,4-diphosphate to inositol-1,4,5-triphosphate:

$$P = -PO_3^{2-}$$

16.33 Cyclic AMP

16.35 Protein Kinase

16.37 Glucagon initiates a series of reactions that eventually activates protein kinase. The protein kinase phosphorylates two key enzymes in the liver, activating one and inhibiting the other. The combination of these effects lowers the level of fructose 2,6-bisphosphate, a key regulator of carbohydrate metabolism. Fructose 2,6-bisphosphate stimulates glycolysis and inhibits gluconeogenesis. Therefore, when fructose 2,6-bisphosphate is decreased, gluconeogenesis is stimulated and glycolysis is inhibited.

16.39 Insulin binds to insulin receptors on liver and muscle cells. The receptor is an example of a protein called a tyrosine kinase. A specific tyrosine residue becomes phosphorylated on the receptor, activating its kinase activity. The target protein called IRS is then phosphorylated by the active tyrosine kinase. The phosphorylated IRS acts as the second messenger. It causes the phosphorylation of many target enzymes in the cell. The effect is to reduce the level of glucose in the blood by increasing the rate of pathways that use glucose and slowing the rate of pathways that make glucose.

16.41 Most receptors for steroid hormones are located in the cell nucleus.

16.43 In the brain, steroid hormones can act as neurotransmitters.

16.45 Calmodulin, a calcium-ion-binding protein, activates protein kinase II, which catalyzes phosphorylation of other proteins. This process transmits the calcium signal to the cell.

16.47 Local injections of the toxin prevent release of acetylcholine in that area.

16.49 The neurofibrillar tangles found in the brains of Alzheimer's patients are composed of tau proteins. Mutated tau proteins, which normally interact with the cytoskeleton, grow into these tangles instead, thus altering normal cell structure.

16.51 Drugs that increase the concentration of the neurotransmitter acetylcholine may be effective in the treatment of Alzheimer's disease. Acetylcholinesterase inhibitors, such as Aricept, inhibit the enzyme that decomposes the neurotransmitter.

16.53 The neurotransmitter dopamine is deficient in Parkinson's disease, but a dopamine pill would not be an effective treatment. Dopamine cannot cross the blood-brain barrier.

16.55 Drugs like Cogentin that block cholinergic receptors are often used to treat the symptoms of Parkinson's disease. These drugs lessen spastic motions and tremors.

16.57 Nitric oxide relaxes the smooth muscle that surrounds blood vessels. This relaxation causes increased blood flow in the brain, which in turn causes headaches.

16.59 Neurons adjacent to those damaged by the stroke begin to release glutamate and NO, which kills other cells in the region.

16.61 Insulin-dependent (type 1) diabetes is caused by insufficient production of insulin by the pancreas. Administration of insulin relieves symptoms of this type of diabetes. Non-insulin-dependent (type 2) diabetes is caused by a deficiency of insulin receptors or by the presence of inactive insulin receptors. Other drugs are used to relieve symptoms.

16.63 Monitoring glucose in the tears relieves the patient of taking many blood samples every day.

16.65 Some possible dangers include enlargement of the prostate, increases in chromosomal abnormalities, breast cancer and early onset of puberty.

16.67 Aldosterone binds to a specific receptor in the nucleus. The aldosterone–receptor complex serves as a transcription factor that regulates gene expression. Proteins for mineral metabolism are produced as a result.

16.69 Large doses of acetylcholine will help. Decamethonium bromide is a competitive inhibitor of acetylcholine esterase. The inhibitor can be removed by increasing substrate concentration.

16.71 Alanine is an α-amino acid in which the amino group is bonded to the same carbon as the carboxyl group. In β-alanine, the amino group is bonded to the carbon adjacent to the one to which the carboxyl group is bonded.

16.73 Effects of NO on smooth muscle are as follows: dilation of blood vessels and increased blood flow; headaches caused by dilation of blood vessels in brain; increased blood flow in the penis, leading to erections.

16.75 Acetylcholine esterase catalyzes the hydrolysis of the neurotransmitter acetylcholine to produce acetate and choline. Acetylcholine transferase catalyzes the synthesis of acetylcholine from acetyl—CoA and choline.

16.77 The reaction shown below is the hydrolysis of GTP:

16.79 Ritalin increases serotonin levels. Serotonin has a calming effect on the brain. One advantage of this drug is that it does not increase levels of the stimulant dopamine.

16.81 Proteins are capable of specific interactions at recognition sites. This ability makes for useful selectivity in receptors.

16.83 Adrenergic messengers, such as dopamine, are derivatives of amino acids. For example, a biochemical pathway exists that produces dopamine from the amino acid tyrosine.

16.85 Insulin is a small protein. It would go through protein digestion if taken orally and would not be taken up as the whole protein.

16.87 Steroid hormones directly affect nucleic acid synthesis.

16.89 Chemical messengers vary in their response times. Those that operate over short distances, such as neurotransmitters, have short response times. Their mode of action frequently consists of opening or closing channels in a membrane or binding to a membrane-bound receptor. Hormones must be transmitted in the bloodstream, which requires a longer time for them to take effect. Some hormones can and do affect protein synthesis, which makes the response time even longer.

16.91 Having two different enzymes for the synthesis and breakdown of acetylcholine means that the rates of formation and breakdown can be controlled independently.

Chapter 17 Nucleotides, Nucleic Acids, and Heredity

17.1

17.3 hemophilia, sickle cell anemia, etc.

17.5 (a) In eukaryotic cells, DNA is located in the cell nucleus and in mitochondria.

(b) RNA is synthesized from DNA in the nucleus, but further use of RNA (protein synthesis) occurs on ribosomes in the cytoplasm.

17.7 DNA has the sugar deoxyribose, while RNA has the sugar ribose. Also, RNA has uracil, while DNA has thymine.

17.9 Thymine and uracil are both based on the pyrimidine ring. However, thymine has a methyl substituent at carbon 5, whereas uracil has a hydrogen. All of the other ring substituents are the same.

17.11

17.13 D-Ribose and 2-deoxy-D-ribose have the same structure except at carbon 2. D-Ribose has a hydroxyl group and hydrogen on carbon 2, whereas deoxyribose has two hydrogens.

D-Ribose   2-Deoxy-D-ribose

17.15 The name "nucleic acid" derives from the fact that the nucleosides are linked by phosphate groups, which are the dissociated form of phosphoric acid.

17.17 anhydride bonds

17.19 In RNA, carbons 3′ and 5′ of the ribose are linked by ester bonds to phosphates. Carbon 1 is linked to the nitrogen base with an *N*-glycosidic bond.

17.21 (a)

(b)

17.23 (a) One end will have a free 5′ phosphate or hydroxyl group that is not in phosphodiester linkage. That end is called the 5′ end. The other end, the 3′ end, will have a 3′ free phosphate or hydroxyl group.
(b) By convention, the end drawn to the left is the 5′ end. A is the 5′ end, and C is the 3′ end.
(c) The complementary strand would be GTATTGCCAT written from 5′ to 3′.

17.25 two

17.27 electrostatic interactions

17.29 The superstructure of chromosomes consists of many elements. DNA and histones combine to form nucleosomes that are wound into chromatin fibers. These fibers are further twisted into loops and minibands to form the chromosome superstructure.

17.31 the double helix

17.33 DNA is wound around histones, collectively forming nucleosomes that are further wound into solenoids, loops, and bands.

17.35 rRNA

17.37 mRNA

17.39 Ribozymes, or catalytic forms of RNA, are involved in post-transcriptional splicing reactions that cleave larger RNA

molecules into smaller, more active forms. For example, tRNA molecules are formed in this way.

17.41 Small nuclear RNA is involved in splicing reactions of other RNA molecules.

17.43 Micro RNAs are 22 bases long and prevent transcription of certain genes. Small interfering RNAs vary from 22 to 30 bases and are involved in the degradation of specific mRNA molecules.

17.45 Immediately after transcription, messenger RNA contains both introns and exons. The introns are cleaved out by the action of ribozymes that catalyze splicing reactions on the mRNA.

17.47 no

17.49 the specificity between the base pairs, A-T and G-C

17.51

Thymine   Adenine
AT pair

17.53 four

17.55 In semiconservative DNA replication, the new daughter DNA helix is composed of one strand from the original (or parent) molecule and one new strand.

17.57

$$\text{Histone}-(CH_2)_4-NH_3^+ + CH_3-COO^- \underset{\text{deacetylation}}{\overset{\text{acetylation}}{\rightleftharpoons}}$$

$$\text{Histone}-(CH_2)_4-NH-\overset{\overset{\displaystyle O}{\|}}{C}-CH_3$$

17.59 Helicases are enzymes that break the hydrogen bonds between the base pairs in double-helix DNA and thus help the helix to unwind. This prepares the DNA for the replication process.

17.61 pyrophosphate

17.63 The leading strand or continuous strand is synthesized in the 5′ to 3′ direction.

17.65 DNA ligase

17.67 From the 5′ to the 3′ direction

17.69 One of the enzymes involved in the DNA base excision repair (BER) pathway is an endonuclease that catalyzes the hydrolytic cleavage of the phosphodiester backbone. The enzyme hydrolyzes on the 5′ side of the AP site.

17.71 a β-*N*-glycosidic bond between the damaged base and the deoxyribose

17.73 Individuals with the inherited disease XP lack an enzyme involved in the NER pathway. They are not able to make repairs in DNA damaged by UV light.

17.75   5′ATGGCAGTAGGC3′

17.77   The anticancer drug fluorouracil causes the inhibition of thymidine synthesis, thereby disrupting replication.

17.79   DNA polymerase, the enzyme that makes the phosphodiester bonds in DNA, is not able to work at the end of linear DNA. This results in the shortening of the telomeres at each replication. The telomere shortening acts as a timer for the cell, allowing it to keep track of the number of divisions.

17.81   Because the genome is circular, even if the 5′ primers are removed, there will always be DNA upstream that can act as a primer for DNA polymerase to use as it synthesizes DNA.

17.83   A DNA fingerprint is made from the DNA of the child, the mother, and any prospective fathers and used to eliminate possible fathers.

17.85   Once a DNA fingerprint is made, each band in the child's DNA must come from one of the parents. Therefore, if the child has a band, and the mother does not, then the father must have that band. In this way, possible fathers are eliminated.

17.87   One example might be that a life insurance company could raise your rates or refuse to give you insurance if your genetic profile had negative indicators. The same could happen with health insurance. Companies could begin to select for people with certain positive traits thereby discriminating against everyone else. Having that information could lead to a new form of discrimination.

17.89   It detoxifies drugs and other synthetic chemicals by adding a hydroxyl group to them.

17.91   A three-dimensional pocket of ribonucleotides where substrate molecules are bound for catalytic reaction. Functional groups for catalysis include the phosphate backbone, ribose hydroxyl groups, and the nitrogen bases.

17.93   (a) The structure of the nitrogen base uracil is shown in Figure 17.1. It is a component of RNA. (b) Uracil with a ribose attached by an N-glycosidic bond is called uridine.

17.95   native DNA

17.97   mol % A = 29.3; mol % T = 29.3; mol % G = 20.7; mol % C = 20.7.

17.99   RNA synthesis is 5′ to 3′

17.101   DNA replication requires a primer, which is RNA. Because RNA synthesis does not require a primer, it makes sense that RNA must have preceded DNA as a genetic material. This, added to the fact that RNA has been shown to be able to catalyze reactions, means that RNA can be both an enzyme and a heredity molecule.

17.103   The guanine-cytosine base pair has three hydrogen bonds, while the adenine-thymine base pair has only two. Therefore it takes more energy to separate DNA strands with more G — C base pairs as it takes more energy to break their three hydrogen bonds.

17.105   DNA is the blueprint for all of the components of an organism. It is important that it have repair mechanisms because if it is wrong, all of its products will always be wrong. If correct DNA leads to incorrect RNA by some mutation, then the products of the RNA may be wrong, but RNA is short lived and the next time the RNA is produced, it will be correct. A good analogy is that of a cookbook. The words on the page are the DNA. How you read them is the RNA. If you misread the words, you may make the recipe wrong once. If the book is printed wrong, however, you will always make the recipe wrong.

Chapter 18 Gene Expression and Protein Synthesis

18.1   First, binding proteins must make the portion of the chromosome where the gene is less condensed and more accessible. Second, the helicase enzyme must unwind the double helix near the gene. Third, the polymerase must recognize the initiation signal on the gene.

18.2   (a) CAU and CAC   (b) GUA and GUG

18.3   valine + ATP + $tRNA_{val}$

18.4   —CCT CGATTG —
        —GGAGC TAAC —

18.5   (c); gene expression refers to both processes—transcription and translation.

18.7   Protein translation occurs on the ribosomes.

18.9   Helicases are enzymes that catalyze the unwinding of the DNA double helix prior to transcription. The helicases break the hydrogen bonds between base pairs.

18.11   The termination signal is at the 5′ end of the template strand that is being transcribed. It can also be said to be at the 3′ end of the coding strand.

18.13   The "guanine cap" methyl group is located on nitrogen number 7 of guanine.

18.15   on the messenger RNA

18.17   The main subunits are the 60S and the 40S ribosomal subunits, although these can be dissociated into even smaller subunits.

18.19   326

18.21   Leucine, arginine, and serine have the most, with six codons. Methionine and tryptophan have the fewest, with one apiece.

18.23   The amino acid for protein translation is linked via an ester bond to the 3′ end of the tRNA. The energy for producing the ester bond comes from breaking two energy-rich phosphate anhydride bonds in ATP (producing AMP and two phosphates).

18.25   (a) The 40S subunit in eukaryotes forms the pre-initiation complex with the mRNA and the Met-tRNA that will become the first amino acid in the protein. (b) The 60S subunit binds to the pre-initiation complex and brings in the next aminoacyl-tRNA. The 60S subunit contains the peptidyl transferase enzyme.

18.27   Elongation factors are proteins that participate in the process of tRNA binding and movement of the ribosome on the mRNA during the elongation process in translation.

18.29   A special tRNA molecule is used for initiating protein synthesis. In prokaryotes, it is $tRNA^{fmet}$, which will carry a formyl-methionine. In eukaryotes, there is a similar molecule, but it carries methionine. However, this tRNA carrying methionine for the initiation of synthesis is different from the tRNA carrying methionine for internal positions.

18.31   There are no amino acids in the vicinity of the nucleophilic attack that leads to peptide bond formation. Therefore, the ribosome must be using its RNA portion to catalyze the reaction, so it is a type of enzyme called a ribozyme.

18.33   Parts of the DNA involved are promoters, enhancers, silencers, and response elements. Molecules that bind to DNA include RNA polymerase, transcription factors, and other proteins that may bind the RNA polymerase and a transcription factor.

18.35   The active site of aminoacyl-tRNA synthases (AARS) contains the sieving portions that ensure that each amino acid is linked to its correct tRNA. The two sieving steps work on the basis of the size of the amino acid.

18.37 Both are DNA sequences that bind to transcription factors. The difference is largely due to our own understanding of the big picture. A response element controls a set of responses in a particular metabolic context. For example, a response element may activate several genes when the organism is challenged metabolically by heavy metals, by heat, or by a reduction in oxygen pressure.

18.39 Proteosomes play a role in post-translational degradation of damaged proteins. Proteins that are damaged by age or proteins that have misfolded are degraded by the proteosomes.

18.41 (a) Silent mutation: Assume the DNA sequence is TAT on the coding strand, which will lead to UAU on the mRNA. Tyrosine is incorporated into the protein. Now assume a mutation in the DNA to TAC. This will lead to UAC in mRNA. Again, the amino acid will be tyrosine. (b) Lethal mutation: the original DNA sequence is GAA on the coding strand, which transcribes into GAA on mRNA. This codes for the amino acid glutamic acid. The DNA mutation TAA will lead to UAA, a stop signal that incorporates no amino acid.

18.43 Yes, a harmful mutation may be carried as a recessive gene from generation to generation, with no individual demonstrating symptoms of the disease. Only when both parents carry recessive genes does an offspring have a 25% chance of inheriting the disease.

18.45 Restriction endonucleases are enzymes that recognize specific sequences on DNA and catalyze the hydrolysis of phosphodiester bonds in that region, thereby cleaving both strands of the DNA.

18.47 Mutation by natural selection is an exceedingly long, slow process that has occurred for centuries. Each natural change in the gene has been ecologically tested and found usually to have a positive effect or the organism is not viable. Genetic engineering, where a DNA mutation is done very fast, does not provide sufficient time to observe all of the possible biological and ecological consequences of the change.

18.49 The discovery of restriction enzymes allowed scientists to cut DNA at specific locations and link different pieces of DNA together. This led to the ability to clone foreign DNA into a host, leading to the ability to both amplify DNA of interest and also have it expressed. Without restriction enzymes scientists would not be able to express a human protein in a bacterial cell, for example, or to create the therapeutic gene used in gene therapy.

18.51 The viral coat is a protective protein covering around a virus particle. All of the components necessary to make the coat—for example, amino acids and lipids—come from the host.

18.53 An invariant site is a location in a protein that has the same amino acid in all species that have been studied. Studies of invariant sites help establish genetic links and evolutionary relationships.

18.55 A silent mutation is a change in the DNA that does not lead to a change in the DNA product. This can happen when there is a base change in the DNA but due to the redundancy of the genetic code the change does not change the amino acid coded for.

18.57 A silent mutation may require a different tRNA molecule even though the same amino acid will be incorporated. The pace of the ribosome movement during translation may be different depending on the tRNA used, leading to the potential for different folding patterns in the protein produced.

18.59 The protein p53 is a tumor suppressor. When its gene is mutated, the protein no longer controls replication and the cell begins to grow at an increased rate.

18.61 The Duffy protein is found on the surface of human red blood cells. It acts as a docking protein for malaria, so mutations that lead to loss of the Duffy protein make the person resistant to malaria.

18.63 Several mutation types could affect production of protein Y. A mutation of gene X might change the protein sequence, such as we see in Duffy or sickle cell anemia. These changes can be minor or could lead to complete loss of function of the protein. Another mutation in gene X might be a silent mutation, but as we saw in Chemical Connection 18D, even a silent mutation could lead to a changed protein. Another possibility is that a mutation affects not the gene X directly but rather the promoter for gene X. If the promoter region is mutated, it might lead to fewer RNA polymerase molecules binding and reduced expression of the protein. Mutations could also affect enhancer or silencer regions, altering the level of expression of protein Y.

18.65 (a) Transcription: The units include the DNA being transcribed, the RNA polymerases, and a variety of transcription factors.
(b) Translation: mRNA, ribosomal subunits, aminoacyl-tRNA, initiation factors, elongation factors.

18.67 Hereditary diseases cannot be prevented, but genetic counseling can help people understand the risks involved in passing a mutated gene to their offspring.

18.69 (a) Plasmid: a small, closed circular piece of DNA found in bacteria. It is replicated in a process independent of the bacterial chromosome. (b) Gene: a section of chromosomal DNA that codes for a particular protein molecule or RNA.

18.71 Each of the amino acids has four codons. All of the codons start with G. The second base is different for each amino acid. The third base may be any of the four possible bases. The distinguishing feature for each amino acid is the second base.

18.73 The hexapeptide is Ala-Glu-Val-Glu-Val-Trp.

Chapter 19 Bioenergetics: How the Body Converts Food to Energy

19.1 ATP

19.3 (a) 2 (b) the outer membrane

19.5 Cristae are folded membranes originating from the inner membrane. They are connected to the inner membrane by tubular channels.

19.7 There are two phosphate anhydride bonds:

19.9 Neither; they yield the same energy.

19.11 It is a phosphate ester bond.

19.13 The two nitrogen atoms that are part of $C=N$ bonds are reduced to form $FADH_2$.

19.15 (a) ATP (b) NAD^+ and FAD (c) acetyl groups

19.17 An amide bond is formed between the amine portion of mercaptoethanolamine and the carboxyl group of pantothenic acid (see Figure 19.7).

19.19 No. The pantothenic acid portion is not the active part. The active part is the —SH group at the end of the molecule.

19.21 Both fats and carbohydrates are degraded to acetyl coenzyme A.

19.23 α-ketoglutarate

19.25 Succinate is oxidized by FAD, and the oxidation product is fumarate.

19.27 Fumarase is a lyase (it adds water across a double bond).

19.29 No, but GTP is produced in Step ⑤.

19.31 It allows the energy to be released in small packets.

19.33 carbon–carbon double bonds occur in *cis*-aconitate and fumarate.

19.35 α-Ketoglutarate transfers its electrons to NAD^+, which becomes $NADH + H^+$.

19.37 Mobile electron carriers of the electron transport chain: cytochrome c and CoQ

19.39 When H^+ passes through the ion channel, the proteins of the channel rotate. The kinetic energy of this rotatory motion is converted to and stored as the chemical energy in ATP.

19.41 This process takes place in the inner membranes of the mitochondria.

19.43 (a) 0.5 (b) 12

19.45 Ions reenter the mitochondrial matrix through the proton-translocating ATPase.

19.47 The F_1 portion of ATPase catalyzes the conversion of ADP to ATP.

19.49 The molecular weight of acetate = 59 g/mol, so 1 g acetate = 1 ÷ 59 = 0.017 mol acetate. Each mole of acetate produces 12 moles of ATP [see Problem 19.43(b)], so 0.017 mol × 12 = 0.204 mol ATP. This gives 0.204 mol ATP × 7.3 kcal/mol = 1.5 kcal.

19.51 (a) Muscles contract by sliding the thick filaments (myosin) and the thin filaments (actin) past each other. (b) The energy comes from the hydrolysis of ATP.

19.53 ATP transfers a phosphate group to the serine residue at the active site of glycogen phosphorylase, thereby activating the enzyme.

19.55 No. It would harm humans because they would not synthesize enough ATP molecules.

19.57 This amount of energy (87.6 kcal) is obtained from 12 mol of ATP (87.6 kcal ÷ 7.3 kcal/mol ATP = 12 mol ATP). Oxidation of 1 mol of acetate yields 12 mol ATP. The molecular weight of CH_3COOH is 60 g/mol, so the answer is 60 g or 1 mol CH_3COOH.

19.59 The energy of motion appears first in the ion channel, where the passage of H^+ causes the proteins lining the channel to rotate.

19.61 They are both hydroxy acids.

19.63 Myosin, the thick filament in muscle, is an enzyme that acts as an ATPase.

19.65 Isocitrate has two stereocenters.

19.67 The ion channel is the F_0 portion of the ATPase; it is made of 12 subunits.

19.69 No, it largely comes from the chemical energy as a result of the breaking of bonds in the O_2 molecule.

19.71 It removes two hydrogens from succinate to produce fumarate.

19.73 The carbon dioxide that we exhale is released by the two oxidative decarboxylation steps in the citric acid cycle.

19.75 Because of the central role of citric acid in metabolism, it can be considered a good nutrient.

19.77 Complex II does not generate enough energy to produce ATP. The rest do.

19.79 Citrate isomerizes to isocitrate to convert a tertiary alcohol to a secondary alcohol. Tertiary alcohols cannot be oxidized, but secondary alcohols can be oxidized to produce a keto group.

19.81 Iron is found in iron-sulfur clusters in proteins and is also part of the heme group of cytochromes.

19.83 Mobile electron carriers transfer electrons on their path from one large, less mobile protein complex to another.

19.85 ATP and reducing agents such as NADH and $FADH_2$, which are generated by the citric acid cycle, are needed for biosynthetic pathways.

19.87 Biosynthetic pathways are likely to feature reduction reactions because their net effect is to reverse catabolism, which is oxidative.

19.89 ATP is not stored in the body. It is hydrolyzed to provide energy for many different kinds of processes and thus turns over rapidly.

19.91 The citric acid cycle generates NADH and $FADH_2$, which are linked to oxygen by the electron transport chain.

Chapter 20 Specific Catabolic Pathways: Carbohydrate, Lipid, and Protein Metabolism

20.1 According to Table 20.2, the ATP yield from stearic acid is 146 ATP. This makes 146/18 = 8.1 ATP/carbon atom. For lauric acid (C_{12}):

Step ① Activation	–2 ATP
Step ② Dehydrogenation five times	10 ATP
Step ③ Dehydrogenation five times	15 ATP
Six C_2 fragments in common pathway	72 ATP
Total	95 ATP

95/12 = 7.9 ATP per carbon atom for lauric acid. Thus stearic acid yields more ATP/C atom.

20.3 They serve as building blocks for the synthesis of proteins.

20.5 The two C_3 fragments are in equilibrium. As the glyceraldehyde phosphate is used up, the equilibrium shifts and converts the other C_3 fragment (dihydroxyacetone phosphate) to glyceraldehyde phosphate.

20.7 (a) Steps ① and ③ (b) Steps ⑥ and ⑨

20.9 ATP inhibition takes place at Step ⑨. It inhibits the pyruvate kinase by feedback regulation.

20.11 NADPH is the compound in question.

20.13 Each mole of glucose produces two moles of lactate, so three moles of glucose give rise to six moles of lactate.

20.15 According to Table 20.1, two moles of ATP are produced directly in the cytoplasm.

20.17 Two net ATP molecules are produced in both cases.

20.19 Enzymes that catalyze the phosphorylation of substrates using ATP are called kinases. Therefore, the enzyme that transforms glycerol to glycerol 1-phosphate is called glycerol kinase.

20.21 (a) The two enzymes are thiokinase and thiolase. (b) "Thio" refers to the presence of a sulfur-containing group, such as —SH. (c) Both enzymes insert a CoA—SH into a compound.

20.23 Each turn of fatty acid β-oxidation yields one acetyl CoA, one $FADH_2$, and one NADH. After three turns, $CH_3(CH_2)_4CO$—CoA remains from the original lauric acid; three acetyl CoA, three $FADH_2$, and three NADH + H^+ are produced.

20.25 Using data from Table 20.2, we obtain a figure of 112 moles of ATP for each mole of myristic acid.

20.27 The body preferentially uses carbohydrates as an energy source.

20.29 (a) The transformation of acetoacetate to β-hydroxybutyrate is a reduction reaction. (b) Acetone is produced by decarboxylation of acetoacetate.

20.31 It enters the citric acid cycle.

20.33 Oxidative deamination of alanine to pyruvate:

$$CH_3-\underset{\underset{NH_3^+}{|}}{CH}-COO^- + NAD^+ + H_2O \longrightarrow$$

$$CH_3-\underset{\underset{O}{\|}}{C}-COO^- + NADH + H^+ + NH_4^+$$

20.35 One of the nitrogens comes from ammonium ion through the intermediate carbamoyl phosphate. The other nitrogen comes from aspartate.

20.37 (a) The toxic product is ammonium ion. (b) The body gets rid of it by converting it to urea.

20.39 Tyrosine is considered a glucogenic amino acid because pyruvate can be converted to glucose when the body needs it.

20.41 It is stored in ferritin and reused.

20.43 Muscle cramps come from lactic acid accumulation.

20.45 The bicarbonate/carbonic acid buffer counteracts the acidic effects of ketone bodies.

20.47 The reaction is a transamination:

20.49 Black and blue are due to the hemoglobin in congealed blood, green to biliverdin, and yellow to bilirubin.

20.51 Production of ethanol in yeast takes place as a result of glycolysis, giving a net yield of two ATP molecules for each mole of glucose metabolized.

20.53 Glucose can be converted to ribose by the pentose phosphate pathway.

20.55 The step in glycolysis in which a phosphate group is transferred from phosphoenolpyruvate (PEP) to ADP to produce ATP indicates that the energy of the phosphate group in PEP is higher than that in ATP.

20.57 Carbamoyl phosphate has an amide group and a phosphate group.

20.59 Pyruvate can be converted to oxaloacetate.

20.61 Table 20.1 takes into account the fact that glucose can be metabolized further by the citric acid cycle, which produces NADH and $FADH_2$. These coenzymes pass electrons to oxygen, giving rise to ATP in the process.

20.63 Lactate plays a key role in regenerating NAD^+.

20.65 Amino acids can be catabolized to yield energy, but usually only under starvation conditions.

20.67 Catabolism, oxidative, energy-yielding; anabolism, reductive, energy-requiring.

20.69 If you look at the balanced chemical equations for the two processes, they are the exact opposite of each other. They differ in that photosynthesis requires energy from the sun and occurs only in some organisms such as plants, whereas aerobic catabolism of glucose releases energy and occurs in organisms of all sorts.

20.71 Sugars are already partially oxidized, so their pathway of complete oxidation is further advanced, producing less energy.

20.73 The reactions of glycolysis take place in the cytosol. Because of their charge the compounds that form a part of the pathway are not as prone to crossing the cell membrane to the exterior as they would be if they were uncharged. The reactions of the citric acid cycle take place in mitochondria, which have double membrane. The intermediates of the citric acid cycle tend to stay within mitochondria even without a charge.

20.75 ATP production takes place in connection with the re-oxidation of the NADH and $FADH_2$ produced in the citric acid cycle.

Chapter 21 Biosynthetic Pathways

21.1 Different pathways allow for flexibility and overcome unfavorable equilibria. Separate control of anabolism and catabolism becomes possible.

21.3 The main biosynthesis of glycogen does not use inorganic phosphate because the presence of a large inorganic phosphate pool would shift the reaction to the degradation process such that no substantial amount of glycogen would be synthesized.

21.5 Photosynthesis is the reverse of respiration:

$$6CO_2 + 6H_2O \longrightarrow C_6H_{12}O_6 + 6O_2 \quad \text{Photosynthesis}$$
$$C_6H_{12}O_6 + 6O_2 \longrightarrow 6CO_2 + 6H_2O \quad \text{Respiration}$$

21.7 A compound that can be used for gluconeogenesis:
(a) from glycolysis: pyruvate
(b) from the citric acid cycle: oxaloacetate
(c) from amino acid oxidation: alanine

21.9 Glucose needs for the brain are met by gluconeogenesis, because the other pathways metabolize glucose, and only gluconeogenesis manufactures it.

21.11 Maltose is a disaccharide that is composed of two glucose units linked by an α-1,4-glycosidic bond.

$$\text{UDP-glucose} + \text{glucose} \longrightarrow \text{maltose} + \text{UDP}$$

21.13 UTP consists of uracil, ribose, and three phosphates.
21.15 (a) Fatty acid biosynthesis occurs primarily in the cytoplasm. (b) No, fatty acid degradation occurs in the mitochondrial matrix.
21.17 In fatty acid biosynthesis, a three-carbon compound, malonyl ACP, is repeatedly added to the synthase.
21.19 Carbon dioxide is released from malonyl ACP, leading to the addition of two carbons to the growing fatty acid chain.
21.21 It is an oxidation step because the substrate is oxidized with concomitant removal of hydrogen. The oxidizing agent is O_2. NADPH is also oxidized during this step.
21.23 NADPH is bulkier than NADH because of its extra phosphate group; it also has two more negative charges.
21.25 No, the body makes other unsaturated fatty acids, such as oleic acid and arachidonic acid.
21.27 The activated components needed are sphingosine, acyl CoA, and UDP-glucose.
21.29 All of the carbons in cholesterol originate in acetyl CoA. A C_5 fragment called isopentenyl pyrophosphate is an important intermediate in steroid biosynthesis.

$$3 \underset{C_2}{\text{Acetyl-CoA}} \longrightarrow \underset{C_6}{\text{mevalonate}}$$
$$\longrightarrow \underset{C_5}{\text{isopentenyl pyrophosphate}} + CO_2$$

21.31 The amino acid product is aspartic acid.
21.33 The products of the transamination reaction shown are valine and α-ketoglutarate.

$$(CH_3)_2CH\overset{\overset{O}{\|}}{-C}-COO^- + {}^-OOC-CH_2-CH_2-\underset{\underset{NH_3^+}{|}}{CH}-COO^- \longrightarrow$$

The keto form of valine   Glutamate

$$(CH_3)_2CH-\underset{\underset{NH_3^+}{|}}{CH}-COO^- + {}^-OOC-CH_2-CH_2-\overset{\overset{O}{\|}}{C}-COO^-$$

Valine α-Ketoglutarate

21.35 NADPH is the reducing agent in the process of carbon dioxide being incorporated into carbohydrates.
21.37 Acetyl-CoA carboxylase (ACC) is a key enzyme in fatty acid biosynthesis. It exists in two forms, one found in liver and one in muscle tissue. The one found in muscle affects weight loss and may become a target for anti-obesity drugs.
21.39 The bonds that connect the nitrogen bases to the ribose units are β-N-glycosidic bonds just like those found in nucleotides.
21.41 The amino acid produced by this transamination is phenylalanine.
21.43 The structure of a lecithin (phosphatidyl choline) is shown in Section 13.6. Synthesis of a molecule of this sort requires activated glycerol, two activated fatty acids, and activated choline. Each activation requires one ATP molecule, for a total number of four ATP molecules.

21.45 The compound that reacts with glutamate in a transamination reaction to form serine is 3-hydroxypyruvate. The reverse of the reaction is shown below:

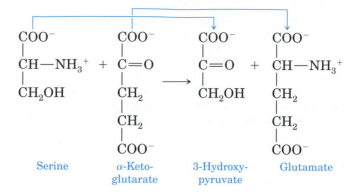

Serine α-Keto-glutarate   3-Hydroxy-pyruvate   Glutamate

21.47 HMG-CoA is hydroxymethylglutaryl CoA. Its structure is shown in Section 21.4. Carbon 1 is the carbonyl group linked to the thio group of CoA.
21.49 Heme is a porphyrin ring with iron at the center. Chlorophyll is a porphyrin ring with magnesium at the center.
21.51 Fatty acid biosynthesis takes place in the cytoplasm, requires NADPH, and uses malonyl CoA. Fatty acid catabolism takes place in the mitochondrial matrix, produces NADH and $FADH_2$, and has no requirement for malonyl CoA.
21.53 Photosynthesis has high requirements for light energy from the Sun.
21.55 Lack of essential amino acids would hinder the synthesis of the protein part. Gluconeogenesis can produce sugars even under starvation conditions.
21.57 Separation of catabolic and anabolic pathways allows for greater efficiency, especially in control of the pathways.
21.59 If laboratory rats are fed all the amino acids but minus one of the essential ones, they will be unable to synthesize protein. Administering the essential amino acid later will not be useful because the other amino acids have already been metabolized.

Chapter 22 Nutrition

22.1 No, nutrient requirements vary from person to person.
22.3 Sodium benzoate is not catabolized by the body; therefore, it does not comply with the definition of a nutrient—components of food that provide growth, replacement, and energy. Calcium propionate enters mainstream metabolism by conversion to succinyl-CoA and catabolism by the citric acid cycle and thus is a nutrient.
22.5 The Nutrition Facts label found on all foods must list the percentage of Daily Values for four important nutrients: vitamins A and C, calcium, and iron.
22.7 Chemically, fiber is cellulose, a polysaccharide that cannot be degraded by humans. It is important for proper operation of dietary processes, especially in the colon.
22.9 The basal caloric requirement is calculated assuming the body is completely at rest. Because most of us perform some activity, we need more calories than this basic minimum.
22.11 1833 Cal
22.13 No. Using diuretics would be a temporary fix at best.

22.15 The product would be different-sized oligosaccharide fragments much smaller than the original amylose molecules.

22.17 No. Dietary maltose, the disaccharide composed of glucose units linked by an α-1,4-glycosidic bond, is rapidly hydrolyzed in the stomach and small intestines. By the time it reaches the blood, it is the monosaccharide glucose.

22.19 linoleic acid

22.21 No. Lipases degrade neither; they degrade triacylglycerols.

22.23 Yes, it is possible for a vegetarian to obtain a sufficient supply of adequate proteins; however, the person must be very knowledgeable about the amino acid content of vegetables, so as to allow for protein complementation.

22.25 Dietary proteins begin degradation in the stomach, which contains HCl in a concentration of about 0.5%. Trypsin is a protease present in the small intestines that continues protein digestion after the stomach. Stomach HCl denatures dietary protein and causes somewhat random hydrolysis of the amide bonds in the protein. Fragments of the protein are produced. Trypsin catalyzes hydrolysis of peptide bonds only on the carboxyl side of the amino acids Arg and Lys.

22.27 It is expected that many of the prisoners will develop deficiency diseases in the near future.

22.29 Limes provided sailors with a supply of vitamin C to prevent scurvy.

22.31 Vitamin K is essential for proper blood clotting.

22.33 The only disease that has been proven scientifically to be prevented by vitamin C is scurvy.

22.35 Vitamins E and C and the carotenoids may have significant effects on respiratory health. This may be due to their activity as antioxidants.

22.37 There is a sulfur atom in biotin and in vitamin B_1 (also called thiamine).

22.39 The original Food Guide Pyramid did not consider the differences between types of nutrients. It assumed that all fats were to be limited and that all carbohydrates were healthy. The new guidelines recognizes that polyunsaturated fats are necessary and that carbohydrates from whole grains are better for you than those from refined sources. The new pyramid also recognizes the importance of exercise, which the original did not.

22.41 All proteins, carbohydrates, and fats in excess have metabolic pathways that lead to increased levels of fatty acids. However, there is no pathway that allows fats to generate a net surplus of carbohydrates. Thus fat stores cannot be used to make carbohydrates when a person's blood glucose is low.

22.43 All effective weight loss is based on increasing activity while limiting caloric intake. However, it is more effective to concentrate on increasing activity than on limiting intake.

22.45 Theoretically speaking, if humans had the glyoxylate pathway, dieting would be easier. By eliminating the two decarboxylation steps of the citric acid cycle, there is no loss of carbon from the acetyl-CoA. Therefore, carbon compounds could be removed from the pathway to form glucose. A person could diet and use fat stores to power the body's systems and maintain blood glucose levels.

22.47 (a) Most studies show that the artificial sweeteners Sucralose and acesulfame-K are not metabolized in any measurable amounts. (b) Digestion of aspartame can lead to high levels of phenylalanine.

22.49 Iron is an important cofactor in many biological compounds. The most obvious is the part iron plays in hemoglobin. It is the iron that directly binds the oxygen that is the source of respiration for our metabolism. Iron must be consumed in the diet to maintain iron levels for hemoglobin and many other compounds.

22.51 Factors that affect absorption include the solubility of the compound of iron, the presence of antacids in the digestive tract, and the source of the iron.

22.53 arginine

22.55 Carbohydrate loading before the event and consuming carbohydrates during the event

22.57 Caffeine acts as a central nervous system stimulant, which provides a feeling of energy that athletes often enjoy. In addition, caffeine reduces insulin levels and stimulates oxidation of fatty acids, which would be beneficial to endurance athletes. However, caffeine is also a diuretic and can lead to dehydration in long-distance events.

22.59 Cost is the most significant downside to organic food, as organic food can be up to 100% more expensive than non-organic. The type of food is also an important consideration as pesticides or other chemicals are transferred from the food to the consumer while others are not. For example, if a pesticide is concentrated in a banana peel, that is not as big a problem as if it were concentrated in the banana itself. Pesticides or other chemicals are more hazardous to children and pregnant women than to others.

22.61 The vitamin pantothenic acid is part of CoA.
(a) Glycolysis: Pyruvate dehydrogenase uses CoA as a coenzyme.
(b) Fatty acid synthesis: The first step involves the enzyme fatty acid synthase.

22.63 Proteins that are ingested in the diet are degraded to free amino acids, which are then used to build proteins that carry out specific functions. Two very important functions include structural integrity and biological catalysis. Our proteins are constantly being turned over—that is, continuously being degraded and rebuilt using free amino acids.

22.65 The very tip of the Food Guide Pyramid displays fats, oils, and sweets, with the cautionary statement, "Use sparingly." We can omit sweets completely from the diet; however, complete omission of fats and oils is dangerous. We must have dietary fats and oils that contain the two essential fatty acids. The essential fatty acids may be present as components in other food groups—that is, the meat, poultry, and fish group.

22.67 Walnuts are not just a tasty snack—they are a healthy one. Walnuts have protein. In fact, nuts are included in a group of the U.S. Department of Agriculture's Food Guide Pyramid. Walnuts are also a good source of vitamins and minerals, including vitamins E and B, biotin, potassium, magnesium, phosphorus, zinc, and manganese.

22.69 No, the lecithin is degraded in the stomach and intestines long before it could get into the blood. The phosphoglyceride is degraded to fatty acids, glycerol, and choline, which are absorbed through the intestinal walls.

22.71 Patients who have undergone ulcer surgery are administered digestive enzymes that may have been lost during the procedure. The enzyme supplement should contain proteases to help break down proteins as well as lipases to assist in fat digestion.

Chapter 23 Immunochemistry

23.1 Examples of external innate immunity include action by the skin, tears, and mucus.

23.3 The skin fights infection by providing a barrier against penetration of pathogens. The skin also secretes lactic acid and fatty acids, both of which create a low pH, thereby inhibiting bacterial growth.

23.5 Innate immunity processes have little ability to change in response to immune dangers. The key features of adaptive (acquired) immunity are specificity and memory. The acquired immune system uses antibody molecules designed for each type of invader. In a second encounter with the same danger, the response is more rapid and more prolonged than the first response.

23.7 T cells originate in the bone marrow, but grow and develop in the thymus gland. B cells originate and grow in the bone marrow.

23.9 Macrophages are the first cells in the blood that encounter potential threats to the system. They attack virtually anything that is not recognized as part of the body, including pathogens, cancer cells, and damaged tissue. Macrophages engulf an invading bacterium or virus and kill it with (NO), nitric oxide, and then digest it.

23.11 protein-based antigens

23.13 Class II MHC molecules pick up damaged antigens. A targeted antigen is first processed in lysosomes, where it is degraded by proteolytic enzymes. An enzyme, GILT, reduces the disulfide bridges of the antigen. The reduced peptide antigens unfold and are further degraded by proteases. The peptide fragments remaining serve as epitopes that are recognized by class II MHC molecules.

23.15 MHC molecules are transmembrane proteins that belong to the immunoglobulin superfamily. They are originally present inside cells until they become associated with antigens and move to the surface membrane.

23.17 If we assume that the rabbit has never been exposed to the antigen, the response will occur 1–2 weeks after the injection of antigen.

23.19 (a) IgE molecules have a carbohydrate content of 10–12%, which is equal to that of IgM molecules. IgE molecules have the lowest concentration in the blood. The blood concentration of IgE is about 0.01–0.1 mg/100 mL of blood.
(b) IgE molecules are involved in the effects of hay fever and other allergies. They also offer protection against parasites.

23.21 The two Fab fragments would be able to bind to an antigen. These fragments contain the variable protein sequence regions and hence are able to change during synthesis against a specific antigen.

23.23 *Immunoglobulin superfamily* refers to all of the proteins that have the standard structure of a heavy chain and a light chain.

23.25 Antibodies and antigens are held together by weak, noncovalent interactions: hydrogen bonds, electrostatic interactions (dipole–dipole), and hydrophobic interactions.

23.27 The DNA for the immunoglobulin superfamily has multiple ways of recombining during cell development. The diversity is a reflection of the number of permutations and ways of combining various constant regions, variable regions, joining regions, and diversity regions.

23.29 T cells carry on their surfaces unique receptor proteins that are specific for antigens. These receptors (TcR), which are members of the immunoglobulin superfamily, have constant and variable regions. They are anchored in the T-cell membrane by hydrophobic interactions. They are not able to bind antigens alone, but rather need additional protein molecules called cluster determinants that act as coreceptors. When TcR molecules combine with cluster determinant proteins, they form T-cell-receptor complexes (TcR complexes).

23.31 The components of the TcR complexes are (1) accessory protein molecules called cluster determinants and (2) the T-cell receptor.

23.33 CD4

23.35 They are adhesion molecules that help dock antigen-presenting cells to T cells. They also act as signal transducers.

23.37 Cytokines are glycoproteins that interact with cytokine receptors on macrophages, B cells, and T cells. They do not recognize and bind antigens.

23.39 Chemokines are a class of cytokines that send messages between cells. They attract leukocytes to the site of injury and bind to specific receptors on the leukocytes.

23.41 All chemokines are low-molecular-weight proteins that have four cysteine residues that are linked in very specific disulfide bonds: Cys1—Cys3 and Cys2—Cys4.

23.43 The T cells mature in the thymus gland. During maturation, those cells that fail to interact with MHC and thus cannot respond to foreign antigens are eliminated by a special selection process. T cells that express receptors that may interact with normal self antigens are eliminated by the same selection process.

23.45 A signaling pathway that controls the maturation of B cells is the phosphorylation pathway activated by tyrosine kinase and deactivated by phosphatase.

23.47 the cytokines and chemokines

23.49 helper T cells

23.51 It is hard to find because the virus mutates quickly. Also, one of its docking proteins changes conformation when it docks, so that antibodies elicited against undocked proteins are ineffective. It binds to several proteins that inhibit antiviral factors, and cloaks its outer membrane with sugars that are very similar to the natural sugars found on host cells.

23.53 Vaccines rely on the immune system's ability to recognize a foreign molecule and make specific antibodies to it. HIV hides from the immune system in a variety of ways, and it changes often. The body makes antibodies, but they are not very effective at finding or neutralizing the virus.

23.55 It had been known since the eighteen hundreds that the mayapple had anticancer properties. It was later found that a chemical found in the mayapple, picropodophyllin, inhibits spindle formation during mitosis in dividing cells. As all chemotherapy agents do, they hinder rapidly dividing cells, like cancer cells, more then regular cells.

23.57 Most cancer cells have specific proteins on their surface that help allow their identification as cancerous. Monoclonal antibodies are very specific for the molecules they will bind to, making them an excellent choice for a weapon against cancer. The antibodies will attack the cancer cell and only the cancer cell if the monoclonal antibody is specific enough.

23.59 Fluorescence labeling studies show that breast cancer cells have elevated levels of the HER2 protein. In addition, drugs designed to attack HER2 are very successful at targeting breast cancer cells.

23.61 Many cancers are linked to dimerization of specific cell receptors. Tyrosine kinase is a type of cell receptor that functions via dimerization. Specific monoclonal antibodies are being designed to block the dimerization of these tyrosine kinases.

23.63 Jenner noticed that milkmaids, who would often be exposed to cowpox, rarely if ever got smallpox

23.65 Allergies to antibiotics can be very potent. A person may show no symptoms with the first exposure, but a second or third may produce severe reactions or even be fatal.

23.67 Sex workers in some countries use constant low doses of antibiotics in an attempt to avoid sexually transmitted diseases. The unfortunate side effect of this practice has been to allow the evolution of strains of gonorrhea that are antibiotic resistant.

23.69 One of the molecules on the streptococcus bacteria resembles a protein found in the valves of the heart. The body's attempt to fight strep throat can lead to antibodies that attack not only the bacteria but also the person's own heart valves. This is the danger in rheumatic fever.

23.71 Stem cells can be transformed into other cell types. Scientists are working to find ways to use stem cells to repair damaged nerve tissue or brain tissue. In some animal models, brain cell function has been restored after a stroke by adding stem cells to the brain in the area of the damage.

23.73 IgA molecules are the first line of defense since they are found in tears and mucous secretions. They can intercept invaders before they get into the bloodstream.

23.75 Chemokines (or, more generally, cytokines) help leukocytes migrate out of a blood vessel to the site of injury. Cytokines help the proliferation of leukocytes.

23.77 A compound called 12:13 dEpoB, a derivative of epothilon B, is being studied as an anticancer vaccine.

23.79 Tumor necrosis factor receptors are located on the surfaces of several cell types, but especially on tumor cells.

Glossary

A site (*Section 18.3*) The site on the large ribosomal subunit where the incoming tRNA molecule binds.

Abzyme (*Section 15.8*) An immunoglobin generated by using a transition-state analog as an antigen.

Acetal (*Section 9.4C*) A molecule containing two —OR groups bonded to the same carbon.

Acetyl group (*Section 19.3*) The group CH_2CO—

Achiral (*Section 6.1*) An object that lacks chirality; an object that is superposable on its mirror image.

Acid–base reaction (*Section 7.3*) A proton-transfer reaction.

Acid ionization constant (K_a) (*Section 7.5*) An equilibrium constant for the ionization of an acid in aqueous solution to H_3O^+ and its conjugate base. K_a is also called an **acid dissociation constant**.

Acidosis (*Chemical Connections 7C*) A condition in which the pH of blood is lower than 7.35.

Acquired immunity (*Section 23.1*) The second line of defense that vertebrates have against invading organisms.

Activating receptor (*Section 23.7*) A receptor on a cell of the innate immune system that triggers activation of the immune cell in response to a foreign antigen.

Activation of an amino acid (*Section 18.5*) The process by which an amino acid is bonded to an AMP molecule and then to the 3′—OH of a tRNA molecule.

Activation of an enzyme (*Section 15.2*) Any process by which an inactive enzyme is transformed into an active enzyme.

Active site (*Section 15.3*) A three-dimensional cavity of an enzyme with specific chemical properties to accommodate the substrate.

Adaptive immunity (*Section 23.1*) Acquired immunity with specificity and memory.

Adenovirus (*Section 18.9*) A common vector used in gene therapy.

Adhesion molecule (*Section 23.5*) A protein that helps to bind an antigen to the T-cell receptor.

Adrenergic neurotransmitter (*Section 16.4*) A monoamine neurotransmitter or hormone, the most common of which are epinephrine (adrenaline), serotonin, histamine, and dopamine.

Advanced glycation end products (*Section 14.7*) A chemical product of sugars and proteins linking together to produce an imine.

Affinity maturation (*Section 23.4*) The process of mutation of T cells and B cells in response to an antigen.

Agonist (*Section 16.1*) A molecule that mimics the structure of a natural neurotransmitter or hormone, binds to the same receptor, and elicits the same response.

AIDS (*Section 23.8*) Acquired *i*mmune *d*eficiency *s*yndrome. The disease caused by the human immunodeficiency virus, which attacks and depletes T cells.

Alcohol (*Section 1.4A*) A compound containing an —OH (hydroxyl) group bonded to a tetrahedral carbon atom.

Aldehyde (*Section 1.4C*) A compound containing a carbonyl group bonded to a hydrogen; a —CHO group.

Alditol (*Section 12.3*) The product formed when the CHO group of a monosaccharide is reduced to a CH_2OH.

Aldose (*Section 12.1*) A monosaccharide containing an aldehyde group.

Aliphatic amine (*Section 8.1*) An amine in which nitrogen is bonded only to alkyl groups.

Aliphatic hydrocarbon (*Section 2.1*) An alkane.

Alkaloid (*Chemical Connections 8B*) A basic nitrogen-containing compound of plant origin, many of which have physiological activity when administered to humans.

Alkalosis (*Chemical Connections 7D*) A condition in which the pH of blood is greater than 7.45.

Alkane (*Section 2.1*) A saturated hydrocarbon whose carbon atoms are arranged in an open chain—that is, not arranged in a ring.

Alkene (*Section 3.1*) An unsaturated hydrocarbon that contains a carbon–carbon double bond.

Alkyl group (*Section 2.3A*) A group derived by removing a hydrogen atom from an alkane; given the symbol R—.

Alkyne (*Section 3.1*) An unsaturated hydrocarbon that contains a carbon–carbon triple bond.

Allosteric protein (*Chemical Connections 14G*) A protein that exhibits a behavior such that binding of one molecule at one site changes the ability of the protein to bind another molecule at a different site.

Allosterism (Allosteric enzyme) (*Section 15.6*) An enzyme regulation in which the binding of a regulator on one site of the enzyme modifies the ability of the enzyme to bind the substrate at the active site. Allosteric enzymes often have multiple polypeptide chains with the possibility of chemical communication between the chains.

Alpha (α-) amino acid (*Section 14.2*) An amino acid in which the amino group is bonded to the carbon atom next to the —COOH carbon.

Alpha helix (*Section 14.9*) A type of repeating secondary structure of a protein in which the peptide chain adopts a helical conformation stabilized by hydrogen bonding between a peptide backbone N—H and the backbone C=O four amino acids farther up the chain.

Amine (*Section 1.4*) An organic compound in which one, two, or three hydrogens of ammonia are replaced by carbon groups; RNH_2, R_2NH, or R_3N.

Amino acid (*Section 14.1*) An organic compound containing an amino group and a carboxyl group.

Amino acid neurotransmitter (*Section 16.5*) A neurotransmitter or hormone that is an amino acid.

Amino acid pool (*Section 20.1*) The free amino acids found both inside and outside cells throughout the body.

Aminoacyl tRNA synthetase (*Section 18.3*) An enzyme that links the correct amino acid to a tRNA molecule.

Amino group (*Section 1.4B*) An —NH$_2$ group.

Amino sugar (*Section 12.1*) A monosaccharide in which an —OH group is replaced by an —NH$_2$ group.

Amphiprotic (*Section 7.3*) A substance that can act as either an acid or a base.

Amphoteric (*Section 7.3*) An alternative for amphiprotic.

Amylase (*Section 22.3*) An enzyme that catalyzes the hydrolysis of α-1,4-glycosidic bonds in dietary starches.

Anabolism (*Section 19.1*) The biochemical process of building up larger molecules from smaller ones.

Aneuploid cell (*Chemical Connections 23F*) A cell with the wrong number of chromosomes.

Anomeric projection (*Section 12.2*) The hemiacetal carbon of the cyclic form of a monosaccharide.

Anomers (*Section 12.2*) Monosaccharides that differ in configuration only at their anomeric carbons.

Antagonist (*Section 16.1*) A molecule that binds to a neurotransmitter receptor but does not elicit the natural response.

Antibody (*Section 23.1*) A defense glycoprotein synthesized by the immune system of vertebrates that interacts with an antigen; also called an immunoglobulin.

Anticodon (*Section 18.3*) A sequence of three nucleotides on tRNA, also called a codon recognition site, complementary to the codon in mRNA.

Antigen (*Sections 23.1 and 23.3*) A substance foreign to the body that triggers an immune response.

Antigen-presenting cells (**APCs**) (*Section 23.2*) Cells that cleave foreign molecules and present them on their surfaces for binding to T cells or B cells.

Antisense strand (*Section 18.2*) The strand of DNA that acts as the template for transcription. Also called the template strand and the (-) strand.

AP site (*Section 17.7*) The ribose and phosphate that are left after a glycolase removes a purine or pyrimidine base during DNA repair.

Apoenzyme (*Section 15.2*) The protein portion of an enzyme that has cofactors or prosthetic groups.

Ar— (*Section 4.1*) The symbol used for an aryl group.

Arene (*Section 4.1*) A compound containing one or more benzene rings.

Aromatic amine (*Section 8.1*) An amine in which nitrogen is bonded to one or more aromatic rings.

Aromatic compound (*Section 4.1*) A term used to classify benzene and its derivatives.

Aromatic sextet (*Section 4.1B*) The closed loop of six electrons (two from the second bond of each double bond) characteristic of a benzene ring.

Aryl group (*Section 4.1*) A group derived from an arene by removal of a hydrogen atom. Given the symbol Ar—.

Autoxidation (*Section 4.4C*) The reaction of a C—H group with oxygen, O$_2$, to form a hydroperoxide, R—OOH.

Axial position (*Section 2.6B*) A position on a chair conformation of a cyclohexane ring that extends from the ring parallel to the imaginary axis of the ring.

Axon (*Section 16.2*) The long part of a nerve cell that comes out of the main cell body and eventually connects with another nerve cell or tissue cell.

B cell (*Section 23.1*) A type of lymphocyte that is produced in and matures in the bone marrow. B cells produce antibody molecules.

Basal caloric requirement (*Section 22.2*) The caloric requirement for an individual at rest, usually given in Cal/day.

Base (*Section 7.1*) An Arrhenius base is a substance that ionizes in aqueous solution to give hydroxide (OH$^-$) ions.

Bases (*Section 17.2*) Purines and pyrimidines, which are components of nucleosides. DNA and RNA.

Beta (β) oxidation (*Section 20.5*) The biochemical pathway that degrades fatty acids to acetyl CoA by removing two carbons at a time and yielding energy.

Beta (β-) pleated sheet (*Section 14.9*) A type of secondary protein structure in which the backbone of two protein chains in the same or different molecules is held together by hydrogen bonds.

Binding protein (*Section 18.2*) A protein that binds to the nucleosomes making the DNA more accessible for transcription.

Brønsted-Lowry acid (*Section 7.3*) A proton donor.

Brønsted-Lowry base (*Section 7.3*) A proton acceptor.

Buffer (*Section 7.10*) A solution that resists change in pH when limited amounts of an acid or a base are added to it; an aqueous solution containing a weak acid and its conjugate base.

Buffer capacity (*Section 7.10*) The extent to which a buffer solution can prevent a significant change in the pH of a solution upon addition of an acid or a base.

Carbocation (*Section 3.6A*) A species containing a carbon atom with only three bonds to it and bearing a positive charge.

Carbohydrate (*Section 12.1*) A polyhydroxyaldehyde or polyhydroxyketone, or a substance that gives these compounds on hydrolysis.

Carboxylic ester (*Section 1.4*) A derivative of a carboxylic acid in which the H of the carboxyl group is replaced by a carbon atom.

Carbonyl group (*Section 1.4C*) A C=O group.

Carboxyl group (*Section 1.4D*) A —COOH group.

Carboxylic acid (*Section 1.4D*) A compound containing a —COOH group.

Carcinogen (*Section 18.7*) A chemical mutagen that can cause cancer.

Catabolism (*Section 19.1*) The biochemical process of breaking down molecules to supply energy.

Cell reprogramming (*Chemical Connections 23F*) A technique used in whole-mammal cloning, in which a somatic cell is reprogrammed to behave like a fertilized egg.

Central dogma (*Section 18.1*) A doctrine stating the basic directionality of heredity when DNA leads to RNA, which leads to protein. This doctrine is true in almost all life forms except certain viruses.

Chair conformation (*Section 2.6*) The most stable conformation of a cyclohexane ring; all bond angles are approximately 109.5°.

Chaperone (*Section 14.10*) A protein molecule that helps other proteins to fold into the biologically active conformation and enables partially denatured proteins to regain their biologically active conformation.

Chemical messenger (*Section 16.1*) Any chemical that is released from one location and travels to another location before acting. It may be a hormone, neurotransmitter, or ion.

Chemiosmotic theory (*Section 19.5*) Mitchell's proposal that electron transport is accompanied by an accumulation of protons in the intermembrane space of the mitochondrion, which in turn creates osmotic pressure; the protons driven back to the mitochondrion under this pressure generate ATP.

Chemokine (*Section 23.6*) A chemotactic cytokine that facilitates the migration of leukocytes from the blood vessels to the site of injury or inflammation.

Chiral (*Section 6.1*) From the Greek *cheir*, meaning "hand"; an object that is not superposable on its mirror image.

Cholinergic neurotransmitter (*Section 16.1*) A neurotransmitter or hormone based on acetylcholine.

Chromatin (*Section 17.6*) A complex of DNA with histones and non-histone proteins that exists in eukaryotic cells between cell divisions.

Chromosomes (*Section 17.1*) Structures within the nucleus of eukaryotes that contain DNA and protein and that are replicated as units during mitosis. Each chromosome is made up of one long DNA molecule that contains many heritable genes.

Cis (*Section 2.7*) A prefix meaning "on the same side."

Cis-trans isomers (*Section 2.7*) Isomers that have the same (1) molecular formula and (2) connectivity of their atoms, but (3) a different arrangement of their atoms in space due to the presence of either a ring or a carbon–carbon double bond.

Cloning (*Section 17.8*) A process whereby DNA is amplified by inserting it into a host and having the host replicate it along with the host's own DNA.

Cluster determinant (*Section 23.5*) A set of membrane proteins on T cells that help the binding of antigens to the T-cell receptors.

Coding strand (*Section 18.2*) The DNA strand that is not used as a template for transcription but that has a sequence identical to the RNA produced. Also called the $(+)^-$strand and the sense strand.

Codon (*Section 18.3*) A three-nucleotide sequence on mRNA that specifies a particular amino acid.

Coenzyme (*Section 15.3*) An organic molecule, frequently a B vitamin, that acts as a cofactor.

Cofactor (*Section 15.3*) The nonprotein part of an enzyme necessary for its catalytic function.

Competitive inhibition (*Section 15.3*) A mechanism of enzyme regulation in which an inhibitor competes with the substrate for the active site.

Complementary base pairs (*Section 17.3*) The combination of a purine and a pyrimidine base that hydrogen bond together in DNA.

Complete protein (*Section 22.5*) A protein source that contains sufficient quantities of all amino acids required for normal growth and development.

Configuration (*Section 2.7*) The arrangement of atoms about a stereocenter—that is, the relative arrangements of the parts of a molecule in space.

Conformation (*Section 2.5A*) Any three-dimensional arrangement of atoms in a molecule that results from rotation about a single bond.

Conjugate acid (*Section 7.3*) In the Brønsted-Lowry theory, a substance formed when a base accepts a proton.

Conjugate acid–base pair (*Section 7.3*) A pair of molecules or ions that are related to one another by the gain or loss of a proton.

Conjugate base (*Section 7.3*) In the Brønsted-Lowry theory, a substance formed when an acid donates a proton to another molecule or ion.

Conjugated protein (*Section 14.11*) A protein that contains a nonprotein part, such as the heme part of hemoglobin.

Consensus sequence (*Section 18.2*) A sequence of DNA in the promoter region that is relatively conserved from species to species.

Constitutional isomers (*Section 2.2*) Compounds with the same molecular formula but a different order of attachment (connectivity) of their atoms.

Control site (*Section 18.6*) A DNA sequence that is part of a prokaryotic operon. This sequence is upstream of the structural gene DNA and plays a role in controlling whether the structural gene is transcribed.

C-terminus (*Section 14.6*) The amino acid at the end of a peptide chain that has a free carboxyl group.

Cyclic AMP response element binding protein (CREB) (*Chemical Connections 18E*) An important transcription factor that binds to the cAMP response element stimulating transcription of many eukaryotic genes.

Cyclic ether (*Section 5.3B*) An ether in which oxygen is one of the atoms of a ring.

Cycloalkane (*Section 2.5*) A saturated hydrocarbon that contains carbon atoms bonded to form a ring.

Cystine (*Section 14.4*) A dimer of cysteine in which the two amino acids are covalently bonded by a disulfide bond between their side chain —SH groups.

Cytokine (*Section 23.6*) A glycoprotein that traffics between cells and alters the function of a target cell.

Debranching enzyme (*Section 22.3*) The enzyme that catalyzes the hydrolysis of the 1,6-glycosidic bonds in starch and glycogen.

Decarboxylation (*Section 10.5E*) The loss of CO_2 from a carboxyl (—COOH) group.

Dehydration (*Section 5.2B*) The elimination of a molecule of water from an alcohol. An OH is removed from one carbon and an H is removed from an adjacent carbon.

Dehydrogenase (*Section 15.2*) A class of enzymes that catalyze oxidation–reduction reactions, often using NAD^+ as the oxidizing agent.

Denaturation (*Section 14.12*) The loss of the secondary, tertiary, and quaternary structure of a protein by a chemical or physical agent that leaves the primary structure intact.

Dendrite (*Section 16.2*) A hair-like projection that extends from the cell body of a nerve cell on the opposite side from the axon.

Dendritic cells (*Sections 23.1 and 23.2*) Important cells in the innate immune system that are often the first cells to defend against invaders.

Deoxyribonucleic acid (*Section 17.2*) The macromolecule of heredity in eukaryotes and prokaryotes. It is composed of chains of nucleotide monomers of a nitrogenous base, 2-deoxy-D-ribose, and phosphate.

Detergent (*Section 10.4D*) A synthetic soap. The most common are the linear alkylbenzene sulfonic acids (LAS).

Dextrorotatory (*Section 6.4B*) The clockwise (to the right) rotation of the plane of polarized light in a polarimeter.

Diastereomers (*Section 6.3A*) Stereoisomers that are not mirror images of each other.

Dietary Reference Intake (DRI) (*Section 22.1*) The current numerical system for reporting nutrient requirements; an average daily requirement for nutrients published by the U.S. Food and Drug Administration.

Diet faddism (*Section 22.1*) An exaggerated belief in the effects of nutrition upon health and disease.

Digestion (*Section 22.1*) The process in which the body breaks down large molecules into smaller ones that can then be absorbed and metabolized.

Disaccharide (*Section 12.4*) A carbohydrate containing two monosaccharide units joined by a glycosidic bond.

Diol (*Section 5.1B*) A compound containing two —OH (hydroxyl) groups.

Dipeptide (*Section 14.6*) A peptide with two amino acids.

Diprotic acid (*Section 7.3*) An acid that can give up two protons.

Discriminatory curtailment diet (*Section 22.1*) A diet that avoids certain food ingredients that are considered harmful to the health of an individual—for example, low-sodium diets for people with high blood pressure.

Disulfide (*Section 5.4D*) A compound containing an —S—S— group.

D–monosaccharide (*Section 12.1*) A monosaccharide that, when written as a Fischer projection, has the —OH group on its penultimate carbon to the right.

DNA (*Section 17.2*) Deoxyribonucleic acid.

DNA fingerprint (*Chemical Connections 17C*) A pattern of DNA fragments generated by electrophoresis that is used in forensic science.

Double helix (*Section 17.3*) The arrangement in which two strands of DNA are coiled around each other in a screw-like fashion.

EGF (*Section 23.6*) Epidermal growth factor; a cytokine that stimulates epidermal cells during healing of wounds.

Electrophoresis (*Chemical Connections 17C*) A laboratory technique involving the separation of molecules in an electric field.

Elongation (*Section 18.2*) The phase of protein synthesis during which activated tRNA molecules deliver new amino acids to ribosomes where they are joined by peptide bonds to form a polypeptide.

Elongation factor (*Section 18.5*) A small protein molecule that is involved in the process of tRNA binding and movement of the ribosome on the mRNA during elongation.

Embryonal carcinoma cell (*Chemical Connections 23D*) A cell that is multipotent and is derived from carcinomas.

Embryonic stem cell (*Chemical Connections 23D*) Stem cells derived from embryonic tissue. Embryonic tissue is the richest source of stem cells.

Enantiomers (*Section 6.1*) Stereoisomers that are nonsuperposable mirror images; refers to a relationship between pairs of objects.

End point (*Section 7.9*). The point in a titration at which a visible change occurs.

Endocrine gland (*Section 16.2*) A gland such as the pancreas, pituitary, and hypothalamus that produces hormones involved in the control of chemical reactions and metabolism.

Enhancer (*Section 18.6*) A DNA sequence that is not part of the promoter that binds a transcription factor enhancing transcription.

Enkephalin (*Section 16.6*) A pentapeptide found in nerve cells of the brain that acts to control the perception of pain.

Enol (*Section 9.5*) A molecule containing an —OH group bonded to a carbon of a carbon–carbon double bond.

Enzyme (*Section 15.1*) A biological catalyst that increases the rate of a chemical reaction by providing an alternative pathway with a lower activation energy.

Enzyme activity (*Section 15.4*) The rate at which an enzyme-catalyzed reaction proceeds, commonly measured as the amount of product produced per minute.

Enzyme specificity (*Section 15.1*) The limitation of an enzyme to catalyze one specific reaction with one specific substrate.

Enzyme-substrate complex (*Section 15.5*) A part of an enzyme reaction mechanism where the enzyme is bound to the substrate.

Epigenetics (*Chemical Connections 23D*) The study of heritable processes that alter gene expression without altering the actual DNA.

Epitope (*Section 23.3*) The smallest number of amino acids on an antigen that elicits an immune response.

Equatorial position (*Section 2.6B*) A position on a chair conformation of a cyclohexane ring that extends from the ring roughly perpendicular to the imaginary axis of the ring.

Equivalence point (*Section 7.9*) The point in an acid–base titration at which there is an equal amount of acid and base.

Ergogenic aid (*Chemical Connections 22D*) A substance that can be consumed to enhance athletic performance.

Essential amino acid (*Section 22.5*) An amino acid that the body cannot synthesize in the required amounts and so must be obtained in the diet.

Essential fatty acid (*Section 22.4*) A fatty acid required in the diet.

Ester (*Section 10.5D*) A compound in which the —OH of a carboxyl group, RCOOH, is replaced by an —OR′ (alkoxy) group or —OAr (aryloxy) group.

Ether (*Section 5.3A*) A compound containing an oxygen atom bonded to two carbon atoms.

Excitatory neurotransmitter (*Section 16.4*) A neurotransmitter that increases the transmission of nerve impulses.

Exon (*Section 17.5*) A nucleotide sequence in mRNA that codes for a protein.

Expression cassette (*Section 18.9*) A gene sequence containing a gene being introduced via gene therapy that is incorporated into a vector replacing some of the vector's own DNA.

Extended helix (*Section 14.9*) A type of helix found in collagen, caused by a repeating sequence.

External innate immunity (*Section 23.1*) The innate protection against foreign invaders characteristic of the skin barrier, tears, and mucus.

Fat (*Section 13.3*) A mixture of triglycerides containing a high proportion of long-chain, saturated fatty acids.

Fatty acid (*Section 10.4A*) A long, unbranched chain carboxylic acid, most commonly with 10–20 carbon atoms, derived from animal fats, vegetable oils, or the phospholipids of biological membranes. The hydrocarbon chain may be saturated or unsaturated. In most unsaturated fatty acids, the *cis* isomer predominates. *Trans* isomers are rare.

Feedback control (*Section 15.6*) A type of enzyme regulation where the product of a series of reactions inhibits the enzyme that catalyzes the first reaction in the series.

Fiber (*Section 22.1*) The cellulosic, non-nutrient component in our food.

Fibrous protein (*Section 14.1*) A protein used for structural purposes. Fibrous proteins are insoluble in water and have a high percentage of secondary structures, such as alpha helices and/or beta-pleated sheets.

Fischer esterification (*Section 10.5D*) The process of forming an ester by refluxing a carboxylic acid and an alcohol in the presence of an acid catalyst, commonly sulfuric acid.

Fischer projection (*Section 12.1B*) A two-dimensional representation of a sugar structure.

Functional group (*Section 1.4*) An atom or group of atoms within a molecule that shows a characteristic set of physical and chemical properties.

Furanose (*Section 12.2*) A five-membered cyclic hemiacetal form of a monosaccharide.

Gene (*Section 17.1*) The unit of heredity; a DNA segment that codes for a protein.

Gene expression (*Section 18.1*) The activation of a gene to produce a specific protein. It involves both transcription and translation.

Gene regulation (*Section 18.6*) The various methods used by organisms to control which genes will be expressed and when.

Gene therapy (*Section 18.9*) The process of treating a disease by introducing a functional copy of a gene to an organism that was lacking it.

General transcription factor (GTF) (*Section 18.6*) Proteins that make a complex with the DNA being transcribed and the RNA polymerase.

Genetic code (*Section 18.4*) The sequence of triplets of nucleotides (codons) that determines the sequence of amino acids in a protein.

Genetic engineering (*Section 18.8*) The process by which genes are inserted into cells.

Genome (*Chemical Connections 17E*) A complete DNA sequence of an organism.

Globular protein (*Section 14.1*) Protein that is used mainly for nonstructural purposes and is largely soluble in water.

Glycogenesis (*Section 21.2*) The conversion of glucose to glycogen.

Glycogenolysis (*Section 20.3*) The biochemical pathway for the breakdown of glycogen to glucose.

Glycol (*Section 5.1B*) A compound with hydroxyl (—OH) groups on adjacent carbons.

Glycolysis (*Section 20.2*) The catabolic pathway in which glucose is broken down to pyruvate.

Glycomics (*Chemical Connections 23D*) The collective knowledge of all the carbohydrates, including glycoproteins and glycolipids, that a cell or a tissue contains and the determination of their functions.

Glycoside (*Section 12.3*) A carbohydrate in which the —OH group on its anomeric carbon is replaced by an —OR group.

Glycosidic bond (*Section 12.3*) The bond from the anomeric carbon of a glycoside to an —OR bond.

Gluconeogenesis (*Section 21.2*) The process by which glucose is synthesized in the body.

Gp120 (*Section 23.5*) A 120,000-molecular-weight glycoprotein on the surface of the human immunodeficiency virus that binds strongly to the CD4 molecules on T cells.

G-Protein (*Section 13.5*) A protein that is either stimulated or inhibited when a hormone binds to a receptor and subsequently alters the activity or another protein such as adenyl cyclase.

Guanosine (*Section 17.2*) A nucleoside made of D-ribose and guanine.

Haber process (*Chemical Connections 7E*) An industrial process by which H_2 and N_2 are converted to NH_3.

Haworth projection (*Section 12.2*) A representation of the cyclic structure of a sugar.

HDPE (*Section 3.7C*) *High-density polyethylene.*

Helicase (*Section 17.6*) An unwinding protein that acts at a replication fork to unwind DNA so that DNA polymerase can synthesize a new DNA strand.

Helix-turn-helix (*Section 18.6*) A common motif for a transcription factor.

Helper T cells (*Section 23.2*) A type of T cell that helps in the response of the acquired immune system against invaders but does not kill infected cells directly.

Hemiacetal (*Section 9.4C*) A molecule containing a carbon bonded to one —OH and one —OR group; the product of adding one molecule of alcohol to the carbonyl group of an aldehyde or ketone.

Henderson-Hasselbalch equation (*Section 7.11*) A mathematical relationship between pH, the pK_a of a weak acid, HA, and the concentrations of the weak acid and its conjugate base.

Heterocyclic aliphatic amine (*Section 8.1*) A heterocyclic amine in which nitrogen is bonded only to alkyl groups.

Heterocyclic amine (*Section 8.1*) An amine in which nitrogen is one of the atoms of a ring.

Heterocyclic aromatic amine (*Section 8.1*) An amine in which nitrogen is one of the atoms of an aromatic ring.

Heterogeneous catalyst (*Section 7.4D*) A catalyst in a separate phase from the reactants—for example, the solid platinum, Pt(s), in the reaction between CO(*g*) and H_2(*g*) to produce CH_3OH(*l*).

Highly active antiretroviral therapy (HAART) (*Section 23.8*) An aggressive treatment against AIDS involving the use of several different drugs.

Histone (*Section 17.6*) A basic protein that is found in complexes with DNA in eukaryotes.

HIV (*Sections 23.4 and 23.8*) Human immunodeficiency virus.

Homogeneous catalyst (*Section 7.4D*) A catalyst in the same phase as the reactants—for example, enzymes in body tissues.

Hormone (*Section 16.2*) A chemical messenger released by an endocrine gland into the bloodstream and transported there to reach its target cell.

Hybridization (*Section 17.8*) A process whereby two strands of nucleic acids or segments thereof form a double-stranded structure through hydrogen bonding of complementary base pairs.

Hybridoma (*Section 23.4*) A combination of a mylenoma cell with a B cell to produce monoclonal antibodies.

Hydration (*Section 3.6C*) The addition of water.

Hydrocarbon (*Section 2.1*) A compound that contains only carbon and hydrogen atoms.

Hydrogenation (*Section 3.6B*) Addition of hydrogen atoms to a double or triple bond using H_2 in the presence of a transition metal catalyst, most commonly Ni, Pd, or Pt. Also called catalytic reduction or catalytic hydrogenation.

Hydrolase (*Section 15.2*) An enzyme that catalyzes a hydrolysis reaction.

Hydronium ion (*Section 7.1*) The H_3O^+ ion.

Hydrophobic interaction (*Section 14.10*) Interaction by London dispersion forces between hydrophobic groups.

Hydroxyl group (*Section 1.4A*) An —OH group bonded to a tetrahedral carbon atom.

Hyperthermophile (*Section 15.4*) An organism that lives at extremely high temperatures.

Immunogen (*Section 23.3*) Another term for antigen.

Immunoglobulin (*Section 23.4*) An antibody protein generated against and capable of binding specifically to an antigen.

Immunoglobulin superfamily (*Section 23.1*) A family of molecules based on a similar structure that includes the immunoglobulins, T-cell receptors, and other membrane proteins that are involved in cell communications. All molecules in this class have a certain portion that can react with antigens.

Indicator, acid–base (*Section 7.8*) A substance that changes color within a given pH range.

Induced-fit model (*Section 15.5*) A model explaining the specificity of enzyme action by comparing the active site to a glove and the substrate to a hand.

Inhibition of enzyme activity (*Section 15.3*) Any reversible or irreversible process that makes an enzyme less active.

Inhibitor (*Section 15.3*) A compound that binds to an enzyme and lowers its activity.

Inhibitory neurotransmitter (*Section 16.4*) A neurotransmitter that decreases the transmission of nerve impulses.

Inhibitory receptor (*Section 23.7*) A receptor on the surface of a cell of the innate immune system that recognizes antigens on healthy cells and prevents activation of the immune system.

Initiation of protein synthesis (*Section 18.5*) The first step in the process whereby the base sequence of a mRNA is translated into the primary structure of a polypeptide.

Initiation signal (*Section 18.2*) A sequence on DNA that identifies the location where transcription is to begin.

Innate immunity (*Section 23.1*) The first line of defense against foreign invaders, which includes skin resistance to penetration, tears, mucus, and nonspecific macrophages that engulf bacteria.

Interleukin (*Section 23.6*) A cytokine that controls and coordinates the action of leukocytes.

Internal innate immunity (*Section 23.1*) The type of innate immunity that is used once a pathogen has already penetrated a tissue.

Intron (*Section 17.5*) A nucleotide sequence in mRNA that does not code for a protein.

Ion product of water, K_w (*Section 7.7*) The concentration of H_3O^+ multiplied by the concentration of OH^-; $[H_3O^+][OH^-] = 1 \times 10^{-14}$.

Isoelectric point (pI) (*Section 14.3*) The pH at which a molecule has no net charge.

Isoenzyme (*Section 15.6*) An enzyme that can be found in multiple forms, each of which catalyzes the same reaction. Also called an **isozyme**.

Isomerase (*Section 15.2*) An enzyme that catalyzes an isomerization reaction.

Isozymes (*Section 15.6*) Two or more enzymes that perform the same functions but have different combinations of subunits—that is, different quaternary structures.

Ketone (*Section 1.4C*) A compound containing a carbonyl group bonded to two carbons.

Ketone body (*Chemical Connections 10C*) One of several ketone-based molecules—for example, acetone, 3-hydroxybutanoic acid (β-hydroxybutyric acid) and acetoacetic acid (3-oxobutanoic acid)—produced in the liver during over-utilization of fatty acids when the supply of carbohydrates is limited.

Ketose (*Section 12.1*) A monosaccharide containing a ketone group.

Killer T cell (*Section 23.2*) A T cell that kills invading foreign cells by cell-to-cell contact. Also called cytotoxic T cell.

Kinase (*Section 15.6*) A class of enzymes that covalently modifies a protein with a phosphate group usually through the —OH group on the side chain of a serine, threonine, or tyrosine.

Kwashiorkor (*Section 22.5*) A disease caused by insufficient protein intake and characterized by a swollen stomach, skin discoloration, and retarded growth.

Lagging strand (*Section 17.6*) A discontinuously synthesized DNA that elongates in a direction away from the replication fork.

LDPE (*Section 3.7B*) Low-density polyethylene.

Leading strand (*Section 17.6*) The continuously synthesized DNA strand that elongates toward the replication fork.

Leucine zipper (*Section 18.6*) A common motif for a transcription factor.

Leukocytes (*Section 23.2*) White blood cells, which are the principal parts of the acquired immunity system and act via phagocytosis or antibody production.

Levorotatory (*Section 6.4B*) The counterclockwise rotation of the plane of polarized light in a polarimeter.

Ligase (*Section 15.2*) A class of enzymes that catalyzes a reaction joining two molecules. They are often called synthetases or synthases.

Line-angle formula (*Section 2.1*) An abbreviated way to draw structural formulas in which each vertex and line terminus represents a carbon atom and each line represents a bond.

Lipase (*Section 22.4*) An enzyme that catalyzes the hydrolysis of an ester bond between a fatty acid and glycerol.

Lipoprotein (*Section 12.9*) Spherically shaped clusters containing both lipid molecules and protein molecules.

L–monosaccharide (*Section 12.1*) A monosaccharide that, when written as a Fischer projection, has the —OH group on its penultimate carbon to the left.

Lock-and-key model (*Section 15.5*) A model for enzyme-substrate interaction based on the postulate that the active site of an enzyme is a perfect fit for the substrate.

Lyase (*Section 15.2*) A class of enzymes that catalyzes the addition of two atoms or groups of atoms to a double bond or their removal to form a double bond.

Lymphocyte (*Sections 23.1 and 23.2*) A white blood cell that spends most of its time in the lymphatic tissues. Those that mature in the bone marrow are B cells. Those that mature in the thymus are T cells.

Lymphoid organs (*Section 23.2*) The main organs of the immune system, such as the lymph nodes, spleen, and thymus, which are connected together by lymphatic capillary vessels.

Macrophage (*Sections 23.1 and 23.2*) An amoeboid white blood cell that moves through tissue fibers, engulfing dead cells and bacteria by phagocytosis, and then displays some of the engulfed antigens on its surface.

Major groove (*Section 17.3*) The wider of the unequal grooves found in a B–DNA double helix.

Major histocompatibility complex (MHC) (*Sections 23.2 and 23.3*) A transmembrane protein complex that brings the epitope of an antigen to the surface of the infected cell to be presented to the T cells.

Maloney murine leukemia virus (MMLV) (*Section 18.9*) A common vector used for gene therapy.

Marasmus (*Section 22.2*) Another term for chronic starvation, whereby the individual does not have adequate caloric intake. It is characterized by arrested growth, muscle wasting, anemia, and general weakness.

Markovnikov's rule (*Section 3.6A*) In the addition of HX or H_2O to an alkene, hydrogen adds to the carbon of the double bond having the greater number of hydrogens.

Memory cell (*Section 23.2*) A type of T cell that stays in the blood after an infection is over and acts as a quick line of defense if the same antigen is encountered again.

Mercaptan (*Section 5.4B*) A common name for any molecule containing an —SH group.

Messenger RNA (*Section 17.4*) The RNA that carries genetic information from DNA to the ribosome and acts as a template for protein synthesis.

Meta (m) (*Section 4.2B*) Refers to groups occupying the 1 and 3 positions on a benzene ring.

Metabolic acidosis (*Chemical Connections 7C*) The lowering of the blood pH due to metabolic effects such as starvation or intense exercise.

Metabolism (*Section 19.1*) The sum of all chemical reactions in a cell.

Metal binding finger (*Section 18.6*) A type of transcription factor containing heavy metal ions, such as Zn^{2+}, that is involved in helping RNA polymerase bind to the DNA being transcribed.

Micelle (*Section 10.4*) A spherical arrangement of molecules in aqueous solution such that their hydrophobic (water-hating) parts are shielded from the aqueous environment and their hydrophilic (water-loving) parts are on the surface of the sphere and in contact with the aqueous environment.

Micro RNA (*Section 17.4*) Small RNA of 22 nucleosides that are involved in regulation of genes and development of an organism.

Minisatellite (*Section 17.5*) A small repetitive DNA sequence that is sometimes associated with cancer when it mutates.

Minor groove (*Section 17.3*) The narrower of the unequal grooves found in a B–DNA double helix.

Mirror image (*Section 6.1*) The reflection of an object in a mirror.

Monoclonal antibody (*Section 23.4*) An antibody produced by clones of a single B cell specific to a single epitope.

Monomer (*Section 3.7A*) From the Greek *mono*, "single," and *meros*, "part"; the simplest nonredundant unit from which a polymer is synthesized.

Monoprotic acid (*Section 7.3*) An acid that can give up only one proton.

Monosaccharide (*Section 12.1*) A carbohydrate that cannot be hydrolyzed to a simpler compound.

Multiclonal antibodies (*Chemical Connections 23B*) The type of antibodies found in the serum after a vertebrate is exposed to an antigen.

Multipotent stem cell (*Chemical Connections 23D*) A stem cell capable of differentiating into many, but not all, cell types.

Mutagen (*Section 18.7*) A chemical substance that induces a base change or mutation in DNA.

Mutarotation (*Section 12.2*) The change in a specific rotation that occurs when an α or β form of a carbohydrate is converted to an equilibrium mixture of the two forms.

Natural killer cell (*Sections 23.1 and 23.2*) A cell of the innate immune system that attacks infected or cancerous cells.

Negative modulation (*Section 15.6*) The process whereby an allosteric regulator inhibits enzyme action.

Neuron (*Section 16.1*) Another name for a nerve cell.

Neuropeptide Y (*Section 16.6*) A brain peptide that affects the hypothalamus and is an appetite-stimulating agent.

Neurotransmitter (*Section 16.2*) A chemical messenger between a neuron and another cell, which may be a neuron, a muscle cell, or a cell of a gland.

Neutralizing antibody (*Section 23.8*) A type of antibody that completely destroys its target antigen.

Noncompetitive inhibition (*Section 15.3*) An enzyme regulation in which an inhibitor binds to the active site, thereby changing the shape of the active site and reducing its catalytic activity.

N-terminus (*Section 14.6*) The amino acid at the end of a peptide chain that has a free amino group.

Nucleic acids (*Section 17.3*) A polymer composed of nucleotides.

Nucleophilic attack (*Section 15.5*) A chemical reaction where an electron-rich atom, such as oxygen or sulfur, bonds to an electron-deficient atom, such as carbonyl carbon.

Nucleoside (*Section 17.2*) The combination of a heterocyclic aromatic amine bonded by a glycosidic bond to either D-ribose or 2-deoxy-D-ribose.

Nucleosome (*Section 17.3*) Combinations of DNA and proteins.

Nucleotide (*Section 17.2*) A phosphoric ester of a nucleoside.

Nutrient (*Section 22.1*) Components of food and drink that provide energy, replacement, and growth.

Oil (*Section 13.2*) A mixture of triglycerides containing a high portion of long-chain, unsaturated fatty acids.

Okazaki fragment (*Section 17.6*) A short segment of DNA made up of about 200 nucleotides in higher organisms and 2000 nucleosides in prokaryotes.

Oligosaccharide (*Section 12.4*) A carbohydrate containing from six to ten monosaccharide units, each joined to the next by a glycosidic bond.

Open complex (*Section 18.6*) The complex of DNA, RNA polymerase, and general transcription factors that must be formed before transcription can take place. In this complex, the DNA is being separated.

Optically active (*Section 6.4A*) Showing that a compound rotates the plane of polarized light.

Organic chemistry (*Section 1.1*) The study of the compounds of carbon.

Origin of replication (*Section 17.6*) The point in a DNA molecule where replication starts.

Ortho (o) (*Section 4.2B*) Refers to groups occupying the 1 and 2 positions on a benzene ring.

Oxidative deamination (*Section 20.8*) The reaction in which the amino group of an amino acid is removed and an α-ketoacid is formed.

Oxidoreductase (*Section 15.2*) A class of enzymes that catalyzes an oxidation–reduction reaction.

Oxonium ion (*Section 3.6B*) An ion in which oxygen is bonded to three other atoms and bears a positive charge.

P site (*Section 18.5*) The site on the large ribosomal subunit where the current peptide is bound before peptidyl transferase links it to the amino acid attached to the A site during elongation.

Para (p) (*Section 4.2B*) Refers to groups occupying the 1 and 4 positions on a benzene ring.

Parenteral nutrition (*Chemical Connections 21A*) The technical term for intravenous feeding.

Pentose phosphate pathway (*Section 20.2*) The biochemical pathway that produces ribose and NADPH from glucose-6-phosphate or, alternatively, releases energy.

Peptide (*Section 14.6*) A short chain of amino acids linked via peptide bonds.

Peptide backbone (*Section 14.7*) The repeating pattern of peptide bonds in a polypeptide or protein.

Peptide bond (*Section 14.6*) An amide bond that links two amino acids. Also called a peptide linkage.

Peptidergic neurotransmitter (*Section 16.6*) A type of neurotransmitter or hormone that is based on a peptide, such as glucagon, insulin, and the enkephalins.

Peptidyl transferase (*Section 18.5*) The enzyme activity of the ribosomal complex that is responsible for formation of peptide bonds between the amino acids of the growing peptide.

Perforin (*Section 23.2*) A protein produced by killer T cells that punches holes in the membrane of target cells.

Peroxide (*Section 3.7B*) A compound that contains an —O—O— bond—for example, hydrogen peroxide, H—O—O—H.

pH (*Section 7.8*) The negative logarithm of the hydronium ion concentration; $pH = -\log[H_3O^+]$

Phagocytosis (*Section 23.4*) The process by which large particulates including bacteria are pulled inside a white cell called a phagocyte.

Pharmacogenomics (*Chemical Connections 17E*) The study of how variations affect drug design.

Phenol (*Section 4.4*) A compound that contains an —OH group bonded to a benzene ring.

Phenyl group (*Section 4.2*) C_6H_5—, the aryl group derived by removing a hydrogen atom from benzene.

Pheromone (*Chemical Connections 3B*) A chemical secreted by an organism to influence the behavior of another member of the same species.

Photosynthesis (*Section 21.2*) The process in which plants synthesize carbohydrates from CO_2 and H_2O with the help of sunlight and chlorophyll.

Plane-polarized light (*Section 6.4A*) Light vibrating in only parallel planes.

Plasma cell (*Section 23.2*) A cell derived from a B cell that has been exposed to an antigen.

Plastic (*Chemical Connections 3D*) A polymer that can be molded when hot and that retains its shape when cooled.

Pluripotent stem cell (*Chemical Connections 23F*) A stem cell that is capable of developing into every cell type.

Plasma cell (*Section 23.2*) A cell derived from a B cell that has been exposed to an antigen.

Plasmids (*Section 18.8*) Small circular DNAs of bacterial origin often used to construct recombinant DNA.

Pluripotent stem cell (*Chemical Connections 23D*) A stem cell that is capable of developing into every cell type.

pOH (*Section 7.8*) The negative logarithm of the hydroxide ion concentration; $pOH = -\log[OH^-]$

Polarimeter (*Section 6.4B*) An instrument for measuring the ability of a compound to rotate the plane of polarized light.

Polymer (*Section 3.7A*) From the Greek *poly*, "many", and *meros*, "parts"; any long-chain molecule synthesized by bonding together many single parts called monomers.

Polymerase chain reaction (PCR) (*Section 17.8*) An automated technique for amplifying DNA using a heat-stable DNA polymerase from thermophilic bacteria.

Polynuclear aromatic hydrocarbon (*Section 4.2D*) A hydrocarbon containing two or more benzene rings, each of which shares two carbon atoms with another benzene ring.

Polypeptide (*Section 14.6*) A long chain of amino acids bonded via peptide bonds.

Polysaccharide (*Section 12.4*) A carbohydrate containing a large number of monosaccharide units, each joined to the next by one or more glycosidic bonds.

Positive cooperativity (*Chemical Connections 14G*) A type of allosterism where binding of one molecule of a protein makes it easier to bind another of the same molecule.

Positive modulation (*Section 15.6*) The process whereby an allosteric regulator increases enzyme action.

Postsynapic membrane (*Section 16.2*) The membrane on the side of the synapse nearest the dendrite of the neuron receiving the transmission.

Post-transcription process (*Section 18.2*) A process, such as splicing or capping, that alters RNA after it is initially made during transcription.

Presynaptic membrane (*Section 16.2*) The membrane on the side of the synapse nearest the dendrite of the axon of the neuron transmitting the signal.

Primary (1°) alcohol (*Section 1.4A*) An alcohol in which the carbon atom bearing the —OH group is bonded to only one other carbon group, a —CH₂OH group.

Primary (1°) amine (*Section 1.4B*) An amine in which nitrogen is bonded to one carbon group and two hydrogens.

Primary structure, of DNA (*Section 17.3*) The order of the bases in DNA.

Primary structure, of proteins (*Section 14.8*) The order of amino acids in a peptide, polypeptide, or protein.

Primer (*Section 17.6*) Short pieces of DNA or RNA that initiate DNA replication.

Proenzyme (*Section 15.6*) An inactive form of an enzyme that must have part of its polypeptide chain cleaved before it becomes active.

Progenitor cells (*Chemical Connections 23D*) Another term for stem cells.

Prokaryote (*Section 17.6*) An organism that has no true nucleus or organelles.

Promoter (*Section 18.2*) An upstream DNA sequence that is used for RNA polymerase recognition and binding to DNA.

Prosthetic group (*Section 14.11*) The non-amino-acid part of a conjugated protein.

Proteasome (*Section 18.6*) A large protein complex that is involved in the degradation of other proteins.

Protein (*Section 14.1*) A long chain of amino acids linked via peptide bonds. There must usually be a minimum of 30 to 50 amino acids in a chain before it is considered a protein.

Protein complementation (*Section 22.5*) A diet that combines proteins of varied sources to arrive at a complete protein.

Protein microarray (*Chemical Connection 14F*) An automated technique used to study proteomics that is based on having thousands of protein samples imprinted on a chip.

Protein modification (*Section 15.6*) The process of affecting the enzyme activity by covalently modifying the enzyme, such as phosphorylation of a particular amino acid.

Proteomics (*Chemical Connections 14F*) The collective knowledge of all the proteins and peptides of a cell or a tissue and their functions.

Pyranose (*Section 12.2*) A six-membered cyclic hemiacetal form of a monosaccharide.

Quaternary structure (*Section 14.11*) The organization of a protein that has multiple polypeptide chains, or subunits; refers principally to the way the multiple chains interact.

R— (*Section 2.3A*) A symbol used to represent an alkyl group.

R— (*Section 6.2*) From the Latin *rectus*, meaning "straight, correct"; used in the *R,S* system to show that, when the lowest-priority group is away from you, the order of priority of groups on a stereocenter is clockwise.

R form (*Section 15.6*) The more active form of an allosteric enzyme.

R,S system (*Section 6.2*) A set of rules for specifying configuration about a stereocenter.

Racemic mixture (*Section 6.1*) A mixture of equal amounts of two enantiomers.

Random coils (*Section 14.9*) Proteins that do not exhibit any repeated pattern.

Reaction mechanism (*Section 3.6A*) A step-by-step description of how a chemical reaction occurs.

Receptor (*Section 16.1*) A membrane protein that can bind a chemical messenger and then perform a function, such as synthesizing a second messenger or opening an ion channel.

Recognition site (*Section 18.3*) The area of the tRNA molecule that recognizes the mRNA codon.

Recombinant DNA (*Section 18.8*) DNAs from two sources that have been combined into one molecule.

Recommended Daily Allowance (RDA) (*Section 22.1*) *also* **Recommended Dietary Allowance**; An average daily requirement for nutrients published by the U.S. Food and Drug Administration.

Reducing sugar (*Section 12.3*) A carbohydrate that reacts with a mild oxidizing agent under basic conditions to give an aldonic acid; the carbohydrate reduces the oxidizing agent.

Regioselective reaction (*Section 3.6A*) A reaction in which one direction of bond forming or bond breaking occurs in preference to all other directions.

Regulator (*Section 15.6*) A molecule that binds to an allosteric enzyme and changes its activity. The change could be positive or negative.

Regulatory site (*Section 15.6*) A site, other than the active site, where a regulator binds to an allosteric site and affects the rate of reaction.

Replication (*Section 17.6*) The process whereby DNA is duplicated to form two exact replicas of the original DNA molecule.

Replication fork (*Section 17.6*) The point on a DNA molecule where replication is proceeding.

Residue (*Section 14.6*) Another term for an amino acid in a peptide chain.

Respiratory acidosis (*Chemical Connections 7C*) The lowering of the blood pH due to difficulty breathing.

Response element (*Section 18.6*) A sequence of DNA upstream from a promoter that interacts with a transcription factor to stimulate transcription in eukaryotes. Response elements may control several similar genes based on a single stimulus.

Restriction endonuclease (*Section 18.8*) An enzyme, usually purified from bacteria, that cuts DNA at a specific base sequence.

Retrovaccination (*Section 23.8*) A process whereby scientists have an antibody they want to use and try to develop molecules to elicit it.

Retrovirus (*Section 18.1*) A virus, such as HIV, that has an RNA genome.

Reuptake (*Section 16.4*) The transport of a neurotransmitter from its receptor back through the presynaptic membrane into the neuron.

Ribonucleic acid (RNA) (*Section 17.5*) A type of nucleic acid consisting of nucleotide monomers of a nitrogenous base, D-ribose and phosphate.

Ribosomal RNA (rRNA) (*Section 17.5*) The type of RNA that is complexed with proteins and makes up the ribosomes used in translation of mRNA into protein.

Ribosome (*Section 17.4*) Small spherical bodies in the cell made of protein and RNA; the site of protein synthesis.

Ribozyme (*Section 15.1*) An enzyme that is made up of ribonucleic acid. The currently recognized ribozymes catalyze cleavage of part of their own sequences in mRNA and tRNA.

RNA (*Section 17.2*) Ribonucleic acid.

S— (*Section 6.2*) From the Latin *sinister*, meaning "left"; used in the *R,S* system to show that, when the lowest-priority group is away from you, the order of priority of groups on a stereocenter is counterclockwise.

Saponification (*Section 10.4B*) The hydrolysis of an ester in aqueous NaOH or KOH to give an alcohol and the sodium or potassium salt of a carboxylic acid.

Satellites (*Section 17.5*) Short sequences of DNA that are repeated hundreds of thousands of times but do not code for any protein in RNA.

Saturated hydrocarbon (*Section 2.1*) A hydrocarbon that contains only carbon–carbon single bonds.

Saturation curve (*Section 15.4*) A graph of enzyme rate versus substrate concentration. At high levels of substrate, the enzyme becomes saturated and the velocity does not increase linearly with increasing substrate.

Secondary (2°) alcohol (*Section 1.4A*) An alcohol in which the carbon atom bearing the —OH group is bonded to two other carbon groups.

Secondary (2°) amine (*Section 1.4B*) An amine in which nitrogen is bonded to two carbons groups and one hydrogen.

Secondary messenger (*Section 16.1*) A molecule that is created or released due to the binding of a hormone or neurotransmitter which then proceeds to carry and amplify the signal inside the cell.

Secondary structure of DNA (*Section 17.3*) Specific forms of DNA due to pairing of complementary bases.

Secondary structure of proteins (*Section 14.9*) Repeating structures within polypeptides that are based solely on interactions of the peptide backbone. Examples are the alpha helix and the beta-pleated sheet.

Semiconservative replication (*Section 17.6*) Replication of DNA strands whereby each daughter molecule has one parental strand and one newly synthesized strand.

Sense strand (*Section 18.2*) The DNA strand that is not used as a template for transcription but that has a sequence identical to the RNA produced. Also called the coding strand and the (+) strand.

Severe Combined Immuno Deficiency (SCID) (*Section 18.9*) A disease caused by several possible missing enzymes that leads to the organism having no immune system.

Shine-Dalgarno Sequence (*Section 18.5*) A sequence on the mRNA that attracts the ribosome for translation.

Side chains (*Section 14.7*) The part of an amino acid that varies one from the other. The side chain is attached to the alpha carbon, and the nature of the side chain determines the characteristics of the amino acid.

Sieving portion (*Section 18.6*) A part of a ribosome that directs only certain tRNA molecules to enter.

Signal transduction (*Section 16.5*) A cascade of events through which the signal of a neurotransmitter or hormone delivered to its receptor is carried inside the target cell and amplified into many signals that can cause protein modification, enzyme activation, or the opening of membrane channels.

Significant figures (*Section 1.3*) Numbers that are known with certainty.

Silencer (*Section 18.6*) A DNA sequence that is not part of the promoter that binds a transcription factor suppressing transcription.

Small interfering RNA (*Section 17.4*) Small RNA molecules that are involved in degradation of specific mRNA molecules.

Small nuclear RNA (*Section 17.4*) Small RNA molecules (100–200 nucleotides) located in the nucleus that are distinct from tRNA and rRNA.

Small nuclear ribonucleoprotein particles (*Section 17.4*) Combinations of RNA and protein that are used in RNA splicing reactions.

Soap (*Section 10.4B*) A sodium or potassium salt of a fatty acid.

Solenoid (*Section 17.3*) A coil wound in the form of a helix.

Specificity (*Section 23.1*) A characteristic of acquired immunity based on the fact that cells make specific antibodies to a wide range of pathogens.

Splicing (*Section 17.4*) The removal of an internal RNA segment and the joining of the remaining ends of the RNA molecule.

Step-growth polymerization (*Section 11.6*) A polymerization in which chain growth occurs in a stepwise manner between difunctional monomers—as, for example, between adipic acid and hexamethylenediamine to form nylon-66.

Stereocenter (*Section 2.7*) An atom, most commonly a tetrahedral carbon atom, at which exchange of two groups produces a stereoisomer.

Stereoisomers (*Section 2.7*) Isomers that have the same connectivity (the same order of attachment of their atoms) but different orientations of their atoms in space.

(–) Strand (*Section 18.2*) The strand of DNA used as a template for transcription. Also called the template strand and the antisense strand.

(+) Strand (*Section 18.2*) The DNA strand that is not used as a template for transcription but that has a sequence identical to the RNA produced. Also called the coding and the sense strand.

Strong acid (*Section 7.2*) An acid that ionizes completely in aqueous solution.

Strong base (*Section 7.2*) A base that ionizes completely in aqueous solution.

Structural genes (*Section 18.2*) Genes that code for the product proteins.

Substrate (*Section 15.3*) The compound or compounds whose reaction an enzyme catalyzes.

Substrate specificity (*Section 15.1*) The limitation of an enzyme to catalyze specific reactions with specific substrates.

Substance P (*Section 16.6*) An 11-amino acid peptidergic neurotransmitter involved in the transmission of pain signals.

Subunit (*Section 15.6*) An individual polypeptide chain of an enzyme that has multiple chains.

Sulfhydryl group (*Section 5.4A*) An —SH group.

Surface presentation (*Section 23.1*) The process whereby a portion of an antigen from a foreign pathogen that infected a cell is brought to the surface of the cell.

Synapse (*Section 16.2*) A small aqueous space between the tip of a neuron and its target cell.

T cell (*Section 23.1*) A type of lymphoid cell that matures in the thymus and that reacts with antigens via bound receptors on its cell surface. T cells can differentiate into memory T cells or killer T cells.

T-cell receptor (*Section 23.1*) A glycoprotein of the immunoglobulin superfamily on the surface of T cells that interacts with the epitope presented by MHC.

T-cell receptor complex (*Section 23.5*) The combination of T-cell receptors, antigens, and cluster determinants (CD) that are all involved in the T cell's ability to bind antigen.

T-form (*Section 15.6*) The form of an allosteric enzyme that is less active.

Tautomers (*Section 9.5*) Constitutional isomers that differ in the location of an H atom.

Template strand (*Section 18.2*) The strand of DNA used as a template for transcription. Also called the (−) strand and the antisense strand.

Termination (*Sections 18.2 and 18.5*) The final stage of translation during which a termination sequence on mRNA tells the ribosomes to dissociate and release the newly synthesized peptide.

Termination sequence (*Section 18.2*) A sequence of DNA that tells the RNA polymerase to terminate synthesis.

Terpene (*Section 3.5*) A compound whose carbon skeleton can be divided into two or more units identical to the carbon skeleton of isoprene.

Tertiary (3°) alcohol (*Section 1.4A*) An alcohol in which the carbon atom bearing the —OH group is bonded to three other carbon groups.

Tertiary (3°) amine (*Section 1.4B*) An amine in which nitrogen is bonded to three carbon groups.

Tertiary structure (*Section 14.10*) The overall conformation of a polypeptide chain, including the interactions of the side chains and the position of every atom in the polypeptide.

Thiol (*Section 5.4A*) A compound containing an —SH (sulfhydryl) group bonded to a tetrahedral carbon atom.

Tissue necrosis factor (**TNF**) (*Section 23.6*) A type of cytokine produced by T cells and macrophages that has the ability to lyse susceptible tumor cells.

Titration (*Section 7.9*) An analytical procedure whereby we react a known volume of a solution of known concentration with a known volume of a solution of unknown concentration.

TNF (*Section 23.6*) Tumor necrosis factor; a type of cytokine produced by T cells and macrophages that has the ability to lyse tumor cells.

Trans (*Section 2.7*) A prefix meaning "across from."

Transamination (*Section 20.8*) The exchange of the amino group of an amino acid and a keto group of an alpha-ketoacid.

Transcription (*Section 17.4*) The process whereby DNA is used as a template for the synthesis of RNA.

Transcription factor (*Section 18.2*) A binding protein that facilitates the binding of RNA polymerase to the DNA to be transcribed, or that binds to a remote location and stimulate transcription.

Transfer RNA (tRNA) (*Section 17.4*) The RNA that transports amino acids to the site of protein synthesis on ribosomes.

Transferase (*Section 15.2*) A class of enzymes that catalyzes a reaction where a group of atoms such as an acyl group or amino group is transferred from one molecule to another.

Transition state (*Section 15.5*) An unstable species formed during a chemical reaction; a maximum on an energy diagram.

Transition-state analog (*Section 15.8*) A molecule constructed to mimic the transition state of an enzyme-catalyzed reaction.

Translation (*Section 18.1*) The process in which information encoded in a mRNA is used to assemble a specific protein.

Translocation (*Section 18.5*) The part of translation where the ribosome moves down the mRNA a distance of three bases, so that the new codon is on the A site.

Transporter (*Section 16.5*) A protein molecule carrying small molecules, such as glucose or glutamic acid, across a membrane.

Triglyceride (*Section 13.7*) A kind of lipid formed by bonding glycerol to three fatty acids by ester bonds.

Triple helix (*Section 14.11*) The collagen triple helix is composed of three peptide chains. Each chain is itself a left-handed helix. These chains are twisted around each other in a right-handed helix.

Triprotic acid (*Section 7.3*) An acid that can give up three protons.

tRNA^{fmet} (*Section 18.5*) The special tRNA molecule that initiates translation.

Tumor suppression factor (*Chemical Connections 18F*) A protein that controls replication of DNA so that cells do not divide constantly. Many cancers are caused by mutated tumor suppression factors.

Unwinding proteins (*Section 17.6*) Special proteins that help unwind DNA so that it can be replicated.

Urea cycle (*Section 20.8*) A cyclic pathway that produces urea from ammonia and carbon dioxide.

Vesicle, synaptic (*Section 16.2*) A compartment containing a neurotransmitter that fuses with a presynaptic membrane and releases its contents when a nerve impulse arrives.

Vitamin (*Section 22.6*) An organic substance required in small quantities in the diet of most species, which generally functions as a cofactor in important metabolic reactions.

Weak acid (*Section 7.2*) An acid that is only partially ionized in aqueous solution.

Weak base (*Section 7.2*) A base that is only partially ionized in aqueous solution.

Zwitterion (*Section 14.3*) A molecule that has equal numbers of positive and negative charges, giving it a net charge of zero.

Zymogen (*Section 15.6*) An inactive form of an enzyme that must have part of its polypeptide chain cleaved before it becomes active; a proenzyme.